AFA
Tables of Houses

Placidus System

compiled by
Astro Numeric Service

AMERICAN FEDERATION OF ASTROLOGERS, INC.
6535 South Rural Road Tempe, Arizona 85283

First Printing 1977
Ninth Printing 1993
ISBN: 0-86690-252-X
Library of Congress Catalog Card Number: 77-77344

Published by:
American Federation of Astrologers, Inc.
PO Box 22040, 6535 S. Rural Road
Tempe, AZ 85285

Printed in the United States of America

Notes

This new compilation of house cusps offers students, teachers, counselors, and investigators of astrology an authoritative reference for the accurate division of the zodiac into twelve houses. Designed to meet the most demanding needs of astrological research, the tables cover the range of latitudes from 0–66° and list all cusps to one tenth of a minute of arc. Great care has been taken to make this book convenient for frequent use – the number of pages has been kept to a minimum, yet the readability of the tables has not been sacrificed. The tables list the cusps at even four minute intervals of sidereal time, making linear interpolation practical and straightforward.

The design, computation, and typesetting of the tables was done by Astro Numeric Service, utilizing the same computer programs used in their well known chart calculation service. A full complement of auxiliary tables were also specially computed for this volume, including a unique table of logarithms able to handle angles to an accuracy of one tenth of one minute of arc. To enhance the value of the book for students, complete instructions for the use of the tables are provided, including examples of three different interpolation techniques.

General Layout

The twelve houses of an astrological chart divide the zodiac into twelve unequal regions. The houses are numbered from 1 to 12, and the exact point marking the beginning of each house is called the *cusp* of that house. The first six houses occupy the part of the zodiac which lies below the horizon, while the second six houses occupy the zodiac above the horizon. The cusp of the first house is called the *ascendant* since it is positioned exactly on the horizon in the east and represents the point on the zodiac which is currently rising. The seventh house cusp is called the *descendant* since it lies on the western horizon and represents the point on the zodiac that is just setting. The tenth house cusp plays a major role in astrology and is called the *midheaven* or Medium Coeli because it marks the point on the zodiac that is exactly midway between rising and setting. In this volume the tenth house cusp is always abbreviated MC.

The exact positions of the twelve house cusps are determined by the geometric orientation of the zodiac to the horizon at a particular time and place on the earth. The purpose of these tables of houses is to provide the cusps of the twelve house without having to compute the trigonometric functions involved. To use the tables, only two numbers are needed – the latitude of the place and the local sidereal time.

Although all twelve house cusps are needed for a chart, only six are listed in the tables. This is not an inconvenience, however, since opposite house cusps such as the ascendant and descendant, the second and eighth houses, etc., always have exactly opposite positions in the zodiac, 6 signs (180°) apart. To find the cusps not listed in the tables it is only necessary to change the zodiacal signs of the cusps that are listed to the opposite side of the zodiac.

This volume actually contains 360 separate tables of house cusps printed two on each page. Each table shows the house cusps for latitudes 0–66° at a single value of sidereal time. The next table shows the cusps for the sidereal time four minutes greater, and so on, until the entire range of 24 hours has been covered.

At the top of each table the sidereal time is printed in hours, minutes, and seconds on the left, and in degrees, minutes and seconds on the right. In the latter form it is called the *right ascension* of the meridian, but it represents only a change of units from sidereal time. Since the MC remains the same for all latitudes, it is printed once at the top of the page under the letters MC, and because of its importance is listed to a precision of one second of arc.

The main body of each table consists of the ascendant and four intermediate house cusps for the various whole degrees of latitude. Notice that for the range of latitudes from 0–15°, cusps are listed every five degrees. For readability, the zodiacal signs are printed only when the cusps change sign in a column. When reading cusps, use the sign that last appears above in the column.

Southern Latitudes

The house cusps for a southern latitude will not be the same as the cusps for the corresponding northern latitude. The general rule for finding southern latitude cusps is to add twelve hours to the sidereal time, look up the cusps as if the latitude were north, and then reverse the zodiacal sign of all the cusps found.

This process is made easier in this volume through the use of separate headings for southern latitudes at the bottom of each table. Instead of adding twelve hours to the sidereal time, use the sidereal time headings at the bottom of the page when locating the proper table. You will notice that this gives you the same table you would have found by adding the twelve hours and using the northern latitude headings. When read according to the southern latitude headings, the columns list the cusps for houses 5, 6, 8, 9, and the descendant. These cusps may be used directly from the table. To find the opposite houses (11, 12, 2, 3, and ascendant), it is necessary only to change the sign of each cusp to the opposite side of the zodiac. Be sure to use the MC listing at the bottom of the page rather than at the top.

Interpolation

It is usually necessary to interpolate to find cusps for an exact time or latitude that falls between two listings in the tables. Interpolation is the process of finding a small correction to be added or subtracted from a cusp in the tables to compensate for the difference between the value of the cusp at the listed sidereal time or latitude and the value of the cusp at the actual sidereal time or latitude. Generally speaking, interpolation must be done twice for each cusp – a correction for the sidereal time and another for the latitude. The MC (10th house cusp) need only be corrected for sidereal time since it remains the same at all latitudes.

It is no surprise that the difficulty of interpolation depends upon the accuracy required. We describe here three techniques, each giving successively more accurate results, and each appropriate under different circumstances. The direct proportions method first described is little more than a simple procedure for determining the closest degree of a cusp, but it does illustrate the basic approach of interpolation. Following this same approach the cusp correction tables and the proportional logarithm tables allow interpolation to the precision generally required in astrological practice.

Two terms which are used frequently in these explanations of interpolation are *excess sidereal time* and *excess latitude*. The excess sidereal time describes the amount that the actual sidereal time exceeds the time that is listed in the closest preceding house table. In the sample problem used in the illustrations, the sidereal time is 3h 22m 16s. Since the closest house table whose sidereal time is smaller is for 3h 20m 0s, the excess sidereal time is 2 minutes 16 seconds.

Excess latitude is determined the same way. It is the amount that the actual latitude exceeds the next smaller latitude in the house tables. In the sample problem the latitude is 32° 8′, and since the house tables list cusps at 32°, the excess latitude is 8′.

Another quantity that occurs frequently in interpolation is the difference between two successive listings of a cusp, such as between the value of a cusp at one latitude and the next, or between one sidereal time and the next. Whenever the word *difference* is used in the explanations that follow, it always refers to the full amount that a cusp changes from one appearance in the house tables to the next.

Sidereal time: 3h 22m 16s (excess time is approx. 2 minutes)
Latitude: 32° 8′ N (excess latitude is ignored)

		MC	11	12	Asc.	2	3
1.	Cusps at 3h 24m 0s *(32° latitude)*	♉ 23° 23′	♊ 26° 32′	♋ 27° 57′	♌ 26° 47′	♍ 21° 41′	♎ 20° 52′
2.	Cusps at 3h 20m 0s *(32° latitude)*	♉ 22° 24′	♊ 25° 36′	♋ 27° 4′	♌ 25° 57′	♍ 20° 46′	♎ 19° 53′
3.	Difference *(subtract line 2 from line 1)*	59′	56′	53′	50′	55′	59′
4.	Fraction *(¼, ½, or ¾)*	½	½	½	½	½	½
5.	Correction *(multiply line 4 times line 3)*	29′	28′	26′	25′	27′	29′
6.	Cusps at 3h 22m *(add line 5 to line 2)*	♉ 22° 53′	♊ 26° 4′	♋ 27° 30′	♌ 26° 22′	♍ 21° 13′	♎ 20° 22′
7.	Rounded	♉ 23°	♊ 26°	♋ 28°	♌ 26°	♍ 21°	♎ 20°

Illustration 1
Interpolation by Direct Proportions

Simple Proportions

This simplest method of interpolation is appropriate when there is an uncertainty of perhaps 1 to 4 minutes in the birth time. Since under these circumstances accuracies of about 1° in the cusps is about the best that can be obtained, seconds and tenths of minutes should be ignored in the house tables. When the final cusps are found, they should be rounded to the nearest whole degree so that the final figures will not mislead others into thinking that more accuracy was available.

First round the local sidereal time to the nearest minute and if this time is listed in the tables, the cusps can be taken directly from the tables. No corrections for exact latitude are needed in this method since they would be much smaller than the inaccuracy caused by the uncertainty of the birth time. Use the cusps for the nearest whole degree of latitude.

If the local sidereal time is not listed in the tables, then write down the cusps from the two tables whose sidereal time is just less and just greater than the actual time (at the closest latitude), and subtract the smaller cusps from the larger to find the difference for each cusp. The size of the correction will be 1/4, 1/2, or 3/4 of the difference between the two tables for that cusp, depending upon whether the excess sidereal time (rounded to the nearest minute) is 1, 2, or 3 minutes. These corrections should then be added to the cusps of the first house table, and the results rounded to the nearest degree. See Illustration 1 for a sample problem worked using the direct proportions method of interpolation.

House Cusp Correction Tables

To facilitate interpolation of house cusps to an accuracy of one minute of arc we have included in this volume auxiliary tables which allow the user to simply look up the corrections for sidereal time and latitude rather than calculate them. These tables are recomended for use when the birth time is known to the exact minute.

Like the direct proportions method, it is first necessary to write down the cusps from the house tables whose sidereal time is just less than and just greater than the true sidereal time of birth. Since the correction tables needs only degrees and minutes, it is best to round the cusps given in the house tables to the nearest minute of arc. Also, the cusps should be taken from the latitude entry in the table that is equal to or smaller than the actual latitude. As before, the smaller cusps should be subtracted from the larger cusps to find the difference between the adjacent tables for each cusp.

Now instead of multiplying the cusp differences by a fraction to find the correction, look in the auxiliary table called HOUSE CORRECTIONS FOR EXACT SIDEREAL TIME under the column headed by the cusp difference. The correction is found in the row corresponding to the excess sidereal time. Add the proper correction to each cusp found originally in the first house table.

The MC needs only this sidereal time correction, but the other cusps must also be corrected for the exact latitude. This procedure is again very similar. Using the house table whose sidereal time is closest to the actual time, take the house cusps for the latitudes just smaller and just greater than the actual latitude and again find the difference between these listings for each cusp. Now using the table HOUSE CORRECTIONS FOR EXACT LATITUDE, find the column corresponding to the difference between latitudes for each cusp. The latitude correction is taken from the row corresponding to the excess latitude. The correction for latitude must be either added to or subtracted from the corrected cusp found above for the exact sidereal time. If the cusps in the house table are increasing as the latitude increases, add the correction – otherwise subtract it. Illustration 2 shows each step in using the correction tables to solve the sample problem.

If the actual latitude is less than 15°, where the house tables list cusps only every 5°, the excess latitude must be divided by 5 before using the correction tables. For example, if the actual latitude exceeds the house table by 2° 34′ (154′), use 31′ (154′÷5=31′) as the excess latitude and find the correction in the usual way.

Sidereal Time: 3h 22m 16s (excess sidereal time is 2m 16s)
Latitude: 32° 8′ N (excess latitude is 8′)

	MC	11	12	Asc.	2	3
1. Cusps at 3h 24m 0s *(32° latitude)*	♉ 23° 23′	♊ 26° 32′	♋ 27° 57′	♌ 26° 48′	♍ 21° 42′	♎ 20° 53′
2. Cusps at 3h 20m 0s *(32° latitude)*	♉ 22° 25′	♊ 25° 37′	♋ 27° 5′	♌ 25° 57′	♍ 20° 47′	♎ 19° 54′
3. Difference *(subtract line 2 from line 1)*	58′	55′	52′	51′	55′	59′
4. Correction *(from correction tables)*	33′	31′	29′	29′	31′	33′
5. Cusps at 3h 22m 16s *(add line 4 to line 2)*	♉ 22° 58′	♊ 26° 8′	♋ 27° 34′	♌ 26° 26′	♍ 21° 18′	♎ 20° 27′
6. Cusps at 33° *(3h 24m 0s)*		♊ 26° 44′	♋ 28° 16′	♌ 27° 4′	♍ 21° 45′	♎ 20° 49′
7. Cusps at 32° *(3h 24m 0s)*		♊ 26° 32′	♋ 27° 57′	♌ 26° 48′	♍ 21° 42′	♎ 20° 53′
8. Difference *(subtract line 7 from line 6)*		12′	19′	16′	3′	4′
9. Correction *(from correction tables)*		2′	3′	2′	0′	−1′
10. Corrected Cusps *(add line 9 to line 5)*	♉ 22° 58′	♊ 26° 10′	♋ 27° 37′	♌ 26° 28′	♍ 21° 18′	♎ 20° 26′

Illustration 2
Interpolation by Correction Tables

Logarithms

The proportional logarithm tables given at the end of this volume allow linear interpolation of the cusps to the full accuracy of the tables – one tenth of one minute of arc. Although this is the greatest accuracy of the three methods of interpolation described here, a word of caution is necessary. Several parts of the house tables, especially high latitudes (above 50°), cannot be interpolated to this accuracy by *any* means of linear interpolation due the uneven change of house cusps between tables. Proportional logarithms do, however, offer a computational method for interpolation that is more accurate than the previous methods discussed.

In principle the method of proportional logarithms proceeds much like the direct proportions method, except that to simplify the multiplication of degrees and minutes (and to obtain greater accuracy), we first convert all such quantities into logarithms (logs). The logarithms can be handled quite easily using only additions and subtractions, and when the computations are finished, the results can then be re-converted back to degrees and minutes.

Illustration 3 shows the same interpolation problem solved earlier, but now done with proportional logs. Because of the large number of steps, our explanation of the method will refer to the line numbers of the illustration.

As in the previous methods of interpolation, the cusps are copied from the house tables whose sidereal time are on either side of the actual sidereal time. These are put in lines 13 and 14, where they are subtracted to obtain the difference between the cusps of the two tables (15). The differences should be written in degrees, minutes, and tenths of minutes. To convert each difference to a logarithm, use the page of the logarithm table corresponding to the correct degree, and locate the row corresponding to the correct number of minutes. The logarithm itself is found in the column headed by the appropriate number of tenths of degrees. The logarithms found for each cusp are put in line 16.

Now that the logs of the cusp differences have been found, it is necessary to compute the log of the fraction similar to that used back in the direct proportions method. This is done in lines 1 through 6. First the sidereal time of the earlier house table (2), is subtracted from the actual sidereal time (1), giving the excess sidereal time (3). The log of the excess time (4) is then found by turning to the page of the log tables for the appropriate minute (listed in the upper left-hand corner of the tables), and then locating the line corresponding to the number of seconds. Since tenths of seconds are generally not known, use the logarithm in the first column indicated by zero tenths. Next the log of 4 minutes in line 5 (the time difference between tables), is subtracted from the log of the excess time (4), giving the log of the fraction in line 6 which can be immediately transferred to line 17 for all the cusps.

Now add to each difference log (16) the log of the fraction (17) to form the log of the correction for each cusp (18). These corrections must be converted back to degrees, minutes, and tenths of minutes so that they can be added to the original cusps from the house tables (14). To do this conversion locate the logarithm in the log tables whose value is closest to the value of the log to be converted. The degrees can then be read from the page heading, the minutes from the number of the line, and the tenths from the column heading. Once the corrections have been changed to degrees and minutes (19) they can be added to the cusps in line 14 to get the corrected cusps for the exact sidereal time in line 20.

1. Exact Sidereal Time	3h 22m 16s	**7.** Exact Latitude	32° 8′
2. Previous Sidereal Time	3h 20m 0s	**8.** Previous Latitude	32° 0′
3. Excess Sidereal Time	2m 16s	**9.** Excess Latitude	8′
4. Log of Excess Sidereal Time	3180	**10.** Log of Excess Latitude	14513
5. Log of 4m	909	**11.** Log of 1°	6453
6. Log of Fraction	2271	**12.** Log of Fraction	8060

	MC	**11**	**12**	**Asc.**	**2**	**3**
13. Cusps at 3h 24m 0s *(32° latitude)*	♉ 23° 23.4′	♊ 26° 32.4′	♋ 27° 57.1′	♌ 26° 47.8′	♍ 21° 41.8′	♎ 20° 52.8′
14. Cusps at 3h 20m 0s *(32° latitude)*	♉ 22° 24.6′	♊ 25° 36.9′	♋ 27° 4.6′	♌ 25° 57.4′	♍ 20° 46.5′	♎ 19° 53.7′
15. Difference *(subtract line 14 from line 13)*	58.8′	55.5′	52.5′	50.4′	55.3′	59.1′
16. Log of Difference *(from log tables)*	6534	6765	6987	7151	6779	6514
17. Log of Fraction *(from line 6)*	2271	2271	2271	2271	2271	2271
18. Correction Log *(add line 16 to line 17)*	8805	9036	9258	9422	9050	8785
19. Correction *(from log tables)*	33.3′	31.5′	29.8′	28.6′	31.3′	33.5′
20. Cusps at 3h 22m 16s *(add line 19 to line 14)*	♉ 22° 57.9′	♊ 26° 8.4′	♋ 27° 34.4′	♌ 26° 26.0′	♍ 21° 17.8′	♎ 20° 27.2′
21. Cusps at 3h 24m 0s		♊ 26° 44.0′	♋ 28° 16.0′	♌ 27° 3.7′	♍ 21° 44.8′	♎ 20° 52.8′
22. Cusps at 3h 24m 0s		♊ 26° 32.4′	♋ 27° 57.1′	♌ 26° 47.8′	♍ 21° 41.8′	♎ 20° 48.8′
23. Difference *(subtract line 22 from line 21)*		11.6′	18.9′	15.9′	3.0′	4.0′
24. Log of Difference *(from log tables)*		13027	11074	11765	18436	17285
25. Log of Fraction *(from line 12)*		8060	8060	8060	8060	8060
26. Log of Correction *(add line 24 to line 25)*		21087	19134	19825	26496	25345
27. Correction *(from log tables)*		1.5′	2.5′	2.1′	0.4′	−0.5′
28. Corrected Cusps *(add line 27 to line 20)*	♉ 22° 57.9′	♊ 26° 9.9′	♋ 27° 36.9′	♌ 26° 28.1′	♍ 21° 18.2′	♎ 20° 26.7′

Illustration 3
Interpolation by Proportional Logarithms

To correct the cusps for the exact latitude, it is again necessary to copy the cusps from the house tables for the latitudes less than and greater than the true latitude. Using the house table whose sidereal time is closest to the actual sidereal time, the larger cusps are placed in line 13 so that the smaller cusps (14) can be subtracted to form the difference (15). The log of the difference for each cusp (16) is the found from the log tables as described above.

The log of the fraction which must be added to the log of the cusp differences is calculated in lines 7–12 in the same way that it was determined for the sidereal time correction. First the smaller latitude of the house tables (8) is subtracted from the actual latitude (7) to yield the excess latitude (9) in degrees and minutes. The log of the excess latitude (10) is found from the log tables in the usual way. Now the log of either 1° or 5° must be subtracted, depending upon whether the latitudes used from the house tables were one or five degrees apart. In this example, the log of 1° is written in line 11 and subtracted from line 10 to form the log of the fraction (12) which is then copied to every column of line 25.

Lines 24 and 25 can now be added to form the exact latitude correction log for each cusp (26). The correction logs should be converted back to degrees, minutes, and tenths of minutes (27) and then either added or subtracted from the cusps which have been previously corrected for the exact sidereal time (20). Add the corrections if the cusps in the house table increase with increasing latitude, but subtract them if the cusps get smaller with increasing latitude. The final corrected cusps are shown in line 28.

GEOCENTRIC CORRECTIONS TO LATITUDE

Subtract from Geographic Latitude

Latitude	Correction	Latitude
0°	0.0′	90°
1	0.4	89
2	0.8	88
3	1.2	87
4	1.6	86
5	2.0	85
6	2.4	84
7	2.8	83
8	3.2	82
9	3.6	81
10	4.0	80
11	4.3	79
12	4.7	78
13	5.1	77
14	5.4	76
15	5.8	75
16	6.1	74
17	6.5	73
18	6.8	72
19	7.1	71
20	7.4	70
21	7.7	69
22	8.0	68
23	8.3	67
24	8.6	66
25	8.9	65
26	9.1	64
27	9.4	63
28	9.6	62
29	9.8	61
30	10.0	60
31	10.2	59
32	10.4	58
33	10.6	57
34	10.7	56
35	10.9	55
36	11.0	54
37	11.1	53
38	11.2	52
39	11.3	51
40	11.4	50
41	11.5	49
42	11.5	48
43	11.6	47
44	11.6	46
45	11.6	45

CONVERSION OF ARC TO TIME

Degrees

°	h m	°	h m	°	h m	°	h m	°	h m	°	h m
1	0:04	**61**	4:04	**121**	8:04	**181**	12:04	**241**	16:04	**301**	20:04
2	0:08	**62**	4:08	**122**	8:08	**182**	12:08	**242**	16:08	**302**	20:08
3	0:12	**63**	4:12	**123**	8:12	**183**	12:12	**243**	16:12	**303**	20:12
4	0:16	**64**	4:16	**124**	8:16	**184**	12:16	**244**	16:16	**304**	20:16
5	0:20	**65**	4:20	**125**	8:20	**185**	12:20	**245**	16:20	**305**	20:20
6	0:24	**66**	4:24	**126**	8:24	**186**	12:24	**246**	16:24	**306**	20:24
7	0:28	**67**	4:28	**127**	8:28	**187**	12:28	**247**	16:28	**307**	20:28
8	0:32	**68**	4:32	**128**	8:32	**188**	12:32	**248**	16:32	**308**	20:32
9	0:36	**69**	4:36	**129**	8:36	**189**	12:36	**249**	16:36	**309**	20:36
10	0:40	**70**	4:40	**130**	8:40	**190**	12:40	**250**	16:40	**310**	20:40
11	0:44	**71**	4:44	**131**	8:44	**191**	12:44	**251**	16:44	**311**	20:44
12	0:48	**72**	4:48	**132**	8:48	**192**	12:48	**252**	16:48	**312**	20:48
13	0:52	**73**	4:52	**133**	8:52	**193**	12:52	**253**	16:52	**313**	20:52
14	0:56	**74**	4:56	**134**	8:56	**194**	12:56	**254**	16:56	**314**	20:56
15	1:00	**75**	5:00	**135**	9:00	**195**	13:00	**255**	17:00	**315**	21:00
16	1:04	**76**	5:04	**136**	9:04	**196**	13:04	**256**	17:04	**316**	21:04
17	1:08	**77**	5:08	**137**	9:08	**197**	13:08	**257**	17:08	**317**	21:08
18	1:12	**78**	5:12	**138**	9:12	**198**	13:12	**258**	17:12	**318**	21:12
19	1:16	**79**	5:16	**139**	9:16	**199**	13:16	**259**	17:16	**319**	21:16
20	1:20	**80**	5:20	**140**	9:20	**200**	13:20	**260**	17:20	**320**	21:20
21	1:24	**81**	5:24	**141**	9:24	**201**	13:24	**261**	17:24	**321**	21:24
22	1:28	**82**	5:28	**142**	9:28	**202**	13:28	**262**	17:28	**322**	21:28
23	1:32	**83**	5:32	**143**	9:32	**203**	13:32	**263**	17:32	**323**	21:32
24	1:36	**84**	5:36	**144**	9:36	**204**	13:36	**264**	17:36	**324**	21:36
25	1:40	**85**	5:40	**145**	9:40	**205**	13:40	**265**	17:40	**325**	21:40
26	1:44	**86**	5:44	**146**	9:44	**206**	13:44	**266**	17:44	**326**	21:44
27	1:48	**87**	5:48	**147**	9:48	**207**	13:48	**267**	17:48	**327**	21:48
28	1:52	**88**	5:52	**148**	9:52	**208**	13:52	**268**	17:52	**328**	21:52
29	1:56	**89**	5:56	**149**	9:56	**209**	13:56	**269**	17:56	**329**	21:56
30	2:00	**90**	6:00	**150**	10:00	**210**	14:00	**270**	18:00	**330**	22:00
31	2:04	**91**	6:04	**151**	10:04	**211**	14:04	**271**	18:04	**331**	22:04
32	2:08	**92**	6:08	**152**	10:08	**212**	14:08	**272**	18:08	**332**	22:08
33	2:12	**93**	6:12	**153**	10:12	**213**	14:12	**273**	18:12	**333**	22:12
34	2:16	**94**	6:16	**154**	10:16	**214**	14:16	**274**	18:16	**334**	22:16
35	2:20	**95**	6:20	**155**	10:20	**215**	14:20	**275**	18:20	**335**	22:20
36	2:24	**96**	6:24	**156**	10:24	**216**	14:24	**276**	18:24	**336**	22:24
37	2:28	**97**	6:28	**157**	10:28	**217**	14:28	**277**	18:28	**337**	22:28
38	2:32	**98**	6:32	**158**	10:32	**218**	14:32	**278**	18:32	**338**	22:32
39	2:36	**99**	6:36	**159**	10:36	**219**	14:36	**279**	18:36	**339**	22:36
40	2:40	**100**	6:40	**160**	10:40	**220**	14:40	**280**	18:40	**340**	22:40
41	2:44	**101**	6:44	**161**	10:44	**221**	14:44	**281**	18:44	**341**	22:44
42	2:48	**102**	6:48	**162**	10:48	**222**	14:48	**282**	18:48	**342**	22:48
43	2:52	**103**	6:52	**163**	10:52	**223**	14:52	**283**	18:52	**343**	22:52
44	2:56	**104**	6:56	**164**	10:56	**224**	14:56	**284**	18:56	**344**	22:56
45	3:00	**105**	7:00	**165**	11:00	**225**	15:00	**285**	19:00	**345**	23:00
46	3:04	**106**	7:04	**166**	11:04	**226**	15:04	**286**	19:04	**346**	23:04
47	3:08	**107**	7:08	**167**	11:08	**227**	15:08	**287**	19:08	**347**	23:08
48	3:12	**108**	7:12	**168**	11:12	**228**	15:12	**288**	19:12	**348**	23:12
49	3:16	**109**	7:16	**169**	11:16	**229**	15:16	**289**	19:16	**349**	23:16
50	3:20	**110**	7:20	**170**	11:20	**230**	15:20	**290**	19:20	**350**	23:20
51	3:24	**111**	7:24	**171**	11:24	**231**	15:24	**291**	19:24	**351**	23:24
52	3:28	**112**	7:28	**172**	11:28	**232**	15:28	**292**	19:28	**352**	23:28
53	3:32	**113**	7:32	**173**	11:32	**233**	15:32	**293**	19:32	**353**	23:32
54	3:36	**114**	7:36	**174**	11:36	**234**	15:36	**294**	19:36	**354**	23:36
55	3:40	**115**	7:40	**175**	11:40	**235**	15:40	**295**	19:40	**355**	23:40
56	3:44	**116**	7:44	**176**	11:44	**236**	15:44	**296**	19:44	**356**	23:44
57	3:48	**117**	7:48	**177**	11:48	**237**	15:48	**297**	19:48	**357**	23:48
58	3:52	**118**	7:52	**178**	11:52	**238**	15:52	**298**	19:52	**358**	23:52
59	3:56	**119**	7:56	**179**	11:56	**239**	15:56	**299**	19:56	**359**	23:56
60	4:00	**120**	8:00	**180**	12:00	**240**	16:00	**300**	20:00	**360**	24:00

CONVERSION OF ARC TO TIME

Minutes and Seconds

′	″	m s	′	″	m s	′	″	m s	′	″	m s
0	**0**	0:00	**15**	**0**	1:00	**30**	**0**	2:00	**45**	**0**	3:00
	15	0:01		**15**	1:01		**15**	2:01		**15**	3:01
	30	0:02		**30**	1:02		**30**	2:02		**30**	3:02
	45	0:03		**45**	1:03		**45**	2:03		**45**	3:03
1	**0**	0:04	**16**	**0**	1:04	**31**	**0**	2:04	**46**	**0**	3:04
	15	0:05		**15**	1:05		**15**	2:05		**15**	3:05
	30	0:06		**30**	1:06		**30**	2:06		**30**	3:06
	45	0:07		**45**	1:07		**45**	2:07		**45**	3:07
2	**0**	0:08	**17**	**0**	1:08	**32**	**0**	2:08	**47**	**0**	3:08
	15	0:09		**15**	1:09		**15**	2:09		**15**	3:09
	30	0:10		**30**	1:10		**30**	2:10		**30**	3:10
	45	0:11		**45**	1:11		**45**	2:11		**45**	3:11
3	**0**	0:12	**18**	**0**	1:12	**33**	**0**	2:12	**48**	**0**	3:12
	15	0:13		**15**	1:13		**15**	2:13		**15**	3:13
	30	0:14		**30**	1:14		**30**	2:14		**30**	3:14
	45	0:15		**45**	1:15		**45**	2:15		**45**	3:15
4	**0**	0:16	**19**	**0**	1:16	**34**	**0**	2:16	**49**	**0**	3:16
	15	0:17		**15**	1:17		**15**	2:17		**15**	3:17
	30	0:18		**30**	1:18		**30**	2:18		**30**	3:18
	45	0:19		**45**	1:19		**45**	2:19		**45**	3:19
5	**0**	0:20	**20**	**0**	1:20	**35**	**0**	2:20	**50**	**0**	3:20
	15	0:21		**15**	1:21		**15**	2:21		**15**	3:21
	30	0:22		**30**	1:22		**30**	2:22		**30**	3:22
	45	0:23		**45**	1:23		**45**	2:23		**45**	3:23
6	**0**	0:24	**21**	**0**	1:24	**36**	**0**	2:24	**51**	**0**	3:24
	15	0:25		**15**	1:25		**15**	2:25		**15**	3:25
	30	0:26		**30**	1:26		**30**	2:26		**30**	3:26
	45	0:27		**45**	1:27		**45**	2:27		**45**	3:27
7	**0**	0:28	**22**	**0**	1:28	**37**	**0**	2:28	**52**	**0**	3:28
	15	0:29		**15**	1:29		**15**	2:29		**15**	3:29
	30	0:30		**30**	1:30		**30**	2:30		**30**	3:30
	45	0:31		**45**	1:31		**45**	2:31		**45**	3:31
8	**0**	0:32	**23**	**0**	1:32	**38**	**0**	2:32	**53**	**0**	3:32
	15	0:33		**15**	1:33		**15**	2:33		**15**	3:33
	30	0:34		**30**	1:34		**30**	2:34		**30**	3:34
	45	0:35		**45**	1:35		**45**	2:35		**45**	3:35
9	**0**	0:36	**24**	**0**	1:36	**39**	**0**	2:36	**54**	**0**	3:36
	15	0:37		**15**	1:37		**15**	2:37		**15**	3:37
	30	0:38		**30**	1:38		**30**	2:38		**30**	3:38
	45	0:39		**45**	1:39		**45**	2:39		**45**	3:39
10	**0**	0:40	**25**	**0**	1:40	**40**	**0**	2:40	**55**	**0**	3:40
	15	0:41		**15**	1:41		**15**	2:41		**15**	3:41
	30	0:42		**30**	1:42		**30**	2:42		**30**	3:42
	45	0:43		**45**	1:43		**45**	2:43		**45**	3:43
11	**0**	0:44	**26**	**0**	1:44	**41**	**0**	2:44	**56**	**0**	3:44
	15	0:45		**15**	1:45		**15**	2:45		**15**	3:45
	30	0:46		**30**	1:46		**30**	2:46		**30**	3:46
	45	0:47		**45**	1:47		**45**	2:47		**45**	3:47
12	**0**	0:48	**27**	**0**	1:48	**42**	**0**	2:48	**57**	**0**	3:48
	15	0:49		**15**	1:49		**15**	2:49		**15**	3:49
	30	0:50		**30**	1:50		**30**	2:50		**30**	3:50
	45	0:51		**45**	1:51		**45**	2:51		**45**	3:51
13	**0**	0:52	**28**	**0**	1:52	**43**	**0**	2:52	**58**	**0**	3:52
	15	0:53		**15**	1:53		**15**	2:53		**15**	3:53
	30	0:54		**30**	1:54		**30**	2:54		**30**	3:54
	45	0:55		**45**	1:55		**45**	2:55		**45**	3:55
14	**0**	0:56	**29**	**0**	1:56	**44**	**0**	2:56	**59**	**0**	3:56
	15	0:57		**15**	1:57		**15**	2:57		**15**	3:57
	30	0:58		**30**	1:58		**30**	2:58		**30**	3:58
	45	0:59		**45**	1:59		**45**	2:59		**45**	3:59

0m	PROPORTIONAL LOGARITHMS FOR FINDING HOUSE CUSPS										0°
	.0	.1	.2	.3	.4	.5	.6	.7	.8	.9	
0	32041	32041	29268	27646	26496	25603	24874	24257	23723	23252	0
1	22831	22449	22101	21781	21485	21209	20951	20708	20479	20263	1
2	20058	19863	19677	19499	19329	19165	19008	18858	18712	18572	2
3	18436	18305	18178	18055	17935	17819	17707	17597	17491	17387	3
4	17285	17187	17090	16996	16904	16814	16726	16640	16556	16474	4
5	16393	16314	16236	16160	16085	16012	15939	15869	15799	15731	5
6	15663	15597	15532	15468	15405	15343	15282	15222	15163	15104	6
7	15047	14990	14934	14879	14825	14771	14718	14666	14614	14563	7
8	14513	14463	14414	14365	14318	14270	14223	14177	14132	14086	8
9	14042	13997	13954	13910	13868	13825	13783	13742	13701	13660	9
10	13620	13580	13541	13502	13463	13425	13387	13350	13312	13275	10
11	13239	13203	13167	13131	13096	13061	13027	12992	12958	12924	11
12	12891	12858	12825	12792	12760	12728	12696	12664	12633	12602	12
13	12571	12540	12510	12479	12450	12420	12390	12361	12332	12303	13
14	12274	12246	12218	12189	12162	12134	12106	12079	12052	12025	14
15	11998	11972	11945	11919	11893	11867	11841	11816	11790	11765	15
16	11740	11715	11690	11666	11641	11617	11593	11569	11545	11521	16
17	11498	11474	11451	11428	11405	11382	11359	11336	11314	11291	17
18	11269	11247	11225	11203	11181	11159	11138	11116	11095	11074	18
19	11053	11032	11011	10990	10969	10949	10928	10908	10888	10868	19
20	10848	10828	10808	10788	10768	10749	10729	10710	10691	10672	20
21	10652	10633	10615	10596	10577	10558	10540	10521	10503	10485	21
22	10466	10448	10430	10412	10394	10376	10359	10341	10323	10306	22
23	10289	10271	10254	10237	10220	10203	10186	10169	10152	10135	23
24	10118	10102	10085	10069	10052	10036	10020	10003	9987	9971	24
25	9955	9939	9923	9907	9892	9876	9860	9845	9829	9814	25
26	9798	9783	9767	9752	9737	9722	9707	9692	9647	9662	26
27	9647	9632	9618	9603	9588	9574	9559	9545	9530	9516	27
28	9502	9487	9473	9459	9445	9431	9417	9403	9389	9375	28
29	9361	9348	9334	9320	9307	9293	9279	9266	9252	9239	29
30	9226	9212	9199	9186	9173	9160	9147	9133	9120	9107	30
31	9095	9082	9069	9056	9043	9031	9018	9005	8993	8980	31
32	8968	8955	8943	8930	8918	8906	8893	8881	8869	8857	32
33	8844	8832	8820	8808	8796	8784	8772	8761	8749	8737	33
34	8725	8713	8702	8690	8678	8667	8655	8644	8632	8621	34
35	8609	8598	8586	8575	8564	8552	8541	8530	8519	8508	35
36	8496	8485	8474	8463	8452	8441	8430	8419	8409	8398	36
37	8387	8376	8365	8355	8344	8333	8323	8312	8301	8291	37
38	8280	8270	8259	8249	8238	8228	8218	8207	8197	8187	38
39	8176	8166	8156	8146	8135	8125	8115	8105	8095	8085	39
40	8075	8065	8055	8045	8035	8025	8015	8006	7996	7986	40
41	7976	7966	7957	7947	7937	7928	7918	7909	7899	7889	41
42	7880	7870	7861	7851	7842	7833	7823	7814	7804	7795	42
43	7786	7776	7767	7758	7749	7739	7730	7721	7712	7703	43
44	7694	7685	7676	7667	7658	7649	7640	7631	7622	7613	44
45	7604	7595	7586	7577	7568	7560	7551	7542	7533	7525	45
46	7516	7507	7499	7490	7481	7473	7464	7456	7447	7438	46
47	7430	7421	7413	7404	7396	7388	7379	7371	7362	7354	47
48	7346	7337	7329	7321	7313	7304	7296	7288	7280	7271	48
49	7263	7255	7247	7239	7231	7223	7215	7207	7198	7190	49
50	7182	7174	7166	7159	7151	7143	7135	7127	7119	7111	50
51	7103	7095	7088	7080	7072	7064	7056	7049	7041	7033	51
52	7026	7018	7010	7003	6995	6987	6980	6972	6964	6957	52
53	6949	6942	6934	6927	6919	6912	6904	6897	6889	6882	53
54	6875	6867	6860	6852	6845	6838	6830	6823	6816	6808	54
55	6801	6794	6787	6779	6772	6765	6758	6751	6743	6736	55
56	6729	6722	6715	6708	6701	6694	6686	6679	6672	6665	56
57	6658	6651	6644	6637	6630	6623	6616	6609	6603	6596	57
58	6589	6582	6575	6568	6561	6554	6548	6541	6534	6527	58
59	6520	6514	6507	6500	6493	6487	6480	6473	6466	6460	59

SECONDS OF TIME

MINUTES OF ARC

Log 4 min. = 909 Log 1° = 6453 Log 5° = 16

PROPORTIONAL LOGARITHMS FOR FINDING HOUSE CUSPS

1^m Seconds of Time	.0	.1	.2	.3	.4	.5	.6	.7	.8	.9	1° Minutes of Arc
0	6453	6446	6440	6433	6427	6420	6413	6407	6400	6394	0
1	6387	6380	6374	6367	6361	6354	6348	6341	6335	6328	1
2	6322	6316	6309	6303	6296	6290	6283	6277	6271	6264	2
3	6258	6252	6245	6239	6233	6226	6220	6214	6208	6201	3
4	6195	6189	6183	6176	6170	6164	6158	6151	6145	6139	4
5	6133	6127	6121	6115	6108	6102	6096	6090	6084	6078	5
6	6072	6066	6060	6054	6048	6042	6036	6030	6024	6018	6
7	6012	6006	6000	5994	5988	5982	5976	5970	5964	5958	7
8	5952	5947	5941	5935	5929	5923	5917	5912	5906	5900	8
9	5894	5888	5883	5877	5871	5865	5859	5854	5848	5842	9
10	5837	5831	5825	5819	5814	5808	5802	5797	5791	5785	10
11	5780	5774	5769	5763	5757	5752	5746	5741	5735	5729	11
12	5724	5718	5713	5707	5702	5696	5691	5685	5680	5674	12
13	5669	5663	5658	5652	5647	5641	5636	5631	5625	5620	13
14	5614	5609	5603	5598	5593	5587	5582	5577	5571	5566	14
15	5561	5555	5550	5545	5539	5534	5529	5523	5518	5513	15
16	5508	5502	5497	5492	5487	5481	5476	5471	5466	5460	16
17	5455	5450	5445	5440	5435	5429	5424	5419	5414	5409	17
18	5404	5399	5393	5388	5383	5378	5373	5368	5363	5358	18
19	5353	5348	5343	5338	5333	5327	5322	5317	5312	5307	19
20	5302	5297	5292	5287	5282	5277	5273	5268	5263	5258	20
21	5253	5248	5243	5238	5233	5228	5223	5218	5213	5209	21
22	5204	5199	5194	5189	5184	5179	5174	5170	5165	5160	22
23	5155	5150	5146	5141	5136	5131	5126	5122	5117	5112	23
24	5107	5102	5098	5093	5088	5084	5079	5074	5069	5065	24
25	5060	5055	5051	5046	5041	5036	5032	5027	5022	5018	25
26	5013	5008	5004	4999	4995	4990	4985	4981	4976	4971	26
27	4967	4962	4958	4953	4949	4944	4939	4935	4930	4926	27
28	4921	4917	4912	4908	4903	4899	4894	4889	4885	4880	28
29	4876	4871	4867	4863	4858	4854	4849	4845	4840	4836	29
30	4831	4827	4822	4818	4814	4809	4805	4800	4796	4791	30
31	4787	4783	4778	4774	4770	4765	4761	4756	4752	4748	31
32	4743	4739	4735	4730	4726	4722	4717	4713	4709	4704	32
33	4700	4696	4692	4687	4683	4679	4674	4670	4666	4662	33
34	4657	4653	4649	4645	4640	4636	4632	4628	4623	4619	34
35	4615	4611	4607	4602	4598	4594	4590	4586	4581	4577	35
36	4573	4569	4565	4561	4556	4552	4548	4544	4540	4536	36
37	4532	4528	4523	4519	4515	4511	4507	4503	4499	4495	37
38	4491	4487	4482	4478	4474	4470	4466	4462	4458	4454	38
39	4450	4446	4442	4438	4434	4430	4426	4422	4418	4414	39
40	4410	4406	4402	4398	4394	4390	4386	4382	4378	4374	40
41	4370	4366	4362	4358	4354	4350	4346	4342	4338	4335	41
42	4331	4327	4323	4319	4315	4311	4307	4303	4299	4375	42
43	4292	4288	4284	4280	4276	4272	4268	4265	4261	4257	43
44	4253	4249	4245	4241	4238	4234	4230	4226	4222	4218	44
45	4215	4211	4207	4203	4199	4196	4192	4188	4184	4181	45
46	4177	4173	4169	4165	4162	4158	4154	4150	4147	4143	46
47	4139	4135	4132	4128	4124	4121	4117	4113	4109	4106	47
48	4102	4098	4095	4091	4087	4084	4080	4076	4072	4069	48
49	4065	4061	4058	4054	4050	4047	4043	4040	4036	4032	49
50	4029	4025	4021	4018	4014	4010	4007	4003	4000	3996	50
51	3992	3989	3985	3982	3978	3974	3971	3967	3964	3960	51
52	3957	3953	3949	3946	3942	3939	3935	3932	3928	3925	52
53	3921	3917	3914	3910	3907	3903	3900	3896	3893	3889	53
54	3886	3882	3879	3875	3872	3868	3865	3861	3858	3854	54
55	3851	3847	3844	3840	3837	3833	3830	3827	3823	3820	55
56	3816	3813	3809	3806	3802	3799	3796	3792	3789	3785	56
57	3782	3778	3775	3772	3768	3765	3761	3758	3755	3751	57
58	3748	3744	3741	3738	3734	3731	3727	3724	3721	3717	58
59	3714	3711	3707	3704	3701	3697	3694	3691	3687	3684	59

Log 4 min. = 909 Log 1° = 6453 Log 5° = 16

2^m	PROPORTIONAL LOGARITHMS FOR FINDING HOUSE CUSPS	2°

SECONDS OF TIME / MINUTES OF ARC

	.0	.1	.2	.3	.4	.5	.6	.7	.8	.9	
0	3681	3677	3674	3671	3667	3664	3661	3657	3654	3651	**0**
1	3647	3644	3641	3637	3634	3631	3628	3624	3621	3618	**1**
2	3614	3611	3608	3605	3601	3598	3595	3592	3588	3585	**2**
3	3582	3579	3575	3572	3569	3566	3562	3559	3556	3553	**3**
4	3549	3546	3543	3540	3537	3533	3530	3527	3524	3520	**4**
5	3517	3514	3511	3508	3504	3501	3498	3495	3492	3489	**5**
6	3485	3482	3479	3476	3473	3470	3466	3463	3460	3457	**6**
7	3454	3451	3447	3444	3441	3438	3435	3432	3429	3426	**7**
8	3422	3419	3416	3413	3410	3407	3404	3401	3397	3394	**8**
9	3391	3388	3385	3382	3379	3376	3373	3370	3366	3363	**9**
10	3360	3357	3354	3351	3348	3345	3342	3339	3336	3333	**10**
11	3330	3327	3324	3321	3318	3314	3311	3308	3305	3302	**11**
12	3299	3296	3293	3290	3287	3284	3281	3278	3275	3272	**12**
13	3269	3266	3263	3260	3257	3254	3251	3248	3245	3242	**13**
14	3239	3236	3233	3230	3227	3224	3221	3218	3215	3212	**14**
15	3209	3206	3204	3201	3198	3195	3192	3189	3186	3183	**15**
16	3180	3177	3174	3171	3168	3165	3162	3159	3156	3154	**16**
17	3151	3148	3145	3142	3139	3136	3133	3130	3127	3124	**17**
18	3122	3119	3116	3113	3110	3107	3104	3101	3098	3096	**18**
19	3093	3090	3087	3084	3081	3078	3075	3073	3070	3067	**19**
20	3064	3061	3058	3055	3053	3050	3047	3044	3041	3038	**20**
21	3035	3033	3030	3027	3024	3021	3018	3016	3013	3010	**21**
22	3007	3004	3002	2999	2996	2993	2990	2988	2985	2982	**22**
23	2979	2976	2974	2971	2968	2965	2962	2960	2957	2954	**23**
24	2951	2948	2946	2943	2940	2937	2935	2932	2929	2926	**24**
25	2924	2921	2918	2915	2913	2910	2907	2904	2902	2899	**25**
26	2896	2893	2891	2888	2885	2882	2880	2877	2874	2872	**26**
27	2869	2866	2863	2861	2858	2855	2852	2850	2847	2844	**27**
28	2842	2839	2836	2834	2831	2828	2825	2823	2820	2817	**28**
29	2815	2812	2809	2807	2804	2801	2799	2796	2793	2791	**29**
30	2788	2785	2783	2780	2777	2775	2772	2769	2767	2764	**30**
31	2761	2759	2756	2753	2751	2748	2746	2743	2740	2738	**31**
32	2735	2732	2730	2727	2724	2722	2719	2717	2714	2711	**32**
33	2709	2706	2704	2701	2698	2696	2693	2691	2688	2685	**33**
34	2683	2680	2678	2675	2672	2670	2667	2665	2662	2659	**34**
35	2657	2654	2652	2649	2647	2644	2641	2639	2636	2634	**35**
36	2631	2629	2626	2623	2621	2618	2616	2613	2611	2608	**36**
37	2606	2603	2600	2598	2595	2593	2590	2588	2585	2583	**37**
38	2580	2578	2575	2573	2570	2568	2565	2562	2560	2557	**38**
39	2555	2552	2550	2547	2545	2542	2540	2537	2535	2532	**39**
40	2530	2527	2525	2522	2520	2517	2515	2512	2510	2507	**40**
41	2505	2502	2500	2497	2495	2493	2490	2488	2485	2483	**41**
42	2480	2478	2475	2473	2470	2468	2465	2463	2460	2458	**42**
43	2456	2453	2451	2448	2446	2443	2441	2438	2436	2433	**43**
44	2431	2429	2426	2424	2421	2419	2416	2414	2412	2409	**44**
45	2407	2404	2402	2399	2397	2395	2392	2390	2387	2385	**45**
46	2383	2380	2378	2375	2373	2371	2368	2366	2363	2361	**46**
47	2359	2356	2354	2351	2349	2347	2344	2342	2339	2337	**47**
48	2335	2332	2330	2328	2325	2323	2320	2318	2316	2313	**48**
49	2311	2309	2306	2304	2301	2299	2297	2294	2292	2290	**49**
50	2287	2285	2283	2280	2278	2276	2273	2271	2269	2266	**50**
51	2264	2262	2259	2257	2255	2252	2250	2248	2245	2243	**51**
52	2241	2238	2236	2234	2231	2229	2227	2224	2222	2220	**52**
53	2217	2215	2213	2210	2208	2206	2204	2201	2199	2197	**53**
54	2194	2192	2190	2187	2185	2183	2181	2178	2176	2174	**54**
55	2171	2169	2167	2165	2162	2160	2158	2155	2153	2151	**55**
56	2149	2146	2144	2142	2140	2137	2135	2133	2130	2128	**56**
57	2126	2124	2121	2119	2117	2115	2112	2110	2108	2106	**57**
58	2103	2101	2099	2097	2094	2092	2090	2088	2085	2083	**58**
59	2081	2079	2077	2074	1672	2070	2068	2065	2063	2061	**59**

Log 4 min. = 909 Log 1° = 6453 Log 5° = 16

3m	PROPORTIONAL LOGARITHMS FOR FINDING HOUSE CUSPS										3°
	.0	.1	.2	.3	.4	.5	.6	.7	.8	.9	
0	2059	2056	2054	2052	2050	2048	2045	2043	2041	2039	**0**
1	2037	2034	2032	2030	2028	2025	2023	2021	2019	2017	**1**
2	2014	2012	2010	2008	2006	2004	2001	1999	1997	1995	**2**
3	1993	1990	1988	1986	1984	1982	1979	1977	1975	1973	**3**
4	1971	1969	1966	1964	1962	1960	1958	1956	1953	1951	**4**
5	1949	1947	1945	1943	1940	1938	1936	1934	1932	1930	**5**
6	1928	1925	1923	1921	1919	1917	1915	1913	1910	1908	**6**
7	1906	1904	1902	1900	1898	1895	1893	1891	1889	1887	**7**
8	1885	1883	1881	1878	1876	1874	1872	1870	1868	1866	**8**
9	1864	1861	1859	1857	1855	1853	1851	1849	1847	1845	**9**
10	1842	1840	1838	1836	1834	1832	1830	1828	1826	1824	**10**
11	1821	1819	1817	1815	1813	1811	1809	1807	1805	1803	**11**
12	1801	1798	1796	1794	1792	1790	1788	1786	1784	1782	**12**
13	1780	1778	1776	1774	1771	1769	1767	1765	1763	1761	**13**
14	1759	1757	1755	1753	1751	1749	1747	1745	1743	1741	**14**
15	1739	1736	1734	1732	1730	1728	1726	1724	1722	1720	**15**
16	1718	1716	1714	1712	1710	1708	1706	1704	1702	1700	**16**
17	1698	1696	1694	1692	1690	1688	1686	1684	1681	1679	**17**
18	1677	1675	1673	1671	1669	1667	1665	1663	1661	1659	**18**
19	1657	1655	1653	1651	1649	1647	1645	1643	1641	1639	**19**
20	1637	1635	1633	1631	1629	1627	1625	1623	1621	1619	**20**
21	1617	1615	1613	1611	1609	1607	1605	1603	1601	1599	**21**
22	1597	1595	1593	1592	1590	1588	1586	1584	1582	1580	**22**
23	1578	1576	1574	1572	1570	1568	1566	1564	1562	1560	**23**
24	1558	1556	1554	1552	1550	1548	1546	1544	1542	1540	**24**
25	1538	1537	1535	1533	1531	1529	1527	1525	1523	1521	**25**
26	1519	1517	1515	1513	1511	1509	1507	1505	1504	1502	**26**
27	1500	1498	1496	1494	1492	1490	1488	1486	1484	1482	**27**
28	1480	1478	1477	1475	1473	1471	1469	1467	1465	1463	**28**
29	1461	1459	1457	1455	1454	1452	1450	1448	1446	1444	**29**
30	1442	1440	1438	1436	1434	1433	1431	1429	1427	1425	**30**
31	1423	1421	1419	1417	1416	1414	1412	1410	1408	1406	**31**
32	1404	1402	1400	1399	1397	1395	1393	1391	1389	1387	**32**
33	1385	1383	1382	1380	1378	1376	1374	1372	1370	1368	**33**
34	1367	1365	1363	1361	1359	1357	1355	1354	1352	1350	**34**
35	1348	1346	1344	1342	1341	1339	1337	1335	1333	1331	**35**
36	1329	1328	1326	1324	1322	1320	1318	1316	1315	1313	**36**
37	1311	1309	1307	1305	1304	1302	1300	1298	1296	1294	**37**
38	1293	1291	1289	1287	1285	1283	1282	1280	1278	1276	**38**
39	1274	1272	1271	1269	1267	1265	1263	1261	1260	1258	**39**
40	1256	1254	1252	1251	1249	1247	1245	1243	1241	1240	**40**
41	1238	1236	1234	1232	1231	1229	1227	1225	1223	1222	**41**
42	1220	1218	1216	1214	1213	1211	1209	1207	1205	1204	**42**
43	1202	1200	1198	1196	1195	1193	1191	1189	1188	1186	**43**
44	1184	1182	1180	1179	1177	1175	1173	1171	1170	1168	**44**
45	1166	1164	1163	1161	1159	1157	1155	1154	1152	1150	**45**
46	1148	1147	1145	1143	1141	1140	1138	1136	1134	1132	**46**
47	1131	1129	1127	1125	1124	1122	1120	1118	1117	1115	**47**
48	1113	1111	1110	1108	1106	1104	1103	1101	1099	1097	**48**
49	1096	1094	1092	1090	1089	1087	1085	1083	1082	1080	**49**
50	1078	1076	1075	1073	1071	1070	1068	1066	1064	1063	**50**
51	1061	1059	1057	1056	1054	1052	1050	1049	1047	1045	**51**
52	1044	1042	1040	1038	1037	1035	1033	1032	1030	1028	**52**
53	1026	1025	1023	1021	1020	1018	1016	1014	1013	1011	**53**
54	1009	1008	1006	1004	1002	1001	999	997	996	994	**54**
55	992	990	989	987	985	984	982	980	979	977	**55**
56	975	973	972	970	968	967	965	963	962	960	**56**
57	958	957	955	953	952	950	948	946	945	943	**57**
58	941	940	938	936	935	933	931	930	928	926	**58**
59	925	923	921	920	918	916	915	913	911	910	**59**

SECONDS OF TIME

MINUTES OF ARC

Log 4 min. = 909 Log 1° = 6453 Log 5° = 16

HOUSE CORRECTIONS FOR EXACT SIDEREAL TIME

Excess Time	House Cusp Difference Between Tables														Excess Time
	36′	37′	38′	39′	40′	41′	42′	43′	44′	45′	46′	47′	48′	49′	
m s	° ′	° ′	° ′	° ′	° ′	° ′	° ′	° ′	° ′	° ′	° ′	° ′	° ′	° ′	m s
0 4	1	1	1	1	1	1	1	1	1	1	1	1	1	1	0 4
8	1	1	1	1	1	1	1	1	1	1	2	2	2	2	8
12	2	2	2	2	2	2	2	2	2	2	2	2	2	2	12
16	2	2	3	3	3	3	3	3	3	3	3	3	3	3	16
0 20	3	3	3	3	3	3	3	4	4	4	4	4	4	4	0 20
24	4	4	4	4	4	4	4	4	4	4	5	5	5	5	24
28	4	4	4	5	5	5	5	5	5	5	5	5	6	6	28
32	5	5	5	5	5	5	6	6	6	6	6	6	6	7	32
36	5	6	6	6	6	6	6	6	7	7	7	7	7	7	36
0 40	6	6	6	6	7	7	7	7	7	7	8	8	8	8	0 40
44	7	7	7	7	7	8	8	8	8	8	8	9	9	9	44
48	7	7	8	8	8	8	8	9	9	9	9	9	10	10	48
52	8	8	8	8	9	9	9	9	10	10	10	10	10	11	52
56	8	9	9	9	9	10	10	10	10	10	11	11	11	11	56
1 0	9	9	9	10	10	10	10	11	11	11	11	12	12	12	1 0
4	10	10	10	10	11	11	11	11	12	12	12	13	13	13	4
8	10	10	11	11	11	12	12	12	12	13	13	13	14	14	8
12	11	11	11	12	12	12	13	13	13	13	14	14	14	15	12
16	11	12	12	12	13	13	13	14	14	14	15	15	15	16	16
1 20	12	12	13	13	13	14	14	14	15	15	15	16	16	16	1 20
24	13	13	13	14	14	14	15	15	15	16	16	16	17	17	24
28	13	14	14	14	15	15	15	16	16	16	17	17	18	18	28
32	14	14	15	15	15	16	16	16	17	17	18	18	18	19	32
36	14	15	15	16	16	16	17	17	18	18	18	19	19	20	36
1 40	15	15	16	16	17	17	17	18	18	19	19	20	20	20	1 40
44	16	16	16	17	17	18	18	19	19	19	20	20	21	21	44
48	16	17	17	18	18	18	19	19	20	20	21	21	22	22	48
52	17	17	18	18	19	19	20	20	21	21	21	22	22	23	52
56	17	18	18	19	19	20	20	21	21	22	22	23	23	24	56
2 0	18	18	19	19	20	20	21	21	22	22	23	23	24	24	2 0
4	19	19	20	20	21	21	22	22	23	23	24	24	25	25	4
8	19	20	20	21	21	22	22	23	23	24	25	25	26	26	8
12	20	20	21	21	22	23	23	24	24	25	25	26	26	27	12
16	20	21	22	22	23	23	24	24	25	25	26	27	27	28	16
2 20	21	22	22	23	23	24	24	25	26	26	27	27	28	29	2 20
24	22	22	23	23	24	25	25	26	26	27	28	28	29	29	24
28	22	23	23	24	25	25	26	27	27	28	28	29	30	30	28
32	23	23	24	25	25	26	27	27	28	28	29	30	30	31	32
36	23	24	25	25	26	27	27	28	29	29	30	31	31	32	36
2 40	24	25	25	26	27	27	28	29	29	30	31	31	32	33	2 40
44	25	25	26	27	27	28	29	29	30	31	31	32	33	33	44
48	25	26	27	27	28	29	29	30	31	31	32	33	34	34	48
52	26	27	27	28	29	29	30	31	32	32	33	34	34	35	52
56	26	27	28	29	29	30	31	32	32	33	34	34	35	36	56
3 0	27	28	28	29	30	31	31	32	33	34	34	35	36	37	3 0
4	28	28	29	30	31	31	32	33	34	34	35	36	37	38	4
8	28	29	30	31	31	32	33	34	34	35	36	37	38	38	8
12	29	30	30	31	32	33	34	34	35	36	37	38	38	39	12
16	29	30	31	32	33	33	34	35	36	37	38	38	39	40	16
3 20	30	31	32	32	33	34	35	36	37	37	38	39	40	41	3 20
24	31	31	32	33	34	35	36	37	37	38	39	40	41	42	24
28	31	32	33	34	35	36	36	37	38	39	40	41	42	42	28
32	32	33	34	34	35	36	37	38	39	40	41	42	42	43	32
36	32	33	34	35	36	37	38	39	40	40	41	42	43	44	36
3 40	33	34	35	36	37	38	38	39	40	41	42	43	44	45	3 40
44	34	35	35	36	37	38	39	40	41	42	43	44	45	46	44
48	34	35	36	37	38	39	40	41	42	43	44	45	46	47	48
52	35	36	37	38	39	40	41	42	43	43	44	45	46	47	52
56	35	36	37	38	39	40	41	42	43	44	45	46	47	48	56

HOUSE CORRECTIONS FOR EXACT SIDEREAL TIME

Excess Time	House Cusp Difference Between Tables														Excess Time
	50′	51′	52′	53′	54′	55′	56′	57′	58′	59′	1° 0′	1° 1′	1° 2′	1° 3′	
m s	° ′	° ′	° ′	° ′	° ′	° ′	° ′	° ′	° ′	° ′	° ′	° ′	° ′	° ′	m s
0 4	1	1	1	1	1	1	1	1	1	1	1	1	1	1	0 4
8	2	2	2	2	2	2	2	2	2	2	2	2	2	2	8
12	2	3	3	3	3	3	3	3	3	3	3	3	3	3	12
16	3	3	3	4	4	4	4	4	4	4	4	4	4	4	16
0 20	4	4	4	4	4	5	5	5	5	5	5	5	5	5	0 20
24	5	5	5	5	5	5	6	6	6	6	6	6	6	6	24
28	6	6	6	6	6	6	7	7	7	7	7	7	7	7	28
32	7	7	7	7	7	7	7	8	8	8	8	8	8	8	32
36	7	8	8	8	8	8	8	9	9	9	9	9	9	9	36
0 40	8	8	9	9	9	9	9	9	10	10	10	10	10	10	0 40
44	9	9	10	10	10	10	10	10	11	11	11	11	11	12	44
48	10	10	10	11	11	11	11	11	12	12	12	12	12	13	48
52	11	11	11	11	12	12	12	12	13	13	13	13	13	14	52
56	12	12	12	12	13	13	13	13	14	14	14	14	14	15	56
1 0	12	13	13	13	13	14	14	14	14	15	15	15	15	16	1 0
4	13	14	14	14	14	15	15	15	15	16	16	16	17	17	4
8	14	14	15	15	15	16	16	16	16	17	17	17	18	18	8
12	15	15	16	16	16	16	17	17	17	18	18	18	19	19	12
16	16	16	16	17	17	17	18	18	18	19	19	19	20	20	16
1 20	17	17	17	18	18	18	19	19	19	20	20	20	21	21	1 20
24	17	18	18	19	19	19	20	20	20	21	21	21	22	22	24
28	18	19	19	19	20	20	21	21	21	22	22	22	23	23	28
32	19	20	20	20	21	21	21	22	22	23	23	23	24	24	32
36	20	20	21	21	22	22	22	23	23	24	24	24	25	25	36
1 40	21	21	22	22	22	23	23	24	24	25	25	25	26	26	1 40
44	22	22	23	23	23	24	24	25	25	26	26	26	27	27	44
48	22	23	23	24	24	25	25	26	26	27	27	27	28	28	48
52	23	24	24	25	25	26	26	27	27	28	28	28	29	29	52
56	24	25	25	26	26	27	27	28	28	29	29	29	30	30	56
2 0	25	25	26	26	27	27	28	28	29	29	30	30	31	31	2 0
4	26	26	27	27	28	28	29	29	30	30	31	32	32	33	4
8	27	27	28	28	29	29	30	30	31	31	32	33	33	34	8
12	27	28	29	29	30	30	31	31	32	32	33	34	34	35	12
16	28	29	29	30	31	31	32	32	33	33	34	35	35	36	16
2 20	29	30	30	31	31	32	33	33	34	34	35	36	36	37	2 20
24	30	31	31	32	32	33	34	34	35	35	36	37	37	38	24
28	31	31	32	33	33	34	35	35	36	36	37	38	38	39	28
32	32	32	33	34	34	35	35	36	37	37	38	39	39	40	32
36	32	33	34	34	35	36	36	37	38	38	39	40	40	41	36
2 40	33	34	35	35	36	37	37	38	39	39	40	41	41	42	2 40
44	34	35	36	36	37	38	38	39	40	40	41	42	42	43	44
48	35	36	36	37	38	38	39	40	41	41	42	43	43	44	48
52	36	37	37	38	39	39	40	41	42	42	43	44	44	45	52
56	37	37	38	39	40	40	41	42	43	43	44	45	45	46	56
3 0	37	38	39	40	40	41	42	43	43	44	45	46	46	47	3 0
4	38	39	40	41	41	42	43	44	44	45	46	47	48	48	4
8	39	40	41	42	42	43	44	45	45	46	47	48	49	49	8
12	40	41	42	42	43	44	45	46	46	47	48	49	50	50	12
16	41	42	42	43	44	45	46	47	47	48	49	50	51	51	16
3 20	42	42	43	44	45	46	47	47	48	49	50	51	52	52	3 20
24	42	43	44	45	46	47	48	48	49	50	51	52	53	54	24
28	43	44	45	46	47	48	49	49	50	51	52	53	54	55	28
32	44	45	46	47	48	49	49	50	51	52	53	54	55	56	32
36	45	46	47	48	49	49	50	51	52	53	54	55	56	57	36
3 40	46	47	48	49	49	50	51	52	53	54	55	56	57	58	3 40
44	47	48	49	49	50	51	52	53	54	55	56	57	58	59	44
48	47	48	49	50	51	52	53	54	55	56	57	58	59	1 0	48
52	48	49	50	51	52	53	54	55	56	57	58	59	1 0	1 1	52
56	49	50	51	52	53	54	55	56	57	58	59	1 0	1 1	1 2	56

HOUSE CORRECTIONS FOR EXACT SIDEREAL TIME

Excess Time	House Cusp Difference Between Tables														Excess Time
	1° 4′	1° 5′	1° 6′	1° 7′	1° 8′	1° 9′	1° 10′	1° 11′	1° 12′	1° 13′	1° 14′	1° 15′	1° 16′	1° 17′	
m s	° ′	° ′	° ′	° ′	° ′	° ′	° ′	° ′	° ′	° ′	° ′	° ′	° ′	° ′	m s
0 4	1	1	1	1	1	1	1	1	1	1	1	1	1	1	0 4
8	2	2	2	2	2	2	2	2	2	2	2	2	3	3	8
12	3	3	3	3	3	3	3	4	4	4	4	4	4	4	12
16	4	4	4	4	5	5	5	5	5	5	5	5	5	5	16
0 20	5	5	5	6	6	6	6	6	6	6	6	6	6	6	0 20
24	6	6	7	7	7	7	7	7	7	7	7	7	8	8	24
28	7	8	8	8	8	8	8	8	8	9	9	9	9	9	28
32	9	9	9	9	9	9	9	9	10	10	10	10	10	10	32
36	10	10	10	10	10	10	10	11	11	11	11	11	11	12	36
0 40	11	11	11	11	11	11	12	12	12	12	12	12	13	13	0 40
44	12	12	12	12	12	13	13	13	13	13	14	14	14	14	44
48	13	13	13	13	14	14	14	14	14	15	15	15	15	15	48
52	14	14	14	15	15	15	15	15	16	16	16	16	16	17	52
56	15	15	15	16	16	16	16	17	17	17	17	17	18	18	56
1 0	16	16	16	17	17	17	17	18	18	18	18	19	19	19	1 0
4	17	17	18	18	18	18	19	19	19	19	20	20	20	21	4
8	18	18	19	19	19	20	20	20	20	21	21	21	22	22	8
12	19	19	20	20	20	21	21	21	22	22	22	22	23	23	12
16	20	21	21	21	22	22	22	22	23	23	23	24	24	24	16
1 20	21	22	22	22	23	23	23	24	24	24	25	25	25	26	1 20
24	22	23	23	23	24	24	24	25	25	26	26	26	27	27	24
28	23	24	24	25	25	25	26	26	26	27	27	27	28	28	28
32	25	25	25	26	26	26	27	27	28	28	28	29	29	30	32
36	26	26	26	27	27	28	28	28	29	29	30	30	30	31	36
1 40	27	27	27	28	28	29	29	30	30	30	31	31	32	32	1 40
44	28	28	29	29	29	30	30	31	31	32	32	32	33	33	44
48	29	29	30	30	31	31	31	32	32	33	33	34	34	35	48
52	30	30	31	31	32	32	33	33	34	34	35	35	35	36	52
56	31	31	32	32	33	33	34	34	35	35	36	36	37	37	56
2 0	32	32	33	33	34	34	35	35	36	36	37	37	38	38	2 0
4	33	34	34	35	35	36	36	37	37	38	38	39	39	40	4
8	34	35	35	36	36	37	37	38	38	39	39	40	41	41	8
12	35	36	36	37	37	38	38	39	40	40	41	41	42	42	12
16	36	37	37	38	39	39	40	40	41	41	42	42	43	44	16
2 20	37	38	38	39	40	40	41	41	42	43	43	44	44	45	2 20
24	38	39	40	40	41	41	42	43	43	44	44	45	46	46	24
28	39	40	41	41	42	43	43	44	44	45	46	46	47	47	28
32	41	41	42	42	43	44	44	45	46	46	47	47	48	49	32
36	42	42	43	44	44	45	45	46	47	47	48	49	49	50	36
2 40	43	43	44	45	45	46	47	47	48	49	49	50	51	51	2 40
44	44	44	45	46	46	47	48	49	49	50	51	51	52	53	44
48	45	45	46	47	48	48	49	50	50	51	52	52	53	54	48
52	46	47	47	48	49	49	50	51	52	52	53	54	54	55	52
56	47	48	48	49	50	51	51	52	53	54	54	55	56	56	56
3 0	48	49	49	50	51	52	52	53	54	55	55	56	57	58	3 0
4	49	50	51	51	52	53	54	54	55	56	57	57	58	59	4
8	50	51	52	52	53	54	55	56	56	57	58	59	1 0	1 0	8
12	51	52	53	54	54	55	56	57	58	58	59	1 0	1 1	1 2	12
16	52	53	54	55	56	56	57	58	59	1 0	1 0	1 1	1 2	1 3	16
3 20	53	54	55	56	57	57	58	59	1 0	1 1	1 2	1 2	1 3	1 4	3 20
24	54	55	56	57	58	59	59	1 0	1 1	1 2	1 3	1 4	1 5	1 5	24
28	55	56	57	58	59	1 0	1 1	1 2	1 2	1 3	1 4	1 5	1 6	1 7	28
32	57	57	58	59	1 0	1 1	1 2	1 3	1 4	1 4	1 5	1 6	1 7	1 8	32
36	58	58	59	1 0	1 1	1 2	1 3	1 4	1 5	1 6	1 7	1 7	1 8	1 9	36
3 40	59	1 0	1 0	1 1	1 2	1 3	1 4	1 5	1 6	1 7	1 8	1 9	1 10	1 11	3 40
44	1 0	1 1	1 2	1 3	1 3	1 4	1 5	1 6	1 7	1 8	1 9	1 10	1 11	1 12	44
48	1 1	1 2	1 3	1 4	1 5	1 6	1 6	1 7	1 8	1 9	1 10	1 11	1 12	1 13	48
52	1 2	1 3	1 4	1 5	1 6	1 7	1 8	1 9	1 10	1 11	1 12	1 12	1 13	1 14	52
56	1 3	1 4	1 5	1 6	1 7	1 8	1 9	1 10	1 11	1 12	1 13	1 14	1 15	1 16	56

HOUSE CORRECTIONS FOR EXACT SIDEREAL TIME

Excess Time	House Cusp Difference Between Tables														Excess Time
	1° 18′	1° 19′	1° 20′	1° 21′	1° 22′	1° 23′	1° 24′	1° 25′	1° 26′	1° 27′	1° 28′	1° 29′	1° 30′	1° 31′	
m s	° ′	° ′	° ′	° ′	° ′	° ′	° ′	° ′	° ′	° ′	° ′	° ′	° ′	° ′	m s
0 4	1	1	1	1	1	1	1	1	1	1	1	1	1	2	0 4
8	3	3	3	3	3	3	3	3	3	3	3	3	3	3	8
12	4	4	4	4	4	4	4	4	4	4	4	4	4	5	12
16	5	5	5	5	5	6	6	6	6	6	6	6	6	6	16
0 20	6	7	7	7	7	7	7	7	7	7	7	7	7	8	0 20
24	8	8	8	8	8	8	8	8	9	9	9	9	9	9	24
28	9	9	9	9	10	10	10	10	10	10	10	10	10	11	28
32	10	11	11	11	11	11	11	11	11	12	12	12	12	12	32
36	12	12	12	12	12	12	13	13	13	13	13	13	13	14	36
0 40	13	13	13	13	14	14	14	14	14	14	15	15	15	15	0 40
44	14	14	15	15	15	15	15	16	16	16	16	16	16	17	44
48	16	16	16	16	16	17	17	17	17	17	18	18	18	18	48
52	17	17	17	18	18	18	18	18	19	19	19	19	19	20	52
56	18	18	19	19	19	19	20	20	20	20	21	21	21	21	56
1 0	19	20	20	20	20	21	21	21	21	22	22	22	22	23	1 0
4	21	21	21	22	22	22	22	23	23	23	23	24	24	24	4
8	22	22	23	23	23	24	24	24	24	25	25	25	25	26	8
12	23	24	24	24	25	25	25	25	26	26	26	27	27	27	12
16	25	25	25	26	26	26	27	27	27	28	28	28	28	29	16
1 20	26	26	27	27	27	28	28	28	29	29	29	30	30	30	1 20
24	27	28	28	28	29	29	29	30	30	30	31	31	31	32	24
28	29	29	29	30	30	30	31	31	32	32	32	33	33	33	28
32	30	30	31	31	31	32	32	33	33	33	34	34	34	35	32
36	31	32	32	32	33	33	34	34	34	35	35	36	36	36	36
1 40	32	33	33	34	34	35	35	35	36	36	37	37	37	38	1 40
44	34	34	35	35	36	36	36	37	37	38	38	39	39	39	44
48	35	36	36	36	37	37	38	38	39	39	40	40	40	41	48
52	36	37	37	38	38	39	39	40	40	41	41	42	42	42	52
56	38	38	39	39	40	40	41	41	42	42	43	43	43	44	56
2 0	39	39	40	40	41	41	42	42	43	43	44	44	45	45	2 0
4	40	41	41	42	42	43	43	44	44	45	45	46	46	47	4
8	42	42	43	43	44	44	45	45	46	46	47	47	48	49	8
12	43	43	44	45	45	46	46	47	47	48	48	49	49	50	12
16	44	45	45	46	46	47	48	48	49	49	50	50	51	52	16
2 20	45	46	47	47	48	48	49	50	50	51	51	52	52	53	2 20
24	47	47	48	49	49	50	50	51	52	52	53	53	54	55	24
28	48	49	49	50	51	51	52	52	53	54	54	55	55	56	28
32	49	50	51	51	52	53	53	54	54	55	56	56	57	58	32
36	51	51	52	53	53	54	55	55	56	57	57	58	58	59	36
2 40	52	53	53	54	55	55	56	57	57	58	59	59	1 0	1 1	2 40
44	53	54	55	55	56	57	57	58	59	59	1 0	1 1	1 1	1 2	44
48	55	55	56	57	57	58	59	59	1 0	1 1	1 2	1 2	1 3	1 4	48
52	56	57	57	58	59	59	1 0	1 1	1 2	1 2	1 3	1 4	1 4	1 5	52
56	57	58	59	59	1 0	1 1	1 2	1 2	1 3	1 4	1 5	1 5	1 6	1 7	56
3 0	58	59	1 0	1 1	1 1	1 2	1 3	1 4	1 4	1 5	1 6	1 7	1 7	1 8	3 0
4	1 0	1 1	1 1	1 2	1 3	1 4	1 4	1 5	1 6	1 7	1 7	1 8	1 9	1 10	4
8	1 1	1 2	1 3	1 3	1 4	1 5	1 6	1 7	1 7	1 8	1 9	1 10	1 10	1 11	8
12	1 2	1 3	1 4	1 5	1 6	1 6	1 7	1 8	1 9	1 10	1 10	1 11	1 12	1 13	12
16	1 4	1 5	1 5	1 6	1 7	1 8	1 9	1 9	1 10	1 11	1 12	1 13	1 13	1 14	16
3 20	1 5	1 6	1 7	1 7	1 8	1 9	1 10	1 11	1 12	1 12	1 13	1 14	1 15	1 16	3 20
24	1 6	1 7	1 8	1 9	1 10	1 11	1 11	1 12	1 13	1 14	1 15	1 16	1 16	1 17	24
28	1 8	1 8	1 9	1 10	1 11	1 12	1 13	1 14	1 15	1 15	1 16	1 17	1 18	1 19	28
32	1 9	1 10	1 11	1 12	1 12	1 13	1 14	1 15	1 16	1 17	1 18	1 19	1 19	1 20	32
36	1 10	1 11	1 12	1 13	1 14	1 15	1 16	1 16	1 17	1 18	1 19	1 20	1 21	1 22	36
3 40	1 11	1 12	1 13	1 14	1 15	1 16	1 17	1 18	1 19	1 20	1 21	1 22	1 22	1 23	3 40
44	1 13	1 14	1 15	1 16	1 17	1 17	1 18	1 19	1 20	1 21	1 22	1 23	1 24	1 25	44
48	1 14	1 15	1 16	1 17	1 18	1 19	1 20	1 21	1 22	1 23	1 24	1 25	1 25	1 26	48
52	1 15	1 16	1 17	1 18	1 19	1 20	1 21	1 22	1 23	1 24	1 25	1 26	1 27	1 28	52
56	1 17	1 18	1 19	1 20	1 21	1 22	1 23	1 24	1 25	1 26	1 27	1 28	1 28	1 29	56

HOUSE CORRECTIONS FOR EXACT SIDEREAL TIME

Excess Time	House Cusp Difference Between Tables														Excess Time
	1° 32′	1° 33′	1° 34′	1° 35′	1° 36′	1° 37′	1° 38′	1° 39′	1° 40′	1° 41′	1° 42′	1° 43′	1° 44′	1° 45′	
m s	° ′	° ′	° ′	° ′	° ′	° ′	° ′	° ′	° ′	° ′	° ′	° ′	° ′	° ′	m s
0 4	2	2	2	2	2	2	2	2	2	2	2	2	2	2	0 4
8	3	3	3	3	3	3	3	3	3	3	3	3	3	3	8
12	5	5	5	5	5	5	5	5	5	5	5	5	5	5	12
16	6	6	6	6	6	6	7	7	7	7	7	7	7	7	16
0 20	8	8	8	8	8	8	8	8	8	8	8	9	9	9	0 20
24	9	9	9	9	10	10	10	10	10	10	10	10	10	10	24
28	11	11	11	11	11	11	11	12	12	12	12	12	12	12	28
32	12	12	13	13	13	13	13	13	13	13	14	14	14	14	32
36	14	14	14	14	14	15	15	15	15	15	15	15	16	16	36
0 40	15	15	16	16	16	16	16	16	17	17	17	17	17	17	0 40
44	17	17	17	17	18	18	18	18	18	19	19	19	19	19	44
48	18	19	19	19	19	19	20	20	20	20	20	21	21	21	48
52	20	20	20	21	21	21	21	21	22	22	22	22	23	23	52
56	21	22	22	22	22	23	23	23	23	24	24	24	24	24	56
1 0	23	23	23	24	24	24	24	25	25	25	25	26	26	26	1 0
4	25	25	25	25	26	26	26	26	27	27	27	27	28	28	4
8	26	26	27	27	27	27	28	28	28	29	29	29	29	30	8
12	28	28	28	28	29	29	29	30	30	30	31	31	31	31	12
16	29	29	30	30	30	31	31	31	32	32	32	33	33	33	16
1 20	31	31	31	32	32	32	33	33	33	34	34	34	35	35	1 20
24	32	33	33	33	34	34	34	35	35	35	36	36	36	37	24
28	34	34	34	35	35	36	36	36	37	37	37	38	38	38	28
32	35	36	36	36	37	37	38	38	38	39	39	39	40	40	32
36	37	37	38	38	38	39	39	40	40	40	41	41	42	42	36
1 40	38	39	39	40	40	40	41	41	42	42	42	43	43	44	1 40
44	40	40	41	41	42	42	42	43	43	44	44	45	45	45	44
48	41	42	42	43	43	44	44	45	45	45	46	46	47	47	48
52	43	43	44	44	45	45	46	46	47	47	48	48	49	49	52
56	44	45	45	46	46	47	47	48	48	49	49	50	50	51	56
2 0	46	46	47	47	48	48	49	49	50	50	51	51	52	52	2 0
4	48	48	49	49	50	50	51	51	52	52	53	53	54	54	4
8	49	50	50	51	51	52	52	53	53	54	54	55	55	56	8
12	51	51	52	52	53	53	54	54	55	56	56	57	57	58	12
16	52	53	53	54	54	55	56	56	57	57	58	58	59	59	16
2 20	54	54	55	55	56	57	57	58	58	59	59	1 0	1 1	1 1	2 20
24	55	56	56	57	58	58	59	59	1 0	1 1	1 1	1 2	1 2	1 3	24
28	57	57	58	59	59	1 0	1 0	1 1	1 2	1 2	1 3	1 4	1 4	1 5	28
32	58	59	1 0	1 0	1 1	1 1	1 2	1 3	1 3	1 4	1 5	1 5	1 6	1 6	32
36	1 0	1 0	1 1	1 2	1 2	1 3	1 4	1 4	1 5	1 6	1 6	1 7	1 8	1 8	36
2 40	1 1	1 2	1 3	1 3	1 4	1 5	1 5	1 6	1 7	1 7	1 8	1 9	1 9	1 10	2 40
44	1 3	1 4	1 4	1 5	1 6	1 6	1 7	1 8	1 8	1 9	1 10	1 10	1 11	1 12	44
48	1 4	1 5	1 6	1 6	1 7	1 8	1 9	1 9	1 10	1 11	1 11	1 12	1 13	1 13	48
52	1 6	1 7	1 7	1 8	1 9	1 10	1 10	1 11	1 12	1 12	1 13	1 14	1 15	1 15	52
56	1 7	1 8	1 9	1 10	1 10	1 11	1 12	1 13	1 13	1 14	1 15	1 16	1 16	1 17	56
3 0	1 9	1 10	1 10	1 11	1 12	1 13	1 13	1 14	1 15	1 16	1 16	1 17	1 18	1 19	3 0
4	1 11	1 11	1 12	1 13	1 14	1 14	1 15	1 16	1 17	1 17	1 18	1 19	1 20	1 20	4
8	1 12	1 13	1 14	1 14	1 15	1 16	1 17	1 18	1 18	1 19	1 20	1 21	1 21	1 22	8
12	1 14	1 14	1 15	1 16	1 17	1 18	1 18	1 19	1 20	1 21	1 22	1 22	1 23	1 24	12
16	1 15	1 16	1 17	1 18	1 18	1 19	1 20	1 21	1 22	1 22	1 23	1 24	1 25	1 26	16
3 20	1 17	1 17	1 18	1 19	1 20	1 21	1 22	1 22	1 23	1 24	1 25	1 26	1 27	1 27	3 20
24	1 18	1 19	1 20	1 21	1 22	1 22	1 23	1 24	1 25	1 26	1 27	1 28	1 28	1 29	24
28	1 20	1 21	21	1 22	1 23	1 24	1 25	1 26	1 27	1 28	1 28	1 29	1 30	1 31	28
32	1 21	1 22	1 23	1 24	1 25	1 26	1 27	1 27	1 28	1 29	1 30	1 31	1 32	1 33	32
36	1 23	1 24	1 25	1 25	1 26	1 27	1 28	1 29	1 30	1 31	1 32	1 33	1 34	1 34	36
3 40	1 24	1 25	1 26	1 27	1 28	1 29	1 30	1 31	1 32	1 33	1 33	1 34	1 35	1 36	3 40
44	1 26	1 27	1 28	1 29	1 30	1 31	1 31	1 32	1 33	1 34	1 35	1 36	1 37	1 38	44
48	1 27	1 28	1 29	1 30	1 31	1 32	1 33	1 34	1 35	1 36	1 37	1 38	1 39	1 40	48
52	1 29	1 30	1 31	1 32	1 33	1 34	1 35	1 36	1 37	1 38	1 39	1 40	1 41	1 41	52
56	1 30	1 31	1 32	1 33	1 34	1 35	1 36	1 37	1 38	1 39	1 40	1 41	1 42	1 43	56

HOUSE CORRECTIONS FOR EXACT SIDEREAL TIME

Excess Time	House Cusp Difference Between Tables														Excess Time
	1° 46′	1° 47′	1° 48′	1° 49′	1° 50′	1° 51′	1° 52′	1° 53′	1° 54′	1° 55′	1° 56′	1° 57′	1° 58′	1° 59′	
m s	° ′	° ′	° ′	° ′	° ′	° ′	° ′	° ′	° ′	° ′	° ′	° ′	° ′	° ′	m s
0 4	2	2	2	2	2	2	2	2	2	2	2	2	2	2	0 4
8	4	4	4	4	4	4	4	4	4	4	4	4	4	4	8
12	5	5	5	5	5	6	6	6	6	6	6	6	6	6	12
16	7	7	7	7	7	7	7	8	8	8	8	8	8	8	16
0 20	9	9	9	9	9	9	9	9	9	10	10	10	10	10	0 20
24	11	11	11	11	11	11	11	11	11	11	12	12	12	12	24
28	12	12	13	13	13	13	13	13	13	13	14	14	14	14	28
32	14	14	14	15	15	15	15	15	15	15	15	16	16	16	32
36	16	16	16	16	16	17	17	17	17	17	17	18	18	18	36
0 40	18	18	18	18	18	18	19	19	19	19	19	19	20	20	0 40
44	19	20	20	20	20	20	21	21	21	21	21	21	22	22	44
48	21	21	22	22	22	22	22	23	23	23	23	23	24	24	48
52	23	23	23	24	24	24	24	24	25	25	25	25	26	26	52
56	25	25	25	25	26	26	26	26	27	27	27	27	28	28	56
1 0	26	27	27	27	27	28	28	28	28	29	29	29	29	30	1 0
4	28	29	29	29	29	30	30	30	30	31	31	31	31	32	4
8	30	30	31	31	31	31	32	32	32	33	33	33	33	34	8
12	32	32	32	33	33	33	34	34	34	34	35	35	35	36	12
16	34	34	34	35	35	35	35	36	36	36	37	37	37	38	16
1 20	35	36	36	36	37	37	37	38	38	38	39	39	39	40	1 20
24	37	37	38	38	38	39	39	40	40	40	41	41	41	42	24
28	39	39	40	40	40	41	41	41	42	42	43	43	43	44	28
32	41	41	41	42	42	43	43	43	44	44	44	45	45	46	32
36	42	43	43	44	44	44	45	45	46	46	46	47	47	48	36
1 40	44	45	45	45	46	46	47	47	47	48	48	49	49	50	1 40
44	46	46	47	47	48	48	49	49	49	50	50	51	51	52	44
48	48	48	49	49	49	50	50	51	51	52	52	53	53	54	48
52	49	50	50	51	51	52	52	53	53	54	54	55	55	56	52
56	51	52	52	53	53	54	54	55	55	56	56	57	57	58	56
2 0	53	53	54	54	55	55	56	56	57	57	58	58	59	59	2 0
4	55	55	56	56	57	57	58	58	59	59	1 0	1 0	1 1	1 1	4
8	57	57	58	58	59	59	1 0	1 0	1 1	1 1	1 2	1 2	1 3	1 3	8
12	58	59	59	1 0	1 0	1 1	1 2	1 2	1 3	1 3	1 4	1 4	1 5	1 5	12
16	1 0	1 1	1 1	1 2	1 2	1 3	1 3	1 4	1 5	1 5	1 6	1 6	1 7	1 7	16
2 20	1 2	1 2	1 3	1 4	1 4	1 5	1 5	1 6	1 6	1 7	1 8	1 8	1 9	1 9	2 20
24	1 4	1 4	1 5	1 5	1 6	1 7	1 7	1 8	1 8	1 9	1 10	1 10	1 11	1 11	24
28	1 5	1 6	1 7	1 7	1 8	1 8	1 9	1 10	1 10	1 11	1 12	1 12	1 13	1 13	28
32	1 7	1 8	1 8	1 9	1 10	1 10	1 11	1 12	1 12	1 13	1 13	1 14	1 15	1 15	32
36	1 9	1 10	1 10	1 11	1 11	1 12	1 13	1 13	1 14	1 15	1 15	1 16	1 17	1 17	36
2 40	1 11	1 11	1 12	1 13	1 13	1 14	1 15	1 15	1 16	1 17	1 17	1 18	1 19	1 19	2 40
44	1 12	1 13	1 14	1 14	1 15	1 16	1 17	1 17	1 18	1 19	1 19	1 20	1 21	1 21	44
48	1 14	1 15	1 16	1 16	1 17	1 18	1 18	1 19	1 20	1 20	1 21	1 22	1 23	1 23	48
52	1 16	1 17	1 17	1 18	1 19	1 20	1 20	1 21	1 22	1 22	1 23	1 24	1 25	1 25	52
56	1 18	1 18	1 19	1 20	1 21	1 21	1 22	1 23	1 24	1 24	1 25	1 26	1 27	1 27	56
3 0	1 19	1 20	1 21	1 22	1 22	1 23	1 24	1 25	1 25	1 26	1 27	1 28	1 28	1 29	3 0
4	1 21	1 22	1 23	1 24	1 24	1 25	1 26	1 27	1 27	1 28	1 29	1 30	1 30	1 31	4
8	1 23	1 24	1 25	1 25	1 26	1 27	1 28	1 29	1 29	1 30	1 31	1 32	1 32	1 33	8
12	1 25	1 26	1 26	1 27	1 28	1 29	1 30	1 30	1 31	1 32	1 33	1 34	1 34	1 35	12
16	1 27	1 27	1 28	1 29	1 30	1 31	1 31	1 32	1 33	1 34	1 35	1 36	1 36	1 37	16
3 20	1 28	1 29	1 30	1 31	1 32	1 32	1 33	1 34	1 35	1 36	1 37	1 37	1 38	1 39	3 20
24	1 30	1 31	1 32	1 33	1 33	1 34	1 35	1 36	1 37	1 38	1 39	1 39	1 40	1 41	24
28	1 32	1 33	1 34	1 34	1 35	1 36	1 37	1 38	1 39	1 40	1 41	1 41	1 42	1 43	28
32	1 34	1 35	1 35	1 36	1 37	1 38	1 39	1 40	1 41	1 42	1 42	1 43	1 44	1 45	32
36	1 35	1 36	1 37	1 38	1 39	1 40	1 41	1 42	1 43	1 43	1 44	1 45	1 46	1 47	36
3 40	1 37	1 38	1 39	1 40	1 41	1 42	1 43	1 44	1 44	1 45	1 46	1 47	1 48	1 49	3 40
44	1 39	1 40	1 41	1 42	1 43	1 44	1 45	1 45	1 46	1 47	1 48	1 49	1 50	1 51	44
48	1 41	1 42	1 43	1 44	1 44	1 45	1 46	1 47	1 48	1 49	1 50	1 51	1 52	1 53	48
52	1 42	1 43	1 44	1 45	1 46	1 47	1 48	1 49	1 50	1 51	1 52	1 53	1 54	1 55	52
56	1 44	1 45	1 46	1 47	1 48	1 49	1 50	1 51	1 52	1 53	1 54	1 55	1 56	1 57	56

HOUSE CORRECTIONS FOR EXACT LATITUDE

Excess Lat.	House Cusp Difference Between Latitudes														Excess Lat.
	1′	2′	3′	4′	5′	6′	7′	8′	9′	10′	11′	12′	13′	14′	
	° ′	° ′	° ′	° ′	° ′	° ′	° ′	° ′	° ′	° ′	° ′	° ′	° ′	° ′	
1′	0	0	0	0	0	0	0	0	0	0	0	0	0	0	1′
2	0	0	0	0	0	0	0	0	0	0	0	0	0	0	2
3	0	0	0	0	0	0	0	0	0	0	1	1	1	1	3
4	0	0	0	0	0	0	0	1	1	1	1	1	1	1	4
5	0	0	0	0	0	0	1	1	1	1	1	1	1	1	5
6	0	0	0	0	0	1	1	1	1	1	1	1	1	1	6
7	0	0	0	0	1	1	1	1	1	1	1	1	2	2	7
8	0	0	0	1	1	1	1	1	1	1	1	2	2	2	8
9	0	0	0	1	1	1	1	1	1	1	2	2	2	2	9
10	0	0	0	1	1	1	1	1	1	2	2	2	2	2	10
11	0	0	1	1	1	1	1	1	2	2	2	2	2	3	11
12	0	0	1	1	1	1	1	2	2	2	2	2	3	3	12
13	0	0	1	1	1	1	2	2	2	2	2	3	3	3	13
14	0	0	1	1	1	1	2	2	2	2	3	3	3	3	14
15	0	0	1	1	1	1	2	2	2	2	3	3	3	3	15
16	0	1	1	1	1	2	2	2	2	3	3	3	3	4	16
17	0	1	1	1	1	2	2	2	3	3	3	3	4	4	17
18	0	1	1	1	1	2	2	2	3	3	3	4	4	4	18
19	0	1	1	1	2	2	2	3	3	3	3	4	4	4	19
20	0	1	1	1	2	2	2	3	3	3	4	4	4	5	20
21	0	1	1	1	2	2	2	3	3	3	4	4	5	5	21
22	0	1	1	1	2	2	3	3	3	4	4	4	5	5	22
23	0	1	1	2	2	2	3	3	3	4	4	5	5	5	23
24	0	1	1	2	2	2	3	3	4	4	4	5	5	6	24
25	0	1	1	2	2	2	3	3	4	4	5	5	5	6	25
26	0	1	1	2	2	3	3	3	4	4	5	5	6	6	26
27	0	1	1	2	2	3	3	4	4	4	5	5	6	6	27
28	0	1	1	2	2	3	3	4	4	5	5	6	6	7	28
29	0	1	1	2	2	3	3	4	4	5	5	6	6	7	29
30	0	1	1	2	2	3	3	4	4	5	5	6	6	7	30
31	1	1	2	2	3	3	4	4	5	5	6	6	7	7	31
32	1	1	2	2	3	3	4	4	5	5	6	6	7	7	32
33	1	1	2	2	3	3	4	4	5	5	6	7	7	8	33
34	1	1	2	2	3	3	4	5	5	6	6	7	7	8	34
35	1	1	2	2	3	3	4	5	5	6	6	7	8	8	35
36	1	1	2	2	3	4	4	5	5	6	7	7	8	8	36
37	1	1	2	2	3	4	4	5	6	6	7	7	8	9	37
38	1	1	2	3	3	4	4	5	6	6	7	8	8	9	38
39	1	1	2	3	3	4	5	5	6	6	7	8	8	9	39
40	1	1	2	3	3	4	5	5	6	7	7	8	9	9	40
41	1	1	2	3	3	4	5	5	6	7	8	8	9	10	41
42	1	1	2	3	3	4	5	6	6	7	8	8	9	10	42
43	1	1	2	3	4	4	5	6	6	7	8	9	9	10	43
44	1	1	2	3	4	4	5	6	7	7	8	9	10	10	44
45	1	1	2	3	4	4	5	6	7	7	8	9	10	10	45
46	1	2	2	3	4	5	5	6	7	8	8	9	10	11	46
47	1	2	2	3	4	5	5	6	7	8	9	9	10	11	47
48	1	2	2	3	4	5	6	6	7	8	9	10	10	11	48
49	1	2	2	3	4	5	6	7	7	8	9	10	11	11	49
50	1	2	2	3	4	5	6	7	7	8	9	10	11	12	50
51	1	2	3	3	4	5	6	7	8	8	9	10	11	12	51
52	1	2	3	3	4	5	6	7	8	9	10	10	11	12	52
53	1	2	3	4	4	5	6	7	8	9	10	11	11	12	53
54	1	2	3	4	4	5	6	7	8	9	10	11	12	13	54
55	1	2	3	4	5	5	6	7	8	9	10	11	12	13	55
56	1	2	3	4	5	6	7	7	8	9	10	11	12	13	56
57	1	2	3	4	5	6	7	8	9	9	10	11	12	13	57
58	1	2	3	4	5	6	7	8	9	10	11	12	13	14	58
59	1	2	3	4	5	6	7	8	9	10	11	12	13	14	59

HOUSE CORRECTIONS FOR EXACT LATITUDE

Excess Lat.	House Cusp Difference Between Latitudes														Excess Lat.
	15′	16′	17′	18′	19′	20′	21′	22′	23′	24′	25′	26′	27′	28′	
	° ′	° ′	° ′	° ′	° ′	° ′	° ′	° ′	° ′	° ′	° ′	° ′	° ′	° ′	
1′	0	0	0	0	0	0	0	0	0	0	0	0	0	0	1′
2	0	1	1	1	1	1	1	1	1	1	1	1	1	1	2
3	1	1	1	1	1	1	1	1	1	1	1	1	1	1	3
4	1	1	1	1	1	1	1	1	2	2	2	2	2	2	4
5	1	1	1	1	2	2	2	2	2	2	2	2	2	2	5
6	1	2	2	2	2	2	2	2	2	2	2	3	3	3	6
7	2	2	2	2	2	2	2	3	3	3	3	3	3	3	7
8	2	2	2	2	3	3	3	3	3	3	3	3	4	4	8
9	2	2	3	3	3	3	3	3	3	4	4	4	4	4	9
10	2	3	3	3	3	3	3	4	4	4	4	4	4	5	10
11	3	3	3	3	3	4	4	4	4	4	5	5	5	5	11
12	3	3	3	4	4	4	4	4	5	5	5	5	5	6	12
13	3	3	4	4	4	4	5	5	5	5	5	6	6	6	13
14	3	4	4	4	4	5	5	5	5	6	6	6	6	7	14
15	4	4	4	4	5	5	5	5	6	6	6	6	7	7	15
16	4	4	5	5	5	5	6	6	6	6	7	7	7	7	16
17	4	5	5	5	5	6	6	6	7	7	7	7	8	8	17
18	4	5	5	5	6	6	6	7	7	7	7	8	8	8	18
19	5	5	5	6	6	6	7	7	7	8	8	8	9	9	19
20	5	5	6	6	6	7	7	7	8	8	8	9	9	9	20
21	5	6	6	6	7	7	7	8	8	8	9	9	9	10	21
22	5	6	6	7	7	7	8	8	8	9	9	10	10	10	22
23	6	6	7	7	7	8	8	8	9	9	10	10	10	11	23
24	6	6	7	7	8	8	8	9	9	10	10	10	11	11	24
25	6	7	7	7	8	8	9	9	10	10	10	11	11	12	25
26	6	7	7	8	8	9	9	10	10	10	11	11	12	12	26
27	7	7	8	8	9	9	9	10	10	11	11	12	12	13	27
28	7	7	8	8	9	9	10	10	11	11	12	12	13	13	28
29	7	8	8	9	9	10	10	11	11	12	12	13	13	14	29
30	7	8	8	9	9	10	10	11	11	12	12	13	13	14	30
31	8	8	9	9	10	10	11	11	12	12	13	13	14	14	31
32	8	9	9	10	10	11	11	12	12	13	13	14	14	15	32
33	8	9	9	10	10	11	12	12	13	13	14	14	15	15	33
34	8	9	10	10	11	11	12	12	13	14	14	15	15	16	34
35	9	9	10	10	11	12	12	13	13	14	15	15	16	16	35
36	9	10	10	11	11	12	13	13	14	14	15	16	16	17	36
37	9	10	10	11	12	12	13	14	14	15	15	16	17	17	37
38	9	10	11	11	12	13	13	14	15	15	16	16	17	18	38
39	10	10	11	12	12	13	14	14	15	16	16	17	18	18	39
40	10	11	11	12	13	13	14	15	15	16	17	17	18	19	40
41	10	11	12	12	13	14	14	15	16	16	17	18	18	19	41
42	10	11	12	13	13	14	15	15	16	17	17	18	19	20	42
43	11	11	12	13	14	14	15	16	16	17	18	19	19	20	43
44	11	12	12	13	14	15	15	16	17	18	18	19	20	21	44
45	11	12	13	13	14	15	16	16	17	18	19	19	20	21	45
46	11	12	13	14	15	15	16	17	18	18	19	20	21	21	46
47	12	13	13	14	15	16	16	17	18	19	20	20	21	22	47
48	12	13	14	14	15	16	17	18	18	19	20	21	22	22	48
49	12	13	14	15	16	16	17	18	19	20	20	21	22	23	49
50	12	13	14	15	16	17	17	18	19	20	21	22	22	23	50
51	13	14	14	15	16	17	18	19	20	20	21	22	23	24	51
52	13	14	15	16	16	17	18	19	20	21	22	23	23	24	52
53	13	14	15	16	17	18	19	19	20	21	22	23	24	25	53
54	13	14	15	16	17	18	19	20	21	22	22	23	24	25	54
55	14	15	16	16	17	18	19	20	21	22	23	24	25	26	55
56	14	15	16	17	18	19	20	21	21	22	23	24	25	26	56
57	14	15	16	17	18	19	20	21	22	23	24	25	26	27	57
58	14	15	16	17	18	19	20	21	22	23	24	25	26	27	58
59	15	16	17	18	19	20	21	22	23	24	25	26	27	28	59

HOUSE CORRECTIONS FOR EXACT LATITUDE

Excess Lat.	House Cusp Difference Between Latitudes														Excess Lat.
	29′	30′	31′	32′	33′	34′	35′	36′	37′	38′	39′	40′	41′	42′	
	° ′	° ′	° ′	° ′	° ′	° ′	° ′	° ′	° ′	° ′	° ′	° ′	° ′	° ′	
1′	0	0	1	1	1	1	1	1	1	1	1	1	1	1	1′
2	1	1	1	1	1	1	1	1	1	1	1	1	1	1	2
3	1	1	2	2	2	2	2	2	2	2	2	2	2	2	3
4	2	2	2	2	2	2	2	2	2	3	3	3	3	3	4
5	2	2	3	3	3	3	3	3	3	3	3	3	3	3	5
6	3	3	3	3	3	3	3	4	4	4	4	4	4	4	6
7	3	3	4	4	4	4	4	4	4	4	5	5	5	5	7
8	4	4	4	4	4	5	5	5	5	5	5	5	5	6	8
9	4	4	5	5	5	5	5	5	6	6	6	6	6	6	9
10	5	5	5	5	5	6	6	6	6	6	6	7	7	7	10
11	5	5	6	6	6	6	6	7	7	7	7	7	8	8	11
12	6	6	6	6	7	7	7	7	7	8	8	8	8	8	12
13	6	6	7	7	7	7	8	8	8	8	8	9	9	9	13
14	7	7	7	7	8	8	8	8	9	9	9	9	10	10	14
15	7	7	8	8	8	8	9	9	9	9	10	10	10	10	15
16	8	8	8	9	9	9	9	10	10	10	10	11	11	11	16
17	8	8	9	9	9	10	10	10	10	11	11	11	12	12	17
18	9	9	9	10	10	10	10	11	11	11	12	12	12	13	18
19	9	9	10	10	10	11	11	11	12	12	12	13	13	13	19
20	10	10	10	11	11	11	12	12	12	13	13	13	14	14	20
21	10	10	11	11	12	12	12	13	13	13	14	14	14	15	21
22	11	11	11	12	12	12	13	13	14	14	14	15	15	15	22
23	11	11	12	12	13	13	13	14	14	15	15	15	16	16	23
24	12	12	12	13	13	14	14	14	15	15	16	16	16	17	24
25	12	12	13	13	14	14	15	15	15	16	16	17	17	17	25
26	13	13	13	14	14	15	15	16	16	16	17	17	18	18	26
27	13	13	14	14	15	15	16	16	17	17	18	18	18	19	27
28	14	14	14	15	15	16	16	17	17	18	18	19	19	20	28
29	14	14	15	15	16	16	17	17	18	18	19	19	20	20	29
30	14	15	15	16	16	17	17	18	18	19	19	20	20	21	30
31	15	15	16	17	17	18	18	19	19	20	20	21	21	22	31
32	15	16	17	17	18	18	19	19	20	20	21	21	22	22	32
33	16	16	17	18	18	19	19	20	20	21	21	22	23	23	33
34	16	17	18	18	19	19	20	20	21	22	22	23	23	24	34
35	17	17	18	19	19	20	20	21	22	22	23	23	24	24	35
36	17	18	19	19	20	20	21	22	22	23	23	24	25	25	36
37	18	18	19	20	20	21	22	22	23	23	24	25	25	26	37
38	18	19	20	20	21	22	22	23	23	24	25	25	26	27	38
39	19	19	20	21	21	22	23	23	24	25	25	26	27	27	39
40	19	20	21	21	22	23	23	24	25	25	26	27	27	28	40
41	20	20	21	22	23	23	24	25	25	26	27	27	28	29	41
42	20	21	22	22	23	24	24	25	26	27	27	28	29	29	42
43	21	21	22	23	24	24	25	26	27	27	28	29	29	30	43
44	21	22	23	23	24	25	26	26	27	28	29	29	30	31	44
45	22	22	23	24	25	25	26	27	28	28	29	30	31	31	45
46	22	23	24	25	25	26	27	28	28	29	30	31	31	32	46
47	23	23	24	25	26	27	27	28	29	30	31	31	32	33	47
48	23	24	25	26	26	27	28	29	30	30	31	32	33	34	48
49	24	24	25	26	27	28	29	29	30	31	32	33	33	34	49
50	24	25	26	27	27	28	29	30	31	32	32	33	34	35	50
51	25	25	26	27	28	29	30	31	31	32	33	34	35	36	51
52	25	26	27	28	29	29	30	31	32	33	34	35	36	36	52
53	26	26	27	28	29	30	31	32	33	34	34	35	36	37	53
54	26	27	28	29	30	31	31	32	33	34	35	36	37	38	54
55	27	27	28	29	30	31	32	33	34	35	36	37	38	38	55
56	27	28	29	30	31	32	33	34	35	35	36	37	38	39	56
57	28	28	29	30	31	32	33	34	35	36	37	38	39	40	57
58	28	29	30	31	32	33	34	35	36	37	38	39	40	41	58
59	29	29	30	31	32	33	34	35	36	37	38	39	40	41	59

HOUSE CORRECTIONS FOR EXACT LATITUDE

Excess Lat.	House Cusp Difference Between Latitudes														Excess Lat.
	43′	44′	45′	46′	47′	48′	49′	50′	51′	52′	53′	54′	55′	56′	
	° ′	° ′	° ′	° ′	° ′	° ′	° ′	° ′	° ′	° ′	° ′	° ′	° ′	° ′	
1′	1	1	1	1	1	1	1	1	1	1	1	1	1	1	1′
2	1	1	1	2	2	2	2	2	2	2	2	2	2	2	2
3	2	2	2	2	2	2	2	2	3	3	3	3	3	3	3
4	3	3	3	3	3	3	3	3	3	3	4	4	4	4	4
5	4	4	4	4	4	4	4	4	4	4	4	4	5	5	5
6	4	4	4	5	5	5	5	5	5	5	5	5	5	6	6
7	5	5	5	5	5	6	6	6	6	6	6	6	6	7	7
8	6	6	6	6	6	6	7	7	7	7	7	7	7	7	8
9	6	7	7	7	7	7	7	7	8	8	8	8	8	8	9
10	7	7	7	8	8	8	8	8	8	9	9	9	9	9	10
11	8	8	8	8	9	9	9	9	9	10	10	10	10	10	11
12	9	9	9	9	9	10	10	10	10	10	11	11	11	11	12
13	9	10	10	10	10	10	11	11	11	11	11	12	12	12	13
14	10	10	10	11	11	11	11	12	12	12	12	13	13	13	14
15	11	11	11	11	12	12	12	12	13	13	13	13	14	14	15
16	11	12	12	12	13	13	13	13	14	14	14	14	15	15	16
17	12	12	13	13	13	14	14	14	14	15	15	15	16	16	17
18	13	13	13	14	14	14	15	15	15	16	16	16	16	17	18
19	14	14	14	15	15	15	16	16	16	16	17	17	17	18	19
20	14	15	15	15	16	16	16	17	17	17	18	18	18	19	20
21	15	15	16	16	16	17	17	17	18	18	19	19	19	20	21
22	16	16	16	17	17	18	18	18	19	19	19	20	20	21	22
23	16	17	17	18	18	18	19	19	20	20	20	21	21	21	23
24	17	18	18	18	19	19	20	20	20	21	21	22	22	22	24
25	18	18	19	19	20	20	20	21	21	22	22	22	23	23	25
26	19	19	19	20	20	21	21	22	22	23	23	23	24	24	26
27	19	20	20	21	21	22	22	22	23	23	24	24	25	25	27
28	20	21	21	21	22	22	23	23	24	24	25	25	26	26	28
29	21	21	22	22	23	23	24	24	25	25	26	26	27	27	29
30	21	22	22	23	23	24	24	25	25	26	26	27	27	28	30
31	22	23	23	24	24	25	25	26	26	27	27	28	28	29	31
32	23	23	24	25	25	26	26	27	27	28	28	29	29	30	32
33	24	24	25	25	26	26	27	27	28	29	29	30	30	31	33
34	24	25	25	26	27	27	28	28	29	29	30	31	31	32	34
35	25	26	26	27	27	28	29	29	30	30	31	31	32	33	35
36	26	26	27	28	28	29	29	30	31	31	32	32	33	34	36
37	27	27	28	28	29	30	30	31	31	32	33	33	34	35	37
38	27	28	28	29	30	30	31	32	32	33	34	34	35	35	38
39	28	29	29	30	31	31	32	32	33	34	34	35	36	36	39
40	29	29	30	31	31	32	33	33	34	35	35	36	37	37	40
41	29	30	31	31	32	33	33	34	35	36	36	37	38	38	41
42	30	31	31	32	33	34	34	35	36	36	37	38	38	39	42
43	31	32	32	33	34	34	35	36	37	37	38	39	39	40	43
44	32	32	33	34	34	35	36	37	37	38	39	40	40	41	44
45	32	33	34	34	35	36	37	37	38	39	40	40	41	42	45
46	33	34	34	35	36	37	38	38	39	40	41	41	42	43	46
47	34	34	35	36	37	38	38	39	40	41	42	42	43	44	47
48	34	35	36	37	38	38	39	40	41	42	42	43	44	45	48
49	35	36	37	38	38	39	40	41	42	42	43	44	45	46	49
50	36	37	37	38	39	40	41	42	42	43	44	45	46	47	50
51	37	37	38	39	40	41	42	42	43	44	45	46	47	48	51
52	37	38	39	40	41	42	42	43	44	45	46	47	48	49	52
53	38	39	40	41	42	42	43	44	45	46	47	48	49	49	53
54	39	40	40	41	42	43	44	45	46	47	48	49	49	50	54
55	39	40	41	42	43	44	45	46	47	48	49	49	50	51	55
56	40	41	42	43	44	45	46	47	48	49	49	50	51	52	56
57	41	42	43	44	45	46	47	47	48	49	50	51	52	53	57
58	42	43	43	44	45	46	47	48	49	50	51	52	53	54	58
59	42	43	44	45	46	47	48	49	50	51	52	53	54	55	59

HOUSE CORRECTIONS FOR EXACT LATITUDE

Excess Lat.	House Cusp Difference Between Latitudes														Excess Lat.
	57′	58′	59′	1° 0′	1° 1′	1° 2′	1° 3′	1° 4′	1° 5′	1° 6′	1° 7′	1° 8′	1° 9′	1° 10′	
	° ′	° ′	° ′	° ′	° ′	° ′	° ′	° ′	° ′	° ′	° ′	° ′	° ′	° ′	
1′	1	1	1	1	1	1	1	1	1	1	1	1	1	1	1′
2	2	2	2	2	2	2	2	2	2	2	2	2	2	2	2
3	3	3	3	3	3	3	3	3	3	3	3	3	3	3	3
4	4	4	4	4	4	4	4	4	4	4	4	5	5	5	4
5	5	5	5	5	5	5	5	5	5	5	6	6	6	6	5
6	6	6	6	6	6	6	6	6	6	7	7	7	7	7	6
7	7	7	7	7	7	7	7	7	8	8	8	8	8	8	7
8	8	8	8	8	8	8	8	9	9	9	9	9	9	9	8
9	9	9	9	9	9	9	9	10	10	10	10	10	10	10	9
10	9	10	10	10	10	10	10	11	11	11	11	11	11	12	10
11	10	11	11	11	11	11	12	12	12	12	12	12	13	13	11
12	11	12	12	12	12	12	13	13	13	13	13	14	14	14	12
13	12	13	13	13	13	13	14	14	14	14	15	15	15	15	13
14	13	14	14	14	14	14	15	15	15	15	16	16	16	16	14
15	14	14	15	15	15	15	16	16	16	16	17	17	17	17	15
16	15	15	16	16	16	17	17	17	17	18	18	18	18	19	16
17	16	16	17	17	17	18	18	18	18	19	19	19	20	20	17
18	17	17	18	18	18	19	19	19	19	20	20	20	21	21	18
19	18	18	19	19	19	20	20	20	21	21	21	22	22	22	19
20	19	19	20	20	20	21	21	21	22	22	22	23	23	23	20
21	20	20	21	21	21	22	22	22	23	23	23	24	24	24	21
22	21	21	22	22	22	23	23	23	24	24	25	25	25	26	22
23	22	22	23	23	23	24	24	25	25	25	26	26	26	27	23
24	23	23	24	24	24	25	25	26	26	26	27	27	28	28	24
25	24	24	25	25	25	26	26	27	27	27	28	28	29	29	25
26	25	25	26	26	26	27	27	28	28	29	29	29	30	30	26
27	26	26	27	27	27	28	28	29	29	30	30	31	31	31	27
28	27	27	28	28	28	29	29	30	30	31	31	32	32	33	28
29	28	28	29	29	29	30	30	31	31	32	32	33	33	34	29
30	28	29	29	30	30	31	31	32	32	33	33	34	34	35	30
31	29	30	30	31	32	32	33	33	34	34	35	35	36	36	31
32	30	31	31	32	33	33	34	34	35	35	36	36	37	37	32
33	31	32	32	33	34	34	35	35	36	36	37	37	38	38	33
34	32	33	33	34	35	35	36	36	37	37	38	39	39	40	34
35	33	34	34	35	36	36	37	37	38	38	39	40	40	41	35
36	34	35	35	36	37	37	38	38	39	40	40	41	41	42	36
37	35	36	36	37	38	38	39	39	40	41	41	42	43	43	37
38	36	37	37	38	39	39	40	41	41	42	42	43	44	44	38
39	37	38	38	39	40	40	41	42	42	43	44	44	45	45	39
40	38	39	39	40	41	41	42	43	43	44	45	45	46	47	40
41	39	40	40	41	42	42	43	44	44	45	46	46	47	48	41
42	40	41	41	42	43	43	44	45	45	46	47	48	48	49	42
43	41	42	42	43	44	44	45	46	47	47	48	49	49	50	43
44	42	43	43	44	45	45	46	47	48	48	49	50	51	51	44
45	43	43	44	45	46	46	47	48	49	49	50	51	52	52	45
46	44	44	45	46	47	48	48	49	50	51	51	52	53	54	46
47	45	45	46	47	48	49	49	50	51	52	52	53	54	55	47
48	46	46	47	48	49	50	50	51	52	53	54	54	55	56	48
49	47	47	48	49	50	51	51	52	53	54	55	56	56	57	49
50	47	48	49	50	51	52	52	53	54	55	56	57	57	58	50
51	48	49	50	51	52	53	54	54	55	56	57	58	59	59	51
52	49	50	51	52	53	54	55	55	56	57	58	59	1 0	1 1	52
53	50	51	52	53	54	55	56	57	57	58	59	1 0	1 1	1 2	53
54	51	52	53	54	55	56	57	58	58	59	1 0	1 1	1 2	1 3	54
55	52	53	54	55	56	57	58	59	1 0	1 0	1 1	1 2	1 3	1 4	55
56	53	54	55	56	57	58	59	1 0	1 1	1 2	1 3	1 3	1 4	1 5	56
57	54	55	56	57	58	59	1 0	1 1	1 2	1 3	1 4	1 5	1 6	1 6	57
58	55	56	57	58	59	1 0	1 1	1 2	1 3	1 4	1 5	1 6	1 7	1 8	58
59	56	57	58	59	1 0	1 1	1 2	1 3	1 4	1 5	1 6	1 7	1 8	1 9	59

0h 0m 0s — MC — 0° 0′ 0″

♈ 0° 0′ 0″

N LAT °	11 ° ′	12 ° ′	Ascendant ° ′	2 ° ′	3 ° ′
0	♉ 2 10.9	♊ 2 5.4	♋ 0 0.0	♋ 27 54.6	♌ 27 49.1
5	2 33.8	3 18.1	1 59.6	29 5.7	28 11.5
10	2 57.5	4 33.7	4 0.8	♌ 0 16.4	28 33.8
15	3 22.5	5 53.3	6 5.1	1 27.6	28 56.3
16	3 27.7	6 9.9	6 30.5	1 42.0	29 0.8
17	3 33.0	6 26.7	6 56.1	1 56.4	29 5.4
18	3 38.4	6 43.8	7 21.9	2 11.0	29 10.0
19	3 43.9	7 1.2	7 48.0	2 25.7	29 14.7
20	3 49.4	7 18.9	8 14.3	2 40.4	29 19.4
21	3 55.1	7 36.8	8 41.0	2 55.3	29 24.1
22	4 0.8	7 55.2	9 7.9	3 10.3	29 28.8
23	4 6.7	8 13.8	9 35.1	3 25.4	29 33.7
24	4 12.7	8 32.9	10 2.7	3 40.7	29 38.5
25	4 18.8	8 52.3	10 30.6	3 56.1	29 43.4
26	4 25.1	9 12.2	10 58.9	4 11.7	29 48.4
27	4 31.5	9 32.6	11 27.5	4 27.5	29 53.4
28	4 38.0	9 53.4	11 56.6	4 43.4	29 58.5
29	4 44.8	10 14.7	12 26.2	4 59.6	♍ 0 3.7
30	4 51.7	10 36.6	12 56.1	5 16.0	0 9.0
31	4 58.7	10 59.0	13 26.6	5 32.6	0 14.3
32	5 6.0	11 22.1	13 57.6	5 49.5	0 19.7
33	5 13.5	11 45.8	14 29.1	6 6.6	0 25.2
34	5 21.3	12 10.3	15 1.3	6 24.0	0 30.8
35	5 29.2	12 35.4	15 34.0	6 41.7	0 36.5
36	5 37.5	13 1.4	16 7.3	6 59.8	0 42.3
37	5 46.0	13 28.3	16 41.3	7 18.1	0 48.3
38	5 54.8	13 56.0	17 16.0	7 36.9	0 54.3
39	6 3.9	14 24.7	17 51.4	7 55.9	1 0.5
40	6 13.4	14 54.5	18 27.6	8 15.4	1 6.8
41	6 23.3	15 25.4	19 4.6	8 35.4	1 13.3
42	6 33.5	15 57.6	19 42.5	8 55.7	1 19.9
43	6 44.2	16 31.0	20 21.2	9 16.5	1 26.7
44	6 55.4	17 5.8	21 0.9	9 37.9	1 33.7
45	7 7.1	17 42.2	21 41.6	9 59.8	1 40.9
46	7 19.3	18 20.2	22 23.4	10 22.2	1 48.3
47	7 32.1	19 0.1	23 6.2	10 45.3	1 55.9
48	7 45.7	19 41.9	23 50.2	11 8.9	2 3.7
49	7 59.9	20 25.8	24 35.5	11 33.3	2 11.8
50	8 15.0	21 12.1	25 22.0	11 58.4	2 20.1
51	8 30.9	22 0.9	26 9.8	12 24.3	2 28.7
52	8 47.8	22 52.5	26 59.1	12 50.9	2 37.7
53	9 5.8	23 47.2	27 49.9	13 18.5	2 46.9
54	9 25.1	24 45.4	28 42.2	13 47.0	2 56.5
55	9 45.7	25 47.4	29 36.2	14 16.5	3 6.5
56	10 7.9	26 53.6	♌ 0 31.9	14 47.0	3 16.9
57	10 31.9	28 4.6	1 29.5	15 18.7	3 27.7
58	10 57.9	29 21.0	2 29.0	15 51.6	3 39.0
59	11 26.2	♋ 0 43.5	3 30.5	16 25.8	3 50.8
60	11 57.3	2 13.0	4 34.2	17 1.4	4 3.2
61	12 31.7	3 50.4	5 40.0	17 38.5	4 16.1
62	13 9.9	5 37.1	6 48.2	18 17.1	4 29.8
63	13 52.9	7 34.7	7 58.9	18 57.6	4 44.1
64	14 41.7	9 44.9	9 12.2	19 39.8	4 59.3
65	15 38.2	12 10.5	10 28.1	20 24.1	5 15.3
66	16 44.8	14 54.8	11 46.9	21 10.6	5 32.2
S LAT	5	6	Descendant	8	9

♎ 0° 0′ 0″

12h 0m 0s — MC — 180° 0′ 0″

0h 4m 0s — MC — 1° 0′ 0″

♈ 1° 5′ 24″

N LAT °	11 ° ′	12 ° ′	Ascendant ° ′	2 ° ′	3 ° ′
0	♉ 3 13.3	♊ 3 2.6	♋ 0 55.0	♋ 28 52.0	♌ 28 51.6
5	3 36.8	4 15.9	2 54.6	♌ 0 2.5	29 13.4
10	4 1.1	5 31.9	4 55.5	1 12.6	29 35.1
15	4 26.8	6 52.0	6 59.5	2 23.1	29 56.9
16	4 32.2	7 8.7	7 24.8	2 37.3	♍ 0 1.3
17	4 37.6	7 25.6	7 50.3	2 51.7	0 5.8
18	4 43.1	7 42.7	8 16.0	3 6.1	0 10.3
19	4 48.7	8 0.2	8 42.0	3 20.6	0 14.8
20	4 54.4	8 17.9	9 8.2	3 35.2	0 19.3
21	5 0.2	8 36.0	9 34.7	3 49.9	0 23.9
22	5 6.1	8 54.4	10 1.5	4 4.7	0 28.6
23	5 12.2	9 13.1	10 28.6	4 19.7	0 33.2
24	5 18.3	9 32.2	10 56.0	4 34.8	0 37.9
25	5 24.6	9 51.8	11 23.7	4 50.0	0 42.7
26	5 31.0	10 11.7	11 51.8	5 5.5	0 47.5
27	5 37.6	10 32.1	12 20.3	5 21.0	0 52.4
28	5 44.3	10 53.0	12 49.2	5 36.8	0 57.4
29	5 51.2	11 14.4	13 18.6	5 52.8	1 2.4
30	5 58.3	11 36.3	13 48.4	6 9.0	1 7.5
31	6 5.6	11 58.8	14 18.6	6 25.4	1 12.7
32	6 13.0	12 21.9	14 49.4	6 42.1	1 17.9
33	6 20.7	12 45.7	15 20.7	6 59.0	1 23.3
34	6 28.6	13 10.1	15 52.5	7 16.2	1 28.7
35	6 36.8	13 35.4	16 25.0	7 33.7	1 34.2
36	6 45.3	14 1.4	16 58.0	7 51.5	1 39.9
37	6 54.0	14 28.2	17 31.7	8 9.6	1 45.6
38	7 3.0	14 56.0	18 6.1	8 28.1	1 51.5
39	7 12.4	15 24.7	18 41.2	8 46.9	1 57.5
40	7 22.1	15 54.5	19 17.0	9 6.1	2 3.6
41	7 32.2	16 25.3	19 53.7	9 25.8	2 9.9
42	7 42.7	16 57.4	20 31.2	9 45.8	2 16.3
43	7 53.7	17 30.8	21 9.5	10 6.4	2 22.9
44	8 5.1	18 5.6	21 48.8	10 27.4	2 29.7
45	8 17.1	18 41.9	22 29.1	10 49.0	2 36.7
46	8 29.7	19 19.9	23 10.4	11 11.1	2 43.8
47	8 42.8	19 59.6	23 52.7	11 33.8	2 51.2
48	8 56.7	20 41.2	24 36.2	11 57.1	2 58.7
49	9 11.3	21 25.0	25 20.9	12 21.1	3 6.6
50	9 26.7	22 11.0	26 6.8	12 45.8	3 14.6
51	9 43.1	22 59.6	26 54.1	13 11.2	3 23.0
52	10 0.4	23 50.9	27 42.7	13 37.5	3 31.6
53	10 18.9	24 45.3	28 32.9	14 4.6	3 40.6
54	10 38.7	25 43.1	29 24.5	14 32.6	3 49.9
55	10 59.8	26 44.6	♌ 0 17.8	15 1.6	3 59.5
56	11 22.6	27 50.3	1 12.7	15 31.6	4 9.6
57	11 47.2	29 0.6	2 9.5	16 2.7	4 20.0
58	12 13.9	♋ 0 16.2	3 8.1	16 35.0	4 31.0
59	12 43.0	1 37.8	4 8.7	17 8.6	4 42.4
60	13 15.0	3 6.0	5 11.4	17 43.6	4 54.3
61	13 50.3	4 42.1	6 16.3	18 20.0	5 6.9
62	14 29.7	6 27.0	7 23.5	18 57.9	5 20.1
63	15 13.9	8 22.3	8 33.1	19 37.6	5 33.9
64	16 4.4	10 29.9	9 45.2	20 19.0	5 48.5
65	17 2.9	12 52.0	11 0.0	21 2.5	6 4.0
66	18 12.3	15 31.5	12 17.5	21 48.0	6 20.3
S LAT	5	6	Descendant	8	9

♎ 1° 5′ 24″

12h 4m 0s — MC — 181° 0′ 0″

0h 8m 0s MC 2° 0′ 0″

♈ 2° 10′ 47″

11	12	Ascendant	2	3	N LAT
° ′	° ′	° ′	° ′	° ′	°
♉ 4 15.5	♊ 3 59.8	♋ 1 50.1	♋ 29 49.5	♌ 29 54.3	0
4 39.6	5 13.5	3 49.5	♌ 0 59.5	♍ 0 15.5	5
5 4.6	6 29.9	5 50.2	2 8.9	0 36.5	10
5 30.9	7 50.5	7 53.8	3 18.7	0 57.7	15
5 36.4	8 7.2	8 19.0	3 32.8	1 2.0	16
5 42.0	8 24.2	8 44.4	3 47.0	1 6.3	17
5 47.6	8 41.4	9 10.0	4 1.2	1 10.7	18
5 53.4	8 59.0	9 35.9	4 15.6	1 15.0	19
5 59.2	9 16.8	10 2.0	4 30.0	1 19.5	20
6 5.1	9 34.9	10 28.4	4 44.6	1 23.9	21
6 11.2	9 53.4	10 55.0	4 59.2	1 28.4	22
6 17.4	10 12.2	11 21.9	5 14.0	1 32.9	23
6 23.7	10 31.3	11 49.2	5 29.0	1 37.5	24
6 30.1	10 50.9	12 16.8	5 44.0	1 42.1	25
6 36.7	11 10.9	12 44.7	5 59.3	1 46.8	26
6 43.4	11 31.4	13 13.0	6 14.7	1 51.5	27
6 50.3	11 52.3	13 41.8	6 30.3	1 56.3	28
6 57.4	12 13.7	14 10.9	6 46.1	2 1.2	29
7 4.6	12 35.7	14 40.5	7 2.0	2 6.1	30
7 12.1	12 58.3	15 10.5	7 18.3	2 11.2	31
7 19.7	13 21.4	15 41.0	7 34.7	2 16.2	32
7 27.6	13 45.2	16 12.1	7 51.4	2 21.4	33
7 35.7	14 9.7	16 43.7	8 8.4	2 26.7	34
7 44.1	14 34.9	17 15.8	8 25.6	2 32.0	35
7 52.8	15 0.9	17 48.6	8 43.2	2 37.5	36
8 1.7	15 27.8	18 22.0	9 1.1	2 43.1	37
8 11.0	15 55.5	18 56.1	9 19.3	2 48.8	38
8 20.5	16 24.3	19 30.8	9 37.9	2 54.6	39
8 30.5	16 54.0	20 6.3	9 56.8	3 0.5	40
8 40.8	17 24.9	20 42.6	10 16.2	3 6.6	41
8 51.6	17 56.9	21 19.7	10 36.0	3 12.8	42
9 2.8	18 30.3	21 57.7	10 56.2	3 19.2	43
9 14.6	19 5.0	22 36.6	11 16.9	3 25.8	44
9 26.8	19 41.2	23 16.4	11 38.2	3 32.5	45
9 39.7	20 19.0	23 57.2	11 59.9	3 39.4	46
9 53.2	20 58.6	24 39.1	12 22.3	3 46.5	47
10 7.3	21 40.1	25 22.1	12 45.2	3 53.9	48
10 22.3	22 23.7	26 6.2	13 8.9	4 1.4	49
10 38.1	23 9.5	26 51.6	13 33.2	4 9.2	50
10 54.8	23 57.8	27 38.2	13 58.2	4 17.3	51
11 12.6	24 48.9	28 26.3	14 24.0	4 25.7	52
11 31.6	25 42.9	29 15.7	14 50.7	4 34.3	53
11 51.8	26 40.2	♌ 0 6.7	15 18.2	4 43.3	54
12 13.5	27 41.2	0 59.2	15 46.7	4 52.6	55
12 36.8	28 46.3	1 53.4	16 16.2	5 2.3	56
13 2.0	29 56.0	2 49.4	16 46.8	5 12.5	57
13 29.4	♋ 1 10.8	3 47.2	17 18.5	5 23.0	58
13 59.3	2 31.3	4 46.9	17 51.5	5 34.0	59
14 32.1	3 58.5	5 48.7	18 25.8	5 45.6	60
15 8.4	5 33.1	6 52.6	19 1.5	5 57.7	61
15 48.8	7 16.3	7 58.8	19 38.7	6 10.4	62
16 34.5	9 9.5	9 7.3	20 17.6	6 23.8	63
17 26.6	11 14.4	10 18.3	20 58.3	6 37.9	64
18 27.1	13 33.0	11 31.9	21 40.8	6 52.8	65
19 39.3	16 8.2	12 48.2	22 25.5	7 8.5	66
5	6	Descendant	8	9	S LAT

♎ 2° 10′ 47″

12h 8m 0s MC 182° 0′ 0″

0h 12m 0s MC 3° 0′ 0″

♈ 3° 16′ 10″

N LAT	11	12	Ascendant	2	3
°	° ′	° ′	° ′	° ′	° ′
0	♉ 5 17.5	♊ 4 56.7	♋ 2 45.2	♌ 0 47.2	♍ 0 57.2
5	5 42.2	6 10.9	4 44.5	1 56.6	1 17.7
10	6 7.8	7 27.8	6 44.9	3 5.3	1 38.1
15	6 34.8	8 48.7	8 48.1	4 14.4	1 58.6
16	6 40.4	9 5.6	9 13.2	4 28.3	2 2.8
17	6 46.1	9 22.6	9 38.5	4 42.4	2 7.0
18	6 51.9	9 39.9	10 4.0	4 56.5	2 11.2
19	6 57.8	9 57.5	10 29.7	5 10.7	2 15.4
20	7 3.8	10 15.4	10 55.7	5 24.9	2 19.7
21	7 9.8	10 33.6	11 21.9	5 39.3	2 24.0
22	7 16.0	10 52.1	11 48.4	5 53.8	2 28.3
23	7 22.4	11 11.0	12 15.2	6 8.4	2 32.7
24	7 28.8	11 30.2	12 42.3	6 23.2	2 37.2
25	7 35.4	11 49.8	13 9.7	6 38.1	2 41.6
26	7 42.1	12 9.9	13 37.5	6 53.1	2 46.2
27	7 49.0	12 30.4	14 5.6	7 8.4	2 50.8
28	7 56.1	12 51.3	14 34.1	7 23.8	2 55.4
29	8 3.3	13 12.8	15 3.1	7 39.3	3 0.1
30	8 10.7	13 34.8	15 32.4	7 55.1	3 4.9
31	8 18.4	13 57.4	16 2.2	8 11.1	3 9.7
32	8 26.2	14 20.6	16 32.5	8 27.4	3 14.7
33	8 34.2	14 44.4	17 3.3	8 43.9	3 19.7
34	8 42.6	15 8.9	17 34.7	9 0.6	3 24.8
35	8 51.1	15 34.1	18 6.5	9 17.6	3 30.0
36	9 0.0	16 0.2	18 39.0	9 35.0	3 35.2
37	9 9.1	16 27.0	19 12.1	9 52.6	3 40.6
38	9 18.6	16 54.8	19 45.9	10 10.5	3 46.1
39	9 28.4	17 23.5	20 20.3	10 28.9	3 51.8
40	9 38.6	17 53.2	20 55.5	10 47.5	3 57.5
41	9 49.1	18 24.0	21 31.4	11 6.6	4 3.4
42	10 0.2	18 56.0	22 8.1	11 26.1	4 9.4
43	10 11.6	19 29.3	22 45.7	11 46.0	4 15.6
44	10 23.6	20 3.9	23 24.1	12 6.5	4 21.9
45	10 36.2	20 40.0	24 3.5	12 27.4	4 28.4
46	10 49.3	21 17.7	24 43.9	12 48.8	4 35.1
47	11 3.1	21 57.2	25 25.3	13 10.8	4 42.0
48	11 17.6	22 38.5	26 7.7	13 33.4	4 49.1
49	11 32.9	23 21.9	26 51.4	13 56.6	4 56.4
50	11 49.1	24 7.5	27 36.2	14 20.5	5 3.9
51	12 6.2	24 55.5	28 22.2	14 45.2	5 11.7
52	12 24.4	25 46.2	29 9.6	15 10.5	5 19.8
53	12 43.8	26 39.9	29 58.4	15 36.7	5 28.1
54	13 4.4	27 36.8	♌ 0 48.7	16 3.8	5 36.8
55	13 26.6	28 37.3	1 40.5	16 31.8	5 45.8
56	13 50.5	29 41.8	2 34.0	17 0.8	5 55.2
57	14 16.3	♋ 0 50.8	3 29.1	17 30.8	6 4.9
58	14 44.4	2 4.8	4 26.1	18 2.0	6 15.1
59	15 15.0	3 24.3	5 25.0	18 34.3	6 25.8
60	15 48.6	4 50.3	6 25.8	19 8.0	6 36.9
61	16 25.9	6 23.5	7 28.8	19 43.0	6 48.6
62	17 7.4	8 5.0	8 34.0	20 19.6	7 0.8
63	17 54.4	9 56.1	9 41.4	20 57.7	7 13.7
64	18 48.1	11 58.4	10 51.3	21 37.6	7 27.3
65	19 50.7	14 13.7	12 3.8	22 19.3	7 41.6
66	21 5.8	16 44.7	13 18.9	23 3.0	7 56.8
S LAT	5	6	Descendant	8	9

♎ 3° 16′ 10″

12h 12m 0s MC 183° 0′ 0″

0h 16m 0s — MC 4° 0′ 0″ — ♈ 4° 21′ 31″

N LAT	11	12	Ascendant	2	3
°	° ′	° ′	° ′	° ′	° ′
0	♉ 6 19.4	♊ 5 53.6	♋ 3 40.2	♌ 1 45.0	♍ 2 0.3
5	6 44.7	7 8.2	5 39.4	2 53.7	2 20.1
10	7 10.9	8 25.5	7 39.5	4 1.8	2 39.8
15	7 38.5	9 46.8	9 42.3	5 10.2	2 59.7
16	7 44.2	10 3.7	10 7.3	5 24.0	3 3.7
17	7 50.0	10 20.8	10 32.5	5 37.8	3 7.8
18	7 55.9	10 38.2	10 57.9	5 51.8	3 11.8
19	8 2.0	10 55.8	11 23.5	6 5.8	3 15.9
20	8 8.1	11 13.8	11 49.3	6 19.9	3 20.1
21	8 14.3	11 32.0	12 15.4	6 34.1	3 24.2
22	8 20.6	11 50.6	12 41.7	6 48.5	3 28.4
23	8 27.1	12 9.5	13 8.4	7 2.9	3 32.7
24	8 33.7	12 28.8	13 35.3	7 17.5	3 36.9
25	8 40.4	12 48.5	14 2.6	7 32.2	3 41.3
26	8 47.3	13 8.6	14 30.1	7 47.1	3 45.7
27	8 54.4	13 29.1	14 58.1	8 2.1	3 50.1
28	9 1.6	13 50.1	15 26.4	8 17.3	3 54.6
29	9 9.0	14 11.6	15 55.1	8 32.7	3 59.1
30	9 16.6	14 33.7	16 24.3	8 48.3	4 3.8
31	9 24.3	14 56.3	16 53.9	9 4.1	4 8.4
32	9 32.3	15 19.5	17 23.9	9 20.1	4 13.2
33	9 40.6	15 43.3	17 54.5	9 36.4	4 18.0
34	9 49.1	16 7.8	18 25.5	9 52.9	4 23.0
35	9 57.8	16 33.1	18 57.1	10 9.7	4 28.0
36	10 6.9	16 59.1	19 29.3	10 26.7	4 33.1
37	10 16.2	17 25.9	20 2.1	10 44.1	4 38.3
38	10 25.9	17 53.6	20 35.6	11 1.8	4 43.6
39	10 35.9	18 22.3	21 9.7	11 19.9	4 49.0
40	10 46.3	18 52.0	21 44.5	11 38.3	4 54.6
41	10 57.1	19 22.8	22 20.1	11 57.1	5 0.2
42	11 8.4	19 54.7	22 56.4	12 16.3	5 6.1
43	11 20.1	20 27.9	23 33.6	12 35.9	5 12.0
44	11 32.4	21 2.4	24 11.6	12 56.0	5 18.1
45	11 45.2	21 38.4	24 50.5	13 16.6	5 24.4
46	11 58.6	22 16.0	25 30.4	13 37.7	5 30.8
47	12 12.7	22 55.3	26 11.3	13 59.3	5 37.5
48	12 27.5	23 36.5	26 53.3	14 21.6	5 44.3
49	12 43.1	24 19.6	27 36.4	14 44.4	5 51.4
50	12 59.6	25 5.0	28 20.6	15 7.9	5 58.6
51	13 17.1	25 52.7	29 6.1	15 32.1	6 6.2
52	13 35.7	26 43.1	29 52.9	15 57.1	6 13.9
53	13 55.5	27 36.4	♌ 0 41.0	16 22.8	6 22.0
54	14 16.7	28 32.9	1 30.6	16 49.4	6 30.4
55	14 39.3	29 32.9	2 21.7	17 16.9	6 39.1
56	15 3.8	♋ 0 36.8	3 14.5	17 45.4	6 48.1
57	15 30.2	1 45.1	4 8.8	18 14.9	6 57.5
58	15 58.9	2 58.2	5 5.0	18 45.5	7 7.3
59	16 30.2	4 16.8	6 3.0	19 17.2	7 17.5
60	17 4.6	5 41.6	7 2.9	19 50.2	7 28.3
61	17 42.8	7 13.4	8 5.0	20 24.6	7 39.5
62	18 25.5	8 53.2	9 9.1	21 0.4	7 51.3
63	19 13.7	10 42.2	10 15.6	21 37.8	8 3.7
64	20 9.0	12 41.9	11 24.4	22 16.9	8 16.8
65	21 13.8	14 54.2	12 35.7	22 57.8	8 30.6
66	22 31.8	17 21.1	13 49.6	23 40.6	8 45.2
S LAT	5	6	Descendant	8	9

♎ 4° 21′ 31″ — 12h 16m 0s — MC 184° 0′ 0″

0h 20m 0s — MC 5° 0′ 0″ — ♈ 5° 26′ 50″

N LAT	11	12	Ascendant	2	3
°	° ′	° ′	° ′	° ′	° ′
0	♉ 7 21.1	♊ 6 50.3	♋ 4 35.3	♌ 2 43.0	♍ 3 3.5
5	7 46.9	8 5.4	6 34.3	3 51.1	3 22.6
10	8 13.7	9 23.0	8 34.1	4 58.4	3 41.7
15	8 41.9	10 44.7	10 36.4	6 6.0	4 0.9
16	8 47.8	11 1.6	11 1.3	6 19.7	4 4.7
17	8 53.7	11 18.8	11 26.4	6 33.4	4 8.7
18	8 59.8	11 36.2	11 51.7	6 47.2	4 12.6
19	9 5.9	11 53.9	12 17.1	7 1.1	4 16.5
20	9 12.2	12 11.9	12 42.8	7 15.0	4 20.5
21	9 18.5	12 30.2	13 8.8	7 29.0	4 24.6
22	9 25.0	12 48.9	13 35.0	7 43.2	4 28.6
23	9 31.6	13 7.8	14 1.5	7 57.5	4 32.7
24	9 38.3	13 27.2	14 28.2	8 11.9	4 36.8
25	9 45.2	13 46.9	14 55.3	8 26.4	4 41.0
26	9 52.2	14 7.0	15 22.7	8 41.1	4 45.2
27	9 59.4	14 27.6	15 50.5	8 55.9	4 49.5
28	10 6.8	14 48.6	16 18.6	9 10.9	4 53.9
29	10 14.4	15 10.2	16 47.1	9 26.1	4 58.3
30	10 22.1	15 32.2	17 16.0	9 41.5	5 2.7
31	10 30.1	15 54.8	17 45.4	9 57.0	5 7.2
32	10 38.2	16 18.0	18 15.2	10 12.8	5 11.8
33	10 46.6	16 41.9	18 45.5	10 28.9	5 16.5
34	10 55.3	17 6.4	19 16.3	10 45.2	5 21.2
35	11 4.2	17 31.6	19 47.6	11 1.7	5 26.1
36	11 13.5	17 57.6	20 19.5	11 18.5	5 31.0
37	11 23.0	18 24.5	20 52.0	11 35.7	5 36.0
38	11 32.9	18 52.2	21 25.1	11 53.1	5 41.1
39	11 43.1	19 20.8	21 58.9	12 10.9	5 46.4
40	11 53.7	19 50.4	22 33.4	12 29.0	5 51.7
41	12 4.8	20 21.2	23 8.6	12 47.5	5 57.2
42	12 16.2	20 53.0	23 44.5	13 6.4	6 2.8
43	12 28.2	21 26.1	24 21.3	13 25.8	6 8.5
44	12 40.7	22 0.6	24 58.9	13 45.5	6 14.4
45	12 53.8	22 36.5	25 37.4	14 5.8	6 20.5
46	13 7.5	23 13.9	26 16.8	14 26.6	6 26.7
47	13 21.9	23 53.0	26 57.2	14 47.9	6 33.1
48	13 37.0	24 34.0	27 38.7	15 9.7	6 39.7
49	13 52.9	25 16.9	28 21.3	15 32.2	6 46.5
50	14 9.8	26 2.0	29 4.9	15 55.3	6 53.5
51	14 27.7	26 49.5	29 49.9	16 19.1	7 0.7
52	14 46.6	27 39.6	♌ 0 36.0	16 43.6	7 8.2
53	15 6.8	28 32.5	1 23.5	17 8.9	7 16.0
54	15 28.4	29 28.5	2 12.5	17 35.0	7 24.0
55	15 51.6	♋ 0 27.9	3 2.9	18 2.0	7 32.4
56	16 16.5	1 31.2	3 54.8	18 30.0	7 41.0
57	16 43.5	2 38.8	4 48.4	18 58.9	7 50.1
58	17 12.8	3 51.1	5 43.8	19 29.0	7 59.5
59	17 44.9	5 8.7	6 40.9	20 0.1	8 9.4
60	18 20.1	6 32.3	7 40.0	20 32.5	8 19.7
61	18 59.2	8 2.7	8 41.1	21 6.2	8 30.5
62	19 42.9	9 40.9	9 44.3	21 41.3	8 41.8
63	20 32.4	11 27.9	10 49.7	22 18.0	8 53.7
64	21 29.4	13 25.1	11 57.4	22 56.3	9 6.3
65	22 36.2	15 34.3	13 7.6	23 36.3	9 19.5
66	23 57.2	17 57.4	14 20.4	24 18.2	9 33.6
S LAT	5	6	Descendant	8	9

♎ 5° 26′ 50″ — 12h 20m 0s — MC 185° 0′ 0″

0ʰ 24ᵐ 0ˢ MC 6° 0′ 0″

♈ 6° 32′ 7″

N LAT	11	12	Ascendant	2	3
°	° ′	° ′	° ′	° ′	° ′
0	♉ 8 22.6	♊ 7 46.9	♋ 5 30.5	♌ 3 41.2	♍ 4 6.8
5	8 49.0	9 2.4	7 29.2	4 48.6	4 25.3
10	9 16.3	10 20.4	9 28.7	5 55.2	4 43.6
15	9 45.1	11 42.4	11 30.5	7 2.0	5 2.2
16	9 51.1	11 59.4	11 55.3	7 15.5	5 5.9
17	9 57.2	12 16.6	12 20.3	7 29.1	5 9.7
18	10 3.4	12 34.1	12 45.4	7 42.7	5 13.5
19	10 9.6	12 51.8	13 10.8	7 56.4	5 17.3
20	10 16.0	13 9.9	13 36.3	8 10.2	5 21.1
21	10 22.5	13 28.2	14 2.1	8 24.0	5 25.0
22	10 29.1	13 46.9	14 28.2	8 38.0	5 28.9
23	10 35.8	14 5.9	14 54.5	8 52.1	5 32.9
24	10 42.7	14 25.3	15 21.1	9 6.3	5 36.8
25	10 49.7	14 45.0	15 48.0	9 20.6	5 40.9
26	10 56.9	15 5.2	16 15.2	9 35.1	5 44.9
27	11 4.3	15 25.8	16 42.7	9 49.7	5 49.1
28	11 11.8	15 46.9	17 10.7	10 4.5	5 53.2
29	11 19.5	16 8.4	17 39.0	10 19.5	5 57.5
30	11 27.4	16 30.5	18 7.6	10 34.7	6 1.8
31	11 35.5	16 53.1	18 36.8	10 50.1	6 6.1
32	11 43.8	17 16.3	19 6.3	11 5.6	6 10.5
33	11 52.4	17 40.2	19 36.4	11 21.4	6 15.0
34	12 1.2	18 4.7	20 6.9	11 37.5	6 19.6
35	12 10.4	18 29.9	20 38.0	11 53.8	6 24.3
36	12 19.8	18 55.9	21 9.6	12 10.4	6 29.0
37	12 29.5	19 22.7	21 41.8	12 27.3	6 33.8
38	12 39.6	19 50.4	22 14.6	12 44.4	6 38.8
39	12 50.0	20 19.0	22 48.0	13 1.9	6 43.8
40	13 0.8	20 48.5	23 22.1	13 19.8	6 48.9
41	13 12.1	21 19.2	23 56.9	13 38.0	6 54.2
42	13 23.8	21 51.0	24 32.5	13 56.6	6 59.6
43	13 36.0	22 24.0	25 8.9	14 15.7	7 5.1
44	13 48.7	22 58.3	25 46.1	14 35.1	7 10.8
45	14 2.1	23 34.1	26 24.1	14 55.0	7 16.6
46	14 16.0	24 11.4	27 3.1	15 15.5	7 22.6
47	14 30.7	24 50.3	27 43.0	15 36.4	7 28.7
48	14 46.1	25 31.1	28 24.0	15 57.9	7 35.1
49	15 2.4	26 13.8	29 6.0	16 20.0	7 41.6
50	15 19.5	26 58.6	29 49.2	16 42.7	7 48.3
51	15 37.8	27 45.8	♌ 0 33.5	17 6.1	7 55.3
52	15 57.1	28 35.5	1 19.1	17 30.2	8 2.5
53	16 17.7	29 28.0	2 5.9	17 55.0	8 10.0
54	16 39.7	♋ 0 23.6	2 54.2	18 20.7	8 17.7
55	17 3.4	1 22.5	3 43.9	18 47.2	8 25.7
56	17 28.8	2 25.2	4 35.1	19 14.6	8 34.1
57	17 56.3	3 32.0	5 28.0	19 43.0	8 42.8
58	18 26.3	4 43.4	6 22.5	20 12.5	8 51.8
59	18 59.0	6 0.1	7 18.8	20 43.0	9 1.3
60	19 35.0	7 22.5	8 17.0	21 14.8	9 11.2
61	20 15.0	8 51.6	9 17.2	21 47.8	9 21.5
62	20 59.8	10 28.1	10 19.4	22 22.3	9 32.4
63	21 50.6	12 13.1	11 23.8	22 58.2	9 43.9
64	22 49.1	14 8.0	12 30.5	23 35.7	9 55.9
65	23 58.1	16 14.2	13 39.6	24 14.9	10 8.6
66	25 22.2	18 33.6	14 51.2	24 55.9	10 22.1
S LAT	5	6	Descendant	8	9

♎ 6° 32′ 7″

12ʰ 24ᵐ 0ˢ MC 186° 0′ 0″

0ʰ 28ᵐ 0ˢ MC 7° 0′ 0″

♈ 7° 37′ 22″

N LAT	11	12	Ascendant	2	3
°	° ′	° ′	° ′	° ′	° ′
0	♉ 9 23.9	♊ 8 43.3	♋ 6 25.6	♌ 4 39.5	♍ 5 10.3
5	9 50.8	9 59.2	8 24.2	5 46.2	5 28.1
10	10 18.7	11 17.6	10 23.3	6 52.0	5 45.8
15	10 48.1	12 39.9	12 24.6	7 58.1	6 3.6
16	10 54.2	12 56.9	12 49.3	8 11.4	6 7.2
17	11 0.4	13 14.2	13 14.1	8 24.8	6 10.8
18	11 6.7	13 31.7	13 39.1	8 38.3	6 14.5
19	11 13.1	13 49.5	14 4.3	8 51.8	6 18.1
20	11 19.6	14 7.6	14 29.7	9 5.4	6 21.8
21	11 26.2	14 26.0	14 55.4	9 19.1	6 25.6
22	11 33.0	14 44.7	15 21.3	9 32.9	6 29.3
23	11 39.9	15 3.8	15 47.4	9 46.8	6 33.1
24	11 46.9	15 23.2	16 13.8	10 0.8	6 37.0
25	11 54.0	15 42.9	16 40.5	10 14.9	6 40.8
26	12 1.3	16 3.1	17 7.6	10 29.2	6 44.7
27	12 8.8	16 23.8	17 34.9	10 43.7	6 48.7
28	12 16.5	16 44.8	18 2.6	10 58.3	6 52.7
29	12 24.3	17 6.4	18 30.7	11 13.0	6 56.8
30	12 32.4	17 28.5	18 59.2	11 28.0	7 0.9
31	12 40.6	17 51.1	19 28.1	11 43.1	7 5.1
32	12 49.1	18 14.3	19 57.4	11 58.5	7 9.3
33	12 57.9	18 38.2	20 27.2	12 14.0	7 13.7
34	13 6.9	19 2.7	20 57.4	12 29.9	7 18.1
35	13 16.2	19 27.9	21 28.2	12 45.9	7 22.5
36	13 25.8	19 53.9	21 59.5	13 2.3	7 27.1
37	13 35.7	20 20.6	22 31.4	13 18.9	7 31.7
38	13 45.9	20 48.2	23 3.9	13 35.8	7 36.5
39	13 56.6	21 16.8	23 37.0	13 53.0	7 41.3
40	14 7.6	21 46.3	24 10.7	14 10.6	7 46.3
41	14 19.0	22 16.9	24 45.2	14 28.5	7 51.3
42	14 31.0	22 48.6	25 20.4	14 46.8	7 56.5
43	14 43.4	23 21.5	25 56.3	15 5.6	8 1.8
44	14 56.4	23 55.7	26 33.1	15 24.7	8 7.2
45	15 10.0	24 31.3	27 10.7	15 44.3	8 12.8
46	15 24.2	25 8.4	27 49.3	16 4.4	8 18.6
47	15 39.1	25 47.2	28 28.7	16 25.0	8 24.5
48	15 54.8	26 27.7	29 9.2	16 46.1	8 30.6
49	16 11.4	27 10.2	29 50.7	17 7.8	8 36.8
50	16 28.9	27 54.8	♌ 0 33.3	17 30.1	8 43.3
51	16 47.4	28 41.6	1 17.0	17 53.1	8 50.0
52	17 7.1	29 31.0	2 2.0	18 16.7	8 56.9
53	17 28.1	♋ 0 23.1	2 48.2	18 41.1	9 4.0
54	17 50.6	1 18.2	3 35.8	19 6.3	9 11.4
55	18 14.6	2 16.6	4 24.8	19 32.3	9 19.1
56	18 40.6	3 18.6	5 15.4	19 59.2	9 27.2
57	19 8.6	4 24.7	6 7.4	20 27.1	9 35.5
58	19 39.2	5 35.3	7 1.2	20 56.0	9 44.2
59	20 12.6	6 51.0	7 56.7	21 26.0	9 53.2
60	20 49.3	8 12.3	8 54.0	21 57.1	10 2.7
61	21 30.2	9 40.0	9 53.2	22 29.5	10 12.7
62	22 16.0	11 14.9	10 54.5	23 3.2	10 23.1
63	23 8.1	12 58.0	11 57.9	23 38.4	10 34.0
64	24 8.2	14 50.5	13 3.5	24 15.1	10 45.6
65	25 19.4	16 53.8	14 11.6	24 53.5	10 57.7
66	26 46.7	19 9.7	15 22.0	25 33.7	11 10.6
S LAT	5	6	Descendant	8	9

♎ 7° 37′ 22″

12ʰ 28ᵐ 0ˢ MC 187° 0′ 0″

0h 32m 0s		MC		8° 0′ 0″	
		♈ 8° 42′ 33″			
11	**12**	**Ascendant**	**2**	**3**	**N LAT**
° ′	° ′	° ′	° ′	° ′	°
♉10 25.0	♊ 9 39.7	♋ 7 20.8	♌ 5 38.0	♍ 6 14.0	**0**
10 52.5	10 55.9	9 19.1	6 43.9	6 31.0	**5**
11 21.0	12 14.6	11 17.9	7 49.0	6 48.0	**10**
11 50.9	13 37.2	13 18.6	8 54.3	7 5.1	**15**
11 57.1	13 54.3	13 43.2	9 7.5	7 8.6	**16**
12 3.5	14 11.6	14 7.9	9 20.7	7 12.1	**17**
12 9.9	14 29.2	14 32.8	9 33.9	7 15.6	**18**
12 16.4	14 47.0	14 57.8	9 47.3	7 19.1	**19**
12 23.0	15 5.2	15 23.1	10 0.7	7 22.7	**20**
12 29.7	15 23.6	15 48.6	10 14.2	7 26.2	**21**
12 36.6	15 42.3	16 14.3	10 27.8	7 29.8	**22**
12 43.6	16 1.4	16 40.3	10 41.5	7 33.5	**23**
12 50.7	16 20.8	17 6.5	10 55.4	7 37.2	**24**
12 58.0	16 40.6	17 33.0	11 9.3	7 40.9	**25**
13 5.5	17 0.8	17 59.9	11 23.4	7 44.6	**26**
13 13.1	17 21.5	18 27.0	11 37.6	7 48.4	**27**
13 20.9	17 42.6	18 54.5	11 52.0	7 52.3	**28**
13 28.9	18 4.2	19 22.4	12 6.6	7 56.2	**29**
13 37.1	18 26.2	19 50.6	12 21.3	8 0.2	**30**
13 45.5	18 48.9	20 19.3	12 36.2	8 4.2	**31**
13 54.2	19 12.1	20 48.3	12 51.4	8 8.2	**32**
14 3.1	19 35.9	21 17.8	13 6.7	8 12.4	**33**
14 12.2	20 0.4	21 47.8	13 22.3	8 16.6	**34**
14 21.7	20 25.6	22 18.3	13 38.1	8 20.9	**35**
14 31.5	20 51.5	22 49.3	13 54.2	8 25.3	**36**
14 41.5	21 18.2	23 20.9	14 10.5	8 29.7	**37**
14 52.0	21 45.8	23 53.1	14 27.2	8 34.3	**38**
15 2.8	22 14.3	24 25.8	14 44.1	8 38.9	**39**
15 14.0	22 43.7	24 59.2	15 1.4	8 43.6	**40**
15 25.7	23 14.2	25 33.3	15 19.1	8 48.5	**41**
15 37.8	23 45.8	26 8.1	15 37.1	8 53.5	**42**
15 50.5	24 18.6	26 43.7	15 55.5	8 58.5	**43**
16 3.7	24 52.7	27 20.1	16 14.3	9 3.8	**44**
16 17.5	25 28.2	27 57.2	16 33.6	9 9.1	**45**
16 32.0	26 5.1	28 35.3	16 53.3	9 14.6	**46**
16 47.2	26 43.7	29 14.3	17 13.5	9 20.3	**47**
17 3.1	27 24.0	29 54.2	17 34.3	9 26.1	**48**
17 20.0	28 6.3	♌ 0 35.2	17 55.6	9 32.1	**49**
17 37.8	28 50.5	1 17.3	18 17.5	9 38.3	**50**
17 56.7	29 37.1	2 0.4	18 40.1	9 44.7	**51**
18 16.7	♋ 0 26.1	2 44.8	19 3.3	9 51.3	**52**
18 38.1	1 17.7	3 30.4	19 27.3	9 58.2	**53**
19 0.9	2 12.3	4 17.4	19 52.0	10 5.3	**54**
19 25.5	3 10.2	5 5.7	20 17.5	10 12.6	**55**
19 51.9	4 11.6	5 55.5	20 43.9	10 20.3	**56**
20 20.5	5 17.0	6 46.8	21 11.2	10 28.3	**57**
20 51.6	6 26.7	7 39.8	21 39.5	10 36.6	**58**
21 25.6	7 41.4	8 34.4	22 8.9	10 45.3	**59**
22 3.1	9 1.6	9 30.9	22 39.4	10 54.3	**60**
22 44.8	10 27.9	10 29.2	23 11.2	11 3.8	**61**
23 31.7	12 1.2	11 29.6	23 44.2	11 13.8	**62**
24 25.0	13 42.4	12 32.0	24 18.7	11 24.3	**63**
25 26.7	15 32.7	13 36.6	24 54.6	11 35.3	**64**
26 40.0	17 33.2	14 43.5	25 32.2	11 46.9	**65**
28 10.6	19 45.7	15 52.9	26 11.5	11 59.2	**66**
5	**6**	**Descendant**	**8**	**9**	**S LAT**
		♎ 8° 42′ 33″			
12h 32m 0s		MC		188° 0′ 0″	

0h 36m 0s		MC		9° 0′ 0″	
		♈ 9° 47′ 41″			
11	**12**	**Ascendant**	**2**	**3**	**N LAT**
° ′	° ′	° ′	° ′	° ′	°
♉11 26.0	♊10 35.9	♋ 8 16.1	♌ 6 36.6	♍ 7 17.7	**0**
11 54.0	11 52.5	10 14.1	7 41.8	7 34.1	**5**
12 23.0	13 11.5	12 12.4	8 46.2	7 50.4	**10**
12 53.5	14 34.3	14 12.6	9 50.6	8 6.8	**15**
12 59.8	14 51.5	14 37.1	10 3.6	8 10.1	**16**
13 6.2	15 8.8	15 1.6	10 16.6	8 13.4	**17**
13 12.8	15 26.5	15 26.4	10 29.7	8 16.8	**18**
13 19.4	15 44.3	15 51.3	10 42.9	8 20.2	**19**
13 26.1	16 2.5	16 16.4	10 56.1	8 23.6	**20**
13 33.0	16 21.0	16 41.7	11 9.4	8 27.0	**21**
13 40.0	16 39.7	17 7.3	11 22.9	8 30.5	**22**
13 47.1	16 58.8	17 33.1	11 36.4	8 33.9	**23**
13 54.4	17 18.3	17 59.1	11 50.0	8 37.5	**24**
14 1.8	17 38.1	18 25.5	12 3.8	8 41.0	**25**
14 9.4	17 58.3	18 52.1	12 17.7	8 44.6	**26**
14 17.1	18 19.0	19 19.1	12 31.7	8 48.3	**27**
14 25.1	18 40.1	19 46.3	12 45.9	8 52.0	**28**
14 33.2	19 1.7	20 14.0	13 0.2	8 55.7	**29**
14 41.5	19 23.7	20 42.0	13 14.7	8 59.5	**30**
14 50.1	19 46.4	21 10.4	13 29.4	9 3.3	**31**
14 58.9	20 9.6	21 39.2	13 44.3	9 7.2	**32**
15 8.0	20 33.4	22 8.4	13 59.4	9 11.2	**33**
15 17.3	20 57.8	22 38.2	14 14.7	9 15.2	**34**
15 26.9	21 23.0	23 8.4	14 30.3	9 19.3	**35**
15 36.8	21 48.9	23 39.1	14 46.1	9 23.5	**36**
15 47.1	22 15.5	24 10.3	15 2.2	9 27.8	**37**
15 57.7	22 43.1	24 42.2	15 18.6	9 32.1	**38**
16 8.7	23 11.5	25 14.6	15 35.3	9 36.6	**39**
16 20.1	23 40.8	25 47.6	15 52.3	9 41.1	**40**
16 32.0	24 11.2	26 21.4	16 9.6	9 45.7	**41**
16 44.3	24 42.7	26 55.8	16 27.4	9 50.5	**42**
16 57.2	25 15.4	27 30.9	16 45.4	9 55.4	**43**
17 10.6	25 49.3	28 6.9	17 3.9	10 0.3	**44**
17 24.7	26 24.7	28 43.6	17 22.9	10 5.5	**45**
17 39.4	27 1.4	29 21.2	17 42.3	10 10.7	**46**
17 54.8	27 39.8	29 59.7	18 2.1	10 16.1	**47**
18 11.1	28 19.9	♌ 0 39.2	18 22.5	10 21.7	**48**
18 28.2	29 1.9	1 19.6	18 43.5	10 27.4	**49**
18 46.3	29 45.9	2 1.2	19 5.0	10 33.4	**50**
19 5.5	♋ 0 32.1	2 43.8	19 27.1	10 39.5	**51**
19 25.9	1 20.7	3 27.5	19 49.9	10 45.8	**52**
19 47.6	2 11.9	4 12.5	20 13.4	10 52.4	**53**
20 10.9	3 6.1	4 58.8	20 37.6	10 59.1	**54**
20 35.8	4 3.3	5 46.5	21 2.7	11 6.2	**55**
21 2.7	5 4.1	6 35.6	21 28.6	11 13.5	**56**
21 31.8	6 8.8	7 26.2	21 55.3	11 21.1	**57**
22 3.4	7 17.7	8 18.3	22 23.1	11 29.1	**58**
22 38.1	8 31.4	9 12.2	22 51.9	11 37.3	**59**
23 16.3	9 50.4	10 7.8	23 21.8	11 46.0	**60**
23 58.9	11 15.4	11 5.2	23 52.9	11 55.1	**61**
24 46.7	12 47.2	12 4.6	24 25.2	12 4.6	**62**
25 41.2	14 26.5	13 6.1	24 59.0	12 14.6	**63**
26 44.5	16 14.6	14 9.7	25 34.2	12 25.1	**64**
28 0.0	18 12.5	15 15.6	26 10.9	12 36.2	**65**
29 34.0	20 21.7	16 23.8	26 49.4	12 47.9	**66**
5	**6**	**Descendant**	**8**	**9**	**S LAT**
		♎ 9° 47′ 41″			
12h 36m 0s		MC		189° 0′ 0″	

$0^h\ 40^m\ 0^s$ MC 10° 0′ 0″

♈ 10° 52′ 45″

11	12	Ascendant	2	3	N LAT
° ′	° ′	° ′	° ′	° ′	°
♉12 26.7	♊11 32.1	♋ 9 11.4	♌ 7 35.4	♍ 8 21.7	**0**
12 55.3	12 49.0	11 9.1	8 39.9	8 37.3	**5**
13 24.8	14 8.3	13 7.0	9 43.4	8 52.9	**10**
13 55.8	15 31.3	15 6.6	10 47.0	9 8.6	**15**
14 2.3	15 48.5	15 30.9	10 59.8	9 11.7	**16**
14 8.8	16 5.9	15 55.3	11 12.7	9 14.9	**17**
14 15.4	16 23.6	16 19.9	11 25.6	9 18.1	**18**
14 22.2	16 41.5	16 44.7	11 38.6	9 21.4	**19**
14 29.0	16 59.6	17 9.7	11 51.6	9 24.6	**20**
14 36.0	17 18.1	17 34.8	12 4.8	9 27.9	**21**
14 43.1	17 36.9	18 0.2	12 18.0	9 31.2	**22**
14 50.4	17 56.0	18 25.8	12 31.3	9 34.5	**23**
14 57.8	18 15.5	18 51.7	12 44.7	9 37.9	**24**
15 5.3	18 35.3	19 17.8	12 58.3	9 41.3	**25**
15 13.0	18 55.6	19 44.3	13 12.0	9 44.7	**26**
15 20.9	19 16.2	20 11.0	13 25.8	9 48.2	**27**
15 29.0	19 37.3	20 38.1	13 39.7	9 51.7	**28**
15 37.2	19 58.9	21 5.5	13 53.9	9 55.3	**29**
15 45.7	20 21.0	21 33.2	14 8.2	9 58.9	**30**
15 54.4	20 43.6	22 1.4	14 22.6	10 2.6	**31**
16 3.4	21 6.8	22 30.0	14 37.3	10 6.3	**32**
16 12.6	21 30.6	22 58.9	14 52.2	10 10.1	**33**
16 22.1	21 55.0	23 28.4	15 7.2	10 13.9	**34**
16 31.8	22 20.1	23 58.3	15 22.6	10 17.9	**35**
16 41.9	22 45.9	24 28.7	15 38.1	10 21.9	**36**
16 52.3	23 12.6	24 59.6	15 54.0	10 25.9	**37**
17 3.1	23 40.0	25 31.1	16 10.1	10 30.1	**38**
17 14.3	24 8.3	26 3.2	16 26.5	10 34.3	**39**
17 25.9	24 37.6	26 35.9	16 43.2	10 38.6	**40**
17 37.9	25 7.9	27 9.3	17 0.2	10 43.1	**41**
17 50.5	25 39.3	27 43.3	17 17.6	10 47.6	**42**
18 3.6	26 11.8	28 18.1	17 35.4	10 52.2	**43**
18 17.2	26 45.6	28 53.6	17 53.6	10 57.0	**44**
18 31.5	27 20.8	29 29.9	18 12.2	11 1.9	**45**
18 46.4	27 57.4	♌ 0 7.1	18 31.2	11 6.9	**46**
19 2.1	28 35.6	0 45.1	18 50.7	11 12.1	**47**
19 18.6	29 15.4	1 24.1	19 10.8	11 17.4	**48**
19 36.0	29 57.1	2 4.0	19 31.3	11 22.9	**49**
19 54.4	♋ 0 40.8	2 45.0	19 52.4	11 28.5	**50**
20 13.9	1 26.7	3 27.0	20 14.2	11 34.3	**51**
20 34.6	2 14.9	4 10.2	20 36.5	11 40.4	**52**
20 56.7	3 5.7	4 54.6	20 59.6	11 46.6	**53**
21 20.3	3 59.4	5 40.2	21 23.3	11 53.1	**54**
21 45.7	4 56.1	6 27.2	21 47.9	11 59.8	**55**
22 13.0	5 56.2	7 15.6	22 13.2	12 6.8	**56**
22 42.5	7 0.2	8 5.4	22 39.5	12 14.0	**57**
23 14.7	8 8.2	8 56.8	23 6.7	12 21.6	**58**
23 50.0	9 21.0	9 49.9	23 34.9	12 29.5	**59**
24 29.0	10 38.9	10 44.6	24 4.2	12 37.7	**60**
25 12.3	12 2.6	11 41.2	24 34.6	12 46.4	**61**
26 1.1	13 32.8	12 39.7	25 6.3	12 55.4	**62**
26 56.8	15 10.3	13 40.2	25 39.3	13 4.9	**63**
28 1.7	16 56.2	14 42.8	26 13.7	13 14.9	**64**
29 19.4	18 51.5	15 47.6	26 49.7	13 25.5	**65**
♊ 0 56.9	20 57.6	16 54.8	27 27.3	13 36.7	**66**
5	6	Descendant	8	9	S LAT

♎ 10° 52′ 45″

$12^h\ 40^m\ 0^s$ MC 190° 0′ 0″

$0^h\ 44^m\ 0^s$ MC 11° 0′ 0″

♈ 11° 57′ 44″

11	12	Ascendant	2	3	N LAT
° ′	° ′	° ′	° ′	° ′	°
♉13 27.3	♊12 28.1	♋10 6.7	♌ 8 34.4	♍ 9 25.7	**0**
13 56.4	13 45.3	12 4.1	9 38.1	9 40.7	**5**
14 26.4	15 4.9	14 1.5	10 40.8	9 55.5	**10**
14 57.9	16 28.1	16 0.6	11 43.5	10 10.4	**15**
15 4.5	16 45.3	16 24.7	11 56.1	10 13.5	**16**
15 11.1	17 2.8	16 49.0	12 8.8	10 16.5	**17**
15 17.9	17 20.5	17 13.5	12 21.6	10 19.6	**18**
15 24.7	17 38.4	17 38.1	12 34.4	10 22.6	**19**
15 31.7	17 56.6	18 2.9	12 47.2	10 25.7	**20**
15 38.8	18 15.1	18 27.9	13 0.2	10 28.8	**21**
15 46.0	18 33.9	18 53.1	13 13.2	10 32.0	**22**
15 53.4	18 53.1	19 18.5	13 26.3	10 35.2	**23**
16 0.9	19 12.5	19 44.2	13 39.6	10 38.4	**24**
16 8.6	19 32.4	20 10.2	13 52.9	10 41.6	**25**
16 16.4	19 52.6	20 36.4	14 6.4	10 44.9	**26**
16 24.4	20 13.3	21 2.9	14 20.0	10 48.2	**27**
16 32.6	20 34.4	21 29.7	14 33.7	10 51.6	**28**
16 41.0	20 55.9	21 56.9	14 47.6	10 55.0	**29**
16 49.6	21 18.0	22 24.4	15 1.7	10 58.4	**30**
16 58.5	21 40.6	22 52.3	15 15.9	11 1.9	**31**
17 7.5	22 3.8	23 20.6	15 30.3	11 5.5	**32**
17 16.9	22 27.5	23 49.4	15 45.0	11 9.1	**33**
17 26.5	22 51.9	24 18.5	15 59.8	11 12.7	**34**
17 36.4	23 17.0	24 48.1	16 14.9	11 16.5	**35**
17 46.7	23 42.7	25 18.2	16 30.2	11 20.3	**36**
17 57.3	24 9.3	25 48.9	16 45.7	11 24.1	**37**
18 8.2	24 36.7	26 20.0	17 1.6	11 28.1	**38**
18 19.6	25 4.9	26 51.8	17 17.7	11 32.1	**39**
18 31.3	25 34.1	27 24.1	17 34.1	11 36.3	**40**
18 43.6	26 4.3	27 57.1	17 50.9	11 40.5	**41**
18 56.3	26 35.6	28 30.8	18 8.0	11 44.8	**42**
19 9.6	27 8.0	29 5.1	18 25.4	11 49.2	**43**
19 23.4	27 41.6	29 40.2	18 43.3	11 53.7	**44**
19 37.9	28 16.6	♌ 0 16.1	19 1.5	11 58.4	**45**
19 53.1	28 53.0	0 52.8	19 20.2	12 3.2	**46**
20 9.0	29 31.0	1 30.3	19 39.4	12 8.1	**47**
20 25.8	♋ 0 10.6	2 8.8	19 59.0	12 13.1	**48**
20 43.4	0 52.0	2 48.2	20 19.2	12 18.3	**49**
21 2.1	1 35.4	3 28.7	20 39.9	12 23.7	**50**
21 21.9	2 20.9	4 10.1	21 1.2	12 29.2	**51**
21 42.9	3 8.8	4 52.7	21 23.1	12 35.0	**52**
22 5.3	3 59.1	5 36.5	21 45.7	12 40.9	**53**
22 29.3	4 52.3	6 21.5	22 9.0	12 47.1	**54**
22 55.0	5 48.4	7 7.8	22 33.1	12 53.4	**55**
23 22.8	6 47.9	7 55.5	22 57.9	13 0.1	**56**
23 52.8	7 51.1	8 44.6	23 23.7	13 7.0	**57**
24 25.6	8 58.4	9 35.3	23 50.3	13 14.2	**58**
25 1.4	10 10.1	10 27.5	24 17.9	13 21.7	**59**
25 41.0	11 26.9	11 21.4	24 46.6	13 29.5	**60**
26 25.2	12 49.3	12 17.1	25 16.4	13 37.7	**61**
27 14.9	14 18.0	13 14.7	25 47.4	13 46.3	**62**
28 11.8	15 53.8	14 14.3	26 19.7	13 55.3	**63**
29 18.2	17 37.5	15 15.9	26 53.4	14 4.8	**64**
♊ 0 38.1	19 30.4	16 19.7	27 28.5	14 14.9	**65**
2 19.2	21 33.5	17 25.7	28 5.3	14 25.5	**66**
5	6	Descendant	8	9	S LAT

♎ 11° 57′ 44″

$12^h\ 44^m\ 0^s$ MC 191° 0′ 0″

0ʰ 48ᵐ 0ˢ — MC 12° 0′ 0″ — ♈ 13° 2′ 39″

N LAT	11	12	Ascendant	2	3
°	° ′	° ′	° ′	° ′	° ′
0	♉14 27.7	♊13 24.0	♋11 2.1	♌ 9 33.6	♍10 29.9
5	14 57.2	14 41.6	12 59.1	10 36.4	10 44.1
10	15 27.8	16 1.4	14 56.1	11 38.3	10 58.2
15	15 59.8	17 24.8	16 54.5	12 40.1	11 12.4
16	16 6.5	17 42.0	17 18.5	12 52.6	11 15.3
17	16 13.2	17 59.5	17 42.7	13 5.1	11 18.2
18	16 20.1	18 17.2	18 7.0	13 17.6	11 21.1
19	16 27.1	18 35.2	18 31.4	13 30.2	11 24.0
20	16 34.1	18 53.4	18 56.1	13 42.9	11 26.9
21	16 41.3	19 11.9	19 20.9	13 55.7	11 29.9
22	16 48.7	19 30.7	19 45.9	14 8.5	11 32.9
23	16 56.2	19 49.9	20 11.2	14 21.4	11 35.9
24	17 3.8	20 9.4	20 36.7	14 34.4	11 39.0
25	17 11.6	20 29.2	21 2.4	14 47.6	11 42.0
26	17 19.5	20 49.5	21 28.4	15 0.8	11 45.2
27	17 27.6	21 10.1	21 54.7	15 14.2	11 48.3
28	17 36.0	21 31.2	22 21.3	15 27.7	11 51.5
29	17 44.5	21 52.7	22 48.3	15 41.4	11 54.7
30	17 53.2	22 14.8	23 15.6	15 55.2	11 58.0
31	18 2.2	22 37.4	23 43.2	16 9.2	12 1.3
32	18 11.4	23 0.5	24 11.3	16 23.4	12 4.7
33	18 20.9	23 24.2	24 39.7	16 37.8	12 8.1
34	18 30.7	23 48.5	25 8.6	16 52.4	12 11.6
35	18 40.8	24 13.6	25 37.9	17 7.2	12 15.1
36	18 51.1	24 39.3	26 7.7	17 22.3	12 18.8
37	19 1.9	25 5.8	26 38.0	17 37.6	12 22.4
38	19 13.0	25 33.1	27 8.9	17 53.1	12 26.2
39	19 24.5	26 1.2	27 40.3	18 9.0	12 30.0
40	19 36.5	26 30.3	28 12.2	18 25.1	12 33.9
41	19 48.9	27 0.4	28 44.9	18 41.5	12 37.9
42	20 1.8	27 31.5	29 18.1	18 58.3	12 42.0
43	20 15.2	28 3.8	29 52.1	19 15.5	12 46.2
44	20 29.3	28 37.3	♌ 0 26.8	19 33.0	12 50.5
45	20 44.0	29 12.1	1 2.2	19 50.9	12 54.9
46	20 59.4	29 48.3	1 38.4	20 9.3	12 59.5
47	21 15.5	♋ 0 26.0	2 15.5	20 28.0	13 4.1
48	21 32.5	1 5.4	2 53.5	20 47.3	13 8.9
49	21 50.4	1 46.6	3 32.4	21 7.1	13 13.9
50	22 9.4	2 29.6	4 12.3	21 27.4	13 18.9
51	22 29.4	3 14.8	4 53.2	21 48.3	13 24.2
52	22 50.8	4 2.2	5 35.2	22 9.8	13 29.6
53	23 13.5	4 52.2	6 18.4	22 31.9	13 35.3
54	23 37.8	5 44.8	7 2.8	22 54.8	13 41.1
55	24 3.9	6 40.4	7 48.4	23 18.3	13 47.2
56	24 32.1	7 39.2	8 35.4	23 42.7	13 53.4
57	25 2.6	8 41.7	9 23.8	24 7.9	14 0.0
58	25 35.8	9 48.1	10 13.7	24 33.9	14 6.8
59	26 12.3	10 58.9	11 5.1	25 1.0	14 13.9
60	26 52.5	12 14.6	11 58.2	25 29.0	14 21.3
61	27 37.4	13 35.7	12 53.1	25 58.2	14 29.1
62	28 28.1	15 2.9	13 49.7	26 28.5	14 37.2
63	29 26.1	16 37.0	14 48.3	27 0.1	14 45.8
64	♊ 0 34.0	18 18.7	15 49.0	27 33.0	14 54.8
65	1 56.1	20 9.1	16 51.7	28 7.4	15 4.3
66	3 41.0	22 9.3	17 56.8	28 43.4	15 14.3
S LAT	5	6	Descendant	8	9

♎ 13° 2′ 39″ — 12ʰ 48ᵐ 0ˢ — MC 192° 0′ 0″

0ʰ 52ᵐ 0ˢ — MC 13° 0′ 0″ — ♈ 14° 7′ 28″

N LAT	11	12	Ascendant	2	3
°	° ′	° ′	° ′	° ′	° ′
0	♉15 28.0	♊14 19.9	♋11 57.6	♌10 32.9	♍11 34.2
5	15 58.0	15 37.7	13 54.2	11 34.9	11 47.7
10	16 28.9	16 57.7	15 50.7	12 35.9	12 1.0
15	17 1.5	18 21.3	17 48.5	13 36.9	12 14.5
16	17 8.3	18 38.6	18 12.3	13 49.1	12 17.2
17	17 15.1	18 56.1	18 36.3	14 1.4	12 20.0
18	17 22.1	19 13.8	19 0.5	14 13.8	12 22.7
19	17 29.1	19 31.8	19 24.7	14 26.2	12 25.5
20	17 36.3	19 50.0	19 49.2	14 38.7	12 28.3
21	17 43.6	20 8.5	20 13.9	14 51.2	12 31.1
22	17 51.1	20 27.4	20 38.7	15 3.9	12 33.9
23	17 58.7	20 46.5	21 3.8	15 16.6	12 36.8
24	18 6.4	21 6.0	21 29.1	15 29.4	12 39.6
25	18 14.3	21 25.8	21 54.6	15 42.3	12 42.6
26	18 22.4	21 46.1	22 20.4	15 55.4	12 45.5
27	18 30.6	22 6.7	22 46.5	16 8.5	12 48.5
28	18 39.1	22 27.8	23 12.9	16 21.8	12 51.5
29	18 47.7	22 49.3	23 39.6	16 35.3	12 54.6
30	18 56.6	23 11.3	24 6.6	16 48.9	12 57.7
31	19 5.7	23 33.9	24 34.0	17 2.6	13 0.8
32	19 15.0	23 57.0	25 1.8	17 16.6	13 4.0
33	19 24.7	24 20.6	25 30.0	17 30.7	13 7.2
34	19 34.6	24 44.9	25 58.5	17 45.1	13 10.5
35	19 44.8	25 9.9	26 27.6	17 59.6	13 13.9
36	19 55.3	25 35.6	26 57.1	18 14.4	13 17.3
37	20 6.2	26 2.0	27 27.1	18 29.4	13 20.8
38	20 17.5	26 29.2	27 57.6	18 44.7	13 24.3
39	20 29.1	26 57.2	28 28.6	19 0.2	13 28.0
40	20 41.2	27 26.2	29 0.3	19 16.1	13 31.7
41	20 53.8	27 56.2	29 32.5	19 32.2	13 35.5
42	21 6.9	28 27.2	♌ 0 5.4	19 48.7	13 39.3
43	21 20.6	28 59.3	0 38.9	20 5.5	13 43.3
44	21 34.8	29 32.6	1 13.2	20 22.7	13 47.4
45	21 49.7	♋ 0 7.2	1 48.2	20 40.3	13 51.5
46	22 5.3	0 43.3	2 24.0	20 58.3	13 55.8
47	22 21.7	1 20.8	3 0.6	21 16.7	14 0.2
48	22 38.9	1 59.9	3 38.1	21 35.6	14 4.8
49	22 57.0	2 40.7	4 16.5	21 55.0	14 9.4
50	23 16.2	3 23.5	4 55.8	22 14.9	14 14.2
51	23 36.6	4 8.3	5 36.2	22 35.4	14 19.2
52	23 58.2	4 55.3	6 17.6	22 56.4	14 24.4
53	24 21.2	5 44.8	7 0.2	23 18.1	14 29.7
54	24 45.9	6 36.9	7 43.9	23 40.5	14 35.2
55	25 12.4	7 32.0	8 28.9	24 3.6	14 40.9
56	25 40.9	8 30.2	9 15.2	24 27.4	14 46.9
57	26 11.8	9 31.9	10 2.9	24 52.1	14 53.0
58	26 45.6	10 37.5	10 52.0	25 17.6	14 59.5
59	27 22.5	11 47.3	11 42.7	25 44.0	15 6.2
60	28 3.4	13 1.9	12 35.0	26 11.5	15 13.2
61	28 49.1	14 21.8	13 29.0	26 40.0	15 20.5
62	29 40.6	15 47.5	14 24.8	27 9.7	15 28.2
63	♊ 0 39.8	17 19.9	15 22.4	27 40.5	15 36.3
64	1 49.2	18 59.6	16 22.1	28 12.7	15 44.8
65	3 13.4	20 47.6	17 23.9	28 46.4	15 53.8
66	5 2.1	22 45.1	18 27.8	29 21.5	16 3.2
S LAT	5	6	Descendant	8	9

♎ 14° 7′ 28″ — 12ʰ 52ᵐ 0ˢ — MC 193° 0′ 0″

$0^h\ 56^m\ 0^s$ MC 14° 0′ 0″

♈ 15° 12′ 13″

N LAT	11	12	Ascendant	2	3
°	° ′	° ′	° ′	° ′	° ′
0	♉16 28.0	♊15 15.6	♋12 53.1	♌11 32.4	♍12 38.6
5	16 58.5	16 33.7	14 49.3	12 33.6	12 51.3
10	17 29.9	17 53.9	16 45.2	13 33.7	13 4.0
15	18 2.9	19 17.6	18 42.4	14 33.7	13 16.7
16	18 9.8	19 34.9	19 6.1	14 45.8	13 19.2
17	18 16.8	19 52.5	19 29.9	14 57.9	13 21.8
18	18 23.8	20 10.2	19 53.9	15 10.1	13 24.4
19	18 31.0	20 28.2	20 18.0	15 22.3	13 27.0
20	18 38.3	20 46.4	20 42.3	15 34.6	13 29.7
21	18 45.7	21 5.0	21 6.8	15 46.9	13 32.3
22	18 53.2	21 23.8	21 31.5	15 59.3	13 35.0
23	19 0.9	21 43.0	21 56.3	16 11.8	13 37.7
24	19 8.8	22 2.4	22 21.4	16 24.5	13 40.4
25	19 16.8	22 22.3	22 46.8	16 37.2	13 43.2
26	19 25.0	22 42.5	23 12.4	16 50.0	13 45.9
27	19 33.3	23 3.1	23 38.2	17 2.9	13 48.7
28	19 41.9	23 24.2	24 4.4	17 16.0	13 51.6
29	19 50.7	23 45.7	24 30.8	17 29.2	13 54.5
30	19 59.6	24 7.7	24 57.6	17 42.6	13 57.4
31	20 8.9	24 30.2	25 24.8	17 56.1	14 0.4
32	20 18.4	24 53.2	25 52.3	18 9.8	14 3.4
33	20 28.1	25 16.9	26 20.2	18 23.7	14 6.4
34	20 38.1	25 41.1	26 48.5	18 37.8	14 9.5
35	20 48.5	26 6.0	27 17.2	18 52.1	14 12.7
36	20 59.2	26 31.6	27 46.4	19 6.6	14 15.9
37	21 10.2	26 57.9	28 16.0	19 21.3	14 19.2
38	21 21.6	27 25.0	28 46.2	19 36.3	14 22.6
39	21 33.4	27 53.0	29 16.9	19 51.6	14 26.0
40	21 45.7	28 21.9	29 48.2	20 7.1	14 29.5
41	21 58.5	28 51.7	♌ 0 20.1	20 23.0	14 33.0
42	22 11.7	29 22.6	0 52.6	20 39.1	14 36.7
43	22 25.5	29 54.5	1 25.7	20 55.6	14 40.4
44	22 40.0	♋ 0 27.7	1 59.6	21 12.5	14 44.3
45	22 55.0	1 2.1	2 34.1	21 29.7	14 48.2
46	23 10.8	1 37.9	3 9.5	21 47.4	14 52.2
47	23 27.4	2 15.2	3 45.6	22 5.4	14 56.4
48	23 44.9	2 54.0	4 22.6	22 24.0	15 0.7
49	24 3.3	3 34.6	5 0.5	22 43.0	15 5.1
50	24 22.7	4 17.0	5 39.3	23 2.5	15 9.6
51	24 43.3	5 1.5	6 19.1	23 22.5	15 14.3
52	25 5.2	5 48.1	7 0.0	23 43.1	15 19.1
53	25 28.5	6 37.1	7 41.9	24 4.4	15 24.1
54	25 53.5	7 28.7	8 25.0	24 26.3	15 29.3
55	26 20.3	8 23.2	9 9.4	24 48.9	15 34.7
56	26 49.2	9 20.7	9 55.0	25 12.2	15 40.3
57	27 20.6	10 21.7	10 41.9	25 36.3	15 46.1
58	27 54.8	11 26.5	11 30.3	26 1.3	15 52.2
59	28 32.3	12 35.4	12 20.2	26 27.1	15 58.5
60	29 13.8	13 48.9	13 11.7	26 54.0	16 5.1
61	♊ 0 0.1	15 7.5	14 4.9	27 21.9	16 12.0
62	0 52.5	16 31.8	14 59.8	27 50.9	16 19.3
63	1 52.8	18 2.5	15 56.5	28 21.0	16 26.9
64	3 3.6	19 40.3	16 55.2	28 52.5	16 34.8
65	4 30.1	21 26.1	17 56.0	29 25.3	16 43.3
66	6 22.7	23 20.9	18 58.9	29 59.7	16 52.2
S LAT	5	6	Descendant	8	9

♎ 15° 12′ 13″

$12^h\ 56^m\ 0^s$ MC 194° 0′ 0″

$1^h\ 0^m\ 0^s$ MC 15° 0′ 0″

♈ 16° 16′ 51″

N LAT	11	12	Ascendant	2	3
°	° ′	° ′	° ′	° ′	° ′
0	♉17 27.9	♊16 11.3	♋13 48.7	♌12 32.1	♍13 43.2
5	17 58.8	17 29.6	15 44.5	13 32.4	13 55.1
10	18 30.7	18 50.0	17 39.9	14 31.6	14 7.0
15	19 4.2	20 13.9	19 36.3	15 30.7	14 18.9
16	19 11.1	20 31.2	19 59.9	15 42.6	14 21.3
17	19 18.2	20 48.7	20 23.6	15 54.5	14 23.8
18	19 25.3	21 6.5	20 47.4	16 6.5	14 26.2
19	19 32.6	21 24.5	21 11.3	16 18.5	14 28.7
20	19 40.0	21 42.7	21 35.4	16 30.5	14 31.1
21	19 47.5	22 1.3	21 59.7	16 42.7	14 33.6
22	19 55.2	22 20.1	22 24.2	16 54.9	14 36.1
23	20 3.0	22 39.2	22 48.9	17 7.2	14 38.7
24	20 10.9	22 58.7	23 13.8	17 19.6	14 41.2
25	20 19.0	23 18.6	23 38.9	17 32.1	14 43.8
26	20 27.3	23 38.8	24 4.3	17 44.7	14 46.4
27	20 35.8	23 59.4	24 29.9	17 57.4	14 49.1
28	20 44.5	24 20.4	24 55.8	18 10.2	14 51.8
29	20 53.3	24 41.9	25 22.0	18 23.2	14 54.5
30	21 2.4	25 3.8	25 48.6	18 36.4	14 57.2
31	21 11.8	25 26.3	26 15.4	18 49.6	15 0.0
32	21 21.4	25 49.3	26 42.7	19 3.1	15 2.8
33	21 31.3	26 12.8	27 10.3	19 16.7	15 5.7
34	21 41.4	26 37.0	27 38.3	19 30.5	15 8.6
35	21 51.9	27 1.8	28 6.7	19 44.6	15 11.6
36	22 2.7	27 27.4	28 35.6	19 58.8	15 14.6
37	22 13.9	27 53.6	29 5.0	20 13.3	15 17.7
38	22 25.5	28 20.6	29 34.8	20 28.0	15 20.8
39	22 37.4	28 48.5	♌ 0 5.2	20 43.0	15 24.1
40	22 49.8	29 17.2	0 36.1	20 58.2	15 27.3
41	23 2.7	29 46.9	1 7.6	21 13.8	15 30.7
42	23 16.2	♋ 0 17.7	1 39.7	21 29.6	15 34.1
43	23 30.2	0 49.5	2 12.4	21 45.8	15 37.6
44	23 44.8	1 22.4	2 45.8	22 2.3	15 41.2
45	24 0.0	1 56.7	3 20.0	22 19.2	15 44.9
46	24 16.0	2 32.2	3 54.8	22 36.5	15 48.7
47	24 32.8	3 9.3	4 30.5	22 54.2	15 52.6
48	24 50.5	3 47.8	5 7.0	23 12.3	15 56.6
49	25 9.1	4 28.1	5 44.4	23 30.9	16 0.7
50	25 28.7	5 10.2	6 22.7	23 50.0	16 5.0
51	25 49.6	5 54.3	7 1.9	24 9.7	16 9.4
52	26 11.7	6 40.5	7 42.2	24 29.8	16 13.9
53	26 35.4	7 29.1	8 23.6	24 50.6	16 18.6
54	27 0.6	8 20.2	9 6.1	25 12.1	16 23.5
55	27 27.8	9 14.1	9 49.8	25 34.2	16 28.6
56	27 57.1	10 11.0	10 34.7	25 57.0	16 33.8
57	28 28.8	11 11.2	11 21.0	26 20.6	16 39.3
58	29 3.5	12 15.2	12 8.6	26 45.0	16 45.0
59	29 41.5	13 23.1	12 57.7	27 10.3	16 50.9
60	♊ 0 23.5	14 35.6	13 48.4	27 36.5	16 57.1
61	1 10.6	15 53.0	14 40.7	28 3.8	17 3.5
62	2 3.8	17 15.9	15 34.8	28 32.1	17 10.3
63	3 5.1	18 44.9	16 30.6	29 1.6	17 17.5
64	4 17.4	20 20.8	17 28.4	29 32.3	17 24.9
65	5 46.0	22 4.4	18 28.2	♍ 0 4.4	17 32.8
66	7 42.6	23 56.6	19 30.1	0 37.9	17 41.2
S LAT	5	6	Descendant	8	9

♎ 16° 16′ 51″

$13^h\ 0^m\ 0^s$ MC 195° 0′ 0″

1ʰ 4ᵐ 0ˢ		MC		16° 0′ 0″
		♈ 17° 21′ 23″		

N LAT	11	12	Ascendant	2	3
°	° ′	° ′	° ′	° ′	° ′
0	♉18 27.6	♊17 6.9	♋14 44.4	♌13 32.0	♍14 47.8
5	18 58.9	18 25.4	16 39.7	14 31.3	14 59.0
10	19 31.2	19 46.0	18 34.5	15 29.6	15 10.1
15	20 5.2	21 10.0	20 30.3	16 27.8	15 21.3
16	20 12.2	21 27.3	20 53.7	16 39.5	15 23.5
17	20 19.4	21 44.8	21 17.2	16 51.2	15 25.8
18	20 26.6	22 2.6	21 40.8	17 2.9	15 28.1
19	20 34.0	22 20.6	22 4.6	17 14.7	15 30.4
20	20 41.5	22 38.9	22 28.5	17 26.6	15 32.7
21	20 49.1	22 57.4	22 52.6	17 38.5	15 35.0
22	20 56.8	23 16.2	23 16.9	17 50.6	15 37.4
23	21 4.7	23 35.4	23 41.4	18 2.6	15 39.8
24	21 12.8	23 54.8	24 6.1	18 14.8	15 42.1
25	21 21.0	24 14.6	24 31.0	18 27.1	15 44.6
26	21 29.4	24 34.8	24 56.1	18 39.5	15 47.0
27	21 38.0	24 55.4	25 21.5	18 51.9	15 49.5
28	21 46.8	25 16.4	25 47.2	19 4.6	15 52.0
29	21 55.7	25 37.8	26 13.2	19 17.3	15 54.5
30	22 5.0	25 59.7	26 39.5	19 30.2	15 57.1
31	22 14.4	26 22.2	27 6.1	19 43.2	15 59.7
32	22 24.1	26 45.1	27 33.0	19 56.4	16 2.3
33	22 34.1	27 8.6	28 0.4	20 9.8	16 5.0
34	22 44.4	27 32.7	28 28.1	20 23.4	16 7.8
35	22 55.0	27 57.5	28 56.2	20 37.1	16 10.5
36	23 6.0	28 22.9	29 24.8	20 51.1	16 13.4
37	23 17.3	28 49.1	29 53.8	21 5.3	16 16.2
38	23 29.0	29 16.0	♌ 0 23.3	21 19.7	16 19.2
39	23 41.1	29 43.7	0 53.3	21 34.4	16 22.2
40	23 53.7	♋ 0 12.4	1 23.9	21 49.3	16 25.3
41	24 6.7	0 41.9	1 55.0	22 4.6	16 28.4
42	24 20.3	1 12.5	2 26.7	22 20.1	16 31.6
43	24 34.4	1 44.1	2 59.0	22 36.0	16 34.9
44	24 49.2	2 16.9	3 32.0	22 52.1	16 38.2
45	25 4.6	2 50.9	4 5.7	23 8.7	16 41.7
46	25 20.8	3 26.3	4 40.2	23 25.6	16 45.2
47	25 37.8	4 3.1	5 15.4	23 42.9	16 48.9
48	25 55.7	4 41.4	5 51.4	24 0.7	16 52.6
49	26 14.5	5 21.3	6 28.2	24 18.9	16 56.5
50	26 34.4	6 3.1	7 6.0	24 37.6	17 0.5
51	26 55.5	6 46.8	7 44.7	24 56.8	17 4.6
52	27 17.9	7 32.6	8 24.4	25 16.6	17 8.8
53	27 41.8	8 20.7	9 5.2	25 36.9	17 13.2
54	28 7.3	9 11.3	9 47.1	25 57.9	17 17.7
55	28 34.8	10 4.6	10 30.1	26 19.5	17 22.5
56	29 4.4	11 0.9	11 14.4	26 41.8	17 27.4
57	29 36.6	12 0.4	11 59.9	27 4.9	17 32.5
58	♊ 0 11.6	13 3.5	12 46.9	27 28.7	17 37.8
59	0 50.1	14 10.5	13 35.2	27 53.4	17 43.3
60	1 32.7	15 21.9	14 25.1	28 19.1	17 49.1
61	2 20.4	16 38.1	15 16.6	28 45.7	17 55.1
62	3 14.4	17 59.7	16 9.8	29 13.3	18 1.4
63	4 16.7	19 27.1	17 4.7	29 42.1	18 8.1
64	5 30.4	21 1.2	18 1.5	♍ 0 12.1	18 15.1
65	7 1.1	22 42.6	19 0.4	0 43.4	18 22.4
66	9 1.8	24 32.3	20 1.2	1 16.1	18 30.2
S LAT	5	6	Descendant	8	9

		♎ 17° 21′ 23″		
13ʰ 4ᵐ 0ˢ		MC		196° 0′ 0″

1ʰ 8ᵐ 0ˢ		MC		17° 0′ 0″
		♈ 18° 25′ 48″		

N LAT	11	12	Ascendant	2	3
°	° ′	° ′	° ′	° ′	° ′
0	♉19 27.1	♊18 2.4	♋15 40.1	♌14 32.0	♍15 52.5
5	19 58.8	19 21.2	17 35.0	15 30.5	16 3.0
10	20 31.6	20 41.9	19 29.2	16 27.8	16 13.3
15	21 6.0	22 5.9	21 24.2	17 25.0	16 23.7
16	21 13.1	22 23.2	21 47.4	17 36.5	16 25.8
17	21 20.4	22 40.8	22 10.8	17 48.0	16 27.9
18	21 27.7	22 58.6	22 34.3	17 59.5	16 30.0
19	21 35.2	23 16.6	22 57.9	18 11.1	16 32.2
20	21 42.7	23 34.8	23 21.6	18 22.8	16 34.3
21	21 50.4	23 53.4	23 45.5	18 34.5	16 36.5
22	21 58.3	24 12.2	24 9.6	18 46.3	16 38.7
23	22 6.3	24 31.3	24 33.9	18 58.2	16 40.9
24	22 14.4	24 50.7	24 58.3	19 10.1	16 43.1
25	22 22.7	25 10.5	25 23.0	19 22.2	16 45.4
26	22 31.2	25 30.7	25 47.9	19 34.3	16 47.6
27	22 39.9	25 51.2	26 13.1	19 46.6	16 49.9
28	22 48.8	26 12.2	26 38.5	19 58.9	16 52.3
29	22 57.9	26 33.6	27 4.3	20 11.5	16 54.6
30	23 7.2	26 55.5	27 30.3	20 24.1	16 57.0
31	23 16.8	27 17.8	27 56.6	20 36.9	16 59.4
32	23 26.6	27 40.7	28 23.3	20 49.8	17 1.9
33	23 36.7	28 4.2	28 50.4	21 2.9	17 4.4
34	23 47.1	28 28.2	29 17.8	21 16.2	17 7.0
35	23 57.9	28 52.9	29 45.6	21 29.7	17 9.5
36	24 8.9	29 18.2	♌ 0 13.9	21 43.4	17 12.2
37	24 20.4	29 44.3	0 42.6	21 57.3	17 14.8
38	24 32.2	♋ 0 11.1	1 11.7	22 11.5	17 17.6
39	24 44.4	0 38.7	1 41.4	22 25.9	17 20.4
40	24 57.2	1 7.2	2 11.6	22 40.5	17 23.2
41	25 10.4	1 36.6	2 42.3	22 55.4	17 26.1
42	25 24.1	2 7.1	3 13.7	23 10.6	17 29.1
43	25 38.4	2 38.5	3 45.6	23 26.2	17 32.2
44	25 53.3	3 11.1	4 18.2	23 42.0	17 35.3
45	26 8.9	3 44.9	4 51.4	23 58.2	17 38.5
46	26 25.3	4 20.0	5 25.4	24 14.8	17 41.8
47	26 42.4	4 56.6	6 0.1	24 31.7	17 45.2
48	27 0.5	5 34.6	6 35.7	24 49.1	17 48.7
49	27 19.5	6 14.3	7 12.0	25 6.9	17 52.3
50	27 39.6	6 55.7	7 49.3	25 25.2	17 56.0
51	28 0.9	7 39.0	8 27.4	25 44.0	17 59.8
52	28 23.6	8 24.4	9 6.6	26 3.3	18 3.7
53	28 47.7	9 12.0	9 46.8	26 23.2	18 7.8
54	29 13.6	10 2.1	10 28.0	26 43.7	18 12.0
55	29 41.3	10 54.8	11 10.4	27 4.8	18 16.4
56	♊ 0 11.3	11 50.5	11 54.0	27 26.6	18 21.0
57	0 43.8	12 49.3	12 38.9	27 49.2	18 25.7
58	1 19.2	13 51.5	13 25.1	28 12.5	18 30.6
59	1 58.2	14 57.7	14 12.7	28 36.6	18 35.7
60	2 41.3	16 8.0	15 1.8	29 1.7	18 41.1
61	3 29.6	17 23.0	15 52.5	29 27.7	18 46.7
62	4 24.4	18 43.2	16 44.8	29 54.7	18 52.6
63	5 27.7	20 9.1	17 38.8	♍ 0 22.8	18 58.8
64	6 42.7	21 41.4	18 34.7	0 52.0	19 5.3
65	8 15.5	23 20.8	19 32.6	1 22.6	19 12.1
66	10 20.3	25 8.1	20 32.5	1 54.5	19 19.3
S LAT	5	6	Descendant	8	9

		♎ 18° 25′ 48″		
13ʰ 8ᵐ 0ˢ		MC		197° 0′ 0″

1ʰ 12ᵐ 0ˢ MC 18° 0′ 0″

♈ 19° 30′ 6″

N LAT	11	12	Ascendant	2	3
°	° ′	° ′	° ′	° ′	° ′
0	♉20 26.4	♊18 57.9	♋16 36.0	♌15 32.3	♍16 57.4
5	20 58.6	20 16.8	18 30.3	16 29.7	17 7.0
10	21 31.8	21 37.7	20 23.9	17 26.1	17 16.6
15	22 6.6	23 1.7	22 18.2	18 22.3	17 26.2
16	22 13.8	23 19.1	22 41.2	18 33.6	17 28.1
17	22 21.1	23 36.6	23 4.4	18 44.9	17 30.1
18	22 28.5	23 54.4	23 27.7	18 56.2	17 32.1
19	22 36.1	24 12.4	23 51.1	19 7.6	17 34.0
20	22 43.7	24 30.7	24 14.7	19 19.1	17 36.0
21	22 51.5	24 49.2	24 38.4	19 30.6	17 38.0
22	22 59.5	25 8.0	25 2.3	19 42.1	17 40.1
23	23 7.6	25 27.1	25 26.3	19 53.8	17 42.1
24	23 15.8	25 46.5	25 50.6	20 5.5	17 44.2
25	23 24.2	26 6.3	26 15.0	20 17.3	17 46.2
26	23 32.8	26 26.4	26 39.7	20 29.2	17 48.4
27	23 41.6	26 46.9	27 4.7	20 41.3	17 50.5
28	23 50.6	27 7.9	27 29.9	20 53.4	17 52.6
29	23 59.8	27 29.2	27 55.3	21 5.7	17 54.8
30	24 9.2	27 51.0	28 21.1	21 18.1	17 57.0
31	24 18.9	28 13.3	28 47.2	21 30.6	17 59.3
32	24 28.8	28 36.2	29 13.6	21 43.3	18 1.5
33	24 39.0	28 59.5	29 40.3	21 56.2	18 3.9
34	24 49.6	29 23.5	♌ 0 7.5	22 9.2	18 6.2
35	25 0.4	29 48.1	0 35.0	22 22.4	18 8.6
36	25 11.6	♋ 0 13.3	1 2.9	22 35.8	18 11.0
37	25 23.2	0 39.3	1 31.3	22 49.4	18 13.5
38	25 35.1	1 6.0	2 0.1	23 3.3	18 16.0
39	25 47.5	1 33.5	2 29.4	23 17.4	18 18.6
40	26 0.3	2 1.9	2 59.3	23 31.7	18 21.2
41	26 13.7	2 31.1	3 29.6	23 46.3	18 23.9
42	26 27.5	3 1.4	4 0.6	24 1.2	18 26.7
43	26 42.0	3 32.7	4 32.1	24 16.4	18 29.5
44	26 57.1	4 5.1	5 4.2	24 31.9	18 32.4
45	27 12.8	4 38.7	5 37.1	24 47.8	18 35.4
46	27 29.4	5 13.5	6 10.6	25 4.0	18 38.4
47	27 46.7	5 49.8	6 44.9	25 20.6	18 41.5
48	28 4.9	6 27.6	7 19.9	25 37.6	18 44.8
49	28 24.2	7 6.9	7 55.8	25 55.0	18 48.1
50	28 44.5	7 48.0	8 32.5	26 12.9	18 51.5
51	29 6.0	8 30.9	9 10.1	26 31.2	18 55.0
52	29 28.9	9 15.9	9 48.7	26 50.1	18 58.7
53	29 53.3	10 3.1	10 28.3	27 9.6	19 2.4
54	♊ 0 19.4	10 52.6	11 8.9	27 29.6	19 6.3
55	0 47.4	11 44.7	11 50.7	27 50.2	19 10.4
56	1 17.7	12 39.7	12 33.6	28 11.5	19 14.6
57	1 50.5	13 37.8	13 17.8	28 33.5	19 18.9
58	2 26.4	14 39.3	14 3.3	28 56.3	19 23.5
59	3 5.7	15 44.5	14 50.1	29 19.9	19 28.2
60	3 49.4	16 53.8	15 38.4	29 44.3	19 33.2
61	4 38.2	18 7.6	16 28.3	♍ 0 9.6	19 38.4
62	5 33.7	19 26.5	17 19.8	0 36.0	19 43.8
63	6 37.9	20 50.9	18 13.0	1 3.4	19 49.5
64	7 54.2	22 21.4	19 7.9	1 32.0	19 55.5
65	9 29.0	23 58.8	20 4.8	2 1.7	20 1.8
66	11 38.1	25 43.8	21 3.7	2 32.9	20 8.4
S LAT	5	6	Descendant	8	9

♎ 19° 30′ 6″

13ʰ 12ᵐ 0ˢ MC 198° 0′ 0″

1ʰ 16ᵐ 0ˢ MC 19° 0′ 0″

♈ 20° 34′ 17″

N LAT	11	12	Ascendant	2	3
°	° ′	° ′	° ′	° ′	° ′
0	♉21 25.6	♊19 53.3	♋17 31.9	♌16 32.7	♍18 2.3
5	21 58.1	21 12.4	19 25.7	17 29.2	18 11.1
10	22 31.7	22 33.3	21 18.6	18 24.6	18 19.9
15	23 6.9	23 57.5	23 12.1	19 19.7	18 28.8
16	23 14.2	24 14.8	23 35.0	19 30.8	18 30.6
17	23 21.6	24 32.4	23 58.0	19 41.9	18 32.4
18	23 29.2	24 50.1	24 21.1	19 53.0	18 34.2
19	23 36.8	25 8.1	24 44.4	20 4.2	18 36.0
20	23 44.5	25 26.4	25 7.7	20 15.4	18 37.8
21	23 52.4	25 44.9	25 31.2	20 26.7	18 39.6
22	24 0.4	26 3.6	25 54.9	20 38.1	18 41.5
23	24 8.6	26 22.7	26 18.8	20 49.5	18 43.4
24	24 17.0	26 42.1	26 42.8	21 1.0	18 45.3
25	24 25.5	27 1.9	27 7.0	21 12.6	18 47.2
26	24 34.1	27 22.0	27 31.5	21 24.3	18 49.1
27	24 43.0	27 42.4	27 56.2	21 36.0	18 51.1
28	24 52.1	28 3.3	28 21.1	21 47.9	18 53.0
29	25 1.4	28 24.6	28 46.4	22 0.0	18 55.0
30	25 10.9	28 46.4	29 11.9	22 12.1	18 57.1
31	25 20.7	29 8.6	29 37.7	22 24.4	18 59.1
32	25 30.8	29 31.4	♌ 0 3.8	22 36.8	19 1.2
33	25 41.1	29 54.7	0 30.3	22 49.4	19 3.3
34	25 51.7	♋ 0 18.6	0 57.1	23 2.2	19 5.5
35	26 2.7	0 43.1	1 24.3	23 15.1	19 7.7
36	26 14.0	1 8.2	1 51.9	23 28.3	19 9.9
37	26 25.6	1 34.1	2 19.9	23 41.6	19 12.2
38	26 37.7	2 0.7	2 48.4	23 55.2	19 14.5
39	26 50.2	2 28.0	3 17.4	24 8.9	19 16.9
40	27 3.2	2 56.3	3 46.9	24 23.0	19 19.3
41	27 16.7	3 25.4	4 16.8	24 37.3	19 21.8
42	27 30.7	3 55.4	4 47.4	24 51.8	19 24.3
43	27 45.3	4 26.6	5 18.5	25 6.7	19 26.9
44	28 0.5	4 58.8	5 50.3	25 21.9	19 29.5
45	28 16.4	5 32.1	6 22.7	25 37.4	19 32.3
46	28 33.1	6 6.8	6 55.7	25 53.2	19 35.1
47	28 50.6	6 42.8	7 29.5	26 9.4	19 37.9
48	29 9.0	7 20.2	8 4.1	26 26.1	19 40.9
49	29 28.4	7 59.3	8 39.4	26 43.1	19 43.9
50	29 48.9	8 40.0	9 15.6	27 0.6	19 47.1
51	♊ 0 10.6	9 22.6	9 52.7	27 18.5	19 50.3
52	0 33.7	10 7.1	10 30.7	27 37.0	19 53.6
53	0 58.4	10 53.8	11 9.7	27 55.9	19 57.1
54	1 24.7	11 42.8	11 49.8	28 15.5	20 0.7
55	1 53.0	12 34.4	12 30.9	28 35.6	20 4.4
56	2 23.6	13 28.7	13 13.2	28 56.4	20 8.2
57	2 56.8	14 26.1	13 56.7	29 17.9	20 12.2
58	3 32.9	15 26.8	14 41.4	29 40.1	20 16.4
59	4 12.7	16 31.1	15 27.5	♍ 0 3.1	20 20.8
60	4 56.8	17 39.3	16 15.1	0 26.9	20 25.3
61	5 46.3	18 52.0	17 4.1	0 51.7	20 30.1
62	6 42.4	20 9.6	17 54.8	1 17.4	20 35.0
63	7 47.5	21 32.5	18 47.1	1 44.1	20 40.2
64	9 5.0	23 1.3	19 41.2	2 11.9	20 45.7
65	10 41.8	24 36.8	20 37.1	2 41.0	20 51.5
66	12 55.0	26 19.5	21 35.0	3 11.3	20 57.6
S LAT	5	6	Descendant	8	9

♎ 20° 34′ 17″

13ʰ 16ᵐ 0ˢ MC 199° 0′ 0″

1h 20m 0s MC 20° 0′ 0″

♈ 21° 38′ 20″

11	12	Ascendant	2	3	N LAT
° ′	° ′	° ′	° ′	° ′	°
♉22 24.6	♊20 48.6	♋18 27.9	♌17 33.3	♍19 7.3	0
22 57.5	22 7.9	20 21.2	18 28.8	19 15.3	5
23 31.5	23 28.9	22 13.4	19 23.2	19 23.3	10
24 7.1	24 53.1	24 6.1	20 17.3	19 31.4	15
24 14.5	25 10.4	24 28.9	20 28.1	19 33.0	16
24 22.0	25 27.9	24 51.7	20 39.0	19 34.7	17
24 29.6	25 45.7	25 14.6	20 49.9	19 36.3	18
24 37.3	26 3.7	25 37.6	21 0.9	19 38.0	19
24 45.1	26 21.9	26 0.8	21 11.9	19 39.6	20
24 53.1	26 40.4	26 24.1	21 23.0	19 41.3	21
25 1.2	26 59.2	26 47.6	21 34.1	19 43.0	22
25 9.4	27 18.2	27 11.2	21 45.3	19 44.7	23
25 17.9	27 37.6	27 35.0	21 56.6	19 46.4	24
25 26.5	27 57.3	27 59.0	22 7.9	19 48.2	25
25 35.2	28 17.4	28 23.2	22 19.4	19 49.9	26
25 44.2	28 37.8	28 47.7	22 30.9	19 51.7	27
25 53.4	28 58.6	29 12.4	22 42.6	19 53.5	28
26 2.8	29 19.9	29 37.4	22 54.3	19 55.3	29
26 12.4	29 41.6	♌ 0 2.6	23 6.2	19 57.2	30
26 22.3	♋ 0 3.8	0 28.1	23 18.3	19 59.1	31
26 32.4	0 26.4	0 54.0	23 30.4	20 1.0	32
26 42.8	0 49.7	1 20.1	23 42.7	20 2.9	33
26 53.6	1 13.5	1 46.7	23 55.2	20 4.8	34
27 4.6	1 37.9	2 13.6	24 7.9	20 6.8	35
27 16.1	2 2.9	2 40.9	24 20.8	20 8.9	36
27 27.8	2 28.6	3 8.6	24 33.8	20 11.0	37
27 40.0	2 55.1	3 36.7	24 47.1	20 13.1	38
27 52.7	3 22.4	4 5.3	25 0.6	20 15.2	39
28 5.7	3 50.4	4 34.4	25 14.3	20 17.4	40
28 19.3	4 19.4	5 4.0	25 28.3	20 19.7	41
28 33.5	4 49.3	5 34.2	25 42.5	20 22.0	42
28 48.2	5 20.2	6 4.9	25 57.0	20 24.3	43
29 3.6	5 52.2	6 36.2	26 11.9	20 26.7	44
29 19.6	6 25.4	7 8.2	26 27.0	20 29.2	45
29 36.5	6 59.8	7 40.8	26 42.5	20 31.8	46
29 54.1	7 35.5	8 14.1	26 58.3	20 34.4	47
♊ 0 12.7	8 12.7	8 48.2	27 14.6	20 37.0	48
0 32.3	8 51.4	9 23.1	27 31.2	20 39.8	49
0 53.0	9 31.7	9 58.7	27 48.3	20 42.7	50
1 14.9	10 13.9	10 35.3	28 5.8	20 45.6	51
1 38.2	10 58.0	11 12.7	28 23.8	20 48.6	52
2 3.0	11 44.2	11 51.1	28 42.3	20 51.8	53
2 29.6	12 32.8	12 30.6	29 1.4	20 55.0	54
2 58.2	13 23.7	13 11.1	29 21.1	20 58.4	55
3 29.0	14 17.4	13 52.7	29 41.3	21 1.9	56
4 2.5	15 14.1	14 35.5	♍ 0 2.3	21 5.6	57
4 39.0	16 14.0	15 19.6	0 24.0	21 9.4	58
5 19.2	17 17.4	16 5.0	0 46.4	21 13.3	59
6 3.7	18 24.6	16 51.7	1 9.6	21 17.4	60
6 53.7	19 36.2	17 40.0	1 33.7	21 21.8	61
7 50.5	20 52.5	18 29.8	1 58.8	21 26.3	62
8 56.4	22 13.9	19 21.2	2 24.8	21 31.0	63
10 15.0	23 41.1	20 14.4	2 52.0	21 36.0	64
11 53.7	25 14.6	21 9.4	3 20.2	21 41.3	65
14 11.1	26 55.2	22 6.3	3 49.8	21 46.8	66
5	6	Descendant	8	9	S LAT

♎ 21° 38′ 20″

13h 20m 0s MC 200° 0′ 0″

1h 24m 0s MC 21° 0′ 0″

♈ 22° 42′ 16″

11	12	Ascendant	2	3	N LAT
° ′	° ′	° ′	° ′	° ′	°
♉23 23.4	♊21 43.9	♋19 24.1	♌18 34.0	♍20 12.3	0
23 56.7	23 3.3	21 16.7	19 28.5	20 19.6	5
24 31.0	24 24.4	23 8.3	20 21.9	20 26.8	10
25 7.0	25 48.6	25 0.2	21 15.0	20 34.1	15
25 14.5	26 5.9	25 22.7	21 25.6	20 35.6	16
25 22.1	26 23.4	25 45.3	21 36.3	20 37.0	17
25 29.7	26 41.2	26 8.1	21 47.0	20 38.5	18
25 37.5	26 59.1	26 30.9	21 57.7	20 40.0	19
25 45.4	27 17.3	26 53.9	22 8.5	20 41.5	20
25 53.5	27 35.8	27 17.0	22 19.3	20 43.0	21
26 1.7	27 54.6	27 40.2	22 30.2	20 44.5	22
26 10.0	28 13.6	28 3.6	22 41.2	20 46.1	23
26 18.5	28 32.9	28 27.2	22 52.2	20 47.6	24
26 27.2	28 52.6	28 51.0	23 3.3	20 49.2	25
26 36.1	29 12.6	29 15.0	23 14.5	20 50.8	26
26 45.1	29 33.0	29 39.2	23 25.8	20 52.4	27
26 54.4	29 53.8	♌ 0 3.6	23 37.2	20 54.0	28
27 3.9	♋ 0 15.0	0 28.3	23 48.8	20 55.7	29
27 13.6	0 36.6	0 53.3	24 0.4	20 57.3	30
27 23.6	0 58.7	1 18.5	24 12.2	20 59.0	31
27 33.8	1 21.3	1 44.1	24 24.1	21 0.7	32
27 44.3	1 44.5	2 10.0	24 36.1	21 2.5	33
27 55.2	2 8.2	2 36.2	24 48.4	21 4.2	34
28 6.3	2 32.5	3 2.8	25 0.7	21 6.0	35
28 17.8	2 57.4	3 29.8	25 13.3	21 7.9	36
28 29.7	3 23.0	3 57.1	25 26.1	21 9.7	37
28 42.0	3 49.4	4 24.9	25 39.0	21 11.6	38
28 54.8	4 16.5	4 53.2	25 52.2	21 13.6	39
29 8.0	4 44.4	5 21.9	26 5.6	21 15.6	40
29 21.7	5 13.2	5 51.1	26 19.3	21 17.6	41
29 36.0	5 42.9	6 20.9	26 33.2	21 19.7	42
29 50.8	6 13.6	6 51.2	26 47.4	21 21.8	43
♊ 0 6.3	6 45.4	7 22.1	27 1.9	21 24.0	44
0 22.5	7 18.3	7 53.6	27 16.7	21 26.2	45
0 39.5	7 52.5	8 25.8	27 31.8	21 28.5	46
0 57.3	8 28.0	8 58.7	27 47.3	21 30.8	47
1 16.0	9 4.8	9 32.3	28 3.1	21 33.2	48
1 35.8	9 43.2	10 6.6	28 19.3	21 35.7	49
1 56.6	10 23.2	10 41.8	28 36.0	21 38.3	50
2 18.7	11 5.0	11 17.8	28 53.1	21 41.0	51
2 42.2	11 48.7	11 54.7	29 10.7	21 43.7	52
3 7.3	12 34.4	12 32.5	29 28.7	21 46.5	53
3 34.1	13 22.4	13 11.3	29 47.4	21 49.5	54
4 2.9	14 12.8	13 51.2	♍ 0 6.5	21 52.5	55
4 34.0	15 5.9	14 32.2	0 26.3	21 55.7	56
5 7.7	16 1.8	15 14.3	0 46.7	21 58.9	57
5 44.6	17 0.9	15 57.7	1 7.9	22 2.3	58
6 25.1	18 3.4	16 42.3	1 29.7	22 5.9	59
7 10.0	19 9.7	17 28.4	1 52.4	22 9.6	60
8 0.5	20 20.2	18 15.8	2 15.8	22 13.5	61
8 57.9	21 35.2	19 4.8	2 40.2	22 17.6	62
10 4.5	22 55.2	19 55.4	3 5.6	22 21.8	63
11 24.3	24 20.8	20 47.7	3 32.0	22 26.3	64
13 4.7	25 52.5	21 41.7	3 59.6	22 31.1	65
15 26.2	27 30.9	22 37.7	4 28.3	22 36.0	66
5	6	Descendant	8	9	S LAT

♎ 22° 42′ 16″

13h 24m 0s MC 201° 0′ 0″

1ʰ 28ᵐ 0ˢ MC 22° 0′ 0″

♈ 23° 46′ 3″

11	12	Ascendant	2	3	N LAT
° ′	° ′	° ′	° ′	° ′	°
♉24 22.0	♊22 39.2	♋20 20.3	♌19 35.0	♍21 17.5	0
24 55.7	23 58.6	22 12.4	20 28.5	21 23.9	5
25 30.4	25 19.8	24 3.2	21 20.8	21 30.4	10
26 6.8	26 44.0	25 54.2	22 12.7	21 36.8	15
26 14.3	27 1.3	26 16.6	22 23.2	21 38.1	16
26 21.9	27 18.8	26 39.0	22 33.6	21 39.5	17
26 29.7	27 36.5	27 1.5	22 44.1	21 40.8	18
26 37.6	27 54.5	27 24.2	22 54.6	21 42.1	19
26 45.6	28 12.7	27 46.9	23 5.2	21 43.4	20
26 53.7	28 31.1	28 9.8	23 15.8	21 44.8	21
27 2.0	28 49.8	28 32.9	23 26.4	21 46.1	22
27 10.4	29 8.8	28 56.1	23 37.2	21 47.5	23
27 19.0	29 28.1	29 19.4	23 48.0	21 48.9	24
27 27.7	29 47.7	29 43.0	23 58.8	21 50.3	25
27 36.7	♋ 0 7.7	♌ 0 6.7	24 9.8	21 51.7	26
27 45.8	0 28.0	0 30.7	24 20.9	21 53.1	27
27 55.2	0 48.8	0 54.8	24 32.0	21 54.6	28
28 4.7	1 9.9	1 19.3	24 43.3	21 56.0	29
28 14.5	1 31.5	1 44.0	24 54.7	21 57.5	30
28 24.6	1 53.5	2 8.9	25 6.2	21 59.0	31
28 34.9	2 16.0	2 34.2	25 17.8	22 0.5	32
28 45.5	2 39.1	2 59.8	25 29.6	22 2.1	33
28 56.5	3 2.7	3 25.7	25 41.5	22 3.7	34
29 7.7	3 26.9	3 52.0	25 53.6	22 5.3	35
29 19.4	3 51.7	4 18.6	26 5.9	22 6.9	36
29 31.4	4 17.2	4 45.7	26 18.4	22 8.6	37
29 43.8	4 43.4	5 13.1	26 31.1	22 10.3	38
29 56.6	5 10.4	5 41.0	26 43.9	22 12.0	39
♊ 0 9.9	5 38.1	6 9.4	26 57.0	22 13.7	40
0 23.7	6 6.8	6 38.2	27 10.4	22 15.5	41
0 38.1	6 36.3	7 7.6	27 23.9	22 17.4	42
0 53.1	7 6.8	7 37.5	27 37.8	22 19.3	43
1 8.7	7 38.4	8 8.0	27 51.9	22 21.2	44
1 25.1	8 11.1	8 39.1	28 6.4	22 23.2	45
1 42.2	8 45.0	9 10.8	28 21.1	22 25.2	46
2 0.1	9 20.2	9 43.2	28 36.2	22 27.3	47
2 19.0	9 56.8	10 16.3	28 51.7	22 29.5	48
2 38.9	10 34.8	10 50.2	29 7.5	22 31.7	49
2 59.9	11 14.5	11 24.8	29 23.8	22 34.0	50
3 22.2	11 55.9	12 0.3	29 40.4	22 36.3	51
3 45.9	12 39.1	12 36.6	29 57.6	22 38.8	52
4 11.1	13 24.4	13 13.9	♍ 0 15.2	22 41.3	53
4 38.1	14 11.8	13 52.1	0 33.3	22 43.9	54
5 7.2	15 1.7	14 31.3	0 52.0	22 46.6	55
5 38.5	15 54.1	15 11.7	1 11.3	22 49.4	56
6 12.5	16 49.3	15 53.1	1 31.2	22 52.3	57
6 49.6	17 47.6	16 35.8	1 51.8	22 55.4	58
7 30.5	18 49.3	17 19.7	2 13.1	22 58.5	59
8 15.8	19 54.6	18 5.0	2 35.1	23 1.8	60
9 6.7	21 3.9	18 51.7	2 58.0	23 5.3	61
10 4.6	22 17.7	19 39.8	3 21.7	23 8.9	62
11 12.0	23 36.3	20 29.6	3 46.4	23 12.7	63
12 32.8	25 0.3	21 21.0	4 12.1	23 16.7	64
14 14.8	26 30.2	22 14.1	4 38.9	23 20.9	65
16 40.2	28 6.6	23 9.1	5 6.9	23 25.3	66
5	6	Descendant	8	9	S LAT

♎ 23° 46′ 3″

13ʰ 28ᵐ 0ˢ MC 202° 0′ 0″

1ʰ 32ᵐ 0ˢ MC 23° 0′ 0″

♈ 24° 49′ 42″

N LAT	11	12	Ascendant	2	3
°	° ′	° ′	° ′	° ′	° ′
0	♉25 20.5	♊23 34.4	♋21 16.7	♌20 36.1	♍22 22.6
5	25 54.5	24 53.9	23 8.1	21 28.5	22 28.3
10	26 29.6	26 15.1	24 58.1	22 19.8	22 34.0
15	27 6.3	27 39.3	26 48.3	23 10.7	22 39.6
16	27 13.9	27 56.6	27 10.5	23 20.9	22 40.8
17	27 21.6	28 14.1	27 32.7	23 31.1	22 41.9
18	27 29.4	28 31.8	27 55.1	23 41.3	22 43.1
19	27 37.4	28 49.7	28 17.5	23 51.6	22 44.2
20	27 45.5	29 7.9	28 40.0	24 2.0	22 45.4
21	27 53.7	29 26.3	29 2.7	24 12.3	22 46.6
22	28 2.0	29 44.9	29 25.5	24 22.8	22 47.8
23	28 10.5	♋ 0 3.9	29 48.5	24 33.2	22 49.0
24	28 19.2	0 23.2	♌ 0 11.6	24 43.8	22 50.2
25	28 28.0	0 42.7	0 34.9	24 54.4	22 51.4
26	28 37.0	1 2.7	0 58.4	25 5.2	22 52.7
27	28 46.2	1 22.9	1 22.1	25 16.0	22 53.9
28	28 55.7	1 43.6	1 46.0	25 26.9	22 55.2
29	29 5.3	2 4.7	2 10.2	25 37.9	22 56.4
30	29 15.2	2 26.2	2 34.6	25 49.0	22 57.7
31	29 25.4	2 48.1	2 59.3	26 0.2	22 59.1
32	29 35.8	3 10.6	3 24.3	26 11.6	23 0.4
33	29 46.5	3 33.5	3 49.6	26 23.1	23 1.8
34	29 57.5	3 57.0	4 15.2	26 34.8	23 3.1
35	♊ 0 8.9	4 21.1	4 41.1	26 46.6	23 4.5
36	0 20.6	4 45.8	5 7.4	26 58.6	23 6.0
37	0 32.7	5 11.2	5 34.1	27 10.8	23 7.4
38	0 45.2	5 37.3	6 1.3	27 23.1	23 8.9
39	0 58.1	6 4.1	6 28.8	27 35.7	23 10.4
40	1 11.6	6 31.7	6 56.8	27 48.5	23 12.0
41	1 25.5	7 0.2	7 25.2	28 1.5	23 13.5
42	1 40.0	7 29.5	7 54.2	28 14.7	23 15.2
43	1 55.1	7 59.8	8 23.7	28 28.3	23 16.8
44	2 10.8	8 31.2	8 53.8	28 42.0	23 18.5
45	2 27.3	9 3.6	9 24.4	28 56.1	23 20.2
46	2 44.5	9 37.3	9 55.7	29 10.5	23 22.0
47	3 2.6	10 12.2	10 27.7	29 25.2	23 23.8
48	3 21.6	10 48.5	11 0.3	29 40.3	23 25.7
49	3 41.7	11 26.2	11 33.6	29 55.7	23 27.7
50	4 2.8	12 5.5	12 7.8	♍ 0 11.6	23 29.7
51	4 25.3	12 46.5	12 42.7	0 27.8	23 31.7
52	4 49.1	13 29.3	13 18.5	0 44.5	23 33.9
53	5 14.5	14 14.1	13 55.2	1 1.7	23 36.1
54	5 41.7	15 1.0	14 32.8	1 19.3	23 38.4
55	6 11.0	15 50.3	15 11.4	1 37.5	23 40.7
56	6 42.5	16 42.1	15 51.1	1 56.3	23 43.2
57	7 16.8	17 36.6	16 31.9	2 15.7	23 45.7
58	7 54.2	18 34.1	17 13.9	2 35.7	23 48.4
59	8 35.3	19 34.9	17 57.1	2 56.4	23 51.2
60	9 21.0	20 39.2	18 41.6	3 17.9	23 54.0
61	10 12.3	21 47.5	19 27.5	3 40.2	23 57.1
62	11 10.7	23 0.0	20 14.8	4 3.3	24 0.2
63	12 18.8	24 17.3	21 3.7	4 27.3	24 3.5
64	13 40.4	25 39.8	21 54.3	4 52.3	24 7.0
65	15 24.0	27 7.9	22 46.5	5 18.3	24 10.7
66	17 53.2	28 42.4	23 40.6	5 45.5	24 14.6
S LAT	5	6	Descendant	8	9

♎ 24° 49′ 42″

13ʰ 32ᵐ 0ˢ MC 203° 0′ 0″

$1^h\ 36^m\ 0^s$		MC		24° 0′ 0″		$1^h\ 40^m\ 0^s$		MC		25° 0′ 0″
		♈ 25° 53′ 12″			N			♈ 26° 56′ 33″		
11	**12**	**Ascendant**	**2**	**3**	**LAT**	**11**	**12**	**Ascendant**	**2**	**3**
° ′	° ′	° ′	° ′	° ′	°	° ′	° ′	° ′	° ′	° ′
♉26 18.8	♊24 29.5	♋22 13.1	♌21 37.4	♍23 27.9	**0**	♉27 17.0	♊25 24.7	♋23 9.7	♌22 38.9	♍24 33.2
26 53.2	25 49.2	24 3.9	22 28.8	23 32.8	**5**	27 51.7	26 44.4	24 59.8	23 29.2	24 37.2
27 28.6	27 10.4	25 53.1	23 18.9	23 37.6	**10**	28 27.4	28 5.6	26 48.2	24 18.2	24 41.3
28 5.6	28 34.5	27 42.4	24 8.7	23 42.5	**15**	29 4.8	29 29.6	28 36.6	25 6.8	24 45.3
28 13.3	28 51.8	28 4.4	24 18.7	23 43.4	**16**	29 12.5	29 46.9	28 58.4	25 16.6	24 46.1
28 21.1	29 9.3	28 26.4	24 28.7	23 44.4	**17**	29 20.4	♋ 0 4.3	29 20.2	25 26.3	24 47.0
28 29.0	29 26.9	28 48.6	24 38.7	23 45.4	**18**	29 28.3	0 22.0	29 42.1	25 36.1	24 47.8
28 37.0	29 44.8	29 10.8	24 48.7	23 46.4	**19**	29 36.4	0 39.9	♌ 0 4.2	25 46.0	24 48.6
28 45.1	♋ 0 3.0	29 33.1	24 58.8	23 47.4	**20**	29 44.6	0 58.0	0 26.3	25 55.8	24 49.5
28 53.4	0 21.3	29 55.6	25 9.0	23 48.4	**21**	29 53.0	1 16.3	0 48.5	26 5.7	24 50.3
29 1.8	0 40.0	♌ 0 18.2	25 19.2	23 49.5	**22**	♊ 0 1.4	1 34.9	1 10.9	26 15.7	24 51.2
29 10.4	0 58.9	0 40.9	25 29.4	23 50.5	**23**	0 10.1	1 53.8	1 33.4	26 25.7	24 52.0
29 19.1	1 18.1	1 3.8	25 39.7	23 51.5	**24**	0 18.9	2 12.9	1 56.0	26 35.7	24 52.9
29 28.1	1 37.6	1 26.9	25 50.1	23 52.6	**25**	0 27.9	2 32.4	2 18.8	26 45.9	24 53.8
29 37.2	1 57.5	1 50.1	26 0.6	23 53.6	**26**	0 37.0	2 52.2	2 41.8	26 56.1	24 54.6
29 46.5	2 17.7	2 13.6	26 11.1	23 54.7	**27**	0 46.4	3 12.3	3 5.0	27 6.4	24 55.5
29 56.0	2 38.3	2 37.2	26 21.8	23 55.8	**28**	0 56.0	3 32.9	3 28.4	27 16.8	24 56.4
♊ 0 5.7	2 59.3	3 1.1	26 32.5	23 56.9	**29**	1 5.8	3 53.8	3 52.0	27 27.3	24 57.4
0 15.7	3 20.7	3 25.3	26 43.4	23 58.0	**30**	1 15.9	4 15.1	4 15.9	27 37.9	24 58.3
0 25.9	3 42.6	3 49.7	26 54.4	23 59.1	**31**	1 26.2	4 36.9	4 40.0	27 48.6	24 59.2
0 36.4	4 5.0	4 14.3	27 5.5	24 0.3	**32**	1 36.7	4 59.2	5 4.4	27 59.4	25 0.2
0 47.2	4 27.8	4 39.3	27 16.7	24 1.4	**33**	1 47.6	5 22.0	5 29.1	28 10.3	25 1.2
0 58.3	4 51.2	5 4.6	27 28.1	24 2.6	**34**	1 58.8	5 45.2	5 54.0	28 21.4	25 2.1
1 9.7	5 15.2	5 30.3	27 39.6	24 3.8	**35**	2 10.3	6 9.1	6 19.4	28 32.7	25 3.2
1 21.5	5 39.8	5 56.2	27 51.3	24 5.1	**36**	2 22.2	6 33.6	6 45.0	28 44.1	25 4.2
1 33.7	6 5.0	6 22.6	28 3.2	24 6.3	**37**	2 34.5	6 58.7	7 11.0	28 55.7	25 5.2
1 46.3	6 31.0	6 49.4	28 15.2	24 7.6	**38**	2 47.2	7 24.5	7 37.4	29 7.4	25 6.3
1 59.4	6 57.6	7 16.5	28 27.5	24 8.9	**39**	3 0.3	7 51.0	8 4.2	29 19.3	25 7.4
2 12.9	7 25.1	7 44.2	28 40.0	24 10.2	**40**	3 13.9	8 18.3	8 31.5	29 31.5	25 8.5
2 26.9	7 53.3	8 12.2	28 52.6	24 11.6	**41**	3 28.1	8 46.3	8 59.2	29 43.9	25 9.6
2 41.5	8 22.5	8 40.8	29 5.6	24 12.9	**42**	3 42.8	9 15.3	9 27.4	29 56.4	25 10.7
2 56.7	8 52.6	9 9.9	29 18.7	24 14.4	**43**	3 58.1	9 45.2	9 56.1	♍ 0 9.3	25 11.9
3 12.6	9 23.8	9 39.6	29 32.2	24 15.8	**44**	4 14.0	10 16.1	10 25.3	0 22.4	25 13.1
3 29.2	9 56.0	10 9.8	29 45.9	24 17.3	**45**	4 30.7	10 48.1	10 55.1	0 35.7	25 14.4
3 46.5	10 29.4	10 40.6	29 59.9	24 18.8	**46**	4 48.2	11 21.2	11 25.5	0 49.4	25 15.7
4 4.7	11 4.0	11 12.1	♍ 0 14.3	24 20.4	**47**	5 6.5	11 55.5	11 56.5	1 3.3	25 17.0
4 23.9	11 39.9	11 44.2	0 29.0	24 22.0	**48**	5 25.8	12 31.2	12 28.1	1 17.6	25 18.3
4 44.0	12 17.3	12 17.1	0 44.0	24 23.7	**49**	5 46.1	13 8.2	13 0.5	1 32.3	25 19.7
5 5.4	12 56.2	12 50.7	0 59.4	24 25.4	**50**	6 7.5	13 46.8	13 33.6	1 47.3	25 21.1
5 28.0	13 36.8	13 25.1	1 15.2	24 27.2	**51**	6 30.3	14 27.0	14 7.4	2 2.7	25 22.6
5 52.0	14 19.2	14 0.3	1 31.5	24 29.0	**52**	6 54.4	15 9.0	14 42.1	2 18.5	25 24.1
6 17.5	15 3.5	14 36.4	1 48.2	24 30.9	**53**	7 20.2	15 52.8	15 17.7	2 34.7	25 25.7
6 44.9	15 50.0	15 13.5	2 5.4	24 32.8	**54**	7 47.7	16 38.7	15 54.1	2 51.5	25 27.3
7 14.4	16 38.6	15 51.5	2 23.1	24 34.9	**55**	8 17.3	17 26.8	16 31.5	3 8.7	25 29.0
7 46.1	17 29.8	16 30.5	2 41.4	24 37.0	**56**	8 49.2	18 17.3	17 9.9	3 26.4	25 30.8
8 20.6	18 23.6	17 10.7	3 0.2	24 39.2	**57**	9 23.9	19 10.5	17 49.4	3 44.8	25 32.6
8 58.2	19 20.4	17 51.9	3 19.7	24 41.4	**58**	10 1.8	20 6.4	18 30.0	4 3.7	25 34.5
9 39.7	20 20.3	18 34.4	3 39.9	24 43.8	**59**	10 43.5	21 5.5	19 11.8	4 23.3	25 36.5
10 25.7	21 23.7	19 18.2	4 0.7	24 46.3	**60**	11 29.7	22 8.0	19 54.8	4 43.6	25 38.5
11 17.3	22 30.9	20 3.3	4 22.4	24 48.9	**61**	12 21.7	23 14.1	20 39.2	5 4.6	25 40.7
12 16.2	23 42.2	20 49.9	4 44.8	24 51.6	**62**	13 21.0	24 24.3	21 24.9	5 26.4	25 43.0
13 24.8	24 58.2	21 37.9	5 8.2	24 54.4	**63**	14 30.1	25 38.9	22 12.2	5 49.1	25 45.3
14 47.3	26 19.1	22 27.6	5 32.5	24 57.4	**64**	15 53.4	26 58.4	23 1.0	6 12.7	25 47.8
16 32.3	27 45.6	23 18.9	5 57.8	25 0.6	**65**	17 39.5	28 23.2	23 51.4	6 37.3	25 50.4
19 4.9	29 18.1	24 12.1	6 24.2	25 3.9	**66**	20 15.4	29 53.9	24 43.6	7 2.9	25 53.2
5	**6**	**Descendant**	**8**	**9**	**S**	**5**	**6**	**Descendant**	**8**	**9**
		♎ 25° 53′ 12″			**LAT**			♎ 26° 56′ 33″		
$13^h\ 36^m\ 0^s$		MC		204° 0′ 0″		$13^h\ 40^m\ 0^s$		MC		205° 0′ 0″

1ʰ 44ᵐ 0ˢ MC 26° 0′ 0″

♈ 27° 59′ 44″

11	12	Ascendant	2	3	N LAT
° ′	° ′	° ′	° ′	° ′	°
♉28 15.0	♊26 19.8	♋24 6.4	♌23 40.6	♍25 38.5	0
28 50.0	27 39.5	25 55.7	24 29.7	25 41.7	5
29 26.0	29 0.7	27 43.4	25 17.6	25 45.0	10
♊ 0 3.7	♋ 0 24.7	29 30.8	26 5.1	25 48.2	15
0 11.5	0 41.9	29 52.4	26 14.6	25 48.9	16
0 19.4	0 59.3	♌ 0 14.0	26 24.1	25 49.5	17
0 27.5	1 17.0	0 35.7	26 33.7	25 50.2	18
0 35.6	1 34.8	0 57.5	26 43.3	25 50.9	19
0 43.9	1 52.8	1 19.4	26 52.9	25 51.5	20
0 52.3	2 11.1	1 41.4	27 2.6	25 52.2	21
1 0.8	2 29.7	2 3.6	27 12.3	25 52.9	22
1 9.5	2 48.5	2 25.8	27 22.0	25 53.6	23
1 18.4	3 7.6	2 48.2	27 31.8	25 54.3	24
1 27.5	3 27.0	3 10.8	27 41.7	25 55.0	25
1 36.7	3 46.8	3 33.5	27 51.7	25 55.7	26
1 46.2	4 6.9	3 56.4	28 1.7	25 56.4	27
1 55.8	4 27.3	4 19.6	28 11.9	25 57.1	28
2 5.7	4 48.2	4 42.9	28 22.1	25 57.9	29
2 15.8	5 9.4	5 6.5	28 32.4	25 58.6	30
2 26.2	5 31.1	5 30.3	28 42.8	25 59.4	31
2 36.8	5 53.3	5 54.4	28 53.4	26 0.1	32
2 47.8	6 15.9	6 18.8	29 4.0	26 0.9	33
2 59.1	6 39.1	6 43.4	29 14.9	26 1.7	34
3 10.7	7 2.9	7 8.4	29 25.8	26 2.5	35
3 22.6	7 27.2	7 33.7	29 36.9	26 3.3	36
3 35.0	7 52.2	7 59.4	29 48.2	26 4.1	37
3 47.8	8 17.8	8 25.5	29 59.6	26 5.0	38
4 1.0	8 44.2	8 51.9	♍ 0 11.3	26 5.9	39
4 14.7	9 11.3	9 18.8	0 23.1	26 6.7	40
4 28.9	9 39.2	9 46.1	0 35.1	26 7.6	41
4 43.7	10 7.9	10 13.9	0 47.4	26 8.6	42
4 59.1	10 37.6	10 42.2	0 59.9	26 9.5	43
5 15.2	11 8.3	11 11.0	1 12.6	26 10.5	44
5 32.0	11 40.0	11 40.3	1 25.6	26 11.5	45
5 49.5	12 12.9	12 10.3	1 38.9	26 12.5	46
6 8.0	12 46.9	12 40.8	1 52.5	26 13.6	47
6 27.3	13 22.3	13 12.0	2 6.4	26 14.6	48
6 47.8	13 59.0	13 43.9	2 20.6	26 15.7	49
7 9.3	14 37.2	14 16.4	2 35.2	26 16.9	50
7 32.2	15 17.0	14 49.8	2 50.1	26 18.1	51
7 56.5	15 58.5	15 23.9	3 5.5	26 19.3	52
8 22.4	16 41.8	15 58.9	3 21.3	26 20.5	53
8 50.1	17 27.2	16 34.7	3 37.6	26 21.8	54
9 19.8	18 14.7	17 11.5	3 54.3	26 23.2	55
9 51.9	19 4.6	17 49.3	4 11.6	26 24.6	56
10 26.8	19 57.1	18 28.1	4 29.4	26 26.1	57
11 4.9	20 52.3	19 8.1	4 47.8	26 27.6	58
11 46.8	21 50.5	19 49.1	5 6.8	26 29.2	59
12 33.3	22 52.1	20 31.4	5 26.5	26 30.8	60
13 25.6	23 57.2	21 15.0	5 46.9	26 32.5	61
14 25.2	25 6.2	22 0.0	6 8.1	26 34.4	62
15 34.8	26 19.5	22 46.4	6 30.1	26 36.2	63
16 58.6	27 37.6	23 34.3	6 53.0	26 38.2	64
18 45.8	29 0.8	24 23.9	7 16.8	26 40.3	65
21 24.4	♌ 0 29.7	25 15.1	7 41.7	26 42.6	66
5	6	Descendant	8	9	S LAT

♎ 27° 59′ 44″

13ʰ 44ᵐ 0ˢ MC 206° 0′ 0″

1ʰ 48ᵐ 0ˢ MC 27° 0′ 0″

♈ 29° 2′ 47″

11	12	Ascendant	2	3	N LAT
° ′	° ′	° ′	° ′	° ′	°
♉29 12.8	♊27 14.8	♋25 3.3	♌24 42.5	♍26 43.8	0
29 48.1	28 34.6	26 51.8	25 30.4	26 46.3	5
♊ 0 24.4	29 55.8	28 38.6	26 17.2	26 48.7	10
1 2.5	♋ 1 19.6	♌ 0 25.0	27 3.5	26 51.1	15
1 10.3	1 36.9	0 46.4	27 12.8	26 51.6	16
1 18.3	1 54.3	1 7.9	27 22.1	26 52.1	17
1 26.4	2 11.8	1 29.4	27 31.4	26 52.6	18
1 34.6	2 29.6	1 50.9	27 40.7	26 53.1	19
1 42.9	2 47.7	2 12.6	27 50.1	26 53.6	20
1 51.4	3 5.9	2 34.4	27 59.5	26 54.1	21
2 0.0	3 24.4	2 56.3	28 9.0	26 54.7	22
2 8.8	3 43.2	3 18.3	28 18.5	26 55.2	23
2 17.7	4 2.2	3 40.4	28 28.0	26 55.7	24
2 26.9	4 21.6	4 2.7	28 37.7	26 56.2	25
2 36.2	4 41.2	4 25.2	28 47.4	26 56.7	26
2 45.7	5 1.3	4 47.9	28 57.1	26 57.3	27
2 55.4	5 21.6	5 10.7	29 7.0	26 57.8	28
3 5.3	5 42.4	5 33.8	29 17.0	26 58.4	29
3 15.5	6 3.6	5 57.1	29 27.0	26 58.9	30
3 26.0	6 25.2	6 20.6	29 37.2	26 59.5	31
3 36.7	6 47.2	6 44.4	29 47.4	27 0.1	32
3 47.7	7 9.8	7 8.5	29 57.8	27 0.7	33
3 59.1	7 32.9	7 32.8	♍ 0 8.3	27 1.3	34
4 10.8	7 56.5	7 57.5	0 19.0	27 1.9	35
4 22.8	8 20.7	8 22.5	0 29.8	27 2.5	36
4 35.2	8 45.5	8 47.8	0 40.8	27 3.1	37
4 48.1	9 11.0	9 13.5	0 51.9	27 3.7	38
5 1.4	9 37.2	9 39.6	1 3.2	27 4.4	39
5 15.2	10 4.1	10 6.1	1 14.7	27 5.0	40
5 29.5	10 31.8	10 33.0	1 26.4	27 5.7	41
5 44.4	11 0.4	11 0.4	1 38.3	27 6.4	42
5 59.8	11 29.9	11 28.3	1 50.5	27 7.1	43
6 16.0	12 0.3	11 56.7	2 2.9	27 7.9	44
6 32.9	12 31.8	12 25.6	2 15.5	27 8.6	45
6 50.6	13 4.4	12 55.1	2 28.4	27 9.4	46
7 9.1	13 38.1	13 25.1	2 41.6	27 10.2	47
7 28.6	14 13.1	13 55.8	2 55.1	27 11.0	48
7 49.1	14 49.5	14 27.2	3 8.9	27 11.8	49
8 10.8	15 27.3	14 59.3	3 23.1	27 12.7	50
8 33.8	16 6.7	15 32.1	3 37.6	27 13.5	51
8 58.2	16 47.8	16 5.7	3 52.6	27 14.5	52
9 24.2	17 30.7	16 40.1	4 7.9	27 15.4	53
9 52.0	18 15.5	17 15.4	4 23.7	27 16.4	54
10 21.9	19 2.5	17 51.5	4 39.9	27 17.4	55
10 54.2	19 51.8	18 28.7	4 56.7	27 18.4	56
11 29.2	20 43.5	19 6.9	5 14.0	27 19.5	57
12 7.5	21 38.0	19 46.1	5 31.8	27 20.7	58
12 49.5	22 35.4	20 26.5	5 50.3	27 21.9	59
13 36.3	23 36.0	21 8.0	6 9.4	27 23.1	60
14 28.8	24 40.1	21 50.9	6 29.2	27 24.4	61
15 28.8	25 48.0	22 35.1	6 49.7	27 25.8	62
16 38.7	27 0.1	23 20.6	7 11.1	27 27.2	63
18 3.0	28 16.7	24 7.7	7 33.3	27 28.7	64
19 51.1	29 38.4	24 56.4	7 56.4	27 30.2	65
22 32.0	♌ 1 5.5	25 46.8	8 20.6	27 31.9	66
5	6	Descendant	8	9	S LAT

♎ 29° 2′ 47″

13ʰ 48ᵐ 0ˢ MC 207° 0′ 0″

1h 52m 0s — MC 28° 0′ 0″

♉ 0° 5′ 39″

N LAT	11	12	Ascendant	2	3
°	° ′	° ′	° ′	° ′	° ′
0	♊ 0 10.5	♊28 9.9	♋26 0.2	♌25 44.5	♍27 49.2
5	0 46.1	29 29.7	27 48.0	26 31.3	27 50.8
10	1 22.7	♋ 0 50.8	29 33.9	27 16.8	27 52.5
15	2 1.0	2 14.6	♌ 1 19.4	28 2.0	27 54.1
16	2 8.9	2 31.7	1 40.5	28 11.0	27 54.4
17	2 17.0	2 49.1	2 1.7	28 20.1	27 54.7
18	2 25.1	3 6.7	2 23.0	28 29.1	27 55.1
19	2 33.4	3 24.4	2 44.4	28 38.2	27 55.4
20	2 41.8	3 42.4	3 5.8	28 47.4	27 55.7
21	2 50.3	4 0.6	3 27.4	28 56.5	27 56.1
22	2 59.0	4 19.0	3 49.0	29 5.7	27 56.4
23	3 7.8	4 37.7	4 10.8	29 15.0	27 56.8
24	3 16.8	4 56.7	4 32.7	29 24.3	27 57.1
25	3 26.0	5 16.0	4 54.7	29 33.7	27 57.5
26	3 35.4	5 35.6	5 16.9	29 43.1	27 57.8
27	3 45.0	5 55.6	5 39.3	29 52.6	27 58.2
28	3 54.7	6 15.8	6 1.9	♍ 0 2.2	27 58.5
29	4 4.7	6 36.5	6 24.7	0 11.9	27 58.9
30	4 15.0	6 57.6	6 47.7	0 21.7	27 59.3
31	4 25.5	7 19.1	7 10.9	0 31.6	27 59.7
32	4 36.3	7 41.1	7 34.4	0 41.5	28 0.0
33	4 47.4	8 3.5	7 58.2	0 51.7	28 0.4
34	4 58.8	8 26.5	8 22.2	1 1.9	28 0.8
35	5 10.6	8 49.9	8 46.5	1 12.2	28 1.2
36	5 22.7	9 14.0	9 11.2	1 22.8	28 1.6
37	5 35.2	9 38.7	9 36.2	1 33.4	28 2.1
38	5 48.1	10 4.0	10 1.5	1 44.2	28 2.5
39	6 1.5	10 30.0	10 27.2	1 55.2	28 2.9
40	6 15.4	10 56.8	10 53.3	2 6.4	28 3.4
41	6 29.8	11 24.3	11 19.9	2 17.8	28 3.8
42	6 44.7	11 52.7	11 46.9	2 29.3	28 4.3
43	7 0.3	12 21.9	12 14.3	2 41.1	28 4.7
44	7 16.5	12 52.1	12 42.3	2 53.2	28 5.2
45	7 33.5	13 23.4	13 10.8	3 5.4	28 5.7
46	7 51.3	13 55.7	13 39.8	3 18.0	28 6.2
47	8 9.9	14 29.1	14 9.4	3 30.8	28 6.8
48	8 29.5	15 3.8	14 39.6	3 43.9	28 7.3
49	8 50.1	15 39.8	15 10.5	3 57.3	28 7.9
50	9 11.9	16 17.3	15 42.1	4 11.1	28 8.4
51	9 35.0	16 56.3	16 14.4	4 25.2	28 9.0
52	9 59.5	17 36.9	16 47.4	4 39.7	28 9.6
53	10 25.6	18 19.3	17 21.3	4 54.5	28 10.3
54	10 53.6	19 3.6	17 55.9	5 9.9	28 10.9
55	11 23.6	19 50.0	18 31.5	5 25.6	28 11.6
56	11 56.0	20 38.7	19 8.0	5 41.9	28 12.3
57	12 31.2	21 29.8	19 45.6	5 58.6	28 13.0
58	13 9.6	22 23.5	20 24.1	6 15.9	28 13.8
59	13 51.8	23 20.1	21 3.8	6 33.8	28 14.6
60	14 38.8	24 19.8	21 44.7	6 52.3	28 15.4
61	15 31.5	25 22.9	22 26.7	7 11.5	28 16.3
62	16 31.7	26 29.7	23 10.1	7 31.5	28 17.2
63	17 41.9	27 40.5	23 54.9	7 52.1	28 18.1
64	19 6.7	28 55.8	24 41.2	8 13.7	28 19.1
65	20 55.4	♌ 0 15.9	25 29.0	8 36.1	28 20.2
66	23 38.0	1 41.3	26 18.4	8 59.4	28 21.3
S LAT	5	6	Descendant	8	9

♏ 0° 5′ 39″

13h 52m 0s — MC 208° 0′ 0″

1h 56m 0s — MC 29° 0′ 0″

♉ 1° 8′ 22″

N LAT	11	12	Ascendant	2	3
°	° ′	° ′	° ′	° ′	° ′
0	♊ 1 8.0	♊29 5.0	♋26 57.4	♌26 46.7	♍28 54.6
5	1 43.9	♋ 0 24.7	28 44.3	27 32.3	28 55.4
10	2 20.8	1 45.8	♌ 0 29.3	28 16.7	28 56.2
15	2 59.4	3 9.4	2 13.7	29 0.6	28 57.0
16	3 7.4	3 26.6	2 34.7	29 9.4	28 57.2
17	3 15.5	3 43.9	2 55.7	29 18.2	28 57.4
18	3 23.7	4 1.4	3 16.7	29 27.0	28 57.5
19	3 32.0	4 19.1	3 37.9	29 35.9	28 57.7
20	3 40.4	4 37.0	3 59.1	29 44.7	28 57.9
21	3 49.0	4 55.2	4 20.4	29 53.7	28 58.0
22	3 57.8	5 13.6	4 41.8	♍ 0 2.6	28 58.2
23	4 6.7	5 32.2	5 3.3	0 11.6	28 58.4
24	4 15.7	5 51.1	5 24.9	0 20.7	28 58.6
25	4 25.0	6 10.3	5 46.7	0 29.8	28 58.7
26	4 34.4	6 29.9	6 8.7	0 39.0	28 58.9
27	4 44.0	6 49.7	6 30.8	0 48.2	28 59.1
28	4 53.9	7 10.0	6 53.1	0 57.5	28 59.3
29	5 3.9	7 30.5	7 15.6	1 6.9	28 59.5
30	5 14.3	7 51.5	7 38.3	1 16.4	28 59.6
31	5 24.8	8 12.9	8 1.2	1 26.0	28 59.8
32	5 35.7	8 34.8	8 24.4	1 35.7	29 0.0
33	5 46.9	8 57.1	8 47.8	1 45.5	29 0.2
34	5 58.3	9 19.9	9 11.6	1 55.5	29 0.4
35	6 10.2	9 43.3	9 35.6	2 5.6	29 0.6
36	6 22.3	10 7.2	9 59.9	2 15.8	29 0.8
37	6 34.9	10 31.7	10 24.5	2 26.1	29 1.0
38	6 47.9	10 56.9	10 49.5	2 36.6	29 1.2
39	7 1.4	11 22.7	11 14.8	2 47.3	29 1.5
40	7 15.3	11 49.3	11 40.6	2 58.1	29 1.7
41	7 29.8	12 16.7	12 6.7	3 9.2	29 1.9
42	7 44.8	12 44.8	12 33.3	3 20.4	29 2.1
43	8 0.4	13 13.8	13 0.4	3 31.8	29 2.4
44	8 16.8	13 43.8	13 27.9	3 43.5	29 2.6
45	8 33.8	14 14.8	13 56.0	3 55.4	29 2.9
46	8 51.7	14 46.8	14 24.5	4 7.6	29 3.1
47	9 10.4	15 19.9	14 53.7	4 20.0	29 3.4
48	9 30.0	15 54.3	15 23.4	4 32.7	29 3.6
49	9 50.8	16 30.0	15 53.8	4 45.7	29 3.9
50	10 12.6	17 7.1	16 24.9	4 59.1	29 4.2
51	10 35.8	17 45.6	16 56.6	5 12.8	29 4.5
52	11 0.4	18 25.8	17 29.1	5 26.8	29 4.8
53	11 26.7	19 7.8	18 2.4	5 41.2	29 5.1
54	11 54.7	19 51.6	18 36.5	5 56.1	29 5.5
55	12 24.9	20 37.4	19 11.5	6 11.3	29 5.8
56	12 57.4	21 25.5	19 47.4	6 27.1	29 6.1
57	13 32.7	22 15.9	20 24.3	6 43.3	29 6.5
58	14 11.2	23 8.8	21 2.2	7 0.1	29 6.9
59	14 53.6	24 4.6	21 41.1	7 17.4	29 7.3
60	15 40.7	25 3.4	22 21.3	7 35.3	29 7.7
61	16 33.6	26 5.5	23 2.6	7 53.9	29 8.1
62	17 34.0	27 11.2	23 45.2	8 13.2	29 8.6
63	18 44.4	28 20.9	24 29.2	8 33.2	29 9.1
64	20 9.5	29 34.8	25 14.6	8 54.0	29 9.6
65	21 58.6	♌ 0 53.4	26 1.6	9 15.7	29 10.1
66	24 42.3	2 17.2	26 50.1	9 38.3	29 10.6
S LAT	5	6	Descendant	8	9

♏ 1° 8′ 22″

13h 56m 0s — MC 209° 0′ 0″

$2^h\ 0^m\ 0^s$ MC 30° 0′ 0″

♉ 2° 10′ 55″

N LAT	11	12	Ascendant	2	3
	° ′	° ′	° ′	° ′	° ′
0	♊ 2 5.4	♋ 0 0.0	♋27 54.6	♌27 49.1	♎ 0 0.0
5	2 41.6	1 19.8	29 40.7	28 33.4	0 0.0
10	3 18.7	2 40.7	♌ 1 24.7	29 16.6	0 0.0
15	3 57.6	4 4.2	3 8.2	29 59.3	0 0.0
16	4 5.6	4 21.3	3 28.9	♍ 0 7.9	0 0.0
17	4 13.8	4 38.6	3 49.7	0 16.4	0 0.0
18	4 22.0	4 56.1	4 10.5	0 25.0	0 0.0
19	4 30.4	5 13.7	4 31.4	0 33.6	0 0.0
20	4 38.9	5 31.6	4 52.3	0 42.2	0 0.0
21	4 47.6	5 49.7	5 13.4	0 50.9	0 0.0
22	4 56.3	6 8.0	5 34.6	0 59.6	0 0.0
23	5 5.3	6 26.6	5 55.8	1 8.3	0 0.0
24	5 14.4	6 45.4	6 17.2	1 17.1	0 0.0
25	5 23.7	7 4.6	6 38.7	1 26.0	0 0.0
26	5 33.2	7 24.0	7 0.4	1 34.9	0 0.0
27	5 42.9	7 43.8	7 22.3	1 43.9	0 0.0
28	5 52.8	8 4.0	7 44.3	1 52.9	0 0.0
29	6 2.9	8 24.4	8 6.5	2 2.0	0 0.0
30	6 13.3	8 45.3	8 28.9	2 11.3	0 0.0
31	6 23.9	9 6.6	8 51.5	2 20.6	0 0.0
32	6 34.8	9 28.4	9 14.4	2 30.0	0 0.0
33	6 46.1	9 50.6	9 37.5	2 39.5	0 0.0
34	6 57.6	10 13.3	10 0.9	2 49.1	0 0.0
35	7 9.5	10 36.5	10 24.6	2 58.9	0 0.0
36	7 21.7	11 0.3	10 48.5	3 8.8	0 0.0
37	7 34.4	11 24.6	11 12.8	3 18.8	0 0.0
38	7 47.4	11 49.6	11 37.5	3 29.0	0 0.0
39	8 0.9	12 15.3	12 2.4	3 39.4	0 0.0
40	8 14.9	12 41.7	12 27.8	3 49.9	0 0.0
41	8 29.5	13 8.8	12 53.6	4 0.6	0 0.0
42	8 44.6	13 36.8	13 19.8	4 11.5	0 0.0
43	9 0.3	14 5.6	13 46.4	4 22.6	0 0.0
44	9 16.7	14 35.3	14 13.5	4 33.9	0 0.0
45	9 33.8	15 6.0	14 41.1	4 45.4	0 0.0
46	9 51.7	15 37.7	15 9.2	4 57.2	0 0.0
47	10 10.5	16 10.6	15 37.9	5 9.3	0 0.0
48	10 30.3	16 44.6	16 7.2	5 21.6	0 0.0
49	10 51.1	17 20.0	16 37.1	5 34.2	0 0.0
50	11 13.0	17 56.7	17 7.6	5 47.1	0 0.0
51	11 36.3	18 34.8	17 38.9	6 0.4	0 0.0
52	12 1.0	19 14.6	18 10.8	6 13.9	0 0.0
53	12 27.3	19 56.1	18 43.5	6 27.9	0 0.0
54	12 55.5	20 39.4	19 17.1	6 42.3	0 0.0
55	13 25.7	21 24.6	19 51.4	6 57.1	0 0.0
56	13 58.4	22 12.1	20 26.7	7 12.3	0 0.0
57	14 33.8	23 1.8	21 3.0	7 28.0	0 0.0
58	15 12.4	23 54.0	21 40.2	7 44.2	0 0.0
59	15 54.9	24 49.0	22 18.5	8 1.0	0 0.0
60	16 42.1	25 46.9	22 57.9	8 18.4	0 0.0
61	17 35.1	26 48.1	23 38.5	8 36.3	0 0.0
62	18 35.7	27 52.7	24 20.3	8 55.0	0 0.0
63	19 46.3	29 1.2	25 3.5	9 14.3	0 0.0
64	21 11.5	♌ 0 13.8	25 48.1	9 34.5	0 0.0
65	23 0.9	1 31.0	26 34.2	9 55.4	0 0.0
66	25 45.0	2 53.1	27 21.8	10 17.3	0 0.0
S LAT	5	6	Descendant	8	9

♏ 2° 10′ 55″

$14^h\ 0^m\ 0^s$ MC 210° 0′ 0″

$2^h\ 4^m\ 0^s$ MC 31° 0′ 0″

♉ 3° 13′ 17″

N LAT	11	12	Ascendant	2	3
	° ′	° ′	° ′	° ′	° ′
0	♊ 3 2.6	♋ 0 55.0	♋28 52.0	♌28 51.6	♎ 1 5.4
5	3 39.1	2 14.8	♌ 0 37.2	29 34.7	1 4.6
10	4 16.5	3 35.6	2 20.3	♍ 0 16.7	1 3.8
15	4 55.6	4 58.9	4 2.6	0 58.2	1 3.0
16	5 3.7	5 16.0	4 23.1	1 6.4	1 2.8
17	5 11.9	5 33.3	4 43.7	1 14.8	1 2.6
18	5 20.2	5 50.7	5 4.3	1 23.1	1 2.5
19	5 28.6	6 8.3	5 24.9	1 31.4	1 2.3
20	5 37.2	6 26.1	5 45.7	1 39.8	1 2.1
21	5 45.9	6 44.1	6 6.5	1 48.2	1 2.0
22	5 54.7	7 2.4	6 27.4	1 56.6	1 1.8
23	6 3.7	7 20.9	6 48.4	2 5.1	1 1.6
24	6 12.9	7 39.7	7 9.5	2 13.6	1 1.4
25	6 22.2	7 58.8	7 30.8	2 22.2	1 1.3
26	6 31.8	8 18.1	7 52.2	2 30.9	1 1.1
27	6 41.5	8 37.8	8 13.7	2 39.6	1 0.9
28	6 51.5	8 57.9	8 35.5	2 48.4	1 0.7
29	7 1.7	9 18.3	8 57.4	2 57.2	1 0.5
30	7 12.1	9 39.0	9 19.5	3 6.1	1 0.4
31	7 22.8	10 0.2	9 41.8	3 15.2	1 0.2
32	7 33.8	10 21.8	10 4.4	3 24.3	1 0.0
33	7 45.0	10 43.9	10 27.2	3 33.5	0 59.8
34	7 56.6	11 6.5	10 50.2	3 42.9	0 59.6
35	8 8.6	11 29.6	11 13.6	3 52.3	0 59.4
36	8 20.9	11 53.2	11 37.2	4 1.9	0 59.2
37	8 33.6	12 17.4	12 1.1	4 11.6	0 59.0
38	8 46.7	12 42.3	12 25.4	4 21.5	0 58.8
39	9 0.3	13 7.7	12 50.0	4 31.5	0 58.5
40	9 14.3	13 33.9	13 15.0	4 41.7	0 58.3
41	9 28.9	14 0.9	13 40.4	4 52.1	0 58.1
42	9 44.1	14 28.6	14 6.2	5 2.6	0 57.9
43	9 59.9	14 57.2	14 32.4	5 13.4	0 57.6
44	10 16.3	15 26.7	14 59.1	5 24.3	0 57.4
45	10 33.5	15 57.1	15 26.2	5 35.5	0 57.1
46	10 51.5	16 28.5	15 53.9	5 46.9	0 56.9
47	11 10.4	17 1.1	16 22.2	5 58.6	0 56.6
48	11 30.2	17 34.8	16 50.9	6 10.5	0 56.4
49	11 51.1	18 9.8	17 20.3	6 22.7	0 56.1
50	12 13.1	18 46.1	17 50.4	6 35.2	0 55.8
51	12 36.5	19 23.9	18 21.1	6 48.0	0 55.5
52	13 1.2	20 3.2	18 52.5	7 1.1	0 55.2
53	13 27.6	20 44.2	19 24.7	7 14.6	0 54.9
54	13 55.9	21 27.0	19 57.6	7 28.5	0 54.5
55	14 26.2	22 11.7	20 31.4	7 42.8	0 54.2
56	14 58.9	22 58.5	21 6.1	7 57.6	0 53.9
57	15 34.4	23 47.6	21 41.6	8 12.7	0 53.5
58	16 13.1	24 39.1	22 18.2	8 28.4	0 53.1
59	16 55.7	25 33.3	22 55.8	8 44.6	0 52.7
60	17 43.0	26 30.3	23 34.5	9 1.4	0 52.3
61	18 36.1	27 30.5	24 14.4	9 18.8	0 51.9
62	19 36.7	28 34.1	24 55.5	9 36.8	0 51.4
63	20 47.4	29 41.4	25 37.9	9 55.5	0 50.9
64	22 12.7	♌ 0 52.7	26 21.6	10 14.9	0 50.4
65	24 2.1	2 8.5	27 6.8	10 35.2	0 49.9
66	26 46.0	3 29.0	27 53.6	10 56.3	0 49.4
S LAT	5	6	Descendant	8	9

♏ 3° 13′ 17″

$14^h\ 4^m\ 0^s$ MC 211° 0′ 0″

2h 8m 0s — MC 32° 0′ 0″ — ♉ 4° 15′ 30″

N LAT	11	12	Ascendant	2	3
0	♊ 3 59.8	♋ 1 50.1	♋ 29 49.5	♌ 29 54.3	♎ 2 10.8
5	4 36.4	3 9.8	♌ 1 33.8	♍ 0 36.2	2 9.2
10	5 14.1	4 30.5	3 15.9	1 16.9	2 7.5
15	5 53.5	5 53.6	4 57.2	1 57.1	2 5.9
16	6 1.6	6 10.7	5 17.5	2 5.1	2 5.6
17	6 9.8	6 27.9	5 37.8	2 13.2	2 5.3
18	6 18.2	6 45.2	5 58.1	2 21.3	2 4.9
19	6 26.7	7 2.8	6 18.6	2 29.3	2 4.6
20	6 35.3	7 20.5	6 39.0	2 37.5	2 4.3
21	6 44.0	7 38.5	6 59.6	2 45.6	2 3.9
22	6 52.9	7 56.7	7 20.2	2 53.8	2 3.6
23	7 2.0	8 15.1	7 41.0	3 2.0	2 3.2
24	7 11.2	8 33.9	8 1.9	3 10.3	2 2.9
25	7 20.6	8 52.8	8 22.8	3 18.6	2 2.5
26	7 30.2	9 12.1	8 44.0	3 26.9	2 2.2
27	7 39.9	9 31.7	9 5.3	3 35.4	2 1.8
28	7 50.0	9 51.7	9 26.7	3 43.9	2 1.5
29	8 0.2	10 12.0	9 48.3	3 52.4	2 1.1
30	8 10.7	10 32.6	10 10.1	4 1.1	2 0.7
31	8 21.4	10 53.7	10 32.2	4 9.8	2 0.3
32	8 32.5	11 15.2	10 54.4	4 18.7	2 0.0
33	8 43.8	11 37.2	11 16.9	4 27.6	1 59.6
34	8 55.4	11 59.6	11 39.6	4 36.7	1 59.2
35	9 7.4	12 22.5	12 2.6	4 45.8	1 58.8
36	9 19.8	12 46.0	12 25.9	4 55.1	1 58.4
37	9 32.5	13 10.1	12 49.4	5 4.5	1 57.9
38	9 45.7	13 34.7	13 13.4	5 14.0	1 57.5
39	9 59.4	14 0.1	13 37.6	5 23.7	1 57.1
40	10 13.5	14 26.1	14 2.2	5 33.6	1 56.6
41	10 28.1	14 52.8	14 27.2	5 43.6	1 56.2
42	10 43.4	15 20.3	14 52.6	5 53.8	1 55.7
43	10 59.2	15 48.6	15 18.4	6 4.2	1 55.3
44	11 15.7	16 17.9	15 44.6	6 14.8	1 54.8
45	11 33.0	16 48.0	16 11.4	6 25.6	1 54.3
46	11 51.0	17 19.2	16 38.6	6 36.6	1 53.8
47	12 9.9	17 51.4	17 6.4	6 47.9	1 53.2
48	12 29.8	18 24.8	17 34.7	6 59.4	1 52.7
49	12 50.8	18 59.5	18 3.6	7 11.2	1 52.1
50	13 12.9	19 35.4	18 33.1	7 23.3	1 51.6
51	13 36.3	20 12.8	19 3.3	7 35.6	1 51.0
52	14 1.1	20 51.6	19 34.2	7 48.4	1 50.4
53	14 27.6	21 32.1	20 5.8	8 1.4	1 49.7
54	14 55.9	22 14.4	20 38.1	8 14.8	1 49.1
55	15 26.3	22 58.6	21 11.3	8 28.6	1 48.4
56	15 59.1	23 44.8	21 45.4	8 42.8	1 47.7
57	16 34.6	24 33.2	22 20.3	8 57.5	1 47.0
58	17 13.4	25 24.0	22 56.2	9 12.7	1 46.2
59	17 56.1	26 17.4	23 33.2	9 28.3	1 45.4
60	18 43.4	27 13.6	24 11.2	9 44.5	1 44.6
61	19 36.6	28 12.8	24 50.3	10 1.3	1 43.7
62	20 37.2	29 15.4	25 30.6	10 18.6	1 42.8
63	21 47.9	♌ 0 21.5	26 12.2	10 36.7	1 41.9
64	23 13.1	1 31.6	26 55.1	10 55.5	1 40.9
65	25 2.3	2 46.0	27 39.5	11 15.0	1 39.8
66	27 45.4	4 5.0	28 25.4	11 35.3	1 38.7
S LAT	5	6	Descendant	8	9

♏ 4° 15′ 30″ — 14h 8m 0s — MC 212° 0′ 0″

2h 12m 0s — MC 33° 0′ 0″ — ♉ 5° 17′ 32″

N LAT	11	12	Ascendant	2	3
0	♊ 4 56.7	♋ 2 45.2	♌ 0 47.2	♍ 0 57.2	♎ 3 16.2
5	5 33.6	4 4.8	2 30.6	1 37.8	3 13.7
10	6 11.6	5 25.4	4 11.6	2 17.2	3 11.3
15	6 51.1	6 48.3	5 51.8	2 56.1	3 8.9
16	6 59.3	7 5.3	6 11.9	3 3.9	3 8.4
17	7 7.6	7 22.4	6 31.9	3 11.7	3 7.9
18	7 16.0	7 39.7	6 52.0	3 19.5	3 7.4
19	7 24.5	7 57.2	7 12.2	3 27.4	3 6.9
20	7 33.2	8 14.9	7 32.4	3 35.2	3 6.4
21	7 42.0	8 32.8	7 52.7	3 43.1	3 5.9
22	7 50.9	8 50.9	8 13.1	3 51.0	3 5.3
23	8 0.0	9 9.3	8 33.6	3 59.0	3 4.8
24	8 9.3	9 27.9	8 54.2	4 6.9	3 4.3
25	8 18.7	9 46.9	9 14.9	4 15.0	3 3.8
26	8 28.3	10 6.1	9 35.8	4 23.1	3 3.3
27	8 38.2	10 25.6	9 56.8	4 31.2	3 2.7
28	8 48.2	10 45.4	10 17.9	4 39.5	3 2.2
29	8 58.5	11 5.6	10 39.3	4 47.8	3 1.6
30	9 9.1	11 26.2	11 0.8	4 56.1	3 1.1
31	9 19.9	11 47.1	11 22.5	5 4.6	3 0.5
32	9 30.9	12 8.5	11 44.4	5 13.1	2 59.9
33	9 42.3	12 30.3	12 6.5	5 21.7	2 59.3
34	9 54.0	12 52.6	12 28.9	5 30.5	2 58.7
35	10 6.0	13 15.4	12 51.6	5 39.3	2 58.1
36	10 18.5	13 38.7	13 14.5	5 48.3	2 57.5
37	10 31.3	14 2.6	13 37.8	5 57.4	2 56.9
38	10 44.5	14 27.1	14 1.3	6 6.6	2 56.3
39	10 58.2	14 52.2	14 25.2	6 16.0	2 55.6
40	11 12.4	15 18.0	14 49.4	6 25.5	2 55.0
41	11 27.1	15 44.6	15 14.0	6 35.2	2 54.3
42	11 42.3	16 11.9	15 38.9	6 45.0	2 53.6
43	11 58.2	16 40.0	16 4.3	6 55.1	2 52.9
44	12 14.8	17 8.9	16 30.2	7 5.3	2 52.1
45	12 32.1	17 38.8	16 56.5	7 15.7	2 51.4
46	12 50.2	18 9.7	17 23.3	7 26.4	2 50.6
47	13 9.2	18 41.6	17 50.5	7 37.2	2 49.8
48	13 29.1	19 14.7	18 18.4	7 48.4	2 49.0
49	13 50.1	19 49.0	18 46.8	7 59.7	2 48.2
50	14 12.3	20 24.5	19 15.8	8 11.4	2 47.3
51	14 35.8	21 1.5	19 45.5	8 23.3	2 46.5
52	15 0.7	21 39.9	20 15.8	8 35.6	2 45.5
53	15 27.2	22 20.0	20 46.9	8 48.2	2 44.6
54	15 55.6	23 1.7	21 18.7	9 1.1	2 43.6
55	16 26.0	23 45.3	21 51.3	9 14.5	2 42.6
56	16 58.8	24 30.9	22 24.7	9 28.2	2 41.6
57	17 34.4	25 18.7	22 59.0	9 42.3	2 40.5
58	18 13.3	26 8.8	23 34.3	9 56.9	2 39.3
59	18 56.0	27 1.4	24 10.5	10 12.0	2 38.1
60	19 43.3	27 56.7	24 47.8	10 27.6	2 36.9
61	20 36.5	28 55.0	25 26.2	10 43.8	2 35.6
62	21 37.1	29 56.6	26 5.8	11 0.5	2 34.2
63	22 47.7	♌ 1 1.6	26 46.6	11 17.9	2 32.8
64	24 12.8	2 10.5	27 28.7	11 36.0	2 31.3
65	26 1.6	3 23.4	28 12.2	11 54.8	2 29.8
66	28 43.2	4 41.0	28 57.2	12 14.4	2 28.1
S LAT	5	6	Descendant	8	9

♏ 5° 17′ 32″ — 14h 12m 0s — MC 213° 0′ 0″

2ʰ 16ᵐ 0ˢ		MC		34° 0′ 0″	
		♉ 6° 19′ 23″			N
11	**12**	**Ascendant**	**2**	**3**	**LAT**
° ′	° ′	° ′	° ′	° ′	°
♊ 5 53.6	♋ 3 40.2	♌ 1 45.0	♍ 2 0.3	♎ 4 21.5	**0**
6 30.7	4 59.8	3 27.4	2 39.5	4 18.3	**5**
7 8.8	6 20.2	5 7.5	3 17.6	4 15.0	**10**
7 48.6	7 42.9	6 46.5	3 55.3	4 11.8	**15**
7 56.9	7 59.9	7 6.3	4 2.8	4 11.1	**16**
8 5.2	8 16.9	7 26.1	4 10.4	4 10.5	**17**
8 13.6	8 34.2	7 46.0	4 17.9	4 9.8	**18**
8 22.2	8 51.6	8 5.9	4 25.5	4 9.1	**19**
8 30.9	9 9.2	8 25.9	4 33.1	4 8.5	**20**
8 39.7	9 27.1	8 45.9	4 40.7	4 7.8	**21**
8 48.7	9 45.1	9 6.1	4 48.3	4 7.1	**22**
8 57.9	10 3.4	9 26.3	4 56.0	4 6.4	**23**
9 7.2	10 22.0	9 46.6	5 3.7	4 5.7	**24**
9 16.7	10 40.8	10 7.1	5 11.5	4 5.0	**25**
9 26.3	10 59.9	10 27.6	5 19.3	4 4.3	**26**
9 36.2	11 19.3	10 48.3	5 27.2	4 3.6	**27**
9 46.3	11 39.1	11 9.2	5 35.1	4 2.9	**28**
9 56.7	11 59.1	11 30.2	5 43.1	4 2.1	**29**
10 7.2	12 19.6	11 51.4	5 51.2	4 1.4	**30**
10 18.1	12 40.4	12 12.8	5 59.4	4 0.6	**31**
10 29.2	13 1.7	12 34.4	6 7.6	3 59.9	**32**
10 40.6	13 23.4	12 56.2	6 15.9	3 59.1	**33**
10 52.4	13 45.5	13 18.3	6 24.4	3 58.3	**34**
11 4.5	14 8.2	13 40.6	6 32.9	3 57.5	**35**
11 16.9	14 31.3	14 3.2	6 41.6	3 56.7	**36**
11 29.8	14 55.1	14 26.1	6 50.3	3 55.9	**37**
11 43.0	15 19.4	14 49.2	6 59.3	3 55.0	**38**
11 56.8	15 44.3	15 12.7	7 8.3	3 54.1	**39**
12 11.0	16 9.9	15 36.5	7 17.5	3 53.3	**40**
12 25.7	16 36.2	16 0.7	7 26.8	3 52.4	**41**
12 41.1	17 3.3	16 25.3	7 36.3	3 51.4	**42**
12 57.0	17 31.1	16 50.3	7 46.0	3 50.5	**43**
13 13.6	17 59.9	17 15.7	7 55.8	3 49.5	**44**
13 31.0	18 29.5	17 41.6	8 5.9	3 48.5	**45**
13 49.2	19 0.1	18 7.9	8 16.2	3 47.5	**46**
14 8.2	19 31.7	18 34.7	8 26.6	3 46.4	**47**
14 28.2	20 4.4	19 2.1	8 37.4	3 45.4	**48**
14 49.2	20 38.4	19 30.0	8 48.3	3 44.3	**49**
15 11.4	21 13.5	19 58.5	8 59.6	3 43.1	**50**
15 34.9	21 50.1	20 27.7	9 11.1	3 41.9	**51**
15 59.9	22 28.1	20 57.5	9 22.9	3 40.7	**52**
16 26.5	23 7.6	21 28.0	9 35.0	3 39.5	**53**
16 54.9	23 48.9	21 59.2	9 47.5	3 38.2	**54**
17 25.3	24 31.9	22 31.2	10 0.3	3 36.8	**55**
17 58.2	25 16.9	23 4.0	10 13.5	3 35.4	**56**
18 33.8	26 4.0	23 37.7	10 27.1	3 33.9	**57**
19 12.7	26 53.4	24 12.3	10 41.2	3 32.4	**58**
19 55.4	27 45.3	24 47.9	10 55.7	3 30.8	**59**
20 42.7	28 39.8	25 24.5	11 10.7	3 29.2	**60**
21 35.9	29 37.2	26 2.1	11 26.3	3 27.5	**61**
22 36.4	♌ 0 37.7	26 40.9	11 42.4	3 25.6	**62**
23 46.9	1 41.6	27 21.0	11 59.2	3 23.8	**63**
25 11.7	2 49.3	28 2.3	12 16.6	3 21.8	**64**
26 59.9	4 0.9	28 45.0	12 34.7	3 19.7	**65**
29 39.5	5 17.0	29 29.1	12 53.5	3 17.4	**66**
5	**6**	**Descendant**	**8**	**9**	**S**
		♏ 6° 19′ 23″			**LAT**
14ʰ 16ᵐ 0ˢ		MC		214° 0′ 0″	

2ʰ 20ᵐ 0ˢ		MC		35° 0′ 0″	
		♉ 7° 21′ 3″			N
11	**12**	**Ascendant**	**2**	**3**	**LAT**
° ′	° ′	° ′	° ′	° ′	°
♊ 6 50.3	♋ 4 35.3	♌ 2 43.0	♍ 3 3.5	♎ 5 26.8	**0**
7 27.6	5 54.8	4 24.4	3 41.4	5 22.8	**5**
8 6.0	7 15.1	6 3.4	4 18.2	5 18.7	**10**
8 46.0	8 37.5	7 41.3	4 54.6	5 14.7	**15**
8 54.3	8 54.4	8 0.8	5 1.8	5 13.9	**16**
9 2.6	9 11.4	8 20.4	5 9.1	5 13.0	**17**
9 11.1	9 28.6	8 40.0	5 16.4	5 12.2	**18**
9 19.7	9 46.0	8 59.7	5 23.7	5 11.4	**19**
9 28.5	10 3.5	9 19.4	5 31.0	5 10.5	**20**
9 37.3	10 21.3	9 39.2	5 38.3	5 9.7	**21**
9 46.4	10 39.3	9 59.1	5 45.7	5 8.8	**22**
9 55.5	10 57.5	10 19.0	5 53.1	5 8.0	**23**
10 4.9	11 15.9	10 39.0	6 0.6	5 7.1	**24**
10 14.4	11 34.7	10 59.2	6 8.1	5 6.2	**25**
10 24.1	11 53.7	11 19.5	6 15.6	5 5.4	**26**
10 34.1	12 13.0	11 39.9	6 23.2	5 4.5	**27**
10 44.2	12 32.6	12 0.5	6 30.9	5 3.6	**28**
10 54.6	12 52.6	12 21.2	6 38.6	5 2.6	**29**
11 5.2	13 12.9	12 42.1	6 46.4	5 1.7	**30**
11 16.1	13 33.7	13 3.1	6 54.2	5 0.8	**31**
11 27.3	13 54.8	13 24.4	7 2.2	4 59.8	**32**
11 38.7	14 16.3	13 45.9	7 10.2	4 58.8	**33**
11 50.5	14 38.3	14 7.6	7 18.3	4 57.8	**34**
12 2.6	15 0.8	14 29.6	7 26.6	4 56.8	**35**
12 15.1	15 23.8	14 51.8	7 34.9	4 55.8	**36**
12 28.0	15 47.4	15 14.3	7 43.4	4 54.8	**37**
12 41.3	16 11.5	15 37.2	7 51.9	4 53.7	**38**
12 55.1	16 36.3	16 0.3	8 0.6	4 52.6	**39**
13 9.4	17 1.7	16 23.7	8 9.5	4 51.5	**40**
13 24.2	17 27.8	16 47.5	8 18.5	4 50.4	**41**
13 39.6	17 54.6	17 11.7	8 27.6	4 49.3	**42**
13 55.5	18 22.2	17 36.2	8 36.9	4 48.1	**43**
14 12.2	18 50.7	18 1.2	8 46.4	4 46.9	**44**
14 29.6	19 20.0	18 26.6	8 56.1	4 45.6	**45**
14 47.8	19 50.3	18 52.5	9 6.0	4 44.3	**46**
15 6.9	20 21.6	19 18.9	9 16.1	4 43.0	**47**
15 26.9	20 54.0	19 45.8	9 26.4	4 41.7	**48**
15 48.0	21 27.6	20 13.2	9 36.9	4 40.3	**49**
16 10.2	22 2.4	20 41.2	9 47.8	4 38.9	**50**
16 33.8	22 38.5	21 9.8	9 58.8	4 37.4	**51**
16 58.8	23 16.1	21 39.1	10 10.2	4 35.9	**52**
17 25.4	23 55.2	22 9.0	10 21.9	4 34.3	**53**
17 53.8	24 35.9	22 39.7	10 33.9	4 32.7	**54**
18 24.3	25 18.4	23 11.1	10 46.2	4 31.0	**55**
18 57.2	26 2.8	23 43.3	10 58.9	4 29.2	**56**
19 32.8	26 49.3	24 16.4	11 12.0	4 27.4	**57**
20 11.7	27 38.0	24 50.3	11 25.5	4 25.5	**58**
20 54.4	28 29.1	25 25.2	11 39.5	4 23.5	**59**
21 41.7	29 22.7	26 1.1	11 53.9	4 21.5	**60**
22 34.8	♌ 0 19.2	26 38.1	12 8.9	4 19.3	**61**
23 35.2	1 18.8	27 16.1	12 24.4	4 17.0	**62**
24 45.4	2 21.6	27 55.4	12 40.5	4 14.7	**63**
26 9.8	3 28.1	28 35.9	12 57.2	4 12.2	**64**
27 57.2	4 38.4	29 17.7	13 14.6	4 9.6	**65**
♋ 0 34.4	5 53.0	♍ 0 1.0	13 32.7	4 6.8	**66**
5	**6**	**Descendant**	**8**	**9**	**S**
		♏ 7° 21′ 3″			**LAT**
14ʰ 20ᵐ 0ˢ		MC		215° 0′ 0″	

$2^h\ 24^m\ 0^s$ MC 36° 0′ 0″

♉ 8° 22′ 33″

11	12	Ascendant	2	3	N LAT
° ′	° ′	° ′	° ′	° ′	°
♊ 7 46.9	♋ 5 30.5	♌ 3 41.2	♍ 4 6.8	♎ 6 32.1	0
8 24.4	6 49.8	5 21.6	4 43.4	6 27.2	5
9 3.0	8 9.9	6 59.4	5 18.9	6 22.4	10
9 43.2	9 32.1	8 36.1	5 53.9	6 17.5	15
9 51.5	9 48.9	8 55.4	6 0.9	6 16.6	16
9 59.9	10 5.9	9 14.8	6 7.9	6 15.6	17
10 8.4	10 23.0	9 34.1	6 14.9	6 14.6	18
10 17.1	10 40.3	9 53.5	6 22.0	6 13.6	19
10 25.9	10 57.8	10 13.0	6 29.0	6 12.6	20
10 34.8	11 15.5	10 32.5	6 36.1	6 11.6	21
10 43.8	11 33.4	10 52.1	6 43.2	6 10.5	22
10 53.1	11 51.5	11 11.8	6 50.3	6 9.5	23
11 2.4	12 9.9	11 31.5	6 57.5	6 8.5	24
11 12.0	12 28.5	11 51.4	7 4.7	6 7.4	25
11 21.8	12 47.4	12 11.4	7 12.0	6 6.4	26
11 31.7	13 6.6	12 31.5	7 19.3	6 5.3	27
11 41.9	13 26.1	12 51.7	7 26.6	6 4.2	28
11 52.3	13 46.0	13 12.2	7 34.1	6 3.1	29
12 3.0	14 6.2	13 32.7	7 41.6	6 2.0	30
12 13.9	14 26.8	13 53.5	7 49.1	6 0.9	31
12 25.1	14 47.8	14 14.5	7 56.8	5 59.7	32
12 36.6	15 9.2	14 35.6	8 4.5	5 58.6	33
12 48.4	15 31.1	14 57.0	8 12.3	5 57.4	34
13 0.6	15 53.4	15 18.6	8 20.3	5 56.2	35
13 13.1	16 16.2	15 40.5	8 28.3	5 54.9	36
13 26.1	16 39.6	16 2.6	8 36.4	5 53.7	37
13 39.4	17 3.6	16 25.1	8 44.6	5 52.4	38
13 53.2	17 28.1	16 47.8	8 53.0	5 51.1	39
14 7.5	17 53.3	17 10.9	9 1.5	5 49.8	40
14 22.4	18 19.2	17 34.3	9 10.2	5 48.4	41
14 37.8	18 45.8	17 58.0	9 19.0	5 47.1	42
14 53.8	19 13.2	18 22.2	9 27.9	5 45.6	43
15 10.5	19 41.4	18 46.7	9 37.0	5 44.2	44
15 28.0	20 10.5	19 11.7	9 46.3	5 42.7	45
15 46.2	20 40.5	19 37.1	9 55.8	5 41.2	46
16 5.3	21 11.5	20 3.1	10 5.5	5 39.6	47
16 25.4	21 43.5	20 29.5	10 15.4	5 38.0	48
16 46.5	22 16.7	20 56.4	10 25.6	5 36.3	49
17 8.7	22 51.2	21 23.9	10 36.0	5 34.6	50
17 32.3	23 26.9	21 52.0	10 46.6	5 32.8	51
17 57.3	24 4.0	22 20.7	10 57.5	5 31.0	52
18 24.0	24 42.6	22 50.1	11 8.7	5 29.1	53
18 52.4	25 22.8	23 20.2	11 20.3	5 27.2	54
19 22.9	26 4.8	23 51.0	11 32.1	5 25.1	55
19 55.8	26 48.6	24 22.6	11 44.3	5 23.0	56
20 31.4	27 34.4	24 55.1	11 56.9	5 20.8	57
21 10.3	28 22.4	25 28.4	12 9.9	5 18.6	58
21 52.9	29 12.7	26 2.6	12 23.3	5 16.2	59
22 40.2	♌ 0 5.6	26 37.8	12 37.1	5 13.7	60
23 33.1	1 1.2	27 14.0	12 51.5	5 11.1	61
24 33.4	1 59.8	27 51.3	13 6.3	5 8.4	62
25 43.4	3 1.6	28 29.8	13 21.8	5 5.6	63
27 7.3	4 6.9	29 9.5	13 37.8	5 2.6	64
28 53.7	5 15.9	29 50.6	13 54.5	4 59.4	65
♋ 1 27.9	6 29.2	♍ 0 33.0	14 11.9	4 56.1	66
5	6	Descendant	8	9	S LAT

♏ 8° 22′ 33″

$14^h\ 24^m\ 0^s$ MC 216° 0′ 0″

$2^h\ 28^m\ 0^s$ MC 37° 0′ 0″

♉ 9° 23′ 53″

11	12	Ascendant	2	3	N LAT
° ′	° ′	° ′	° ′	° ′	°
♊ 8 43.3	♋ 6 25.6	♌ 4 39.5	♍ 5 10.3	♎ 7 37.4	0
9 21.1	7 44.8	6 18.8	5 45.5	7 31.7	5
9 59.9	9 4.7	7 55.5	6 19.7	7 26.0	10
10 40.2	10 26.7	9 31.0	6 53.4	7 20.4	15
10 48.6	10 43.4	9 50.1	7 0.1	7 19.2	16
10 57.0	11 0.3	10 9.2	7 6.9	7 18.1	17
11 5.6	11 17.4	10 28.3	7 13.6	7 16.9	18
11 14.3	11 34.6	10 47.4	7 20.4	7 15.8	19
11 23.1	11 52.0	11 6.6	7 27.1	7 14.6	20
11 32.0	12 9.6	11 25.8	7 33.9	7 13.4	21
11 41.1	12 27.4	11 45.2	7 40.8	7 12.2	22
11 50.4	12 45.4	12 4.5	7 47.6	7 11.0	23
11 59.8	13 3.7	12 24.0	7 54.5	7 9.8	24
12 9.4	13 22.3	12 43.6	8 1.4	7 8.6	25
12 19.2	13 41.1	13 3.3	8 8.4	7 7.3	26
12 29.2	14 0.2	13 23.1	8 15.4	7 6.1	27
12 39.4	14 19.6	13 43.1	8 22.5	7 4.8	28
12 49.9	14 39.3	14 3.2	8 29.6	7 3.6	29
13 0.6	14 59.4	14 23.4	8 36.8	7 2.3	30
13 11.5	15 19.9	14 43.9	8 44.1	7 0.9	31
13 22.8	15 40.7	15 4.5	8 51.5	6 59.6	32
13 34.3	16 2.0	15 25.3	8 58.9	6 58.2	33
13 46.2	16 23.7	15 46.4	9 6.4	6 56.9	34
13 58.4	16 45.9	16 7.7	9 14.0	6 55.5	35
14 10.9	17 8.6	16 29.2	9 21.7	6 54.0	36
14 23.9	17 31.8	16 51.0	9 29.5	6 52.6	37
14 37.3	17 55.5	17 13.0	9 37.4	6 51.1	38
14 51.1	18 19.9	17 35.4	9 45.4	6 49.6	39
15 5.5	18 44.9	17 58.0	9 53.6	6 48.0	40
15 20.3	19 10.5	18 21.0	10 1.9	6 46.5	41
15 35.8	19 36.9	18 44.4	10 10.3	6 44.8	42
15 51.9	20 4.0	19 8.1	10 18.9	6 43.2	43
16 8.6	20 32.0	19 32.2	10 27.7	6 41.5	44
16 26.0	21 0.8	19 56.8	10 36.6	6 39.8	45
16 44.3	21 30.5	20 21.8	10 45.7	6 38.0	46
17 3.4	22 1.2	20 47.2	10 55.0	6 36.2	47
17 23.5	22 32.9	21 13.1	11 4.5	6 34.3	48
17 44.7	23 5.7	21 39.6	11 14.3	6 32.3	49
18 7.0	23 39.8	22 6.6	11 24.2	6 30.3	50
18 30.6	24 15.1	22 34.2	11 34.4	6 28.3	51
18 55.6	24 51.8	23 2.4	11 44.9	6 26.1	52
19 22.2	25 29.9	23 31.2	11 55.6	6 23.9	53
19 50.7	26 9.6	24 0.7	12 6.7	6 21.6	54
20 21.2	26 51.0	24 30.9	12 18.0	6 19.3	55
20 54.1	27 34.3	25 1.9	12 29.7	6 16.8	56
21 29.7	28 19.4	25 33.8	12 41.8	6 14.3	57
22 8.4	29 6.7	26 6.4	12 54.2	6 11.6	58
22 51.0	29 56.3	26 40.0	13 7.1	6 8.8	59
23 38.2	♌ 0 48.4	27 14.5	13 20.3	6 6.0	60
24 31.0	1 43.1	27 50.0	13 34.1	6 2.9	61
25 31.0	2 40.7	28 26.6	13 48.3	5 59.8	62
26 40.7	3 41.5	29 4.3	14 3.1	5 56.5	63
28 4.0	4 45.6	29 43.2	14 18.5	5 53.0	64
29 49.3	5 53.5	♍ 0 23.4	14 34.5	5 49.3	65
♋ 2 20.3	7 5.3	1 4.9	14 51.1	5 45.4	66
5	6	Descendant	8	9	S LAT

♏ 9° 23′ 53″

$14^h\ 28^m\ 0^s$ MC 217° 0′ 0″

$2^h\ 32^m\ 0^s$ MC 38° 0′ 0″

♉ 10° 25′ 1″

11	12	Ascendant	2	3	N LAT
° ′	° ′	° ′	° ′	° ′	°
♊ 9 39.7	♋ 7 20.8	♌ 5 38.0	♍ 6 14.0	♎ 8 42.5	**0**
10 17.7	8 39.8	7 16.2	6 47.8	8 36.1	**5**
10 56.6	9 59.5	8 51.7	7 20.6	8 29.6	**10**
11 37.1	11 21.2	10 26.0	7 52.9	8 23.2	**15**
11 45.5	11 37.9	10 44.8	7 59.4	8 21.9	**16**
11 54.0	11 54.7	11 3.6	8 5.9	8 20.5	**17**
12 2.6	12 11.7	11 22.5	8 12.3	8 19.2	**18**
12 11.3	12 28.8	11 41.4	8 18.8	8 17.9	**19**
12 20.2	12 46.2	12 0.3	8 25.3	8 16.6	**20**
12 29.1	13 3.7	12 19.2	8 31.8	8 15.2	**21**
12 38.3	13 21.4	12 38.3	8 38.4	8 13.9	**22**
12 47.6	13 39.4	12 57.4	8 45.0	8 12.5	**23**
12 57.0	13 57.6	13 16.6	8 51.6	8 11.1	**24**
13 6.7	14 16.0	13 35.9	8 58.2	8 9.7	**25**
13 16.5	14 34.7	13 55.3	9 4.9	8 8.3	**26**
13 26.5	14 53.7	14 14.8	9 11.6	8 6.9	**27**
13 36.8	15 13.0	14 34.4	9 18.4	8 5.4	**28**
13 47.2	15 32.6	14 54.2	9 25.3	8 4.0	**29**
13 58.0	15 52.6	15 14.2	9 32.2	8 2.5	**30**
14 9.0	16 12.9	15 34.3	9 39.1	8 1.0	**31**
14 20.2	16 33.6	15 54.6	9 46.2	7 59.5	**32**
14 31.8	16 54.7	16 15.1	9 53.3	7 57.9	**33**
14 43.7	17 16.3	16 35.8	10 0.5	7 56.3	**34**
14 55.9	17 38.3	16 56.7	10 7.8	7 54.7	**35**
15 8.5	18 0.8	17 17.9	10 15.2	7 53.1	**36**
15 21.5	18 23.8	17 39.3	10 22.6	7 51.4	**37**
15 34.9	18 47.4	18 1.0	10 30.2	7 49.7	**38**
15 48.8	19 11.6	18 22.9	10 37.9	7 48.0	**39**
16 3.2	19 36.3	18 45.2	10 45.7	7 46.3	**40**
16 18.1	20 1.8	19 7.8	10 53.7	7 44.5	**41**
16 33.6	20 27.9	19 30.8	11 1.8	7 42.6	**42**
16 49.6	20 54.8	19 54.1	11 10.0	7 40.7	**43**
17 6.4	21 22.5	20 17.8	11 18.4	7 38.8	**44**
17 23.9	21 51.0	20 41.8	11 26.9	7 36.8	**45**
17 42.2	22 20.4	21 6.4	11 35.7	7 34.8	**46**
18 1.3	22 50.8	21 31.3	11 44.6	7 32.7	**47**
18 21.4	23 22.1	21 56.8	11 53.7	7 30.5	**48**
18 42.6	23 54.6	22 22.8	12 3.0	7 28.3	**49**
19 4.9	24 28.3	22 49.3	12 12.5	7 26.0	**50**
19 28.5	25 3.2	23 16.3	12 22.3	7 23.7	**51**
19 53.5	25 39.4	23 44.0	12 32.3	7 21.2	**52**
20 20.2	26 17.1	24 12.3	12 42.6	7 18.7	**53**
20 48.6	26 56.3	24 41.2	12 53.1	7 16.1	**54**
21 19.1	27 37.2	25 10.9	13 4.0	7 13.4	**55**
21 51.9	28 19.8	25 41.3	13 15.2	7 10.6	**56**
22 27.5	29 4.4	26 12.4	13 26.7	7 7.7	**57**
23 6.2	29 51.0	26 44.5	13 38.6	7 4.6	**58**
23 48.7	♌ 0 39.8	27 17.4	13 50.9	7 1.5	**59**
24 35.8	1 31.1	27 51.2	14 3.6	6 58.2	**60**
25 28.4	2 25.0	28 26.0	14 16.7	6 54.7	**61**
26 28.2	3 21.6	29 1.8	14 30.4	6 51.1	**62**
27 37.4	4 21.4	29 38.8	14 44.5	6 47.3	**63**
29 0.1	5 24.4	♍ 0 16.9	14 59.2	6 43.3	**64**
♋ 0 44.1	6 31.0	0 56.3	15 14.5	6 39.1	**65**
3 11.6	7 41.5	1 37.0	15 30.4	6 34.7	**66**
5	**6**	**Descendant**	**8**	**9**	**S LAT**

♏ 10° 25′ 1″

$14^h\ 32^m\ 0^s$ MC 218° 0′ 0″

$2^h\ 36^m\ 0^s$ MC 39° 0′ 0″

♉ 11° 25′ 58″

11	12	Ascendant	2	3	N LAT
° ′	° ′	° ′	° ′	° ′	°
♊ 10 35.9	♋ 8 16.1	♌ 6 36.6	♍ 7 17.7	♎ 9 47.7	**0**
11 14.1	9 34.9	8 13.7	7 50.2	9 40.4	**5**
11 53.2	10 54.4	9 48.1	8 21.6	9 33.2	**10**
12 33.9	12 15.7	11 21.1	8 52.6	9 25.9	**15**
12 42.3	12 32.4	11 39.6	8 58.8	9 24.4	**16**
12 50.8	12 49.1	11 58.2	9 5.0	9 23.0	**17**
12 59.5	13 6.0	12 16.8	9 11.2	9 21.5	**18**
13 8.2	13 23.1	12 35.4	9 17.4	9 20.0	**19**
13 17.1	13 40.3	12 54.0	9 23.6	9 18.5	**20**
13 26.1	13 57.8	13 12.7	9 29.8	9 17.0	**21**
13 35.3	14 15.4	13 31.5	9 36.1	9 15.5	**22**
13 44.6	14 33.3	13 50.3	9 42.4	9 13.9	**23**
13 54.1	14 51.3	14 9.2	9 48.7	9 12.4	**24**
14 3.7	15 9.7	14 28.2	9 55.1	9 10.8	**25**
14 13.6	15 28.3	14 47.3	10 1.5	9 9.2	**26**
14 23.6	15 47.2	15 6.5	10 7.9	9 7.6	**27**
14 33.9	16 6.3	15 25.8	10 14.4	9 6.0	**28**
14 44.4	16 25.8	15 45.3	10 21.0	9 4.3	**29**
14 55.2	16 45.7	16 4.9	10 27.6	9 2.7	**30**
15 6.2	17 5.9	16 24.7	10 34.2	9 1.0	**31**
15 17.5	17 26.4	16 44.7	10 41.0	8 59.3	**32**
15 29.1	17 47.4	17 4.8	10 47.8	8 57.5	**33**
15 41.0	18 8.8	17 25.2	10 54.7	8 55.8	**34**
15 53.3	18 30.6	17 45.7	11 1.6	8 54.0	**35**
16 5.9	18 53.0	18 6.5	11 8.7	8 52.1	**36**
16 18.9	19 15.8	18 27.6	11 15.8	8 50.3	**37**
16 32.4	19 39.2	18 48.9	11 23.1	8 48.4	**38**
16 46.3	20 3.1	19 10.5	11 30.4	8 46.4	**39**
17 0.7	20 27.7	19 32.4	11 37.9	8 44.4	**40**
17 15.6	20 52.9	19 54.6	11 45.5	8 42.4	**41**
17 31.1	21 18.8	20 17.1	11 53.2	8 40.3	**42**
17 47.2	21 45.4	20 40.0	12 1.1	8 38.2	**43**
18 4.0	22 12.8	21 3.3	12 9.1	8 36.0	**44**
18 21.5	22 41.1	21 26.9	12 17.3	8 33.8	**45**
18 39.8	23 10.2	21 51.0	12 25.6	8 31.5	**46**
18 59.0	23 40.2	22 15.5	12 34.1	8 29.2	**47**
19 19.1	24 11.3	22 40.5	12 42.8	8 26.8	**48**
19 40.2	24 43.4	23 5.9	12 51.7	8 24.3	**49**
20 2.6	25 16.7	23 31.9	13 0.8	8 21.7	**50**
20 26.2	25 51.2	23 58.5	13 10.2	8 19.0	**51**
20 51.2	26 27.0	24 25.6	13 19.7	8 16.3	**52**
21 17.8	27 4.2	24 53.3	13 29.5	8 13.5	**53**
21 46.2	27 42.9	25 21.7	13 39.6	8 10.5	**54**
22 16.7	28 23.2	25 50.8	13 50.0	8 7.5	**55**
22 49.5	29 5.3	26 20.6	14 0.7	8 4.3	**56**
23 25.0	29 49.2	26 51.1	14 11.7	8 1.1	**57**
24 3.6	♌ 0 35.2	27 22.5	14 23.0	7 57.7	**58**
24 46.0	1 23.3	27 54.7	14 34.8	7 54.1	**59**
25 32.9	2 13.7	28 27.9	14 46.9	7 50.4	**60**
26 25.3	3 6.7	29 2.0	14 59.4	7 46.5	**61**
27 24.8	4 2.5	29 37.1	15 12.4	7 42.4	**62**
28 33.5	5 1.2	♍ 0 13.3	15 25.9	7 38.2	**63**
29 55.5	6 3.1	0 50.6	15 39.9	7 33.7	**64**
♋ 1 38.2	7 8.5	1 29.2	15 54.5	7 28.9	**65**
4 2.0	8 17.7	2 9.0	16 9.7	7 24.0	**66**
5	**6**	**Descendant**	**8**	**9**	**S LAT**

♏ 11° 25′ 58″

$14^h\ 36^m\ 0^s$ MC 219° 0′ 0″

2h 40m 0s — MC — 40° 0′ 0″

♉ 12° 26′ 45″

11	12	Ascendant	2	3	N LAT
° ′	° ′	° ′	° ′	° ′	°
♊11 32.1	♋ 9 11.4	♌ 7 35.4	♍ 8 21.7	♎10 52.7	0
12 10.4	10 30.0	9 11.4	8 52.7	10 44.7	5
12 49.7	11 49.2	10 44.5	9 22.7	10 36.7	10
13 30.5	13 10.3	12 16.2	9 52.3	10 28.6	15
13 39.0	13 26.8	12 34.5	9 58.2	10 27.0	16
13 47.5	13 43.5	12 52.8	10 4.1	10 25.3	17
13 56.2	14 0.3	13 11.1	10 10.1	10 23.7	18
14 4.9	14 17.3	13 29.4	10 16.0	10 22.0	19
14 13.9	14 34.5	13 47.8	10 21.9	10 20.4	20
14 22.9	14 51.8	14 6.2	10 27.9	10 18.7	21
14 32.1	15 9.4	14 24.7	10 33.9	10 17.0	22
14 41.4	15 27.1	14 43.2	10 39.9	10 15.3	23
14 50.9	15 45.1	15 1.8	10 45.9	10 13.6	24
15 0.6	16 3.3	15 20.5	10 52.0	10 11.8	25
15 10.5	16 21.8	15 39.3	10 58.1	10 10.1	26
15 20.6	16 40.6	15 58.2	11 4.3	10 8.3	27
15 30.9	16 59.6	16 17.2	11 10.5	10 6.5	28
15 41.5	17 19.0	16 36.4	11 16.7	10 4.7	29
15 52.2	17 38.7	16 55.7	11 23.0	10 2.8	30
16 3.3	17 58.8	17 15.1	11 29.4	10 0.9	31
16 14.6	18 19.2	17 34.8	11 35.8	9 59.0	32
16 26.2	18 40.0	17 54.6	11 42.3	9 57.1	33
16 38.2	19 1.2	18 14.6	11 48.9	9 55.2	34
16 50.5	19 22.9	18 34.8	11 55.5	9 53.2	35
17 3.1	19 45.1	18 55.2	12 2.2	9 51.1	36
17 16.2	20 7.7	19 15.9	12 9.1	9 49.0	37
17 29.6	20 30.9	19 36.8	12 16.0	9 46.9	38
17 43.5	20 54.7	19 58.1	12 23.0	9 44.8	39
17 58.0	21 19.0	20 19.5	12 30.1	9 42.6	40
18 12.9	21 44.0	20 41.3	12 37.4	9 40.3	41
18 28.4	22 9.6	21 3.5	12 44.7	9 38.0	42
18 44.5	22 36.0	21 25.9	12 52.2	9 35.7	43
19 1.3	23 3.1	21 48.8	12 59.9	9 33.3	44
19 18.8	23 31.1	22 12.0	13 7.7	9 30.8	45
19 37.2	23 59.9	22 35.6	13 15.6	9 28.2	46
19 56.3	24 29.6	22 59.6	13 23.7	9 25.6	47
20 16.5	25 0.3	23 24.1	13 32.0	9 23.0	48
20 37.6	25 32.1	23 49.1	13 40.5	9 20.2	49
20 59.9	26 5.0	24 14.6	13 49.2	9 17.3	50
21 23.5	26 39.1	24 40.6	13 58.1	9 14.4	51
21 48.5	27 14.4	25 7.2	14 7.2	9 11.4	52
22 15.1	27 51.2	25 34.4	14 16.5	9 8.2	53
22 43.5	28 29.4	26 2.2	14 26.1	9 5.0	54
23 14.0	29 9.2	26 30.7	14 36.0	9 1.6	55
23 46.7	29 50.6	26 59.9	14 46.2	8 58.1	56
24 22.1	♌ 0 34.0	27 29.9	14 56.7	8 54.4	57
25 0.7	1 19.2	28 0.6	15 7.5	8 50.6	58
25 42.9	2 6.6	28 32.2	15 18.6	8 46.7	59
26 29.6	2 56.3	29 4.6	15 30.2	8 42.6	60
27 21.8	3 48.5	29 38.0	15 42.1	8 38.2	61
28 20.9	4 43.3	♍ 0 12.4	15 54.5	8 33.7	62
29 29.2	5 41.0	0 47.8	16 7.4	8 29.0	63
♋ 0 50.3	6 41.9	1 24.4	16 20.7	8 24.0	64
2 31.5	7 46.1	2 2.1	16 34.6	8 18.7	65
4 51.5	8 54.0	2 41.1	16 49.0	8 13.2	66
5	6	Descendant	8	9	S LAT

♏ 12° 26′ 45″

14h 40m 0s — MC — 220° 0′ 0″

2h 44m 0s — MC — 41° 0′ 0″

♉ 13° 27′ 20″

11	12	Ascendant	2	3	N LAT
° ′	° ′	° ′	° ′	° ′	°
♊12 28.1	♋10 6.7	♌ 8 34.4	♍ 9 25.7	♎11 57.7	0
13 6.6	11 25.1	10 9.2	9 55.3	11 48.9	5
13 46.0	12 44.1	11 41.1	10 23.9	11 40.1	10
14 27.0	14 4.8	13 11.5	10 52.1	11 31.2	15
14 35.5	14 21.3	13 29.5	10 57.8	11 29.4	16
14 44.1	14 37.9	13 47.5	11 3.4	11 27.6	17
14 52.7	14 54.6	14 5.5	11 9.0	11 25.8	18
15 1.5	15 11.5	14 23.6	11 14.7	11 24.0	19
15 10.5	15 28.6	14 41.7	11 20.4	11 22.2	20
15 19.6	15 45.8	14 59.8	11 26.0	11 20.4	21
15 28.8	16 3.3	15 18.0	11 31.7	11 18.5	22
15 38.1	16 21.0	15 36.2	11 37.5	11 16.6	23
15 47.7	16 38.8	15 54.5	11 43.2	11 14.7	24
15 57.4	16 57.0	16 12.9	11 49.0	11 12.8	25
16 7.3	17 15.3	16 31.4	11 54.8	11 10.9	26
16 17.4	17 34.0	16 50.0	12 0.7	11 8.9	27
16 27.7	17 52.9	17 8.7	12 6.6	11 7.0	28
16 38.3	18 12.1	17 27.5	12 12.5	11 5.0	29
16 49.1	18 31.7	17 46.5	12 18.5	11 2.9	30
17 0.2	18 51.6	18 5.6	12 24.6	11 0.9	31
17 11.5	19 11.9	18 24.9	12 30.7	10 58.8	32
17 23.2	19 32.5	18 44.4	12 36.9	10 56.7	33
17 35.1	19 53.6	19 4.0	12 43.1	10 54.5	34
17 47.4	20 15.1	19 23.9	12 49.4	10 52.3	35
18 0.1	20 37.1	19 44.0	12 55.8	10 50.1	36
18 13.2	20 59.6	20 4.3	13 2.3	10 47.8	37
18 26.7	21 22.5	20 24.8	13 8.9	10 45.5	38
18 40.6	21 46.1	20 45.6	13 15.6	10 43.1	39
18 55.0	22 10.2	21 6.7	13 22.4	10 40.7	40
19 10.0	22 35.0	21 28.1	13 29.3	10 38.2	41
19 25.5	23 0.4	21 49.8	13 36.3	10 35.7	42
19 41.6	23 26.5	22 11.9	13 43.4	10 33.1	43
19 58.4	23 53.3	22 34.3	13 50.7	10 30.5	44
20 16.0	24 21.0	22 57.0	13 58.1	10 27.7	45
20 34.3	24 49.5	23 20.2	14 5.6	10 24.9	46
20 53.5	25 18.9	23 43.8	14 13.3	10 22.1	47
21 13.6	25 49.3	24 7.8	14 21.2	10 19.1	48
21 34.8	26 20.7	24 32.3	14 29.3	10 16.1	49
21 57.1	26 53.2	24 57.3	14 37.5	10 12.9	50
22 20.6	27 26.9	25 22.8	14 46.0	10 9.7	51
22 45.6	28 1.8	25 48.8	14 54.6	10 6.4	52
23 12.2	28 38.1	26 15.5	15 3.5	10 2.9	53
23 40.5	29 15.8	26 42.7	15 12.7	9 59.3	54
24 10.9	29 55.0	27 10.6	15 22.1	9 55.6	55
24 43.6	♌ 0 35.9	27 39.2	15 31.7	9 51.8	56
25 18.9	1 18.6	28 8.6	15 41.7	9 47.8	57
25 57.3	2 3.2	28 38.7	15 52.0	9 43.6	58
26 39.4	2 49.9	29 9.6	16 2.6	9 39.2	59
27 25.9	3 38.8	29 41.3	16 13.5	9 34.7	60
28 17.8	4 30.2	♍ 0 14.0	16 24.9	9 29.9	61
29 16.5	5 24.1	0 47.7	16 36.6	9 25.0	62
♋ 0 24.2	6 20.8	1 22.4	16 48.8	9 19.8	63
1 44.5	7 20.6	1 58.2	17 1.5	9 14.3	64
3 24.1	8 23.7	2 35.1	17 14.7	9 8.5	65
5 40.2	9 30.3	3 13.3	17 28.4	9 2.4	66
5	6	Descendant	8	9	S LAT

♏ 13° 27′ 20″

14h 44m 0s — MC — 221° 0′ 0″

2h 48m 0s — MC 42° 0′ 0″ — ♉ 14° 27′ 45″

N LAT	11	12	Ascendant	2	3
0	♊13 24.0	♋11 2.1	♌ 9 33.6	♍10 29.9	♎13 2.6
5	14 2.7	12 20.3	11 7.1	10 58.0	12 53.0
10	14 42.2	13 39.0	12 37.7	11 25.2	12 43.4
15	15 23.4	14 59.4	14 6.8	11 52.1	12 33.8
16	15 31.9	15 15.7	14 24.6	11 57.4	12 31.9
17	15 40.5	15 32.3	14 42.3	12 2.8	12 29.9
18	15 49.2	15 48.9	15 0.0	12 8.1	12 27.9
19	15 58.0	16 5.7	15 17.8	12 13.5	12 26.0
20	16 7.0	16 22.7	15 35.6	12 18.8	12 24.0
21	16 16.1	16 39.9	15 53.4	12 24.2	12 22.0
22	16 25.3	16 57.2	16 11.3	12 29.7	12 19.9
23	16 34.7	17 14.8	16 29.2	12 35.1	12 17.9
24	16 44.3	17 32.5	16 47.3	12 40.6	12 15.8
25	16 54.0	17 50.5	17 5.3	12 46.0	12 13.8
26	17 3.9	18 8.8	17 23.5	12 51.6	12 11.6
27	17 14.1	18 27.3	17 41.8	12 57.1	12 9.5
28	17 24.4	18 46.1	18 0.2	13 2.7	12 7.4
29	17 35.0	19 5.2	18 18.7	13 8.4	12 5.2
30	17 45.8	19 24.7	18 37.3	13 14.1	12 3.0
31	17 56.9	19 44.4	18 56.1	13 19.8	12 0.7
32	18 8.3	20 4.5	19 15.1	13 25.6	11 58.5
33	18 19.9	20 25.0	19 34.2	13 31.5	11 56.1
34	18 31.9	20 45.9	19 53.5	13 37.4	11 53.8
35	18 44.2	21 7.3	20 13.0	13 43.4	11 51.4
36	18 56.9	21 29.1	20 32.7	13 49.5	11 49.0
37	19 10.0	21 51.3	20 52.6	13 55.6	11 46.5
38	19 23.5	22 14.1	21 12.8	14 1.9	11 44.0
39	19 37.5	22 37.5	21 33.2	14 8.2	11 41.4
40	19 51.9	23 1.4	21 53.9	14 14.7	11 38.8
41	20 6.9	23 25.9	22 14.9	14 21.2	11 36.1
42	20 22.4	23 51.0	22 36.2	14 27.8	11 33.3
43	20 38.5	24 16.9	22 57.8	14 34.6	11 30.5
44	20 55.4	24 43.5	23 19.8	14 41.5	11 27.6
45	21 12.9	25 10.9	23 42.1	14 48.5	11 24.6
46	21 31.2	25 39.1	24 4.8	14 55.7	11 21.6
47	21 50.4	26 8.1	24 27.9	15 3.0	11 18.5
48	22 10.5	26 38.2	24 51.5	15 10.5	11 15.2
49	22 31.6	27 9.2	25 15.5	15 18.1	11 11.9
50	22 53.9	27 41.3	25 39.9	15 25.9	11 8.5
51	23 17.5	28 14.6	26 4.9	15 33.9	11 5.0
52	23 42.4	28 49.1	26 30.5	15 42.1	11 1.3
53	24 9.0	29 24.9	26 56.6	15 50.6	10 57.6
54	24 37.3	♌ 0 2.1	27 23.3	15 59.2	10 53.7
55	25 7.6	0 40.8	27 50.6	16 8.1	10 49.6
56	25 40.1	1 21.1	28 18.6	16 17.3	10 45.4
57	26 15.4	2 3.2	28 47.3	16 26.7	10 41.1
58	26 53.7	2 47.2	29 16.7	16 36.4	10 36.5
59	27 35.6	3 33.1	29 47.0	16 46.5	10 31.8
60	28 21.8	4 21.3	♍ 0 18.1	16 56.9	10 26.8
61	29 13.4	5 11.8	0 50.1	17 7.6	10 21.6
62	♋ 0 11.7	6 4.8	1 23.0	17 18.8	10 16.2
63	1 18.8	7 0.6	1 57.0	17 30.3	10 10.5
64	2 38.1	7 59.4	2 32.0	17 42.3	10 4.5
65	4 16.0	9 1.3	3 8.1	17 54.8	9 58.2
66	6 28.3	10 6.7	3 45.4	18 7.8	9 51.6
S LAT	5	6	Descendant	8	9

♏ 14° 27′ 45″ — 14h 48m 0s — MC 222° 0′ 0″

2h 52m 0s — MC 43° 0′ 0″ — ♉ 15° 27′ 59″

N LAT	11	12	Ascendant	2	3
0	♊14 19.9	♋11 57.6	♌10 32.9	♍11 34.2	♎14 7.5
5	14 58.7	13 15.5	12 5.2	12 0.8	13 57.0
10	15 38.4	14 33.9	13 34.5	12 26.6	13 46.7
15	16 19.6	15 53.9	15 2.2	12 52.0	13 36.3
16	16 28.2	16 10.2	15 19.7	12 57.1	13 34.2
17	16 36.8	16 26.7	15 37.2	13 2.2	13 32.1
18	16 45.5	16 43.2	15 54.6	13 7.2	13 30.0
19	16 54.4	16 59.9	16 12.1	13 12.3	13 27.8
20	17 3.3	17 16.8	16 29.6	13 17.4	13 25.7
21	17 12.4	17 33.9	16 47.1	13 22.5	13 23.5
22	17 21.7	17 51.1	17 4.7	13 27.6	13 21.3
23	17 31.1	18 8.6	17 22.3	13 32.8	13 19.1
24	17 40.7	18 26.2	17 40.0	13 38.0	13 16.9
25	17 50.5	18 44.1	17 57.8	13 43.1	13 14.6
26	18 0.4	19 2.3	18 15.7	13 48.4	13 12.4
27	18 10.6	19 20.6	18 33.6	13 53.6	13 10.1
28	18 20.9	19 39.3	18 51.7	13 58.9	13 7.7
29	18 31.5	19 58.3	19 9.9	14 4.3	13 5.4
30	18 42.4	20 17.6	19 28.2	14 9.7	13 3.0
31	18 53.5	20 37.2	19 46.6	14 15.1	13 0.5
32	19 4.9	20 57.1	20 5.2	14 20.6	12 58.1
33	19 16.5	21 17.5	20 24.0	14 26.1	12 55.6
34	19 28.5	21 38.2	20 43.0	14 31.8	12 53.0
35	19 40.9	21 59.4	21 2.1	14 37.4	12 50.5
36	19 53.6	22 21.0	21 21.4	14 43.2	12 47.8
37	20 6.7	22 43.1	21 41.0	14 49.0	12 45.2
38	20 20.2	23 5.6	22 0.8	14 54.9	12 42.4
39	20 34.1	23 28.8	22 20.8	15 0.9	12 39.6
40	20 48.6	23 52.4	22 41.1	15 7.0	12 36.8
41	21 3.6	24 16.7	23 1.7	15 13.2	12 33.9
42	21 19.1	24 41.6	23 22.6	15 19.4	12 30.9
43	21 35.2	25 7.2	23 43.8	15 25.8	12 27.8
44	21 52.0	25 33.5	24 5.3	15 32.3	12 24.7
45	22 9.6	26 0.6	24 27.2	15 39.0	12 21.5
46	22 27.9	26 28.5	24 49.4	15 45.8	12 18.2
47	22 47.0	26 57.3	25 12.1	15 52.7	12 14.8
48	23 7.1	27 27.0	25 35.1	15 59.7	12 11.3
49	23 28.3	27 57.6	25 58.6	16 6.9	12 7.7
50	23 50.5	28 29.4	26 22.6	16 14.3	12 4.0
51	24 14.1	29 2.2	26 47.1	16 21.9	12 0.2
52	24 39.0	29 36.3	27 12.1	16 29.7	11 56.3
53	25 5.4	♌ 0 11.6	27 37.6	16 37.6	11 52.2
54	25 33.7	0 48.3	28 3.8	16 45.8	11 48.0
55	26 3.9	1 26.5	28 30.5	16 54.2	11 43.6
56	26 36.4	2 6.3	28 57.9	17 2.9	11 39.0
57	27 11.5	2 47.7	29 26.0	17 11.8	11 34.3
58	27 49.6	3 31.0	29 54.8	17 21.0	11 29.4
59	28 31.3	4 16.3	♍ 0 24.4	17 30.5	11 24.3
60	29 17.3	5 3.7	0 54.9	17 40.3	11 18.9
61	♋ 0 8.6	5 53.4	1 26.2	17 50.4	11 13.3
62	1 6.4	6 45.6	1 58.4	18 0.9	11 7.4
63	2 12.9	7 40.4	2 31.6	18 11.8	11 1.2
64	3 31.2	8 38.1	3 5.8	18 23.1	10 54.7
65	5 7.4	9 38.9	3 41.1	18 34.9	10 47.9
66	7 15.7	10 43.1	4 17.6	18 47.2	10 40.7
S LAT	5	6	Descendant	8	9

♏ 15° 27′ 59″ — 14h 52m 0s — MC 223° 0′ 0″

2h 56m 0s MC 44° 0′ 0″

♉ 16° 28′ 1″

N LAT	11	12	Ascendant	2	3
°	° ′	° ′	° ′	° ′	° ′
0	♊15 15.6	♋12 53.1	♌11 32.4	♍12 38.6	♎15 12.2
5	15 54.6	14 10.8	13 3.4	13 3.8	15 1.0
10	16 34.4	15 28.9	14 31.4	13 28.1	14 49.9
15	17 15.8	16 48.5	15 57.7	13 52.1	14 38.7
16	17 24.3	17 4.7	16 14.9	13 56.9	14 36.5
17	17 32.9	17 21.1	16 32.1	14 1.7	14 34.2
18	17 41.7	17 37.5	16 49.3	14 6.4	14 31.9
19	17 50.6	17 54.2	17 6.4	14 11.2	14 29.6
20	17 59.6	18 10.9	17 23.7	14 16.0	14 27.3
21	18 8.7	18 27.9	17 40.9	14 20.9	14 25.0
22	18 18.0	18 45.0	17 58.2	14 25.7	14 22.6
23	18 27.4	19 2.4	18 15.5	14 30.5	14 20.2
24	18 37.0	19 19.9	18 32.9	14 35.4	14 17.9
25	18 46.8	19 37.7	18 50.4	14 40.3	14 15.4
26	18 56.8	19 55.7	19 7.9	14 45.2	14 13.0
27	19 6.9	20 14.0	19 25.5	14 50.2	14 10.5
28	19 17.3	20 32.5	19 43.3	14 55.2	14 8.0
29	19 27.9	20 51.3	20 1.1	15 0.2	14 5.5
30	19 38.8	21 10.5	20 19.1	15 5.3	14 2.9
31	19 49.9	21 29.9	20 37.2	15 10.4	14 0.3
32	20 1.3	21 49.7	20 55.5	15 15.6	13 57.7
33	20 13.0	22 9.9	21 13.9	15 20.8	13 55.0
34	20 25.0	22 30.5	21 32.5	15 26.1	13 52.2
35	20 37.3	22 51.4	21 51.2	15 31.5	13 49.5
36	20 50.0	23 12.8	22 10.2	15 36.9	13 46.6
37	21 3.1	23 34.7	22 29.4	15 42.4	13 43.8
38	21 16.7	23 57.1	22 48.8	15 48.0	13 40.8
39	21 30.6	24 20.0	23 8.4	15 53.6	13 37.8
40	21 45.1	24 43.5	23 28.3	15 59.3	13 34.7
41	22 0.1	25 7.5	23 48.5	16 5.2	13 31.6
42	22 15.6	25 32.2	24 9.0	16 11.1	13 28.4
43	22 31.7	25 57.5	24 29.7	16 17.1	13 25.1
44	22 48.5	26 23.5	24 50.8	16 23.2	13 21.8
45	23 6.1	26 50.3	25 12.2	16 29.5	13 18.3
46	23 24.4	27 17.9	25 34.0	16 35.9	13 14.8
47	23 43.5	27 46.3	25 56.2	16 42.4	13 11.1
48	24 3.6	28 15.7	26 18.8	16 49.0	13 7.4
49	24 24.7	28 46.0	26 41.8	16 55.8	13 3.5
50	24 46.9	29 17.3	27 5.3	17 2.8	12 59.5
51	25 10.4	29 49.8	27 29.3	17 9.9	12 55.4
52	25 35.3	♌ 0 23.4	27 53.7	17 17.2	12 51.2
53	26 1.7	0 58.2	28 18.7	17 24.7	12 46.8
54	26 29.8	1 34.5	28 44.3	17 32.4	12 42.3
55	27 0.0	2 12.1	29 10.5	17 40.3	12 37.5
56	27 32.3	2 51.3	29 37.3	17 48.5	12 32.6
57	28 7.3	3 32.2	♍ 0 4.8	17 56.8	12 27.5
58	28 45.3	4 14.9	0 33.0	18 5.5	12 22.2
59	29 26.8	4 59.4	1 1.9	18 14.4	12 16.7
60	♋ 0 12.5	5 46.1	1 31.7	18 23.7	12 10.9
61	1 3.4	6 35.0	2 2.3	18 33.2	12 4.9
62	2 0.7	7 26.3	2 33.8	18 43.1	11 58.6
63	3 6.5	8 20.2	3 6.2	18 53.4	11 51.9
64	4 23.7	9 16.9	3 39.7	19 4.0	11 44.9
65	5 58.2	10 16.6	4 14.2	19 15.1	11 37.6
66	8 2.7	11 19.6	4 49.9	19 26.6	11 29.8
S LAT	5	6	Descendant	8	9

♏ 16° 28′ 1″

14h 56m 0s MC 224° 0′ 0″

3h 0m 0s MC 45° 0′ 0″

♉ 17° 27′ 54″

N LAT	11	12	Ascendant	2	3
°	° ′	° ′	° ′	° ′	° ′
0	♊16 11.3	♋13 48.7	♌12 32.1	♍13 43.2	♎16 16.8
5	16 50.4	15 6.1	14 1.8	14 6.8	16 4.9
10	17 30.3	16 23.9	15 28.4	14 29.7	15 53.0
15	18 11.8	17 43.1	16 53.3	14 52.2	15 41.1
16	18 20.3	17 59.2	17 10.2	14 56.7	15 38.7
17	18 29.0	18 15.5	17 27.1	15 1.2	15 36.2
18	18 37.8	18 31.8	17 44.0	15 5.7	15 33.8
19	18 46.7	18 48.4	18 0.9	15 10.2	15 31.3
20	18 55.7	19 5.1	18 17.8	15 14.7	15 28.9
21	19 4.8	19 21.9	18 34.7	15 19.2	15 26.4
22	19 14.1	19 38.9	18 51.7	15 23.8	15 23.9
23	19 23.6	19 56.2	19 8.7	15 28.3	15 21.3
24	19 33.2	20 13.6	19 25.8	15 32.9	15 18.8
25	19 43.0	20 31.2	19 42.9	15 37.5	15 16.2
26	19 53.0	20 49.1	20 0.2	15 42.2	15 13.6
27	20 3.1	21 7.2	20 17.5	15 46.8	15 10.9
28	20 13.5	21 25.6	20 34.9	15 51.5	15 8.2
29	20 24.2	21 44.3	20 52.4	15 56.2	15 5.5
30	20 35.0	22 3.3	21 10.0	16 1.0	15 2.8
31	20 46.2	22 22.6	21 27.8	16 5.8	15 0.0
32	20 57.6	22 42.3	21 45.7	16 10.7	14 57.2
33	21 9.3	23 2.3	22 3.8	16 15.6	14 54.3
34	21 21.3	23 22.7	22 22.0	16 20.6	14 51.4
35	21 33.6	23 43.4	22 40.4	16 25.6	14 48.4
36	21 46.4	24 4.7	22 59.0	16 30.7	14 45.4
37	21 59.5	24 26.4	23 17.8	16 35.8	14 42.3
38	22 13.0	24 48.5	23 36.8	16 41.0	14 39.2
39	22 27.0	25 11.2	23 56.0	16 46.3	14 35.9
40	22 41.4	25 34.4	24 15.5	16 51.7	14 32.7
41	22 56.4	25 58.2	24 35.3	16 57.2	14 29.3
42	23 11.9	26 22.6	24 55.3	17 2.7	14 25.9
43	23 28.0	26 47.7	25 15.7	17 8.4	14 22.4
44	23 44.8	27 13.5	25 36.3	17 14.1	14 18.8
45	24 2.3	27 39.9	25 57.3	17 20.0	14 15.1
46	24 20.6	28 7.2	26 18.7	17 26.0	14 11.3
47	24 39.7	28 35.3	26 40.4	17 32.1	14 7.4
48	24 59.8	29 4.3	27 2.5	17 38.3	14 3.4
49	25 20.9	29 34.3	27 25.0	17 44.7	13 59.3
50	25 43.0	♌ 0 5.2	27 48.0	17 51.2	13 55.0
51	26 6.5	0 37.2	28 11.4	17 57.9	13 50.6
52	26 31.3	1 10.4	28 35.4	18 4.8	13 46.1
53	26 57.6	1 44.8	28 59.8	18 11.8	13 41.4
54	27 25.7	2 20.6	29 24.8	18 19.0	13 36.5
55	27 55.7	2 57.7	29 50.4	18 26.4	13 31.4
56	28 28.0	3 36.3	♍ 0 16.6	18 34.1	13 26.2
57	29 2.8	4 16.6	0 43.5	18 41.9	13 20.7
58	29 40.6	4 58.6	1 11.1	18 50.1	13 15.0
59	♋ 0 21.8	5 42.5	1 39.4	18 58.4	13 9.1
60	1 7.3	6 28.4	2 8.5	19 7.1	13 2.9
61	1 57.8	7 16.5	2 38.4	19 16.0	12 56.5
62	2 54.6	8 6.9	3 9.1	19 25.3	12 49.7
63	3 59.7	8 59.9	3 40.9	19 34.9	12 42.5
64	5 15.8	9 55.6	4 13.6	19 44.9	12 35.1
65	6 48.4	10 54.3	4 47.3	19 55.3	12 27.2
66	8 49.2	11 56.1	5 22.2	20 6.1	12 18.8
S LAT	5	6	Descendant	8	9

♏ 17° 27′ 54″

15h 0m 0s MC 225° 0′ 0″

3^h 4^m 0^s	MC		46° 0′ 0″		
	♉ 18° 27′ 35″				
N LAT	**11**	**12**	**Ascendant**	**2**	**3**
°	° ′	° ′	° ′	° ′	° ′
0	♊17 6.9	♋14 44.4	♌13 32.0	♍14 47.8	♎17 21.4
5	17 46.1	16 1.5	15 0.3	15 9.9	17 8.7
10	18 26.1	17 18.9	16 25.5	15 31.4	16 56.0
15	19 7.7	18 37.7	17 49.0	15 52.4	16 43.3
16	19 16.3	18 53.7	18 5.6	15 56.6	16 40.8
17	19 24.9	19 9.9	18 22.2	16 0.8	16 38.2
18	19 33.7	19 26.2	18 38.8	16 5.0	16 35.6
19	19 42.6	19 42.6	18 55.4	16 9.2	16 33.0
20	19 51.7	19 59.2	19 12.0	16 13.5	16 30.3
21	20 0.8	20 15.9	19 28.6	16 17.7	16 27.7
22	20 10.1	20 32.8	19 45.3	16 21.9	16 25.0
23	20 19.6	20 49.9	20 2.0	16 26.2	16 22.3
24	20 29.2	21 7.2	20 18.7	16 30.5	16 19.6
25	20 39.0	21 24.8	20 35.6	16 34.8	16 16.8
26	20 49.0	21 42.5	20 52.5	16 39.1	16 14.1
27	20 59.2	22 0.5	21 9.4	16 43.5	16 11.3
28	21 9.6	22 18.8	21 26.5	16 47.9	16 8.4
29	21 20.3	22 37.3	21 43.7	16 52.3	16 5.5
30	21 31.2	22 56.1	22 1.0	16 56.7	16 2.6
31	21 42.3	23 15.3	22 18.4	17 1.2	15 59.6
32	21 53.7	23 34.8	22 36.0	17 5.8	15 56.6
33	22 5.4	23 54.6	22 53.7	17 10.4	15 53.6
34	22 17.4	24 14.8	23 11.5	17 15.0	15 50.5
35	22 29.8	24 35.4	23 29.6	17 19.7	15 47.3
36	22 42.5	24 56.5	23 47.8	17 24.5	15 44.1
37	22 55.6	25 17.9	24 6.2	17 29.3	15 40.8
38	23 9.1	25 39.9	24 24.8	17 34.1	15 37.4
39	23 23.1	26 2.3	24 43.7	17 39.1	15 34.0
40	23 37.5	26 25.3	25 2.8	17 44.1	15 30.5
41	23 52.5	26 48.9	25 22.1	17 49.2	15 27.0
42	24 8.0	27 13.0	25 41.8	17 54.4	15 23.3
43	24 24.1	27 37.8	26 1.7	17 59.7	15 19.6
44	24 40.9	28 3.3	26 21.9	18 5.1	15 15.7
45	24 58.4	28 29.5	26 42.4	18 10.6	15 11.8
46	25 16.7	28 56.5	27 3.3	18 16.1	15 7.8
47	25 35.8	29 24.3	27 24.5	18 21.8	15 3.6
48	25 55.8	29 52.9	27 46.2	18 27.7	14 59.3
49	26 16.8	♌ 0 22.5	28 8.2	18 33.6	14 54.9
50	26 39.0	0 53.0	28 30.7	18 39.7	14 50.4
51	27 2.3	1 24.7	28 53.6	18 46.0	14 45.7
52	27 27.1	1 57.4	29 17.0	18 52.4	14 40.9
53	27 53.3	2 31.4	29 40.9	18 58.9	14 35.9
54	28 21.3	3 6.6	♍ 0 5.4	19 5.7	14 30.7
55	28 51.2	3 43.2	0 30.4	19 12.6	14 25.3
56	29 23.4	4 21.3	0 56.0	19 19.7	14 19.7
57	29 58.0	5 1.0	1 22.3	19 27.1	14 13.9
58	♋ 0 35.6	5 42.3	1 49.2	19 34.6	14 7.8
59	1 16.6	6 25.5	2 16.9	19 42.4	14 1.5
60	2 1.7	7 10.7	2 45.3	19 50.5	13 54.9
61	2 51.8	7 58.0	3 14.5	19 58.9	13 48.0
62	3 48.1	8 47.6	3 44.6	20 7.5	13 40.7
63	4 52.4	9 39.7	4 15.5	20 16.5	13 33.1
64	6 7.4	10 34.4	4 47.5	20 25.8	13 25.2
65	7 38.2	11 32.0	5 20.4	20 35.5	13 16.7
66	9 35.3	12 32.7	5 54.5	20 45.6	13 7.8
S LAT	**5**	**6**	**Descendant**	**8**	**9**
	♏ 18° 27′ 35″				
15^h 4^m 0^s	MC		226° 0′ 0″		

3^h 8^m 0^s	MC		47° 0′ 0″		
	♉ 19° 27′ 5″				
N LAT	**11**	**12**	**Ascendant**	**2**	**3**
°	° ′	° ′	° ′	° ′	° ′
0	♊18 2.4	♋15 40.1	♌14 32.0	♍15 52.5	♎18 25.8
5	18 41.7	16 57.0	15 58.9	16 13.1	18 12.3
10	19 21.8	18 14.0	17 22.8	16 33.1	17 59.0
15	20 3.5	19 32.4	18 44.8	16 52.7	17 45.5
16	20 12.1	19 48.3	19 1.1	16 56.6	17 42.8
17	20 20.8	20 4.4	19 17.4	17 0.5	17 40.0
18	20 29.6	20 20.5	19 33.7	17 4.4	17 37.3
19	20 38.5	20 36.9	19 50.0	17 8.3	17 34.5
20	20 47.6	20 53.3	20 6.3	17 12.3	17 31.7
21	20 56.7	21 10.0	20 22.6	17 16.2	17 28.9
22	21 6.1	21 26.8	20 38.9	17 20.2	17 26.1
23	21 15.5	21 43.7	20 55.3	17 24.1	17 23.2
24	21 25.2	22 0.9	21 11.8	17 28.1	17 20.4
25	21 35.0	22 18.3	21 28.3	17 32.1	17 17.4
26	21 45.0	22 35.9	21 44.8	17 36.1	17 14.5
27	21 55.2	22 53.8	22 1.5	17 40.2	17 11.5
28	22 5.6	23 11.9	22 18.2	17 44.3	17 8.5
29	22 16.3	23 30.3	22 35.1	17 48.4	17 5.4
30	22 27.1	23 49.0	22 52.0	17 52.5	17 2.3
31	22 38.3	24 8.0	23 9.1	17 56.7	16 59.2
32	22 49.7	24 27.3	23 26.3	18 0.9	16 56.0
33	23 1.4	24 46.9	23 43.6	18 5.2	16 52.8
34	23 13.4	25 7.0	24 1.1	18 9.5	16 49.5
35	23 25.8	25 27.4	24 18.8	18 13.9	16 46.1
36	23 38.5	25 48.2	24 36.6	18 18.3	16 42.7
37	23 51.6	26 9.5	24 54.6	18 22.8	16 39.2
38	24 5.1	26 31.2	25 12.9	18 27.3	16 35.7
39	24 19.1	26 53.4	25 31.3	18 31.9	16 32.0
40	24 33.5	27 16.2	25 50.0	18 36.6	16 28.3
41	24 48.5	27 39.5	26 9.0	18 41.3	16 24.5
42	25 4.0	28 3.4	26 28.2	18 46.1	16 20.7
43	25 20.1	28 27.9	26 47.6	18 51.0	16 16.7
44	25 36.8	28 53.1	27 7.4	18 56.0	16 12.6
45	25 54.3	29 19.0	27 27.5	19 1.1	16 8.5
46	26 12.5	29 45.7	27 47.9	19 6.3	16 4.2
47	26 31.6	♌ 0 13.1	28 8.7	19 11.6	15 59.8
48	26 51.6	0 41.4	28 29.9	19 17.0	15 55.2
49	27 12.6	1 10.6	28 51.4	19 22.6	15 50.6
50	27 34.6	1 40.8	29 13.4	19 28.2	15 45.8
51	27 58.0	2 12.0	29 35.8	19 34.0	15 40.8
52	28 22.6	2 44.3	29 58.7	19 40.0	15 35.6
53	28 48.8	3 17.8	♍ 0 22.0	19 46.1	15 30.3
54	29 16.7	3 52.6	0 45.9	19 52.3	15 24.8
55	29 46.5	4 28.6	1 10.4	19 58.8	15 19.1
56	♋ 0 18.4	5 6.2	1 35.4	20 5.4	15 13.1
57	0 52.9	5 45.3	2 1.0	20 12.2	15 7.0
58	1 30.3	6 26.0	2 27.4	20 19.2	15 0.5
59	2 11.0	7 8.5	2 54.4	20 26.5	14 53.8
60	2 55.8	7 53.0	3 22.1	20 34.0	14 46.8
61	3 45.5	8 39.5	3 50.6	20 41.7	14 39.5
62	4 41.2	9 28.3	4 20.0	20 49.8	14 31.8
63	5 44.7	10 19.4	4 50.2	20 58.1	14 23.7
64	6 58.6	11 13.2	5 21.4	21 6.7	14 15.2
65	8 27.6	12 9.8	5 53.6	21 15.7	14 6.2
66	10 21.1	13 9.3	6 26.8	21 25.1	13 56.8
S LAT	**5**	**6**	**Descendant**	**8**	**9**
	♏ 19° 27′ 5″				
15^h 8^m 0^s	MC		227° 0′ 0″		

3h 12m 0s — MC 48° 0′ 0″ — ♉ 20° 26′ 26″

11	12	Ascendant	2	3	N LAT
° ′	° ′	° ′	° ′	° ′	°
♊18 57.9	♋16 36.0	♌15 32.3	♍16 57.4	♎19 30.1	**0**
19 37.3	17 52.5	16 57.7	17 16.4	19 15.9	**5**
20 17.5	19 9.1	18 20.1	17 34.9	19 1.8	**10**
20 59.2	20 27.0	19 40.7	17 53.0	18 47.6	**15**
21 7.8	20 42.9	19 56.7	17 56.6	18 44.7	**16**
21 16.5	20 58.8	20 12.7	18 0.3	18 41.8	**17**
21 25.3	21 14.9	20 28.6	18 3.9	18 38.9	**18**
21 34.3	21 31.1	20 44.6	18 7.5	18 36.0	**19**
21 43.3	21 47.5	21 0.6	18 11.1	18 33.1	**20**
21 52.5	22 4.0	21 16.6	18 14.8	18 30.1	**21**
22 1.9	22 20.7	21 32.6	18 18.4	18 27.1	**22**
22 11.3	22 37.5	21 48.7	18 22.1	18 24.1	**23**
22 21.0	22 54.6	22 4.8	18 25.8	18 21.0	**24**
22 30.8	23 11.9	22 21.0	18 29.5	18 18.0	**25**
22 40.8	23 29.3	22 37.2	18 33.2	18 14.8	**26**
22 51.0	23 47.0	22 53.6	18 36.9	18 11.7	**27**
23 1.5	24 5.0	23 10.0	18 40.7	18 8.5	**28**
23 12.1	24 23.3	23 26.4	18 44.5	18 5.3	**29**
23 23.0	24 41.8	23 43.0	18 48.3	18 2.0	**30**
23 34.1	25 0.6	23 59.8	18 52.2	17 58.7	**31**
23 45.5	25 19.7	24 16.6	18 56.1	17 55.3	**32**
23 57.3	25 39.2	24 33.6	19 0.0	17 51.9	**33**
24 9.3	25 59.1	24 50.7	19 4.0	17 48.4	**34**
24 21.6	26 19.3	25 8.0	19 8.1	17 44.9	**35**
24 34.3	26 39.9	25 25.5	19 12.1	17 41.2	**36**
24 47.4	27 1.0	25 43.1	19 16.3	17 37.6	**37**
25 1.0	27 22.5	26 1.0	19 20.5	17 33.8	**38**
25 14.9	27 44.5	26 19.0	19 24.7	17 30.0	**39**
25 29.3	28 7.0	26 37.3	19 29.0	17 26.1	**40**
25 44.3	28 30.1	26 55.8	19 33.4	17 22.1	**41**
25 59.8	28 53.7	27 14.6	19 37.9	17 18.0	**42**
26 15.8	29 18.0	27 33.6	19 42.4	17 13.8	**43**
26 32.6	29 42.9	27 53.0	19 47.0	17 9.5	**44**
26 50.0	♌ 0 8.5	28 12.6	19 51.7	17 5.1	**45**
27 8.2	0 34.8	28 32.6	19 56.5	17 0.5	**46**
27 27.2	1 2.0	28 52.9	20 1.4	16 55.9	**47**
27 47.2	1 29.9	29 13.6	20 6.4	16 51.1	**48**
28 8.1	1 58.7	29 34.6	20 11.5	16 46.1	**49**
28 30.1	2 28.5	29 56.1	20 16.8	16 41.1	**50**
28 53.4	2 59.3	♍ 0 18.0	20 22.1	16 35.8	**51**
29 17.9	3 31.2	0 40.3	20 27.6	16 30.4	**52**
29 44.0	4 4.2	1 3.1	20 33.2	16 24.7	**53**
♋ 0 11.8	4 38.5	1 26.5	20 39.0	16 18.9	**54**
0 41.5	5 14.0	1 50.3	20 44.9	16 12.8	**55**
1 13.3	5 51.0	2 14.8	20 51.0	16 6.6	**56**
1 47.6	6 29.5	2 39.8	20 57.3	16 0.0	**57**
2 24.7	7 9.6	3 5.5	21 3.8	15 53.2	**58**
3 5.2	7 51.5	3 31.9	21 10.5	15 46.1	**59**
3 49.6	8 35.2	3 59.0	21 17.4	15 38.7	**60**
4 38.8	9 21.0	4 26.8	21 24.6	15 30.9	**61**
5 34.0	10 8.9	4 55.5	21 32.0	15 22.8	**62**
6 36.7	10 59.2	5 25.0	21 39.7	15 14.2	**63**
7 49.4	11 52.0	5 55.4	21 47.7	15 5.2	**64**
9 16.5	12 47.5	6 26.8	21 56.0	14 55.7	**65**
11 6.6	13 46.0	6 59.2	22 4.6	14 45.7	**66**
5	6	Descendant	8	9	S LAT

♏ 20° 26′ 26″ — 15h 12m 0s — MC 228° 0′ 0″

3h 16m 0s — MC 49° 0′ 0″ — ♉ 21° 25′ 35″

11	12	Ascendant	2	3	N LAT
° ′	° ′	° ′	° ′	° ′	°
♊19 53.3	♋17 31.9	♌16 32.7	♍18 2.3	♎20 34.3	**0**
20 32.8	18 48.1	17 56.7	18 19.8	20 19.3	**5**
21 13.0	20 4.3	19 17.6	18 36.7	20 4.5	**10**
21 54.9	21 21.8	20 36.7	18 53.4	19 49.6	**15**
22 3.5	21 37.5	20 52.4	18 56.7	19 46.5	**16**
22 12.2	21 53.4	21 8.0	19 0.0	19 43.5	**17**
22 21.0	22 9.3	21 23.7	19 3.4	19 40.4	**18**
22 29.9	22 25.4	21 39.3	19 6.7	19 37.4	**19**
22 39.0	22 41.7	21 55.0	19 10.0	19 34.3	**20**
22 48.2	22 58.1	22 10.7	19 13.4	19 31.2	**21**
22 57.5	23 14.6	22 26.4	19 16.7	19 28.0	**22**
23 7.0	23 31.4	22 42.2	19 20.1	19 24.8	**23**
23 16.7	23 48.3	22 58.0	19 23.5	19 21.6	**24**
23 26.5	24 5.4	23 13.8	19 26.9	19 18.4	**25**
23 36.5	24 22.7	23 29.7	19 30.3	19 15.1	**26**
23 46.8	24 40.3	23 45.7	19 33.7	19 11.8	**27**
23 57.2	24 58.1	24 1.7	19 37.2	19 8.4	**28**
24 7.8	25 16.2	24 17.9	19 40.7	19 5.0	**29**
24 18.7	25 34.6	24 34.1	19 44.2	19 1.6	**30**
24 29.9	25 53.2	24 50.5	19 47.7	18 58.1	**31**
24 41.3	26 12.2	25 7.0	19 51.3	18 54.5	**32**
24 53.0	26 31.5	25 23.6	19 54.9	18 50.9	**33**
25 5.0	26 51.2	25 40.3	19 58.6	18 47.3	**34**
25 17.3	27 11.2	25 57.3	20 2.3	18 43.5	**35**
25 30.1	27 31.6	26 14.3	20 6.0	18 39.7	**36**
25 43.1	27 52.5	26 31.6	20 9.8	18 35.9	**37**
25 56.6	28 13.8	26 49.1	20 13.7	18 31.9	**38**
26 10.6	28 35.6	27 6.7	20 17.6	18 27.9	**39**
26 25.0	28 57.8	27 24.6	20 21.5	18 23.7	**40**
26 39.9	29 20.6	27 42.7	20 25.5	18 19.5	**41**
26 55.4	29 44.0	28 1.0	20 29.6	18 15.2	**42**
27 11.4	♌ 0 8.0	28 19.7	20 33.8	18 10.8	**43**
27 28.1	0 32.6	28 38.6	20 38.0	18 6.3	**44**
27 45.5	0 57.9	28 57.7	20 42.3	18 1.6	**45**
28 3.7	1 23.9	29 17.2	20 46.7	17 56.8	**46**
28 22.7	1 50.7	29 37.1	20 51.2	17 51.9	**47**
28 42.6	2 18.3	29 57.3	20 55.8	17 46.9	**48**
29 3.4	2 46.8	♍ 0 17.8	21 0.5	17 41.7	**49**
29 25.4	3 16.2	0 38.8	21 5.3	17 36.3	**50**
29 48.5	3 46.6	1 0.2	21 10.2	17 30.8	**51**
♋ 0 13.0	4 18.0	1 22.0	21 15.2	17 25.0	**52**
0 39.0	4 50.6	1 44.3	21 20.4	17 19.1	**53**
1 6.7	5 24.3	2 7.0	21 25.7	17 12.9	**54**
1 36.2	5 59.4	2 30.3	21 31.1	17 6.6	**55**
2 7.8	6 35.8	2 54.2	21 36.7	16 59.9	**56**
2 41.9	7 13.7	3 18.6	21 42.5	16 53.0	**57**
3 18.8	7 53.2	3 43.7	21 48.5	16 45.8	**58**
3 59.0	8 34.4	4 9.4	21 54.6	16 38.3	**59**
4 43.1	9 17.5	4 35.8	22 0.9	16 30.5	**60**
5 31.8	10 2.5	5 3.0	22 7.5	16 22.3	**61**
6 26.4	10 49.6	5 30.9	22 14.3	16 13.7	**62**
7 28.2	11 39.0	5 59.7	22 21.3	16 4.7	**63**
8 39.8	12 30.8	6 29.4	22 28.7	15 55.2	**64**
10 5.0	13 25.4	7 0.0	22 36.3	15 45.1	**65**
11 51.8	14 22.8	7 31.6	22 44.2	15 34.5	**66**
5	6	Descendant	8	9	S LAT

♏ 21° 25′ 35″ — 15h 16m 0s — MC 229° 0′ 0″

$3^h\ 20^m\ 0^s$ MC 50° 0′ 0″

♉ 22° 24′ 34″

N LAT	11	12	Ascendant	2	3
	° ′	° ′	° ′	° ′	° ′
0	♊20 48.6	♋18 27.9	♌17 33.3	♍19 7.3	♎21 38.3
5	21 28.2	19 43.7	18 55.8	19 23.2	21 22.7
10	22 8.5	20 59.6	20 15.2	19 38.7	21 7.1
15	22 50.4	22 16.5	21 32.7	19 53.8	20 51.4
16	22 59.0	22 32.2	21 48.1	19 56.9	20 48.3
17	23 7.7	22 47.9	22 3.5	19 59.9	20 45.1
18	23 16.6	23 3.8	22 18.8	20 2.9	20 41.9
19	23 25.5	23 19.8	22 34.2	20 5.9	20 38.6
20	23 34.6	23 35.9	22 49.5	20 9.0	20 35.4
21	23 43.8	23 52.2	23 4.9	20 12.0	20 32.1
22	23 53.1	24 8.6	23 20.3	20 15.1	20 28.8
23	24 2.6	24 25.2	23 35.7	20 18.1	20 25.5
24	24 12.3	24 42.0	23 51.1	20 21.2	20 22.1
25	24 22.1	24 59.0	24 6.6	20 24.3	20 18.7
26	24 32.2	25 16.2	24 22.2	20 27.4	20 15.3
27	24 42.4	25 33.6	24 37.8	20 30.5	20 11.8
28	24 52.8	25 51.3	24 53.6	20 33.7	20 8.3
29	25 3.4	26 9.2	25 9.4	20 36.9	20 4.7
30	25 14.3	26 27.4	25 25.3	20 40.1	20 1.1
31	25 25.5	26 45.9	25 41.3	20 43.3	19 57.4
32	25 36.9	27 4.6	25 57.4	20 46.5	19 53.7
33	25 48.6	27 23.8	26 13.6	20 49.8	19 49.9
34	26 0.6	27 43.2	26 30.0	20 53.2	19 46.1
35	26 12.9	28 3.1	26 46.5	20 56.5	19 42.1
36	26 25.6	28 23.3	27 3.2	20 59.9	19 38.1
37	26 38.7	28 43.9	27 20.1	21 3.4	19 34.1
38	26 52.2	29 5.0	27 37.2	21 6.9	19 29.9
39	27 6.1	29 26.6	27 54.4	21 10.4	19 25.7
40	27 20.5	29 48.6	28 11.9	21 14.0	19 21.4
41	27 35.4	♌ 0 11.2	28 29.6	21 17.7	19 16.9
42	27 50.8	0 34.3	28 47.5	21 21.4	19 12.4
43	28 6.9	0 58.0	29 5.7	21 25.2	19 7.8
44	28 23.5	1 22.3	29 24.1	21 29.1	19 3.0
45	28 40.9	1 47.3	29 42.9	21 33.0	18 58.1
46	28 59.0	2 13.0	♍ 0 1.9	21 37.0	18 53.1
47	29 17.9	2 39.5	0 21.3	21 41.1	18 47.9
48	29 37.8	3 6.7	0 41.0	21 45.2	18 42.6
49	29 58.6	3 34.8	1 1.1	21 49.5	18 37.1
50	♋ 0 20.5	4 3.8	1 21.5	21 53.8	18 31.5
51	0 43.5	4 33.8	1 42.4	21 58.3	18 25.7
52	1 7.9	5 4.8	2 3.7	22 2.9	18 19.6
53	1 33.8	5 36.9	2 25.4	22 7.6	18 13.4
54	2 1.3	6 10.2	2 47.6	22 12.4	18 6.9
55	2 30.7	6 44.7	3 10.3	22 17.3	18 0.2
56	3 2.2	7 20.6	3 33.6	22 22.4	17 53.2
57	3 36.0	7 57.9	3 57.4	22 27.7	17 46.0
58	4 12.7	8 36.8	4 21.9	22 33.1	17 38.4
59	4 52.5	9 17.3	4 47.0	22 38.7	17 30.5
60	5 36.2	9 59.7	5 12.7	22 44.4	17 22.3
61	6 24.5	10 43.9	5 39.2	22 50.4	17 13.6
62	7 18.4	11 30.2	6 6.4	22 56.6	17 4.6
63	8 19.4	12 18.8	6 34.5	23 3.0	16 55.1
64	9 29.8	13 9.7	7 3.4	23 9.6	16 45.1
65	10 53.3	14 3.2	7 33.2	23 16.5	16 34.5
66	12 36.8	14 59.6	8 4.0	23 23.7	16 23.3
S LAT	5	6	Descendant	8	9

♏ 22° 24′ 34″

$15^h\ 20^m\ 0^s$ MC 230° 0′ 0″

$3^h\ 24^m\ 0^s$ MC 51° 0′ 0″

♉ 23° 23′ 23″

N LAT	11	12	Ascendant	2	3
	° ′	° ′	° ′	° ′	° ′
0	♊21 43.9	♋19 24.1	♌18 34.0	♍20 12.3	♎22 42.3
5	22 23.5	20 39.5	19 55.1	20 26.7	22 25.9
10	23 4.0	21 54.9	21 12.9	20 40.6	22 9.6
15	23 45.9	23 11.4	22 28.9	20 54.3	21 53.2
16	23 54.5	23 26.9	22 44.0	20 57.0	21 49.9
17	24 3.2	23 42.5	22 59.0	20 59.8	21 46.6
18	24 12.1	23 58.3	23 14.0	21 2.5	21 43.2
19	24 21.0	24 14.1	23 29.0	21 5.2	21 39.8
20	24 30.1	24 30.1	23 44.1	21 8.0	21 36.4
21	24 39.3	24 46.3	23 59.1	21 10.7	21 33.0
22	24 48.6	25 2.6	24 14.2	21 13.4	21 29.5
23	24 58.1	25 19.0	24 29.3	21 16.2	21 26.1
24	25 7.8	25 35.7	24 44.4	21 19.0	21 22.5
25	25 17.6	25 52.5	24 59.5	21 21.8	21 19.0
26	25 27.7	26 9.6	25 14.8	21 24.6	21 15.4
27	25 37.9	26 26.9	25 30.1	21 27.4	21 11.7
28	25 48.3	26 44.4	25 45.4	21 30.2	21 8.0
29	25 58.9	27 2.1	26 0.9	21 33.1	21 4.3
30	26 9.8	27 20.2	26 16.4	21 36.0	21 0.5
31	26 21.0	27 38.5	26 32.1	21 38.9	20 56.7
32	26 32.4	27 57.1	26 47.8	21 41.8	20 52.8
33	26 44.0	28 16.0	27 3.7	21 44.8	20 48.8
34	26 56.0	28 35.3	27 19.7	21 47.8	20 44.8
35	27 8.4	28 54.9	27 35.9	21 50.8	20 40.7
36	27 21.1	29 15.0	27 52.2	21 53.9	20 36.5
37	27 34.1	29 35.4	28 8.7	21 57.0	20 32.2
38	27 47.6	29 56.2	28 25.3	22 0.1	20 27.9
39	28 1.5	♌ 0 17.5	28 42.2	22 3.3	20 23.4
40	28 15.9	0 39.3	28 59.2	22 6.6	20 18.9
41	28 30.7	1 1.6	29 16.5	22 9.9	20 14.3
42	28 46.1	1 24.5	29 34.0	22 13.2	20 9.5
43	29 2.1	1 47.9	29 51.7	22 16.6	20 4.6
44	29 18.8	2 12.0	♍ 0 9.7	22 20.1	19 59.7
45	29 36.1	2 36.6	0 28.0	22 23.6	19 54.5
46	29 54.2	3 2.0	0 46.6	22 27.2	19 49.3
47	♋ 0 13.0	3 28.1	1 5.5	22 30.9	19 43.9
48	0 32.8	3 55.0	1 24.7	22 34.7	19 38.3
49	0 53.5	4 22.8	1 44.3	22 38.5	19 32.6
50	1 15.3	4 51.4	2 4.2	22 42.4	19 26.6
51	1 38.3	5 21.0	2 24.6	22 46.4	19 20.5
52	2 2.6	5 51.5	2 45.3	22 50.6	19 14.2
53	2 28.4	6 23.2	3 6.5	22 54.8	19 7.6
54	2 55.8	6 56.0	3 28.2	22 59.1	19 0.9
55	3 25.0	7 30.0	3 50.4	23 3.6	18 53.8
56	3 56.3	8 5.3	4 13.0	23 8.2	18 46.5
57	4 29.9	8 42.1	4 36.3	23 12.9	18 38.9
58	5 6.3	9 20.3	5 0.1	23 17.7	18 30.9
59	5 45.8	10 0.2	5 24.5	23 22.8	18 22.7
60	6 29.1	10 41.9	5 49.6	23 28.0	18 14.0
61	7 16.9	11 25.3	6 15.4	23 33.3	18 4.9
62	8 10.2	12 10.9	6 42.0	23 38.9	17 55.4
63	9 10.3	12 58.5	7 9.3	23 44.6	17 45.4
64	10 19.5	13 48.6	7 37.4	23 50.6	17 34.9
65	11 41.1	14 41.1	8 6.5	23 56.8	17 23.8
66	13 21.5	15 36.4	8 36.5	24 3.3	17 12.1
S LAT	5	6	Descendant	8	9

♏ 23° 23′ 23″

$15^h\ 24^m\ 0^s$ MC 231° 0′ 0″

3h 28m 0s MC 52° 0′ 0″

♉ 24° 22′ 2″

11	12	Ascendant	2	3	N LAT
° ′	° ′	° ′	° ′	° ′	°
♊22 39.2	♋20 20.3	♌19 35.0	♍21 17.5	♎23 46.0	0
23 18.9	21 35.3	20 54.4	21 30.3	23 29.0	5
23 59.3	22 50.3	22 10.8	21 42.7	23 12.0	10
24 41.3	24 6.2	23 25.1	21 54.8	22 54.9	15
24 49.9	24 21.6	23 39.9	21 57.3	22 51.4	16
24 58.6	24 37.2	23 54.6	21 59.7	22 47.9	17
25 7.5	24 52.8	24 9.3	22 2.1	22 44.4	18
25 16.4	25 8.5	24 24.0	22 4.5	22 40.9	19
25 25.5	25 24.4	24 38.7	22 7.0	22 37.3	20
25 34.7	25 40.4	24 53.4	22 9.4	22 33.8	21
25 44.1	25 56.6	25 8.1	22 11.9	22 30.2	22
25 53.6	26 12.9	25 22.9	22 14.3	22 26.5	23
26 3.2	26 29.4	25 37.7	22 16.8	22 22.8	24
26 13.1	26 46.1	25 52.5	22 19.3	22 19.1	25
26 23.1	27 3.0	26 7.4	22 21.8	22 15.4	26
26 33.3	27 20.2	26 22.3	22 24.3	22 11.6	27
26 43.7	27 37.5	26 37.3	22 26.8	22 7.7	28
26 54.3	27 55.1	26 52.4	22 29.3	22 3.8	29
27 5.2	28 13.0	27 7.6	22 31.9	21 59.8	30
27 16.3	28 31.1	27 22.9	22 34.5	21 55.8	31
27 27.7	28 49.5	27 38.3	22 37.1	21 51.8	32
27 39.4	29 8.3	27 53.8	22 39.7	21 47.6	33
27 51.4	29 27.4	28 9.4	22 42.4	21 43.4	34
28 3.7	29 46.8	28 25.2	22 45.1	21 39.1	35
28 16.4	♌ 0 6.6	28 41.1	22 47.8	21 34.7	36
28 29.4	0 26.8	28 57.2	22 50.6	21 30.3	37
28 42.9	0 47.4	29 13.5	22 53.4	21 25.7	38
28 56.7	1 8.5	29 29.9	22 56.2	21 21.1	39
29 11.1	1 30.0	29 46.5	22 59.1	21 16.4	40
29 25.9	1 52.1	♍ 0 3.4	23 2.1	21 11.5	41
29 41.3	2 14.7	0 20.5	23 5.0	21 6.5	42
29 57.3	2 37.8	0 37.8	23 8.1	21 1.5	43
♋ 0 13.9	3 1.6	0 55.3	23 11.2	20 56.2	44
0 31.1	3 26.0	1 13.2	23 14.3	20 50.9	45
0 49.1	3 51.0	1 31.3	23 17.5	20 45.4	46
1 8.0	4 16.8	1 49.7	23 20.8	20 39.7	47
1 27.7	4 43.3	2 8.5	23 24.1	20 33.9	48
1 48.3	5 10.7	2 27.5	23 27.5	20 27.9	49
2 10.0	5 38.9	2 47.0	23 31.0	20 21.7	50
2 32.9	6 8.1	3 6.8	23 34.6	20 15.3	51
2 57.1	6 38.2	3 27.0	23 38.2	20 8.7	52
3 22.7	7 9.4	3 47.7	23 42.0	20 1.8	53
3 50.0	7 41.7	4 8.8	23 45.9	19 54.7	54
4 19.0	8 15.2	4 30.4	23 49.8	19 47.4	55
4 50.1	8 50.0	4 52.5	23 53.9	19 39.7	56
5 23.5	9 26.2	5 15.1	23 58.1	19 31.7	57
5 59.6	10 3.9	5 38.3	24 2.4	19 23.4	58
6 38.8	10 43.1	6 2.1	24 6.9	19 14.7	59
7 21.7	11 24.0	6 26.5	24 11.5	19 5.7	60
8 9.0	12 6.8	6 51.7	24 16.3	18 56.2	61
9 1.6	12 51.5	7 17.5	24 21.2	18 46.2	62
10 0.9	13 38.4	7 44.1	24 26.3	18 35.7	63
11 8.9	14 27.5	8 11.5	24 31.6	18 24.7	64
12 28.7	15 19.1	8 39.8	24 37.2	18 13.1	65
14 6.2	16 13.3	9 9.0	24 42.9	18 0.8	66
5	6	Descendant	8	9	S LAT

♏ 24° 22′ 2″

15h 28m 0s MC 232° 0′ 0″

3h 32m 0s MC 53° 0′ 0″

♉ 25° 20′ 30″

11	12	Ascendant	2	3	N LAT
° ′	° ′	° ′	° ′	° ′	°
♊23 34.4	♋21 16.7	♌20 36.1	♍22 22.6	♎24 49.7	0
24 14.1	22 31.3	21 54.0	22 33.9	24 31.9	5
24 54.6	23 45.7	23 8.7	22 44.7	24 14.2	10
25 36.6	25 1.2	24 21.5	22 55.4	23 56.4	15
25 45.2	25 16.5	24 35.9	22 57.5	23 52.8	16
25 53.9	25 31.9	24 50.3	22 59.6	23 49.2	17
26 2.8	25 47.4	25 4.7	23 1.8	23 45.5	18
26 11.7	26 3.0	25 19.1	23 3.9	23 41.9	19
26 20.8	26 18.7	25 33.4	23 6.0	23 38.2	20
26 30.0	26 34.6	25 47.8	23 8.2	23 34.4	21
26 39.4	26 50.6	26 2.2	23 10.3	23 30.7	22
26 48.9	27 6.8	26 16.6	23 12.5	23 26.9	23
26 58.5	27 23.2	26 31.0	23 14.6	23 23.0	24
27 8.4	27 39.8	26 45.5	23 16.8	23 19.2	25
27 18.4	27 56.5	27 0.1	23 19.0	23 15.3	26
27 28.6	28 13.5	27 14.7	23 21.2	23 11.3	27
27 39.0	28 30.7	27 29.3	23 23.4	23 7.3	28
27 49.6	28 48.1	27 44.1	23 25.6	23 3.2	29
28 0.5	29 5.8	27 58.9	23 27.9	22 59.1	30
28 11.6	29 23.7	28 13.8	23 30.1	22 54.9	31
28 23.0	29 42.0	28 28.8	23 32.4	22 50.7	32
28 34.7	♌ 0 0.5	28 43.9	23 34.7	22 46.3	33
28 46.6	0 19.4	28 59.2	23 37.1	22 41.9	34
28 58.9	0 38.6	29 14.6	23 39.4	22 37.5	35
29 11.6	0 58.2	29 30.1	23 41.8	22 32.9	36
29 24.6	1 18.2	29 45.8	23 44.2	22 28.3	37
29 38.0	1 38.6	♍ 0 1.7	23 46.7	22 23.5	38
29 51.9	1 59.4	0 17.7	23 49.2	22 18.7	39
♋ 0 6.2	2 20.7	0 33.9	23 51.7	22 13.7	40
0 21.0	2 42.5	0 50.3	23 54.3	22 8.7	41
0 36.3	3 4.8	1 7.0	23 56.9	22 3.5	42
0 52.2	3 27.7	1 23.9	23 59.5	21 58.2	43
1 8.8	3 51.2	1 41.0	24 2.2	21 52.8	44
1 26.0	4 15.2	1 58.4	24 5.0	21 47.2	45
1 44.0	4 40.0	2 16.0	24 7.8	21 41.4	46
2 2.7	5 5.4	2 34.0	24 10.6	21 35.5	47
2 22.3	5 31.6	2 52.2	24 13.6	21 29.4	48
2 42.9	5 58.6	3 10.8	24 16.5	21 23.2	49
3 4.6	6 26.4	3 29.7	24 19.6	21 16.7	50
3 27.4	6 55.2	3 49.1	24 22.7	21 10.0	51
3 51.4	7 24.9	4 8.7	24 25.9	21 3.1	52
4 16.9	7 55.6	4 28.9	24 29.2	20 56.0	53
4 44.0	8 27.4	4 49.4	24 32.6	20 48.6	54
5 12.9	9 0.5	5 10.4	24 36.1	20 40.9	55
5 43.7	9 34.7	5 31.9	24 39.6	20 32.8	56
6 16.9	10 10.3	5 53.9	24 43.3	20 24.5	57
6 52.7	10 47.4	6 16.5	24 47.1	20 15.8	58
7 31.6	11 25.9	6 39.7	24 51.0	20 6.8	59
8 14.1	12 6.2	7 3.5	24 55.0	19 57.3	60
9 0.8	12 48.2	7 27.9	24 59.2	19 47.3	61
9 52.8	13 32.2	7 53.1	25 3.5	19 36.9	62
10 51.2	14 18.2	8 18.9	25 8.0	19 26.0	63
11 58.0	15 6.4	8 45.6	25 12.7	19 14.4	64
13 16.1	15 57.0	9 13.1	25 17.5	19 2.3	65
14 50.6	16 50.3	9 41.5	25 22.5	18 49.4	66
5	6	Descendant	8	9	S LAT

♏ 25° 20′ 30″

15h 32m 0s MC 233° 0′ 0″

3h 36m 0s MC 54° 0′ 0″
♉ 26° 18′ 49″

N LAT	11	12	Ascendant	2	3
°	° ′	° ′	° ′	° ′	° ′
0	♊24 29.5	♋22 13.1	♌21 37.4	♍23 27.9	♎25 53.2
5	25 9.3	23 27.3	22 53.7	23 37.5	25 34.7
10	25 49.8	24 41.3	24 6.8	23 46.8	25 16.4
15	26 31.8	25 56.1	25 17.9	23 56.0	24 57.8
16	26 40.5	26 11.3	25 32.0	23 57.8	24 54.1
17	26 49.2	26 26.6	25 46.1	23 59.6	24 50.3
18	26 58.1	26 42.0	26 0.1	24 1.5	24 46.5
19	27 7.0	26 57.5	26 14.2	24 3.3	24 42.7
20	27 16.1	27 13.1	26 28.2	24 5.1	24 38.9
21	27 25.3	27 28.8	26 42.2	24 7.0	24 35.0
22	27 34.6	27 44.7	26 56.3	24 8.8	24 31.1
23	27 44.1	28 0.8	27 10.4	24 10.6	24 27.1
24	27 53.8	28 17.0	27 24.5	24 12.5	24 23.2
25	28 3.6	28 33.4	27 38.6	24 14.4	24 19.1
26	28 13.6	28 50.0	27 52.8	24 16.2	24 15.1
27	28 23.8	29 6.8	28 7.0	24 18.1	24 10.9
28	28 34.2	29 23.8	28 21.3	24 20.0	24 6.8
29	28 44.8	29 41.1	28 35.7	24 21.9	24 2.5
30	28 55.7	29 58.6	28 50.2	24 23.8	23 58.2
31	29 6.8	♌ 0 16.4	29 4.7	24 25.8	23 53.9
32	29 18.2	0 34.4	29 19.4	24 27.7	23 49.5
33	29 29.8	0 52.8	29 34.1	24 29.7	23 45.0
34	29 41.8	1 11.5	29 49.0	24 31.7	23 40.4
35	29 54.0	1 30.5	♍ 0 4.0	24 33.8	23 35.7
36	♋ 0 6.7	1 49.9	0 19.1	24 35.8	23 31.0
37	0 19.7	2 9.6	0 34.4	24 37.9	23 26.2
38	0 33.1	2 29.8	0 49.9	24 40.0	23 21.2
39	0 46.9	2 50.4	1 5.5	24 42.1	23 16.2
40	1 1.2	3 11.4	1 21.3	24 44.3	23 11.1
41	1 15.9	3 33.0	1 37.3	24 46.5	23 5.8
42	1 31.2	3 55.0	1 53.5	24 48.7	23 0.4
43	1 47.1	4 17.6	2 9.9	24 51.0	22 54.9
44	2 3.6	4 40.7	2 26.6	24 53.3	22 49.2
45	2 20.8	5 4.5	2 43.5	24 55.7	22 43.4
46	2 38.6	5 28.9	3 0.7	24 58.1	22 37.4
47	2 57.3	5 54.0	3 18.2	25 0.5	22 31.3
48	3 16.9	6 19.9	3 36.0	25 3.0	22 24.9
49	3 37.4	6 46.5	3 54.1	25 5.6	22 18.4
50	3 58.9	7 13.9	4 12.5	25 8.2	22 11.7
51	4 21.6	7 42.2	4 31.3	25 10.9	22 4.7
52	4 45.5	8 11.5	4 50.5	25 13.6	21 57.5
53	5 10.9	8 41.8	5 10.0	25 16.5	21 50.0
54	5 37.8	9 13.1	5 30.0	25 19.3	21 42.3
55	6 6.5	9 45.6	5 50.5	25 22.3	21 34.3
56	6 37.2	10 19.4	6 11.4	25 25.4	21 25.9
57	7 10.1	10 54.4	6 32.8	25 28.5	21 17.2
58	7 45.6	11 30.8	6 54.8	25 31.8	21 8.2
59	8 24.1	12 8.8	7 17.3	25 35.1	20 58.7
60	9 6.2	12 48.4	7 40.4	25 38.6	20 48.8
61	9 52.4	13 29.7	8 4.2	25 42.2	20 38.5
62	10 43.7	14 12.8	8 28.6	25 45.9	20 27.6
63	11 41.2	14 58.0	8 53.8	25 49.7	20 16.1
64	12 46.8	15 45.4	9 19.7	25 53.7	20 4.1
65	14 3.2	16 35.1	9 46.5	25 57.8	19 51.4
66	15 34.9	17 27.3	10 14.1	26 2.1	19 37.9
S LAT	5	6	Descendant	8	9

♏ 26° 18′ 49″
15h 36m 0s MC 234° 0′ 0″

3h 40m 0s MC 55° 0′ 0″
♉ 27° 16′ 58″

N LAT	11	12	Ascendant	2	3
°	° ′	° ′	° ′	° ′	° ′
0	♊25 24.7	♋23 9.7	♌22 38.9	♍24 33.2	♎26 56.5
5	26 4.5	24 23.4	23 53.5	24 41.2	26 37.4
10	26 45.0	25 36.9	25 5.0	24 49.0	26 18.3
15	27 27.0	26 51.2	26 14.5	24 56.6	25 59.1
16	27 35.7	27 6.3	26 28.2	24 58.1	25 55.3
17	27 44.4	27 21.4	26 42.0	24 59.7	25 51.3
18	27 53.2	27 36.7	26 55.7	25 1.2	25 47.4
19	28 2.2	27 52.0	27 9.4	25 2.7	25 43.5
20	28 11.3	28 7.5	27 23.1	25 4.2	25 39.5
21	28 20.5	28 23.1	27 36.8	25 5.8	25 35.4
22	28 29.8	28 38.9	27 50.5	25 7.3	25 31.4
23	28 39.3	28 54.8	28 4.2	25 8.8	25 27.3
24	28 49.0	29 10.8	28 17.9	25 10.4	25 23.2
25	28 58.8	29 27.1	28 31.7	25 11.9	25 19.0
26	29 8.8	29 43.5	28 45.6	25 13.5	25 14.8
27	29 19.0	♌ 0 0.2	28 59.4	25 15.1	25 10.5
28	29 29.3	0 17.0	29 13.4	25 16.6	25 6.1
29	29 40.0	0 34.1	29 27.4	25 18.2	25 1.7
30	29 50.8	0 51.4	29 41.5	25 19.8	24 57.3
31	♋ 0 1.9	1 9.0	29 55.7	25 21.5	24 52.8
32	0 13.2	1 26.9	♍ 0 9.9	25 23.1	24 48.2
33	0 24.9	1 45.1	0 24.3	25 24.8	24 43.5
34	0 36.8	2 3.5	0 38.8	25 26.4	24 38.8
35	0 49.1	2 22.3	0 53.4	25 28.1	24 33.9
36	1 1.7	2 41.5	1 8.2	25 29.8	24 29.0
37	1 14.6	3 1.0	1 23.1	25 31.5	24 24.0
38	1 28.0	3 21.0	1 38.1	25 33.3	24 18.9
39	1 41.8	3 41.3	1 53.3	25 35.1	24 13.6
40	1 56.0	4 2.1	2 8.7	25 36.9	24 8.3
41	2 10.7	4 23.4	2 24.3	25 38.7	24 2.8
42	2 26.0	4 45.1	2 40.1	25 40.6	23 57.2
43	2 41.8	5 7.4	2 56.1	25 42.5	23 51.5
44	2 58.3	5 30.3	3 12.3	25 44.4	23 45.6
45	3 15.4	5 53.7	3 28.7	25 46.4	23 39.5
46	3 33.2	6 17.8	3 45.5	25 48.4	23 33.3
47	3 51.8	6 42.6	4 2.5	25 50.4	23 26.9
48	4 11.3	7 8.1	4 19.8	25 52.5	23 20.3
49	4 31.7	7 34.3	4 37.4	25 54.6	23 13.5
50	4 53.1	8 1.4	4 55.3	25 56.8	23 6.5
51	5 15.7	8 29.3	5 13.6	25 59.1	22 59.3
52	5 39.5	8 58.1	5 32.2	26 1.4	22 51.8
53	6 4.7	9 28.0	5 51.2	26 3.7	22 44.0
54	6 31.5	9 58.8	6 10.7	26 6.1	22 36.0
55	6 59.9	10 30.8	6 30.5	26 8.6	22 27.6
56	7 30.4	11 4.0	6 50.9	26 11.1	22 19.0
57	8 3.1	11 38.5	7 11.7	26 13.8	22 9.9
58	8 38.3	12 14.3	7 33.0	26 16.5	22 0.5
59	9 16.4	12 51.6	7 54.9	26 19.3	21 50.6
60	9 58.0	13 30.5	8 17.4	26 22.1	21 40.3
61	10 43.7	14 11.1	8 40.5	26 25.1	21 29.5
62	11 34.3	14 53.5	9 4.2	26 28.2	21 18.2
63	12 31.0	15 37.9	9 28.7	26 31.4	21 6.3
64	13 35.4	16 24.4	9 53.9	26 34.7	20 53.7
65	14 50.0	17 13.1	10 19.9	26 38.2	20 40.5
66	16 19.2	18 4.3	10 46.7	26 41.8	20 26.4
S LAT	5	6	Descendant	8	9

♏ 27° 16′ 58″
15h 40m 0s MC 235° 0′ 0″

3h 44m 0s — MC 56° 0′ 0″ — ♉ 28° 14′ 58″

N LAT	11	12	Ascendant	2	3
°	° ′	° ′	° ′	° ′	° ′
0	♊26 19.8	♋24 6.4	♌23 40.6	♍25 38.5	♎27 59.7
5	26 59.6	25 19.7	24 53.5	25 44.9	27 39.9
10	27 40.2	26 32.6	26 3.3	25 51.2	27 20.2
15	28 22.2	27 46.3	27 11.1	25 57.3	27 0.3
16	28 30.8	28 1.3	27 24.5	25 58.5	26 56.3
17	28 39.5	28 16.3	27 37.9	25 59.7	26 52.2
18	28 48.4	28 31.4	27 51.3	26 0.9	26 48.2
19	28 57.3	28 46.6	28 4.6	26 2.1	26 44.1
20	29 6.4	29 2.0	28 18.0	26 3.4	26 39.9
21	29 15.6	29 17.4	28 31.3	26 4.6	26 35.8
22	29 24.9	29 33.0	28 44.7	26 5.8	26 31.6
23	29 34.4	29 48.8	28 58.1	26 7.0	26 27.3
24	29 44.1	♌ 0 4.7	29 11.5	26 8.3	26 23.1
25	29 53.9	0 20.8	29 24.9	26 9.5	26 18.7
26	♋ 0 3.8	0 37.1	29 38.4	26 10.8	26 14.3
27	0 14.0	0 53.6	29 51.9	26 12.0	26 9.9
28	0 24.4	1 10.2	♍ 0 5.5	26 13.3	26 5.4
29	0 35.0	1 27.1	0 19.1	26 14.6	26 0.9
30	0 45.8	1 44.3	0 32.9	26 15.9	25 56.2
31	0 56.9	2 1.7	0 46.7	26 17.2	25 51.6
32	1 8.2	2 19.4	1 0.6	26 18.5	25 46.8
33	1 19.8	2 37.3	1 14.6	26 19.8	25 42.0
34	1 31.7	2 55.6	1 28.7	26 21.1	25 37.0
35	1 44.0	3 14.2	1 42.9	26 22.5	25 32.0
36	1 56.5	3 33.1	1 57.2	26 23.8	25 26.9
37	2 9.5	3 52.4	2 11.7	26 25.2	25 21.7
38	2 22.8	4 12.1	2 26.4	26 26.6	25 16.4
39	2 36.5	4 32.2	2 41.2	26 28.1	25 11.0
40	2 50.7	4 52.8	2 56.1	26 29.5	25 5.4
41	3 5.4	5 13.8	3 11.3	26 31.0	24 59.8
42	3 20.6	5 35.3	3 26.6	26 32.5	24 53.9
43	3 36.4	5 57.3	3 42.2	26 34.0	24 48.0
44	3 52.8	6 19.8	3 58.0	26 35.5	24 41.9
45	4 9.8	6 43.0	4 14.0	26 37.1	24 35.6
46	4 27.6	7 6.7	4 30.2	26 38.7	24 29.2
47	4 46.1	7 31.1	4 46.8	26 40.3	24 22.5
48	5 5.5	7 56.3	5 3.6	26 42.0	24 15.7
49	5 25.8	8 22.1	5 20.7	26 43.7	24 8.6
50	5 47.1	8 48.8	5 38.1	26 45.5	24 1.4
51	6 9.6	9 16.3	5 55.8	26 47.2	23 53.8
52	6 33.3	9 44.7	6 13.9	26 49.1	23 46.1
53	6 58.3	10 14.1	6 32.4	26 51.0	23 38.0
54	7 24.9	10 44.5	6 51.3	26 52.9	23 29.6
55	7 53.2	11 16.0	7 10.6	26 54.9	23 20.9
56	8 23.4	11 48.6	7 30.4	26 56.9	23 11.9
57	8 55.8	12 22.5	7 50.6	26 59.0	23 2.5
58	9 30.7	12 57.8	8 11.3	27 1.2	22 52.7
59	10 8.5	13 34.5	8 32.6	27 3.4	22 42.5
60	10 49.7	14 12.7	8 54.4	27 5.7	22 31.7
61	11 34.8	14 52.5	9 16.8	27 8.1	22 20.5
62	12 24.7	15 34.2	9 39.9	27 10.6	22 8.7
63	13 20.5	16 17.8	10 3.6	27 13.1	21 56.3
64	14 23.7	17 3.4	10 28.1	27 15.8	21 43.2
65	15 36.7	17 51.2	10 53.3	27 18.5	21 29.4
66	17 3.3	18 41.5	11 19.3	27 21.4	21 14.8
S LAT	5	6	Descendant	8	9

♏ 28° 14′ 58″ — 15h 44m 0s — MC 236° 0′ 0″

3h 48m 0s — MC 57° 0′ 0″ — ♉ 29° 12′ 48″

N LAT	11	12	Ascendant	2	3
°	° ′	° ′	° ′	° ′	° ′
0	♊27 14.8	♋25 3.3	♌24 42.5	♍26 43.8	♎29 2.8
5	27 54.7	26 16.0	25 53.6	26 48.7	28 42.3
10	28 35.3	27 28.4	27 1.7	26 53.3	28 21.9
15	29 17.3	28 41.5	28 7.8	26 57.9	28 1.4
16	29 25.9	28 56.3	28 20.9	26 58.8	27 57.2
17	29 34.6	29 11.2	28 34.0	26 59.8	27 53.0
18	29 43.5	29 26.2	28 47.0	27 0.7	27 48.8
19	29 52.4	29 41.3	29 0.0	27 1.6	27 44.6
20	♋ 0 1.5	29 56.5	29 13.0	27 2.5	27 40.3
21	0 10.7	♌ 0 11.8	29 26.0	27 3.4	27 36.0
22	0 20.0	0 27.3	29 39.0	27 4.3	27 31.7
23	0 29.5	0 42.9	29 52.0	27 5.3	27 27.3
24	0 39.1	0 58.6	♍ 0 5.1	27 6.2	27 22.8
25	0 48.9	1 14.6	0 18.2	27 7.1	27 18.4
26	0 58.9	1 30.7	0 31.3	27 8.1	27 13.8
27	1 9.0	1 47.0	0 44.4	27 9.0	27 9.2
28	1 19.4	2 3.5	0 57.7	27 10.0	27 4.6
29	1 30.0	2 20.2	1 10.9	27 10.9	26 59.9
30	1 40.8	2 37.2	1 24.3	27 11.9	26 55.1
31	1 51.8	2 54.4	1 37.7	27 12.9	26 50.3
32	2 3.1	3 11.9	1 51.2	27 13.8	26 45.3
33	2 14.7	3 29.6	2 4.8	27 14.8	26 40.3
34	2 26.6	3 47.7	2 18.5	27 15.8	26 35.2
35	2 38.8	4 6.1	2 32.4	27 16.8	26 30.0
36	2 51.3	4 24.8	2 46.3	27 17.9	26 24.8
37	3 4.2	4 43.8	3 0.4	27 18.9	26 19.4
38	3 17.5	5 3.3	3 14.6	27 20.0	26 13.9
39	3 31.2	5 23.2	3 29.0	27 21.0	26 8.2
40	3 45.4	5 43.4	3 43.6	27 22.1	26 2.5
41	4 0.0	6 4.2	3 58.3	27 23.2	25 56.6
42	4 15.2	6 25.4	4 13.2	27 24.3	25 50.6
43	4 30.9	6 47.1	4 28.3	27 25.5	25 44.4
44	4 47.2	7 9.3	4 43.6	27 26.6	25 38.1
45	5 4.2	7 32.2	4 59.2	27 27.8	25 31.6
46	5 21.9	7 55.6	5 15.0	27 29.0	25 24.9
47	5 40.3	8 19.7	5 31.0	27 30.2	25 18.0
48	5 59.6	8 44.5	5 47.4	27 31.5	25 10.9
49	6 19.8	9 10.0	6 4.0	27 32.8	25 3.6
50	6 41.0	9 36.2	6 20.9	27 34.1	24 56.1
51	7 3.4	10 3.3	6 38.1	27 35.4	24 48.3
52	7 26.9	10 31.3	6 55.7	27 36.8	24 40.2
53	7 51.8	11 0.2	7 13.6	27 38.2	24 31.9
54	8 18.2	11 30.1	7 32.0	27 39.7	24 23.2
55	8 46.3	12 1.1	7 50.7	27 41.1	24 14.2
56	9 16.3	12 33.2	8 9.9	27 42.7	24 4.8
57	9 48.4	13 6.6	8 29.5	27 44.2	23 55.1
58	10 23.0	13 41.2	8 49.6	27 45.9	23 44.9
59	11 0.4	14 17.3	9 10.2	27 47.5	23 34.2
60	11 41.1	14 54.8	9 31.4	27 49.3	23 23.1
61	12 25.7	15 34.0	9 53.1	27 51.1	23 11.4
62	13 14.9	16 14.9	10 15.5	27 52.9	22 59.2
63	14 9.8	16 57.7	10 38.5	27 54.8	22 46.3
64	15 11.9	17 42.5	11 2.3	27 56.8	22 32.7
65	16 23.2	18 29.4	11 26.7	27 58.9	22 18.4
66	17 47.4	19 18.7	11 52.0	28 1.1	22 3.2
S LAT	5	6	Descendant	8	9

♏ 29° 12′ 48″ — 15h 48m 0s — MC 237° 0′ 0″

3ʰ 52ᵐ 0ˢ MC 58° 0′ 0″

♊ 0° 10′ 29″

11	12	Ascendant	2	3	N LAT
° ′	° ′	° ′	° ′	° ′	°
♊28 9.9	♋26 0.2	♌25 44.5	♍27 49.2	♏ 0 5.7	0
28 49.8	27 12.5	26 53.9	27 52.4	♎29 44.5	5
29 30.4	28 24.3	28 0.3	27 55.6	29 23.5	10
♋ 0 12.3	29 36.8	29 4.7	27 58.6	29 2.3	15
0 21.0	29 51.5	29 17.4	27 59.2	28 58.0	16
0 29.7	♌ 0 6.2	29 30.1	27 59.8	28 53.7	17
0 38.5	0 21.1	29 42.8	28 0.4	28 49.3	18
0 47.4	0 36.0	29 55.4	28 1.1	28 45.0	19
0 56.5	0 51.1	♍ 0 8.1	28 1.7	28 40.5	20
1 5.7	1 6.2	0 20.7	28 2.3	28 36.1	21
1 15.0	1 21.5	0 33.4	28 2.9	28 31.6	22
1 24.5	1 37.0	0 46.1	28 3.5	28 27.1	23
1 34.1	1 52.6	0 58.7	28 4.1	28 22.5	24
1 43.8	2 8.4	1 11.5	28 4.7	28 17.9	25
1 53.8	2 24.3	1 24.2	28 5.4	28 13.2	26
2 3.9	2 40.4	1 37.0	28 6.0	28 8.5	27
2 14.3	2 56.8	1 49.9	28 6.6	28 3.7	28
2 24.8	3 13.3	2 2.8	28 7.3	27 58.8	29
2 35.6	3 30.1	2 15.7	28 7.9	27 53.9	30
2 46.7	3 47.1	2 28.8	28 8.6	27 48.8	31
2 57.9	4 4.4	2 41.9	28 9.2	27 43.8	32
3 9.5	4 21.9	2 55.1	28 9.9	27 38.6	33
3 21.3	4 39.8	3 8.5	28 10.5	27 33.3	34
3 33.5	4 57.9	3 21.9	28 11.2	27 28.0	35
3 46.0	5 16.4	3 35.5	28 11.9	27 22.5	36
3 58.9	5 35.3	3 49.1	28 12.6	27 16.9	37
4 12.1	5 54.5	4 3.0	28 13.3	27 11.2	38
4 25.8	6 14.1	4 16.9	28 14.0	27 5.4	39
4 39.9	6 34.1	4 31.0	28 14.7	26 59.5	40
4 54.5	6 54.6	4 45.3	28 15.5	26 53.4	41
5 9.6	7 15.5	4 59.8	28 16.2	26 47.2	42
5 25.3	7 36.9	5 14.5	28 17.0	26 40.8	43
5 41.5	7 58.9	5 29.4	28 17.8	26 34.2	44
5 58.4	8 21.4	5 44.5	28 18.5	26 27.5	45
6 16.0	8 44.5	5 59.8	28 19.3	26 20.6	46
6 34.4	9 8.2	6 15.4	28 20.2	26 13.5	47
6 53.6	9 32.6	6 31.2	28 21.0	26 6.1	48
7 13.7	9 57.7	6 47.3	28 21.8	25 58.6	49
7 34.8	10 23.6	7 3.7	28 22.7	25 50.8	50
7 57.0	10 50.3	7 20.4	28 23.6	25 42.7	51
8 20.4	11 17.9	7 37.5	28 24.5	25 34.3	52
8 45.1	11 46.3	7 54.9	28 25.5	25 25.7	53
9 11.3	12 15.8	8 12.6	28 26.4	25 16.7	54
9 39.2	12 46.2	8 30.8	28 27.4	25 7.4	55
10 8.9	13 17.8	8 49.4	28 28.4	24 57.7	56
10 40.8	13 50.6	9 8.4	28 29.5	24 47.5	57
11 15.1	14 24.7	9 27.9	28 30.6	24 37.0	58
11 52.1	15 0.1	9 47.9	28 31.7	24 26.0	59
12 32.3	15 37.0	10 8.4	28 32.8	24 14.4	60
13 16.3	16 15.5	10 29.5	28 34.0	24 2.3	61
14 4.9	16 55.6	10 51.2	28 35.3	23 49.6	62
14 58.9	17 37.6	11 13.5	28 36.6	23 36.2	63
15 59.8	18 21.5	11 36.5	28 37.9	23 22.1	64
17 9.6	19 7.6	12 0.2	28 39.3	23 7.2	65
18 31.4	19 55.9	12 24.6	28 40.7	22 51.5	66
5	6	Descendant	8	9	S LAT

♐ 0° 10′ 29″

15ʰ 52ᵐ 0ˢ MC 238° 0′ 0″

3ʰ 56ᵐ 0ˢ MC 59° 0′ 0″

♊ 1° 8′ 1″

11	12	Ascendant	2	3	N LAT
° ′	° ′	° ′	° ′	° ′	°
♊29 5.0	♋26 57.4	♌26 46.7	♍28 54.6	♏ 1 8.4	0
29 44.8	28 9.0	27 54.3	28 56.2	0 46.6	5
♋ 0 25.4	29 20.3	28 58.9	28 57.8	0 24.9	10
1 7.4	♌ 0 32.1	♍ 0 1.6	28 59.3	0 3.1	15
1 16.0	0 46.7	0 14.0	28 59.6	♎29 58.7	16
1 24.7	1 1.3	0 26.3	28 59.9	29 54.2	17
1 33.5	1 16.0	0 38.6	29 0.2	29 49.7	18
1 42.4	1 30.8	0 50.9	29 0.5	29 45.2	19
1 51.5	1 45.7	1 3.2	29 0.8	29 40.7	20
2 0.6	2 0.7	1 15.5	29 1.1	29 36.1	21
2 9.9	2 15.9	1 27.8	29 1.4	29 31.4	22
2 19.4	2 31.2	1 40.1	29 1.8	29 26.8	23
2 29.0	2 46.6	1 52.5	29 2.1	29 22.1	24
2 38.8	3 2.2	2 4.8	29 2.4	29 17.3	25
2 48.7	3 18.0	2 17.2	29 2.7	29 12.5	26
2 58.8	3 33.9	2 29.6	29 3.0	29 7.6	27
3 9.1	3 50.1	2 42.1	29 3.3	29 2.6	28
3 19.7	4 6.4	2 54.6	29 3.6	28 57.6	29
3 30.4	4 23.0	3 7.2	29 4.0	28 52.5	30
3 41.4	4 39.8	3 19.9	29 4.3	28 47.3	31
3 52.7	4 56.9	3 32.7	29 4.6	28 42.1	32
4 4.2	5 14.3	3 45.5	29 4.9	28 36.7	33
4 16.0	5 31.9	3 58.4	29 5.3	28 31.3	34
4 28.2	5 49.8	4 11.5	29 5.6	28 25.8	35
4 40.6	6 8.1	4 24.6	29 6.0	28 20.1	36
4 53.5	6 26.7	4 37.9	29 6.3	28 14.4	37
5 6.7	6 45.7	4 51.3	29 6.7	28 8.5	38
5 20.3	7 5.0	5 4.8	29 7.0	28 2.5	39
5 34.4	7 24.8	5 18.5	29 7.4	27 56.4	40
5 48.9	7 44.9	5 32.4	29 7.7	27 50.1	41
6 3.9	8 5.6	5 46.4	29 8.1	27 43.7	42
6 19.5	8 26.7	6 0.7	29 8.5	27 37.1	43
6 35.7	8 48.4	6 15.1	29 8.9	27 30.3	44
6 52.5	9 10.6	6 29.7	29 9.3	27 23.3	45
7 10.1	9 33.3	6 44.6	29 9.7	27 16.2	46
7 28.3	9 56.7	6 59.7	29 10.1	27 8.8	47
7 47.4	10 20.8	7 15.0	29 10.5	27 1.3	48
8 7.4	10 45.5	7 30.6	29 10.9	26 53.4	49
8 28.4	11 11.0	7 46.5	29 11.4	26 45.4	50
8 50.5	11 37.3	8 2.7	29 11.8	26 37.0	51
9 13.7	12 4.4	8 19.3	29 12.3	26 28.4	52
9 38.3	12 32.4	8 36.1	29 12.7	26 19.4	53
10 4.3	13 1.4	8 53.3	29 13.2	26 10.1	54
10 32.0	13 31.4	9 10.9	29 13.7	26 0.5	55
11 1.4	14 2.4	9 28.9	29 14.2	25 50.4	56
11 33.0	14 34.7	9 47.4	29 14.7	25 40.0	57
12 6.9	15 8.1	10 6.2	29 15.3	25 29.0	58
12 43.6	15 42.9	10 25.6	29 15.8	25 17.6	59
13 23.4	16 19.2	10 45.5	29 16.4	25 5.7	60
14 6.8	16 57.0	11 5.9	29 17.0	24 53.1	61
14 54.7	17 36.4	11 26.9	29 17.6	24 39.9	62
15 47.8	18 17.6	11 48.5	29 18.3	24 26.1	63
16 47.6	19 0.7	12 10.7	29 18.9	24 11.5	64
17 55.8	19 45.8	12 33.7	29 19.6	23 56.0	65
19 15.3	20 33.2	12 57.3	29 20.3	23 39.7	66
5	6	Descendant	8	9	S LAT

♐ 1° 8′ 1″

15ʰ 56ᵐ 0ˢ MC 239° 0′ 0″

4h 0m 0s — MC 60° 0′ 0″ — ♊ 2° 5′ 24″

N LAT	11	12	Ascendant	2	3
0	♋ 0 0.0	♋27 54.6	♌27 49.1	♎ 0 0.0	♏ 2 10.9
5	0 39.9	29 5.7	28 54.9	0 0.0	1 48.5
10	1 20.4	♌ 0 16.4	29 57.7	0 0.0	1 26.2
15	2 2.3	1 27.6	♍ 0 58.6	0 0.0	1 3.7
16	2 11.0	1 42.0	1 10.6	0 0.0	0 59.2
17	2 19.7	1 56.4	1 22.6	0 0.0	0 54.6
18	2 28.5	2 11.0	1 34.6	0 0.0	0 50.0
19	2 37.4	2 25.7	1 46.5	0 0.0	0 45.3
20	2 46.4	2 40.4	1 58.4	0 0.0	0 40.6
21	2 55.6	2 55.3	2 10.4	0 0.0	0 35.9
22	3 4.9	3 10.3	2 22.3	0 0.0	0 31.2
23	3 14.3	3 25.4	2 34.3	0 0.0	0 26.3
24	3 23.9	3 40.7	2 46.2	0 0.0	0 21.5
25	3 33.6	3 56.1	2 58.2	0 0.0	0 16.6
26	3 43.5	4 11.7	3 10.2	0 0.0	0 11.6
27	3 53.6	4 27.5	3 22.3	0 0.0	0 6.6
28	4 3.9	4 43.4	3 34.4	0 0.0	0 1.5
29	4 14.4	4 59.6	3 46.6	0 0.0	♎29 56.3
30	4 25.2	5 16.0	3 58.8	0 0.0	29 51.0
31	4 36.1	5 32.6	4 11.1	0 0.0	29 45.7
32	4 47.4	5 49.5	4 23.4	0 0.0	29 40.3
33	4 58.9	6 6.6	4 35.9	0 0.0	29 34.8
34	5 10.7	6 24.0	4 48.4	0 0.0	29 29.2
35	5 22.8	6 41.7	5 1.0	0 0.0	29 23.5
36	5 35.2	6 59.8	5 13.8	0 0.0	29 17.7
37	5 48.0	7 18.1	5 26.7	0 0.0	29 11.7
38	6 1.1	7 36.9	5 39.6	0 0.0	29 5.7
39	6 14.7	7 55.9	5 52.8	0 0.0	28 59.5
40	6 28.7	8 15.4	6 6.1	0 0.0	28 53.2
41	6 43.2	8 35.4	6 19.5	0 0.0	28 46.7
42	6 58.2	8 55.7	6 33.1	0 0.0	28 40.1
43	7 13.7	9 16.5	6 46.9	0 0.0	28 33.3
44	7 29.8	9 37.9	7 0.8	0 0.0	28 26.3
45	7 46.6	9 59.8	7 15.0	0 0.0	28 19.1
46	8 4.0	10 22.2	7 29.4	0 0.0	28 11.7
47	8 22.2	10 45.3	7 44.0	0 0.0	28 4.1
48	8 41.2	11 8.9	7 58.9	0 0.0	27 56.3
49	9 1.0	11 33.3	8 14.0	0 0.0	27 48.2
50	9 21.9	11 58.4	8 29.4	0 0.0	27 39.9
51	9 43.8	12 24.3	8 45.1	0 0.0	27 31.3
52	10 6.9	12 50.9	9 1.1	0 0.0	27 22.3
53	10 31.3	13 18.5	9 17.4	0 0.0	27 13.1
54	10 57.1	13 47.0	9 34.0	0 0.0	27 3.5
55	11 24.6	14 16.5	9 51.1	0 0.0	26 53.5
56	11 53.8	14 47.0	10 8.5	0 0.0	26 43.1
57	12 25.1	15 18.7	10 26.3	0 0.0	26 32.3
58	12 58.7	15 51.6	10 44.6	0 0.0	26 21.0
59	13 34.9	16 25.8	11 3.3	0 0.0	26 9.2
60	14 14.2	17 1.4	11 22.5	0 0.0	25 56.8
61	14 57.1	17 38.5	11 42.3	0 0.0	25 43.9
62	15 44.2	18 17.1	12 2.6	0 0.0	25 30.2
63	16 36.5	18 57.6	12 23.5	0 0.0	25 15.9
64	17 35.2	19 39.8	12 45.0	0 0.0	25 0.7
65	18 41.8	20 24.1	13 7.2	0 0.0	24 44.7
66	19 59.2	21 10.6	13 30.1	0 0.0	24 27.8
S LAT	5	6	Descendant	8	9

♐ 2° 5′ 24″ — 16h 0m 0s — MC 240° 0′ 0″

4h 4m 0s — MC 61° 0′ 0″ — ♊ 3° 2′ 39″

N LAT	11	12	Ascendant	2	3
0	♋ 0 55.0	♋28 52.0	♌28 51.6	♎ 1 5.4	♏ 3 13.3
5	1 34.9	♌ 0 2.5	29 55.6	1 3.8	2 50.3
10	2 15.4	1 12.6	♍ 0 56.6	1 2.2	2 27.4
15	2 57.3	2 23.1	1 55.7	1 0.7	2 4.3
16	3 5.9	2 37.3	2 7.3	1 0.4	1 59.6
17	3 14.6	2 51.7	2 19.0	1 0.1	1 54.9
18	3 23.4	3 6.1	2 30.6	0 59.8	1 50.1
19	3 32.3	3 20.6	2 42.2	0 59.5	1 45.3
20	3 41.3	3 35.2	2 53.7	0 59.2	1 40.5
21	3 50.5	3 49.9	3 5.3	0 58.9	1 35.7
22	3 59.7	4 4.7	3 16.9	0 58.6	1 30.8
23	4 9.1	4 19.7	3 28.5	0 58.2	1 25.8
24	4 18.7	4 34.8	3 40.1	0 57.9	1 20.8
25	4 28.4	4 50.0	3 51.7	0 57.6	1 15.7
26	4 38.3	5 5.5	4 3.3	0 57.3	1 10.6
27	4 48.4	5 21.0	4 15.0	0 57.0	1 5.4
28	4 58.7	5 36.8	4 26.8	0 56.7	1 0.2
29	5 9.2	5 52.8	4 38.5	0 56.4	0 54.9
30	5 19.9	6 9.0	4 50.4	0 56.0	0 49.5
31	5 30.8	6 25.4	5 2.3	0 55.7	0 44.0
32	5 42.0	6 42.1	5 14.2	0 55.4	0 38.4
33	5 53.5	6 59.0	5 26.3	0 55.1	0 32.7
34	6 5.2	7 16.2	5 38.4	0 54.7	0 27.0
35	6 17.3	7 33.7	5 50.7	0 54.4	0 21.1
36	6 29.7	7 51.5	6 3.0	0 54.0	0 15.1
37	6 42.4	8 9.6	6 15.5	0 53.7	0 9.0
38	6 55.5	8 28.1	6 28.0	0 53.3	0 2.8
39	7 9.0	8 46.9	6 40.7	0 53.0	♎29 56.4
40	7 23.0	9 6.1	6 53.6	0 52.6	29 49.9
41	7 37.4	9 25.8	7 6.6	0 52.3	29 43.2
42	7 52.3	9 45.8	7 19.8	0 51.9	29 36.4
43	8 7.8	10 6.4	7 33.1	0 51.5	29 29.4
44	8 23.8	10 27.4	7 46.6	0 51.1	29 22.2
45	8 40.5	10 49.0	8 0.3	0 50.7	29 14.8
46	8 57.8	11 11.1	8 14.2	0 50.3	29 7.2
47	9 15.9	11 33.8	8 28.4	0 49.9	28 59.3
48	9 34.8	11 57.1	8 42.7	0 49.5	28 51.3
49	9 54.6	12 21.1	8 57.4	0 49.1	28 42.9
50	10 15.3	12 45.8	9 12.2	0 48.6	28 34.3
51	10 37.0	13 11.2	9 27.4	0 48.2	28 25.4
52	11 0.0	13 37.5	9 42.9	0 47.7	28 16.2
53	11 24.2	14 4.6	9 58.6	0 47.3	28 6.7
54	11 49.8	14 32.6	10 14.7	0 46.8	27 56.8
55	12 17.0	15 1.6	10 31.2	0 46.3	27 46.5
56	12 46.0	15 31.6	10 48.0	0 45.8	27 35.8
57	13 17.0	16 2.7	11 5.3	0 45.3	27 24.6
58	13 50.2	16 35.0	11 22.9	0 44.7	27 12.9
59	14 26.1	17 8.6	11 41.0	0 44.2	27 0.7
60	15 4.9	17 43.6	11 59.6	0 43.6	26 47.9
61	15 47.2	18 20.0	12 18.7	0 43.0	26 34.5
62	16 33.7	18 57.9	12 38.3	0 42.4	26 20.4
63	17 25.1	19 37.6	12 58.5	0 41.7	26 5.6
64	18 22.6	20 19.0	13 19.3	0 41.1	25 49.9
65	19 27.8	21 2.5	13 40.7	0 40.4	25 33.4
66	20 43.1	21 48.0	14 2.8	0 39.7	25 15.8
S LAT	5	6	Descendant	8	9

♐ 3° 2′ 39″ — 16h 4m 0s — MC 241° 0′ 0″

$4^h\ 8^m\ 0^s$ MC 62° 0′ 0″

♊ 3° 59′ 45″

N LAT	11	12	Ascendant	2	3
°	° ′	° ′	° ′	° ′	° ′
0	♋ 1 50.1	♋29 49.5	♌29 54.3	♎ 2 10.8	♏ 4 15.5
5	2 30.0	♌ 0 59.5	♍ 0 56.4	2 7.6	3 51.9
10	3 10.4	2 8.9	1 55.6	2 4.4	3 28.4
15	3 52.3	3 18.7	2 52.8	2 1.4	3 4.6
16	4 0.8	3 32.8	3 4.1	2 0.8	2 59.8
17	4 9.5	3 47.0	3 15.4	2 0.2	2 55.0
18	4 18.3	4 1.2	3 26.7	1 59.6	2 50.1
19	4 27.2	4 15.6	3 37.9	1 58.9	2 45.2
20	4 36.2	4 30.0	3 49.1	1 58.3	2 40.2
21	4 45.3	4 44.6	4 0.3	1 57.7	2 35.3
22	4 54.6	4 59.2	4 11.5	1 57.1	2 30.2
23	5 3.9	5 14.0	4 22.7	1 56.5	2 25.1
24	5 13.5	5 29.0	4 34.0	1 55.9	2 20.0
25	5 23.2	5 44.0	4 45.2	1 55.3	2 14.8
26	5 33.1	5 59.3	4 56.5	1 54.6	2 9.5
27	5 43.1	6 14.7	5 7.8	1 54.0	2 4.2
28	5 53.4	6 30.3	5 19.1	1 53.4	1 58.8
29	6 3.8	6 46.1	5 30.5	1 52.7	1 53.3
30	6 14.5	7 2.0	5 42.0	1 52.1	1 47.8
31	6 25.4	7 18.3	5 53.5	1 51.4	1 42.2
32	6 36.6	7 34.7	6 5.1	1 50.8	1 36.4
33	6 48.0	7 51.4	6 16.7	1 50.1	1 30.6
34	6 59.7	8 8.4	6 28.5	1 49.5	1 24.7
35	7 11.7	8 25.6	6 40.3	1 48.8	1 18.6
36	7 24.1	8 43.2	6 52.2	1 48.1	1 12.5
37	7 36.8	9 1.1	7 4.3	1 47.4	1 6.2
38	7 49.8	9 19.3	7 16.4	1 46.7	0 59.8
39	8 3.3	9 37.9	7 28.7	1 46.0	0 53.2
40	8 17.2	9 56.8	7 41.2	1 45.3	0 46.5
41	8 31.5	10 16.2	7 53.7	1 44.5	0 39.6
42	8 46.4	10 36.0	8 6.4	1 43.8	0 32.6
43	9 1.8	10 56.2	8 19.3	1 43.0	0 25.4
44	9 17.7	11 16.9	8 32.4	1 42.2	0 18.0
45	9 34.3	11 38.2	8 45.6	1 41.5	0 10.4
46	9 51.6	11 59.9	8 59.1	1 40.7	0 2.5
47	10 9.5	12 22.3	9 12.7	1 39.8	♎29 54.5
48	10 28.3	12 45.2	9 26.6	1 39.0	29 46.2
49	10 47.9	13 8.9	9 40.7	1 38.2	29 37.6
50	11 8.5	13 33.2	9 55.1	1 37.3	29 28.7
51	11 30.2	13 58.2	10 9.8	1 36.4	29 19.5
52	11 52.9	14 24.0	10 24.7	1 35.5	29 10.1
53	12 16.9	14 50.7	10 39.9	1 34.5	29 0.2
54	12 42.4	15 18.2	10 55.5	1 33.6	28 50.0
55	13 9.3	15 46.7	11 11.4	1 32.6	28 39.4
56	13 38.0	16 16.2	11 27.6	1 31.6	28 28.3
57	14 8.7	16 46.8	11 44.2	1 30.5	28 16.8
58	14 41.6	17 18.5	12 1.3	1 29.4	28 4.7
59	15 17.1	17 51.5	12 18.8	1 28.3	27 52.2
60	15 55.4	18 25.8	12 36.7	1 27.2	27 39.0
61	16 37.2	19 1.5	12 55.1	1 26.0	27 25.1
62	17 22.9	19 38.7	13 14.0	1 24.7	27 10.6
63	18 13.5	20 17.6	13 33.5	1 23.4	26 55.2
64	19 9.9	20 58.3	13 53.6	1 22.1	26 39.1
65	20 13.6	21 40.8	14 14.2	1 20.7	26 21.9
66	21 26.9	22 25.5	14 35.6	1 19.3	26 3.8
S LAT	5	6	Descendant	8	9

♐ 3° 59′ 45″

$16^h\ 8^m\ 0^s$ MC 242° 0′ 0″

$4^h\ 12^m\ 0^s$ MC 63° 0′ 0″

♊ 4° 56′ 43″

N LAT	11	12	Ascendant	2	3
°	° ′	° ′	° ′	° ′	° ′
0	♋ 2 45.2	♌ 0 47.2	♍ 0 57.2	♎ 3 16.2	♏ 5 17.5
5	3 25.0	1 56.6	1 57.4	3 11.3	4 53.3
10	4 5.4	3 5.3	2 54.7	3 6.7	4 29.2
15	4 47.2	4 14.4	3 50.1	3 2.1	4 4.9
16	4 55.8	4 28.3	4 1.0	3 1.2	3 59.9
17	5 4.4	4 42.4	4 12.0	3 0.2	3 55.0
18	5 13.2	4 56.5	4 22.8	2 59.3	3 50.0
19	5 22.1	5 10.7	4 33.7	2 58.4	3 44.9
20	5 31.0	5 24.9	4 44.5	2 57.5	3 39.8
21	5 40.1	5 39.3	4 55.4	2 56.6	3 34.7
22	5 49.4	5 53.8	5 6.2	2 55.7	3 29.6
23	5 58.7	6 8.4	5 17.1	2 54.7	3 24.3
24	6 8.3	6 23.2	5 27.9	2 53.8	3 19.1
25	6 17.9	6 38.1	5 38.8	2 52.9	3 13.7
26	6 27.8	6 53.1	5 49.7	2 51.9	3 8.3
27	6 37.8	7 8.4	6 0.6	2 51.0	3 2.9
28	6 48.0	7 23.8	6 11.6	2 50.0	2 57.3
29	6 58.4	7 39.3	6 22.6	2 49.1	2 51.7
30	7 9.1	7 55.1	6 33.7	2 48.1	2 46.0
31	7 20.0	8 11.1	6 44.8	2 47.1	2 40.2
32	7 31.1	8 27.4	6 56.0	2 46.2	2 34.3
33	7 42.5	8 43.9	7 7.2	2 45.2	2 28.3
34	7 54.1	9 0.6	7 18.6	2 44.2	2 22.2
35	8 6.1	9 17.6	7 30.0	2 43.2	2 16.0
36	8 18.4	9 35.0	7 41.5	2 42.1	2 9.7
37	8 31.1	9 52.6	7 53.1	2 41.1	2 3.3
38	8 44.1	10 10.5	8 4.9	2 40.0	1 56.7
39	8 57.5	10 28.9	8 16.7	2 39.0	1 49.9
40	9 11.3	10 47.5	8 28.7	2 37.9	1 43.0
41	9 25.6	11 6.6	8 40.9	2 36.8	1 36.0
42	9 40.4	11 26.1	8 53.1	2 35.7	1 28.7
43	9 55.7	11 46.0	9 5.6	2 34.5	1 21.3
44	10 11.6	12 6.5	9 18.2	2 33.4	1 13.7
45	10 28.0	12 27.4	9 31.0	2 32.2	1 5.9
46	10 45.2	12 48.8	9 43.9	2 31.0	0 57.8
47	11 3.1	13 10.8	9 57.1	2 29.8	0 49.5
48	11 21.7	13 33.4	10 10.5	2 28.5	0 41.0
49	11 41.2	13 56.6	10 24.1	2 27.2	0 32.1
50	12 1.7	14 20.5	10 38.0	2 25.9	0 23.0
51	12 23.2	14 45.2	10 52.1	2 24.6	0 13.6
52	12 45.7	15 10.5	11 6.5	2 23.2	0 3.8
53	13 9.6	15 36.7	11 21.2	2 21.8	♎29 53.7
54	13 34.8	16 3.8	11 36.2	2 20.3	29 43.1
55	14 1.5	16 31.8	11 51.5	2 18.9	29 32.2
56	14 30.0	17 0.8	12 7.2	2 17.3	29 20.8
57	15 0.3	17 30.8	12 23.2	2 15.8	29 8.9
58	15 32.9	18 2.0	12 39.7	2 14.1	28 56.5
59	16 7.9	18 34.3	12 56.5	2 12.5	28 43.5
60	16 45.8	19 8.0	13 13.8	2 10.7	28 29.9
61	17 27.0	19 43.0	13 31.5	2 8.9	28 15.6
62	18 12.0	20 19.6	13 49.8	2 7.1	28 0.6
63	19 1.7	20 57.7	14 8.6	2 5.2	27 44.8
64	19 57.1	21 37.6	14 27.9	2 3.2	27 28.1
65	20 59.4	22 19.3	14 47.8	2 1.1	27 10.4
66	22 10.8	23 3.0	15 8.4	1 58.9	26 51.7
S LAT	5	6	Descendant	8	9

♐ 4° 56′ 43″

$16^h\ 12^m\ 0^s$ MC 243° 0′ 0″

4h 16m 0s — MC 64° 0′ 0″ — ♊ 5° 53′ 34″

N LAT °	11 ° ′	12 ° ′	Ascendant ° ′	2 ° ′	3 ° ′
0	♋ 3 40.2	♌ 1 45.0	♍ 2 0.3	♎ 4 21.5	♏ 6 19.4
5	4 20.0	2 53.7	2 58.5	4 15.1	5 54.6
10	5 0.4	4 1.8	3 53.9	4 8.8	5 29.9
15	5 42.1	5 10.2	4 47.5	4 2.7	5 5.0
16	5 50.7	5 24.0	4 58.0	4 1.5	4 59.9
17	5 59.3	5 37.8	5 8.6	4 0.3	4 54.8
18	6 8.1	5 51.8	5 19.1	3 59.1	4 49.7
19	6 16.9	6 5.8	5 29.6	3 57.9	4 44.5
20	6 25.9	6 19.9	5 40.0	3 56.6	4 39.3
21	6 34.9	6 34.1	5 50.5	3 55.4	4 34.1
22	6 44.2	6 48.5	6 1.0	3 54.2	4 28.8
23	6 53.5	7 2.9	6 11.4	3 53.0	4 23.4
24	7 3.0	7 17.5	6 21.9	3 51.7	4 18.0
25	7 12.6	7 32.2	6 32.4	3 50.5	4 12.5
26	7 22.5	7 47.1	6 42.9	3 49.2	4 7.0
27	7 32.5	8 2.1	6 53.5	3 48.0	4 1.4
28	7 42.6	8 17.3	7 4.1	3 46.7	3 55.7
29	7 53.0	8 32.7	7 14.7	3 45.4	3 49.9
30	8 3.6	8 48.3	7 25.4	3 44.1	3 44.1
31	8 14.5	9 4.1	7 36.1	3 42.8	3 38.2
32	8 25.6	9 20.1	7 46.9	3 41.5	3 32.1
33	8 36.9	9 36.4	7 57.7	3 40.2	3 26.0
34	8 48.5	9 52.9	8 8.7	3 38.9	3 19.7
35	9 0.5	10 9.7	8 19.7	3 37.5	3 13.4
36	9 12.7	10 26.7	8 30.8	3 36.2	3 6.9
37	9 25.3	10 44.1	8 42.0	3 34.8	3 0.2
38	9 38.3	11 1.8	8 53.3	3 33.4	2 53.5
39	9 51.6	11 19.9	9 4.8	3 31.9	2 46.5
40	10 5.4	11 38.3	9 16.3	3 30.5	2 39.5
41	10 19.6	11 57.1	9 28.0	3 29.0	2 32.2
42	10 34.3	12 16.3	9 39.9	3 27.5	2 24.8
43	10 49.5	12 35.9	9 51.9	3 26.0	2 17.2
44	11 5.3	12 56.0	10 4.0	3 24.5	2 9.3
45	11 21.7	13 16.6	10 16.3	3 22.9	2 1.3
46	11 38.7	13 37.7	10 28.8	3 21.3	1 53.0
47	11 56.5	13 59.3	10 41.5	3 19.7	1 44.5
48	12 15.1	14 21.6	10 54.4	3 18.0	1 35.7
49	12 34.4	14 44.4	11 7.5	3 16.3	1 26.6
50	12 54.7	15 7.9	11 20.9	3 14.5	1 17.2
51	13 16.0	15 32.1	11 34.5	3 12.8	1 7.5
52	13 38.5	15 57.1	11 48.4	3 10.9	0 57.5
53	14 2.1	16 22.8	12 2.5	3 9.0	0 47.1
54	14 27.1	16 49.4	12 17.0	3 7.1	0 36.2
55	14 53.6	17 16.9	12 31.7	3 5.1	0 25.0
56	15 21.8	17 45.4	12 46.8	3 3.1	0 13.2
57	15 51.8	18 14.9	13 2.2	3 1.0	0 1.0
58	16 24.0	18 45.5	13 18.1	2 58.8	♎ 29 48.2
59	16 58.6	19 17.2	13 34.3	2 56.6	29 34.8
60	17 36.0	19 50.2	13 50.9	2 54.3	29 20.8
61	18 16.6	20 24.6	14 8.0	2 51.9	29 6.1
62	19 1.0	21 0.4	14 25.6	2 49.4	28 50.6
63	19 49.9	21 37.8	14 43.6	2 46.9	28 34.3
64	20 44.1	22 16.9	15 2.2	2 44.2	28 17.1
65	21 45.1	22 57.8	15 21.4	2 41.5	27 58.8
66	22 54.6	23 40.6	15 41.2	2 38.6	27 39.5
S LAT	5	6	Descendant	8	9

♐ 5° 53′ 34″ — 16h 16m 0s — MC 244° 0′ 0″

4h 20m 0s — MC 65° 0′ 0″ — ♊ 6° 50′ 16″

N LAT °	11 ° ′	12 ° ′	Ascendant ° ′	2 ° ′	3 ° ′
0	♋ 4 35.3	♌ 2 43.0	♍ 3 3.5	♎ 5 26.8	♏ 7 21.1
5	5 15.1	3 51.1	3 59.7	5 18.8	6 55.7
10	5 55.4	4 58.4	4 53.2	5 11.0	6 27.7
15	6 37.0	6 6.0	5 44.9	5 3.4	6 4.9
16	6 45.6	6 19.7	5 55.1	5 1.9	5 59.7
17	6 54.2	6 33.4	6 5.2	5 0.3	5 54.5
18	7 2.9	6 47.2	6 15.4	4 58.8	5 49.3
19	7 11.7	7 1.1	6 25.5	4 57.3	5 44.0
20	7 20.7	7 15.0	6 35.6	4 55.8	5 38.7
21	7 29.7	7 29.0	6 45.7	4 54.2	5 33.3
22	7 38.9	7 43.2	6 55.8	4 52.7	5 27.8
23	7 48.2	7 57.5	7 5.9	4 51.2	5 22.4
24	7 57.7	8 11.9	7 16.0	4 49.6	5 16.8
25	8 7.3	8 26.4	7 26.1	4 48.1	5 11.2
26	8 17.1	8 41.1	7 36.2	4 46.5	5 5.6
27	8 27.1	8 55.9	7 46.4	4 44.9	4 59.8
28	8 37.2	9 10.9	7 56.6	4 43.4	4 54.0
29	8 47.6	9 26.1	8 6.8	4 41.8	4 48.1
30	8 58.2	9 41.5	8 17.1	4 40.2	4 42.1
31	9 9.0	9 57.0	8 27.4	4 38.5	4 36.0
32	9 20.0	10 12.8	8 37.8	4 36.9	4 29.8
33	9 31.3	10 28.9	8 48.3	4 35.2	4 23.5
34	9 42.9	10 45.2	8 58.8	4 33.6	4 17.1
35	9 54.8	11 1.7	9 9.4	4 31.9	4 10.6
36	10 7.0	11 18.5	9 20.1	4 30.2	4 3.9
37	10 19.5	11 35.7	9 30.9	4 28.5	3 57.1
38	10 32.4	11 53.1	9 41.8	4 26.7	3 50.2
39	10 45.7	12 10.9	9 52.8	4 24.9	3 43.1
40	10 59.4	12 29.0	10 4.0	4 23.1	3 35.8
41	11 13.5	12 47.5	10 15.2	4 21.3	3 28.4
42	11 28.1	13 6.4	10 26.6	4 19.4	3 20.7
43	11 43.3	13 25.8	10 38.1	4 17.5	3 12.9
44	11 59.0	13 45.5	10 49.8	4 15.6	3 4.9
45	12 15.3	14 5.8	11 1.7	4 13.6	2 56.6
46	12 32.2	14 26.6	11 13.7	4 11.6	2 48.1
47	12 49.9	14 47.9	11 25.9	4 9.6	2 39.4
48	13 8.3	15 9.7	11 38.3	4 7.5	2 30.3
49	13 27.5	15 32.2	11 51.0	4 5.4	2 21.0
50	13 47.7	15 55.3	12 3.8	4 3.2	2 11.4
51	14 8.8	16 19.1	12 16.9	4 0.9	2 1.4
52	14 31.1	16 43.6	12 30.2	3 58.6	1 51.1
53	14 54.5	17 8.9	12 43.8	3 56.3	1 40.4
54	15 19.3	17 35.0	12 57.7	3 53.9	1 29.2
55	15 45.6	18 2.0	13 11.9	3 51.4	1 17.6
56	16 13.4	18 30.0	13 26.4	3 48.9	1 5.6
57	16 43.2	18 58.9	13 41.3	3 46.2	0 53.0
58	17 15.0	19 29.0	13 56.5	3 43.5	0 39.8
59	17 49.2	20 0.1	14 12.1	3 40.7	0 26.1
60	18 24.5	20 32.5	14 28.0	3 37.9	0 11.6
61	19 6.2	21 6.2	14 44.5	3 34.9	♎ 29 56.5
62	19 49.8	21 41.3	15 1.3	3 31.8	29 40.5
63	20 37.9	22 18.0	15 18.7	3 28.6	29 23.7
64	21 31.1	22 56.3	15 36.6	3 25.3	29 6.0
65	22 30.7	23 36.3	15 55.0	3 21.8	28 47.1
66	23 38.5	24 18.2	16 14.1	3 18.2	28 27.2
S LAT	5	6	Descendant	8	9

♐ 6° 50′ 16″ — 16h 20m 0s — MC 245° 0′ 0″

4h 24m 0s — MC 66° 0′ 0″ — ♊ 7° 46′ 52″

11	12	Ascendant	2	3	N LAT
° ′	° ′	° ′	° ′	° ′	°
♋ 5 30.5	♌ 3 41.2	♍ 4 6.8	♎ 6 32.1	♏ 8 22.6	0
6 10.2	4 48.6	5 1.1	6 22.5	7 56.7	5
6 50.4	5 55.2	5 52.6	6 13.2	7 30.8	10
7 31.9	7 2.0	6 42.4	6 4.0	7 4.7	15
7 40.5	7 15.5	6 52.2	6 2.2	6 59.4	16
7 49.1	7 29.1	7 2.0	6 0.4	6 54.1	17
7 57.8	7 42.7	7 11.8	5 58.5	6 48.7	18
8 6.6	7 56.4	7 21.5	5 56.7	6 43.3	19
8 15.5	8 10.2	7 31.2	5 54.9	6 37.9	20
8 24.5	8 24.0	7 40.9	5 53.0	6 32.4	21
8 33.7	8 38.0	7 50.6	5 51.2	6 26.8	22
8 43.0	8 52.1	8 0.4	5 49.4	6 21.2	23
8 52.4	9 6.3	8 10.1	5 47.5	6 15.5	24
9 2.0	9 20.6	8 19.8	5 45.6	6 9.8	25
9 11.7	9 35.1	8 29.6	5 43.8	6 4.0	26
9 21.7	9 49.7	8 39.3	5 41.9	5 58.1	27
9 31.8	10 4.5	8 49.1	5 40.0	5 52.1	28
9 42.1	10 19.5	8 59.0	5 38.1	5 46.1	29
9 52.6	10 34.7	9 8.9	5 36.2	5 40.0	30
10 3.4	10 50.1	9 18.8	5 34.2	5 33.7	31
10 14.4	11 5.6	9 28.8	5 32.3	5 27.4	32
10 25.7	11 21.4	9 38.9	5 30.3	5 20.9	33
10 37.2	11 37.5	9 49.0	5 28.3	5 14.4	34
10 49.0	11 53.8	9 59.2	5 26.2	5 7.7	35
11 1.2	12 10.4	10 9.5	5 24.2	5 0.8	36
11 13.6	12 27.3	10 19.9	5 22.1	4 53.9	37
11 26.5	12 44.4	10 30.3	5 20.0	4 46.8	38
11 39.7	13 1.9	10 40.9	5 17.9	4 39.5	39
11 53.3	13 19.8	10 51.6	5 15.7	4 32.0	40
12 7.4	13 38.0	11 2.4	5 13.5	4 24.4	41
12 21.9	13 56.6	11 13.4	5 11.3	4 16.6	42
12 37.0	14 15.7	11 24.5	5 9.0	4 8.6	43
12 52.6	14 35.1	11 35.7	5 6.7	4 0.3	44
13 8.8	14 55.0	11 47.1	5 4.3	3 51.9	45
13 25.6	15 15.5	11 58.6	5 1.9	3 43.1	46
13 43.2	15 36.4	12 10.4	4 59.5	3 34.2	47
14 1.5	15 57.9	12 22.3	4 57.0	3 24.9	48
14 20.6	16 20.0	12 34.4	4 54.4	3 15.3	49
14 40.6	16 42.7	12 46.7	4 51.8	3 5.4	50
15 1.6	17 6.1	12 59.3	4 49.1	2 55.2	51
15 23.6	17 30.2	13 12.1	4 46.4	2 44.6	52
15 46.9	17 55.0	13 25.2	4 43.5	2 33.6	53
16 11.4	18 20.7	13 38.5	4 40.7	2 22.2	54
16 37.4	18 47.2	13 52.1	4 37.7	2 10.3	55
17 5.0	19 14.6	14 6.1	4 34.6	1 57.9	56
17 34.4	19 43.0	14 20.3	4 31.5	1 44.9	57
18 5.9	20 12.5	14 34.9	4 28.2	1 31.4	58
18 39.7	20 43.0	14 49.9	4 24.9	1 17.2	59
19 16.1	21 14.8	15 5.2	4 21.4	1 2.4	60
19 55.6	21 47.8	15 21.0	4 17.8	0 46.8	61
20 38.6	22 22.3	15 37.2	4 14.1	0 30.4	62
21 25.8	22 58.2	15 53.8	4 10.3	0 13.1	63
22 18.0	23 35.7	16 11.0	4 6.3	♎29 54.8	64
23 16.3	24 14.9	16 28.7	4 2.2	29 35.4	65
24 22.3	24 55.9	16 46.9	3 57.9	29 14.8	66
5	6	Descendant	8	9	S LAT

♐ 7° 46′ 52″ — 16h 24m 0s — MC 246° 0′ 0″

4h 28m 0s — MC 67° 0′ 0″ — ♊ 8° 43′ 20″

N LAT	11	12	Ascendant	2	3
°	° ′	° ′	° ′	° ′	° ′
0	♋ 6 25.6	♌ 4 39.5	♍ 5 10.3	♎ 7 37.4	♏ 9 23.9
5	7 5.3	5 46.2	6 2.5	7 26.1	8 57.4
10	7 45.5	6 52.0	6 52.1	7 15.3	8 31.1
15	8 26.9	7 58.1	7 40.0	7 4.6	8 4.4
16	8 35.4	8 11.4	7 49.4	7 2.5	7 59.0
17	8 43.9	8 24.8	7 58.8	7 0.4	7 53.5
18	8 52.6	8 38.3	8 8.2	6 58.2	7 48.0
19	9 1.4	8 51.8	8 17.6	6 56.1	7 42.5
20	9 10.3	9 5.4	8 26.9	6 54.0	7 36.9
21	9 19.3	9 19.1	8 36.2	6 51.8	7 31.3
22	9 28.4	9 32.9	8 45.6	6 49.7	7 25.6
23	9 37.7	9 46.8	8 54.9	6 47.5	7 19.9
24	9 47.1	10 0.8	9 4.2	6 45.4	7 14.1
25	9 56.6	10 14.9	9 13.6	6 43.2	7 8.2
26	10 6.4	10 29.2	9 23.0	6 41.0	7 2.3
27	10 16.3	10 43.7	9 32.3	6 38.8	6 56.3
28	10 26.3	10 58.3	9 41.8	6 36.6	6 50.2
29	10 36.6	11 13.0	9 51.2	6 34.4	6 44.0
30	10 47.1	11 28.0	10 0.7	6 32.1	6 37.7
31	10 57.8	11 43.1	10 10.2	6 29.9	6 31.3
32	11 8.8	11 58.5	10 19.8	6 27.6	6 24.8
33	11 20.0	12 14.0	10 29.5	6 25.3	6 18.2
34	11 31.5	12 29.9	10 39.2	6 22.9	6 11.5
35	11 43.2	12 45.9	10 49.0	6 20.6	6 4.7
36	11 55.3	13 2.3	10 58.9	6 18.2	5 57.7
37	12 7.7	13 18.9	11 8.8	6 15.8	5 50.5
38	12 20.5	13 35.8	11 18.9	6 13.3	5 43.3
39	12 33.7	13 53.0	11 29.0	6 10.8	5 35.8
40	12 47.2	14 10.6	11 39.3	6 8.3	5 28.2
41	13 1.2	14 28.5	11 49.7	6 5.7	5 20.4
42	13 15.7	14 46.8	12 0.2	6 3.1	5 12.4
43	13 30.6	15 5.6	12 10.8	6 0.5	5 4.2
44	13 46.1	15 24.7	12 21.6	5 57.8	4 55.7
45	14 2.2	15 44.3	12 32.5	5 55.0	4 47.0
46	14 19.0	16 4.4	12 43.6	5 52.2	4 38.1
47	14 36.4	16 25.0	12 54.8	5 49.4	4 28.9
48	14 54.5	16 46.1	13 6.2	5 46.4	4 19.4
49	15 13.5	17 7.8	13 17.9	5 43.5	4 9.6
50	15 33.4	17 30.1	13 29.7	5 40.4	3 59.4
51	15 54.2	17 53.1	13 41.7	5 37.3	3 48.9
52	16 16.1	18 16.7	13 54.0	5 34.1	3 38.1
53	16 39.1	18 41.1	14 6.5	5 30.8	3 26.8
54	17 3.4	19 6.3	14 19.3	5 27.4	3 15.0
55	17 29.2	19 32.3	14 32.4	5 23.9	3 2.8
56	17 56.5	19 59.2	14 45.7	5 20.4	2 50.1
57	18 25.6	20 27.1	14 59.4	5 16.7	2 36.8
58	18 56.7	20 56.0	15 13.3	5 12.9	2 22.9
59	19 30.0	21 26.0	15 27.7	5 9.0	2 8.3
60	20 6.0	21 57.1	15 42.4	5 5.0	1 53.1
61	20 44.9	22 29.5	15 57.5	5 0.8	1 37.0
62	21 27.2	23 3.2	16 13.0	4 56.5	1 20.2
63	22 13.6	23 38.4	16 28.9	4 52.0	1 2.3
64	23 4.8	24 15.1	16 45.4	4 47.3	0 43.5
65	24 1.9	24 53.5	17 2.3	4 42.5	0 23.5
66	25 6.2	25 33.7	17 19.8	4 37.5	0 2.4
S LAT	5	6	Descendant	8	9

♐ 8° 43′ 20″ — 16h 28m 0s — MC 247° 0′ 0″

4h 32m 0s — MC — 68° 0′ 0″

♊ 9° 39′ 41″

11	12	Ascendant	2	3	N LAT
° ′	° ′	° ′	° ′	° ′	°
♋ 7 20.8	♌ 5 38.0	♍ 6 14.0	♎ 8 42.5	♏ 10 25.0	0
8 0.4	6 43.9	7 4.1	8 29.7	9 58.0	5
8 40.5	7 49.0	7 51.7	8 17.3	9 31.1	10
9 21.8	8 54.3	8 37.7	8 5.2	9 3.9	15
9 30.3	9 7.5	8 46.7	8 2.7	8 58.4	16
9 38.8	9 20.7	8 55.7	8 0.3	8 52.8	17
9 47.5	9 33.9	9 4.7	7 57.9	8 47.2	18
9 56.2	9 47.3	9 13.7	7 55.5	8 41.5	19
10 5.1	10 0.7	9 22.7	7 53.0	8 35.8	20
10 14.1	10 14.2	9 31.6	7 50.6	8 30.1	21
10 23.2	10 27.8	9 40.6	7 48.1	8 24.3	22
10 32.4	10 41.5	9 49.5	7 45.7	8 18.4	23
10 41.8	10 55.4	9 58.4	7 43.2	8 12.5	24
10 51.3	11 9.3	10 7.4	7 40.7	8 6.5	25
11 1.0	11 23.4	10 16.4	7 38.2	8 0.5	26
11 10.8	11 37.6	10 25.4	7 35.7	7 54.3	27
11 20.9	11 52.0	10 34.4	7 33.2	7 48.1	28
11 31.1	12 6.6	10 43.5	7 30.7	7 41.8	29
11 41.6	12 21.3	10 52.6	7 28.1	7 35.3	30
11 52.2	12 36.2	11 1.7	7 25.5	7 28.8	31
12 3.1	12 51.4	11 10.9	7 22.9	7 22.2	32
12 14.3	13 6.7	11 20.1	7 20.3	7 15.4	33
12 25.7	13 22.3	11 29.5	7 17.6	7 8.6	34
12 37.4	13 38.1	11 38.8	7 14.9	7 1.6	35
12 49.5	13 54.2	11 48.3	7 12.2	6 54.4	36
13 1.8	14 10.5	11 57.8	7 9.4	6 47.1	37
13 14.5	14 27.2	12 7.4	7 6.6	6 39.7	38
13 27.6	14 44.1	12 17.2	7 3.8	6 32.0	39
13 41.1	15 1.4	12 27.0	7 0.9	6 24.2	40
13 55.0	15 19.1	12 36.9	6 57.9	6 16.3	41
14 9.4	15 37.1	12 47.0	6 55.0	6 8.1	42
14 24.2	15 55.5	12 57.1	6 51.9	5 59.6	43
14 39.6	16 14.3	13 7.4	6 48.8	5 51.0	44
14 55.6	16 33.6	13 17.9	6 45.7	5 42.1	45
15 12.3	16 53.3	13 28.5	6 42.5	5 33.0	46
15 29.5	17 13.5	13 39.3	6 39.2	5 23.5	47
15 47.6	17 34.3	13 50.2	6 35.9	5 13.8	48
16 6.4	17 55.6	14 1.3	6 32.5	5 3.7	49
16 26.1	18 17.5	14 12.6	6 29.0	4 53.3	50
16 46.7	18 40.1	14 24.2	6 25.4	4 42.6	51
17 8.4	19 3.3	14 35.9	6 21.8	4 31.4	52
17 31.2	19 27.3	14 47.9	6 18.0	4 19.8	53
17 55.3	19 52.0	15 0.1	6 14.1	4 7.8	54
18 20.8	20 17.5	15 12.6	6 10.2	3 55.3	55
18 47.9	20 43.9	15 25.4	6 6.1	3 42.2	56
19 16.6	21 11.2	15 38.4	6 1.9	3 28.6	57
19 47.4	21 39.5	15 51.8	5 57.6	3 14.3	58
20 20.3	22 8.9	16 5.5	5 53.1	2 59.4	59
20 55.7	22 39.4	16 19.5	5 48.5	2 43.7	60
21 34.1	23 11.2	16 34.0	5 43.7	2 27.2	61
22 15.7	23 44.2	16 48.8	5 38.8	2 9.9	62
23 1.3	24 18.7	17 4.1	5 33.7	1 51.5	63
23 51.5	24 54.6	17 19.8	5 28.4	1 32.2	64
24 47.4	25 32.2	17 36.0	5 22.8	1 11.6	65
25 50.1	26 11.5	17 52.7	5 17.1	0 49.8	66
5	6	Descendant	8	9	S LAT

♐ 9° 39′ 41″

16h 32m 0s — MC — 248° 0′ 0″

4h 36m 0s — MC — 69° 0′ 0″

♊ 10° 35′ 56″

11	12	Ascendant	2	3	N LAT
° ′	° ′	° ′	° ′	° ′	°
♋ 8 16.1	♌ 6 36.6	♍ 7 17.7	♎ 9 47.7	♏ 11 26.0	0
8 55.6	7 41.8	8 5.8	9 33.3	10 58.5	5
9 35.6	8 46.2	8 51.4	9 19.4	10 31.0	10
10 16.8	9 50.6	9 35.4	9 5.7	10 3.2	15
10 25.2	10 3.6	9 44.1	9 3.0	9 57.6	16
10 33.7	10 16.6	9 52.7	9 0.2	9 51.9	17
10 42.4	10 29.7	10 1.3	8 57.5	9 46.2	18
10 51.1	10 42.9	10 9.9	8 54.8	9 40.4	19
10 59.9	10 56.1	10 18.5	8 52.0	9 34.6	20
11 8.9	11 9.4	10 27.0	8 49.3	9 28.8	21
11 17.9	11 22.9	10 35.6	8 46.6	9 22.9	22
11 27.1	11 36.4	10 44.1	8 43.8	9 16.9	23
11 36.5	11 50.0	10 52.7	8 41.0	9 10.8	24
11 45.9	12 3.8	11 1.3	8 38.2	9 4.7	25
11 55.6	12 17.7	11 9.9	8 35.4	8 58.5	26
12 5.4	12 31.7	11 18.5	8 32.6	8 52.2	27
12 15.4	12 45.9	11 27.1	8 29.8	8 45.9	28
12 25.6	13 0.2	11 35.8	8 26.9	8 39.4	29
12 36.0	13 14.7	11 44.5	8 24.0	8 32.9	30
12 46.6	13 29.4	11 53.2	8 21.1	8 26.2	31
12 57.5	13 44.3	12 2.0	8 18.2	8 19.4	32
13 8.6	13 59.4	12 10.8	8 15.2	8 12.5	33
13 19.9	14 14.7	12 19.7	8 12.2	8 5.5	34
13 31.6	14 30.3	12 28.7	8 9.2	7 58.4	35
13 43.6	14 46.1	12 37.7	8 6.1	7 51.0	36
13 55.9	15 2.2	12 46.8	8 3.0	7 43.6	37
14 8.5	15 18.6	12 56.0	7 59.9	7 36.0	38
14 21.5	15 35.3	13 5.3	7 56.7	7 28.2	39
14 34.9	15 52.3	13 14.7	7 53.4	7 20.2	40
14 48.7	16 9.6	13 24.2	7 50.1	7 12.0	41
15 3.0	16 27.4	13 33.8	7 46.8	7 3.6	42
15 17.8	16 45.4	13 43.5	7 43.4	6 55.0	43
15 33.1	17 3.9	13 53.3	7 39.9	6 46.2	44
15 49.0	17 22.9	14 3.3	7 36.4	6 37.1	45
16 5.5	17 42.3	14 13.4	7 32.8	6 27.7	46
16 22.6	18 2.1	14 23.7	7 29.1	6 18.1	47
16 40.5	18 22.5	14 34.2	7 25.3	6 8.1	48
16 59.2	18 43.5	14 44.8	7 21.5	5 57.8	49
17 18.7	19 5.0	14 55.6	7 17.6	5 47.2	50
17 39.2	19 27.1	15 6.6	7 13.6	5 36.2	51
18 0.7	19 49.9	15 17.8	7 9.4	5 24.7	52
18 23.3	20 13.4	15 29.2	7 5.2	5 12.9	53
18 47.2	20 37.6	15 40.9	7 0.9	5 0.5	54
19 12.4	21 2.7	15 52.8	6 56.4	4 47.7	55
19 39.1	21 28.6	16 5.0	6 51.8	4 34.3	56
20 7.6	21 55.3	16 17.5	6 47.1	4 20.3	57
20 38.0	22 23.1	16 30.2	6 42.3	4 5.6	58
21 10.5	22 51.9	16 43.3	6 37.2	3 50.3	59
21 45.4	23 21.8	16 56.7	6 32.0	3 34.2	60
22 23.2	23 52.9	17 10.5	6 26.7	3 17.3	61
23 4.2	24 25.2	17 24.7	6 21.1	2 59.5	62
23 49.0	24 59.0	17 39.2	6 15.4	2 40.6	63
24 38.2	25 34.2	17 54.2	6 9.4	2 20.7	64
25 32.8	26 10.9	18 9.7	6 3.2	1 59.6	65
26 34.1	26 49.4	18 25.6	5 56.7	1 37.2	66
5	6	Descendant	8	9	S LAT

♐ 10° 35′ 56″

16h 36m 0s — MC — 249° 0′ 0″

4^h 40^m 0^s MC 70° 0′ 0″

♊ 11° 32′ 4″

11	12	Ascendant	2	3	N LAT
° ′	° ′	° ′	° ′	° ′	°
♋ 9 11.4	♌ 7 35.4	♍ 8 21.7	♎ 10 52.7	♏ 12 26.7	**0**
9 50.8	8 39.9	9 7.6	10 36.8	11 58.7	**5**
10 30.7	9 43.4	9 51.2	10 21.3	11 30.8	**10**
11 11.7	10 47.0	10 33.2	10 6.2	11 2.5	**15**
11 20.1	10 59.8	10 41.5	10 3.1	10 56.7	**16**
11 28.7	11 12.7	10 49.7	10 0.1	10 50.9	**17**
11 37.2	11 25.6	10 58.0	9 57.1	10 45.1	**18**
11 45.9	11 38.6	11 6.2	9 54.1	10 39.2	**19**
11 54.7	11 51.6	11 14.3	9 51.0	10 33.3	**20**
12 3.6	12 4.8	11 22.5	9 48.0	10 27.3	**21**
12 12.7	12 18.0	11 30.7	9 44.9	10 21.3	**22**
12 21.8	12 31.3	11 38.8	9 41.9	10 15.2	**23**
12 31.1	12 44.7	11 47.0	9 38.8	10 9.0	**24**
12 40.6	12 58.3	11 55.2	9 35.7	10 2.8	**25**
12 50.2	13 12.0	12 3.4	9 32.6	9 56.4	**26**
13 0.0	13 25.8	12 11.6	9 29.5	9 50.0	**27**
13 9.9	13 39.7	12 19.8	9 26.3	9 43.6	**28**
13 20.1	13 53.9	12 28.1	9 23.1	9 37.0	**29**
13 30.4	14 8.2	12 36.4	9 19.9	9 30.3	**30**
13 41.0	14 22.6	12 44.7	9 16.7	9 23.5	**31**
13 51.8	14 37.3	12 53.1	9 13.5	9 16.6	**32**
14 2.8	14 52.2	13 1.5	9 10.2	9 9.5	**33**
14 14.2	15 7.2	13 10.0	9 6.8	9 2.3	**34**
14 25.8	15 22.6	13 18.6	9 3.5	8 55.0	**35**
14 37.7	15 38.1	13 27.2	9 0.1	8 47.6	**36**
14 49.9	15 54.0	13 35.9	8 56.6	8 40.0	**37**
15 2.4	16 10.1	13 44.6	8 53.1	8 32.2	**38**
15 15.4	16 26.5	13 53.5	8 49.6	8 24.2	**39**
15 28.7	16 43.2	14 2.4	8 46.0	8 16.1	**40**
15 42.4	17 0.2	14 11.5	8 42.3	8 7.7	**41**
15 56.6	17 17.6	14 20.6	8 38.6	7 59.1	**42**
16 11.3	17 35.4	14 29.9	8 34.8	7 50.3	**43**
16 26.5	17 53.6	14 39.3	8 30.9	7 41.3	**44**
16 42.3	18 12.2	14 48.8	8 27.0	7 32.0	**45**
16 58.7	18 31.2	14 58.4	8 23.0	7 22.4	**46**
17 15.7	18 50.7	15 8.2	8 18.9	7 12.6	**47**
17 33.4	19 10.8	15 18.2	8 14.8	7 2.4	**48**
17 52.0	19 31.3	15 28.3	8 10.5	6 51.8	**49**
18 11.3	19 52.4	15 38.6	8 6.2	6 40.9	**50**
18 31.6	20 14.2	15 49.1	8 1.7	6 29.7	**51**
18 52.9	20 36.5	15 59.7	7 57.1	6 17.9	**52**
19 15.3	20 59.6	16 10.6	7 52.4	6 5.8	**53**
19 38.9	21 23.3	16 21.7	7 47.6	5 53.2	**54**
20 3.9	21 47.9	16 33.1	7 42.7	5 40.0	**55**
20 30.3	22 13.2	16 44.7	7 37.6	5 26.3	**56**
20 58.5	22 39.5	16 56.6	7 32.3	5 11.9	**57**
21 28.5	23 6.7	17 8.7	7 26.9	4 56.9	**58**
22 0.5	23 34.9	17 21.2	7 21.3	4 41.2	**59**
22 35.0	24 4.2	17 34.0	7 15.6	4 24.7	**60**
23 12.2	24 34.6	17 47.1	7 9.6	4 7.3	**61**
23 52.5	25 6.3	18 0.5	7 3.4	3 49.0	**62**
24 36.5	25 39.3	18 14.4	6 57.0	3 29.7	**63**
25 24.8	26 13.7	18 28.7	6 50.4	3 9.2	**64**
26 18.3	26 49.7	18 43.4	6 43.5	2 47.5	**65**
27 18.0	27 27.3	18 58.6	6 36.3	2 24.4	**66**
5	6	Descendant	8	9	S LAT

♐ 11° 32′ 4″

16^h 40^m 0^s MC 250° 0′ 0″

4^h 44^m 0^s MC 71° 0′ 0″

♊ 12° 28′ 6″

11	12	Ascendant	2	3	N LAT
° ′	° ′	° ′	° ′	° ′	°
♋10 6.7	♌ 8 34.4	♍ 9 25.7	♎11 57.7	♏13 27.3	**0**
10 46.0	9 38.1	10 9.5	11 40.2	12 58.8	**5**
11 25.8	10 40.8	10 51.1	11 23.3	12 30.4	**10**
12 6.7	11 43.5	11 31.1	11 6.6	12 1.5	**15**
12 15.1	11 56.1	11 39.0	11 3.3	11 55.7	**16**
12 23.6	12 8.8	11 46.8	11 0.0	11 49.8	**17**
12 32.2	12 21.6	11 54.7	10 56.6	11 43.8	**18**
12 40.8	12 34.4	12 2.5	10 53.3	11 37.8	**19**
12 49.6	12 47.2	12 10.3	10 50.0	11 31.8	**20**
12 58.5	13 0.2	12 18.0	10 46.6	11 25.7	**21**
13 7.5	13 13.2	12 25.8	10 43.3	11 19.6	**22**
13 16.6	13 26.3	12 33.6	10 39.9	11 13.4	**23**
13 25.8	13 39.6	12 41.4	10 36.5	11 7.1	**24**
13 35.2	13 52.9	12 49.1	10 33.1	11 0.7	**25**
13 44.8	14 6.4	12 56.9	10 29.7	10 54.3	**26**
13 54.5	14 20.0	13 4.8	10 26.3	10 47.7	**27**
14 4.5	14 33.7	13 12.6	10 22.8	10 41.1	**28**
14 14.6	14 47.6	13 20.5	10 19.3	10 34.4	**29**
14 24.9	15 1.7	13 28.3	10 15.8	10 27.6	**30**
14 35.4	15 15.9	13 36.3	10 12.3	10 20.6	**31**
14 46.1	15 30.3	13 44.2	10 8.7	10 13.6	**32**
14 57.1	15 45.0	13 52.3	10 5.1	10 6.4	**33**
15 8.4	15 59.8	14 0.3	10 1.4	9 59.1	**34**
15 19.9	16 14.9	14 8.5	9 57.7	9 51.6	**35**
15 31.7	16 30.2	14 16.7	9 54.0	9 44.0	**36**
15 43.9	16 45.7	14 24.9	9 50.2	9 36.2	**37**
15 56.4	17 1.6	14 33.3	9 46.3	9 28.3	**38**
16 9.2	17 17.7	14 41.7	9 42.4	9 20.2	**39**
16 22.4	17 34.1	14 50.2	9 38.5	9 11.8	**40**
16 36.1	17 50.9	14 58.8	9 34.5	9 3.3	**41**
16 50.2	18 8.0	15 7.5	9 30.4	8 54.6	**42**
17 4.8	18 25.4	15 16.3	9 26.2	8 45.6	**43**
17 19.9	18 43.3	15 25.2	9 22.0	8 36.3	**44**
17 35.5	19 1.5	15 34.2	9 17.7	8 26.8	**45**
17 51.8	19 20.2	15 43.4	9 13.3	8 17.0	**46**
18 8.7	19 39.4	15 52.7	9 8.8	8 6.9	**47**
18 26.3	19 59.0	16 2.2	9 4.2	7 56.5	**48**
18 44.7	20 19.2	16 11.8	8 59.5	7 45.8	**49**
19 3.9	20 39.9	16 21.6	8 54.7	7 34.6	**50**
19 24.0	21 1.2	16 31.5	8 49.8	7 23.1	**51**
19 45.1	21 23.1	16 41.7	8 44.8	7 11.1	**52**
20 7.2	21 45.7	16 52.0	8 39.6	6 58.7	**53**
20 30.6	22 9.0	17 2.6	8 34.3	6 45.7	**54**
20 55.3	22 33.1	17 13.4	8 28.9	6 32.2	**55**
21 21.5	22 57.9	17 24.4	8 23.3	6 18.2	**56**
21 49.3	23 23.7	17 35.7	8 17.5	6 3.5	**57**
22 18.9	23 50.3	17 47.2	8 11.5	5 48.1	**58**
22 50.5	24 17.9	17 59.0	8 5.4	5 32.0	**59**
23 24.5	24 46.6	18 11.2	7 59.1	5 15.1	**60**
24 1.2	25 16.4	18 23.6	7 52.5	4 57.3	**61**
24 40.8	25 47.4	18 36.4	7 45.7	4 38.5	**62**
25 24.1	26 19.7	18 49.6	7 38.7	4 18.7	**63**
26 11.4	26 53.4	19 3.2	7 31.3	3 57.6	**64**
27 3.7	27 28.5	19 17.1	7 23.7	3 35.3	**65**
28 2.0	28 5.3	19 31.5	7 15.8	3 11.6	**66**
5	6	Descendant	8	9	S LAT

♐ 12° 28′ 6″

16^h 44^m 0^s MC 251° 0′ 0″

4^h 48^m 0^s MC 72° 0′ 0″

♊ 13° 24′ 2″

11	12	Ascendant	2	3	N LAT
° ′	° ′	° ′	° ′	° ′	°
♋11 2.1	♌ 9 33.6	♍10 29.9	♎13 2.6	♏14 27.7	0
11 41.3	10 36.4	11 11.6	12 43.6	13 58.8	5
12 21.0	11 38.3	11 51.1	12 25.1	13 29.8	10
13 1.8	12 40.1	12 29.1	12 7.0	13 0.4	15
13 10.1	12 52.6	12 36.6	12 3.4	12 54.5	16
13 18.6	13 5.1	12 44.0	11 59.7	12 48.5	17
13 27.1	13 17.6	12 51.4	11 56.1	12 42.4	18
13 35.7	13 30.2	12 58.8	11 52.5	12 36.3	19
13 44.4	13 42.9	13 6.2	11 48.9	12 30.2	20
13 53.3	13 55.7	13 13.6	11 45.2	12 24.0	21
14 2.2	14 8.5	13 21.0	11 41.6	12 17.7	22
14 11.3	14 21.4	13 28.4	11 37.9	12 11.4	23
14 20.5	14 34.4	13 35.8	11 34.2	12 5.0	24
14 29.9	14 47.6	13 43.1	11 30.5	11 58.5	25
14 39.4	15 0.8	13 50.5	11 26.8	11 51.9	26
14 49.1	15 14.2	13 58.0	11 23.1	11 45.3	27
14 59.0	15 27.7	14 5.4	11 19.3	11 38.5	28
15 9.0	15 41.4	14 12.9	11 15.5	11 31.7	29
15 19.3	15 55.2	14 20.3	11 11.7	11 24.7	30
15 29.7	16 9.2	14 27.9	11 7.8	11 17.7	31
15 40.4	16 23.4	14 35.4	11 3.9	11 10.5	32
15 51.4	16 37.8	14 43.0	11 0.0	11 3.2	33
16 2.6	16 52.4	14 50.7	10 56.0	10 55.7	34
16 14.0	17 7.2	14 58.4	10 51.9	10 48.1	35
16 25.8	17 22.3	15 6.2	10 47.9	10 40.3	36
16 37.9	17 37.6	15 14.0	10 43.7	10 32.4	37
16 50.3	17 53.1	15 21.9	10 39.5	10 24.3	38
17 3.0	18 9.0	15 29.9	10 35.3	10 16.0	39
17 16.2	18 25.1	15 37.9	10 31.0	10 7.5	40
17 29.7	18 41.5	15 46.1	10 26.6	9 58.8	41
17 43.8	18 58.3	15 54.3	10 22.1	9 49.9	42
17 58.2	19 15.5	16 2.7	10 17.6	9 40.7	43
18 13.2	19 33.0	16 11.1	10 13.0	9 31.3	44
18 28.8	19 50.9	16 19.7	10 8.3	9 21.6	45
18 44.9	20 9.3	16 28.4	10 3.5	9 11.6	46
19 1.6	20 28.0	16 37.2	9 58.6	9 1.3	47
19 19.1	20 47.3	16 46.2	9 53.6	8 50.6	48
19 37.3	21 7.1	16 55.3	9 48.5	8 39.6	49
19 56.4	21 27.4	17 4.6	9 43.2	8 28.2	50
20 16.3	21 48.3	17 14.0	9 37.9	8 16.4	51
20 37.2	22 9.8	17 23.6	9 32.4	8 4.2	52
20 59.1	22 31.9	17 33.4	9 26.8	7 51.5	53
21 22.3	22 54.8	17 43.4	9 21.0	7 38.2	54
21 46.7	23 18.3	17 53.6	9 15.1	7 24.4	55
22 12.5	23 42.7	18 4.1	9 9.0	7 10.0	56
22 40.0	24 7.9	18 14.8	9 2.7	6 55.0	57
23 9.2	24 33.9	18 25.7	8 56.2	6 39.3	58
23 40.5	25 1.0	18 36.9	8 49.5	6 22.8	59
24 14.0	25 29.0	18 48.4	8 42.6	6 5.4	60
24 50.0	25 58.2	19 0.2	8 35.4	5 47.2	61
25 29.1	26 28.5	19 12.3	8 28.0	5 27.9	62
26 11.5	27 0.1	19 24.8	8 20.3	5 7.5	63
26 58.0	27 33.0	19 37.6	8 12.3	4 46.0	64
27 49.2	28 7.4	19 50.9	8 4.0	4 23.1	65
28 46.1	28 43.4	20 4.5	7 55.4	3 58.7	66
5	6	Descendant	8	9	S LAT

♐ 13° 24′ 2″

16^h 48^m 0^s MC 252° 0′ 0″

4^h 52^m 0^s MC 73° 0′ 0″

♊ 14° 19′ 53″

11	12	Ascendant	2	3	N LAT
° ′	° ′	° ′	° ′	° ′	°
♋11 57.6	♌10 32.9	♍11 34.2	♎14 7.5	♏15 28.0	0
12 36.6	11 34.9	12 13.7	13 46.9	14 58.5	5
13 16.2	12 35.9	12 51.1	13 26.9	14 29.1	10
13 56.8	13 36.9	13 27.1	13 7.3	13 59.2	15
14 5.2	13 49.1	13 34.2	13 3.4	13 53.2	16
14 13.6	14 1.4	13 41.2	12 59.5	13 47.0	17
14 22.1	14 13.8	13 48.3	12 55.6	13 40.9	18
14 30.6	14 26.2	13 55.3	12 51.7	13 34.7	19
14 39.3	14 38.7	14 2.3	12 47.7	13 28.4	20
14 48.1	14 51.2	14 9.3	12 43.8	13 22.1	21
14 57.1	15 3.9	14 16.2	12 39.8	13 15.7	22
15 6.1	15 16.6	14 23.2	12 35.9	13 9.3	23
15 15.3	15 29.4	14 30.2	12 31.9	13 2.8	24
15 24.6	15 42.3	14 37.2	12 27.9	12 56.2	25
15 34.1	15 55.4	14 44.2	12 23.9	12 49.5	26
15 43.7	16 8.5	14 51.2	12 19.8	12 42.7	27
15 53.5	16 21.8	14 58.2	12 15.7	12 35.9	28
16 3.5	16 35.3	15 5.3	12 11.6	12 28.9	29
16 13.7	16 48.9	15 12.4	12 7.5	12 21.8	30
16 24.1	17 2.6	15 19.5	12 3.3	12 14.6	31
16 34.8	17 16.6	15 26.6	11 59.1	12 7.3	32
16 45.6	17 30.7	15 33.8	11 54.8	11 59.8	33
16 56.7	17 45.1	15 41.1	11 50.5	11 52.2	34
17 8.1	17 59.6	15 48.3	11 46.1	11 44.5	35
17 19.8	18 14.4	15 55.7	11 41.7	11 36.5	36
17 31.8	18 29.4	16 3.1	11 37.2	11 28.5	37
17 44.2	18 44.7	16 10.6	11 32.7	11 20.2	38
17 56.8	19 0.2	16 18.1	11 28.1	11 11.7	39
18 9.9	19 16.1	16 25.7	11 23.4	11 3.1	40
18 23.4	19 32.2	16 33.4	11 18.7	10 54.2	41
18 37.3	19 48.7	16 41.2	11 13.9	10 45.1	42
18 51.7	20 5.5	16 49.1	11 9.0	10 35.7	43
19 6.5	20 22.7	16 57.1	11 4.0	10 26.1	44
19 21.9	20 40.3	17 5.2	10 58.9	10 16.2	45
19 37.9	20 58.3	17 13.4	10 53.7	10 6.0	46
19 54.6	21 16.7	17 21.8	10 48.4	9 55.5	47
20 11.9	21 35.6	17 30.2	10 43.0	9 44.6	48
20 29.9	21 55.0	17 38.8	10 37.4	9 33.4	49
20 48.8	22 14.9	17 47.6	10 31.8	9 21.7	50
21 8.5	22 35.4	17 56.5	10 26.0	9 9.7	51
21 29.2	22 56.4	18 5.6	10 20.0	8 57.2	52
21 50.9	23 18.1	18 14.8	10 13.9	8 44.2	53
22 13.8	23 40.5	18 24.3	10 7.7	8 30.6	54
22 38.0	24 3.6	18 33.9	10 1.2	8 16.5	55
23 3.5	24 27.4	18 43.8	9 54.6	8 1.8	56
23 30.7	24 52.1	18 53.9	9 47.8	7 46.4	57
23 59.5	25 17.6	19 4.2	9 40.8	7 30.3	58
24 30.3	25 44.0	19 14.8	9 33.5	7 13.4	59
25 3.4	26 11.5	19 25.6	9 26.0	6 55.7	60
25 38.9	26 40.0	19 36.8	9 18.3	6 37.0	61
26 17.3	27 9.7	19 48.2	9 10.2	6 17.2	62
26 59.0	27 40.5	20 0.0	9 1.9	5 56.3	63
27 44.5	28 12.7	20 12.1	8 53.3	5 34.2	64
28 34.6	28 46.4	20 24.6	8 44.3	5 10.7	65
29 30.2	29 21.5	20 37.5	8 34.9	4 45.7	66
5	6	Descendant	8	9	S LAT

♐ 14° 19′ 53″

16^h 52^m 0^s MC 253° 0′ 0″

4h 56m 0s | MC | 74° 0′ 0″

♊ 15° 15′ 39″

N LAT	11	12	Ascendant	2	3
°	° ′	° ′	° ′	° ′	° ′
0	♋12 53.1	♌11 32.4	♍12 38.6	♎15 12.2	♏16 28.0
5	13 32.0	12 33.6	13 15.9	14 50.1	15 58.1
10	14 11.4	13 33.7	13 51.2	14 28.6	15 28.2
15	14 51.9	14 33.7	14 25.2	14 7.6	14 57.8
16	15 0.2	14 45.8	14 31.8	14 3.4	14 51.7
17	15 8.6	14 57.9	14 38.5	13 59.2	14 45.5
18	15 17.1	15 10.1	14 45.1	13 55.0	14 39.2
19	15 25.6	15 22.3	14 51.7	13 50.8	14 32.9
20	15 34.3	15 34.6	14 58.3	13 46.5	14 26.5
21	15 43.0	15 46.9	15 4.9	13 42.3	14 20.1
22	15 51.9	15 59.3	15 11.5	13 38.1	14 13.6
23	16 0.9	16 11.8	15 18.1	13 33.8	14 7.1
24	16 10.0	16 24.5	15 24.7	13 29.5	14 0.5
25	16 19.3	16 37.2	15 31.3	13 25.2	13 53.7
26	16 28.7	16 50.0	15 37.9	13 20.9	13 46.9
27	16 38.3	17 2.9	15 44.5	13 16.5	13 40.0
28	16 48.1	17 16.0	15 51.1	13 12.1	13 33.1
29	16 58.0	17 29.2	15 57.7	13 7.7	13 26.0
30	17 8.2	17 42.6	16 4.4	13 3.3	13 18.7
31	17 18.5	17 56.1	16 11.1	12 58.8	13 11.4
32	17 29.1	18 9.8	16 17.9	12 54.2	13 3.9
33	17 39.9	18 23.7	16 24.6	12 49.6	12 56.4
34	17 50.9	18 37.8	16 31.5	12 45.0	12 48.6
35	18 2.3	18 52.1	16 38.3	12 40.3	12 40.7
36	18 13.9	19 6.6	16 45.2	12 35.5	12 32.7
37	18 25.8	19 21.3	16 52.2	12 30.7	12 24.4
38	18 38.1	19 36.3	16 59.3	12 25.9	12 16.0
39	18 50.7	19 51.6	17 6.4	12 20.9	12 7.4
40	19 3.6	20 7.1	17 13.5	12 15.9	11 58.6
41	19 17.0	20 23.0	17 20.8	12 10.8	11 49.5
42	19 30.8	20 39.1	17 28.1	12 5.6	11 40.2
43	19 45.1	20 55.6	17 35.6	12 0.3	11 30.7
44	19 59.8	21 12.5	17 43.1	11 54.9	11 20.9
45	20 15.1	21 29.7	17 50.7	11 49.4	11 10.8
46	20 31.0	21 47.4	17 58.4	11 43.9	11 0.4
47	20 47.5	22 5.4	18 6.3	11 38.2	10 49.6
48	21 4.6	22 24.0	18 14.3	11 32.3	10 38.5
49	21 22.5	22 43.0	18 22.4	11 26.4	10 27.1
50	21 41.2	23 2.5	18 30.6	11 20.3	10 15.2
51	22 0.7	23 22.5	18 39.0	11 14.0	10 2.9
52	22 21.2	23 43.1	18 47.6	11 7.6	9 50.1
53	22 42.7	24 4.4	18 56.3	11 1.1	9 36.8
54	23 5.4	24 26.3	19 5.2	10 54.3	9 23.0
55	23 29.2	24 48.9	19 14.2	10 47.4	9 8.6
56	23 54.5	25 12.2	19 23.5	10 40.3	8 53.5
57	24 21.3	25 36.3	19 33.0	10 32.9	8 37.8
58	24 49.8	26 1.3	19 42.7	10 25.4	8 21.3
59	25 20.1	26 27.1	19 52.7	10 17.6	8 4.0
60	25 52.7	26 54.0	20 2.9	10 9.5	7 45.9
61	26 27.7	27 21.9	20 13.4	10 1.1	7 26.7
62	27 5.4	27 50.9	20 24.2	9 52.5	7 6.5
63	27 46.3	28 21.0	20 35.2	9 43.5	6 45.1
64	28 31.0	28 52.5	20 46.6	9 34.2	6 22.4
65	29 20.0	29 25.3	20 58.4	9 24.5	5 58.3
66	♌ 0 14.3	29 59.7	21 10.5	9 14.4	5 32.6
S LAT	5	6	Descendant	8	9

♐ 15° 15′ 39″

16h 56m 0s | MC | 254° 0′ 0″

5h 0m 0s | MC | 75° 0′ 0″

♊ 16° 11′ 19″

N LAT	11	12	Ascendant	2	3
°	° ′	° ′	° ′	° ′	° ′
0	♋13 48.7	♌12 32.1	♍13 43.2	♎16 16.8	♏17 27.9
5	14 27.5	13 32.4	14 18.2	15 53.2	16 57.5
10	15 6.8	14 31.6	14 51.4	15 30.3	16 27.2
15	15 47.1	15 30.7	15 23.3	15 7.8	15 56.3
16	15 55.3	15 42.6	15 29.6	15 3.3	15 50.1
17	16 3.7	15 54.5	15 35.8	14 58.8	15 43.7
18	16 12.1	16 6.5	15 42.1	14 54.3	15 37.4
19	16 20.6	16 18.5	15 48.3	14 49.8	15 31.0
20	16 29.2	16 30.5	15 54.5	14 45.3	15 24.5
21	16 37.9	16 42.7	16 0.7	14 40.8	15 18.0
22	16 46.8	16 54.9	16 6.8	14 36.2	15 11.4
23	16 55.7	17 7.2	16 13.0	14 31.7	15 4.7
24	17 4.8	17 19.6	16 19.2	14 27.1	14 58.0
25	17 14.0	17 32.1	16 25.4	14 22.5	14 51.2
26	17 23.4	17 44.7	16 31.6	14 17.8	14 44.3
27	17 33.0	17 57.4	16 37.8	14 13.2	14 37.2
28	17 42.7	18 10.2	16 44.0	14 8.5	14 30.1
29	17 52.6	18 23.2	16 50.2	14 3.8	14 22.9
30	18 2.6	18 36.4	16 56.5	13 59.0	14 15.6
31	18 12.9	18 49.6	17 2.8	13 54.2	14 8.1
32	18 23.4	19 3.1	17 9.1	13 49.3	14 0.5
33	18 34.2	19 16.7	17 15.5	13 44.4	13 52.8
34	18 45.1	19 30.5	17 21.9	13 39.4	13 44.9
35	18 56.4	19 44.6	17 28.3	13 34.4	13 36.9
36	19 7.9	19 58.8	17 34.8	13 29.3	13 28.7
37	19 19.8	20 13.3	17 41.4	13 24.2	13 20.3
38	19 31.9	20 28.0	17 48.0	13 19.0	13 11.7
39	19 44.5	20 43.0	17 54.6	13 13.7	13 2.9
40	19 57.3	20 58.2	18 1.4	13 8.3	12 54.0
41	20 10.6	21 13.8	18 8.2	13 2.8	12 44.7
42	20 24.3	21 29.6	18 15.0	12 57.3	12 35.3
43	20 38.5	21 45.8	18 22.0	12 51.6	12 25.6
44	20 53.1	22 2.3	18 29.1	12 45.9	12 15.6
45	21 8.3	22 19.2	18 36.2	12 40.0	12 5.3
46	21 24.0	22 36.5	18 43.5	12 34.0	11 54.6
47	21 40.3	22 54.2	18 50.8	12 27.9	11 43.7
48	21 57.3	23 12.3	18 58.3	12 21.7	11 32.4
49	22 15.1	23 30.9	19 5.9	12 15.3	11 20.7
50	22 33.6	23 50.0	19 13.6	12 8.8	11 8.6
51	22 52.9	24 9.7	19 21.5	12 2.1	10 56.0
52	23 13.2	24 29.8	19 29.5	11 55.2	10 43.0
53	23 34.5	24 50.6	19 37.7	11 48.2	10 29.4
54	23 56.8	25 12.1	19 46.0	11 41.0	10 15.3
55	24 20.4	25 34.2	19 54.6	11 33.6	10 0.6
56	24 45.4	25 57.0	20 3.3	11 25.9	9 45.2
57	25 11.8	26 20.6	20 12.2	11 18.1	9 29.1
58	25 39.9	26 45.0	20 21.3	11 9.9	9 12.3
59	26 9.9	27 10.3	20 30.6	11 1.6	8 54.6
60	26 42.0	27 36.5	20 40.2	10 52.9	8 36.0
61	27 16.4	28 3.8	20 50.0	10 44.0	8 16.4
62	27 53.5	28 32.1	21 0.1	10 34.7	7 55.7
63	28 33.7	29 1.6	21 10.5	10 25.1	7 33.7
64	29 17.5	29 32.3	21 21.2	10 15.1	7 10.5
65	♌ 0 5.5	♍ 0 4.4	21 32.2	10 4.7	6 45.8
66	0 58.5	0 37.9	21 43.5	9 53.9	6 19.4
S LAT	5	6	Descendant	8	9

♐ 16° 11′ 19″

17h 0m 0s | MC | 255° 0′ 0″

$5^h\ 4^m\ 0^s$		MC		$76°\ 0'\ 0''$	
		♊ 17° 6′ 55″			
11	**12**	**Ascendant**	**2**	**3**	**N LAT**
° ′	° ′	° ′	° ′	° ′	°
♋14 44.4	♌13 32.0	♍14 47.8	♎17 21.4	♏18 27.6	**0**
15 23.1	14 31.3	15 20.6	16 56.2	17 56.8	**5**
16 2.1	15 29.6	15 51.7	16 31.9	17 26.0	**10**
16 42.3	16 27.8	16 21.5	16 7.9	16 54.7	**15**
16 50.5	16 39.5	16 27.4	16 3.1	16 48.3	**16**
16 58.8	16 51.2	16 33.2	15 58.3	16 41.9	**17**
17 7.2	17 2.9	16 39.0	15 53.6	16 35.4	**18**
17 15.7	17 14.7	16 44.8	15 48.8	16 28.9	**19**
17 24.2	17 26.6	16 50.6	15 44.0	16 22.4	**20**
17 32.9	17 38.5	16 56.4	15 39.1	16 15.7	**21**
17 41.7	17 50.6	17 2.2	15 34.3	16 9.0	**22**
17 50.6	18 2.6	17 8.0	15 29.5	16 2.3	**23**
17 59.6	18 14.8	17 13.7	15 24.6	15 55.4	**24**
18 8.8	18 27.1	17 19.5	15 19.7	15 48.5	**25**
18 18.1	18 39.5	17 25.3	15 14.8	15 41.4	**26**
18 27.6	18 51.9	17 31.1	15 9.8	15 34.3	**27**
18 37.3	19 4.6	17 36.9	15 4.8	15 27.1	**28**
18 47.1	19 17.3	17 42.8	14 59.8	15 19.7	**29**
18 57.1	19 30.2	17 48.6	14 54.7	15 12.3	**30**
19 7.3	19 43.2	17 54.5	14 49.6	15 4.7	**31**
19 17.8	19 56.4	18 0.4	14 44.4	14 57.0	**32**
19 28.4	20 9.8	18 6.3	14 39.2	14 49.1	**33**
19 39.4	20 23.4	18 12.3	14 33.9	14 41.1	**34**
19 50.5	20 37.1	18 18.3	14 28.5	14 32.9	**35**
20 2.0	20 51.1	18 24.4	14 23.1	14 24.6	**36**
20 13.8	21 5.3	18 30.5	14 17.6	14 16.1	**37**
20 25.8	21 19.7	18 36.7	14 12.0	14 7.3	**38**
20 38.3	21 34.4	18 42.9	14 6.4	13 58.4	**39**
20 51.0	21 49.3	18 49.2	14 0.7	13 49.3	**40**
21 4.2	22 4.6	18 55.5	13 54.8	13 39.9	**41**
21 17.8	22 20.1	19 2.0	13 48.9	13 30.2	**42**
21 31.8	22 36.0	19 8.5	13 42.9	13 20.3	**43**
21 46.4	22 52.1	19 15.1	13 36.8	13 10.2	**44**
22 1.4	23 8.7	19 21.8	13 30.5	12 59.7	**45**
22 17.0	23 25.6	19 28.5	13 24.1	12 48.9	**46**
22 33.2	23 42.9	19 35.4	13 17.6	12 37.7	**47**
22 50.0	24 0.7	19 42.4	13 11.0	12 26.2	**48**
23 7.6	24 18.9	19 49.5	13 4.2	12 14.2	**49**
23 25.9	24 37.6	19 56.7	12 57.2	12 1.9	**50**
23 45.1	24 56.8	20 4.0	12 50.1	11 49.1	**51**
24 5.1	25 16.6	20 11.5	12 42.8	11 35.8	**52**
24 26.2	25 36.9	20 19.1	12 35.3	11 21.9	**53**
24 48.3	25 57.9	20 26.9	12 27.6	11 7.5	**54**
25 11.6	26 19.5	20 34.9	12 19.7	10 52.5	**55**
25 36.3	26 41.8	20 43.0	12 11.5	10 36.8	**56**
26 2.4	27 4.9	20 51.3	12 3.2	10 20.4	**57**
26 30.1	27 28.7	20 59.8	11 54.5	10 3.1	**58**
26 59.6	27 53.4	21 8.5	11 45.6	9 45.0	**59**
27 31.2	28 19.1	21 17.4	11 36.3	9 26.0	**60**
28 5.1	28 45.7	21 26.6	11 26.8	9 6.0	**61**
28 41.6	29 13.3	21 36.0	11 16.9	8 44.8	**62**
29 21.1	29 42.1	21 45.7	11 6.6	8 22.3	**63**
♌ 0 4.0	♍ 0 12.1	21 55.7	10 56.0	7 58.5	**64**
0 51.0	0 43.4	22 6.0	10 44.9	7 33.2	**65**
1 42.7	1 16.1	22 16.6	10 33.4	7 6.1	**66**
5	**6**	**Descendant**	**8**	**9**	**S LAT**
		♐ 17° 6′ 55″			
$17^h\ 4^m\ 0^s$		MC		$256°\ 0'\ 0''$	

$5^h\ 8^m\ 0^s$		MC		$77°\ 0'\ 0''$	
		♊ 18° 2′ 27″			
N LAT	**11**	**12**	**Ascendant**	**2**	**3**
°	° ′	° ′	° ′	° ′	° ′
0	♋15 40.1	♌14 32.0	♍15 52.5	♎18 25.8	♏19 27.1
5	16 18.7	15 30.5	16 23.1	17 59.2	18 55.8
10	16 57.6	16 27.8	16 52.0	17 33.4	18 24.6
15	17 37.6	17 25.0	17 19.7	17 8.0	17 52.8
16	17 45.7	17 36.5	17 25.2	17 2.9	17 46.4
17	17 54.0	17 48.0	17 30.6	16 57.8	17 39.9
18	18 2.3	17 59.5	17 36.1	16 52.8	17 33.3
19	18 10.8	18 11.1	17 41.5	16 47.7	17 26.7
20	18 19.3	18 22.8	17 46.8	16 42.6	17 20.1
21	18 27.9	18 34.5	17 52.2	16 37.5	17 13.3
22	18 36.7	18 46.3	17 57.6	16 32.4	17 6.5
23	18 45.5	18 58.2	18 3.0	16 27.2	16 59.7
24	18 54.5	19 10.1	18 8.3	16 22.0	16 52.7
25	19 3.6	19 22.2	18 13.7	16 16.9	16 45.7
26	19 12.9	19 34.3	18 19.1	16 11.6	16 38.5
27	19 22.3	19 46.6	18 24.5	16 6.4	16 31.3
28	19 31.9	19 58.9	18 29.9	16 1.1	16 23.9
29	19 41.7	20 11.5	18 35.3	15 55.7	16 16.5
30	19 51.6	20 24.1	18 40.7	15 50.3	16 8.9
31	20 1.8	20 36.9	18 46.2	15 44.9	16 1.2
32	20 12.2	20 49.8	18 51.7	15 39.4	15 53.3
33	20 22.8	21 2.9	18 57.2	15 33.9	15 45.4
34	20 33.6	21 16.2	19 2.8	15 28.2	15 37.2
35	20 44.7	21 29.7	19 8.4	15 22.6	15 28.9
36	20 56.1	21 43.4	19 14.0	15 16.8	15 20.4
37	21 7.8	21 57.3	19 19.7	15 11.0	15 11.7
38	21 19.7	22 11.5	19 25.4	15 5.1	15 2.9
39	21 32.1	22 25.9	19 31.2	14 59.1	14 53.8
40	21 44.7	22 40.5	19 37.0	14 53.0	14 44.5
41	21 57.8	22 55.4	19 42.9	14 46.8	14 34.9
42	22 11.3	23 10.6	19 48.9	14 40.6	14 25.1
43	22 25.2	23 26.2	19 55.0	14 34.2	14 15.0
44	22 39.6	23 42.0	20 1.1	14 27.7	14 4.7
45	22 54.5	23 58.2	20 7.3	14 21.0	13 54.0
46	23 10.0	24 14.8	20 13.6	14 14.2	13 43.0
47	23 26.0	24 31.7	20 20.0	14 7.3	13 31.6
48	23 42.7	24 49.1	20 26.4	14 0.3	13 19.9
49	24 0.1	25 6.9	20 33.0	13 53.1	13 7.7
50	24 18.2	25 25.2	20 39.7	13 45.7	12 55.1
51	24 37.2	25 44.0	20 46.6	13 38.1	12 42.1
52	24 57.0	26 3.3	20 53.5	13 30.3	12 28.5
53	25 17.8	26 23.2	21 0.6	13 22.4	12 14.4
54	25 39.7	26 43.7	21 7.8	13 14.2	11 59.7
55	26 2.8	27 4.8	21 15.2	13 5.8	11 44.3
56	26 27.1	27 26.6	21 22.7	12 57.1	11 28.3
57	26 52.8	27 49.2	21 30.5	12 48.2	11 11.5
58	27 20.2	28 12.5	21 38.4	12 39.0	10 53.9
59	27 49.3	28 36.6	21 46.4	12 29.5	10 35.4
60	28 20.4	29 1.7	21 54.7	12 19.7	10 16.0
61	28 53.7	29 27.7	22 3.2	12 9.6	9 55.5
62	29 29.6	29 54.7	22 12.0	11 59.1	9 33.8
63	♌ 0 8.4	♍ 0 22.8	22 21.0	11 48.2	9 10.8
64	0 50.5	0 52.0	22 30.2	11 36.9	8 46.4
65	1 36.5	1 22.6	22 39.8	11 25.1	8 20.5
66	2 27.1	1 54.5	22 49.6	11 12.8	7 52.7
S LAT	**5**	**6**	**Descendant**	**8**	**9**
			♐ 18° 2′ 27″		
	$17^h\ 8^m\ 0^s$		MC		$257°\ 0'\ 0''$

5ʰ 12ᵐ 0ˢ		MC		78° 0′ 0″	N
		♊ 18° 57′ 54″			
11	**12**	**Ascendant**	**2**	**3**	**LAT**
° ′	° ′	° ′	° ′	° ′	°
♋16 36.0	♌15 32.3	♍16 57.4	♎19 30.1	♏20 26.4	**0**
17 14.4	16 29.7	17 25.6	19 2.0	19 54.8	**5**
17 53.1	17 26.1	17 52.4	18 34.8	19 23.1	**10**
18 32.9	18 22.3	18 18.0	18 7.9	18 50.9	**15**
18 41.0	18 33.6	18 23.1	18 2.6	18 44.3	**16**
18 49.2	18 44.9	18 28.1	17 57.2	18 37.7	**17**
18 57.5	18 56.2	18 33.1	17 51.9	18 31.1	**18**
19 5.9	19 7.6	18 38.1	17 46.5	18 24.4	**19**
19 14.4	19 19.1	18 43.1	17 41.2	18 17.6	**20**
19 23.0	19 30.6	18 48.1	17 35.8	18 10.8	**21**
19 31.7	19 42.1	18 53.0	17 30.3	18 3.9	**22**
19 40.5	19 53.8	18 58.0	17 24.9	17 56.9	**23**
19 49.4	20 5.5	19 3.0	17 19.4	17 49.9	**24**
19 58.5	20 17.3	19 7.9	17 14.0	17 42.7	**25**
20 7.7	20 29.2	19 12.9	17 8.4	17 35.5	**26**
20 17.0	20 41.3	19 17.9	17 2.9	17 28.1	**27**
20 26.6	20 53.4	19 22.9	16 57.3	17 20.7	**28**
20 36.3	21 5.7	19 27.9	16 51.6	17 13.1	**29**
20 46.2	21 18.1	19 32.9	16 45.9	17 5.4	**30**
20 56.3	21 30.6	19 37.9	16 40.2	16 57.6	**31**
21 6.6	21 43.3	19 43.0	16 34.4	16 49.6	**32**
21 17.1	21 56.2	19 48.1	16 28.5	16 41.5	**33**
21 27.9	22 9.2	19 53.3	16 22.6	16 33.2	**34**
21 38.9	22 22.4	19 58.4	16 16.6	16 24.8	**35**
21 50.2	22 35.8	20 3.6	16 10.5	16 16.1	**36**
22 1.8	22 49.4	20 8.9	16 4.4	16 7.3	**37**
22 13.7	23 3.3	20 14.2	15 58.1	15 58.3	**38**
22 25.9	23 17.4	20 19.5	15 51.8	15 49.1	**39**
22 38.5	23 31.7	20 24.9	15 45.3	15 39.6	**40**
22 51.4	23 46.3	20 30.4	15 38.8	15 29.9	**41**
23 4.8	24 1.2	20 35.9	15 32.2	15 19.9	**42**
23 18.6	24 16.4	20 41.5	15 25.4	15 9.6	**43**
23 32.9	24 31.9	20 47.1	15 18.5	14 59.1	**44**
23 47.6	24 47.8	20 52.8	15 11.5	14 48.2	**45**
24 2.9	25 4.0	20 58.7	15 4.3	14 37.0	**46**
24 18.8	25 20.6	21 4.5	14 57.0	14 25.4	**47**
24 35.4	25 37.6	21 10.5	14 49.5	14 13.5	**48**
24 52.6	25 55.0	21 16.6	14 41.9	14 1.1	**49**
25 10.5	26 12.9	21 22.8	14 34.1	13 48.3	**50**
25 29.3	26 31.2	21 29.1	14 26.1	13 35.0	**51**
25 48.9	26 50.1	21 35.5	14 17.9	13 21.1	**52**
26 9.5	27 9.6	21 42.1	14 9.4	13 6.8	**53**
26 31.1	27 29.6	21 48.7	14 0.8	12 51.8	**54**
26 53.9	27 50.2	21 55.5	13 51.9	12 36.1	**55**
27 17.9	28 11.5	22 2.5	13 42.7	12 19.8	**56**
27 43.3	28 33.5	22 9.6	13 33.3	12 2.6	**57**
28 10.3	28 56.3	22 16.9	13 23.6	11 44.7	**58**
28 39.0	29 19.9	22 24.4	13 13.5	11 25.8	**59**
29 9.6	29 44.3	22 32.0	13 3.1	11 5.9	**60**
29 42.4	♍ 0 9.6	22 39.9	12 52.4	10 44.9	**61**
♌ 0 17.6	0 36.0	22 48.0	12 41.2	10 22.8	**62**
0 55.7	1 3.4	22 56.3	12 29.7	9 59.3	**63**
1 37.0	1 32.0	23 4.8	12 17.7	9 34.3	**64**
2 22.0	2 1.7	23 13.6	12 5.2	9 7.7	**65**
3 11.4	2 32.9	23 22.7	11 52.2	8 39.3	**66**
5	**6**	**Descendant**	**8**	**9**	**S**
		♐ 18° 57′ 54″			**LAT**
17ʰ 12ᵐ 0ˢ		MC		258° 0′ 0″	

5ʰ 16ᵐ 0ˢ		MC		79° 0′ 0″	N
		♊ 19° 53′ 18″			
11	**12**	**Ascendant**	**2**	**3**	**LAT**
° ′	° ′	° ′	° ′	° ′	°
♋17 31.9	♌16 32.7	♍18 2.3	♎20 34.3	♏21 25.6	**0**
18 10.1	17 29.2	18 28.2	20 4.7	20 53.5	**5**
18 48.7	18 24.6	18 52.8	19 36.1	20 21.4	**10**
19 28.3	19 19.7	19 16.4	19 7.9	19 48.8	**15**
19 36.4	19 30.8	19 21.0	19 2.2	19 42.2	**16**
19 44.5	19 41.9	19 25.6	18 56.6	19 35.5	**17**
19 52.8	19 53.0	19 30.2	18 51.0	19 28.7	**18**
20 1.1	20 4.2	19 34.8	18 45.3	19 21.9	**19**
20 9.5	20 15.4	19 39.4	18 39.6	19 15.1	**20**
20 18.1	20 26.7	19 43.9	18 34.0	19 8.2	**21**
20 26.7	20 38.1	19 48.5	18 28.3	19 1.2	**22**
20 35.5	20 49.5	19 53.0	18 22.5	18 54.1	**23**
20 44.3	21 1.0	19 57.6	18 16.8	18 46.9	**24**
20 53.4	21 12.6	20 2.2	18 11.0	18 39.7	**25**
21 2.5	21 24.3	20 6.7	18 5.2	18 32.3	**26**
21 11.8	21 36.0	20 11.3	17 59.3	18 24.8	**27**
21 21.3	21 47.9	20 15.9	17 53.4	18 17.3	**28**
21 30.9	22 0.0	20 20.5	17 47.5	18 9.6	**29**
21 40.8	22 12.1	20 25.1	17 41.5	18 1.8	**30**
21 50.8	22 24.4	20 29.7	17 35.4	17 53.8	**31**
22 1.0	22 36.8	20 34.4	17 29.3	17 45.7	**32**
22 11.5	22 49.4	20 39.0	17 23.1	17 37.5	**33**
22 22.1	23 2.2	20 43.8	17 16.9	17 29.1	**34**
22 33.1	23 15.1	20 48.5	17 10.6	17 20.5	**35**
22 44.3	23 28.3	20 53.3	17 4.2	17 11.7	**36**
22 55.8	23 41.6	20 58.1	16 57.7	17 2.8	**37**
23 7.6	23 55.2	21 2.9	16 51.1	16 53.6	**38**
23 19.7	24 8.9	21 7.8	16 44.4	16 44.2	**39**
23 32.2	24 23.0	21 12.8	16 37.6	16 34.6	**40**
23 45.0	24 37.3	21 17.8	16 30.7	16 24.7	**41**
23 58.3	24 51.8	21 22.8	16 23.7	16 14.6	**42**
24 12.0	25 6.7	21 28.0	16 16.6	16 4.2	**43**
24 26.1	25 21.9	21 33.1	16 9.3	15 53.4	**44**
24 40.8	25 37.4	21 38.4	16 1.9	15 42.4	**45**
24 55.9	25 53.2	21 43.7	15 54.4	15 31.0	**46**
25 11.7	26 9.4	21 49.1	15 46.7	15 19.2	**47**
25 28.0	26 26.1	21 54.6	15 38.8	15 7.0	**48**
25 45.1	26 43.1	22 0.2	15 30.7	14 54.4	**49**
26 2.8	27 0.6	22 5.9	15 22.5	14 41.4	**50**
26 21.4	27 18.5	22 11.6	15 14.0	14 27.8	**51**
26 40.8	27 37.0	22 17.5	15 5.4	14 13.7	**52**
27 1.1	27 55.9	22 23.5	14 56.5	13 59.1	**53**
27 22.5	28 15.5	22 29.6	14 47.3	13 43.8	**54**
27 44.9	28 35.6	22 35.9	14 37.9	13 27.8	**55**
28 8.7	28 56.4	22 42.3	14 28.3	13 11.1	**56**
28 33.7	29 17.9	22 48.8	14 18.3	12 53.7	**57**
29 0.3	29 40.1	22 55.5	14 8.0	12 35.3	**58**
29 28.6	♍ 0 3.1	23 2.3	13 57.4	12 16.1	**59**
29 58.7	0 26.9	23 9.3	13 46.5	11 55.8	**60**
♌ 0 31.0	0 51.7	23 16.5	13 35.1	11 34.3	**61**
1 5.7	1 17.4	23 23.9	13 23.4	11 11.7	**62**
1 43.0	1 44.1	23 31.5	13 11.2	10 47.6	**63**
2 23.5	2 11.9	23 39.4	12 58.5	10 22.1	**64**
3 7.6	2 41.0	23 47.4	12 45.3	9 54.8	**65**
3 55.9	3 11.3	23 55.8	12 31.6	9 25.7	**66**
5	**6**	**Descendant**	**8**	**9**	**S**
		♐ 19° 53′ 18″			**LAT**
17ʰ 16ᵐ 0ˢ		MC		259° 0′ 0″	

5h 20m 0s		MC		80° 0′ 0″	
		♊ 20° 48′ 39″			N
11	12	Ascendant	2	3	LAT
° ′	° ′	° ′	° ′	° ′	°
♋18 27.9	♌17 33.3	♍19 7.3	♎21 38.3	♏22 24.6	**0**
19 6.0	18 28.8	19 30.9	21 7.3	21 52.1	**5**
19 44.4	19 23.2	19 53.3	20 37.3	21 19.6	**10**
20 23.7	20 17.3	20 14.7	20 7.7	20 46.6	**15**
20 31.8	20 28.1	20 19.0	20 1.8	20 39.8	**16**
20 39.9	20 39.0	20 23.2	19 55.9	20 33.1	**17**
20 48.1	20 49.9	20 27.3	19 49.9	20 26.2	**18**
20 56.4	21 0.9	20 31.5	19 44.0	20 19.3	**19**
21 4.8	21 11.9	20 35.7	19 38.1	20 12.4	**20**
21 13.2	21 23.0	20 39.8	19 32.1	20 5.4	**21**
21 21.8	21 34.1	20 44.0	19 26.1	19 58.3	**22**
21 30.5	21 45.3	20 48.1	19 20.1	19 51.1	**23**
21 39.3	21 56.6	20 52.3	19 14.1	19 43.8	**24**
21 48.3	22 7.9	20 56.4	19 8.0	19 36.5	**25**
21 57.4	22 19.4	21 0.6	19 1.9	19 29.0	**26**
22 6.6	22 30.9	21 4.7	18 55.7	19 21.5	**27**
22 16.0	22 42.6	21 8.9	18 49.5	19 13.8	**28**
22 25.6	22 54.3	21 13.1	18 43.3	19 6.0	**29**
22 35.4	23 6.2	21 17.3	18 37.0	18 58.1	**30**
22 45.3	23 18.3	21 21.5	18 30.6	18 50.0	**31**
22 55.5	23 30.4	21 25.7	18 24.2	18 41.8	**32**
23 5.8	23 42.7	21 30.0	18 17.7	18 33.4	**33**
23 16.5	23 55.2	21 34.3	18 11.1	18 24.9	**34**
23 27.3	24 7.9	21 38.6	18 4.5	18 16.2	**35**
23 38.4	24 20.8	21 42.9	17 57.8	18 7.3	**36**
23 49.8	24 33.8	21 47.3	17 50.9	17 58.2	**37**
24 1.5	24 47.1	21 51.7	17 44.0	17 48.9	**38**
24 13.6	25 0.6	21 56.2	17 37.0	17 39.3	**39**
24 25.9	25 14.3	22 0.7	17 29.9	17 29.6	**40**
24 38.7	25 28.3	22 5.2	17 22.6	17 19.5	**41**
24 51.8	25 42.5	22 9.8	17 15.3	17 9.2	**42**
25 5.4	25 57.0	22 14.5	17 7.8	16 58.6	**43**
25 19.4	26 11.9	22 19.2	17 0.1	16 47.7	**44**
25 33.9	26 27.0	22 24.0	16 52.3	16 36.5	**45**
25 48.9	26 42.5	22 28.8	16 44.4	16 24.9	**46**
26 4.5	26 58.3	22 33.7	16 36.3	16 12.9	**47**
26 20.7	27 14.6	22 38.7	16 28.0	16 0.5	**48**
26 37.5	27 31.2	22 43.8	16 19.5	15 47.7	**49**
26 55.1	27 48.3	22 49.0	16 10.8	15 34.4	**50**
27 13.4	28 5.8	22 54.2	16 1.9	15 20.6	**51**
27 32.6	28 23.8	22 59.5	15 52.8	15 6.3	**52**
27 52.7	28 42.3	23 5.0	15 43.5	14 51.3	**53**
28 13.8	29 1.4	23 10.6	15 33.9	14 35.7	**54**
28 36.0	29 21.1	23 16.2	15 24.0	14 19.5	**55**
28 59.4	29 41.3	23 22.0	15 13.8	14 2.5	**56**
29 24.1	♍ 0 2.3	23 28.0	15 3.3	13 44.7	**57**
29 50.4	0 24.0	23 34.0	14 52.5	13 26.0	**58**
♌ 0 18.2	0 46.4	23 40.3	14 41.4	13 6.3	**59**
0 47.9	1 9.6	23 46.6	14 29.8	12 45.6	**60**
1 19.6	1 33.7	23 53.2	14 17.9	12 23.7	**61**
1 53.7	1 58.8	23 59.9	14 5.5	12 0.5	**62**
2 30.4	2 24.8	24 6.8	13 52.6	11 35.9	**63**
3 10.0	2 52.0	24 13.9	13 39.3	11 9.8	**64**
3 53.2	3 20.2	24 21.3	13 25.4	10 41.9	**65**
4 40.4	3 49.8	24 28.8	13 11.0	10 12.0	**66**
5	6	Descendant	8	9	S
		♐ 20° 48′ 39″			LAT
17h 20m 0s		MC		260° 0′ 0″	

5h 24m 0s		MC		81° 0′ 0″	
		♊ 21° 43′ 56″			N
11	12	Ascendant	2	3	LAT
° ′	° ′	° ′	° ′	° ′	°
♋19 24.1	♌18 34.0	♍20 12.3	♎22 42.3	♏23 23.4	**0**
20 1.9	19 28.5	20 33.7	22 9.8	22 50.5	**5**
20 40.1	20 21.9	20 53.8	21 38.4	22 17.7	**10**
21 19.3	21 15.0	21 13.2	21 7.4	21 44.2	**15**
21 27.3	21 25.6	21 17.0	21 1.2	21 37.4	**16**
21 35.3	21 36.3	21 20.7	20 55.0	21 30.5	**17**
21 43.5	21 47.0	21 24.5	20 48.8	21 23.6	**18**
21 51.7	21 57.7	21 28.3	20 42.6	21 16.6	**19**
22 0.0	22 8.5	21 32.0	20 36.4	21 9.6	**20**
22 8.5	22 19.3	21 35.8	20 30.2	21 2.5	**21**
22 17.0	22 30.2	21 39.5	20 23.9	20 55.3	**22**
22 25.6	22 41.2	21 43.2	20 17.6	20 48.0	**23**
22 34.4	22 52.2	21 47.0	20 11.3	20 40.6	**24**
22 43.3	23 3.3	21 50.7	20 4.9	20 33.2	**25**
22 52.3	23 14.5	21 54.5	19 58.5	20 25.6	**26**
23 1.5	23 25.8	21 58.2	19 52.1	20 18.0	**27**
23 10.8	23 37.2	22 2.0	19 45.6	20 10.2	**28**
23 20.4	23 48.8	22 5.7	19 39.0	20 2.3	**29**
23 30.0	24 0.4	22 9.5	19 32.4	19 54.2	**30**
23 39.9	24 12.2	22 13.3	19 25.8	19 46.0	**31**
23 50.0	24 24.1	22 17.1	19 19.0	19 37.7	**32**
24 0.3	24 36.1	22 20.9	19 12.2	19 29.2	**33**
24 10.8	24 48.4	22 24.8	19 5.3	19 20.6	**34**
24 21.6	25 0.7	22 28.7	18 58.4	19 11.7	**35**
24 32.6	25 13.3	22 32.6	18 51.3	19 2.7	**36**
24 43.9	25 26.1	22 36.5	18 44.2	18 53.5	**37**
24 55.5	25 39.0	22 40.5	18 36.9	18 44.0	**38**
25 7.4	25 52.2	22 44.5	18 29.6	18 34.3	**39**
25 19.7	26 5.6	22 48.6	18 22.1	18 24.4	**40**
25 32.3	26 19.3	22 52.7	18 14.5	18 14.2	**41**
25 45.4	26 33.2	22 56.8	18 6.8	18 3.8	**42**
25 58.8	26 47.4	23 1.0	17 58.9	17 53.0	**43**
26 12.7	27 1.9	23 5.2	17 50.9	17 41.9	**44**
26 27.0	27 16.7	23 9.5	17 42.7	17 30.5	**45**
26 41.9	27 31.8	23 13.9	17 34.4	17 18.7	**46**
26 57.3	27 47.3	23 18.3	17 25.9	17 6.5	**47**
27 13.3	28 3.1	23 22.8	17 17.2	16 53.9	**48**
27 30.0	28 19.3	23 27.4	17 8.3	16 40.9	**49**
27 47.4	28 36.0	23 32.0	16 59.2	16 27.4	**50**
28 5.5	28 53.1	23 36.8	16 49.8	16 13.3	**51**
28 24.5	29 10.7	23 41.6	16 40.3	15 58.7	**52**
28 44.3	29 28.7	23 46.5	16 30.5	15 43.5	**53**
29 5.1	29 47.4	23 51.5	16 20.4	15 27.7	**54**
29 27.1	♍ 0 6.5	23 56.6	16 10.0	15 11.1	**55**
29 50.1	0 26.3	24 1.8	15 59.3	14 53.8	**56**
♌ 0 14.5	0 46.7	24 7.2	15 48.3	14 35.6	**57**
0 40.4	1 7.9	24 12.6	15 37.0	14 16.5	**58**
1 7.8	1 29.7	24 18.2	15 25.2	13 56.4	**59**
1 37.0	1 52.4	24 24.0	15 13.1	13 35.3	**60**
2 8.2	2 15.8	24 29.8	15 0.6	13 12.9	**61**
2 41.7	2 40.2	24 35.9	14 47.6	12 49.2	**62**
3 17.7	3 5.6	24 42.1	14 34.1	12 24.1	**63**
3 56.6	3 32.0	24 48.5	14 20.1	11 57.4	**64**
4 38.9	3 59.6	24 55.1	14 5.5	11 28.8	**65**
5 25.0	4 28.3	25 1.9	13 50.3	10 58.3	**66**
5	6	Descendant	8	9	S
		♐ 21° 43′ 56″			LAT
17h 24m 0s		MC		261° 0′ 0″	

5ʰ 28ᵐ 0ˢ MC 82° 0′ 0″

♊ 22° 39′ 10″

11	12	Ascendant	2	3	N LAT
° ′	° ′	° ′	° ′	° ′	°
♋20 20.3	♌19 35.0	♍21 17.5	♎23 46.0	♏24 22.0	0
20 58.0	20 28.5	21 36.5	23 12.2	23 48.8	5
21 36.0	21 20.8	21 54.4	22 39.4	23 15.5	10
22 14.9	22 12.7	22 11.6	22 7.1	22 41.7	15
22 22.8	22 23.2	22 15.0	22 0.6	22 34.8	16
22 30.8	22 33.6	22 18.4	21 54.1	22 27.8	17
22 38.9	22 44.1	22 21.7	21 47.7	22 20.8	18
22 47.1	22 54.6	22 25.1	21 41.2	22 13.8	19
22 55.4	23 5.2	22 28.4	21 34.7	22 6.6	20
23 3.8	23 15.8	22 31.7	21 28.2	21 59.4	21
23 12.2	23 26.4	22 35.1	21 21.6	21 52.2	22
23 20.8	23 37.2	22 38.4	21 15.0	21 44.8	23
23 29.5	23 48.0	22 41.7	21 8.4	21 37.3	24
23 38.4	23 58.8	22 45.0	21 1.8	21 29.8	25
23 47.3	24 9.8	22 48.3	20 55.1	21 22.1	26
23 56.4	24 20.9	22 51.7	20 48.4	21 14.3	27
24 5.7	24 32.0	22 55.0	20 41.6	21 6.5	28
24 15.1	24 43.3	22 58.4	20 34.7	20 58.4	29
24 24.7	24 54.7	23 1.7	20 27.8	20 50.3	30
24 34.5	25 6.2	23 5.1	20 20.9	20 42.0	31
24 44.5	25 17.8	23 8.5	20 13.8	20 33.6	32
24 54.8	25 29.6	23 11.9	20 6.7	20 25.0	33
25 5.2	25 41.5	23 15.3	19 59.5	20 16.2	34
25 15.9	25 53.6	23 18.8	19 52.2	20 7.2	35
25 26.8	26 5.9	23 22.3	19 44.8	19 58.1	36
25 38.0	26 18.4	23 25.8	19 37.4	19 48.7	37
25 49.5	26 31.1	23 29.3	19 29.8	19 39.1	38
26 1.4	26 43.9	23 32.9	19 22.1	19 29.3	39
26 13.5	26 57.0	23 36.5	19 14.3	19 19.2	40
26 26.0	27 10.4	23 40.1	19 6.3	19 8.9	41
26 38.9	27 23.9	23 43.8	18 58.2	18 58.2	42
26 52.2	27 37.8	23 47.5	18 50.0	18 47.3	43
27 6.0	27 51.9	23 51.3	18 41.6	18 36.0	44
27 20.2	28 6.4	23 55.1	18 33.1	18 24.4	45
27 34.9	28 21.1	23 59.0	18 24.3	18 12.5	46
27 50.1	28 36.2	24 2.9	18 15.4	18 0.1	47
28 6.0	28 51.7	24 6.9	18 6.3	17 47.3	48
28 22.5	29 7.5	24 11.0	17 57.0	17 34.0	49
28 39.7	29 23.8	24 15.1	17 47.5	17 20.3	50
28 57.6	29 40.4	24 19.3	17 37.7	17 6.0	51
29 16.3	29 57.6	24 23.6	17 27.7	16 51.1	52
29 35.9	♍ 0 15.2	24 28.0	17 17.4	16 35.6	53
29 56.5	0 33.3	24 32.4	17 6.9	16 19.5	54
♌ 0 18.1	0 52.0	24 37.0	16 56.0	16 2.6	55
0 40.9	1 11.3	24 41.6	16 44.8	15 45.0	56
1 4.9	1 31.2	24 46.3	16 33.3	15 26.5	57
1 30.4	1 51.8	24 51.2	16 21.4	15 7.0	58
1 57.4	2 13.1	24 56.2	16 9.1	14 46.5	59
2 26.1	2 35.1	25 1.3	15 56.4	14 25.0	60
2 56.8	2 58.0	25 6.5	15 43.3	14 2.1	61
3 29.7	3 21.7	25 11.9	15 29.6	13 37.9	62
4 5.0	3 46.4	25 17.4	15 15.5	13 12.3	63
4 43.2	4 12.1	25 23.1	15 0.8	12 44.9	64
5 24.5	4 38.9	25 29.0	14 45.5	12 15.7	65
6 9.6	5 6.9	25 35.0	14 29.6	11 44.4	66
5	6	Descendant	8	9	S LAT

♐ 22° 39′ 10″

17ʰ 28ᵐ 0ˢ MC 262° 0′ 0″

5ʰ 32ᵐ 0ˢ MC 83° 0′ 0″

♊ 23° 34′ 22″

11	12	Ascendant	2	3	N LAT
° ′	° ′	° ′	° ′	° ′	°
♋21 16.7	♌20 36.1	♍22 22.6	♎24 49.7	♏25 20.5	0
21 54.1	21 28.5	22 39.3	24 14.5	24 46.9	5
22 31.9	22 19.8	22 55.0	23 40.3	24 13.3	10
23 10.6	23 10.7	23 10.1	23 6.6	23 39.0	15
23 18.5	23 20.9	23 13.1	22 59.9	23 32.1	16
23 26.4	23 31.1	23 16.0	22 53.1	23 25.0	17
23 34.5	23 41.3	23 18.9	22 46.4	23 17.9	18
23 42.6	23 51.6	23 21.9	22 39.6	23 10.8	19
23 50.8	24 2.0	23 24.8	22 32.9	23 3.6	20
23 59.1	24 12.3	23 27.7	22 26.1	22 56.3	21
24 7.5	24 22.8	23 30.6	22 19.2	22 48.9	22
24 16.1	24 33.2	23 33.5	22 12.4	22 41.5	23
24 24.7	24 43.8	23 36.4	22 5.5	22 33.9	24
24 33.5	24 54.4	23 39.4	21 58.6	22 26.2	25
24 42.4	25 5.2	23 42.3	21 51.6	22 18.5	26
24 51.4	25 16.0	23 45.2	21 44.6	22 10.6	27
25 0.6	25 26.9	23 48.1	21 37.5	22 2.6	28
25 10.0	25 37.9	23 51.0	21 30.4	21 54.5	29
25 19.5	25 49.0	23 54.0	21 23.2	21 46.3	30
25 29.2	26 0.2	23 56.9	21 15.9	21 37.9	31
25 39.2	26 11.6	23 59.9	21 8.5	21 29.3	32
25 49.3	26 23.1	24 2.9	21 1.1	21 20.6	33
25 59.6	26 34.8	24 5.9	20 53.6	21 11.7	34
26 10.2	26 46.6	24 8.9	20 46.0	21 2.6	35
26 21.1	26 58.6	24 12.0	20 38.3	20 53.3	36
26 32.2	27 10.8	24 15.0	20 30.5	20 43.8	37
26 43.6	27 23.1	24 18.1	20 22.6	20 34.1	38
26 55.3	27 35.7	24 21.3	20 14.6	20 24.1	39
27 7.3	27 48.5	24 24.4	20 6.4	20 13.9	40
27 19.7	28 1.5	24 27.6	19 58.1	20 3.4	41
27 32.5	28 14.7	24 30.8	19 49.7	19 52.6	42
27 45.7	28 28.3	24 34.1	19 41.1	19 41.5	43
27 59.3	28 42.0	24 37.4	19 32.3	19 30.1	44
28 13.3	28 56.1	24 40.7	19 23.4	19 18.3	45
28 27.9	29 10.5	24 44.1	19 14.3	19 6.1	46
28 43.0	29 25.2	24 47.6	19 5.0	18 53.6	47
28 58.7	29 40.3	24 51.1	18 55.5	18 40.6	48
29 15.0	29 55.7	24 54.6	18 45.7	18 27.1	49
29 31.9	♍ 0 11.6	24 58.2	18 35.8	18 13.1	50
29 49.7	0 27.8	25 1.9	18 25.6	17 58.6	51
♌ 0 8.2	0 44.5	25 5.6	18 15.1	17 43.5	52
0 27.5	1 1.7	25 9.5	18 4.4	17 27.7	53
0 47.8	1 19.3	25 13.4	17 53.3	17 11.3	54
1 9.1	1 37.5	25 17.3	17 42.0	16 54.1	55
1 31.6	1 56.3	25 21.4	17 30.3	16 36.1	56
1 55.3	2 15.7	25 25.5	17 18.2	16 17.3	57
2 20.4	2 35.7	25 29.8	17 5.8	15 57.5	58
2 47.0	2 56.4	25 34.1	16 52.9	15 36.6	59
3 15.3	3 17.9	25 38.6	16 39.7	15 14.6	60
3 45.4	3 40.2	25 43.2	16 25.9	14 51.3	61
4 17.7	4 3.3	25 47.9	16 11.7	14 26.6	62
4 52.4	4 27.3	25 52.7	15 56.9	14 0.3	63
5 29.8	4 52.3	25 57.7	15 41.5	13 32.4	64
6 10.3	5 18.3	26 2.8	15 25.5	13 2.5	65
6 54.4	5 45.5	26 8.1	15 8.9	12 30.5	66
5	6	Descendant	8	9	S LAT

♐ 23° 34′ 22″

17ʰ 32ᵐ 0ˢ MC 263° 0′ 0″

5ʰ 36ᵐ 0ˢ — MC 84° 0′ 0″ — ♊ 24° 29′ 31″

N LAT	11	12	Ascendant	2	3
°	° ′	° ′	° ′	° ′	° ′
0	♋22 13.1	♌21 37.4	♍23 27.9	♎25 53.2	♏26 18.8
5	22 50.4	22 28.8	23 42.2	25 16.6	25 44.9
10	23 27.9	23 18.9	23 55.7	24 41.1	25 10.9
15	24 6.3	24 8.7	24 8.6	24 6.1	24 36.3
16	24 14.2	24 18.7	24 11.1	23 59.1	24 29.2
17	24 22.1	24 28.7	24 13.7	23 52.1	24 22.1
18	24 30.1	24 38.7	24 16.2	23 45.1	24 14.9
19	24 38.2	24 48.7	24 18.7	23 38.0	24 7.7
20	24 46.3	24 58.8	24 21.2	23 31.0	24 0.4
21	24 54.6	25 9.0	24 23.7	23 23.9	23 53.0
22	25 2.9	25 19.2	24 26.2	23 16.8	23 45.5
23	25 11.4	25 29.4	24 28.7	23 9.7	23 38.0
24	25 20.0	25 39.7	24 31.2	23 2.5	23 30.4
25	25 28.7	25 50.1	24 33.7	22 55.3	23 22.6
26	25 37.5	26 0.6	24 36.2	22 48.0	23 14.8
27	25 46.5	26 11.1	24 38.7	22 40.7	23 6.8
28	25 55.6	26 21.8	24 41.2	22 33.4	22 58.7
29	26 4.9	26 32.5	24 43.7	22 25.9	22 50.5
30	26 14.3	26 43.4	24 46.2	22 18.4	22 42.1
31	26 24.0	26 54.4	24 48.8	22 10.9	22 33.6
32	26 33.8	27 5.5	24 51.3	22 3.2	22 24.9
33	26 43.8	27 16.7	24 53.9	21 55.5	22 16.1
34	26 54.1	27 28.1	24 56.5	21 47.7	22 7.1
35	27 4.6	27 39.6	24 59.1	21 39.7	21 57.9
36	27 15.4	27 51.3	25 1.7	21 31.7	21 48.5
37	27 26.4	28 3.2	25 4.3	21 23.6	21 38.8
38	27 37.7	28 15.2	25 7.0	21 15.4	21 29.0
39	27 49.3	28 27.5	25 9.6	21 7.0	21 18.9
40	28 1.2	28 40.0	25 12.3	20 58.5	21 8.5
41	28 13.5	28 52.6	25 15.1	20 49.8	20 57.9
42	28 26.1	29 5.6	25 17.8	20 41.0	20 46.9
43	28 39.1	29 18.7	25 20.6	20 32.1	20 35.7
44	28 52.6	29 32.2	25 23.5	20 23.0	20 24.1
45	29 6.5	29 45.9	25 26.3	20 13.7	20 12.1
46	29 20.9	29 59.9	25 29.2	20 4.2	19 59.8
47	29 35.9	♍ 0 14.3	25 32.2	19 54.5	19 47.0
48	29 51.4	0 29.0	25 35.2	19 44.6	19 33.8
49	♌ 0 7.5	0 44.0	25 38.2	19 34.4	19 20.1
50	0 24.2	0 59.4	25 41.3	19 24.0	19 5.9
51	0 41.7	1 15.2	25 44.5	19 13.4	18 51.1
52	1 0.0	1 31.5	25 47.7	19 2.5	18 35.8
53	1 19.1	1 48.2	25 51.0	18 51.3	18 19.7
54	1 39.2	2 5.4	25 54.3	18 39.7	18 3.0
55	2 0.2	2 23.1	25 57.7	18 27.9	17 45.6
56	2 22.3	2 41.4	26 1.2	18 15.7	17 27.3
57	2 45.7	3 0.2	26 4.7	18 3.1	17 8.0
58	3 10.4	3 19.7	26 8.4	17 50.1	16 47.9
59	3 36.6	3 39.9	26 12.1	17 36.7	16 26.6
60	4 4.4	4 0.7	26 15.9	17 22.9	16 4.1
61	4 34.0	4 22.4	26 19.9	17 8.5	15 40.4
62	5 5.8	4 44.8	26 23.9	16 53.7	15 15.2
63	5 39.8	5 8.2	26 28.0	16 38.2	14 48.4
64	6 16.4	5 32.5	26 32.3	16 22.2	14 19.8
65	6 56.1	5 57.8	26 36.7	16 5.5	13 49.2
66	7 39.2	6 24.2	26 41.2	15 48.1	13 16.4
S LAT	5	6	Descendant	8	9

♐ 24° 29′ 31″ — 17ʰ 36ᵐ 0ˢ — MC 264° 0′ 0″

5ʰ 40ᵐ 0ˢ — MC 85° 0′ 0″ — ♊ 25° 24′ 39″

N LAT	11	12	Ascendant	2	3
°	° ′	° ′	° ′	° ′	° ′
0	♋23 9.7	♌22 38.9	♍24 33.2	♎26 56.5	♏27 17.0
5	23 46.8	23 29.2	24 45.1	26 18.6	26 42.7
10	24 24.0	24 18.2	24 56.3	25 41.8	26 8.4
15	25 2.2	25 6.8	25 7.1	25 5.4	25 33.4
16	25 10.0	25 16.6	25 9.2	24 58.2	25 26.2
17	25 17.9	25 26.3	25 11.4	24 50.9	25 19.0
18	25 25.8	25 36.1	25 13.5	24 43.6	25 11.8
19	25 33.8	25 46.0	25 15.6	24 36.3	25 4.5
20	25 41.9	25 55.8	25 17.6	24 29.0	24 57.1
21	25 50.1	26 5.7	25 19.7	24 21.7	24 49.6
22	25 58.4	26 15.7	25 21.8	24 14.3	24 42.1
23	26 6.8	26 25.7	25 23.9	24 6.9	24 34.4
24	26 15.3	26 35.7	25 26.0	23 59.4	24 26.7
25	26 23.9	26 45.9	25 28.1	23 51.9	24 18.9
26	26 32.7	26 56.1	25 30.1	23 44.4	24 10.9
27	26 41.6	27 6.4	25 32.2	23 36.8	24 2.9
28	26 50.6	27 16.8	25 34.3	23 29.1	23 54.7
29	26 59.8	27 27.3	25 36.4	23 21.4	23 46.4
30	27 9.2	27 37.9	25 38.5	23 13.6	23 37.9
31	27 18.8	27 48.6	25 40.6	23 5.8	23 29.3
32	27 28.5	27 59.4	25 42.8	22 57.8	23 20.5
33	27 38.5	28 10.3	25 44.9	22 49.8	23 11.5
34	27 48.6	28 21.4	25 47.0	22 41.7	23 2.4
35	27 59.0	28 32.7	25 49.2	22 33.4	22 53.1
36	28 9.7	28 44.1	25 51.4	22 25.1	22 43.6
37	28 20.6	28 55.7	25 53.6	22 16.6	22 33.8
38	28 31.8	29 7.4	25 55.8	22 8.1	22 23.8
39	28 43.3	29 19.3	25 58.0	21 59.4	22 13.6
40	28 55.1	29 31.5	26 0.3	21 50.5	22 3.1
41	29 7.2	29 43.9	26 2.5	21 41.5	21 52.3
42	29 19.7	29 56.4	26 4.9	21 32.4	21 41.2
43	29 32.6	♍ 0 9.3	26 7.2	21 23.1	21 29.8
44	29 46.0	0 22.4	26 9.5	21 13.6	21 18.0
45	29 59.7	0 35.7	26 11.9	21 3.9	21 5.9
46	♌ 0 14.0	0 49.4	26 14.4	20 54.0	20 53.3
47	0 28.7	1 3.3	26 16.8	20 43.9	20 40.4
48	0 44.1	1 17.6	26 19.3	20 33.6	20 26.9
49	1 0.0	1 32.3	26 21.9	20 23.1	20 13.0
50	1 16.6	1 47.3	26 24.4	20 12.2	19 58.6
51	1 33.8	2 2.7	26 27.1	20 1.2	19 43.6
52	1 51.9	2 18.5	26 29.7	19 49.8	19 28.0
53	2 10.7	2 34.7	26 32.5	19 38.1	19 11.7
54	2 30.5	2 51.5	26 35.2	19 26.1	18 54.7
55	2 51.2	3 8.7	26 38.1	19 13.8	18 36.9
56	3 13.1	3 26.4	26 41.0	19 1.1	18 18.3
57	3 36.1	3 44.8	26 43.9	18 48.0	17 58.8
58	4 0.4	4 3.7	26 47.0	18 34.5	17 38.2
59	4 26.2	4 23.3	26 50.1	18 20.5	17 16.5
60	4 53.5	4 43.6	26 53.3	18 6.1	16 53.6
61	5 22.7	5 4.6	26 56.5	17 51.1	16 29.4
62	5 53.8	5 26.4	26 59.9	17 35.6	16 3.7
63	6 27.2	5 49.1	27 3.4	17 19.5	15 36.3
64	7 3.1	6 12.7	27 6.9	17 2.8	15 7.1
65	7 41.9	6 37.3	27 10.6	16 45.4	14 35.8
66	8 24.1	7 2.9	27 14.4	16 27.3	14 2.3
S LAT	5	6	Descendant	8	9

♐ 25° 24′ 39″ — 17ʰ 40ᵐ 0ˢ — MC 265° 0′ 0″

5h 44m 0s — MC 86° 0′ 0″ — ♊ 26° 19′ 45″

11 ° ′	12 ° ′	Ascendant ° ′	2 ° ′	3 ° ′	N LAT °
♋24 6.4	♌23 40.6	♍25 38.5	♎27 59.7	♏28 15.0	0
24 43.2	24 29.7	25 48.0	27 20.5	27 40.4	5
25 20.3	25 17.6	25 57.0	26 42.4	27 5.7	10
25 58.2	26 5.1	26 5.7	26 4.7	26 30.3	15
26 5.9	26 14.6	26 7.4	25 57.2	26 23.1	16
26 13.7	26 24.1	26 9.1	25 49.6	26 15.8	17
26 21.6	26 33.7	26 10.7	25 42.1	26 8.5	18
26 29.5	26 43.3	26 12.4	25 34.5	26 1.1	19
26 37.6	26 52.9	26 14.1	25 26.9	25 53.7	20
26 45.7	27 2.6	26 15.8	25 19.3	25 46.1	21
26 53.9	27 12.3	26 17.4	25 11.7	25 38.5	22
27 2.2	27 22.0	26 19.1	25 4.0	25 30.8	23
27 10.7	27 31.8	26 20.8	24 56.3	25 22.9	24
27 19.2	27 41.7	26 22.4	24 48.5	25 15.0	25
27 27.9	27 51.7	26 24.1	24 40.7	25 7.0	26
27 36.8	28 1.7	26 25.8	24 32.8	24 58.8	27
27 45.7	28 11.9	26 27.4	24 24.9	24 50.5	28
27 54.9	28 22.1	26 29.1	24 16.9	24 42.1	29
28 4.2	28 32.4	26 30.8	24 8.8	24 33.6	30
28 13.6	28 42.8	26 32.5	24 0.6	24 24.8	31
28 23.3	28 53.4	26 34.2	23 52.4	24 16.0	32
28 33.2	29 4.0	26 35.9	23 44.1	24 6.9	33
28 43.2	29 14.9	26 37.6	23 35.6	23 57.7	34
28 53.5	29 25.8	26 39.4	23 27.1	23 48.2	35
29 4.1	29 36.9	26 41.1	23 18.4	23 38.6	36
29 14.9	29 48.2	26 42.8	23 9.7	23 28.7	37
29 26.0	29 59.6	26 44.6	23 0.7	23 18.6	38
29 37.3	♍ 0 11.3	26 46.4	22 51.7	23 8.2	39
29 49.0	0 23.1	26 48.2	22 42.5	22 57.5	40
♌ 0 1.0	0 35.1	26 50.0	22 33.2	22 46.6	41
0 13.4	0 47.4	26 51.9	22 23.7	22 35.3	42
0 26.2	0 59.9	26 53.7	22 14.0	22 23.8	43
0 39.4	1 12.6	26 55.6	22 4.2	22 11.8	44
0 53.0	1 25.6	26 57.5	21 54.1	21 59.5	45
1 7.1	1 38.9	26 59.5	21 43.8	21 46.8	46
1 21.7	1 52.5	27 1.4	21 33.4	21 33.7	47
1 36.8	2 6.4	27 3.4	21 22.6	21 20.1	48
1 52.5	2 20.6	27 5.5	21 11.7	21 5.9	49
2 8.9	2 35.2	27 7.5	21 0.4	20 51.3	50
2 25.9	2 50.1	27 9.6	20 48.9	20 36.0	51
2 43.7	3 5.5	27 11.8	20 37.1	20 20.2	52
3 2.4	3 21.3	27 14.0	20 25.0	20 3.6	53
3 21.9	3 37.6	27 16.2	20 12.5	19 46.4	54
3 42.3	3 54.3	27 18.5	19 59.7	19 28.3	55
4 3.8	4 11.6	27 20.8	19 46.5	19 9.3	56
4 26.5	4 29.4	27 23.2	19 32.9	18 49.4	57
4 50.4	4 47.8	27 25.6	19 18.8	18 28.5	58
5 15.8	5 6.8	27 28.1	19 4.3	18 6.4	59
5 42.7	5 26.5	27 30.6	18 49.3	17 43.1	60
6 11.3	5 46.9	27 33.2	18 33.7	17 18.4	61
6 41.9	6 8.1	27 35.9	18 17.6	16 52.1	62
7 14.7	6 30.1	27 38.7	18 0.8	16 24.2	63
7 49.9	6 53.0	27 41.5	17 43.4	15 54.4	64
8 27.9	7 16.8	27 44.5	17 25.3	15 22.4	65
9 9.1	7 41.7	27 47.5	17 6.5	14 48.0	66
5	6	Descendant	8	9	S LAT

♐ 26° 19′ 45″ — 17h 44m 0s — MC 266° 0′ 0″

5h 48m 0s — MC 87° 0′ 0″ — ♊ 27° 14′ 50″

N LAT °	11 ° ′	12 ° ′	Ascendant ° ′	2 ° ′	3 ° ′
0	♋25 3.3	♌24 42.5	♍26 43.8	♎29 2.8	♏29 12.8
5	25 39.8	25 30.4	26 51.0	28 22.2	28 37.9
10	26 16.6	26 17.2	26 57.8	27 42.8	28 2.9
15	26 54.2	27 3.5	27 4.2	27 3.9	27 27.2
16	27 1.9	27 12.8	27 5.5	26 56.1	27 19.9
17	27 9.6	27 22.1	27 6.8	26 48.3	27 12.5
18	27 17.5	27 31.4	27 8.1	26 40.5	27 5.1
19	27 25.3	27 40.7	27 9.3	26 32.6	26 57.7
20	27 33.3	27 50.1	27 10.6	26 24.8	26 50.1
21	27 41.4	27 59.5	27 11.8	26 16.9	26 42.5
22	27 49.5	28 9.0	27 13.1	26 9.0	26 34.8
23	27 57.8	28 18.5	27 14.3	26 1.0	26 27.0
24	28 6.2	28 28.0	27 15.6	25 53.1	26 19.1
25	28 14.6	28 37.7	27 16.8	25 45.0	26 11.1
26	28 23.3	28 47.4	27 18.1	25 36.9	26 2.9
27	28 32.0	28 57.1	27 19.3	25 28.8	25 54.7
28	28 40.9	29 7.0	27 20.6	25 20.5	25 46.3
29	28 50.0	29 17.0	27 21.8	25 12.2	25 37.8
30	28 59.2	29 27.0	27 23.1	25 3.9	25 29.1
31	29 8.6	29 37.2	27 24.4	24 55.4	25 20.3
32	29 18.1	29 47.4	27 25.6	24 46.9	25 11.3
33	29 27.9	29 57.8	27 26.9	24 38.3	25 2.2
34	29 37.9	♍ 0 8.3	27 28.2	24 29.5	24 52.8
35	29 48.1	0 19.0	27 29.5	24 20.7	24 43.2
36	29 58.5	0 29.8	27 30.8	24 11.7	24 33.5
37	♌ 0 9.2	0 40.8	27 32.1	24 2.6	24 23.5
38	0 20.2	0 51.9	27 33.5	23 53.4	24 13.2
39	0 31.4	1 3.2	27 34.8	23 44.0	24 2.7
40	0 43.0	1 14.7	27 36.2	23 34.5	23 51.9
41	0 54.9	1 26.4	27 37.5	23 24.8	23 40.8
42	1 7.2	1 38.3	27 38.9	23 15.0	23 29.5
43	1 19.8	1 50.5	27 40.3	23 4.9	23 17.7
44	1 32.8	2 2.9	27 41.7	22 54.7	23 5.6
45	1 46.3	2 15.5	27 43.2	22 44.3	22 53.1
46	2 0.2	2 28.4	27 44.6	22 33.6	22 40.3
47	2 14.6	2 41.6	27 46.1	22 22.8	22 26.9
48	2 29.6	2 55.1	27 47.6	22 11.6	22 13.1
49	2 45.1	3 8.9	27 49.1	22 0.3	21 58.8
50	3 1.3	3 23.1	27 50.7	21 48.6	21 43.9
51	3 18.1	3 37.6	27 52.2	21 36.7	21 28.4
52	3 35.6	3 52.6	27 53.8	21 24.4	21 12.3
53	3 54.0	4 7.9	27 55.5	21 11.8	20 55.5
54	4 13.2	4 23.7	27 57.1	20 58.9	20 38.0
55	4 33.4	4 39.9	27 58.8	20 45.5	20 19.6
56	4 54.6	4 56.7	28 0.6	20 31.8	20 0.3
57	5 16.9	5 14.0	28 2.4	20 17.7	19 40.1
58	5 40.5	5 31.8	28 4.2	20 3.1	19 18.7
59	6 5.4	5 50.3	28 6.0	19 48.0	18 56.3
60	6 31.9	6 9.4	28 8.0	19 32.4	18 32.5
61	7 0.0	6 29.2	28 9.9	19 16.2	18 7.3
62	7 30.0	6 49.7	28 11.9	18 59.5	17 40.5
63	8 2.1	7 11.1	28 14.0	18 42.1	17 12.0
64	8 36.6	7 33.3	28 16.1	18 24.0	16 41.5
65	9 13.8	7 56.4	28 18.3	18 5.2	16 8.9
66	9 54.1	8 20.6	28 20.6	17 45.6	15 33.7
S LAT	5	6	Descendant	8	9

♐ 27° 14′ 50″ — 17h 48m 0s — MC 267° 0′ 0″

5h 52m 0s — MC 88° 0′ 0″ — ♊ 28° 9′ 54″

N LAT	11	12	Ascendant	2	3
	° ′	° ′	° ′	° ′	° ′
0	♋26 0.2	♌25 44.5	♍27 49.2	♏ 0 5.7	♐ 0 10.5
5	26 36.6	26 31.3	27 54.0	♎29 23.8	♏29 35.3
10	27 13.1	27 16.8	27 58.5	28 43.1	28 59.9
15	27 50.4	28 2.0	28 2.8	28 2.9	28 23.9
16	27 58.0	28 11.0	28 3.7	27 54.9	28 16.5
17	28 5.7	28 20.1	28 4.5	27 46.8	28 9.1
18	28 13.4	28 29.1	28 5.4	27 38.7	28 1.6
19	28 21.2	28 38.2	28 6.2	27 30.7	27 54.1
20	28 29.2	28 47.4	28 7.0	27 22.5	27 46.5
21	28 37.1	28 56.5	28 7.9	27 14.4	27 38.8
22	28 45.2	29 5.7	28 8.7	27 6.2	27 31.0
23	28 53.4	29 15.0	28 9.5	26 58.0	27 23.1
24	29 1.7	29 24.3	28 10.4	26 49.7	27 15.1
25	29 10.1	29 33.7	28 11.2	26 41.4	27 7.0
26	29 18.7	29 43.1	28 12.0	26 33.1	26 58.8
27	29 27.3	29 52.6	28 12.9	26 24.6	26 50.5
28	29 36.2	♍ 0 2.2	28 13.7	26 16.1	26 42.0
29	29 45.1	0 11.9	28 14.6	26 7.6	26 33.4
30	29 54.3	0 21.7	28 15.4	25 58.9	26 24.6
31	♌ 0 3.6	0 31.6	28 16.2	25 50.2	26 15.7
32	0 13.0	0 41.5	28 17.1	25 41.3	26 6.6
33	0 22.7	0 51.7	28 17.9	25 32.4	25 57.3
34	0 32.6	1 1.9	28 18.8	25 23.3	25 47.9
35	0 42.7	1 12.2	28 19.7	25 14.2	25 38.2
36	0 53.0	1 22.8	28 20.5	25 4.9	25 28.3
37	1 3.6	1 33.4	28 21.4	24 55.5	25 18.2
38	1 14.5	1 44.2	28 22.3	24 46.0	25 7.8
39	1 25.6	1 55.2	28 23.2	24 36.3	24 57.2
40	1 37.0	2 6.4	28 24.1	24 26.4	24 46.3
41	1 48.8	2 17.8	28 25.0	24 16.4	24 35.0
42	2 0.9	2 29.3	28 25.9	24 6.2	24 23.5
43	2 13.4	2 41.1	28 26.9	23 55.8	24 11.6
44	2 26.3	2 53.2	28 27.8	23 45.2	23 59.4
45	2 39.6	3 5.4	28 28.8	23 34.4	23 46.7
46	2 53.3	3 18.0	28 29.7	23 23.4	23 33.6
47	3 7.6	3 30.8	28 30.7	23 12.1	23 20.1
48	3 22.3	3 43.9	28 31.7	23 0.6	23 6.1
49	3 37.7	3 57.3	28 32.7	22 48.8	22 51.6
50	3 53.6	4 11.1	28 33.8	22 36.7	22 36.5
51	4 10.2	4 25.2	28 34.8	22 24.4	22 20.8
52	4 27.6	4 39.7	28 35.9	22 11.6	22 4.4
53	4 45.7	4 54.5	28 37.0	21 58.6	21 47.4
54	5 4.6	5 9.9	28 38.1	21 45.2	21 29.5
55	5 24.5	5 25.6	28 39.2	21 31.4	21 10.8
56	5 45.4	5 41.9	28 40.4	21 17.2	20 51.2
57	6 7.3	5 58.6	28 41.6	21 2.5	20 30.6
58	6 30.5	6 15.9	28 42.8	20 47.3	20 9.0
59	6 55.1	6 33.8	28 44.0	20 31.7	19 46.1
60	7 21.1	6 52.3	28 45.3	20 15.5	19 21.9
61	7 48.7	7 11.5	28 46.6	19 58.7	18 56.2
62	8 18.2	7 31.5	28 48.0	19 41.4	18 28.9
63	8 49.7	7 52.1	28 49.3	19 23.3	17 59.8
64	9 23.5	8 13.7	28 50.8	19 4.5	17 28.7
65	9 59.9	8 36.1	28 52.2	18 45.0	16 55.3
66	10 39.3	8 59.4	28 53.7	18 24.7	16 19.3
S LAT	5	6	Descendant	8	9

♐ 28° 9′ 54″ — 17h 52m 0s — MC 268° 0′ 0″

5h 56m 0s — MC 89° 0′ 0″ — ♊ 29° 4′ 57″

N LAT	11	12	Ascendant	2	3
	° ′	° ′	° ′	° ′	° ′
0	♋26 57.4	♌26 46.7	♍28 54.6	♏ 1 8.4	♐ 1 8.0
5	27 33.4	27 32.3	28 57.0	0 25.3	0 32.5
10	28 9.6	28 16.7	28 59.2	♎29 43.3	♏29 56.9
15	28 46.6	29 0.6	29 1.4	29 1.8	29 20.5
16	28 54.2	29 9.4	29 1.8	28 53.6	29 13.1
17	29 1.8	29 18.2	29 2.3	28 45.2	29 5.6
18	29 9.5	29 27.0	29 2.7	28 36.9	28 58.0
19	29 17.3	29 35.9	29 3.1	28 28.6	28 50.4
20	29 25.1	29 44.7	29 3.5	28 20.2	28 42.7
21	29 33.0	29 53.7	29 3.9	28 11.8	28 34.9
22	29 41.0	♍ 0 2.6	29 4.4	28 3.4	28 27.1
23	29 49.2	0 11.6	29 4.8	27 54.9	28 19.1
24	29 57.4	0 20.7	29 5.2	27 46.4	28 11.0
25	♌ 0 5.7	0 29.8	29 5.6	27 37.8	28 2.9
26	0 14.2	0 39.0	29 6.0	27 29.1	27 54.6
27	0 22.8	0 48.2	29 6.4	27 20.4	27 46.1
28	0 31.5	0 57.5	29 6.9	27 11.6	27 37.6
29	0 40.4	1 6.9	29 7.3	27 2.8	27 28.9
30	0 49.4	1 16.4	29 7.7	26 53.9	27 20.0
31	0 58.6	1 26.0	29 8.1	26 44.8	27 11.0
32	1 8.0	1 35.7	29 8.5	26 35.7	27 1.8
33	1 17.6	1 45.5	29 9.0	26 26.5	26 52.4
34	1 27.4	1 55.5	29 9.4	26 17.1	26 42.9
35	1 37.4	2 5.6	29 9.8	26 7.7	26 33.1
36	1 47.6	2 15.8	29 10.3	25 58.1	26 23.1
37	1 58.1	2 26.1	29 10.7	25 48.4	26 12.8
38	2 8.8	2 36.6	29 11.2	25 38.5	26 2.3
39	2 19.8	2 47.3	29 11.6	25 28.5	25 51.6
40	2 31.1	2 58.1	29 12.0	25 18.3	25 40.5
41	2 42.8	3 9.2	29 12.5	25 7.9	25 29.2
42	2 54.7	3 20.4	29 13.0	24 57.4	25 17.5
43	3 7.1	3 31.8	29 13.4	24 46.6	25 5.4
44	3 19.8	3 43.5	29 13.9	24 35.7	24 53.0
45	3 32.9	3 55.4	29 14.4	24 24.5	24 40.2
46	3 46.5	4 7.6	29 14.9	24 13.1	24 27.0
47	4 0.6	4 20.0	29 15.4	24 1.4	24 13.3
48	4 15.2	4 32.7	29 15.9	23 49.5	23 59.1
49	4 30.3	4 45.7	29 16.4	23 37.3	23 44.3
50	4 46.1	4 59.1	29 16.9	23 24.8	23 29.0
51	5 2.4	5 12.8	29 17.4	23 12.0	23 13.1
52	5 19.5	5 26.8	29 17.9	22 58.9	22 56.5
53	5 37.4	5 41.2	29 18.5	22 45.4	22 39.1
54	5 56.0	5 56.1	29 19.0	22 31.5	22 21.0
55	6 15.6	6 11.3	29 19.6	22 17.2	22 2.0
56	6 36.2	6 27.1	29 20.2	22 2.4	21 42.1
57	6 57.8	6 43.3	29 20.8	21 47.3	21 21.2
58	7 20.6	7 0.1	29 21.4	21 31.6	20 59.1
59	7 44.7	7 17.4	29 22.0	21 15.4	20 35.8
60	8 10.3	7 35.3	29 22.7	20 58.6	20 11.2
61	8 37.4	7 53.9	29 23.3	20 41.2	19 45.0
62	9 6.3	8 13.2	29 24.0	20 23.2	19 17.2
63	9 37.2	8 33.2	29 24.7	20 4.5	18 47.5
64	10 10.3	8 54.0	29 25.4	19 45.1	18 15.7
65	10 46.0	9 15.7	29 26.1	19 24.8	17 41.6
66	11 24.5	9 38.3	29 26.9	19 3.7	17 4.8
S LAT	5	6	Descendant	8	9

♐ 29° 4′ 57″ — 17h 56m 0s — MC 269° 0′ 0″

6ʰ 0ᵐ 0ˢ MC 90° 0′ 0″

♋ 0° 0′ 0″

N LAT	11	12	Ascendant	2	3
°	° ′	° ′	° ′	° ′	° ′
0	♋27 54.6	♌27 49.1	♎ 0 0.0	♏ 2 10.9	♐ 2 5.4
5	28 30.4	28 33.4	0 0.0	1 26.6	1 29.6
10	29 6.3	29 16.6	0 0.0	0 43.4	0 53.7
15	29 43.0	29 59.3	0 0.0	0 0.7	0 17.0
16	29 50.5	♍ 0 7.9	0 0.0	♎29 52.1	0 9.5
17	29 58.1	0 16.4	0 0.0	29 43.6	0 1.9
18	♌ 0 5.7	0 25.0	0 0.0	29 35.0	♏29 54.3
19	0 13.4	0 33.6	0 0.0	29 26.4	29 46.6
20	0 21.1	0 42.2	0 0.0	29 17.8	29 38.9
21	0 29.0	0 50.9	0 0.0	29 9.1	29 31.0
22	0 36.9	0 59.6	0 0.0	29 0.4	29 23.1
23	0 45.0	1 8.3	0 0.0	28 51.7	29 15.0
24	0 53.1	1 17.1	0 0.0	28 42.9	29 6.9
25	1 1.4	1 26.0	0 0.0	28 34.0	28 58.6
26	1 9.8	1 34.9	0 0.0	28 25.1	28 50.2
27	1 18.3	1 43.9	0 0.0	28 16.1	28 41.7
28	1 26.9	1 52.9	0 0.0	28 7.1	28 33.1
29	1 35.7	2 2.0	0 0.0	27 58.0	28 24.3
30	1 44.6	2 11.3	0 0.0	27 48.7	28 15.4
31	1 53.8	2 20.6	0 0.0	27 39.4	28 6.2
32	2 3.1	2 30.0	0 0.0	27 30.0	27 56.9
33	2 12.5	2 39.5	0 0.0	27 20.5	27 47.5
34	2 22.2	2 49.1	0 0.0	27 10.9	27 37.8
35	2 32.1	2 58.9	0 0.0	27 1.1	27 27.9
36	2 42.2	3 8.8	0 0.0	26 51.2	27 17.8
37	2 52.6	3 18.8	0 0.0	26 41.2	27 7.4
38	3 3.2	3 29.0	0 0.0	26 31.0	26 56.8
39	3 14.1	3 39.4	0 0.0	26 20.6	26 45.9
40	3 25.3	3 49.9	0 0.0	26 10.1	26 34.7
41	3 36.8	4 0.6	0 0.0	25 59.4	26 23.2
42	3 48.6	4 11.5	0 0.0	25 48.5	26 11.4
43	4 0.8	4 22.6	0 0.0	25 37.4	25 59.2
44	4 13.4	4 33.9	0 0.0	25 26.1	25 46.6
45	4 26.3	4 45.4	0 0.0	25 14.6	25 33.7
46	4 39.8	4 57.2	0 0.0	25 2.8	25 20.2
47	4 53.6	5 9.3	0 0.0	24 50.7	25 6.4
48	5 8.0	5 21.6	0 0.0	24 38.4	24 52.0
49	5 23.0	5 34.2	0 0.0	24 25.8	24 37.0
50	5 38.5	5 47.1	0 0.0	24 12.9	24 21.5
51	5 54.7	6 0.4	0 0.0	23 59.6	24 5.3
52	6 11.5	6 13.9	0 0.0	23 46.1	23 48.5
53	6 29.1	6 27.9	0 0.0	23 32.1	23 30.9
54	6 47.5	6 42.3	0 0.0	23 17.7	23 12.5
55	7 6.8	6 57.1	0 0.0	23 2.9	22 53.2
56	7 27.0	7 12.3	0 0.0	22 47.7	22 33.0
57	7 48.3	7 28.0	0 0.0	22 32.0	22 11.7
58	8 10.7	7 44.2	0 0.0	22 15.8	21 49.3
59	8 34.4	8 1.0	0 0.0	21 59.0	21 25.6
60	8 59.5	8 18.4	0 0.0	21 41.6	21 0.5
61	9 26.2	8 36.3	0 0.0	21 23.7	20 33.8
62	9 54.5	8 55.0	0 0.0	21 5.0	20 5.5
63	10 24.8	9 14.3	0 0.0	20 45.7	19 35.2
64	10 57.3	9 34.5	0 0.0	20 25.5	19 2.7
65	11 32.2	9 55.4	0 0.0	20 4.6	18 27.8
66	12 9.8	10 17.3	0 0.0	19 42.7	17 50.2
S LAT	5	6	Descendant	8	9

♑ 0° 0′ 0″

18ʰ 0ᵐ 0ˢ MC 270° 0′ 0″

6ʰ 4ᵐ 0ˢ MC 91° 0′ 0″

♋ 0° 55′ 3″

N LAT	11	12	Ascendant	2	3
°	° ′	° ′	° ′	° ′	° ′
0	♋28 52.0	♌28 51.6	♎ 1 5.4	♏ 3 13.3	♐ 3 2.6
5	29 27.5	29 34.7	1 3.0	2 27.7	2 26.6
10	♌ 0 3.1	♍ 0 16.7	1 0.8	1 43.3	1 50.4
15	0 39.5	0 58.2	0 58.6	0 59.4	1 13.4
16	0 46.9	1 6.4	0 58.2	0 50.6	1 5.8
17	0 54.4	1 14.8	0 57.7	0 41.8	0 58.2
18	1 2.0	1 23.1	0 57.3	0 33.0	0 50.5
19	1 9.6	1 31.4	0 56.9	0 24.1	0 42.7
20	1 17.3	1 39.8	0 56.5	0 15.3	0 34.9
21	1 25.1	1 48.2	0 56.1	0 6.3	0 27.0
22	1 32.9	1 56.6	0 55.6	♎29 57.4	0 19.0
23	1 40.9	2 5.1	0 55.2	29 48.4	0 10.8
24	1 49.0	2 13.6	0 54.8	29 39.3	0 2.6
25	1 57.1	2 22.2	0 54.4	29 30.2	♏29 54.3
26	2 5.4	2 30.9	0 54.0	29 21.0	29 45.8
27	2 13.9	2 39.6	0 53.6	29 11.8	29 37.2
28	2 22.4	2 48.4	0 53.1	29 2.5	29 28.5
29	2 31.1	2 57.2	0 52.7	28 53.1	29 19.6
30	2 40.0	3 6.1	0 52.3	28 43.6	29 10.6
31	2 49.0	3 15.2	0 51.9	28 34.0	29 1.4
32	2 58.2	3 24.3	0 51.5	28 24.3	28 52.0
33	3 7.6	3 33.5	0 51.0	28 14.5	28 42.4
34	3 17.1	3 42.9	0 50.6	28 4.5	28 32.6
35	3 26.9	3 52.3	0 50.2	27 54.4	28 22.6
36	3 36.9	4 1.9	0 49.7	27 44.2	28 12.4
37	3 47.2	4 11.6	0 49.3	27 33.9	28 1.9
38	3 57.7	4 21.5	0 48.8	27 23.4	27 51.2
39	4 8.4	4 31.5	0 48.4	27 12.7	27 40.2
40	4 19.5	4 41.7	0 48.0	27 1.9	27 28.9
41	4 30.8	4 52.1	0 47.5	26 50.8	27 17.2
42	4 42.5	5 2.6	0 47.0	26 39.6	27 5.3
43	4 54.6	5 13.4	0 46.6	26 28.2	26 52.9
44	5 7.0	5 24.3	0 46.1	26 16.5	26 40.2
45	5 19.8	5 35.5	0 45.6	26 4.6	26 27.1
46	5 33.0	5 46.9	0 45.1	25 52.4	26 13.5
47	5 46.7	5 58.6	0 44.6	25 40.0	25 59.4
48	6 0.9	6 10.5	0 44.1	25 27.3	25 44.8
49	6 15.7	6 22.7	0 43.6	25 14.3	25 29.7
50	6 31.0	6 35.2	0 43.1	25 0.9	25 13.9
51	6 46.9	6 48.0	0 42.6	24 47.2	24 57.6
52	7 3.5	7 1.1	0 42.1	24 33.2	24 40.5
53	7 20.9	7 14.6	0 41.5	24 18.8	24 22.6
54	7 39.0	7 28.5	0 41.0	24 3.9	24 4.0
55	7 58.0	7 42.8	0 40.4	23 48.7	23 44.4
56	8 17.9	7 57.6	0 39.8	23 32.9	23 23.8
57	8 38.8	8 12.7	0 39.2	23 16.7	23 2.2
58	9 0.9	8 28.4	0 38.6	22 59.9	22 39.4
59	9 24.2	8 44.6	0 38.0	22 42.6	22 15.3
60	9 48.8	9 1.4	0 37.3	22 24.7	21 49.7
61	10 15.0	9 18.8	0 36.7	22 6.1	21 22.6
62	10 42.8	9 36.8	0 36.0	21 46.8	20 53.7
63	11 12.5	9 55.5	0 35.3	21 26.8	20 22.8
64	11 44.3	10 14.9	0 34.6	21 6.0	19 49.7
65	12 18.4	10 35.2	0 33.9	20 44.3	19 14.0
66	12 55.2	10 56.3	0 33.1	20 21.7	18 35.5
S LAT	5	6	Descendant	8	9

♑ 0° 55′ 3″

18ʰ 4ᵐ 0ˢ MC 271° 0′ 0″

$6^h\ 8^m\ 0^s$ MC 92° 0′ 0″

♋ 1° 50′ 6″

N LAT	11	12	Ascendant	2	3
°	° ′	° ′	° ′	° ′	° ′
0	♋29 49.5	♌29 54.3	♎ 2 10.8	♏ 4 15.5	♐ 3 59.8
5	♌ 0 24.7	♍ 0 36.2	2 6.0	3 28.7	3 23.4
10	1 0.1	1 16.9	2 1.5	2 43.2	2 46.9
15	1 36.1	1 57.1	1 57.2	1 58.0	2 9.6
16	1 43.5	2 5.1	1 56.3	1 49.0	2 2.0
17	1 50.9	2 13.2	1 55.5	1 39.9	1 54.3
18	1 58.4	2 21.3	1 54.6	1 30.9	1 46.6
19	2 5.9	2 29.3	1 53.8	1 21.8	1 38.8
20	2 13.5	2 37.5	1 53.0	1 12.6	1 30.8
21	2 21.2	2 45.6	1 52.1	1 3.5	1 22.9
22	2 29.0	2 53.8	1 51.3	0 54.3	1 14.8
23	2 36.9	3 2.0	1 50.5	0 45.0	1 6.6
24	2 44.9	3 10.3	1 49.6	0 35.7	0 58.3
25	2 53.0	3 18.6	1 48.8	0 26.3	0 49.9
26	3 1.2	3 26.9	1 48.0	0 16.9	0 41.3
27	3 9.5	3 35.4	1 47.1	0 7.4	0 32.7
28	3 18.0	3 43.9	1 46.3	♎29 57.8	0 23.8
29	3 26.6	3 52.4	1 45.4	29 48.1	0 14.9
30	3 35.4	4 1.1	1 44.6	29 38.3	0 5.7
31	3 44.3	4 9.8	1 43.8	29 28.4	♏29 56.4
32	3 53.4	4 18.7	1 42.9	29 18.5	29 47.0
33	4 2.7	4 27.6	1 42.1	29 8.3	29 37.3
34	4 12.1	4 36.7	1 41.2	28 58.1	29 27.4
35	4 21.8	4 45.8	1 40.3	28 47.8	29 17.3
36	4 31.7	4 55.1	1 39.5	28 37.2	29 7.0
37	4 41.8	5 4.5	1 38.6	28 26.6	28 56.4
38	4 52.2	5 14.0	1 37.7	28 15.8	28 45.5
39	5 2.8	5 23.7	1 36.8	28 4.8	28 34.4
40	5 13.7	5 33.6	1 35.9	27 53.6	28 23.0
41	5 25.0	5 43.6	1 35.0	27 42.2	28 11.2
42	5 36.5	5 53.8	1 34.1	27 30.7	27 59.1
43	5 48.4	6 4.2	1 33.1	27 18.9	27 46.6
44	6 0.6	6 14.8	1 32.2	27 6.8	27 33.7
45	6 13.3	6 25.6	1 31.2	26 54.6	27 20.4
46	6 26.4	6 36.6	1 30.3	26 42.0	27 6.7
47	6 39.9	6 47.9	1 29.3	26 29.2	26 52.4
48	6 53.9	6 59.4	1 28.3	26 16.1	26 37.7
49	7 8.4	7 11.2	1 27.3	26 2.7	26 22.3
50	7 23.5	7 23.3	1 26.2	25 48.9	26 6.4
51	7 39.2	7 35.6	1 25.2	25 34.8	25 49.8
52	7 55.6	7 48.4	1 24.1	25 20.3	25 32.4
53	8 12.6	8 1.4	1 23.0	25 5.5	25 14.3
54	8 30.5	8 14.8	1 21.9	24 50.1	24 55.4
55	8 49.2	8 28.6	1 20.8	24 34.4	24 35.5
56	9 8.8	8 42.8	1 19.6	24 18.1	24 14.6
57	9 29.4	8 57.5	1 18.4	24 1.4	23 52.7
58	9 51.0	9 12.7	1 17.2	23 44.1	23 29.5
59	10 13.9	9 28.3	1 16.0	23 26.2	23 4.9
60	10 38.1	9 44.5	1 14.7	23 7.7	22 38.9
61	11 3.8	10 1.3	1 13.4	22 48.5	22 11.3
62	11 31.1	10 18.6	1 12.0	22 28.5	21 41.8
63	12 0.2	10 36.7	1 10.7	22 7.9	21 10.3
64	12 31.3	10 55.5	1 9.2	21 46.3	20 36.5
65	13 4.7	11 15.0	1 7.8	21 23.9	20 0.1
66	13 40.7	11 35.3	1 6.3	21 0.6	19 20.7
S LAT	5	6	Descendant	8	9

♑ 1° 50′ 6″

$18^h\ 8^m\ 0^s$ MC 272° 0′ 0″

$6^h\ 12^m\ 0^s$ MC 93° 0′ 0″

♋ 2° 45′ 10″

N LAT	11	12	Ascendant	2	3
°	° ′	° ′	° ′	° ′	° ′
0	♌ 0 47.2	♍ 0 57.2	♎ 3 16.2	♏ 5 17.5	♐ 4 56.7
5	1 22.1	1 37.8	3 9.0	4 29.6	4 20.2
10	1 57.1	2 17.2	3 2.2	3 42.8	3 43.4
15	2 32.8	2 56.1	2 55.8	2 56.5	3 5.8
16	2 40.1	3 3.9	2 54.5	2 47.2	2 58.1
17	2 47.5	3 11.7	2 53.2	2 37.9	2 50.4
18	2 54.9	3 19.5	2 51.9	2 28.6	2 42.5
19	3 2.3	3 27.4	2 50.7	2 19.3	2 34.7
20	3 9.9	3 35.2	2 49.4	2 9.9	2 26.7
21	3 17.5	3 43.1	2 48.2	2 0.5	2 18.6
22	3 25.2	3 51.0	2 46.9	1 51.0	2 10.5
23	3 33.0	3 59.0	2 45.7	1 41.5	2 2.2
24	3 40.9	4 6.9	2 44.4	1 32.0	1 53.8
25	3 48.9	4 15.0	2 43.2	1 22.3	1 45.4
26	3 57.1	4 23.1	2 41.9	1 12.6	1 36.7
27	4 5.3	4 31.2	2 40.7	1 2.9	1 28.0
28	4 13.7	4 39.5	2 39.4	0 53.0	1 19.1
29	4 22.2	4 47.8	2 38.2	0 43.0	1 10.0
30	4 30.9	4 56.1	2 36.9	0 33.0	1 0.8
31	4 39.7	5 4.6	2 35.6	0 22.8	0 51.4
32	4 48.7	5 13.1	2 34.4	0 12.6	0 41.9
33	4 57.8	5 21.7	2 33.1	0 2.2	0 32.1
34	5 7.2	5 30.5	2 31.8	♎29 51.7	0 22.1
35	5 16.8	5 39.3	2 30.5	29 41.0	0 11.9
36	5 26.5	5 48.3	2 29.2	29 30.2	0 1.5
37	5 36.5	5 57.4	2 27.9	29 19.2	♏29 50.8
38	5 46.8	6 6.6	2 26.5	29 8.1	29 39.8
39	5 57.3	6 16.0	2 25.2	28 56.8	29 28.6
40	6 8.1	6 25.5	2 23.8	28 45.3	29 17.0
41	6 19.2	6 35.2	2 22.5	28 33.6	29 5.1
42	6 30.5	6 45.0	2 21.1	28 21.7	28 52.8
43	6 42.3	6 55.1	2 19.7	28 9.5	28 40.2
44	6 54.4	7 5.3	2 18.3	27 57.1	28 27.2
45	7 6.9	7 15.7	2 16.8	27 44.5	28 13.7
46	7 19.7	7 26.4	2 15.4	27 31.6	27 59.8
47	7 33.1	7 37.2	2 13.9	27 18.4	27 45.4
48	7 46.9	7 48.4	2 12.4	27 4.9	27 30.4
49	8 1.2	7 59.7	2 10.9	26 51.1	27 14.9
50	8 16.1	8 11.4	2 9.3	26 36.9	26 58.7
51	8 31.6	8 23.3	2 7.8	26 22.4	26 41.9
52	8 47.7	8 35.6	2 6.2	26 7.4	26 24.4
53	9 4.5	8 48.2	2 4.5	25 52.1	26 6.0
54	9 22.0	9 1.1	2 2.9	25 36.3	25 46.8
55	9 40.4	9 14.5	2 1.2	25 20.1	25 26.6
56	9 59.7	9 28.2	1 59.4	25 3.3	25 5.4
57	10 19.9	9 42.3	1 57.6	24 46.0	24 43.1
58	10 41.3	9 56.9	1 55.8	24 28.2	24 19.5
59	11 3.7	10 12.0	1 54.0	24 9.7	23 54.6
60	11 27.5	10 27.6	1 52.0	23 50.6	23 28.1
61	11 52.7	10 43.8	1 50.1	23 30.8	23 0.0
62	12 19.5	11 0.5	1 48.1	23 10.3	22 30.0
63	12 48.0	11 17.9	1 46.0	22 48.9	21 57.9
64	13 18.5	11 36.0	1 43.9	22 26.7	21 23.4
65	13 51.1	11 54.8	1 41.7	22 3.6	20 46.2
66	14 26.3	12 14.4	1 39.4	21 39.4	20 5.9
S LAT	5	6	Descendant	8	9

♑ 2° 45′ 10″

$18^h\ 12^m\ 0^s$ MC 273° 0′ 0″

$6^h\ 16^m\ 0^s$ MC 94° 0′ 0″

♋ 3° 40′ 15″

N LAT °	11 ° ′	12 ° ′	Ascendant ° ′	2 ° ′	3 ° ′
0	♌ 1 45.0	♍ 2 0.3	♎ 4 21.5	♏ 6 19.4	♐ 5 53.6
5	2 19.6	2 39.5	4 12.0	5 30.3	5 16.8
10	2 54.3	3 17.6	4 3.0	4 42.4	4 39.7
15	3 29.7	3 55.3	3 54.3	3 54.9	4 1.8
16	3 36.9	4 2.8	3 52.6	3 45.4	3 54.1
17	3 44.2	4 10.4	3 50.9	3 35.9	3 46.3
18	3 51.5	4 17.9	3 49.3	3 26.3	3 38.4
19	3 58.9	4 25.5	3 47.6	3 16.7	3 30.5
20	4 6.3	4 33.1	3 45.9	3 7.1	3 22.4
21	4 13.9	4 40.7	3 44.2	2 57.4	3 14.3
22	4 21.5	4 48.3	3 42.6	2 47.7	3 6.1
23	4 29.2	4 56.0	3 40.9	2 38.0	2 57.8
24	4 37.1	5 3.7	3 39.2	2 28.2	2 49.3
25	4 45.0	5 11.5	3 37.6	2 18.3	2 40.8
26	4 53.0	5 19.3	3 35.9	2 8.3	2 32.1
27	5 1.2	5 27.2	3 34.2	1 58.3	2 23.2
28	5 9.5	5 35.1	3 32.6	1 48.1	2 14.3
29	5 17.9	5 43.1	3 30.9	1 37.9	2 5.1
30	5 26.4	5 51.2	3 29.2	1 27.6	1 55.8
31	5 35.2	5 59.4	3 27.5	1 17.2	1 46.4
32	5 44.0	6 7.6	3 25.8	1 6.6	1 36.7
33	5 53.1	6 15.9	3 24.1	0 56.0	1 26.8
34	6 2.3	6 24.4	3 22.4	0 45.1	1 16.8
35	6 11.8	6 32.9	3 20.6	0 34.2	1 6.5
36	6 21.4	6 41.6	3 18.9	0 23.1	0 55.9
37	6 31.3	6 50.3	3 17.2	0 11.8	0 45.1
38	6 41.4	6 59.3	3 15.4	0 0.4	0 34.0
39	6 51.8	7 8.3	3 13.6	♎29 48.7	0 22.7
40	7 2.5	7 17.5	3 11.8	29 36.9	0 11.0
41	7 13.4	7 26.8	3 10.0	29 24.9	♏29 59.0
42	7 24.7	7 36.3	3 8.1	29 12.6	29 46.6
43	7 36.2	7 46.0	3 6.3	29 0.1	29 33.8
44	7 48.2	7 55.8	3 4.4	28 47.4	29 20.6
45	8 0.5	8 5.9	3 2.5	28 34.4	29 7.0
46	8 13.2	8 16.2	3 0.5	28 21.1	28 52.9
47	8 26.3	8 26.6	2 58.6	28 7.5	28 38.3
48	8 39.9	8 37.4	2 56.6	27 53.6	28 23.2
49	8 54.1	8 48.3	2 54.5	27 39.4	28 7.5
50	9 8.7	8 59.6	2 52.5	27 24.8	27 51.1
51	9 24.0	9 11.1	2 50.4	27 9.9	27 34.1
52	9 39.8	9 22.9	2 48.2	26 54.5	27 16.3
53	9 56.4	9 35.0	2 46.0	26 38.7	26 57.6
54	10 13.6	9 47.5	2 43.8	26 22.4	26 38.1
55	10 31.7	10 0.3	2 41.5	26 5.7	26 17.7
56	10 50.7	10 13.5	2 39.2	25 48.4	25 56.2
57	11 10.6	10 27.1	2 36.8	25 30.6	25 33.5
58	11 31.5	10 41.2	2 34.4	25 12.2	25 9.6
59	11 53.6	10 55.7	2 31.9	24 53.2	24 44.2
60	12 16.9	11 10.7	2 29.4	24 33.5	24 17.3
61	12 41.6	11 26.3	2 26.8	24 13.1	23 48.7
62	13 7.9	11 42.4	2 24.1	23 51.9	23 18.1
63	13 35.8	11 59.2	2 21.3	23 29.9	22 45.3
64	14 5.6	12 16.6	2 18.5	23 7.0	22 10.1
65	14 37.6	12 34.7	2 15.5	22 43.2	21 32.1
66	15 12.0	12 53.5	2 12.5	22 18.3	20 50.9
S LAT	5	6	Descendant	8	9

♑ 3° 40′ 15″

$18^h\ 16^m\ 0^s$ MC 274° 0′ 0″

$6^h\ 20^m\ 0^s$ MC 95° 0′ 0″

♋ 4° 35′ 21″

N LAT °	11 ° ′	12 ° ′	Ascendant ° ′	2 ° ′	3 ° ′
0	♌ 2 43.0	♍ 3 3.5	♎ 5 26.8	♏ 7 21.1	♐ 6 50.3
5	3 17.3	3 41.4	5 14.9	6 30.8	6 13.2
10	3 51.6	4 18.2	5 3.7	5 41.8	5 36.0
15	4 26.6	4 54.6	4 52.9	4 53.2	4 57.8
16	4 33.8	5 1.8	4 50.8	4 43.4	4 50.0
17	4 41.0	5 9.1	4 48.6	4 33.7	4 42.1
18	4 48.2	5 16.4	4 46.5	4 23.9	4 34.2
19	4 55.5	5 23.7	4 44.4	4 14.0	4 26.2
20	5 2.9	5 31.0	4 42.4	4 4.2	4 18.1
21	5 10.4	5 38.3	4 40.3	3 54.3	4 9.9
22	5 17.9	5 45.7	4 38.2	3 44.3	4 1.6
23	5 25.6	5 53.1	4 36.1	3 34.3	3 53.2
24	5 33.3	6 0.6	4 34.0	3 24.3	3 44.7
25	5 41.1	6 8.1	4 31.9	3 14.1	3 36.1
26	5 49.1	6 15.6	4 29.9	3 3.9	3 27.3
27	5 57.1	6 23.2	4 27.8	2 53.6	3 18.4
28	6 5.3	6 30.9	4 25.7	2 43.2	3 9.4
29	6 13.6	6 38.6	4 23.6	2 32.7	3 0.2
30	6 22.1	6 46.4	4 21.5	2 22.1	2 50.8
31	6 30.7	6 54.2	4 19.4	2 11.4	2 41.2
32	6 39.5	7 2.2	4 17.2	2 0.6	2 31.5
33	6 48.5	7 10.2	4 15.1	1 49.7	2 21.5
34	6 57.6	7 18.3	4 13.0	1 38.6	2 11.4
35	7 6.9	7 26.6	4 10.8	1 27.3	2 1.0
36	7 16.4	7 34.9	4 8.6	1 15.9	1 50.3
37	7 26.2	7 43.4	4 6.4	1 4.3	1 39.4
38	7 36.2	7 51.9	4 4.2	0 52.6	1 28.2
39	7 46.4	8 0.6	4 2.0	0 40.7	1 16.7
40	7 56.9	8 9.5	3 59.7	0 28.5	1 4.9
41	8 7.7	8 18.5	3 57.5	0 16.1	0 52.8
42	8 18.8	8 27.6	3 55.1	0 3.6	0 40.3
43	8 30.2	8 36.9	3 52.8	♎29 50.7	0 27.4
44	8 42.0	8 46.4	3 50.5	29 37.6	0 14.0
45	8 54.1	8 56.1	3 48.1	29 24.3	0 0.3
46	9 6.7	9 6.0	3 45.6	29 10.6	♏29 46.0
47	9 19.6	9 16.1	3 43.2	28 56.7	29 31.3
48	9 33.1	9 26.4	3 40.7	28 42.4	29 15.9
49	9 47.0	9 36.9	3 38.1	28 27.7	29 0.0
50	10 1.4	9 47.8	3 35.6	28 12.7	28 43.4
51	10 16.4	9 58.8	3 32.9	27 57.3	28 26.2
52	10 32.0	10 10.2	3 30.3	27 41.5	28 8.1
53	10 48.3	10 21.9	3 27.5	27 25.3	27 49.3
54	11 5.3	10 33.9	3 24.8	27 8.5	27 29.5
55	11 23.1	10 46.2	3 21.9	26 51.3	27 8.8
56	11 41.7	10 58.9	3 19.0	26 33.6	26 46.9
57	12 1.2	11 12.0	3 16.1	26 15.2	26 23.9
58	12 21.8	11 25.5	3 13.0	25 56.3	25 59.6
59	12 43.5	11 39.5	3 9.9	25 36.7	25 33.8
60	13 6.4	11 53.9	3 6.7	25 16.4	25 6.5
61	13 30.6	12 8.9	3 3.5	24 55.4	24 37.3
62	13 56.3	12 24.4	3 0.1	24 33.6	24 6.2
63	14 23.7	12 40.5	2 56.6	24 10.9	23 32.8
64	14 52.9	12 57.2	2 53.1	23 47.3	22 56.9
65	15 24.2	13 14.6	2 49.4	23 22.7	22 18.1
66	15 57.7	13 32.7	2 45.6	22 57.1	21 35.9
S LAT	5	6	Descendant	8	9

♑ 4° 35′ 21″

u h18 m20s 0 MC 275° 0′ 0″

6h 24m 0s — MC 96° 0′ 0″ — ♋ 5° 30′ 29″

11	12	Ascendant	2	3	N LAT
° ′	° ′	° ′	° ′	° ′	°
♌ 3 41.2	♍ 4 6.8	♎ 6 32.1	♏ 8 22.6	♐ 7 46.9	0
4 15.1	4 43.4	6 17.8	7 31.2	7 9.6	5
4 49.1	5 18.9	6 4.3	6 41.1	6 32.1	10
5 23.7	5 53.9	5 51.4	5 51.3	5 53.7	15
5 30.8	6 0.9	5 48.9	5 41.3	5 45.8	16
5 37.9	6 7.9	5 46.3	5 31.3	5 37.9	17
5 45.1	6 14.9	5 43.8	5 21.3	5 29.9	18
5 52.3	6 22.0	5 41.3	5 11.3	5 21.8	19
5 59.6	6 29.0	5 38.8	5 1.2	5 13.7	20
6 7.0	6 36.1	5 36.3	4 51.0	5 5.4	21
6 14.5	6 43.2	5 33.8	4 40.8	4 57.1	22
6 22.0	6 50.3	5 31.3	4 30.6	4 48.6	23
6 29.6	6 57.5	5 28.8	4 20.3	4 40.0	24
6 37.4	7 4.7	5 26.3	4 9.9	4 31.3	25
6 45.2	7 12.0	5 23.8	3 59.4	4 22.5	26
6 53.2	7 19.3	5 21.3	3 48.9	4 13.5	27
7 1.3	7 26.6	5 18.8	3 38.2	4 4.4	28
7 9.5	7 34.1	5 16.3	3 27.5	3 55.1	29
7 17.9	7 41.6	5 13.8	3 16.6	3 45.7	30
7 26.4	7 49.1	5 11.2	3 5.6	3 36.0	31
7 35.1	7 56.8	5 8.7	2 54.5	3 26.2	32
6 15.0	8 4.5	5 6.1	2 43.3	3 16.2	33
7 52.9	8 12.3	5 3.5	2 31.9	3 5.9	34
8 2.1	8 20.3	5 0.9	2 20.4	2 55.4	35
8 11.5	8 28.3	4 58.3	2 8.7	2 44.6	36
8 21.2	8 36.4	4 55.7	1 56.8	2 33.6	37
8 31.0	8 44.6	4 53.0	1 44.8	2 22.3	38
8 41.1	8 53.0	4 50.4	1 32.5	2 10.7	39
8 51.5	9 1.5	4 47.7	1 20.0	1 58.8	40
9 2.1	9 10.2	4 44.9	1 7.4	1 46.5	41
9 13.1	9 19.0	4 42.2	0 54.4	1 33.9	42
9 24.3	9 27.9	4 39.4	0 41.3	1 20.9	43
9 35.9	9 37.0	4 36.5	0 27.8	1 7.4	44
9 47.9	9 46.3	4 33.7	0 14.1	0 53.5	45
10 0.2	9 55.8	4 30.8	0 0.1	0 39.1	46
10 13.0	10 5.5	4 27.8	♎29 45.7	0 24.1	47
10 26.2	10 15.4	4 24.8	29 31.0	0 8.6	48
10 39.9	10 25.6	4 21.8	29 16.0	♏29 52.5	49
10 54.1	10 36.0	4 18.7	29 0.6	29 35.8	50
11 8.9	10 46.6	4 15.5	28 44.8	29 18.3	51
11 24.2	10 57.5	4 12.3	28 28.5	29 0.0	52
11 40.3	11 8.7	4 9.0	28 11.8	28 40.9	53
11 57.0	11 20.3	4 5.7	27 54.6	28 20.8	54
12 14.4	11 32.1	4 2.3	27 36.9	27 59.8	55
12 32.7	11 44.3	3 58.8	27 18.6	27 37.7	56
12 52.0	11 56.9	3 55.3	26 59.8	27 14.3	57
13 12.1	12 9.9	3 51.6	26 40.3	26 49.6	58
13 33.4	12 23.3	3 47.9	26 20.1	26 23.4	59
13 55.9	12 37.1	3 44.1	25 59.3	25 55.6	60
14 19.6	12 51.5	3 40.1	25 37.6	25 26.0	61
14 44.8	13 6.3	3 36.1	25 15.2	24 54.2	62
15 11.6	13 21.8	3 32.0	24 51.8	24 20.2	63
15 40.2	13 37.8	3 27.7	24 27.5	23 43.6	64
16 10.8	13 54.5	3 23.3	24 2.2	23 3.9	65
16 43.6	14 11.9	3 18.8	23 35.8	22 20.8	66
5	6	Descendant	8	9	S LAT

♑ 5° 30′ 29″ — 18h 24m 0s — MC 276° 0′ 0″

6h 28m 0s — MC 97° 0′ 0″ — ♋ 6° 25′ 38″

11	12	Ascendant	2	3	N LAT
° ′	° ′	° ′	° ′	° ′	°
♌ 4 39.5	♍ 5 10.3	♎ 7 37.4	♏ 9 23.9	♐ 8 43.3	0
5 13.1	5 45.5	7 20.7	8 31.5	8 5.9	5
5 46.7	6 19.7	7 5.0	7 40.2	7 28.1	10
6 21.0	6 53.4	6 49.9	6 49.3	6 49.4	15
6 27.9	7 0.1	6 46.9	6 39.1	6 41.5	16
6 35.0	7 6.9	6 44.0	6 28.9	6 33.6	17
6 42.1	7 13.6	6 41.1	6 18.7	6 25.5	18
6 49.2	7 20.4	6 38.1	6 8.4	6 17.4	19
6 56.4	7 27.1	6 35.2	5 58.0	6 9.2	20
7 3.7	7 33.9	6 32.3	5 47.7	6 0.9	21
7 11.1	7 40.8	6 29.4	5 37.2	5 52.5	22
7 18.5	7 47.6	6 26.5	5 26.8	5 43.9	23
7 26.1	7 54.5	6 23.6	5 16.2	5 35.3	24
7 33.8	8 1.4	6 20.6	5 5.6	5 26.5	25
7 41.5	8 8.4	6 17.7	4 54.8	5 17.6	26
7 49.4	8 15.4	6 14.8	4 44.0	5 8.6	27
7 57.4	8 22.5	6 11.9	4 33.1	4 59.4	28
8 5.5	8 29.6	6 9.0	4 22.1	4 50.0	29
8 13.7	8 36.8	6 6.0	4 11.0	4 40.5	30
8 22.1	8 44.1	6 3.1	3 59.8	4 30.8	31
8 30.7	8 51.5	6 0.1	3 48.4	4 20.8	32
8 39.4	8 58.9	5 57.1	3 36.9	4 10.7	33
8 48.3	9 6.4	5 54.1	3 25.2	4 0.4	34
8 57.4	9 14.0	5 51.1	3 13.4	3 49.8	35
9 6.7	9 21.7	5 48.0	3 1.4	3 38.9	36
9 16.2	9 29.5	5 45.0	2 49.2	3 27.8	37
9 25.9	9 37.4	5 41.9	2 36.9	3 16.4	38
9 35.9	9 45.4	5 38.7	2 24.3	3 4.7	39
9 46.1	9 53.6	5 35.6	2 11.5	2 52.7	40
9 56.6	10 1.9	5 32.4	1 58.5	2 40.3	41
10 7.4	10 10.3	5 29.2	1 45.3	2 27.5	42
10 18.5	10 18.9	5 25.9	1 31.7	2 14.3	43
10 29.9	10 27.7	5 22.6	1 18.0	2 0.7	44
10 41.7	10 36.6	5 19.3	1 3.9	1 46.7	45
10 53.9	10 45.7	5 15.9	0 49.5	1 32.1	46
11 6.4	10 55.0	5 12.4	0 34.8	1 17.0	47
11 19.4	11 4.5	5 8.9	0 19.7	1 1.3	48
11 32.9	11 14.3	5 5.4	0 4.3	0 45.0	49
11 46.9	11 24.2	5 1.8	♎29 48.4	0 28.1	50
12 1.4	11 34.4	4 58.1	29 32.2	0 10.3	51
12 16.5	11 44.9	4 54.4	29 15.5	♏29 51.8	52
12 32.3	11 55.6	4 50.5	28 58.3	29 32.5	53
12 48.7	12 6.7	4 46.6	28 40.7	29 12.2	54
13 5.9	12 18.0	4 42.7	28 22.5	28 50.9	55
13 23.9	12 29.7	4 38.6	28 3.7	28 28.4	56
13 42.7	12 41.8	4 34.5	27 44.3	28 4.7	57
14 2.5	12 54.2	4 30.2	27 24.3	27 39.6	58
14 23.4	13 7.1	4 25.9	27 3.6	27 13.0	59
14 45.4	13 20.3	4 21.4	26 42.1	26 44.7	60
15 8.7	13 34.1	4 16.8	26 19.8	26 14.6	61
15 33.4	13 48.3	4 12.1	25 56.7	25 42.3	62
15 59.7	14 3.1	4 7.3	25 32.7	25 7.6	63
16 27.6	14 18.5	4 2.3	25 7.7	24 30.2	64
16 57.5	14 34.5	3 57.2	24 41.7	23 49.7	65
17 29.5	14 51.1	3 51.9	24 14.5	23 5.6	66
5	6	Descendant	8	9	S LAT

♑ 6° 25′ 38″ — 18h 28m 0s — MC 277° 0′ 0″

6h 32m 0s		MC		98° 0′ 0″	
		♋ 7° 20′ 50″			N
11	12	Ascendant	2	3	LAT
° ′	° ′	° ′	° ′	° ′	°
♌ 5 38.0	♍ 6 14.0	♎ 8 42.5	♏10 25.0	♐ 9 39.7	0
6 11.2	6 47.8	8 23.5	9 31.5	9 2.0	5
6 44.5	7 20.6	8 5.6	8 39.2	8 24.0	10
7 18.3	7 52.9	7 48.4	7 47.3	7 45.1	15
7 25.2	7 59.4	7 45.0	7 36.8	7 37.2	16
7 32.2	8 5.9	7 41.6	7 26.4	7 29.2	17
7 39.2	8 12.3	7 38.3	7 15.9	7 21.1	18
7 46.2	8 18.8	7 34.9	7 5.4	7 12.9	19
7 53.4	8 25.3	7 31.6	6 54.8	7 4.6	20
8 0.6	8 31.8	7 28.3	6 44.2	6 56.2	21
8 7.8	8 38.4	7 24.9	6 33.6	6 47.8	22
8 15.2	8 45.0	7 21.6	6 23.5	6 39.2	23
8 22.7	8 51.6	7 18.3	6 12.0	6 30.5	24
8 30.2	8 58.2	7 15.0	6 1.2	6 21.6	25
8 37.9	9 4.9	7 11.7	5 50.2	6 12.7	26
8 45.7	9 11.6	7 8.3	5 39.1	6 3.6	27
8 53.5	9 18.4	7 5.0	5 28.0	5 54.3	28
9 1.6	9 25.3	7 1.6	5 16.7	5 44.9	29
9 9.7	9 32.2	6 58.3	5 5.3	5 35.3	30
9 18.0	9 39.1	6 54.9	4 53.8	5 25.5	31
9 26.4	9 46.2	6 51.5	4 42.2	5 15.5	32
9 35.0	9 53.3	6 48.1	4 30.4	5 5.2	33
9 43.8	10 0.5	6 44.7	4 18.5	4 54.8	34
9 52.8	10 7.8	6 41.2	4 6.4	4 44.1	35
10 1.9	10 15.2	6 37.7	3 54.1	4 33.2	36
10 11.3	10 22.6	6 34.2	3 41.6	4 22.0	37
10 20.9	10 30.2	6 30.7	3 28.9	4 10.5	38
10 30.7	10 37.9	6 27.1	3 16.1	3 58.6	39
10 40.8	10 45.7	6 23.5	3 3.0	3 46.5	40
10 51.1	10 53.7	6 19.9	2 49.6	3 34.0	41
11 1.8	11 1.8	6 16.2	2 36.1	3 21.1	42
11 12.7	11 10.0	6 12.5	2 22.2	3 7.8	43
11 24.0	11 18.4	6 8.7	2 8.1	2 54.0	44
11 35.6	11 26.9	6 4.9	1 53.6	2 39.8	45
11 47.5	11 35.7	6 1.0	1 38.9	2 25.1	46
11 59.9	11 44.6	5 57.1	1 23.8	2 9.9	47
12 12.7	11 53.7	5 53.1	1 8.3	1 54.0	48
12 26.0	12 3.0	5 49.0	0 52.5	1 37.5	49
12 39.7	12 12.5	5 44.9	0 36.2	1 20.3	50
12 54.0	12 22.3	5 40.7	0 19.6	1 2.4	51
13 8.9	12 32.3	5 36.4	0 2.4	0 43.7	52
13 24.4	12 42.6	5 32.0	♎29 44.8	0 24.1	53
13 40.5	12 53.1	5 27.6	29 26.7	0 3.5	54
13 57.4	13 4.0	5 23.0	29 8.0	♏29 41.9	55
14 15.0	13 15.2	5 18.4	28 48.7	29 19.1	56
14 33.5	13 26.7	5 13.7	28 28.8	28 55.1	57
14 53.0	13 38.6	5 8.8	28 8.2	28 29.6	58
15 13.5	13 50.9	5 3.8	27 46.9	28 2.6	59
15 35.0	14 3.6	4 58.7	27 24.9	27 33.9	60
15 57.9	14 16.7	4 53.5	27 2.0	27 3.2	61
16 22.1	14 30.4	4 48.1	26 38.3	26 30.3	62
16 47.7	14 44.5	4 42.6	26 13.6	25 55.0	63
17 15.1	14 59.2	4 36.9	25 47.9	25 16.8	64
17 44.3	15 14.5	4 31.0	25 21.1	24 35.5	65
18 15.6	15 30.4	4 25.0	24 53.1	23 50.4	66
5	6	Descendant	8	9	S
		♑ 7° 20′ 50″			LAT
18h 32m 0s		MC		278° 0′ 0″	

6h 36m 0s		MC		99° 0′ 0″	
		♋ 8° 16′ 4″			N
11	12	Ascendant	2	3	LAT
° ′	° ′	° ′	° ′	° ′	°
♌ 7 35.2	♍ 7 17.7	♎ 9 47.7	♏11 26.0	♐10 35.9	0
7 9.5	7 50.2	9 26.3	10 31.5	9 58.1	5
7 42.3	8 21.6	9 6.2	9 38.1	9 19.9	10
8 15.8	8 52.6	8 46.8	8 45.0	8 40.7	15
8 22.6	8 58.8	8 43.0	8 34.4	8 32.7	16
8 29.5	9 5.0	8 39.3	8 23.7	8 24.7	17
8 36.4	9 11.2	8 35.5	8 13.0	8 16.5	18
8 43.4	9 17.4	8 31.7	8 2.3	8 8.3	19
8 50.4	9 23.6	8 28.0	7 51.5	8 0.0	20
8 57.5	9 29.8	8 24.2	7 40.7	7 51.5	21
9 4.7	9 36.1	8 20.5	7 29.8	7 43.0	22
9 12.0	9 42.4	8 16.8	7 18.8	7 34.4	23
9 19.4	9 48.7	8 13.0	7 7.8	7 25.6	24
9 26.8	9 55.1	8 9.3	6 56.7	7 16.7	25
9 34.4	10 1.5	8 5.5	6 45.5	7 7.7	26
9 42.0	10 7.9	8 1.8	6 34.2	6 58.5	27
9 49.8	10 14.4	7 58.0	6 22.8	6 49.2	28
9 57.7	10 21.0	7 54.3	6 11.2	6 39.6	29
10 5.8	10 27.6	7 50.5	5 59.6	6 30.0	30
10 14.0	10 34.2	7 46.7	5 47.8	6 20.1	31
10 22.3	10 41.0	7 42.9	5 35.9	6 10.0	32
10 30.8	10 47.8	7 39.1	5 23.9	5 59.7	33
10 39.4	10 54.7	7 35.2	5 11.6	5 49.2	34
10 48.3	11 1.6	7 31.3	4 59.3	5 38.4	35
10 57.3	11 8.7	7 27.4	4 46.7	5 27.4	36
11 6.5	11 15.8	7 23.5	4 33.9	5 16.1	37
11 16.0	11 23.1	7 19.5	4 21.0	5 4.5	38
11 25.7	11 30.4	7 15.5	4 7.8	4 52.6	39
11 35.6	11 37.9	7 11.4	3 54.4	4 40.3	40
11 45.8	11 45.5	7 7.3	3 40.7	4 27.7	41
11 56.2	11 53.2	7 3.2	3 26.8	4 14.6	42
12 7.0	12 1.1	6 59.0	3 12.6	4 1.2	43
12 18.1	12 9.1	6 54.8	2 58.1	3 47.3	44
12 29.5	12 17.3	6 50.5	2 43.3	3 33.0	45
12 41.3	12 25.6	6 46.1	2 28.2	3 18.1	46
12 53.5	12 34.1	6 41.7	2 12.7	3 2.7	47
13 6.1	12 42.8	6 37.2	1 56.9	2 46.7	48
13 19.1	12 51.7	6 32.6	1 40.7	2 30.0	49
13 32.6	13 0.8	6 28.0	1 24.0	2 12.6	50
13 46.7	13 10.2	6 23.2	1 6.9	1 54.5	51
14 1.3	13 19.7	6 18.4	0 49.3	1 35.5	52
14 16.5	13 29.5	6 13.5	0 31.3	1 15.7	53
14 32.3	13 39.6	6 8.5	0 12.6	0 54.9	54
14 48.9	13 50.0	6 3.4	♎29 53.5	0 32.9	55
15 6.2	14 0.7	5 58.2	29 33.7	0 9.9	56
15 24.4	14 11.7	5 52.8	29 13.3	♏29 45.5	57
15 43.5	14 23.0	5 47.4	28 52.1	29 19.6	58
16 3.6	14 34.8	5 41.8	28 30.3	28 52.2	59
16 24.7	14 46.9	5 36.0	28 7.6	28 23.0	60
16 47.1	14 59.4	5 30.2	27 44.2	27 51.8	61
17 10.8	15 12.4	5 24.1	27 19.8	27 18.3	62
17 35.9	15 25.9	5 17.9	26 54.4	26 42.3	63
18 2.6	15 39.9	5 11.5	26 28.0	26 3.4	64
18 31.2	15 54.5	5 4.9	26 0.4	25 21.1	65
19 1.7	16 9.7	4 58.1	25 31.7	24 35.0	66
5	6	Descendant	8	9	S
		♑ 8° 16′ 4″			LAT
18h 36m 0s		MC		279° 0′ 0″	

6h 40m 0s		MC		100° 0′ 0″
		♋ 9° 11′ 21″		

N LAT	11	12	Ascendant	2	3
0	♌ 7 35.4	♍ 8 21.7	♎ 10 52.7	♏ 12 26.7	♐ 11 32.1
5	8 7.9	8 52.7	10 29.1	11 31.2	10 54.0
10	8 40.4	9 22.7	10 6.7	10 36.8	10 15.6
15	9 13.4	9 52.3	9 45.3	9 42.7	9 36.3
16	9 20.2	9 58.2	9 41.0	9 31.9	9 28.2
17	9 26.9	10 4.1	9 36.8	9 21.0	9 20.1
18	9 33.8	10 10.1	9 32.7	9 10.1	9 11.9
19	9 40.7	10 16.0	9 28.5	8 59.1	9 3.6
20	9 47.6	10 21.9	9 24.3	8 48.1	8 55.2
21	9 54.6	10 27.9	9 20.2	8 37.0	8 46.8
22	10 1.7	10 33.9	9 16.0	8 25.9	8 38.2
23	10 8.9	10 39.9	9 11.9	8 14.7	8 29.5
24	10 16.2	10 45.9	9 7.7	8 3.4	8 20.7
25	10 23.5	10 52.0	9 3.6	7 52.1	8 11.7
26	10 31.0	10 58.1	8 59.4	7 40.6	8 2.6
27	10 38.5	11 4.3	8 55.3	7 29.1	7 53.4
28	10 46.2	11 10.5	8 51.1	7 17.4	7 44.0
29	10 54.0	11 16.7	8 46.9	7 5.7	7 34.4
30	11 1.9	11 23.0	8 42.7	6 53.8	7 24.6
31	11 10.0	11 29.4	8 38.5	6 41.7	7 14.7
32	11 18.2	11 35.8	8 34.3	6 29.6	7 4.5
33	11 26.6	11 42.3	8 30.0	6 17.3	6 54.2
34	11 35.1	11 48.9	8 25.7	6 4.8	6 43.5
35	11 43.8	11 55.5	8 21.4	5 52.1	6 32.7
36	11 52.7	12 2.2	8 17.1	5 39.2	6 21.6
37	12 1.8	12 9.1	8 12.7	5 26.2	6 10.2
38	12 11.1	12 16.0	8 8.3	5 12.9	5 58.5
39	12 20.7	12 23.0	8 3.8	4 59.4	5 46.4
40	12 30.4	12 30.1	7 59.3	4 45.7	5 34.1
41	12 40.5	12 37.4	7 54.8	4 31.7	5 21.3
42	12 50.8	12 44.7	7 50.2	4 17.5	5 8.2
43	13 1.4	12 52.2	7 45.5	4 3.0	4 54.6
44	13 12.3	12 59.9	7 40.8	3 48.1	4 40.6
45	13 23.5	13 7.7	7 36.0	3 33.0	4 26.1
46	13 35.1	13 15.6	7 31.2	3 17.5	4 11.1
47	13 47.1	13 23.7	7 26.3	3 1.7	3 55.5
48	13 59.5	13 32.0	7 21.3	2 45.4	3 39.3
49	14 12.3	13 40.5	7 16.2	2 28.8	3 22.5
50	14 25.6	13 49.2	7 11.0	2 11.7	3 4.9
51	14 39.4	13 58.1	7 5.8	1 54.2	2 46.6
52	14 53.7	14 7.2	7 0.5	1 36.2	2 27.4
53	15 8.7	14 16.5	6 55.0	1 17.7	2 7.3
54	15 24.3	14 26.1	6 49.4	0 58.6	1 46.2
55	15 40.5	14 36.0	6 43.8	0 38.9	1 24.0
56	15 57.5	14 46.2	6 38.0	0 18.7	1 0.6
57	16 15.3	14 56.7	6 32.0	♎ 29 57.7	0 35.9
58	16 34.0	15 7.5	6 26.0	29 36.0	0 9.6
59	16 53.7	15 18.6	6 19.7	29 13.6	♏ 29 41.8
60	17 14.4	15 30.2	6 13.4	28 50.4	29 12.1
61	17 36.3	15 42.1	6 6.8	28 26.3	28 40.4
62	17 59.5	15 54.5	6 0.1	28 1.2	28 6.3
63	18 24.1	16 7.4	5 53.2	27 35.2	27 29.6
64	18 50.2	16 20.7	5 46.1	27 8.0	26 50.0
65	19 18.1	16 34.6	5 38.7	26 39.8	26 6.8
66	19 48.0	16 49.0	5 31.2	26 10.2	25 19.6
S LAT	5	6	Descendant	8	9

		♑ 9° 11′ 21″		
18h 40m 0s		MC		280° 0′ 0″

6h 44m 0s		MC		101° 0′ 0″
		♋ 10° 6′ 42″		

N LAT	11	12	Ascendant	2	3
0	♌ 8 34.4	♍ 9 25.7	♎ 11 57.7	♏ 13 27.3	♐ 12 28.1
5	9 6.5	9 55.3	11 31.8	12 30.8	11 49.9
10	9 38.6	10 23.9	11 7.2	11 35.4	11 11.3
15	10 11.2	10 52.1	10 43.6	10 40.3	10 31.7
16	10 17.8	10 57.8	10 39.0	10 29.2	10 23.6
17	10 24.5	11 3.4	10 34.4	10 18.1	10 15.5
18	10 31.3	11 9.0	10 29.8	10 7.0	10 7.2
19	10 38.1	11 14.7	10 25.2	9 55.8	9 58.9
20	10 44.6	11 20.4	10 20.6	9 44.6	9 50.5
21	10 51.8	11 26.0	10 16.1	9 33.3	9 41.9
22	10 58.8	11 31.7	10 11.5	9 21.9	9 33.3
23	11 5.9	11 37.5	10 7.0	9 10.5	9 24.5
24	11 13.1	11 43.2	10 2.4	8 59.0	9 15.7
25	11 20.3	11 49.0	9 57.8	8 47.4	9 6.6
26	11 27.7	11 54.8	9 53.3	8 35.7	8 57.5
27	11 35.2	12 0.7	9 48.7	8 24.0	8 48.2
28	11 42.7	12 6.6	9 44.1	8 12.1	8 38.7
29	11 50.4	12 12.5	9 39.5	8 0.0	8 29.1
30	11 58.2	12 18.5	9 34.9	7 47.9	8 19.2
31	12 6.2	12 24.6	9 30.3	7 35.6	8 9.2
32	12 14.3	12 30.7	9 25.6	7 23.2	7 59.0
33	12 22.5	12 36.9	9 21.0	7 10.6	7 48.5
34	12 30.9	12 43.1	9 16.2	6 57.8	7 37.9
35	12 39.5	12 49.4	9 11.5	6 44.9	7 26.9
36	12 48.3	12 55.8	9 6.7	6 31.7	7 15.7
37	12 57.2	13 2.3	9 1.9	6 18.4	7 4.2
38	13 6.4	13 8.9	8 57.1	6 4.8	6 52.4
39	13 15.8	13 15.6	8 52.2	5 51.1	6 40.3
40	13 25.4	13 22.4	8 47.2	5 37.0	6 27.8
41	13 35.3	13 29.3	8 42.2	5 22.7	6 15.0
42	13 45.4	13 36.3	8 37.2	5 8.2	6 1.7
43	13 55.8	13 43.4	8 32.0	4 53.3	5 48.0
44	14 6.6	13 50.7	8 26.9	4 38.1	5 33.9
45	14 17.6	13 58.1	8 21.6	4 22.6	5 19.2
46	14 29.0	14 5.6	8 16.3	4 6.8	5 4.1
47	14 40.8	14 13.3	8 10.9	3 50.6	4 48.3
48	14 53.0	14 21.2	8 5.4	3 33.9	4 32.0
49	15 5.6	14 29.3	7 59.8	3 16.9	4 14.9
50	15 18.6	14 37.5	7 54.1	2 59.4	3 57.2
51	15 32.2	14 46.0	7 48.4	2 41.5	3 38.6
52	15 46.3	14 54.6	7 42.5	2 23.0	3 19.2
53	16 0.9	15 3.5	7 36.5	2 4.1	2 58.9
54	16 16.2	15 12.7	7 30.4	1 44.5	2 37.5
55	16 32.2	15 22.1	7 24.1	1 24.4	2 15.1
56	16 48.9	15 31.7	7 17.7	1 3.6	1 51.3
57	17 6.3	15 41.7	7 11.2	0 42.1	1 26.3
58	17 24.7	15 52.0	7 4.5	0 19.9	0 59.7
59	17 43.9	16 2.6	6 57.7	♎ 29 56.9	0 31.4
60	18 4.2	16 13.5	6 50.7	29 33.1	0 1.3
61	18 25.7	16 24.9	6 43.5	29 8.3	♏ 29 29.0
62	18 48.3	16 36.6	6 36.1	28 42.6	28 54.3
63	19 12.4	16 48.8	6 28.5	28 15.9	28 17.0
64	19 37.9	17 1.5	6 20.6	27 48.1	27 36.5
65	20 5.2	17 14.7	6 12.6	27 19.0	26 52.4
66	20 34.3	17 28.4	6 4.2	26 48.7	26 4.1
S LAT	5	6	Descendant	8	9

		♑ 10° 6′ 42″		
18h 44m 0s		MC		281° 0′ 0″

$6^h\ 48^m\ 0^s$	MC	102° 0′ 0″				$6^h\ 52^m\ 0^s$	MC	103° 0′ 0″		
	♋ 11° 2′ 6″				N		♋ 11° 57′ 33″			
11	**12**	**Ascendant**	**2**	**3**	**LAT**	**11**	**12**	**Ascendant**	**2**	**3**
° ′	° ′	° ′	° ′	° ′	°	° ′	° ′	° ′	° ′	° ′
♌ 9 33.6	♍10 29.9	♎13 2.6	♏14 27.7	♐13 24.0	**0**	♌10 32.9	♍11 34.2	♎14 7.5	♏15 28.0	♐14 19.9
10 5.2	10 58.0	12 34.4	13 30.3	12 45.6	**5**	11 4.2	12 0.8	13 36.9	14 29.5	13 41.3
10 36.9	11 25.2	12 7.6	12 33.9	12 6.9	**10**	11 35.4	12 26.6	13 8.0	13 32.2	13 2.4
11 9.1	11 52.1	11 42.0	11 37.7	11 27.1	**15**	12 7.2	12 52.0	12 40.3	12 35.0	12 22.4
11 15.7	11 57.4	11 36.9	11 26.4	11 19.0	**16**	12 13.6	12 57.1	12 34.8	12 23.5	12 14.3
11 22.3	12 2.8	11 31.9	11 15.1	11 10.8	**17**	12 20.1	13 2.2	12 29.4	12 12.0	12 6.0
11 28.9	12 8.1	11 26.9	11 3.8	11 2.5	**18**	12 26.7	13 7.2	12 23.9	12 0.5	11 57.7
11 35.6	12 13.5	11 21.9	10 52.4	10 54.1	**19**	12 33.3	13 12.3	12 18.5	11 48.9	11 49.2
11 42.4	12 18.8	11 16.9	10 40.9	10 45.6	**20**	12 39.9	13 17.4	12 13.2	11 37.2	11 40.7
11 49.2	12 24.2	11 11.9	10 29.4	10 37.0	**21**	12 46.7	13 22.5	12 7.8	11 25.5	11 32.1
11 56.1	12 29.7	11 7.0	10 17.9	10 28.3	**22**	12 53.5	13 27.6	12 2.4	11 13.7	11 23.3
12 3.1	12 35.1	11 2.0	10 6.2	10 19.5	**23**	13 0.3	13 32.8	11 57.0	11 1.8	11 14.5
12 10.1	12 40.6	10 57.0	9 54.5	10 10.6	**24**	13 7.3	13 38.0	11 51.7	10 49.9	11 5.5
12 17.3	12 46.0	10 52.1	9 42.7	10 1.5	**25**	13 14.3	13 43.1	11 46.3	10 37.8	10 56.4
12 24.5	12 51.6	10 47.1	9 30.8	9 52.3	**26**	13 21.5	13 48.4	11 40.9	10 25.7	10 47.1
12 31.9	12 57.1	10 42.1	9 18.7	9 43.0	**27**	13 28.7	13 53.6	11 35.5	10 13.4	10 37.7
12 39.3	13 2.7	10 37.1	9 6.6	9 33.4	**28**	13 36.1	13 58.9	11 30.1	10 1.1	10 28.1
12 46.9	13 8.4	10 32.1	8 54.3	9 23.7	**29**	13 43.5	14 4.3	11 24.7	9 48.5	10 18.3
12 54.6	13 14.1	10 27.1	8 41.9	9 13.8	**30**	13 51.1	14 9.7	11 19.3	9 35.9	10 8.4
13 2.4	13 19.8	10 22.1	8 29.4	9 3.7	**31**	13 58.8	14 15.1	11 13.8	9 23.1	9 58.2
13 10.4	13 25.6	10 17.0	8 16.7	8 53.4	**32**	14 6.7	14 20.6	11 8.3	9 10.2	9 47.8
13 18.5	13 31.5	10 11.9	8 3.8	8 42.9	**33**	14 14.6	14 26.1	11 2.8	8 57.1	9 37.2
13 26.8	13 37.4	10 6.7	7 50.8	8 32.1	**34**	14 22.8	14 31.8	10 57.2	8 43.8	9 26.4
13 35.2	13 43.4	10 1.6	7 37.6	8 21.1	**35**	14 31.1	14 37.4	10 51.6	8 30.3	9 15.3
13 43.9	13 49.5	9 56.4	7 24.2	8 9.8	**36**	14 39.6	14 43.2	10 46.0	8 16.6	9 3.9
13 52.7	13 55.6	9 51.1	7 10.6	7 58.2	**37**	14 48.3	14 49.0	10 40.3	8 2.7	8 52.2
14 1.7	14 1.9	9 45.8	6 56.7	7 46.3	**38**	14 57.1	14 54.9	10 34.6	7 48.5	8 40.3
14 10.9	14 8.2	9 40.5	6 42.6	7 34.1	**39**	15 6.2	15 0.9	10 28.8	7 34.1	8 27.9
14 20.4	14 14.7	9 35.1	6 28.3	7 21.5	**40**	15 15.5	15 7.0	10 23.0	7 19.5	8 15.3
14 30.1	14 21.2	9 29.6	6 13.7	7 8.6	**41**	15 25.1	15 13.2	10 17.1	7 4.6	8 2.2
14 40.1	14 27.8	9 24.1	5 58.8	6 55.2	**42**	15 34.9	15 19.4	10 11.1	6 49.4	7 48.7
14 50.4	14 34.6	9 18.5	5 43.6	6 41.4	**43**	15 45.0	15 25.8	10 5.0	6 33.8	7 34.8
15 0.9	14 41.5	9 12.9	5 28.1	6 27.1	**44**	15 55.3	15 32.3	9 58.9	6 18.0	7 20.4
15 11.8	14 48.5	9 7.2	5 12.2	6 12.4	**45**	16 6.0	15 39.0	9 52.7	6 1.8	7 5.5
15 23.0	14 55.7	9 1.3	4 56.0	5 57.1	**46**	16 17.0	15 45.8	9 46.4	5 45.2	6 50.0
15 34.6	15 3.0	8 55.5	4 39.4	5 41.2	**47**	16 28.4	15 52.7	9 40.0	5 28.3	6 34.0
15 46.5	15 10.5	8 49.5	4 22.4	5 24.6	**48**	16 40.1	15 59.7	9 33.6	5 10.9	6 17.3
15 58.9	15 18.1	8 43.4	4 5.0	5 7.4	**49**	16 52.3	16 6.9	9 27.0	4 53.1	5 59.9
16 11.7	15 25.9	8 37.2	3 47.1	4 49.5	**50**	17 4.9	16 14.3	9 20.3	4 34.8	5 41.8
16 25.0	15 33.9	8 30.9	3 28.8	4 30.7	**51**	17 17.9	16 21.9	9 13.4	4 16.0	5 22.8
16 38.9	15 42.1	8 24.5	3 9.9	4 11.1	**52**	17 31.5	16 29.7	9 6.5	3 56.7	5 3.0
16 53.2	15 50.6	8 17.9	2 50.4	3 50.5	**53**	17 45.6	16 37.6	8 59.4	3 36.8	4 42.2
17 8.2	15 59.2	8 11.3	2 30.4	3 28.9	**54**	18 0.3	16 45.8	8 52.2	3 16.3	4 20.3
17 23.9	16 8.1	8 4.5	2 9.8	3 6.1	**55**	18 15.7	16 54.2	8 44.8	2 55.2	3 57.2
17 40.2	16 17.3	7 57.5	1 48.5	2 42.1	**56**	18 31.7	17 2.9	8 37.3	2 33.4	3 32.9
17 57.4	16 26.7	7 50.4	1 26.5	2 16.7	**57**	18 48.5	17 11.8	8 29.5	2 10.8	3 7.2
18 15.3	16 36.4	7 43.1	1 3.7	1 49.7	**58**	19 6.1	17 21.0	8 21.6	1 47.5	2 39.8
18 34.2	16 46.5	7 35.6	0 40.1	1 21.0	**59**	19 24.6	17 30.5	8 13.6	1 23.4	2 10.7
18 54.1	16 56.9	7 28.0	0 15.7	0 50.4	**60**	19 44.0	17 40.3	8 5.3	0 58.3	1 39.6
19 15.1	17 7.6	7 20.1	♎29 50.4	0 17.6	**61**	20 4.5	17 50.4	7 56.8	0 32.3	1 6.3
19 37.2	17 18.8	7 12.0	29 24.0	♏29 42.4	**62**	20 26.2	18 0.9	7 48.0	0 5.3	0 30.4
20 0.7	17 30.3	7 3.7	28 56.6	29 4.3	**63**	20 49.2	18 11.8	7 39.0	♎29 37.2	♏29 51.6
20 25.7	17 42.3	6 55.2	28 28.0	28 23.0	**64**	21 13.6	18 23.1	7 29.8	29 8.0	29 9.5
20 52.3	17 54.8	6 46.4	27 58.3	27 38.0	**65**	21 39.5	18 34.9	7 20.2	28 37.4	28 23.5
21 20.7	18 7.8	6 37.3	27 27.1	26 48.6	**66**	22 7.3	18 47.2	7 10.4	28 5.5	27 32.9
5	**6**	**Descendant**	**8**	**9**	**S**	**5**	**6**	**Descendant**	**8**	**9**
	♑ 11° 2′ 6″				**LAT**		♑ 11° 57′ 33″			
$18^h\ 48^m\ 0^s$	MC	282° 0′ 0″				$18^h\ 52^m\ 0^s$	MC	283° 0′ 0″		

6h 56m 0s		MC		104° 0′ 0″
		♋ 12° 53′ 5″		

N LAT	11	12	Ascendant	2	3
°	° ′	° ′	° ′	° ′	° ′
0	♌11 32.4	♍12 38.6	♎15 12.2	♏16 28.0	♐15 15.6
5	12 3.2	13 3.8	14 39.4	15 28.7	14 36.9
10	12 34.0	13 28.1	14 8.3	14 30.4	13 57.9
15	13 5.3	13 52.1	13 38.5	13 32.2	13 17.7
16	13 11.7	13 56.9	13 32.6	13 20.5	13 9.5
17	13 18.1	14 1.7	13 26.8	13 8.8	13 1.2
18	13 24.6	14 6.4	13 21.0	12 57.1	12 52.8
19	13 31.1	14 11.2	13 15.2	12 45.3	12 44.3
20	13 37.6	14 16.0	13 9.4	12 33.4	12 35.8
21	13 44.3	14 20.9	13 3.6	12 21.5	12 27.1
22	13 51.0	14 25.7	12 57.8	12 9.4	12 18.3
23	13 57.7	14 30.5	12 52.0	11 57.4	12 9.4
24	14 4.6	14 35.4	12 46.3	11 45.2	12 0.4
25	14 11.5	14 40.3	12 40.5	11 32.9	11 51.2
26	14 18.6	14 45.2	12 34.7	11 20.5	11 41.9
27	14 25.7	14 50.2	12 28.9	11 8.1	11 32.4
28	14 32.9	14 55.2	12 23.1	10 55.4	11 22.7
29	14 40.3	15 0.2	12 17.2	10 42.7	11 12.9
30	14 47.7	15 5.3	12 11.4	10 29.8	11 2.9
31	14 55.3	15 10.4	12 5.5	10 16.8	10 52.7
32	15 3.0	15 15.6	11 59.6	10 3.6	10 42.2
33	15 10.9	15 20.8	11 53.7	9 50.2	10 31.6
34	15 18.9	15 26.1	11 47.7	9 36.6	10 20.6
35	15 27.1	15 31.5	11 41.7	9 22.9	10 9.5
36	15 35.4	15 36.9	11 35.6	9 8.9	9 58.0
37	15 43.9	15 42.4	11 29.5	8 54.7	9 46.2
38	15 52.7	15 48.0	11 23.3	8 40.3	9 34.2
39	16 1.6	15 53.6	11 17.1	8 25.6	9 21.7
40	16 10.7	15 59.3	11 10.8	8 10.7	9 9.0
41	16 20.1	16 5.2	11 4.5	7 55.4	8 55.8
42	16 29.8	16 11.1	10 58.0	7 39.9	8 42.2
43	16 39.7	16 17.1	10 51.5	7 24.0	8 28.2
44	16 49.8	16 23.2	10 44.9	7 7.9	8 13.6
45	17 0.3	16 29.5	10 38.2	6 51.3	7 58.6
46	17 11.1	16 35.9	10 31.5	6 34.4	7 43.0
47	17 22.3	16 42.4	10 24.6	6 17.1	7 26.8
48	17 33.8	16 49.0	10 17.6	5 59.3	7 10.0
49	17 45.8	16 55.8	10 10.5	5 41.1	6 52.4
50	17 58.1	17 2.8	10 3.3	5 22.4	6 34.1
51	18 10.9	17 9.9	9 56.0	5 3.2	6 14.9
52	18 24.2	17 17.2	9 48.5	4 43.4	5 54.9
53	18 38.1	17 24.7	9 40.9	4 23.1	5 33.8
54	18 52.5	17 32.4	9 33.1	4 2.1	5 11.7
55	19 7.5	17 40.3	9 25.1	3 40.5	4 48.4
56	19 23.2	17 48.5	9 17.0	3 18.2	4 23.7
57	19 39.6	17 56.8	9 8.7	2 55.1	3 57.6
58	19 56.9	18 5.5	9 0.2	2 31.3	3 29.9
59	20 15.0	18 14.4	8 51.5	2 6.6	3 0.4
60	20 34.0	18 23.7	8 42.6	1 40.9	2 28.8
61	20 54.0	18 33.2	8 33.4	1 14.3	1 54.9
62	21 15.2	18 43.1	8 24.0	0 46.7	1 18.4
63	21 37.7	18 53.4	8 14.3	0 17.9	0 38.9
64	22 1.5	19 4.0	8 4.3	♎29 47.9	♏29 56.0
65	22 26.8	19 15.1	7 54.0	29 16.6	29 9.0
66	22 53.9	19 26.6	7 43.4	28 43.9	28 17.3
S LAT	5	6	Descendant	8	9

		♑ 12° 53′ 5″		
18h 56m 0s		MC		284° 0′ 0″

7h 0m 0s		MC		105° 0′ 0″
		♋ 13° 48′ 41″		

N LAT	11	12	Ascendant	2	3
°	° ′	° ′	° ′	° ′	° ′
0	♌12 32.1	♍13 43.2	♎16 16.8	♏17 27.9	♐16 11.3
5	13 2.5	14 6.8	15 41.8	16 27.6	15 32.5
10	13 32.8	14 29.7	15 8.6	15 28.4	14 53.2
15	14 3.7	14 52.2	14 36.7	14 29.3	14 12.9
16	14 9.9	14 56.7	14 30.4	14 17.4	14 4.7
17	14 16.3	15 1.2	14 24.2	14 5.5	13 56.3
18	14 22.6	15 5.7	14 17.9	13 53.5	13 47.9
19	14 29.0	15 10.2	14 11.7	13 41.5	13 39.4
20	14 35.5	15 14.7	14 5.5	13 29.5	13 30.8
21	14 42.0	15 19.2	13 59.3	13 17.3	13 22.1
22	14 48.6	15 23.8	13 53.2	13 5.1	13 13.2
23	14 55.3	15 28.3	13 47.0	12 52.8	13 4.3
24	15 2.0	15 32.9	13 40.8	12 40.4	12 55.2
25	15 8.8	15 37.5	13 34.6	12 27.9	12 46.0
26	15 15.7	15 42.2	13 28.4	12 15.3	12 36.6
27	15 22.8	15 46.8	13 22.2	12 2.6	12 27.0
28	15 29.9	15 51.5	13 16.0	11 49.8	12 17.3
29	15 37.1	15 56.2	13 9.8	11 36.8	12 7.4
30	15 44.4	16 1.0	13 3.5	11 23.6	11 57.4
31	15 51.9	16 5.8	12 57.2	11 10.4	11 47.1
32	15 59.5	16 10.7	12 50.9	10 56.9	11 36.6
33	16 7.2	16 15.6	12 44.5	10 43.3	11 25.8
34	16 15.1	16 20.6	12 38.1	10 29.5	11 14.9
35	16 23.1	16 25.6	12 31.7	10 15.4	11 3.6
36	16 31.3	16 30.7	12 25.2	10 1.2	10 52.1
37	16 39.7	16 35.8	12 18.6	9 46.7	10 40.2
38	16 48.3	16 41.0	12 12.0	9 32.0	10 28.1
39	16 57.1	16 46.3	12 5.4	9 17.0	10 15.5
40	17 6.0	16 51.7	11 58.6	9 1.8	10 2.7
41	17 15.3	16 57.2	11 51.8	8 46.2	9 49.4
42	17 24.7	17 2.7	11 45.0	8 30.4	9 35.7
43	17 34.4	17 8.4	11 38.0	8 14.2	9 21.5
44	17 44.4	17 14.1	11 30.9	7 57.7	9 6.9
45	17 54.7	17 20.0	11 23.8	7 40.8	8 51.7
46	18 5.4	17 26.0	11 16.5	7 23.5	8 36.0
47	18 16.3	17 32.1	11 9.2	7 5.8	8 19.7
48	18 27.6	17 38.3	11 1.7	6 47.7	8 2.7
49	18 39.3	17 44.7	10 54.1	6 29.1	7 44.9
50	18 51.4	17 51.2	10 46.4	6 10.0	7 26.4
51	19 4.0	17 57.9	10 38.5	5 50.3	7 7.1
52	19 17.0	18 4.8	10 30.5	5 30.2	6 46.8
53	19 30.6	18 11.8	10 22.3	5 9.4	6 25.5
54	19 44.7	18 19.0	10 14.0	4 47.9	6 3.2
55	19 59.4	18 26.4	10 5.4	4 25.8	5 39.6
56	20 14.8	18 34.1	9 56.7	4 3.0	5 14.6
57	20 30.9	18 41.9	9 47.8	3 39.4	4 48.2
58	20 47.7	18 50.1	9 38.7	3 15.0	4 20.1
59	21 5.4	18 58.4	9 29.4	2 49.7	3 50.1
60	21 24.0	19 7.1	9 19.8	2 23.5	3 18.0
61	21 43.6	19 16.0	9 10.0	1 56.2	2 43.6
62	22 4.3	19 25.3	8 59.9	1 27.9	2 6.5
63	22 26.3	19 34.9	8 49.5	0 58.4	1 26.3
64	22 49.5	19 44.9	8 38.8	0 27.7	0 42.5
65	23 14.2	19 55.3	8 27.8	♎29 55.6	♏29 54.5
66	23 40.6	20 6.1	8 16.5	29 22.1	29 1.5
S LAT	5	6	Descendant	8	9

		♑ 13° 48′ 41″		
19h 0m 0s		MC		285° 0′ 0″

7h 4m 0s — MC — 106° 0′ 0″

♋ 14° 44′ 21″

N LAT	11	12	Ascendant	2	3
°	° ′	° ′	° ′	° ′	° ′
0	♌13 32.0	♍14 47.8	♎17 21.4	♏18 27.6	♐17 6.9
5	14 1.9	15 9.9	16 44.1	17 26.4	16 28.0
10	14 31.8	15 31.4	16 8.8	16 26.3	15 48.6
15	15 2.2	15 52.4	15 34.8	15 26.3	15 8.1
16	15 8.3	15 56.6	15 28.2	15 14.2	14 59.8
17	15 14.5	16 0.8	15 21.5	15 2.1	14 51.4
18	15 20.8	16 5.0	15 14.9	14 49.9	14 42.9
19	15 27.1	16 9.2	15 8.3	14 37.7	14 34.4
20	15 33.5	16 13.5	15 1.7	14 25.4	14 25.7
21	15 39.9	16 17.7	14 55.1	14 13.1	14 17.0
22	15 46.4	16 21.9	14 48.5	14 0.7	14 8.1
23	15 52.9	16 26.2	14 41.9	13 48.2	13 59.1
24	15 59.5	16 30.5	14 35.3	13 35.5	13 50.0
25	16 6.3	16 34.8	14 28.7	13 22.8	13 40.7
26	16 13.1	16 39.1	14 22.1	13 10.0	13 31.3
27	16 20.0	16 43.5	14 15.5	12 57.1	13 21.7
28	16 26.9	16 47.9	14 8.9	12 44.0	13 11.9
29	16 34.0	16 52.3	14 2.3	12 30.8	13 2.0
30	16 41.3	16 56.7	13 55.6	12 17.4	12 51.8
31	16 48.6	17 1.2	13 48.9	12 3.9	12 41.5
32	16 56.1	17 5.8	13 42.1	11 50.2	12 30.9
33	17 3.6	17 10.4	13 35.4	11 36.3	12 20.1
34	17 11.4	17 15.0	13 28.5	11 22.2	12 9.1
35	17 19.3	17 19.7	13 21.7	11 7.9	11 57.7
36	17 27.3	17 24.5	13 14.8	10 53.4	11 46.1
37	17 35.6	17 29.3	13 7.8	10 38.7	11 34.2
38	17 44.0	17 34.1	13 0.7	10 23.7	11 21.9
39	17 52.6	17 39.1	12 53.6	10 8.4	11 9.3
40	18 1.4	17 44.1	12 46.5	9 52.9	10 56.4
41	18 10.5	17 49.2	12 39.2	9 37.0	10 43.0
42	18 19.8	17 54.4	12 31.9	9 20.9	10 29.2
43	18 29.3	17 59.7	12 24.4	9 4.4	10 14.9
44	18 39.1	18 5.1	12 16.9	8 47.5	10 0.2
45	18 49.2	18 10.6	12 9.3	8 30.3	9 44.9
46	18 59.6	18 16.1	12 1.6	8 12.6	9 29.0
47	19 10.4	18 21.8	11 53.7	7 54.6	9 12.5
48	19 21.5	18 27.7	11 45.7	7 36.0	8 55.4
49	19 32.9	18 33.6	11 37.6	7 17.0	8 37.5
50	19 44.8	18 39.7	11 29.4	6 57.5	8 18.8
51	19 57.1	18 46.0	11 21.0	6 37.5	7 59.3
52	20 9.9	18 52.4	11 12.4	6 16.9	7 38.8
53	20 23.2	18 58.9	11 3.7	5 55.6	7 17.3
54	20 37.0	19 5.7	10 54.8	5 33.7	6 54.6
55	20 51.4	19 12.6	10 45.8	5 11.1	6 30.8
56	21 6.5	19 19.7	10 36.5	4 47.8	6 5.5
57	21 22.2	19 27.1	10 27.0	4 23.7	5 38.7
58	21 38.7	19 34.6	10 17.3	3 58.7	5 10.2
59	21 56.0	19 42.4	10 7.3	3 32.9	4 39.9
60	22 14.1	19 50.5	9 57.1	3 6.0	4 7.3
61	22 33.3	19 58.9	9 46.6	2 38.1	3 32.3
62	22 53.5	20 7.5	9 35.8	2 9.1	2 54.6
63	23 14.9	20 16.5	9 24.8	1 39.0	2 13.7
64	23 37.6	20 25.8	9 13.4	1 7.5	1 29.0
65	24 1.7	20 35.5	9 1.6	0 34.7	0 40.0
66	24 27.4	20 45.6	8 49.5	0 0.3	♏29 45.7
S LAT	5	6	Descendant	8	9

♑ 14° 44′ 21″

19h 4m 0s — MC — 286° 0′ 0″

7h 8m 0s — MC — 107° 0′ 0″

♋ 15° 40′ 7″

N LAT	11	12	Ascendant	2	3
°	° ′	° ′	° ′	° ′	° ′
0	♌14 32.0	♍15 52.5	♎18 25.8	♏19 27.1	♐18 2.4
5	15 1.5	16 13.1	17 46.3	18 25.1	17 23.4
10	15 30.9	16 33.1	17 8.9	17 24.1	16 43.8
15	16 0.8	16 52.7	16 32.9	16 23.1	16 3.2
16	16 6.8	16 56.6	16 25.8	16 10.9	15 54.8
17	16 13.0	17 0.5	16 18.8	15 58.6	15 46.4
18	16 19.1	17 4.4	16 11.7	15 46.2	15 37.9
19	16 25.3	17 8.3	16 4.7	15 33.8	15 29.4
20	16 31.6	17 12.3	15 57.7	15 21.3	15 20.7
21	16 37.9	17 16.2	15 50.7	15 8.8	15 11.9
22	16 44.3	17 20.2	15 43.8	14 56.1	15 2.9
23	16 50.7	17 24.1	15 36.8	14 43.4	14 53.9
24	16 57.2	17 28.1	15 29.8	14 30.6	14 44.7
25	17 3.8	17 32.1	15 22.8	14 17.7	14 35.4
26	17 10.5	17 36.1	15 15.8	14 4.6	14 25.9
27	17 17.3	17 40.2	15 8.8	13 51.5	14 16.3
28	17 24.1	17 44.3	15 1.8	13 38.2	14 6.5
29	17 31.1	17 48.4	14 54.7	13 24.7	13 56.5
30	17 38.2	17 52.5	14 47.6	13 11.1	13 46.3
31	17 45.4	17 56.7	14 40.5	12 57.4	13 35.9
32	17 52.7	18 0.9	14 33.4	12 43.4	13 25.2
33	18 0.2	18 5.2	14 26.2	12 29.3	13 14.4
34	18 7.8	18 9.5	14 18.9	12 14.9	13 3.3
35	18 15.5	18 13.9	14 11.7	12 0.4	12 51.9
36	18 23.5	18 18.3	14 4.3	11 45.6	12 40.2
37	18 31.5	18 22.8	13 56.9	11 30.6	12 28.2
38	18 39.8	18 27.3	13 49.4	11 15.3	12 15.8
39	18 48.3	18 31.9	13 41.9	10 59.8	12 3.2
40	18 56.9	18 36.6	13 34.3	10 43.9	11 50.1
41	19 5.8	18 41.3	13 26.6	10 27.8	11 36.6
42	19 14.9	18 46.1	13 18.8	10 11.3	11 22.7
43	19 24.3	18 51.0	13 10.9	9 54.5	11 8.3
44	19 33.9	18 56.0	13 2.9	9 37.3	10 53.5
45	19 43.8	19 1.1	12 54.8	9 19.7	10 38.1
46	19 54.0	19 6.3	12 46.6	9 1.7	10 22.1
47	20 4.5	19 11.6	12 38.2	8 43.3	10 5.4
48	20 15.4	19 17.0	12 29.8	8 24.4	9 48.1
49	20 26.6	19 22.6	12 21.2	8 5.0	9 30.1
50	20 38.3	19 28.2	12 12.4	7 45.1	9 11.2
51	20 50.3	19 34.0	12 3.5	7 24.6	8 51.5
52	21 2.8	19 40.0	11 54.4	7 3.6	8 30.8
53	21 15.8	19 46.1	11 45.2	6 41.9	8 9.1
54	21 29.4	19 52.3	11 35.7	6 19.5	7 46.2
55	21 43.5	19 58.8	11 26.1	5 56.4	7 22.0
56	21 58.2	20 5.4	11 16.2	5 32.6	6 56.5
57	22 13.6	20 12.2	11 6.1	5 7.9	6 29.3
58	22 29.7	20 19.2	10 55.8	4 42.4	6 0.5
59	22 46.6	20 26.5	10 45.2	4 16.0	5 29.7
60	23 4.3	20 34.0	10 34.4	3 48.5	4 56.6
61	23 23.0	20 41.7	10 23.2	3 20.0	4 21.1
62	23 42.8	20 49.8	10 11.8	2 50.3	3 42.7
63	24 3.7	20 58.1	10 0.0	2 19.5	3 1.0
64	24 25.8	21 6.7	9 47.9	1 47.3	2 15.5
65	24 49.3	21 15.7	9 35.4	1 13.6	1 25.4
66	25 14.3	21 25.1	9 22.5	0 38.5	0 29.8
S LAT	5	6	Descendant	8	9

♑ 15° 40′ 7″

19h 8m 0s — MC — 287° 0′ 0″

7ʰ 12ᵐ 0ˢ		MC		108° 0′ 0″
		♋ 16° 35′ 58″		

11	12	Ascendant	2	3	N LAT
° ′	° ′	° ′	° ′	° ′	°
♌15 32.3	♍16 57.4	♎19 30.1	♏20 26.4	♐18 57.9	0
16 1.2	17 16.4	18 48.4	19 23.6	18 18.7	5
16 30.2	17 34.9	18 8.9	18 21.7	17 39.0	10
16 59.6	17 53.0	17 30.9	17 19.9	16 58.2	15
17 5.5	17 56.6	17 23.4	17 7.4	16 49.9	16
17 11.5	18 0.3	17 16.0	16 54.9	16 41.4	17
17 17.6	18 3.9	17 8.6	16 42.4	16 32.9	18
17 23.7	18 7.5	17 1.2	16 29.8	16 24.3	19
17 29.8	18 11.1	16 53.8	16 17.1	16 15.6	20
17 36.0	18 14.8	16 46.4	16 4.3	16 6.7	21
17 42.3	18 18.4	16 39.0	15 51.5	15 57.8	22
17 48.6	18 22.1	16 31.6	15 38.6	15 48.7	23
17 55.0	18 25.8	16 24.2	15 25.6	15 39.5	24
18 1.5	18 29.5	16 16.9	15 12.4	15 30.1	25
18 8.1	18 33.2	16 9.5	14 59.2	15 20.6	26
18 14.7	18 36.9	16 2.0	14 45.8	15 10.9	27
18 21.5	18 40.7	15 54.6	14 32.3	15 1.0	28
18 28.3	18 44.5	15 47.1	14 18.6	14 51.0	29
18 35.3	18 48.3	15 39.7	14 4.8	14 40.7	30
18 42.3	18 52.2	15 32.1	13 50.8	14 30.3	31
18 49.5	18 56.1	15 24.6	13 36.6	14 19.6	32
18 56.8	19 0.0	15 17.0	13 22.2	14 8.6	33
19 4.3	19 4.0	15 9.3	13 7.6	13 57.4	34
19 11.9	19 8.1	15 1.6	12 52.8	13 46.0	35
19 19.7	19 12.1	14 53.8	12 37.7	13 34.2	36
19 27.6	19 16.3	14 46.0	12 22.4	13 22.1	37
19 35.7	19 20.5	14 38.1	12 6.9	13 9.7	38
19 44.0	19 24.7	14 30.1	11 51.0	12 57.0	39
19 52.5	19 29.0	14 22.1	11 34.9	12 43.8	40
20 1.2	19 33.4	14 13.9	11 18.5	12 30.3	41
20 10.1	19 37.9	14 5.7	11 1.7	12 16.2	42
20 19.3	19 42.4	13 57.3	10 44.5	12 1.8	43
20 28.7	19 47.0	13 48.9	10 27.0	11 46.8	44
20 38.4	19 51.7	13 40.3	10 9.1	11 31.2	45
20 48.4	19 56.5	13 31.6	9 50.7	11 15.1	46
20 58.7	20 1.4	13 22.8	9 32.0	10 58.4	47
21 9.4	20 6.4	13 13.8	9 12.7	10 40.9	48
21 20.4	20 11.5	13 4.7	8 52.9	10 22.7	49
21 31.8	20 16.8	12 55.4	8 32.6	10 3.6	50
21 43.6	20 22.1	12 46.0	8 11.7	9 43.7	51
21 55.8	20 27.6	12 36.4	7 50.2	9 22.8	52
22 8.5	20 33.2	12 26.6	7 28.1	9 0.9	53
22 21.8	20 39.0	12 16.6	7 5.2	8 37.7	54
22 35.6	20 44.9	12 6.4	6 41.7	8 13.3	55
22 50.0	20 51.0	11 55.9	6 17.3	7 47.5	56
23 5.0	20 57.3	11 45.2	5 52.1	7 20.0	57
23 20.7	21 3.8	11 34.3	5 26.1	6 50.8	58
23 37.2	21 10.5	11 23.1	4 59.0	6 19.5	59
23 54.6	21 17.4	11 11.6	4 31.0	5 46.0	60
24 12.8	21 24.6	10 59.8	4 1.8	5 10.0	61
24 32.1	21 32.0	10 47.7	3 31.5	4 30.9	62
24 52.5	21 39.7	10 35.2	2 59.9	3 48.5	63
25 14.0	21 47.7	10 22.4	2 27.0	3 2.0	64
25 36.9	21 56.0	10 9.1	1 52.6	2 10.8	65
26 1.3	22 4.6	9 55.5	1 16.6	1 13.9	66
5	6	Descendant	8	9	S LAT

		♑ 16° 35′ 58″		
19ʰ 12ᵐ 0ˢ		MC		288° 0′ 0″

7ʰ 16ᵐ 0ˢ		MC		109° 0′ 0″
		♋ 17° 31′ 54″		

N LAT	11	12	Ascendant	2	3
°	° ′	° ′	° ′	° ′	° ′
0	♌16 32.7	♍18 2.3	♎20 34.3	♏21 25.6	♐19 53.3
5	17 1.2	18 19.8	19 50.5	20 21.9	19 14.0
10	17 29.6	18 36.7	19 8.9	19 19.2	18 34.2
15	17 58.5	18 53.4	18 28.9	18 16.5	17 53.3
16	18 4.3	18 56.7	18 21.0	18 3.9	17 44.9
17	18 10.2	19 0.0	18 13.2	17 51.2	17 36.4
18	18 16.2	19 3.4	18 5.3	17 38.4	17 27.8
19	18 22.2	19 6.7	17 57.5	17 25.6	17 19.2
20	18 28.2	19 10.0	17 49.7	17 12.8	17 10.4
21	18 34.3	19 13.4	17 42.0	16 59.8	17 1.5
22	18 40.4	19 16.7	17 34.2	16 46.8	16 52.5
23	18 46.6	19 20.1	17 26.4	16 33.7	16 43.4
24	18 52.9	19 23.5	17 18.6	16 20.4	16 34.2
25	18 59.3	19 26.9	17 10.9	16 7.1	16 24.8
26	19 5.7	19 30.3	17 3.1	15 53.6	16 15.2
27	19 12.3	19 33.7	16 55.2	15 40.0	16 5.5
28	19 18.9	19 37.2	16 47.4	15 26.3	15 55.5
29	19 25.6	19 40.7	16 39.5	15 12.4	15 45.4
30	19 32.4	19 44.2	16 31.7	14 58.3	15 35.1
31	19 39.4	19 47.7	16 23.7	14 44.1	15 24.6
32	19 46.4	19 51.3	16 15.8	14 29.7	15 13.9
33	19 53.6	19 54.9	16 7.7	14 15.0	15 2.9
34	20 0.9	19 58.6	15 59.7	14 0.2	14 51.6
35	20 8.4	20 2.3	15 51.5	13 45.1	14 40.1
36	20 16.0	20 6.0	15 43.3	13 29.8	14 28.3
37	20 23.8	20 9.8	15 35.1	13 14.3	14 16.1
38	20 31.7	20 13.7	15 26.7	12 58.4	14 3.6
39	20 39.8	20 17.6	15 18.3	12 42.3	13 50.8
40	20 48.2	20 21.5	15 9.8	12 25.9	13 37.6
41	20 56.7	20 25.5	15 1.2	12 9.1	13 23.9
42	21 5.4	20 29.6	14 52.5	11 52.0	13 9.8
43	21 14.4	20 33.8	14 43.7	11 34.6	12 55.2
44	21 23.7	20 38.0	14 34.8	11 16.7	12 40.1
45	21 33.2	20 42.3	14 25.8	10 58.5	12 24.5
46	21 43.0	20 46.7	14 16.6	10 39.8	12 8.2
47	21 53.1	20 51.2	14 7.3	10 20.6	11 51.3
48	22 3.5	20 55.8	13 57.8	10 1.0	11 33.7
49	22 14.2	21 0.5	13 48.2	9 40.8	11 15.3
50	22 25.4	21 5.3	13 38.4	9 20.1	10 56.1
51	22 36.9	21 10.2	13 28.5	8 58.8	10 36.0
52	22 48.9	21 15.2	13 18.3	8 36.9	10 14.9
53	23 1.3	21 20.4	13 8.0	8 14.3	9 52.8
54	23 14.3	21 25.7	12 57.4	7 51.0	9 29.4
55	23 27.8	21 31.1	12 46.6	7 26.9	9 4.7
56	23 41.8	21 36.7	12 35.6	7 2.1	8 38.5
57	23 56.5	21 42.5	12 24.3	6 36.3	8 10.7
58	24 11.9	21 48.5	12 12.8	6 9.7	7 41.1
59	24 28.0	21 54.6	12 1.0	5 42.1	7 9.5
60	24 44.9	22 0.9	11 48.8	5 13.4	6 35.5
61	25 2.7	22 7.5	11 36.4	4 43.6	5 58.8
62	25 21.5	22 14.3	11 23.6	4 12.6	5 19.2
63	25 41.3	22 21.3	11 10.4	3 40.3	4 35.9
64	26 2.4	22 28.7	10 56.8	3 6.6	3 48.6
65	26 24.7	22 36.3	10 42.9	2 31.5	2 56.3
66	26 48.4	22 44.2	10 28.5	1 54.7	1 58.0
S LAT	5	6	Descendant	8	9

		♑ 17° 31′ 54″		
19ʰ 16ᵐ 0ˢ		MC		289° 0′ 0″

$7^h\ 20^m\ 0^s$	MC		110° 0′ 0″		
		♋ 18° 27′ 56″			
11	**12**	**Ascendant**	**2**	**3**	**N LAT**
° ′	° ′	° ′	° ′	° ′	°
♌17 33.3	♍19 7.3	♎21 38.3	♏22 24.6	♐20 48.6	**0**
18 1.3	19 23.2	20 52.4	21 20.1	20 9.2	**5**
18 29.2	19 38.7	20 8.8	20 16.6	19 29.3	**10**
18 57.5	19 53.8	19 26.8	19 13.0	18 48.3	**15**
19 3.3	19 56.9	19 18.5	19 0.2	18 39.9	**16**
19 9.1	19 59.9	19 10.3	18 47.3	18 31.3	**17**
19 14.9	20 2.9	19 2.0	18 34.4	18 22.8	**18**
19 20.8	20 5.9	18 53.8	18 21.4	18 14.1	**19**
19 26.7	20 9.0	18 45.7	18 8.4	18 5.3	**20**
19 32.7	20 12.0	18 37.5	17 55.2	17 56.4	**21**
19 38.7	20 15.1	18 29.3	17 42.0	17 47.3	**22**
19 44.8	20 18.1	18 21.2	17 28.7	17 38.2	**23**
19 51.0	20 21.2	18 13.0	17 15.3	17 28.9	**24**
19 57.2	20 24.3	18 4.8	17 1.7	17 19.4	**25**
20 3.6	20 27.4	17 56.6	16 48.0	17 9.8	**26**
20 10.0	20 30.5	17 48.4	16 34.2	17 0.0	**27**
20 16.4	20 33.7	17 40.2	16 20.3	16 50.1	**28**
20 35.3	20 42.7	17 16.7	15 39.9	16 20.9	**29**
20 29.7	20 40.1	17 23.6	15 51.8	16 29.6	**30**
20 36.5	20 43.3	17 15.3	15 37.4	16 19.0	**31**
20 43.4	20 46.5	17 6.9	15 22.7	16 8.2	**32**
20 50.5	20 49.8	16 58.5	15 7.8	15 57.2	**33**
20 57.7	20 53.2	16 50.0	14 52.8	15 45.8	**34**
21 5.0	20 56.5	16 41.4	14 37.4	15 34.2	**35**
21 12.4	20 59.9	16 32.8	14 21.9	15 22.3	**36**
21 20.0	21 3.4	16 24.1	14 6.0	15 10.1	**37**
21 27.8	21 6.9	16 15.4	13 49.9	14 57.6	**38**
21 35.8	21 10.4	16 6.5	13 33.5	14 44.6	**39**
21 43.9	21 14.0	15 57.6	13 16.8	14 31.3	**40**
21 52.3	21 17.7	15 48.5	12 59.8	14 17.6	**41**
22 0.9	21 21.4	15 39.4	12 42.4	14 3.4	**42**
22 9.7	21 25.2	15 30.1	12 24.6	13 48.7	**43**
22 18.7	21 29.1	15 20.7	12 6.4	13 33.5	**44**
22 28.0	21 33.0	15 11.2	11 47.8	13 17.7	**45**
22 37.6	21 37.0	15 1.6	11 28.8	13 1.3	**46**
22 47.4	21 41.1	14 51.8	11 9.3	12 44.3	**47**
22 57.6	21 45.2	14 41.8	10 49.2	12 26.6	**48**
23 8.2	21 49.5	14 31.7	10 28.7	12 8.0	**49**
23 19.1	21 53.8	14 21.4	10 7.6	11 48.7	**50**
23 30.3	21 58.3	14 10.9	9 45.8	11 28.4	**51**
23 42.1	22 2.9	14 0.3	9 23.5	11 7.1	**52**
23 54.2	22 7.6	13 49.4	9 0.4	10 44.7	**53**
24 6.8	22 12.4	13 38.3	8 36.7	10 21.1	**54**
24 20.0	22 17.3	13 26.9	8 12.1	9 56.1	**55**
24 33.7	22 22.4	13 15.3	7 46.8	9 29.7	**56**
24 48.1	22 27.7	13 3.4	7 20.5	9 1.5	**57**
25 3.1	22 33.1	12 51.3	6 53.3	8 31.5	**58**
25 18.8	22 38.7	12 38.8	6 25.1	7 59.5	**59**
25 35.3	22 44.4	12 26.0	5 55.8	7 25.0	**60**
25 52.7	22 50.4	12 12.9	5 25.4	6 47.8	**61**
26 11.0	22 56.6	11 59.5	4 53.7	6 7.5	**62**
26 30.3	23 3.0	11 45.6	4 20.7	5 23.5	**63**
26 50.8	23 9.6	11 31.3	3 46.3	4 35.2	**64**
27 12.5	23 16.5	11 16.6	3 10.3	3 41.7	**65**
27 35.6	23 23.7	11 1.4	2 32.7	2 42.0	**66**
5	**6**	**Descendant**	**8**	**9**	**S LAT**
		♑ 18° 27′ 56″			
$19^h\ 20^m\ 0^s$	MC		290° 0′ 0″		

$7^h\ 24^m\ 0^s$	MC		111° 0′ 0″		
		♋ 19° 24′ 4″			
11	**12**	**Ascendant**	**2**	**3**	**N LAT**
° ′	° ′	° ′	° ′	° ′	°
♌18 34.0	♍20 12.3	♎22 42.3	♏23 23.4	♐21 43.9	**0**
19 1.5	20 26.7	21 54.2	22 18.2	21 4.4	**5**
19 29.0	20 40.6	21 8.6	21 13.8	20 24.4	**10**
19 56.8	20 54.3	20 24.6	20 9.4	19 43.2	**15**
20 2.4	20 57.0	20 15.9	19 56.4	19 34.8	**16**
20 8.1	20 59.8	20 7.3	19 43.4	19 26.3	**17**
20 13.8	21 2.5	19 58.7	19 30.3	19 17.6	**18**
20 19.6	21 5.2	19 50.1	19 17.1	19 8.9	**19**
20 25.4	21 8.0	19 41.5	19 3.9	19 0.1	**20**
20 31.2	21 10.7	19 33.0	18 50.6	18 51.1	**21**
20 37.1	21 13.4	19 24.4	18 37.1	18 42.1	**22**
20 43.1	21 16.2	19 15.9	18 23.6	18 32.9	**23**
20 49.2	21 19.0	19 7.3	18 10.0	18 23.5	**24**
20 55.3	21 21.8	18 58.7	17 56.2	18 14.1	**25**
21 1.5	21 24.6	18 50.1	17 42.3	18 4.4	**26**
21 7.8	21 27.4	18 41.5	17 28.3	17 54.6	**27**
21 14.1	21 30.2	18 32.9	17 14.1	17 44.6	**28**
21 32.6	21 38.4	18 8.3	16 33.2	17 15.3	**29**
21 27.1	21 36.0	18 15.5	16 45.3	17 24.0	**30**
21 33.8	21 38.9	18 6.8	16 30.6	17 13.4	**31**
21 40.6	21 41.8	17 58.0	16 15.7	17 2.5	**32**
21 47.5	21 44.8	17 49.2	16 0.6	16 51.4	**33**
21 54.5	21 47.8	17 40.3	15 45.3	16 40.1	**34**
22 1.6	21 50.8	17 31.3	15 29.7	16 28.4	**35**
22 9.0	21 53.9	17 22.3	15 13.9	16 16.4	**36**
22 16.4	21 57.0	17 13.2	14 57.8	16 4.1	**37**
22 24.0	22 0.1	17 4.0	14 41.4	15 51.5	**38**
22 31.8	22 3.3	16 54.7	14 24.7	15 38.5	**39**
21 18.3	22 6.6	16 45.3	14 7.7	15 25.1	**40**
22 48.0	22 9.9	16 35.8	13 50.4	15 11.3	**41**
22 56.4	22 13.2	16 26.2	13 32.6	14 57.0	**42**
23 5.0	22 16.6	16 16.5	13 14.6	14 42.2	**43**
23 13.8	22 20.1	16 6.7	12 56.1	14 26.9	**44**
23 22.9	22 23.6	15 56.7	12 37.1	14 11.0	**45**
23 32.3	22 27.2	15 46.6	12 17.7	13 54.5	**46**
23 41.9	22 30.9	15 36.3	11 57.9	13 37.4	**47**
23 51.9	22 34.7	15 25.8	11 37.5	13 19.5	**48**
24 2.2	22 38.5	15 15.2	11 16.5	13 0.8	**49**
24 12.8	22 42.4	15 4.4	10 55.0	12 41.3	**50**
24 23.8	22 46.4	14 53.4	10 32.9	12 20.8	**51**
24 35.3	22 50.6	14 42.2	10 10.1	11 59.3	**52**
24 47.1	22 54.8	14 30.8	9 46.6	11 36.7	**53**
24 59.5	22 59.1	14 19.1	9 22.4	11 12.8	**54**
25 12.3	23 3.6	14 7.2	8 57.3	10 47.6	**55**
25 25.7	23 8.2	13 55.0	8 31.4	10 20.9	**56**
25 39.7	23 12.9	13 42.5	8 4.7	9 52.4	**57**
25 54.4	23 17.7	13 29.8	7 36.9	9 22.0	**58**
26 9.7	23 22.8	13 16.7	7 8.1	8 49.5	**59**
26 25.8	23 28.0	13 3.3	6 38.2	8 14.6	**60**
26 42.7	23 33.3	12 49.5	6 7.1	7 36.8	**61**
27 0.5	23 38.9	12 35.3	5 34.8	6 55.8	**62**
27 19.4	23 44.6	12 20.8	5 1.0	6 11.0	**63**
27 39.3	23 50.6	12 5.8	4 25.8	5 21.8	**64**
28 0.4	23 56.8	11 50.3	3 49.1	4 27.2	**65**
28 22.8	24 3.3	11 34.4	3 10.6	3 25.9	**66**
5	**6**	**Descendant**	**8**	**9**	**S LAT**
		♑ 19° 24′ 4″			
$19^h\ 24^m\ 0^s$	MC		291° 0′ 0″		

7h 28m 0s		MC ♋ 20° 20′ 19″		112° 0′ 0″	
11	12	Ascendant	2	3	N LAT
° ′	° ′	° ′	° ′	° ′	°
♌19 35.0	♍21 17.5	♎23 46.0	♏24 22.0	♐22 39.2	0
20 2.0	21 30.3	22 55.9	23 16.1	21 59.6	5
20 28.5	21 42.7	22 8.3	22 11.0	21 19.5	10
20 56.1	21 54.8	21 22.3	21 5.7	20 38.2	15
21 1.6	21 57.3	21 13.3	20 52.5	20 29.7	16
21 7.2	21 59.7	21 4.3	20 39.3	20 21.2	17
21 12.8	22 2.1	20 55.3	20 26.1	20 12.5	18
21 18.5	22 4.5	20 46.3	20 12.7	20 3.8	19
21 24.2	22 7.0	20 37.3	19 59.3	19 54.9	20
21 29.9	22 9.4	20 28.4	19 45.8	19 45.9	21
21 35.7	22 11.9	20 19.4	19 32.2	19 36.8	22
21 41.6	22 14.3	20 10.5	19 18.5	19 27.6	23
21 47.5	22 16.8	20 1.6	19 4.6	19 18.2	24
21 53.5	22 19.3	19 52.6	18 50.7	19 8.7	25
21 59.5	22 21.8	19 43.6	18 36.6	18 59.0	26
22 5.7	22 24.3	19 34.6	18 22.4	18 49.2	27
22 11.9	22 26.8	19 25.6	18 8.0	18 39.1	28
22 18.2	22 29.3	19 16.5	17 53.4	18 28.9	29
22 24.7	22 31.9	19 7.4	17 38.7	18 18.4	30
22 31.2	22 34.5	18 58.3	17 23.8	18 7.8	31
22 37.8	22 37.1	18 49.1	17 8.6	17 56.9	32
22 44.6	22 39.7	18 39.9	16 53.3	17 45.7	33
22 51.4	22 42.4	18 30.5	16 37.7	17 34.3	34
22 58.4	22 45.1	18 21.2	16 21.9	17 22.6	35
23 5.6	22 47.8	18 11.7	16 5.8	17 10.5	36
23 12.9	22 50.6	18 2.2	15 49.5	16 58.2	37
23 20.3	22 53.4	17 52.6	15 32.8	16 45.5	38
23 28.0	22 56.2	17 42.8	15 15.9	16 32.4	39
23 35.8	22 59.1	17 33.0	14 58.6	16 18.9	40
23 43.7	23 2.1	17 23.1	14 40.9	16 5.0	41
23 51.9	23 5.0	17 13.0	14 22.9	15 50.6	42
24 0.4	23 8.1	17 2.9	14 4.5	15 35.8	43
24 9.0	23 11.2	16 52.6	13 45.7	15 20.4	44
24 17.9	23 14.3	16 42.1	13 26.4	15 4.4	45
24 27.0	23 17.5	16 31.5	13 6.7	14 47.7	46
24 36.5	23 20.8	16 20.7	12 46.5	14 30.5	47
24 46.2	23 24.1	16 9.8	12 25.7	14 12.4	48
24 56.3	23 27.5	15 58.7	12 4.4	13 53.6	49
25 6.7	23 31.0	15 47.4	11 42.5	13 33.9	50
25 17.4	23 34.6	15 35.8	11 19.9	13 13.3	51
25 28.6	23 38.2	15 24.1	10 56.7	12 51.6	52
25 40.2	23 42.0	15 12.1	10 32.7	12 28.8	53
25 52.2	23 45.9	14 59.9	10 8.0	12 4.7	54
26 4.7	23 49.8	14 47.4	9 42.5	11 39.2	55
26 17.8	23 53.9	14 34.6	9 16.1	11 12.1	56
26 31.4	23 58.1	14 21.6	8 48.8	10 43.4	57
26 45.7	24 2.4	14 8.2	8 20.5	10 12.6	58
27 0.6	24 6.9	13 54.5	7 51.1	9 39.7	59
27 16.3	24 11.5	13 40.5	7 20.6	9 4.3	60
27 32.8	24 16.3	13 26.0	6 48.8	8 25.9	61
27 50.1	24 21.2	13 11.2	6 15.8	7 44.3	62
28 8.5	24 26.3	12 55.9	5 41.3	6 58.7	63
28 27.8	24 31.6	12 40.2	5 5.4	6 8.5	64
28 48.4	24 37.2	12 24.0	4 27.8	5 12.6	65
29 10.2	24 42.9	12 7.3	3 48.5	4 9.9	66
5	6	Descendant	8	9	S LAT
19h 28m 0s		MC ♑ 20° 20′ 19″		292° 0′ 0″	

7h 32m 0s		MC ♋ 21° 16′ 40″		113° 0′ 0″	
11	12	Ascendant	2	3	N LAT
° ′	° ′	° ′	° ′	° ′	°
♌20 36.1	♍22 22.6	♎24 49.7	♏25 20.5	♐23 34.4	0
21 2.6	22 33.9	23 57.5	24 13.8	22 54.7	5
21 28.9	22 44.7	23 7.9	23 8.0	22 14.5	10
21 55.6	22 55.4	22 20.0	22 1.9	21 33.1	15
22 1.0	22 57.5	22 10.6	21 48.6	21 24.6	16
22 6.5	22 59.6	22 1.2	21 35.2	21 16.1	17
22 12.0	23 1.8	21 51.8	21 21.7	21 7.4	18
22 17.5	23 3.9	21 42.4	21 8.2	20 58.6	19
22 23.1	23 6.0	21 33.1	20 54.6	20 49.7	20
22 28.7	23 8.2	21 23.8	20 40.9	20 40.7	21
22 34.4	23 10.3	21 14.4	20 27.1	20 31.6	22
22 40.1	23 12.5	21 5.1	20 13.2	20 22.3	23
22 45.9	23 14.6	20 55.8	19 59.2	20 12.9	24
22 51.8	23 16.8	20 46.4	19 45.1	20 3.4	25
22 57.7	23 19.0	20 37.0	19 30.8	19 53.6	26
23 3.7	23 21.2	20 27.7	19 16.3	19 43.7	27
23 9.8	23 23.4	20 18.2	19 1.7	19 33.7	28
23 16.0	23 25.6	20 8.8	18 47.0	19 23.4	29
23 22.3	23 27.9	19 59.3	18 32.0	19 12.9	30
23 28.7	23 30.1	19 49.8	18 16.9	19 2.2	31
23 35.2	23 32.4	19 40.2	18 1.5	18 51.2	32
23 41.8	23 34.7	19 30.5	17 46.0	18 40.0	33
23 48.5	23 37.1	19 20.8	17 30.1	18 28.5	34
23 55.3	23 39.4	19 11.0	17 14.1	18 16.8	35
24 2.3	23 41.8	19 1.1	16 57.7	18 4.7	36
24 9.4	23 44.2	18 51.2	16 41.1	17 52.3	37
24 16.7	23 46.7	18 41.1	16 24.2	17 39.5	38
24 24.2	23 49.2	18 31.0	16 7.0	17 26.3	39
24 31.8	23 51.7	18 20.7	15 49.4	17 12.8	40
24 39.6	23 54.3	18 10.3	15 31.5	16 58.8	41
24 47.6	23 56.9	17 59.8	15 13.2	16 44.3	42
24 55.8	23 59.5	17 49.2	14 54.4	16 29.4	43
25 4.3	24 2.2	17 38.4	14 35.3	16 13.9	44
25 13.0	24 5.0	17 27.5	14 15.7	15 57.8	45
25 21.9	24 7.8	17 16.4	13 55.6	15 41.0	46
25 31.1	24 10.6	17 5.2	13 35.0	15 23.6	47
25 40.6	24 13.6	16 53.8	13 13.9	15 5.5	48
25 50.4	24 16.5	16 42.1	12 52.2	14 46.5	49
26 0.6	24 19.6	16 30.3	12 29.9	14 26.6	50
26 11.1	24 22.7	16 18.3	12 6.9	14 5.8	51
26 21.9	24 25.9	16 6.0	11 43.3	13 43.9	52
26 33.2	24 29.2	15 53.5	11 18.9	13 20.9	53
26 45.0	24 32.6	15 40.7	10 53.7	12 56.6	54
26 57.2	24 36.1	15 27.6	10 27.7	12 30.8	55
27 9.9	24 39.6	15 14.3	10 0.8	12 3.5	56
27 23.2	24 43.3	15 0.6	9 32.9	11 34.4	57
27 37.1	24 47.1	14 46.7	9 4.0	11 3.3	58
27 51.7	24 51.0	14 32.3	8 34.0	10 30.0	59
28 6.9	24 55.0	14 17.6	8 2.9	9 54.0	60
28 23.0	24 59.2	14 2.5	7 30.5	9 15.1	61
28 39.8	25 3.5	13 47.0	6 56.8	8 32.8	62
28 57.7	25 8.0	13 31.1	6 21.6	7 46.4	63
29 16.5	25 12.7	13 14.6	5 44.9	6 55.2	64
29 36.5	25 17.5	12 57.7	5 6.5	5 58.1	65
29 57.6	25 22.5	12 40.2	4 26.3	4 53.8	66
5	6	Descendant	8	9	S LAT
19h 32m 0s		MC ♑ 21° 16′ 40″		293° 0′ 0″	

7ʰ 36ᵐ 0ˢ MC 114° 0′ 0″

♋ 22° 13′ 8″

N LAT °	11 ° ′	12 ° ′	Ascendant ° ′	2 ° ′	3 ° ′
0	♌21 37.4	♍23 27.9	♎25 53.2	♏26 18.8	♐24 29.5
5	22 3.3	23 37.5	24 58.9	25 11.4	23 49.8
10	22 29.2	23 46.8	24 7.4	24 4.8	23 9.6
15	22 55.3	23 56.0	23 17.6	22 58.0	22 28.1
16	23 0.6	23 57.8	23 7.8	22 44.5	22 19.5
17	23 5.9	23 59.6	22 58.0	22 30.9	22 10.9
18	23 11.3	24 1.5	22 48.2	22 17.3	22 2.2
19	23 16.7	24 3.3	22 38.5	22 3.6	21 53.4
20	23 22.1	24 5.1	22 28.8	21 49.8	21 44.5
21	23 27.6	24 7.0	22 19.1	21 36.0	21 35.5
22	23 33.2	24 8.8	22 9.4	21 22.0	21 26.3
23	23 38.8	24 10.6	21 59.6	21 7.9	21 17.0
24	23 44.5	24 12.5	21 49.9	20 53.7	21 7.6
25	23 50.2	24 14.4	21 40.2	20 39.4	20 58.0
26	23 56.0	24 16.2	21 30.4	20 24.9	20 48.3
27	24 1.9	24 18.1	21 20.7	20 10.3	20 38.3
28	24 7.9	24 20.0	21 10.9	19 55.5	20 28.2
29	24 13.9	24 21.9	21 1.0	19 40.5	20 17.9
30	24 20.0	24 23.8	20 51.1	19 25.3	20 7.4
31	24 26.3	24 25.8	20 41.2	19 9.9	19 56.6
32	24 32.6	24 27.7	20 31.2	18 54.4	19 45.6
33	24 39.1	24 29.7	20 21.1	18 38.6	19 34.3
34	24 45.6	24 31.7	20 11.0	18 22.5	19 22.8
35	24 52.3	24 33.8	20 0.8	18 6.2	19 11.0
36	24 59.2	24 35.8	19 50.5	17 49.6	18 58.8
37	25 6.1	24 37.9	19 40.1	17 32.7	18 46.4
38	25 13.2	24 40.0	19 29.7	17 15.6	18 33.5
39	25 20.5	24 42.1	19 19.1	16 58.1	18 20.3
40	25 28.0	24 44.3	19 8.4	16 40.2	18 6.7
41	25 35.6	24 46.5	18 57.6	16 22.0	17 52.6
42	25 43.4	24 48.7	18 46.6	16 3.4	17 38.1
43	25 51.4	24 51.0	18 35.5	15 44.3	17 23.0
44	25 59.7	24 53.3	18 24.3	15 24.9	17 7.4
45	26 8.1	24 55.7	18 12.9	15 5.0	16 51.2
46	26 16.9	24 58.1	18 1.4	14 44.5	16 34.4
47	26 25.8	25 0.5	17 49.6	14 23.6	16 16.8
48	26 35.1	25 3.0	17 37.7	14 2.1	15 58.5
49	26 44.7	25 5.6	17 25.6	13 40.0	15 39.4
50	26 54.6	25 8.2	17 13.3	13 17.3	15 19.4
51	27 4.8	25 10.9	17 0.7	12 53.9	14 58.4
52	27 15.4	25 13.6	16 47.9	12 29.8	14 36.4
53	27 26.4	25 16.5	16 34.8	12 5.0	14 13.1
54	27 37.8	25 19.3	16 21.5	11 39.3	13 48.6
55	27 49.7	25 22.3	16 7.9	11 12.8	13 22.6
56	28 2.1	25 25.4	15 53.9	10 45.4	12 55.0
57	28 15.1	25 28.5	15 39.7	10 17.0	12 25.6
58	28 28.6	25 31.8	15 25.1	9 47.5	11 54.1
59	28 42.8	25 35.1	15 10.1	9 17.0	11 20.3
60	28 57.6	25 38.6	14 54.8	8 45.2	10 43.9
61	29 13.2	25 42.2	14 39.0	8 12.2	10 4.4
62	29 29.6	25 45.9	14 22.8	7 37.7	9 21.4
63	29 46.9	25 49.7	14 6.2	7 1.8	8 34.2
64	♍ 0 5.2	25 53.7	13 49.0	6 24.3	7 42.0
65	0 24.6	25 57.8	13 31.3	5 45.1	6 43.7
66	0 45.2	26 2.1	13 13.1	5 4.1	5 37.7
S LAT	5	6	Descendant	8	9

♑ 22° 13′ 8″

19ʰ 36ᵐ 0ˢ MC 294° 0′ 0″

7ʰ 40ᵐ 0ˢ MC 115° 0′ 0″

♋ 23° 9′ 44″

N LAT °	11 ° ′	12 ° ′	Ascendant ° ′	2 ° ′	3 ° ′
0	♌22 38.9	♍24 33.2	♎26 56.5	♏27 17.0	♐25 24.7
5	23 4.3	24 41.2	26 0.3	26 8.9	24 44.9
10	23 29.5	24 49.0	25 6.8	25 1.6	24 4.6
15	23 55.1	24 56.6	24 15.1	23 54.0	23 23.0
16	24 0.3	24 58.1	24 4.9	23 40.3	23 14.4
17	24 5.5	24 59.7	23 54.8	23 26.6	23 5.8
18	24 10.7	25 1.2	23 44.6	23 12.8	22 57.1
19	24 16.0	25 2.7	23 34.5	22 58.9	22 48.3
20	24 21.3	25 4.2	23 24.4	22 45.0	22 39.3
21	24 26.7	25 5.8	23 14.3	22 31.0	22 30.3
22	24 32.2	25 7.3	23 4.2	22 16.8	22 21.1
23	24 37.6	25 8.8	22 54.1	22 2.5	22 11.8
24	24 43.2	25 10.4	22 44.0	21 48.1	22 2.3
25	24 48.8	25 11.9	22 33.9	21 33.6	21 52.7
26	24 54.4	25 13.5	22 23.8	21 18.9	21 42.9
27	25 0.2	25 15.1	22 13.6	21 4.1	21 32.9
28	25 6.0	25 16.6	22 3.4	20 49.1	21 22.8
29	25 11.9	25 18.2	21 53.2	20 33.9	21 12.4
30	25 17.9	25 19.8	21 42.9	20 18.5	21 1.8
31	25 24.0	25 21.5	21 32.6	20 3.0	20 51.0
32	25 30.2	25 23.1	21 22.2	19 47.2	20 40.0
33	25 36.5	25 24.8	21 11.7	19 31.1	20 28.7
34	25 42.9	25 26.4	21 1.2	19 14.8	20 17.1
35	25 49.4	25 28.1	20 50.6	18 58.3	20 5.2
36	25 56.1	25 29.8	20 39.9	18 41.5	19 53.0
37	26 2.9	25 31.5	20 29.1	18 24.3	19 40.5
38	26 9.8	25 33.3	20 18.2	18 6.9	19 27.6
39	26 16.9	25 35.1	20 7.2	17 49.1	19 14.3
40	26 24.2	25 36.9	19 56.0	17 31.0	19 0.6
41	26 31.6	25 38.7	19 44.8	17 12.5	18 46.5
42	26 39.3	25 40.6	19 33.4	16 53.6	18 31.9
43	26 47.1	25 42.5	19 21.9	16 34.2	18 16.7
44	26 55.1	25 44.4	19 10.2	16 14.5	18 1.0
45	27 3.4	25 46.4	18 58.3	15 54.2	17 44.7
46	27 11.9	25 48.4	18 46.3	15 33.4	17 27.8
47	27 20.6	25 50.4	18 34.1	15 12.1	17 10.1
48	27 29.7	25 52.5	18 21.7	14 50.3	16 51.7
49	27 39.0	25 54.6	18 9.0	14 27.8	16 32.5
50	27 48.6	25 56.8	17 56.2	14 4.7	16 12.3
51	27 58.6	25 59.1	17 43.1	13 40.9	15 51.2
52	28 8.9	26 1.4	17 29.8	13 16.4	15 28.9
53	28 19.6	26 3.7	17 16.2	12 51.1	15 5.5
54	28 30.8	26 6.1	17 2.3	12 25.0	14 40.7
55	28 42.4	26 8.6	16 48.1	11 58.0	14 14.4
56	28 54.4	26 11.1	16 33.6	11 30.0	13 46.6
57	29 7.0	26 13.8	16 18.7	11 1.1	13 16.8
58	29 20.2	26 16.5	16 3.5	10 31.0	12 45.0
59	29 33.9	26 19.3	15 47.9	9 59.9	12 10.8
60	29 48.4	26 22.1	15 32.0	9 27.5	11 33.9
61	♍ 0 3.5	26 25.1	15 15.5	8 53.8	10 53.8
62	0 19.5	26 28.2	14 58.7	8 18.7	10 10.2
63	0 36.3	26 31.4	14 41.3	7 42.0	9 22.1
64	0 54.0	26 34.7	14 23.4	7 3.7	8 28.9
65	1 12.9	26 38.2	14 5.0	6 23.7	7 29.3
66	1 32.8	26 41.8	13 45.9	5 41.8	6 21.5
S LAT	5	6	Descendant	8	9

♑ 23° 9′ 44″

19ʰ 40ᵐ 0ˢ MC 295° 0′ 0″

7h 44m 0s	MC		116° 0′ 0″		
	♋ 24° 6′ 26″				
11	12	Ascendant	2	3	N LAT
° ′	° ′	° ′	° ′	° ′	°
♌23 40.6	♍25 38.5	♎27 59.7	♏28 15.0	♐26 19.8	0
24 5.4	25 44.9	27 1.5	27 6.3	25 40.0	5
24 30.1	25 51.2	26 6.1	25 58.2	24 59.6	10
24 55.0	25 57.3	25 12.5	24 49.8	24 17.9	15
25 0.1	25 58.5	25 2.0	24 36.0	24 9.3	16
25 5.2	25 59.7	24 51.4	24 22.2	24 0.7	17
25 10.3	26 0.9	24 40.9	24 8.2	23 51.9	18
25 15.5	26 2.1	24 30.4	23 54.2	23 43.1	19
25 20.7	26 3.4	24 20.0	23 40.1	23 34.1	20
25 25.9	26 4.6	24 9.5	23 25.9	23 25.1	21
25 31.2	26 5.8	23 59.0	23 11.5	23 15.8	22
25 36.6	26 7.0	23 48.6	22 57.1	23 6.5	23
25 42.0	26 8.3	23 38.1	22 42.5	22 57.0	24
25 47.5	26 9.5	23 27.6	22 27.8	22 47.4	25
25 53.0	26 10.8	23 17.1	22 12.9	22 37.5	26
25 58.6	26 12.0	23 6.5	21 57.9	22 27.5	27
26 4.3	26 13.3	22 55.9	21 42.7	22 17.4	28
26 10.1	26 14.6	22 45.3	21 27.3	22 7.0	29
26 15.9	26 15.9	22 34.6	21 11.7	21 56.4	30
26 21.8	26 17.2	22 23.9	20 55.9	21 45.5	31
26 27.9	26 18.5	22 13.1	20 39.9	21 34.4	32
26 34.0	26 19.8	22 2.3	20 23.6	21 23.1	33
26 40.3	26 21.1	21 51.3	20 7.1	21 11.5	34
26 46.6	26 22.5	21 40.3	19 50.3	20 59.5	35
26 53.1	26 23.8	21 29.2	19 33.3	20 47.3	36
26 59.8	26 25.2	21 18.0	19 15.9	20 34.7	37
27 6.5	26 26.6	21 6.7	18 58.2	20 21.7	38
27 13.5	26 28.1	20 55.2	18 40.1	20 8.4	39
27 20.5	26 29.5	20 43.7	18 21.7	19 54.6	40
27 27.8	26 31.0	20 32.0	18 2.9	19 40.4	41
27 35.2	26 32.5	20 20.1	17 43.7	19 25.7	42
27 42.8	26 34.0	20 8.1	17 24.1	19 10.5	43
27 50.7	26 35.5	19 56.0	17 4.0	18 54.7	44
27 58.7	26 37.1	19 43.7	16 43.4	18 38.3	45
28 7.0	26 38.7	19 31.2	16 22.3	18 21.3	46
28 15.5	26 40.3	19 18.5	16 0.7	18 3.5	47
28 24.3	26 42.0	19 5.6	15 38.4	17 44.9	48
28 33.4	26 43.7	18 52.5	15 15.6	17 25.6	49
28 42.8	26 45.5	18 39.1	14 52.1	17 5.3	50
28 52.5	26 47.2	18 25.5	14 27.9	16 44.0	51
29 2.5	26 49.1	18 11.6	14 2.9	16 21.5	52
29 12.9	26 51.0	17 57.5	13 37.2	15 57.9	53
29 23.8	26 52.9	17 43.0	13 10.6	15 32.9	54
29 35.0	26 54.9	17 28.3	12 43.1	15 6.4	55
29 46.8	26 56.9	17 13.2	12 14.6	14 38.2	56
29 59.0	26 59.0	16 57.8	11 45.1	14 8.2	57
♍ 0 11.8	27 1.2	16 41.9	11 14.5	13 36.0	58
0 25.2	27 3.4	16 25.7	10 42.8	13 1.4	59
0 39.2	27 5.7	16 9.1	10 9.8	12 24.0	60
0 53.9	27 8.1	15 52.0	9 35.4	11 43.4	61
1 9.4	27 10.6	15 34.4	8 59.6	10 59.0	62
1 25.7	27 13.1	15 16.4	8 22.2	10 10.1	63
1 42.9	27 15.8	14 57.8	7 43.1	9 15.9	64
2 1.2	27 18.5	14 38.6	7 2.2	8 14.9	65
2 20.5	27 21.4	14 18.8	6 19.4	7 5.4	66
5	6	Descendant	8	9	S LAT
	♑ 24° 6′ 26″				
19h 44m 0s	MC		296° 0′ 0″		

7h 48m 0s	MC		117° 0′ 0″		
	♋ 25° 3′ 17″				
11	12	Ascendant	2	3	N LAT
° ′	° ′	° ′	° ′	° ′	°
♌24 42.5	♍26 43.8	♎29 2.8	♏29 12.8	♐27 14.8	0
25 6.7	26 48.7	28 2.6	28 3.4	26 35.0	5
25 30.8	26 53.3	27 5.3	26 54.7	25 54.6	10
25 55.1	26 57.9	26 9.9	25 45.6	25 12.8	15
26 0.1	26 58.8	25 59.0	25 31.7	25 4.2	16
26 5.0	26 59.8	25 48.0	25 17.6	24 55.6	17
26 10.0	27 0.7	25 37.2	25 3.5	24 46.8	18
26 15.1	27 1.6	25 26.3	24 49.3	24 37.9	19
26 20.2	27 2.5	25 15.5	24 35.1	24 29.0	20
26 25.3	27 3.4	25 4.6	24 20.7	24 19.9	21
26 30.4	27 4.3	24 53.8	24 6.2	24 10.6	22
26 35.7	27 5.3	24 42.9	23 51.6	24 1.3	23
26 40.9	27 6.2	24 32.1	23 36.8	23 51.7	24
26 46.3	27 7.1	24 21.2	23 21.9	23 42.1	25
26 51.7	27 8.1	24 10.3	23 6.9	23 32.2	26
26 57.1	27 9.0	23 59.4	22 51.6	23 22.2	27
27 2.7	27 10.0	23 48.4	22 36.2	23 12.0	28
27 8.3	27 10.9	23 37.4	22 20.7	23 1.6	29
27 14.0	27 11.9	23 26.3	22 4.9	22 50.9	30
27 19.8	27 12.9	23 15.2	21 48.9	22 40.0	31
27 25.7	27 13.8	23 4.0	21 32.6	22 28.9	32
27 31.7	27 14.8	22 52.8	21 16.1	22 17.5	33
27 37.8	27 15.8	22 41.4	20 59.4	22 5.9	34
27 44.0	27 16.8	22 30.0	20 42.4	21 53.9	35
27 50.3	27 17.9	22 18.5	20 25.0	21 41.6	36
27 56.7	27 18.9	22 6.9	20 7.4	21 28.9	37
28 3.3	27 20.0	21 55.1	19 49.5	21 15.9	38
28 10.1	27 21.0	21 43.3	19 31.1	21 2.5	39
28 17.0	27 22.1	21 31.3	19 12.5	20 48.7	40
28 24.0	27 23.2	21 19.1	18 53.4	20 34.4	41
28 31.3	27 24.3	21 6.9	18 33.9	20 19.6	42
28 38.7	27 25.5	20 54.4	18 14.0	20 4.3	43
28 46.3	27 26.6	20 41.8	17 53.5	19 48.4	44
28 54.1	27 27.8	20 29.0	17 32.6	19 32.0	45
29 2.2	27 29.0	20 16.1	17 11.2	19 14.8	46
29 10.5	27 30.2	20 2.9	16 49.2	18 56.9	47
29 19.0	27 31.5	19 49.5	16 26.6	18 38.3	48
29 27.9	27 32.8	19 35.9	16 3.4	18 18.8	49
29 37.0	27 34.1	19 22.0	15 39.5	17 58.3	50
29 46.4	27 35.4	19 7.9	15 14.8	17 36.8	51
29 56.2	27 36.8	18 53.5	14 49.5	17 14.3	52
♍ 0 6.3	27 38.2	18 38.8	14 23.3	16 50.4	53
0 16.9	27 39.7	18 23.8	13 56.2	16 25.2	54
0 27.8	27 41.1	18 8.5	13 28.2	15 58.5	55
0 39.2	27 42.7	17 52.8	12 59.2	15 30.0	56
0 51.1	27 44.2	17 36.8	12 29.2	14 59.7	57
1 3.5	27 45.9	17 20.3	11 58.0	14 27.1	58
1 16.5	27 47.5	17 3.5	11 25.7	13 52.1	59
1 30.1	27 49.3	16 46.2	10 52.0	13 14.2	60
1 44.4	27 51.1	16 28.5	10 17.0	12 33.0	61
1 59.4	27 52.9	16 10.2	9 40.4	11 48.0	62
2 15.2	27 54.8	15 51.4	9 2.3	10 58.3	63
2 31.9	27 56.8	15 32.1	8 22.4	10 2.9	64
2 49.6	27 58.9	15 12.2	7 40.7	9 0.6	65
3 8.3	28 1.1	14 51.6	6 57.0	7 49.2	66
5	6	Descendant	8	9	S LAT
	♑ 25° 3′ 17″				
19h 48m 0s	MC		297° 0′ 0″		

$7^h\ 52^m\ 0^s$	MC 118° 0′ 0″ ♋ 26° 0′ 15″				N	$7^h\ 56^m\ 0^s$	MC 119° 0′ 0″ ♋ 26° 57′ 21″			
11	**12**	**Ascendant**	**2**	**3**	**LAT**	**11**	**12**	**Ascendant**	**2**	**3**
° ′	° ′	° ′	° ′	° ′	°	° ′	° ′	° ′	° ′	° ′
♌25 44.5	♍27 49.2	♏ 0 5.7	♐ 0 10.5	♐28 9.9	**0**	♌26 46.7	♍28 54.6	♏ 1 8.4	♐ 1 8.0	♐29 5.0
26 8.1	27 52.4	♎29 3.6	♏29 0.5	27 30.0	**5**	27 9.7	28 56.2	0 4.4	♏29 57.5	28 25.1
26 31.6	27 55.6	28 4.4	27 51.1	26 49.6	**10**	27 32.6	28 57.8	♎29 3.4	28 47.4	27 44.6
26 55.4	27 58.6	27 7.2	26 41.3	26 7.7	**15**	27 55.7	28 59.3	28 4.3	27 36.9	27 2.7
27 0.2	27 59.2	26 55.9	26 27.2	25 59.2	**16**	28 0.4	28 59.6	27 52.7	27 22.7	26 54.1
27 5.0	27 59.8	26 44.6	26 13.0	25 50.5	**17**	28 5.1	28 59.9	27 41.0	27 8.3	26 45.4
27 9.9	28 0.4	26 33.3	25 58.8	25 41.7	**18**	28 9.9	29 0.2	27 29.4	26 53.9	26 36.6
27 14.8	28 1.1	26 22.1	25 44.4	25 32.8	**19**	28 14.7	29 0.5	27 17.8	26 39.4	26 27.7
27 19.8	28 1.7	26 10.9	25 30.0	25 23.8	**20**	28 19.5	29 0.8	27 6.3	26 24.8	26 18.7
27 24.7	28 2.3	25 59.7	25 15.4	25 14.7	**21**	28 24.3	29 1.1	26 54.7	26 10.1	26 9.5
27 29.8	28 2.9	25 48.5	25 0.8	25 5.4	**22**	28 29.2	29 1.4	26 43.1	25 55.3	26 0.3
27 34.9	28 3.5	25 37.3	24 46.0	24 56.1	**23**	28 34.2	29 1.8	26 31.5	25 40.3	25 50.9
27 40.0	28 4.1	25 26.0	24 31.0	24 46.5	**24**	28 39.2	29 2.1	26 19.9	25 25.2	25 41.3
27 45.2	28 4.7	25 14.8	24 16.0	24 36.8	**25**	28 44.3	29 2.4	26 8.3	25 10.0	25 31.6
27 50.5	28 5.4	25 3.5	24 0.7	24 26.9	**26**	28 49.4	29 2.7	25 56.7	24 54.5	25 21.7
27 55.8	28 6.0	24 52.2	23 45.3	24 16.9	**27**	28 54.6	29 3.0	25 45.0	24 39.0	25 11.6
28 1.2	28 6.6	24 40.9	23 29.7	24 6.6	**28**	28 59.8	29 3.3	25 33.2	24 23.2	25 1.3
28 6.7	28 7.3	24 29.5	23 13.9	23 56.2	**29**	29 5.1	29 3.6	25 21.5	24 7.2	24 50.8
28 12.2	28 7.9	24 18.0	22 58.0	23 45.5	**30**	29 10.5	29 4.0	25 9.6	23 51.0	24 40.1
28 17.8	28 8.6	24 6.5	22 41.7	23 34.6	**31**	29 16.0	29 4.3	24 57.7	23 34.6	24 29.2
28 23.6	28 9.2	23 54.9	22 25.3	23 23.4	**32**	29 21.6	29 4.6	24 45.8	23 17.9	24 18.0
28 29.4	28 9.9	23 43.3	22 8.6	23 12.0	**33**	29 27.3	29 4.9	24 33.7	23 1.0	24 6.5
28 35.3	28 10.5	23 31.5	21 51.6	23 0.3	**34**	29 33.0	29 5.3	24 21.6	22 43.8	23 54.8
28 41.4	28 11.2	23 19.7	21 34.4	22 48.3	**35**	29 38.9	29 5.6	24 9.3	22 26.3	23 42.7
28 47.5	28 11.9	23 7.8	21 16.8	22 35.9	**36**	29 44.9	29 6.0	23 57.0	22 8.5	23 30.3
28 53.8	28 12.6	22 55.7	20 58.9	22 23.2	**37**	29 51.0	29 6.3	23 44.5	21 50.4	23 17.6
29 0.2	28 13.3	22 43.6	20 40.7	22 10.2	**38**	29 57.2	29 6.7	23 32.0	21 31.9	23 4.5
29 6.8	28 14.0	22 31.3	20 22.1	21 56.7	**39**	♍ 0 3.6	29 7.0	23 19.3	21 13.1	22 51.0
29 13.5	28 14.7	22 18.8	20 3.2	21 42.8	**40**	0 10.1	29 7.4	23 6.4	20 53.9	22 37.0
29 20.4	28 15.5	22 6.3	19 43.8	21 28.5	**41**	0 16.8	29 7.7	22 53.4	20 34.2	22 22.6
29 27.4	28 16.2	21 53.6	19 24.0	21 13.6	**42**	0 23.6	29 8.1	22 40.2	20 14.2	22 7.7
29 34.6	28 17.0	21 40.7	19 3.8	20 58.2	**43**	0 30.6	29 8.5	22 26.9	19 53.6	21 52.2
29 42.0	28 17.8	21 27.6	18 43.1	20 42.3	**44**	0 37.8	29 8.9	22 13.4	19 32.6	21 36.2
29 49.6	28 18.5	21 14.4	18 21.8	20 25.7	**45**	0 45.2	29 9.3	21 59.7	19 11.0	21 19.5
29 57.5	28 19.3	21 0.9	18 0.1	20 8.4	**46**	0 52.8	29 9.7	21 45.8	18 48.9	21 2.2
♍ 0 5.5	28 20.2	20 47.3	17 37.7	19 50.5	**47**	1 0.7	29 10.1	21 31.6	18 26.2	20 44.1
0 13.8	28 21.0	20 33.4	17 14.8	19 31.7	**48**	1 8.7	29 10.5	21 17.3	18 2.9	20 25.2
0 22.4	28 21.8	20 19.3	16 51.1	19 12.1	**49**	1 17.1	29 10.9	21 2.6	17 38.9	20 5.4
0 31.3	28 22.7	20 4.9	16 26.8	18 51.5	**50**	1 25.7	29 11.4	20 47.8	17 14.2	19 44.7
0 40.5	28 23.6	19 50.2	16 1.8	18 29.8	**51**	1 34.6	29 11.8	20 32.6	16 48.8	19 23.0
0 49.9	28 24.5	19 35.3	15 36.0	18 7.1	**52**	1 43.8	29 12.3	20 17.1	16 22.5	19 0.0
0 59.8	28 25.5	19 20.1	15 9.3	17 43.1	**53**	1 53.3	29 12.7	20 1.4	15 55.4	18 35.8
1 10.0	28 26.4	19 4.5	14 41.8	17 17.6	**54**	2 3.2	29 13.2	19 45.3	15 27.4	18 10.2
1 20.6	28 27.4	18 48.6	14 13.3	16 50.7	**55**	2 13.5	29 13.7	19 28.8	14 58.4	17 43.0
1 31.7	28 28.4	18 32.4	13 43.8	16 22.0	**56**	2 24.2	29 14.2	19 12.0	14 28.4	17 14.0
1 43.2	28 29.5	18 15.8	13 13.2	15 51.3	**57**	2 35.4	29 14.7	18 54.7	13 57.3	16 43.0
1 55.3	28 30.6	17 58.7	12 41 5	15 18.4	**58**	2 47.1	29 15.3	18 37.1	13 25.0	16 9.8
2 7.8	28 31.7	17 41.2	12 8.5	14 42.9	**59**	2 59.3	29 15.8	18 19.0	12 51.4	15 33.9
2 21.0	28 32.8	17 23.3	11 34.2	14 4.6	**60**	3 12.1	29 16.4	18 0.4	12 16.4	14 55.1
2 34.9	28 34.0	17 4.9	10 58.5	13 22.8	**61**	3 25.5	29 17.0	17 41.3	11 40.0	14 12.8
2 49.4	28 35.3	16 46.0	10 21.3	12 37.1	**62**	3 39.6	29 17.6	17 21.7	11 2.1	13 26.3
3 4.8	28 36.6	16 26.5	9 42.4	11 46.5	**63**	3 54.4	29 18.3	17 1.5	10 22.4	12 34.9
3 20.9	28 37.9	16 6.4	9 1.7	10 50.1	**64**	4 10.1	29 18.9	16 40.7	9 41.0	11 37.4
3 38.1	28 39.3	15 45.8	8 19.2	9 46.4	**65**	4 26.6	29 19.6	16 19.3	8 57.5	10 32.2
3 56.2	28 40.7	15 24.4	7 34.5	8 33.1	**66**	4 44.2	29 20.3	15 57.2	8 12.0	9 16.9
5	**6**	**Descendant**	**8**	**9**	**S LAT**	**5**	**6**	**Descendant**	**8**	**9**
$19^h\ 52^m\ 0^s$	♑ 26° 0′ 15″ MC 298° 0′ 0″					$19^h\ 56^m\ 0^s$	♑ 26° 57′ 21″ MC 299° 0′ 0″			

8ʰ 0ᵐ 0ˢ MC 120° 0′ 0″

♋ 27° 54′ 36″

11	12	Ascendant	2	3	N LAT
° ′	° ′	° ′	° ′	° ′	°
♌27 49.1	♎ 0 0.0	♏ 2 10.9	♐ 2 5.4	♑ 0 0.0	0
28 11.5	0 0.0	1 5.1	0 54.3	♐29 20.1	5
28 33.8	0 0.0	0 2.3	♏29 43.6	28 39.6	10
28 56.3	0 0.0	♎29 1.4	28 32.4	27 57.7	15
29 0.8	0 0.0	28 49.4	28 18.0	27 49.0	16
29 5.4	0 0.0	28 37.4	28 3.6	27 40.3	17
29 10.0	0 0.0	28 25.4	27 49.0	27 31.5	18
29 14.7	0 0.0	28 13.5	27 34.3	27 22.6	19
29 19.4	0 0.0	28 1.6	27 19.6	27 13.6	20
29 24.1	0 0.0	27 49.6	27 4.7	27 4.4	21
29 28.8	0 0.0	27 37.7	26 49.7	26 55.1	22
29 33.7	0 0.0	27 25.7	26 34.6	26 45.7	23
29 38.5	0 0.0	27 13.8	26 19.3	26 36.1	24
29 43.4	0 0.0	27 1.8	26 3.9	26 26.4	25
29 48.4	0 0.0	26 49.8	25 48.3	26 16.5	26
29 53.4	0 0.0	26 37.7	25 32.5	26 6.4	27
29 58.5	0 0.0	26 25.6	25 16.6	25 56.1	28
♍ 0 3.7	0 0.0	26 13.4	25 0.4	25 45.6	29
0 9.0	0 0.0	26 1.2	24 44.0	25 34.8	30
0 14.3	0 0.0	25 48.9	24 27.4	25 23.9	31
0 19.7	0 0.0	25 36.6	24 10.5	25 12.6	32
0 25.2	0 0.0	25 24.1	23 53.4	25 1.1	33
0 30.8	0 0.0	25 11.6	23 36.0	24 49.3	34
0 36.5	0 0.0	24 59.0	23 18.3	24 37.2	35
0 42.3	0 0.0	24 46.2	23 0.2	24 24.8	36
0 48.3	0 0.0	24 33.3	22 41.9	24 12.0	37
0 54.3	0 0.0	24 20.4	22 23.1	23 58.9	38
1 0.5	0 0.0	24 7.2	22 4.1	23 45.3	39
1 6.8	0 0.0	23 53.9	21 44.6	23 31.3	40
1 13.3	0 0.0	23 40.5	21 24.6	23 16.8	41
1 19.9	0 0.0	23 26.9	21 4.3	23 1.8	42
1 26.7	0 0.0	23 13.1	20 43.5	22 46.3	43
1 33.7	0 0.0	22 59.2	20 22.1	22 30.2	44
1 40.9	0 0.0	22 45.0	20 0.2	22 13.4	45
1 48.3	0 0.0	22 30.6	19 37.8	21 56.0	46
1 55.9	0 0.0	22 16.0	19 14.7	21 37.8	47
2 3.7	0 0.0	22 1.1	18 51.1	21 18.8	48
2 11.8	0 0.0	21 46.0	18 26.7	20 59.0	49
2 20.1	0 0.0	21 30.6	18 1.6	20 38.1	50
2 28.7	0 0.0	21 14.9	17 35.7	20 16.2	51
2 37.7	0 0.0	20 58.9	17 9.1	19 53.1	52
2 46.9	0 0.0	20 42.6	16 41.5	19 28.7	53
2 56.5	0 0.0	20 26.0	16 13.0	19 2.9	54
3 6.5	0 0.0	20 8.9	15 43.5	18 35.4	55
3 16.9	0 0.0	19 51.5	15 13.0	18 6.2	56
3 27.7	0 0.0	19 33.7	14 41.3	17 34.9	57
3 39.0	0 0.0	19 15.4	14 8.4	17 1.3	58
3 50.8	0 0.0	18 56.7	13 34.2	16 25.1	59
4 3.2	0 0.0	18 37.5	12 58.6	15 45.8	60
4 16.1	0 0.0	18 17.7	12 21.5	15 2.9	61
4 29.8	0 0.0	17 57.4	11 42.9	14 15.8	62
4 44.1	0 0.0	17 36.5	11 2.4	13 23.5	63
4 59.3	0 0.0	17 15.0	10 20.2	12 24.8	64
5 15.3	0 0.0	16 52.8	9 35.9	11 18.2	65
5 32.2	0 0.0	16 29.9	8 49.4	10 0.8	66
5	6	Descendant	8	9	S LAT

♑ 27° 54′ 36″

20ʰ 0ᵐ 0ˢ MC 300° 0′ 0″

8ʰ 4ᵐ 0ˢ MC 121° 0′ 0″

♋ 28° 51′ 59″

11	12	Ascendant	2	3	N LAT
° ′	° ′	° ′	° ′	° ′	°
♌28 51.6	♎ 1 5.4	♏ 3 13.3	♐ 3 2.6	♑ 0 55.0	0
29 13.4	1 3.8	2 5.7	1 51.0	0 15.2	5
29 35.1	1 2.2	1 1.1	0 39.7	♐29 34.6	10
29 56.9	1 0.7	♎29 58.4	♏29 27.9	28 52.6	15
♍ 0 1.3	1 0.4	29 46.0	29 13.3	28 44.0	16
0 5.8	1 0.1	29 33.7	28 58.7	28 35.3	17
0 10.3	0 59.8	29 21.4	28 44.0	28 26.5	18
0 14.8	0 59.5	29 9.1	28 29.2	28 17.6	19
0 19.3	0 59.2	28 56.8	28 14.3	28 8.5	20
0 23.9	0 58.9	28 44.5	27 59.3	27 59.4	21
0 28.6	0 58.6	28 32.2	27 44.1	27 50.1	22
0 33.2	0 58.2	28 19.9	27 28.8	27 40.6	23
0 37.9	0 57.9	28 7.5	27 13.4	27 31.0	24
0 42.7	0 57.6	27 55.2	26 57.8	27 21.2	25
0 47.5	0 57.3	27 42.8	26 42.0	27 11.3	26
0 52.4	0 57.0	27 30.4	26 26.1	27 1.2	27
0 57.4	0 56.7	27 17.9	26 9.9	26 50.9	28
1 2.4	0 56.4	27 5.4	25 53.6	26 40.3	29
1 7.5	0 56.0	26 52.8	25 37.0	26 29.6	30
1 12.7	0 55.7	26 40.1	25 20.2	26 18.6	31
1 17.9	0 55.4	26 27.3	25 3.1	26 7.3	32
1 23.3	0 55.1	26 14.5	24 45.7	25 55.8	33
1 28.7	0 54.7	26 1.6	24 28.1	25 44.0	34
1 34.2	0 54.4	25 48.5	24 10.2	25 31.8	35
1 39.9	0 54.0	25 35.4	23 51.9	25 19.4	36
1 45.6	0 53.7	25 22.1	23 33.3	25 6.5	37
1 51.5	0 53.3	25 8.7	23 14.3	24 53.3	38
1 57.5	0 53.0	24 55.2	22 55.0	24 39.7	39
2 3.6	0 52.6	24 41.5	22 35.2	24 25.6	40
2 9.9	0 52.3	24 27.6	22 15.1	24 11.1	41
2 16.3	0 51.9	24 13.6	21 54.4	23 56.1	42
2 22.9	0 51.5	23 59.3	21 33.3	23 40.5	43
2 29.7	0 51.1	23 44.9	21 11.6	23 24.3	44
2 36.7	0 50.7	23 30.3	20 49.4	23 7.5	45
2 43.8	0 50.3	23 15.4	20 26.7	22 49.9	46
2 51.2	0 49.9	23 0.3	20 3.3	22 31.7	47
2 58.7	0 49.5	22 45.0	19 39.2	22 12.6	48
3 6.6	0 49.1	22 29.4	19 14.5	21 52.6	49
3 14.6	0 48.6	22 13.5	18 49.0	21 31.6	50
3 23.0	0 48.2	21 57.3	18 22.7	21 9.5	51
3 31.6	0 47.7	21 40.7	17 55.6	20 46.3	52
3 40.6	0 47.3	21 23.9	17 27.6	20 21.7	53
3 49.9	0 46.8	21 6.7	16 58.6	19 55.7	54
3 59.5	0 46.3	20 49.1	16 28.6	19 28.0	55
4 9.6	0 45.8	20 31.1	15 57.6	18 58.6	56
4 20.0	0 45.3	20 12.6	15 25.3	18 27.0	57
4 31.0	0 44.7	19 53.8	14 51.9	17 53.1	58
4 42.4	0 44.2	19 34.4	14 17.1	17 16.4	59
4 54.3	0 43.6	19 14.5	13 40.8	16 36.6	60
5 6.9	0 43.0	18 54.1	13 3.0	15 53.2	61
5 20.1	0 42.4	18 33.1	12 23.6	15 5.3	62
5 33.9	0 41.7	18 11.5	11 42.4	14 12.2	63
5 48.5	0 41.1	17 49.3	10 59.3	13 12.4	64
6 4.0	0 40.4	17 26.3	10 14.2	12 4.2	65
6 20.3	0 39.7	17 2.7	9 26.8	10 44.7	66
5	6	Descendant	8	9	S LAT

♑ 28° 51′ 59″

20ʰ 4ᵐ 0ˢ MC 301° 0′ 0″

8h 8m 0s — MC 122° 0′ 0″ — ♋ 29° 49′ 31″

N LAT	11	12	Ascendant	2	3
°	° ′	° ′	° ′	° ′	° ′
0	♌29 54.3	♎ 2 10.8	♏ 4 15.5	♐ 3 59.8	♑ 1 50.1
5	♍ 0 15.5	2 7.6	3 6.1	2 47.5	1 10.2
10	0 36.5	2 4.4	1 59.7	1 35.7	0 29.6
15	0 57.7	2 1.4	0 55.3	0 23.2	♐29 47.7
16	1 2.0	2 0.8	0 42.6	0 8.5	29 39.0
17	1 6.3	2 0.2	0 29.9	♏29 53.8	29 30.3
18	1 10.7	1 59.6	0 17.2	29 38.9	29 21.5
19	1 15.0	1 58.9	0 4.6	29 24.0	29 12.6
20	1 19.5	1 58.3	♎29 51.9	29 8.9	29 3.5
21	1 23.9	1 57.7	29 39.3	28 53.8	28 54.3
22	1 28.4	1 57.1	29 26.6	28 38.5	28 45.0
23	1 32.9	1 56.5	29 13.9	28 23.0	28 35.5
24	1 37.5	1 55.9	29 1.3	28 7.4	28 25.9
25	1 42.1	1 55.3	28 48.5	27 51.6	28 16.2
26	1 46.8	1 54.6	28 35.8	27 35.7	28 6.2
27	1 51.5	1 54.0	28 23.0	27 19.6	27 56.1
28	1 56.3	1 53.4	28 10.1	27 3.2	27 45.7
29	2 1.2	1 52.7	27 57.2	26 46.7	27 35.2
30	2 6.1	1 52.1	27 44.3	26 29.9	27 24.4
31	2 11.2	1 51.4	27 31.2	26 12.9	27 13.3
32	2 16.2	1 50.8	27 18.1	25 55.6	27 2.1
33	2 21.4	1 50.1	27 4.9	25 38.1	26 50.5
34	2 26.7	1 49.5	26 51.5	25 20.2	26 38.7
35	2 32.0	1 48.8	26 38.1	25 2.1	26 26.5
36	2 37.5	1 48.1	26 24.5	24 43.6	26 14.0
37	2 43.1	1 47.4	26 10.9	24 24.7	26 1.1
38	2 48.8	1 46.7	25 57.0	24 5.5	25 47.9
39	2 54.6	1 46.0	25 43.1	23 45.9	25 34.2
40	3 0.5	1 45.3	25 29.0	23 25.9	25 20.1
41	3 6.6	1 44.5	25 14.7	23 5.4	25 5.5
42	3 12.8	1 43.8	25 0.2	22 44.5	24 50.4
43	3 19.2	1 43.0	24 45.5	22 23.1	24 34.7
44	3 25.8	1 42.2	24 30.6	22 1.1	24 18.5
45	3 32.5	1 41.5	24 15.5	21 38.6	24 1.6
46	3 39.4	1 40.7	24 0.2	21 15.5	23 44.0
47	3 46.5	1 39.8	23 44.6	20 51.8	23 25.6
48	3 53.9	1 39.0	23 28.8	20 27.4	23 6.4
49	4 1.4	1 38.2	23 12.7	20 2.3	22 46.3
50	4 9.2	1 37.3	22 56.3	19 36.4	22 25.2
51	4 17.3	1 36.4	22 39.6	19 9.7	22 3.0
52	4 25.7	1 35.5	22 22.5	18 42.1	21 39.6
53	4 34.3	1 34.5	22 5.1	18 13.7	21 14.9
54	4 43.3	1 33.6	21 47.4	17 44.2	20 48.7
55	4 52.6	1 32.6	21 29.2	17 13.8	20 20.8
56	5 2.3	1 31.6	21 10.6	16 42.2	19 51.1
57	5 12.5	1 30.5	20 51.6	16 9.4	19 19.2
58	5 23.0	1 29.4	20 32.1	15 35.3	18 44.9
59	5 34.0	1 28.3	20 12.1	14 59.9	18 7.9
60	5 45.6	1 27.2	19 51.6	14 23.0	17 27.7
61	5 57.7	1 26.0	19 30.5	13 44.5	16 43.7
62	6 10.4	1 24.7	19 8.8	13 4.4	15 55.1
63	6 23.8	1 23.4	18 46.5	12 22.4	15 1.1
64	6 37.9	1 22.1	18 23.5	11 38.5	14 0.2
65	6 52.8	1 20.7	17 59.8	10 52.4	12 50.4
66	7 8.5	1 19.3	17 35.4	10 4.1	11 28.6
S LAT	5	6	Descendant	8	9

♑ 29° 49′ 31″ — 20h 8m 0s — MC 302° 0′ 0″

8h 12m 0s — MC 123° 0′ 0″ — ♌ 0° 47′ 12″

N LAT	11	12	Ascendant	2	3
°	° ′	° ′	° ′	° ′	° ′
0	♍ 0 57.2	♎ 3 16.2	♏ 5 17.5	♐ 4 56.7	♑ 2 45.2
5	1 17.7	3 11.3	4 6.4	3 44.0	2 5.3
10	1 38.1	3 6.7	2 58.3	2 31.6	1 24.7
15	1 58.6	3 2.1	1 52.2	1 18.5	0 42.7
16	2 2.8	3 1.2	1 39.1	1 3.7	0 34.1
17	2 7.0	3 0.2	1 26.0	0 48.8	0 25.4
18	2 11.2	2 59.3	1 13.0	0 33.8	0 16.5
19	2 15.4	2 58.4	1 0.0	0 18.7	0 7.6
20	2 19.7	2 57.5	0 47.0	0 3.5	♐29 58.5
21	2 24.0	2 56.6	0 34.0	♏29 48.2	29 49.3
22	2 28.3	2 55.7	0 21.0	29 32.7	29 40.0
23	2 32.7	2 54.7	0 8.0	29 17.1	29 30.5
24	2 37.2	2 53.8	♎29 54.9	29 1.4	29 20.9
25	2 41.6	2 52.9	29 41.8	28 45.4	29 11.1
26	2 46.2	2 51.9	29 28.7	28 29.3	29 1.1
27	2 50.8	2 51.0	29 15.6	28 13.0	28 51.0
28	2 55.4	2 50.0	29 2.3	27 56.5	28 40.6
29	3 0.1	2 49.1	28 49.1	27 39.8	28 30.0
30	3 4.9	2 48.1	28 35.7	27 22.8	28 19.2
31	3 9.7	2 47.1	28 22.3	27 5.6	28 8.2
32	3 14.7	2 46.2	28 8.8	26 48.1	27 56.9
33	3 19.7	2 45.2	27 55.2	26 30.4	27 45.3
34	3 24.8	2 44.2	27 41.5	26 12.3	27 33.4
35	3 30.0	2 43.2	27 27.6	25 53.9	27 21.2
36	3 35.2	2 42.1	27 13.7	25 35.2	27 8.7
37	3 40.6	2 41.1	26 59.6	25 16.2	26 55.8
38	3 46.1	2 40.0	26 45.4	24 56.7	26 42.5
39	3 51.8	2 39.0	26 31.0	24 36.8	26 28.8
40	3 57.5	2 37.9	26 16.4	24 16.6	26 14.6
41	4 3.4	2 36.8	26 1.7	23 55.8	26 0.0
42	4 9.4	2 35.7	25 46.8	23 34.6	25 44.8
43	4 15.6	2 34.5	25 31.7	23 12.9	25 29.1
44	4 21.9	2 33.4	25 16.4	22 50.7	25 12.8
45	4 28.4	2 32.2	25 0.8	22 27.8	24 55.8
46	4 35.1	2 31.0	24 45.0	22 4.4	24 38.1
47	4 42.0	2 29.8	24 29.0	21 40.3	24 19.7
48	4 49.1	2 28.5	24 12.6	21 15.5	24 0.4
49	4 56.4	2 27.2	23 56.0	20 50.0	23 40.2
50	5 3.9	2 25.9	23 39.1	20 23.8	23 19.0
51	5 11.7	2 24.6	23 21.9	19 56.7	22 56.6
52	5 19.8	2 23.2	23 4.3	19 28.7	22 33.1
53	5 28.1	2 21.8	22 46.4	18 59.8	22 8.2
54	5 36.8	2 20.3	22 28.0	18 29.9	21 41.8
55	5 45.8	2 18.9	22 9.3	17 58.9	21 13.7
56	5 55.2	2 17.3	21 50.1	17 26.8	20 43.7
57	6 4.9	2 15.8	21 30.5	16 53.4	20 11.6
58	6 15.1	2 14.1	21 10.4	16 18.8	19 37.0
59	6 25.8	2 12.5	20 49.8	15 42.7	18 59.6
60	6 36.9	2 10.7	20 28.6	15 5.2	18 18.9
61	6 48.6	2 8.9	20 6.9	14 26.0	17 34.3
62	7 0.8	2 7.1	19 44.5	13 45.1	16 45.1
63	7 13.7	2 5.2	19 21.5	13 2.3	15 50.2
64	7 27.3	2 3.2	18 57.7	12 17.5	14 48.1
65	7 41.6	2 1.1	18 33.3	11 30.6	13 36.8
66	7 56.8	1 58.9	18 8.0	10 41.3	12 12.6
S LAT	5	6	Descendant	8	9

♒ 0° 47′ 12″ — 20h 12m 0s — MC 303° 0′ 0″

8h 16m 0s MC 124° 0′ 0″

♌ 1° 45′ 2″

N LAT	11	12	Ascendant	2	3
0°	♍ 2° 0.3′	♎ 4° 21.5′	♏ 6° 19.4′	♐ 5° 53.6′	♑ 3° 40.2′
5	2 20.1	4 15.1	5 6.5	4 40.3	3 0.4
10	2 39.8	4 8.8	3 56.7	3 27.4	2 19.8
15	2 59.7	4 2.7	2 48.9	2 13.7	1 37.8
16	3 3.7	4 1.5	2 35.5	1 58.7	1 29.2
17	3 7.8	4 0.3	2 22.1	1 43.7	1 20.5
18	3 11.8	3 59.1	2 8.7	1 28.6	1 11.6
19	3 15.9	3 57.9	1 55.4	1 13.4	1 2.7
20	3 20.1	3 56.6	1 42.0	0 58.0	0 53.6
21	3 24.2	3 55.4	1 28.7	0 42.6	0 44.4
22	3 28.4	3 54.2	1 15.3	0 27.0	0 35.1
23	3 32.7	3 53.0	1 1.9	0 11.2	0 25.6
24	3 36.9	3 51.7	0 48.5	♏29 55.3	0 15.9
25	3 41.3	3 50.5	0 35.1	29 39.2	0 6.1
26	3 45.7	3 49.2	0 21.6	29 22.9	♐29 56.2
27	3 50.1	3 48.0	0 8.1	29 6.4	29 46.0
28	3 54.6	3 46.7	♎29 54.5	28 49.8	29 35.6
29	3 59.1	3 45.4	29 40.9	28 32.9	29 25.0
30	4 3.8	3 44.1	29 27.1	28 15.7	29 14.2
31	4 8.4	3 42.8	29 13.3	27 58.3	29 3.1
32	4 13.2	3 41.5	28 59.4	27 40.6	28 51.8
33	4 18.0	3 40.2	28 45.4	27 22.7	28 40.2
34	4 23.0	3 38.9	28 31.3	27 4.4	28 28.3
35	4 28.0	3 37.5	28 17.1	26 45.8	28 16.0
36	4 33.1	3 36.2	28 2.8	26 26.9	28 3.5
37	4 38.3	3 34.8	27 48.3	26 7.6	27 50.5
38	4 43.6	3 33.4	27 33.6	25 47.9	27 37.2
39	4 49.0	3 31.9	27 18.8	25 27.8	27 23.5
40	4 54.6	3 30.5	27 3.9	25 7.2	27 9.3
41	5 0.2	3 29.0	26 48.7	24 46.2	26 54.6
42	5 6.1	3 27.5	26 33.4	24 24.7	26 39.4
43	5 12.0	3 26.0	26 17.8	24 2.7	26 23.6
44	5 18.1	3 24.5	26 2.0	23 40.2	26 7.2
45	5 24.4	3 22.9	25 46.0	23 17.0	25 50.2
46	5 30.8	3 21.3	25 29.8	22 53.3	25 32.4
47	5 37.5	3 19.7	25 13.2	22 28.9	25 13.9
48	5 44.3	3 18.0	24 56.4	22 3.7	24 54.5
49	5 51.4	3 16.3	24 39.3	21 37.9	24 34.2
50	5 58.6	3 14.5	24 21.9	21 11.2	24 12.9
51	6 6.2	3 12.8	24 4.2	20 43.7	23 50.4
52	6 13.9	3 10.9	23 46.1	20 15.3	23 26.7
53	6 22.0	3 9.0	23 27.6	19 45.9	23 1.7
54	6 30.4	3 7.1	23 8.7	19 15.5	22 35.1
55	6 39.1	3 5.1	22 49.4	18 44.0	22 6.8
56	6 48.1	3 3.1	22 29.6	18 11.4	21 36.6
57	6 57.5	3 1.0	22 9.4	17 37.5	21 4.2
58	7 7.3	2 58.8	21 48.7	17 2.2	20 29.3
59	7 17.5	2 56.6	21 27.4	16 25.5	19 51.5
60	7 28.3	2 54.3	21 5.6	15 47.3	19 10.3
61	7 39.5	2 51.9	20 43.2	15 7.5	18 25.2
62	7 51.3	2 49.4	20 20.1	14 25.8	17 35.3
63	8 3.7	2 46.9	19 56.4	13 42.2	16 39.5
64	8 16.8	2 44.2	19 31.9	12 56.6	15 36.3
65	8 30.6	2 41.5	19 6.7	12 8.8	14 23.3
66	8 45.2	2 38.6	18 40.7	11 18.5	12 56.7
S LAT	5	6	Descendant	8	9

♒ 1° 45′ 2″

20h 16m 0s MC 304° 0′ 0″

8h 20m 0s MC 125° 0′ 0″

♌ 2° 43′ 2″

N LAT	11	12	Ascendant	2	3
0°	♍ 3° 3.5′	♎ 5° 26.8′	♏ 7° 21.1′	♐ 6° 50.3′	♑ 4° 35.3′
5	3 22.6	5 18.8	6 6.5	5 36.6	3 55.5
10	3 41.7	5 11.0	4 55.0	4 23.1	3 15.0
15	4 0.9	5 3.4	3 45.5	3 8.8	2 33.0
16	4 4.7	5 1.9	3 31.8	2 53.7	2 24.3
17	4 8.7	5 0.3	3 18.0	2 38.6	2 15.6
18	4 12.6	4 58.8	3 4.3	2 23.3	2 6.8
19	4 16.5	4 57.3	2 50.6	2 8.0	1 57.8
20	4 20.5	4 55.8	2 36.9	1 52.5	1 48.7
21	4 24.6	4 54.2	2 23.2	1 36.9	1 39.5
22	4 28.6	4 52.7	2 9.5	1 21.1	1 30.2
23	4 32.7	4 51.2	1 55.8	1 5.2	1 20.7
24	4 36.8	4 49.6	1 42.1	0 49.2	1 11.0
25	4 41.0	4 48.1	1 28.3	0 32.9	1 1.2
26	4 45.2	4 46.5	1 14.4	0 16.5	0 51.2
27	4 49.5	4 44.9	1 0.6	♏29 59.8	0 41.0
28	4 53.9	4 43.4	0 46.6	29 43.0	0 30.7
29	4 58.3	4 41.8	0 32.6	29 25.9	0 20.0
30	5 2.7	4 40.2	0 18.5	29 8.6	0 9.2
31	5 7.2	4 38.5	0 4.3	28 51.0	♐29 58.1
32	5 11.8	4 36.9	♎29 50.1	28 33.1	29 46.8
33	5 16.5	4 35.2	29 35.7	28 14.9	29 35.1
34	5 21.2	4 33.6	29 21.2	27 56.5	29 23.2
35	5 26.1	4 31.9	29 6.6	27 37.7	29 10.9
36	5 31.0	4 30.2	28 51.8	27 18.5	28 58.3
37	5 36.0	4 28.5	28 36.9	26 59.0	28 45.4
38	5 41.1	4 26.7	28 21.9	26 39.0	28 32.0
39	5 46.4	4 24.9	28 6.7	26 18.7	28 18.2
40	5 51.7	4 23.1	27 51.3	25 57.9	28 4.0
41	5 57.2	4 21.3	27 35.7	25 36.6	27 49.3
42	6 2.8	4 19.4	27 19.9	25 14.9	27 34.0
43	6 8.5	4 17.5	27 3.9	24 52.6	27 18.2
44	6 14.4	4 15.6	26 47.7	24 29.7	27 1.7
45	6 20.5	4 13.6	26 31.3	24 6.3	26 44.6
46	6 26.7	4 11.6	26 14.5	23 42.2	26 26.8
47	6 33.1	4 9.6	25 57.5	23 17.4	26 8.2
48	6 39.7	4 7.5	25 40.2	22 51.9	25 48.7
49	6 46.5	4 5.4	25 22.6	22 25.7	25 28.3
50	6 53.5	4 3.2	25 4.7	21 58.6	25 6.9
51	7 0.7	4 0.9	24 46.4	21 30.7	24 44.3
52	7 8.2	3 58.6	24 27.8	21 1.9	24 20.5
53	7 16.0	3 56.3	24 8.8	20 32.0	23 55.3
54	7 24.0	3 53.9	23 49.3	20 1.2	23 28.5
55	7 32.4	3 51.4	23 29.5	19 29.2	23 0.1
56	7 41.0	3 48.9	23 9.1	18 56.0	22 29.6
57	7 50.1	3 46.2	22 48.3	18 21.5	21 56.9
58	7 59.5	3 43.5	22 27.0	17 45.7	21 21.7
59	8 9.4	3 40.7	22 5.1	17 8.4	20 43.6
60	8 19.7	3 37.9	21 42.6	16 29.5	20 2.0
61	8 30.5	3 34.9	21 19.5	15 48.9	19 16.3
62	8 41.8	3 31.8	20 55.8	15 6.5	18 25.7
63	8 53.7	3 28.6	20 31.3	14 22.1	17 29.0
64	9 6.3	3 25.3	20 6.1	13 35.6	16 24.6
65	9 19.5	3 21.8	19 40.1	12 46.9	15 10.0
66	9 33.6	3 18.2	19 13.3	11 55.7	13 40.8
S LAT	5	6	Descendant	8	9

♒ 2° 43′ 2″

20h 20m 0s MC 305° 0′ 0″

$8^h\ 24^m\ 0^s$ MC 126° 0′ 0″

♌ 3° 41′ 11″

11	12	Ascendant	2	3	N LAT
° ′	° ′	° ′	° ′	° ′	°
♍ 4 6.8	♎ 6 32.1	♏ 8 22.6	♐ 7 46.9	♑ 5 30.5	0
4 25.3	6 22.5	7 6.3	6 32.7	4 50.7	5
4 43.6	6 13.2	5 53.2	5 18.7	4 10.2	10
5 2.2	6 4.0	4 42.1	4 3.9	3 28.2	15
5 5.9	6 2.2	4 28.0	3 48.7	3 19.5	16
5 9.7	6 0.4	4 13.9	3 33.4	3 10.8	17
5 13.5	5 58.5	3 59.9	3 18.0	3 1.9	18
5 17.3	5 56.7	3 45.8	3 2.5	2 53.0	19
5 21.1	5 54.9	3 31.8	2 46.9	2 43.9	20
5 25.0	5 53.0	3 17.8	2 31.2	2 34.7	21
5 28.9	5 51.2	3 3.7	2 15.3	2 25.4	22
5 32.9	5 49.4	2 49.6	1 59.2	2 15.9	23
5 36.8	5 47.5	2 35.5	1 43.0	2 6.2	24
5 40.9	5 45.6	2 21.4	1 26.6	1 56.4	25
5 44.9	5 43.8	2 7.2	1 10.0	1 46.4	26
5 49.1	5 41.9	1 53.0	0 53.2	1 36.2	27
5 53.2	5 40.0	1 38.7	0 36.2	1 25.8	28
5 57.5	5 38.1	1 24.3	0 18.9	1 15.2	29
6 1.8	5 36.2	1 9.8	0 1.4	1 4.3	30
6 6.1	5 34.2	0 55.3	♏29 43.6	0 53.2	31
6 10.5	5 32.3	0 40.6	29 25.6	0 41.8	32
6 15.0	5 30.3	0 25.9	29 7.2	0 30.2	33
6 19.6	5 28.3	0 11.0	28 48.5	0 18.2	34
6 24.3	5 26.2	♎29 56.0	28 29.5	0 6.0	35
6 29.0	5 24.2	29 40.9	28 10.1	♐29 53.3	36
6 33.8	5 22.1	29 25.6	27 50.4	29 40.3	37
6 38.8	5 20.0	29 10.1	27 30.2	29 26.9	38
6 43.8	5 17.9	28 54.5	27 9.6	29 13.1	39
6 48.9	5 15.7	28 38.7	26 48.6	28 58.8	40
6 54.2	5 13.5	28 22.7	26 27.0	28 44.1	41
6 59.6	5 11.3	28 6.5	26 5.0	28 28.8	42
7 5.1	5 9.0	27 50.1	25 42.4	28 12.9	43
7 10.8	5 6.7	27 33.4	25 19.3	27 56.4	44
7 16.6	5 4.3	27 16.5	24 55.5	27 39.2	45
7 22.6	5 1.9	26 59.3	24 31.1	27 21.4	46
7 28.7	4 59.5	26 41.8	24 6.0	27 2.7	47
7 35.1	4 57.0	26 24.0	23 40.1	26 43.1	48
7 41.6	4 54.4	26 5.9	23 13.5	26 22.6	49
7 48.3	4 51.8	25 47.5	22 46.1	26 1.1	50
7 55.3	4 49.1	25 28.7	22 17.8	25 38.4	51
8 2.5	4 46.4	25 9.5	21 48.5	25 14.5	52
8 10.0	4 43.5	24 50.0	21 18.2	24 49.1	53
8 17.7	4 40.7	24 30.0	20 46.9	24 22.2	54
8 25.7	4 37.7	24 9.5	20 14.4	23 53.5	55
8 34.1	4 34.6	23 48.6	19 40.6	23 22.8	56
8 42.8	4 31.5	23 27.2	19 5.6	22 49.9	57
8 51.8	4 28.2	23 5.2	18 29.2	22 14.4	58
9 1.3	4 24.9	22 42.7	17 51.2	21 35.9	59
9 11.2	4 21.4	22 19.6	17 11.6	20 53.8	60
9 21.5	4 17.8	21 55.8	16 30.3	20 7.6	61
9 32.4	4 14.1	21 31.4	15 47.2	19 16.3	62
9 43.9	4 10.3	21 6.2	15 2.0	18 18.8	63
9 55.9	4 6.3	20 40.3	14 14.6	17 13.2	64
10 8.6	4 2.2	20 13.5	13 24.9	15 56.8	65
10 22.1	3 57.9	19 45.9	12 32.7	14 25.1	66
5	6	Descendant	8	9	S LAT

♒ 3° 41′ 11″

$20^h\ 24^m\ 0^s$ MC 306° 0′ 0″

$8^h\ 28^m\ 0^s$ MC 127° 0′ 0″

♌ 4° 39′ 30″

11	12	Ascendant	2	3	N LAT
° ′	° ′	° ′	° ′	° ′	°
♍ 5 10.3	♎ 7 37.4	♏ 9 23.9	♐ 8 43.3	♑ 6 25.6	0
5 28.1	7 26.1	8 6.0	7 28.7	5 45.9	5
5 45.8	7 15.3	6 51.3	6 14.3	5 5.4	10
6 3.6	7 4.6	5 38.5	4 58.8	4 23.4	15
6 7.2	7 2.5	5 24.1	4 43.5	4 14.8	16
6 10.8	7 0.4	5 9.7	4 28.1	4 6.1	17
6 14.5	6 58.2	4 55.3	4 12.6	3 57.2	18
6 18.1	6 56.1	4 40.9	3 57.0	3 48.3	19
6 21.8	6 54.0	4 26.6	3 41.3	3 39.2	20
6 25.6	6 51.8	4 12.2	3 25.4	3 30.0	21
6 29.3	6 49.7	3 57.8	3 9.3	3 20.6	22
6 33.1	6 47.5	3 43.4	2 53.2	3 11.1	23
6 37.0	6 45.4	3 29.0	2 36.8	3 1.5	24
6 40.8	6 43.2	3 14.5	2 20.2	2 51.6	25
6 44.7	6 41.0	2 59.9	2 3.5	2 41.6	26
6 48.7	6 38.8	2 45.3	1 46.5	2 31.4	27
6 52.7	6 36.6	2 30.7	1 29.3	2 21.0	28
6 56.8	6 34.4	2 15.9	1 11.9	2 10.4	29
7 0.9	6 32.1	2 1.1	0 54.2	1 59.5	30
7 5.1	6 29.9	1 46.2	0 36.3	1 48.4	31
7 9.3	6 27.6	1 31.2	0 18.0	1 37.0	32
7 13.7	6 25.3	1 16.1	♏29 59.5	1 25.3	33
7 18.1	6 22.9	1 0.8	29 40.6	1 13.4	34
7 22.5	6 20.6	0 45.4	29 21.4	1 1.1	35
7 27.1	6 18.2	0 29.9	29 1.8	0 48.4	36
7 31.7	6 15.8	0 14.2	28 41.8	0 35.4	37
7 36.5	6 13.3	♎29 58.3	28 21.4	0 22.0	38
7 41.3	6 10.8	29 42.3	28 0.6	0 8.1	39
7 46.3	6 8.3	29 26.1	27 39.3	♐29 53.8	40
7 51.3	6 5.7	29 9.7	27 17.5	29 39.0	41
7 56.5	6 3.1	28 53.0	26 55.2	29 23.7	42
8 1.8	6 0.5	28 36.1	26 32.3	29 7.8	43
8 7.2	5 57.8	28 19.0	26 8.8	28 51.2	44
8 12.8	5 55.0	28 1.6	25 44.8	28 34.0	45
8 18.6	5 52.2	27 44.0	25 20.0	28 16.0	46
8 24.5	5 49.4	27 26.0	24 54.6	27 57.3	47
8 30.6	5 46.4	27 7.8	24 28.4	27 37.7	48
8 36.8	5 43.5	26 49.2	24 1.4	27 17.1	49
8 43.3	5 40.4	26 30.3	23 33.6	26 55.4	50
8 50.0	5 37.3	26 10.9	23 4.8	26 32.6	51
8 56.9	5 34.1	25 51.3	22 35.1	26 8.6	52
9 4.0	5 30.8	25 31.1	22 4.4	25 43.1	53
9 11.4	5 27.4	25 10.6	21 32.6	25 16.0	54
9 19.1	5 23.9	24 49.6	20 59.5	24 47.1	55
9 27.2	5 20.4	24 28.1	20 25.3	24 16.3	56
9 35.5	5 16.7	24 6.1	19 49.7	23 43.1	57
9 44.2	5 12.9	23 43.5	19 12.6	23 7.3	58
9 53.2	5 9.0	23 20.3	18 34.1	22 28.4	59
10 2.7	5 5.0	22 56.5	17 53.8	21 45.9	60
10 12.7	5 0.8	22 32.1	17 11.8	20 59.2	61
10 23.1	4 56.5	22 6.9	16 27.8	20 7.2	62
10 34.0	4 52.0	21 41.1	15 41.8	19 8.8	63
10 45.6	4 47.3	21 14.4	14 53.6	18 2.0	64
10 57.7	4 42.5	20 46.9	14 3.0	16 43.9	65
11 10.6	4 37.5	20 18.5	13 9.7	15 9.4	66
5	6	Descendant	8	9	S LAT

♒ 4° 39′ 30″

$20^h\ 28^m\ 0^s$ MC 307° 0′ 0″

8h 32m 0s — MC 128° 0′ 0″ — ♌ 5° 37′ 58″

N LAT	11	12	Ascendant	2	3
0°	♍ 6 14.0	♎ 8 42.5	♏ 10 25.0	♐ 9 39.7	♑ 7 20.8
5	6 31.0	8 29.7	9 5.6	8 24.7	6 41.1
10	6 48.0	8 17.3	7 49.2	7 9.7	6 0.7
15	7 5.1	8 5.2	6 34.9	5 53.8	5 18.7
16	7 8.6	8 2.7	6 20.1	5 38.4	5 10.1
17	7 12.1	8 0.3	6 5.4	5 22.8	5 1.4
18	7 15.6	7 57.9	5 50.7	5 7.2	4 52.5
19	7 19.1	7 55.5	5 36.0	4 51.5	4 43.6
20	7 22.7	7 53.0	5 21.3	4 35.6	4 34.5
21	7 26.2	7 50.6	5 6.6	4 19.6	4 25.3
22	7 29.8	7 48.1	4 51.9	4 3.4	4 15.9
23	7 33.5	7 45.7	4 37.1	3 47.1	4 6.4
24	7 37.2	7 43.2	4 22.3	3 30.6	3 56.8
25	7 40.9	7 40.7	4 7.5	3 13.9	3 46.9
26	7 44.6	7 38.2	3 52.6	2 57.0	3 36.9
27	7 48.4	7 35.7	3 37.7	2 39.8	3 26.7
28	7 52.3	7 33.2	3 22.7	2 22.5	3 16.3
29	7 56.2	7 30.7	3 7.6	2 4.9	3 5.7
30	8 0.2	7 28.1	2 52.4	1 47.0	2 54.8
31	8 4.2	7 25.5	2 37.1	1 28.9	2 43.7
32	8 8.2	7 22.9	2 21.7	1 10.5	2 32.3
33	8 12.4	7 20.3	2 6.2	0 51.7	2 20.6
34	8 16.6	7 17.6	1 50.6	0 32.6	2 8.6
35	8 20.9	7 14.9	1 34.8	0 13.2	1 56.3
36	8 25.3	7 12.2	1 18.9	♏ 29 53.4	1 43.6
37	8 29.7	7 9.4	1 2.8	29 33.2	1 30.6
38	8 34.3	7 6.6	0 46.5	29 12.6	1 17.1
39	8 38.9	7 3.8	0 30.1	28 51.5	1 3.3
40	8 43.6	7 0.9	0 13.5	28 30.0	0 48.9
41	8 48.5	6 57.9	♎ 29 56.6	28 7.9	0 34.1
42	8 53.5	6 55.0	29 39.5	27 45.3	0 18.7
43	8 58.5	6 51.9	29 22.2	27 22.2	0 2.7
44	9 3.8	6 48.8	29 4.7	26 58.4	♐ 29 46.1
45	9 9.1	6 45.7	28 46.8	26 34.0	29 28.9
46	9 14.6	6 42.5	28 28.7	26 9.0	29 10.9
47	9 20.3	6 39.2	28 10.3	25 43.2	28 52.0
48	9 26.1	6 35.9	27 51.5	25 16.7	28 32.3
49	9 32.1	6 32.5	27 32.5	24 49.3	28 11.7
50	9 38.3	6 29.0	27 13.0	24 21.1	27 50.0
51	9 44.7	6 25.4	26 53.2	23 51.9	27 27.1
52	9 51.3	6 21.8	26 33.0	23 21.8	27 2.9
53	9 58.2	6 18.0	26 12.3	22 50.6	26 37.3
54	10 5.3	6 14.1	25 51.2	22 18.3	26 10.0
55	10 12.6	6 10.2	25 29.6	21 44.8	25 41.0
56	10 20.3	6 6.1	25 7.5	21 10.0	25 9.9
57	10 28.3	6 1.9	24 44.9	20 33.8	24 36.5
58	10 36.6	5 57.6	24 21.7	19 56.1	24 0.4
59	10 45.3	5 53.1	23 57.9	19 16.9	23 21.2
60	10 54.3	5 48.5	23 33.5	18 36.0	22 38.3
61	11 3.8	5 43.7	23 8.3	17 53.2	21 51.0
62	11 13.8	5 38.8	22 42.5	17 8.5	20 58.4
63	11 24.3	5 33.7	22 15.9	16 21.6	19 59.1
64	11 35.3	5 28.4	21 48.5	15 32.5	18 51.1
65	11 46.9	5 22.8	21 20.2	14 40.9	17 31.3
66	11 59.2	5 17.1	20 51.0	13 46.7	15 53.8
S LAT	5	6	Descendant	8	9

♒ 5° 37′ 58″ — 20h 32m 0s — MC 308° 0′ 0″

8h 36m 0s — MC 129° 0′ 0″ — ♌ 6° 36′ 37″

N LAT	11	12	Ascendant	2	3
0°	♍ 7 17.7	♎ 9 47.7	♏ 11 26.0	♐ 10 35.9	♑ 8 16.1
5	7 34.1	9 33.3	10 4.9	9 20.5	7 36.5
10	7 50.4	9 19.4	8 47.1	8 5.1	6 56.0
15	8 6.8	9 5.7	7 31.1	6 48.6	6 14.1
16	8 10.1	9 3.0	7 16.0	6 33.1	6 5.5
17	8 13.4	9 0.2	7 1.0	6 17.5	5 56.8
18	8 16.8	8 57.5	6 46.0	6 1.7	5 47.9
19	8 20.2	8 54.8	6 31.0	5 45.9	5 39.0
20	8 23.6	8 52.0	6 15.9	5 29.9	5 29.9
21	8 27.0	8 49.3	6 0.9	5 13.7	5 20.7
22	8 30.5	8 46.6	5 45.8	4 57.4	5 11.4
23	8 33.9	8 43.8	5 30.7	4 41.0	5 1.9
24	8 37.5	8 41.0	5 15.6	4 24.3	4 52.2
25	8 41.0	8 38.2	5 0.5	4 7.5	4 42.4
26	8 44.6	8 35.4	4 45.2	3 50.4	4 32.3
27	8 48.3	8 32.6	4 29.9	3 33.1	4 22.1
28	8 52.0	8 29.8	4 14.6	3 15.6	4 11.7
29	8 55.7	8 26.9	3 59.1	2 57.9	4 1.1
30	8 59.5	8 24.0	3 43.6	2 39.8	3 50.2
31	9 3.3	8 21.1	3 27.9	2 21.5	3 39.0
32	9 7.2	8 18.2	3 12.2	2 2.9	3 27.6
33	9 11.2	8 15.2	2 56.3	1 44.0	3 16.0
34	9 15.2	8 12.2	2 40.3	1 24.7	3 4.0
35	9 19.3	8 9.2	2 24.1	1 5.1	2 51.6
36	9 23.5	8 6.1	2 7.8	0 45.0	2 38.9
37	9 27.8	8 3.0	1 51.3	0 24.6	2 25.9
38	9 32.1	7 59.9	1 34.7	0 3.8	2 12.4
39	9 36.6	7 56.7	1 17.8	♏ 29 42.5	1 58.5
40	9 41.1	7 53.4	1 0.8	29 20.7	1 44.1
41	9 45.7	7 50.1	0 43.5	28 58.4	1 29.3
42	9 50.5	7 46.8	0 26.0	28 35.5	1 13.9
43	9 55.4	7 43.4	0 8.3	28 12.1	0 57.9
44	10 0.3	7 39.9	♎ 29 50.3	27 48.0	0 41.2
45	10 5.5	7 36.4	29 32.0	27 23.4	0 23.9
46	10 10.7	7 32.8	29 13.4	26 58.0	0 5.8
47	10 16.1	7 29.1	28 54.5	26 31.9	♐ 29 47.0
48	10 21.7	7 25.3	28 35.3	26 5.0	29 27.2
49	10 27.4	7 21.5	28 15.7	25 37.2	29 6.5
50	10 33.4	7 17.6	27 55.8	25 8.6	28 44.7
51	10 39.5	7 13.6	27 35.4	24 39.0	28 21.7
52	10 45.8	7 9.4	27 14.7	24 8.5	27 57.4
53	10 52.4	7 5.2	26 53.5	23 36.8	27 31.6
54	10 59.1	7 0.9	26 31.8	23 4.0	27 4.2
55	11 6.2	6 56.4	26 9.6	22 30.0	26 35.0
56	11 13.5	6 51.8	25 47.0	21 54.7	26 3.7
57	11 21.1	6 47.1	25 23.7	21 17.9	25 30.1
58	11 29.1	6 42.3	24 59.9	20 39.7	24 53.7
59	11 37.3	6 37.2	24 35.5	19 59.8	24 14.2
60	11 46.0	6 32.0	24 10.4	19 18.1	23 30.9
61	11 55.1	6 26.7	23 44.6	18 34.7	22 43.1
62	12 4.6	6 21.1	23 18.0	17 49.1	21 49.8
63	12 14.6	6 15.4	22 50.7	17 1.5	20 49.7
64	12 25.1	6 9.4	22 22.6	16 11.4	19 40.5
65	12 36.2	6 3.2	21 53.5	15 18.9	18 18.9
66	12 47.9	5 56.7	21 23.5	14 23.6	16 38.5
S LAT	5	6	Descendant	8	9

♒ 6° 36′ 37″ — 20h 36m 0s — MC 309° 0′ 0″

8ʰ 40ᵐ 0ˢ		MC		130° 0′ 0″	
		♌ 7° 35′ 26″			N
11	12	Ascendant	2	3	LAT
° ′	° ′	° ′	° ′	° ′	°
♍ 8 21.7	♎10 52.7	♏12 26.7	♐11 32.1	♑ 9 11.4	0
8 37.3	10 36.8	11 4.2	10 16.3	8 31.8	5
8 52.9	10 21.3	9 44.8	9 0.4	7 51.5	10
9 8.6	10 6.2	8 27.3	7 43.5	7 9.6	15
9 11.7	10 3.1	8 11.9	7 27.8	7 1.0	16
9 14.9	10 0.1	7 56.5	7 12.1	6 52.3	17
9 18.1	9 57.1	7 41.2	6 56.2	6 43.4	18
9 21.4	9 54.1	7 25.8	6 40.2	6 34.5	19
9 24.6	9 51.0	7 10.5	6 24.1	6 25.4	20
9 27.9	9 48.0	6 55.1	6 7.8	6 16.2	21
9 31.2	9 44.9	6 39.7	5 51.4	6 6.9	22
9 34.5	9 41.9	6 24.3	5 34.8	5 57.4	23
9 37.9	9 38.8	6 8.9	5 18.0	5 47.7	24
9 41.3	9 35.7	5 53.4	5 1.0	5 37.9	25
9 44.7	9 32.6	5 37.8	4 43.8	5 27.8	26
9 48.2	9 29.5	5 22.2	4 26.4	5 17.6	27
9 51.7	9 26.3	5 6.4	4 8.7	5 7.2	28
9 55.3	9 23.1	4 50.6	3 50.8	4 56.6	29
9 58.9	9 19.9	4 34.7	3 32.6	4 45.7	30
10 2.6	9 16.7	4 18.7	3 14.1	4 34.5	31
10 6.3	9 13.5	4 2.6	2 55.4	4 23.1	32
10 10.1	9 10.2	3 46.4	2 36.2	4 11.4	33
10 13.9	9 6.8	3 30.0	2 16.8	3 59.4	34
10 17.9	9 3.5	3 13.5	1 56.9	3 47.1	35
10 21.9	9 0.1	2 56.8	1 36.7	3 34.4	36
10 25.9	8 56.6	2 39.9	1 16.1	3 21.3	37
10 30.1	8 53.1	2 22.8	0 55.0	3 7.8	38
10 34.3	8 49.6	2 5.6	0 33.4	2 53.9	39
10 38.6	8 46.0	1 48.1	0 11.4	2 39.5	40
10 43.1	8 42.3	1 30.4	♏29 48.8	2 24.6	41
10 47.6	8 38.6	1 12.5	29 25.7	2 9.2	42
10 52.2	8 34.8	0 54.3	29 2.0	1 53.1	43
10 57.0	8 30.9	0 35.9	28 37.7	1 36.5	44
11 1.9	8 27.0	0 17.1	28 12.7	1 19.1	45
11 6.9	8 23.0	♎29 58.1	27 47.0	1 1.0	46
11 12.1	8 18.9	29 38.7	27 20.5	0 42.1	47
11 17.4	8 14.8	29 19.0	26 53.3	0 22.2	48
11 22.9	8 10.5	28 58.9	26 25.2	0 1.4	49
11 28.5	8 6.2	28 38.5	25 56.2	♐29 39.5	50
11 34.3	8 1.7	28 17.6	25 26.2	29 16.5	51
11 40.4	7 57.1	27 56.3	24 55.2	28 52.1	52
11 46.6	7 52.4	27 34.6	24 23.1	28 26.2	53
11 53.1	7 47.6	27 12.4	23 49.8	27 58.7	54
11 59.8	7 42.7	26 49.7	23 15.3	27 29.3	55
12 6.8	7 37.6	26 26.4	22 39.4	26 57.8	56
12 14.0	7 32.3	26 2.6	22 2.1	26 24.0	57
12 21.6	7 26.9	25 38.1	21 23.2	25 47.3	58
12 29.5	7 21.3	25 13.0	20 42.7	25 7.5	59
12 37.7	7 15.6	24 47.3	20 0.3	24 23.8	60
12 46.4	7 9.6	24 20.8	19 16.1	23 35.5	61
12 55.4	7 3.4	23 53.6	18 29.8	22 41.6	62
13 4.9	6 57.0	23 25.5	17 41.2	21 40.6	63
13 14.9	6 50.4	22 56.6	16 50.3	20 30.2	64
13 25.5	6 43.5	22 26.8	15 56.8	19 6.7	65
13 36.7	6 36.3	21 56.0	15 0.4	17 23.2	66
5	6	Descendant	8	9	S
		♒ 7° 35′ 26″			LAT
20ʰ 40ᵐ 0ˢ		MC		310° 0′ 0″	

8ʰ 44ᵐ 0ˢ		MC		131° 0′ 0″	
		♌ 8° 34′ 25″			N
11	12	Ascendant	2	3	LAT
° ′	° ′	° ′	° ′	° ′	°
♍ 9 25.7	♎11 57.7	♏13 27.3	♐12 28.1	♑10 6.7	0
9 40.7	11 40.2	12 3.3	11 11.9	9 27.2	5
9 55.5	11 23.3	10 42.4	9 55.7	8 47.0	10
10 10.4	11 6.6	9 23.3	8 38.2	8 5.1	15
10 13.5	11 3.3	9 7.6	8 22.5	7 56.5	16
10 16.5	11 0.0	8 52.0	8 6.6	7 47.8	17
10 19.6	10 56.6	8 36.3	7 50.7	7 39.0	18
10 22.6	10 53.3	8 20.7	7 34.6	7 30.1	19
10 25.7	10 50.0	8 5.0	7 18.3	7 21.0	20
10 28.8	10 46.6	7 49.3	7 1.9	7 11.8	21
10 32.0	10 43.3	7 33.6	6 45.4	7 2.5	22
10 35.2	10 39.9	7 17.8	6 28.6	6 53.0	23
10 38.4	10 36.5	7 2.0	6 11.7	6 43.3	24
10 41.6	10 33.1	6 46.2	5 54.6	6 33.5	25
10 44.9	10 29.7	6 30.3	5 37.3	6 23.5	26
10 48.2	10 26.3	6 14.3	5 19.7	6 13.2	27
10 51.6	10 22.8	5 58.3	5 1.9	6 2.8	28
10 55.0	10 19.3	5 42.1	4 43.8	5 52.2	29
10 58.4	10 15.8	5 25.9	4 25.4	5 41.3	30
11 1.9	10 12.3	5 9.5	4 6.8	5 30.1	31
11 5.5	10 8.7	4 53.0	3 47.8	5 18.7	32
11 9.1	10 5.1	4 36.4	3 28.5	5 7.0	33
11 12.7	10 1.4	4 19.7	3 8.8	4 55.0	34
11 16.5	9 57.7	4 2.7	2 48.8	4 42.7	35
11 20.3	9 54.0	3 45.7	2 28.4	4 29.9	36
11 24.1	9 50.2	3 28.4	2 7.5	4 16.9	37
11 28.1	9 46.3	3 10.9	1 46.2	4 3.4	38
11 32.1	9 42.4	2 53.3	1 24.4	3 49.4	39
11 36.3	9 38.5	2 35.4	1 2.2	3 35.0	40
11 40.5	9 34.5	2 17.3	0 39.4	3 20.1	41
11 44.8	9 30.4	1 59.0	0 16.0	3 4.6	42
11 49.2	9 26.2	1 40.3	♏29 52.0	2 48.6	43
11 53.7	9 22.0	1 21.4	29 27.4	2 31.9	44
11 58.4	9 17.7	1 2.3	29 2.1	2 14.5	45
12 3.2	9 13.3	0 42.8	28 36.1	1 56.3	46
12 8.1	9 8.8	0 22.9	28 9.3	1 37.3	47
12 13.1	9 4.2	0 2.7	27 41.7	1 17.4	48
12 18.3	8 59.5	♎29 42.2	27 13.2	0 56.6	49
12 23.7	8 54.7	29 21.2	26 43.8	0 34.6	50
12 29.2	8 49.8	28 59.8	26 13.4	0 11.4	51
12 35.0	8 44.8	28 38.0	25 42.0	♐29 47.0	52
12 40.9	8 39.6	28 15.7	25 9.4	29 21.0	53
12 47.1	8 34.3	27 53.0	24 35.7	28 53.3	54
12 53.4	8 28.9	27 29.7	24 0.6	28 23.8	55
13 0.1	8 23.3	27 5.8	23 24.2	27 52.2	56
13 7.0	8 17.5	26 41.4	22 46.3	27 18.1	57
13 14.2	8 11.5	26 16.3	22 6.8	26 41.2	58
13 21.7	8 5.4	25 50.6	21 25.6	26 1.0	59
13 29.5	7 59.1	25 24.2	20 42.5	25 16.9	60
13 37.7	7 52.5	24 57.0	19 57.5	24 28.2	61
13 46.3	7 45.7	24 29.1	19 10.4	23 33.6	62
13 55.3	7 38.7	24 0.3	18 21.0	22 31.8	63
14 4.8	7 31.3	23 30.6	17 29.2	21 20.2	64
14 14.9	7 23.7	23 0.0	16 34.6	19 55.0	65
14 25.5	7 15.8	22 28.4	15 37.2	18 8.2	66
5	6	Descendant	8	9	S
		♒ 8° 34′ 25″			LAT
20ʰ 44ᵐ 0ˢ		MC		311° 0′ 0″	

8ʰ 48ᵐ 0ˢ MC 132° 0′ 0″

♌ 9° 33′ 34″

11	12	Ascendant	2	3	N LAT
° ′	° ′	° ′	° ′	° ′	°
♍10 29.9	♎13 2.6	♏14 27.7	♐13 24.0	♑11 2.1	**0**
10 44.1	12 43.6	13 2.3	12 7.5	10 22.7	**5**
10 58.2	12 25.1	11 39.9	10 50.9	9 42.5	**10**
11 12.4	12 7.0	10 19.3	9 33.0	9 0.8	**15**
11 15.3	12 3.4	10 3.3	9 17.1	8 52.2	**16**
11 18.2	11 59.7	9 47.3	9 1.2	8 43.5	**17**
11 21.1	11 56.1	9 31.4	8 45.1	8 34.7	**18**
11 24.0	11 52.5	9 15.4	8 28.9	8 25.7	**19**
11 26.9	11 48.9	8 59.4	8 12.5	8 16.7	**20**
11 29.9	11 45.2	8 43.4	7 56.0	8 7.5	**21**
11 32.9	11 41.6	8 27.4	7 39.3	7 58.1	**22**
11 35.9	11 37.9	8 11.3	7 22.5	7 48.7	**23**
11 39.0	11 34.2	7 55.2	7 5.4	7 39.0	**24**
11 42.0	11 30.5	7 39.0	6 48.1	7 29.2	**25**
11 45.2	11 26.8	7 22.8	6 30.7	7 19.2	**26**
11 48.3	11 23.1	7 6.4	6 13.0	7 9.0	**27**
11 51.5	11 19.3	6 50.0	5 55.0	6 58.5	**28**
11 54.7	11 15.5	6 33.6	5 36.7	6 47.9	**29**
11 58.0	11 11.7	6 17.0	5 18.2	6 37.0	**30**
12 1.3	11 7.8	6 0.2	4 59.4	6 25.9	**31**
12 4.7	11 3.9	5 43.4	4 40.3	6 14.5	**32**
12 8.1	11 0.0	5 26.4	4 20.8	6 2.7	**33**
12 11.6	10 56.0	5 9.3	4 0.9	5 50.7	**34**
12 15.1	10 51.9	4 52.0	3 40.7	5 38.4	**35**
12 18.8	10 47.9	4 34.5	3 20.1	5 25.7	**36**
12 22.4	10 43.7	4 16.9	2 59.0	5 12.6	**37**
12 26.2	10 39.5	3 59.0	2 37.5	4 59.0	**38**
12 30.0	10 35.3	3 41.0	2 15.5	4 45.1	**39**
12 33.9	10 31.0	3 22.7	1 53.0	4 30.7	**40**
12 37.9	10 26.6	3 4.2	1 29.9	4 15.7	**41**
12 42.0	10 22.1	2 45.4	1 6.3	4 0.2	**42**
12 46.2	10 17.6	2 26.4	0 42.0	3 44.2	**43**
12 50.5	10 13.0	2 7.0	0 17.1	3 27.4	**44**
12 54.9	10 8.3	1 47.4	♏29 51.5	3 10.0	**45**
12 59.5	10 3.5	1 27.4	29 25.2	2 51.8	**46**
13 4.1	9 58.6	1 7.1	28 58.0	2 32.8	**47**
13 8.9	9 53.6	0 46.4	28 30.1	2 12.8	**48**
13 13.9	9 48.5	0 25.4	28 1.3	1 51.9	**49**
13 18.9	9 43.2	0 3.9	27 31.5	1 29.9	**50**
13 24.2	9 37.9	♎29 42.0	27 0.7	1 6.6	**51**
13 29.6	9 32.4	29 19.7	26 28.8	0 42.1	**52**
13 35.3	9 26.8	28 56.9	25 55.8	0 16.0	**53**
13 41.1	9 21.0	28 33.5	25 21.5	♐29 48.2	**54**
13 47.2	9 15.1	28 9.7	24 46.0	29 18.5	**55**
13 53.4	9 9.0	27 45.2	24 9.0	28 46.7	**56**
14 0.0	9 2.7	27 20.2	23 30.5	28 12.4	**57**
14 6.8	8 56.2	26 54.5	22 50.4	27 35.3	**58**
14 13.9	8 49.5	26 28.1	22 8.5	26 54.8	**59**
14 21.3	8 42.6	26 1.0	21 24.8	26 10.4	**60**
14 29.1	8 35.4	25 33.2	20 39.0	25 21.2	**61**
14 37.2	8 28.0	25 4.5	19 51.1	24 26.0	**62**
14 45.8	8 20.3	24 35.0	19 0.8	23 23.3	**63**
14 54.8	8 12.3	24 4.6	18 8.0	22 10.6	**64**
15 4.3	8 4.0	23 33.2	17 12.5	20 43.5	**65**
15 14.3	7 55.4	23 0.8	16 14.0	18 53.4	**66**
5	6	Descendant	8	9	S LAT

♒ 9° 33′ 34″

20ʰ 48ᵐ 0ˢ MC 312° 0′ 0″

8ʰ 52ᵐ 0ˢ MC 133° 0′ 0″

♌ 10° 32′ 55″

N LAT	11	12	Ascendant	2	3
°	° ′	° ′	° ′	° ′	° ′
0	♍11 34.2	♎14 7.5	♏15 28.0	♐14 19.9	♑11 57.6
5	11 47.7	13 46.9	14 1.1	13 3.0	11 18.3
10	12 1.0	13 26.9	12 37.2	11 46.0	10 38.2
15	12 14.5	13 7.3	11 15.2	10 27.6	9 56.5
16	12 17.2	13 3.4	10 58.9	10 11.7	9 47.9
17	12 20.0	12 59.5	10 42.6	9 55.6	9 39.2
18	12 22.7	12 55.6	10 26.3	9 39.5	9 30.4
19	12 25.5	12 51.7	10 10.0	9 23.1	9 21.5
20	12 28.3	12 47.7	9 53.7	9 6.7	9 12.4
21	12 31.1	12 43.8	9 37.4	8 50.0	9 3.3
22	12 33.9	12 39.8	9 21.1	8 33.2	8 53.9
23	12 36.8	12 35.9	9 4.7	8 16.3	8 44.5
24	12 39.6	12 31.9	8 48.2	7 59.1	8 34.8
25	12 42.6	12 27.9	8 31.7	7 41.7	8 25.0
26	12 45.5	12 23.9	8 15.2	7 24.1	8 15.0
27	12 48.5	12 19.8	7 58.5	7 6.2	8 4.8
28	12 51.5	12 15.7	7 41.8	6 48.1	7 54.4
29	12 54.6	12 11.6	7 24.9	6 29.7	7 43.7
30	12 57.7	12 7.5	7 8.0	6 11.0	7 32.9
31	13 0.8	12 3.3	6 50.9	5 52.0	7 21.7
32	13 4.0	11 59.1	6 33.7	5 32.7	7 10.3
33	13 7.2	11 54.8	6 16.4	5 13.1	6 58.6
34	13 10.5	11 50.5	5 58.9	4 53.0	6 46.6
35	13 13.9	11 46.1	5 41.2	4 32.6	6 34.2
36	13 17.3	11 41.7	5 23.4	4 11.8	6 21.5
37	13 20.8	11 37.2	5 5.4	3 50.5	6 8.4
38	13 24.3	11 32.7	4 47.1	3 28.8	5 54.9
39	13 28.0	11 28.1	4 28.7	3 6.6	5 40.9
40	13 31.7	11 23.4	4 10.0	2 43.8	5 26.5
41	13 35.5	11 18.7	3 51.0	2 20.5	5 11.5
42	13 39.3	11 13.9	3 31.8	1 56.6	4 56.0
43	13 43.3	11 9.0	3 12.4	1 32.1	4 39.9
44	13 47.4	11 4.0	2 52.6	1 6.9	4 23.2
45	13 51.5	10 58.9	2 32.5	0 41.0	4 5.7
46	13 55.8	10 53.7	2 12.1	0 14.3	3 47.5
47	14 0.2	10 48.4	1 51.3	♏29 46.9	3 28.4
48	14 4.8	10 43.0	1 30.1	29 18.6	3 8.4
49	14 9.4	10 37.4	1 8.6	28 49.4	2 47.4
50	14 14.2	10 31.8	0 46.6	28 19.2	2 25.4
51	14 19.2	10 26.0	0 24.2	27 48.0	2 2.0
52	14 24.4	10 20.0	0 1.3	27 15.7	1 37.4
53	14 29.7	10 13.9	♎29 38.0	26 42.2	1 11.2
54	14 35.2	10 7.7	29 14.1	26 7.4	0 43.3
55	14 40.9	10 1.2	28 49.6	25 31.4	0 13.5
56	14 46.9	9 54.6	28 24.6	24 53.8	♐29 41.6
57	14 53.0	9 47.8	27 59.0	24 14.7	29 7.1
58	14 59.5	9 40.8	27 32.6	23 34.0	28 29.7
59	15 6.2	9 33.5	27 5.6	22 51.5	27 49.0
60	15 13.2	9 26.0	26 37.9	22 7.0	27 4.2
61	15 20.5	9 18.3	26 9.4	21 20.5	26 14.5
62	15 28.2	9 10.2	25 40.0	20 31.7	25 18.8
63	15 36.3	9 1.9	25 9.8	19 40.6	24 15.3
64	15 44.8	8 53.3	24 38.6	18 46.8	23 1.4
65	15 53.8	8 44.3	24 6.4	17 50.2	21 32.4
66	16 3.2	8 34.9	23 33.2	16 50.7	19 38.9
S LAT	5	6	Descendant	8	9

♒ 10° 32′ 55″

20ʰ 52ᵐ 0ˢ MC 313° 0′ 0″

$8^h\ 56^m\ 0^s$ MC 134° 0′ 0″

♌ 11° 32′ 25″

N LAT	11	12	Ascendant	2	3
°	° ′	° ′	° ′	° ′	° ′
0	♍12 38.6	♎15 12.2	♏16 28.0	♐15 15.6	♑12 53.1
5	12 51.3	14 50.1	14 59.7	13 58.5	12 13.9
10	13 4.0	14 28.6	13 34.5	12 41.1	11 33.9
15	13 16.7	14 7.6	12 11.0	11 22.3	10 52.3
16	13 19.2	14 3.4	11 54.4	11 6.3	10 43.7
17	13 21.8	13 59.2	11 37.8	10 50.1	10 35.1
18	13 24.4	13 55.0	11 21.2	10 33.8	10 26.3
19	13 27.0	13 50.8	11 4.6	10 17.4	10 17.4
20	13 29.7	13 46.5	10 48.0	10 0.8	10 8.3
21	13 32.3	13 42.3	10 31.4	9 44.1	9 59.2
22	13 35.0	13 38.1	10 14.7	9 27.2	9 49.9
23	13 37.7	13 33.8	9 58.0	9 10.1	9 40.4
24	13 40.4	13 29.5	9 41.3	8 52.8	9 30.8
25	13 43.2	13 25.2	9 24.4	8 35.2	9 21.0
26	13 45.9	13 20.9	9 7.5	8 17.5	9 11.0
27	13 48.7	13 16.5	8 50.6	7 59.5	9 0.8
28	13 51.6	13 12.1	8 33.5	7 41.2	8 50.4
29	13 54.5	13 7.7	8 16.3	7 22.7	8 39.7
30	13 57.4	13 3.3	7 59.0	7 3.9	8 28.8
31	14 0.4	12 58.8	7 41.6	6 44.7	8 17.7
32	14 3.4	12 54.2	7 24.0	6 25.2	8 6.3
33	14 6.4	12 49.6	7 6.3	6 5.4	7 54.6
34	14 9.5	12 45.0	6 48.5	5 45.2	7 42.6
35	14 12.7	12 40.3	6 30.4	5 24.6	7 30.2
36	14 15.9	12 35.5	6 12.2	5 3.5	7 17.5
37	14 19.2	12 30.7	5 53.8	4 42.1	7 4.4
38	14 22.6	12 25.9	5 35.2	4 20.1	6 50.9
39	14 26.0	12 20.9	5 16.3	3 57.7	6 36.9
40	14 29.5	12 15.9	4 57.2	3 34.7	6 22.5
41	14 33.0	12 10.8	4 37.9	3 11.1	6 7.5
42	14 36.7	12 5.6	4 18.2	2 47.0	5 52.0
43	14 40.4	12 0.3	3 58.3	2 22.2	5 35.9
44	14 44.3	11 54.9	3 38.1	1 56.7	5 19.1
45	14 48.2	11 49.4	3 17.6	1 30.5	5 1.6
46	14 52.2	11 43.9	2 56.7	1 3.5	4 43.3
47	14 56.4	11 38.2	2 35.5	0 35.7	4 24.2
48	15 0.7	11 32.3	2 13.8	0 7.1	4 4.2
49	15 5.1	11 26.4	1 51.8	♏29 37.5	3 43.2
50	15 9.6	11 20.3	1 29.3	29 7.0	3 21.0
51	15 14.3	11 14.0	1 6.4	28 35.3	2 57.7
52	15 19.1	11 7.6	0 43.0	28 2.6	2 32.9
53	15 24.1	11 1.1	0 19.1	27 28.6	2 6.7
54	15 29.3	10 54.3	♎29 54.6	26 53.4	1 38.7
55	15 34.7	10 47.4	29 29.6	26 16.8	1 8.8
56	15 40.3	10 40.3	29 4.0	25 38.7	0 36.6
57	15 46.1	10 32.9	28 37.7	24 59.0	0 2.0
58	15 52.2	10 25.4	28 10.8	24 17.7	♐29 24.4
59	15 58.5	10 17.6	27 43.1	23 34.5	28 43.4
60	16 5.1	10 9.5	27 14.7	22 49.3	27 58.3
61	16 12.0	10 1.1	26 45.5	22 2.0	27 8.2
62	16 19.3	9 52.5	26 15.4	21 12.4	26 11.9
63	16 26.9	9 43.5	25 44.5	20 20.3	25 7.6
64	16 34.8	9 34.2	25 12.5	19 25.6	23 52.6
65	16 43.3	9 24.5	24 39.6	18 28.0	22 21.8
66	16 52.2	9 14.4	24 5.5	17 27.3	20 24.7
S LAT	5	6	Descendant	8	9

♒ 11° 32′ 25″

$20^h\ 56^m\ 0^s$ MC 314° 0′ 0″

$9^h\ 0^m\ 0^s$ MC 135° 0′ 0″

♌ 12° 32′ 6″

N LAT	11	12	Ascendant	2	3
°	° ′	° ′	° ′	° ′	° ′
0	♍13 43.2	♎16 16.8	♏17 27.9	♐16 11.3	♑13 48.7
5	13 55.1	15 53.2	15 58.2	14 53.9	13 9.6
10	14 7.0	15 30.3	14 31.6	13 36.1	12 29.7
15	14 18.9	15 7.8	13 6.7	12 16.9	11 48.2
16	14 21.3	15 3.3	12 49.8	12 0.8	11 39.7
17	14 23.8	14 58.8	12 32.9	11 44.5	11 31.0
18	14 26.2	14 54.3	12 16.0	11 28.2	11 22.2
19	14 28.7	14 49.8	11 59.1	11 11.6	11 13.3
20	14 31.1	14 45.3	11 42.2	10 54.9	11 4.3
21	14 33.6	14 40.8	11 25.3	10 38.1	10 55.2
22	14 36.1	14 36.2	11 8.3	10 21.1	10 45.9
23	14 38.7	14 31.7	10 51.3	10 3.8	10 36.4
24	14 41.2	14 27.1	10 34.2	9 46.4	10 26.8
25	14 43.8	14 22.5	10 17.1	9 28.8	10 17.0
26	14 46.4	14 17.8	9 59.8	9 10.9	10 7.0
27	14 49.1	14 13.2	9 42.5	8 52.8	9 56.9
28	14 51.8	14 8.5	9 25.1	8 34.4	9 46.5
29	14 54.5	14 3.8	9 7.6	8 15.7	9 35.8
30	14 57.2	13 59.0	8 50.0	7 56.7	9 25.0
31	15 0.0	13 54.2	8 32.2	7 37.4	9 13.8
32	15 2.8	13 49.3	8 14.3	7 17.7	9 2.4
33	15 5.7	13 44.4	7 56.2	6 57.7	8 50.7
34	15 8.6	13 39.4	7 38.0	6 37.3	8 38.7
35	15 11.6	13 34.4	7 19.6	6 16.6	8 26.4
36	15 14.6	13 29.3	7 1.0	5 55.3	8 13.6
37	15 17.7	13 24.2	6 42.2	5 33.6	8 0.5
38	15 20.8	13 19.0	6 23.2	5 11.5	7 47.0
39	15 24.1	13 13.7	6 4.0	4 48.8	7 33.0
40	15 27.3	13 8.3	5 44.5	4 25.6	7 18.6
41	15 30.7	13 2.8	5 24.7	4 1.8	7 3.6
42	15 34.1	12 57.3	5 4.7	3 37.4	6 48.1
43	15 37.6	12 51.6	4 44.3	3 12.3	6 32.0
44	15 41.2	12 45.9	4 23.7	2 46.5	6 15.2
45	15 44.9	12 40.0	4 2.7	2 20.1	5 57.7
46	15 48.7	12 34.0	3 41.3	1 52.8	5 39.4
47	15 52.6	12 27.9	3 19.6	1 24.7	5 20.3
48	15 56.6	12 21.7	2 57.5	0 55.7	5 0.2
49	16 0.7	12 15.3	2 35.0	0 25.7	4 39.1
50	16 5.0	12 8.8	2 12.0	♏29 54.8	4 17.0
51	16 9.4	12 2.1	1 48.6	29 22.8	3 53.5
52	16 13.9	11 55.2	1 24.6	28 49.6	3 28.7
53	16 18.6	11 48.2	1 0.2	28 15.2	3 2.4
54	16 23.5	11 41.0	0 35.2	27 39.4	2 34.3
55	16 28.6	11 33.6	0 9.6	27 2.3	2 4.3
56	16 33.8	11 25.9	♎29 43.4	26 23.7	1 32.0
57	16 39.3	11 18.1	29 16.5	25 43.4	0 57.2
58	16 45.0	11 9.9	28 48.9	25 1.4	0 19.4
59	16 50.9	11 1.6	28 20.6	24 17.5	♐29 38.2
60	16 57.1	10 52.9	27 51.5	23 31.6	28 52.7
61	17 3.5	10 44.0	27 21.6	22 43.5	28 2.2
62	17 10.3	10 34.7	26 50.9	21 53.1	27 5.4
63	17 17.5	10 25.1	26 19.1	21 0.1	26 0.3
64	17 24.9	10 15.1	25 46.4	20 4.4	24 44.2
65	17 32.8	10 4.7	25 12.7	19 5.7	23 11.6
66	17 41.2	9 53.9	24 37.8	18 3.9	21 10.8
S LAT	5	6	Descendant	8	9

♒ 12° 32′ 6″

$21^h\ 0^m\ 0^s$ MC 315° 0′ 0″

9h 4m 0s		MC		136° 0′ 0″	
		♌ 13° 31′ 59″			
11	12	Ascendant	2	3	N LAT
♍14 47.8	♎17 21.4	♏18 27.6	♐17 6.9	♑14 44.4	0
14 59.0	16 56.2	16 56.6	15 49.2	14 5.4	5
15 10.1	16 31.9	15 28.6	14 31.1	13 25.6	10
15 21.3	16 7.9	14 2.3	13 11.5	12 44.2	15
15 23.5	16 3.1	13 45.1	12 55.3	12 35.7	16
15 25.8	15 58.3	13 27.9	12 38.9	12 27.1	17
15 28.1	15 53.6	13 10.7	12 22.5	12 18.3	18
15 30.4	15 48.8	12 53.6	12 5.8	12 9.4	19
15 32.7	15 44.0	12 36.3	11 49.1	12 0.4	20
15 35.0	15 39.1	12 19.1	11 32.1	11 51.3	21
15 37.4	15 34.3	12 1.8	11 15.0	11 42.0	22
15 39.8	15 29.5	11 44.5	10 57.6	11 32.6	23
15 42.1	15 24.6	11 27.1	10 40.1	11 23.0	24
15 44.6	15 19.7	11 9.6	10 22.3	11 13.2	25
15 47.0	15 14.8	10 52.1	10 4.3	11 3.2	26
15 49.5	15 9.8	10 34.5	9 46.0	10 53.1	27
15 52.0	15 4.8	10 16.7	9 27.5	10 42.7	28
15 54.5	14 59.8	9 58.9	9 8.7	10 32.1	29
15 57.1	14 54.7	9 40.9	8 49.5	10 21.2	30
15 59.7	14 49.6	9 22.8	8 30.1	10 10.1	31
16 2.3	14 44.4	9 4.5	8 10.3	9 58.7	32
16 5.0	14 39.2	8 46.1	7 50.1	9 47.0	33
16 7.8	14 33.9	8 27.5	7 29.5	9 35.0	34
16 10.5	14 28.5	8 8.8	7 8.6	9 22.7	35
16 13.4	14 23.1	7 49.8	6 47.2	9 10.0	36
16 16.2	14 17.6	7 30.6	6 25.3	8 56.9	37
16 19.2	14 12.0	7 11.2	6 2.9	8 43.3	38
16 22.2	14 6.4	6 51.6	5 40.0	8 29.4	39
16 25.3	14 0.7	6 31.7	5 16.5	8 14.9	40
16 28.4	13 54.8	6 11.5	4 52.5	7 59.9	41
16 31.6	13 48.9	5 51.0	4 27.8	7 44.4	42
16 34.9	13 42.9	5 30.3	4 2.5	7 28.3	43
16 38.2	13 36.8	5 9.2	3 36.5	7 11.5	44
16 41.7	13 30.5	4 47.8	3 9.7	6 53.9	45
16 45.2	13 24.1	4 26.0	2 42.1	6 35.6	46
16 48.9	13 17.6	4 3.8	2 13.7	6 16.5	47
16 52.6	13 11.0	3 41.2	1 44.3	5 56.4	48
16 56.5	13 4.2	3 18.2	1 14.0	5 35.3	49
17 0.5	12 57.2	2 54.7	0 42.7	5 13.1	50
17 4.6	12 50.1	2 30.7	0 10.2	4 49.6	51
17 8.8	12 42.8	2 6.3	♏29 36.6	4 24.7	52
17 13.2	12 35.3	1 41.3	29 1.8	3 58.3	53
17 17.7	12 27.6	1 15.7	28 25.5	3 30.2	54
17 22.5	12 19.7	0 49.5	27 47.9	3 0.0	55
17 27.4	12 11.5	0 22.7	27 8.7	2 27.7	56
17 32.5	12 3.2	♎29 55.2	26 27.8	1 52.7	57
17 37.8	11 54.5	29 27.0	25 45.1	1 14.7	58
17 43.3	11 45.6	28 58.1	25 0.6	0 33.2	59
17 49.1	11 36.3	28 28.3	24 13.9	♐29 47.5	60
17 55.1	11 26.8	27 57.7	23 25.0	28 56.6	61
18 1.4	11 16.9	27 26.2	22 33.7	27 59.3	62
18 8.1	11 6.6	26 53.8	21 39.8	26 53.5	63
18 15.1	10 56.0	26 20.3	20 43.1	25 36.3	64
18 22.4	10 44.9	25 45.8	19 43.4	24 1.8	65
18 30.2	10 33.4	25 10.1	18 40.4	21 57.3	66
5	6	Descendant	8	9	S LAT
		♒ 13° 31′ 59″			
21h 4m 0s		MC		316° 0′ 0″	

9h 8m 0s		MC		137° 0′ 0″	
		♌ 14° 32′ 1″			
N LAT	11	12	Ascendant	2	3
0	♍15 52.5	♎18 25.8	♏19 27.1	♐18 2.4	♑15 40.1
5	16 3.0	17 59.2	17 54.8	16 44.5	15 1.3
10	16 13.3	17 33.4	16 25.5	15 26.1	14 21.6
15	16 23.7	17 8.0	14 57.8	14 6.1	13 40.4
16	16 25.8	17 2.9	14 40.3	13 49.8	13 31.8
17	16 27.9	16 57.8	14 22.8	13 33.3	13 23.2
18	16 30.0	16 52.8	14 5.4	13 16.8	13 14.5
19	16 32.2	16 47.7	13 47.9	13 0.1	13 5.6
20	16 34.3	16 42.6	13 30.4	12 43.2	12 56.7
21	16 36.5	16 37.5	13 12.9	12 26.1	12 47.6
22	16 38.7	16 32.4	12 55.3	12 8.9	12 38.3
23	16 40.9	16 27.2	12 37.7	11 51.4	12 28.9
24	16 43.1	16 22.0	12 20.0	11 33.8	12 19.3
25	16 45.4	16 16.9	12 2.2	11 15.9	12 9.5
26	16 47.6	16 11.6	11 44.3	10 57.7	11 59.6
27	16 49.9	16 6.4	11 26.4	10 39.4	11 49.4
28	16 52.3	16 1.1	11 8.3	10 20.7	11 39.1
29	16 54.6	15 55.7	10 50.1	10 1.7	11 28.5
30	16 57.0	15 50.3	10 31.8	9 42.4	11 17.6
31	16 59.4	15 44.9	10 13.4	9 22.8	11 6.5
32	17 1.9	15 39.4	9 54.8	9 2.9	10 55.1
33	17 4.4	15 33.9	9 36.0	8 42.5	10 43.5
34	17 7.0	15 28.2	9 17.0	8 21.8	10 31.5
35	17 9.5	15 22.6	8 57.9	8 0.6	10 19.1
36	17 12.2	15 16.8	8 38.6	7 39.0	10 6.4
37	17 14.8	15 11.0	8 19.0	7 16.9	9 53.3
38	17 17.6	15 5.1	7 59.2	6 54.4	9 39.8
39	17 20.4	14 59.1	7 39.2	6 31.2	9 25.9
40	17 23.2	14 53.0	7 18.9	6 7.6	9 11.4
41	17 26.1	14 46.8	6 58.3	5 43.3	8 56.4
42	17 29.1	14 40.6	6 37.4	5 18.4	8 40.9
43	17 32.2	14 34.2	6 16.2	4 52.8	8 24.8
44	17 35.3	14 27.7	5 54.7	4 26.5	8 8.0
45	17 38.5	14 21.0	5 32.8	3 59.4	7 50.4
46	17 41.8	14 14.2	5 10.6	3 31.5	7 32.1
47	17 45.2	14 7.3	4 47.9	3 2.7	7 13.0
48	17 48.7	14 0.3	4 24.9	2 33.0	6 52.9
49	17 52.3	13 53.1	4 1.4	2 2.4	6 31.7
50	17 56.0	13 45.7	3 37.4	1 30.6	6 9.5
51	17 59.8	13 38.1	3 12.9	0 57.8	5 45.9
52	18 3.7	13 30.3	2 47.9	0 23.7	5 21.0
53	18 7.8	13 22.4	2 22.4	♏29 48.4	4 54.6
54	18 12.0	13 14.2	1 56.2	29 11.7	4 26.3
55	18 16.4	13 5.8	1 29.5	28 33.5	3 56.1
56	18 21.0	12 57.1	1 2.1	27 53.7	3 23.6
57	18 25.7	12 48.2	0 34.0	27 12.3	2 48.5
58	18 30.6	12 39.0	0 5.2	26 29.0	2 10.4
59	18 35.7	12 29.5	♎29 35.6	25 43.7	1 28.7
60	18 41.1	12 19.7	29 5.1	24 56.3	0 42.7
61	18 46.7	12 9.6	28 33.8	24 6.6	♐29 51.4
62	18 52.6	11 59.1	28 1.6	23 14.4	28 53.6
63	18 58.8	11 48.2	27 28.4	22 19.6	27 47.1
64	19 5.3	11 36.9	26 54.2	21 21.9	26 28.8
65	19 12.1	11 25.1	26 18.9	20 21.1	24 52.6
66	19 19.3	11 12.8	25 42.4	19 16.9	22 44.3
S LAT	5	6	Descendant	8	9
			♒ 14° 32′ 1″		
	21h 8m 0s		MC		317° 0′ 0″

9^h 12^m 0^s MC 138° 0′ 0″

♌ 15° 32′ 15″

N LAT	11	12	Ascendant	2	3
°	° ′	° ′	° ′	° ′	° ′
0	♍16 57.4	♎19 30.1	♏20 26.4	♐18 57.9	♑16 36.0
5	17 7.0	19 2.0	18 52.9	17 39.7	15 57.3
10	17 16.6	18 34.8	17 22.3	16 21.0	15 17.8
15	17 26.2	18 7.9	15 53.2	15 0.6	14 36.6
16	17 28.1	18 2.6	15 35.4	14 44.3	14 28.1
17	17 30.1	17 57.2	15 17.7	14 27.7	14 19.5
18	17 32.1	17 51.9	15 0.0	14 11.1	14 10.8
19	17 34.0	17 46.5	14 42.2	13 54.3	14 2.0
20	17 36.0	17 41.2	14 24.4	13 37.3	13 53.0
21	17 38.0	17 35.8	14 6.6	13 20.1	13 43.9
22	17 40.1	17 30.3	13 48.7	13 2.8	13 34.7
23	17 42.1	17 24.9	13 30.8	12 45.2	13 25.3
24	17 44.2	17 19.4	13 12.7	12 27.5	13 15.7
25	17 46.2	17 14.0	12 54.7	12 9.5	13 6.0
26	17 48.4	17 8.4	12 36.5	11 51.2	12 56.1
27	17 50.5	17 2.9	12 18.2	11 32.7	12 45.9
28	17 52.6	16 57.3	11 59.8	11 13.9	12 35.6
29	17 54.8	16 51.6	11 41.3	10 54.8	12 25.0
30	17 57.0	16 45.9	11 22.7	10 35.3	12 14.2
31	17 59.3	16 40.2	11 3.9	10 15.6	12 3.1
32	18 1.5	16 34.4	10 44.9	9 55.5	11 51.7
33	18 3.9	16 28.5	10 25.8	9 35.0	11 40.1
34	18 6.2	16 22.6	10 6.5	9 14.1	11 28.1
35	18 8.6	16 16.6	9 47.0	8 52.7	11 15.8
36	18 11.0	16 10.5	9 27.3	8 30.9	11 3.1
37	18 13.5	16 4.4	9 7.4	8 8.7	10 50.0
38	18 16.0	15 58.1	8 47.2	7 45.9	10 36.5
39	18 18.6	15 51.8	8 26.8	7 22.5	10 22.5
40	18 21.2	15 45.3	8 6.1	6 58.6	10 8.1
41	18 23.9	15 38.8	7 45.1	6 34.1	9 53.1
42	18 26.7	15 32.2	7 23.8	6 9.0	9 37.6
43	18 29.5	15 25.4	7 2.2	5 43.1	9 21.5
44	18 32.4	15 18.5	6 40.2	5 16.5	9 4.6
45	18 35.4	15 11.5	6 17.9	4 49.1	8 47.1
46	18 38.4	15 4.3	5 55.2	4 20.9	8 28.8
47	18 41.5	14 57.0	5 32.1	3 51.9	8 9.6
48	18 44.8	14 49.5	5 8.5	3 21.8	7 49.5
49	18 48.1	14 41.9	4 44.5	2 50.8	7 28.4
50	18 51.5	14 34.1	4 20.1	2 18.7	7 6.1
51	18 55.0	14 26.1	3 55.1	1 45.4	6 42.5
52	18 58.7	14 17.9	3 29.5	1 10.9	6 17.6
53	19 2.4	14 9.4	3 3.4	0 35.1	5 51.0
54	19 6.3	14 0.8	2 36.7	♏29 57.9	5 22.7
55	19 10.4	13 51.9	2 9.4	29 19.2	4 52.4
56	19 14.6	13 42.7	1 41.4	28 38.9	4 19.9
57	19 18.9	13 33.3	1 12.7	27 56.8	3 44.6
58	19 23.5	13 23.6	0 43.3	27 12.8	3 6.3
59	19 28.2	13 13.5	0 13.0	26 26.9	2 24.4
60	19 33.2	13 3.1	♎29 41.9	25 38.7	1 38.2
61	19 38.4	12 52.4	29 9.9	24 48.2	0 46.6
62	19 43.8	12 41.2	28 37.0	23 55.2	♐29 48.3
63	19 49.5	12 29.7	28 3.0	22 59.4	28 41.2
64	19 55.5	12 17.7	27 28.0	22 0.6	27 21.9
65	20 1.8	12 5.2	26 51.9	20 58.7	25 44.0
66	20 8.4	11 52.2	26 14.6	19 53.3	23 31.7
S LAT	5	6	Descendant	8	9

♒ 15° 32′ 15″

21^h 12^m 0^s MC 318° 0′ 0″

9^h 16^m 0^s MC 139° 0′ 0″

♌ 16° 32′ 40″

N LAT	11	12	Ascendant	2	3
°	° ′	° ′	° ′	° ′	° ′
0	♍18 2.3	♎20 34.3	♏21 25.6	♐19 53.3	♑17 31.9
5	18 11.1	20 4.7	19 50.8	18 34.9	16 53.4
10	18 19.9	19 36.1	18 18.9	17 15.9	16 14.0
15	18 28.8	19 7.9	16 48.5	15 55.2	15 33.0
16	18 30.6	19 2.2	16 30.5	15 38.7	15 24.5
17	18 32.4	18 56.6	16 12.5	15 22.1	15 15.9
18	18 34.2	18 51.0	15 54.5	15 5.4	15 7.3
19	18 36.0	18 45.3	15 36.4	14 48.5	14 58.5
20	18 37.8	18 39.6	15 18.3	14 31.4	14 49.5
21	18 39.6	18 34.0	15 0.2	14 14.2	14 40.4
22	18 41.5	18 28.3	14 42.0	13 56.7	14 31.2
23	18 43.4	18 22.5	14 23.8	13 39.0	14 21.9
24	18 45.3	18 16.8	14 5.5	13 21.2	14 12.3
25	18 47.2	18 11.0	13 47.1	13 3.0	14 2.6
26	18 49.1	18 5.2	13 28.6	12 44.7	13 52.7
27	18 51.1	17 59.3	13 10.0	12 26.0	13 42.6
28	18 53.0	17 53.4	12 51.3	12 7.1	13 32.3
29	18 55.0	17 47.5	12 32.5	11 47.9	13 21.7
30	18 57.1	17 41.5	12 13.5	11 28.3	13 10.9
31	18 59.1	17 35.4	11 54.4	11 8.4	12 59.8
32	19 1.2	17 29.3	11 35.1	10 48.1	12 48.5
33	19 3.3	17 23.1	11 15.6	10 27.5	12 36.8
34	19 5.5	17 16.9	10 56.0	10 6.4	12 24.9
35	19 7.7	17 10.6	10 36.1	9 44.9	12 12.6
36	19 9.9	17 4.2	10 16.0	9 22.9	11 59.9
37	19 12.2	16 57.7	9 55.7	9 0.4	11 46.8
38	19 14.5	16 51.1	9 35.2	8 37.5	11 33.3
39	19 16.9	16 44.4	9 14.4	8 13.9	11 19.4
40	19 19.3	16 37.6	8 53.3	7 49.8	11 5.0
41	19 21.8	16 30.7	8 31.9	7 25.0	10 50.0
42	19 24.3	16 23.7	8 10.2	6 59.6	10 34.5
43	19 26.9	16 16.6	7 48.1	6 33.5	10 18.4
44	19 29.5	16 9.3	7 25.7	6 6.7	10 1.6
45	19 32.3	16 1.9	7 3.0	5 39.0	9 44.0
46	19 35.1	15 54.4	6 39.8	5 10.5	9 25.7
47	19 37.9	15 46.7	6 16.2	4 41.1	9 6.5
48	19 40.9	15 38.8	5 52.2	4 10.7	8 46.4
49	19 43.9	15 30.7	5 27.7	3 39.3	8 25.2
50	19 47.1	15 22.5	5 2.7	3 6.8	8 2.9
51	19 50.3	15 14.0	4 37.2	2 33.1	7 39.4
52	19 53.6	15 5.4	4 11.2	1 58.2	7 14.4
53	19 57.1	14 56.5	3 44.5	1 21.9	6 47.8
54	20 0.7	14 47.3	3 17.3	0 44.2	6 19.5
55	20 4.4	14 37.9	2 49.4	0 5.0	5 49.1
56	20 8.2	14 28.3	2 20.8	♏29 24.1	5 16.4
57	20 12.2	14 18.3	1 51.4	28 41.4	4 41.1
58	20 16.4	14 8.0	1 21.3	27 56.8	4 2.7
59	20 20.8	13 57.4	0 50.4	27 10.1	3 20.6
60	20 25.3	13 46.5	0 18.7	26 21.2	2 34.1
61	20 30.1	13 35.1	♎29 46.0	25 29.8	1 42.2
62	20 35.0	13 23.4	29 12.3	24 35.9	0 43.5
63	20 40.2	13 11.2	28 37.6	23 39.2	♐29 35.8
64	20 45.7	12 58.5	28 1.8	22 39.4	28 15.5
65	20 51.5	12 45.3	27 24.9	21 36.3	26 35.9
66	20 57.6	12 31.6	26 46.7	20 29.7	24 19.8
S LAT	5	6	Descendant	8	9

♒ 16° 32′ 40″

21^h 16^m 0^s MC 319° 0′ 0″

9ʰ 20ᵐ 0ˢ — MC — 140° 0′ 0″

♌ 17° 33′ 15″

11	12	Ascendant	2	3	N LAT
° ′	° ′	° ′	° ′	° ′	
♍19 7.3	♎21 38.3	♏22 24.6	♐20 48.6	♑18 27.9	0°
19 15.3	21 7.3	20 48.6	19 30.0	17 49.6	5
19 23.3	20 37.3	19 15.5	18 10.8	17 10.3	10
19 31.4	20 7.7	17 43.8	16 49.7	16 29.5	15
19 33.0	20 1.8	17 25.5	16 33.2	16 21.0	16
19 34.7	19 55.9	17 7.2	16 16.5	16 12.5	17
19 36.3	19 49.9	16 48.9	15 59.7	16 3.8	18
19 38.0	19 44.0	16 30.6	15 42.7	15 55.1	19
19 39.6	19 38.1	16 12.2	15 25.5	15 46.1	20
19 41.3	19 32.1	15 53.8	15 8.2	15 37.1	21
19 43.0	19 26.1	15 35.3	14 50.6	15 27.9	22
19 44.7	19 20.1	15 16.8	14 32.9	15 18.6	23
19 46.4	19 14.1	14 58.2	14 14.9	15 9.1	24
19 48.2	19 8.0	14 39.5	13 56.7	14 59.4	25
19 49.9	19 1.9	14 20.7	13 38.2	14 49.5	26
19 51.7	18 55.7	14 1.8	13 19.4	14 39.4	27
19 53.5	18 49.5	13 42.8	13 0.4	14 29.1	28
19 55.3	18 43.3	13 23.6	12 41.0	14 18.5	29
19 57.2	18 37.0	13 4.3	12 21.3	14 7.8	30
19 59.1	18 30.6	12 44.9	12 1.2	13 56.7	31
20 1.0	18 24.2	12 25.2	11 40.8	13 45.4	32
20 2.9	18 17.7	12 5.4	11 20.0	13 33.8	33
20 4.8	18 11.1	11 45.4	10 58.8	13 21.8	34
20 6.8	18 4.5	11 25.2	10 37.1	13 9.5	35
20 8.9	17 57.8	11 4.8	10 14.9	12 56.9	36
20 11.0	17 50.9	10 44.1	9 52.3	12 43.8	37
20 13.1	17 44.0	10 23.2	9 29.1	12 30.4	38
20 15.2	17 37.0	10 1.9	9 5.3	12 16.5	39
20 17.4	17 29.9	9 40.5	8 41.0	12 2.0	40
20 19.7	17 22.6	9 18.7	8 16.0	11 47.1	41
20 22.0	17 15.3	8 56.5	7 50.4	11 31.6	42
20 24.3	17 7.8	8 34.1	7 24.0	11 15.5	43
20 26.7	17 0.1	8 11.2	6 56.9	10 58.7	44
20 29.2	16 52.3	7 48.0	6 28.9	10 41.2	45
20 31.8	16 44.4	7 24.4	6 0.1	10 22.8	46
20 34.4	16 36.3	7 0.4	5 30.4	10 3.7	47
20 37.0	16 28.0	6 35.9	4 59.7	9 43.5	48
20 39.8	16 19.5	6 10.9	4 27.9	9 22.4	49
20 42.7	16 10.8	5 45.4	3 55.0	9 0.1	50
20 45.6	16 1.9	5 19.4	3 20.9	8 36.5	51
20 48.6	15 52.8	4 52.8	2 45.6	8 11.5	52
20 51.8	15 43.5	4 25.6	2 8.8	7 44.9	53
20 55.0	15 33.9	3 57.8	1 30.6	7 16.5	54
20 58.4	15 24.0	3 29.3	0 50.8	6 46.0	55
21 1.9	15 13.8	3 0.1	0 9.4	6 13.3	56
21 5.6	15 3.3	2 30.1	♏29 26.0	5 37.9	57
21 9.4	14 52.5	1 59.4	28 40.8	4 59.3	58
21 13.3	14 41.4	1 27.8	27 53.4	4 17.1	59
21 17.4	14 29.8	0 55.4	27 3.7	3 30.4	60
21 21.8	14 17.9	0 22.0	26 11.5	2 38.2	61
21 26.3	14 5.5	♎29 47.6	25 16.7	1 39.1	62
21 31.0	13 52.6	29 12.2	24 19.0	0 30.8	63
21 36.0	13 39.3	28 35.6	23 18.1	♐29 9.7	64
21 41.3	13 25.4	27 57.9	22 13.9	27 28.5	65
21 46.8	13 11.0	27 18.9	21 6.0	25 8.5	66
5	6	Descendant	8	9	S LAT

♒ 17° 33′ 15″

21ʰ 20ᵐ 0ˢ — MC — 320° 0′ 0″

9ʰ 24ᵐ 0ˢ — MC — 141° 0′ 0″

♌ 18° 34′ 2″

11	12	Ascendant	2	3	N LAT
° ′	° ′	° ′	° ′	° ′	
♍20 12.3	♎22 42.3	♏23 23.4	♐21 43.9	♑19 24.1	0°
20 19.6	22 9.8	21 46.3	20 25.1	18 45.9	5
20 26.8	21 38.4	20 11.9	19 5.6	18 6.8	10
20 34.1	21 7.4	18 38.9	17 44.3	17 26.1	15
20 35.6	21 1.2	18 20.4	17 27.6	17 17.7	16
20 37.0	20 55.0	18 1.8	17 10.9	17 9.2	17
20 38.5	20 48.8	17 43.2	16 54.0	17 0.5	18
20 40.0	20 42.6	17 24.6	16 36.9	16 51.8	19
20 41.5	20 36.4	17 6.0	16 19.7	16 42.9	20
20 43.0	20 30.2	16 47.3	16 2.2	16 33.9	21
20 44.5	20 23.9	16 28.5	15 44.6	16 24.7	22
20 46.1	20 17.6	16 9.7	15 26.7	16 15.4	23
20 47.6	20 11.3	15 50.8	15 8.7	16 5.9	24
20 49.2	20 4.9	15 31.8	14 50.3	15 56.3	25
20 50.8	19 58.5	15 12.7	14 31.7	15 46.4	26
20 52.4	19 52.1	14 53.5	14 12.8	15 36.4	27
20 54.0	19 45.6	14 34.2	13 53.7	15 26.1	28
20 55.7	19 39.0	14 14.7	13 34.2	15 15.6	29
20 57.3	19 32.4	13 55.1	13 14.3	15 4.8	30
20 59.0	19 25.8	13 35.3	12 54.1	14 53.8	31
21 0.7	19 19.0	13 15.3	12 33.6	14 42.5	32
21 2.5	19 12.2	12 55.2	12 12.6	14 30.9	33
21 4.2	19 5.3	12 34.8	11 51.2	14 19.0	34
21 6.0	18 58.4	12 14.3	11 29.4	14 6.7	35
21 7.9	18 51.3	11 53.5	11 7.0	13 54.1	36
21 9.7	18 44.2	11 32.4	10 44.2	13 41.1	37
21 11.6	18 36.9	11 11.1	10 20.8	13 27.6	38
21 13.6	18 29.6	10 49.5	9 56.9	13 13.7	39
21 15.6	18 22.1	10 27.6	9 32.3	12 59.3	40
21 17.6	18 14.5	10 5.4	9 7.1	12 44.4	41
21 19.7	18 6.8	9 42.9	8 41.2	12 28.9	42
21 21.8	17 58.9	9 20.0	8 14.6	12 12.8	43
21 24.0	17 50.9	8 56.7	7 47.2	11 56.0	44
21 26.2	17 42.7	8 33.1	7 18.9	11 38.5	45
21 28.5	17 34.4	8 9.0	6 49.8	11 20.2	46
21 30.8	17 25.9	7 44.5	6 19.8	11 1.0	47
21 33.2	17 17.2	7 19.5	5 48.7	10 40.9	48
21 35.7	17 8.3	6 54.1	5 16.6	10 19.8	49
21 38.3	16 59.2	6 28.1	4 43.3	9 57.4	50
21 41.0	16 49.8	6 1.5	4 8.8	9 33.8	51
21 43.7	16 40.3	5 34.4	3 33.0	9 8.8	52
21 46.5	16 30.5	5 6.7	2 55.8	8 42.2	53
21 49.5	16 20.4	4 38.3	2 17.1	8 13.8	54
21 52.5	16 10.0	4 9.2	1 36.8	7 43.3	55
21 55.7	15 59.3	3 39.4	0 54.7	7 10.5	56
21 58.9	15 48.3	3 8.9	0 10.8	6 35.0	57
22 2.3	15 37.0	2 37.5	♏29 24.8	5 56.4	58
22 5.9	15 25.2	2 5.3	28 36.7	5 14.0	59
22 9.6	15 13.1	1 32.1	27 46.3	4 27.1	60
22 13.5	15 0.6	0 58.0	26 53.3	3 34.7	61
22 17.6	14 47.6	0 22.9	25 57.5	2 35.2	62
22 21.8	14 34.1	♎29 46.7	24 58.8	1 26.5	63
22 26.3	14 20.1	29 9.4	23 56.9	0 4.5	64
22 31.1	14 5.5	28 30.8	22 51.5	♐28 21.8	65
22 36.0	13 50.3	27 51.0	21 42.3	25 58.0	66
5	6	Descendant	8	9	S LAT

♒ 18° 34′ 2″

21ʰ 24ᵐ 0ˢ — MC — 321° 0′ 0″

9h 28m 0s — MC 142° 0′ 0″ — ♌ 19° 34′ 59″

N LAT	11	12	Ascendant	2	3
°	° ′	° ′	° ′	° ′	° ′
0	♍21 17.5	♎23 46.0	♏24 22.0	♐22 39.2	♑20 20.3
5	21 23.9	23 12.2	22 43.8	21 20.2	19 42.3
10	21 30.4	22 39.4	21 8.3	20 0.5	19 3.4
15	21 36.8	22 7.1	19 34.0	18 38.8	18 22.9
16	21 38.1	22 0.6	19 15.2	18 22.1	18 14.5
17	21 39.5	21 54.1	18 56.4	18 5.3	18 6.0
18	21 40.8	21 47.7	18 37.5	17 48.3	17 57.4
19	21 42.1	21 41.2	18 18.6	17 31.2	17 48.7
20	21 43.4	21 34.7	17 59.7	17 13.8	17 39.8
21	21 44.8	21 28.2	17 40.8	16 56.3	17 30.9
22	21 46.1	21 21.6	17 21.7	16 38.6	17 21.7
23	21 47.5	21 15.0	17 2.6	16 20.6	17 12.4
24	21 48.9	21 8.4	16 43.4	16 2.4	17 3.0
25	21 50.3	21 1.8	16 24.1	15 44.0	16 53.3
26	21 51.7	20 55.1	16 4.7	15 25.3	16 43.5
27	21 53.1	20 48.4	15 45.2	15 6.3	16 33.5
28	21 54.6	20 41.6	15 25.6	14 47.0	16 23.2
29	21 56.0	20 34.7	15 5.8	14 27.4	16 12.8
30	21 57.5	20 27.8	14 45.8	14 7.4	16 2.0
31	21 59.0	20 20.9	14 25.7	13 47.1	15 51.0
32	22 0.5	20 13.8	14 5.4	13 26.4	15 39.8
33	22 2.1	20 6.7	13 44.9	13 5.3	15 28.2
34	22 3.7	19 59.5	13 24.2	12 43.7	15 16.3
35	22 5.3	19 52.2	13 3.3	12 21.7	15 4.1
36	22 6.9	19 44.8	12 42.1	11 59.2	14 51.5
37	22 8.6	19 37.4	12 20.7	11 36.2	14 38.5
38	22 10.3	19 29.8	11 59.0	11 12.6	14 25.1
39	22 12.0	19 22.1	11 37.1	10 48.4	14 11.2
40	22 13.7	19 14.3	11 14.8	10 23.7	13 56.8
41	22 15.5	19 6.3	10 52.2	9 58.2	13 41.9
42	22 17.4	18 58.2	10 29.2	9 32.1	13 26.4
43	22 19.3	18 50.0	10 5.9	9 5.2	13 10.4
44	22 21.2	18 41.6	9 42.2	8 37.5	12 53.6
45	22 23.2	18 33.1	9 18.2	8 9.0	12 36.1
46	22 25.2	18 24.3	8 53.6	7 39.6	12 17.8
47	22 27.3	18 15.4	8 28.7	7 9.2	11 58.7
48	22 29.5	18 6.3	8 3.2	6 37.9	11 38.6
49	22 31.7	17 57.0	7 37.2	6 5.4	11 17.4
50	22 34.0	17 47.5	7 10.7	5 31.7	10 55.1
51	22 36.3	17 37.7	6 43.7	4 56.8	10 31.5
52	22 38.8	17 27.7	6 16.0	4 20.6	10 6.5
53	22 41.3	17 17.4	5 47.7	3 42.9	9 39.8
54	22 43.9	17 6.9	5 18.8	3 3.7	9 11.4
55	22 46.6	16 56.0	4 49.1	2 22.8	8 40.9
56	22 49.4	16 44.8	4 18.7	1 40.2	8 8.1
57	22 52.3	16 33.3	3 47.6	0 55.6	7 32.5
58	22 55.4	16 21.4	3 15.5	0 9.0	6 53.8
59	22 58.5	16 9.1	2 42.6	♏29 20.2	6 11.3
60	23 1.8	15 56.4	2 8.8	28 28.9	5 24.2
61	23 5.3	15 43.3	1 34.0	27 35.0	4 31.6
62	23 8.9	15 29.6	0 58.2	26 38.4	3 31.8
63	23 12.7	15 15.5	0 21.2	25 38.6	2 22.6
64	23 16.7	15 0.8	♎29 43.1	24 35.6	0 59.9
65	23 20.9	14 45.5	29 3.7	23 29.0	♐29 15.9
66	23 25.3	14 29.6	28 23.0	22 18.5	26 48.4
S LAT	5	6	Descendant	8	9

♒ 19° 34′ 59″ — 21h 28m 0s — MC 322° 0′ 0″

9h 32m 0s — MC 143° 0′ 0″ — ♌ 20° 36′ 7″

N LAT	11	12	Ascendant	2	3
°	° ′	° ′	° ′	° ′	° ′
0	♍22 22.6	♎24 49.7	♏25 20.5	♐23 34.4	♑21 16.7
5	22 28.3	24 14.5	23 41.2	22 15.2	20 38.9
10	22 34.0	23 40.3	22 4.5	20 55.3	20 0.1
15	22 39.6	23 6.6	20 29.0	19 33.3	19 19.8
16	22 40.8	22 59.9	20 9.9	19 16.6	19 11.4
17	22 41.9	22 53.1	19 50.8	18 59.7	19 3.0
18	22 43.1	22 46.4	19 31.7	18 42.6	18 54.4
19	22 44.2	22 39.6	19 12.6	18 25.4	18 45.7
20	22 45.4	22 32.9	18 53.4	18 8.0	18 36.9
21	22 46.6	22 26.1	18 34.2	17 50.4	18 28.0
22	22 47.8	22 19.2	18 14.8	17 32.6	18 18.9
23	22 49.0	22 12.4	17 55.5	17 14.6	18 9.6
24	22 50.2	22 5.5	17 36.0	16 56.3	18 0.2
25	20 35.2	21 58.6	17 16.4	16 37.7	17 50.6
26	22 52.7	21 51.6	16 56.7	16 18.9	17 40.8
27	22 53.9	21 44.6	16 36.9	15 59.8	17 30.8
28	22 55.2	21 37.5	16 16.9	15 40.4	17 20.6
29	22 56.4	21 30.4	15 56.8	15 20.7	17 10.1
30	22 57.7	21 23.2	15 36.6	15 0.6	16 59.4
31	22 59.1	21 15.9	15 16.1	14 40.1	16 48.5
32	23 0.4	21 8.5	14 55.5	14 19.3	16 37.2
33	23 1.8	21 1.1	14 34.7	13 58.0	16 25.7
34	23 3.1	20 53.6	14 13.6	13 36.3	16 13.8
35	23 4.5	20 46.0	13 52.3	13 14.1	16 1.6
36	23 6.0	20 38.3	13 30.8	12 51.4	15 49.1
37	23 7.4	20 30.5	13 9.0	12 28.2	15 36.1
38	23 8.9	20 22.6	12 47.0	12 4.5	15 22.7
39	23 10.4	20 14.6	12 24.6	11 40.1	15 8.9
40	23 12.0	20 6.4	12 2.0	11 15.1	14 54.5
41	23 13.5	19 58.1	11 39.0	10 49.5	14 39.7
42	23 15.2	19 49.7	11 15.6	10 23.1	14 24.2
43	23 16.8	19 41.1	10 51.9	9 56.0	14 8.1
44	23 18.5	19 32.3	10 27.8	9 28.0	13 51.4
45	23 20.2	19 23.4	10 3.2	8 59.2	13 34.0
46	23 22.0	19 14.3	9 38.2	8 29.5	13 15.7
47	23 23.8	19 5.0	9 12.8	7 58.8	12 56.6
48	23 25.7	18 55.5	8 46.9	7 27.1	12 36.5
49	23 27.7	18 45.7	8 20.4	6 54.3	12 15.3
50	23 29.7	18 35.8	7 53.4	6 20.2	11 53.0
51	23 31.7	18 25.6	7 25.8	5 44.9	11 29.4
52	23 33.9	18 15.1	6 57.6	5 8.2	11 4.4
53	23 36.1	18 4.4	6 28.8	4 30.1	10 37.8
54	23 38.4	17 53.3	5 59.3	3 50.4	10 9.3
55	23 40.7	17 42.0	5 29.1	3 9.0	9 38.8
56	23 43.2	17 30.3	4 58.1	2 25.7	9 5.9
57	23 45.7	17 18.2	4 26.2	1 40.6	8 30.3
58	23 48.4	17 5.8	3 53.6	0 53.3	7 51.6
59	23 51.2	16 52.9	3 20.0	0 3.7	7 9.0
60	23 54.0	16 39.7	2 45.5	♏29 11.6	6 21.8
61	23 57.1	16 25.9	2 10.0	28 16.9	5 29.0
62	24 0.2	16 11.7	1 33.4	27 19.3	4 29.0
63	24 3.5	15 56.9	0 55.7	26 18.5	3 19.3
64	24 7.0	15 41.5	0 16.8	25 14.4	1 56.0
65	24 10.7	15 25.5	♎29 36.6	24 6.5	0 10.7
66	24 14.6	15 8.9	28 55.1	22 54.7	♐27 39.7
S LAT	5	6	Descendant	8	9

♒ 20° 36′ 7″ — 21h 32m 0s — MC 323° 0′ 0″

9h 36m 0s — MC 144° 0′ 0″ — ♌ 21° 37′ 27″

N LAT	11	12	Ascendant	2	3
°	° ′	° ′	° ′	° ′	° ′
0	♍23 27.9	♎25 53.2	♏26 18.8	♐24 29.5	♑22 13.1
5	23 32.8	25 16.6	24 38.4	23 10.2	21 35.6
10	23 37.6	24 41.1	23 0.6	21 50.1	20 57.0
15	23 42.5	24 6.1	21 23.9	20 27.9	20 16.8
16	23 43.4	23 59.1	21 4.6	20 11.1	20 8.5
17	23 44.4	23 52.1	20 45.2	19 54.1	20 0.1
18	23 45.4	23 45.1	20 25.9	19 37.0	19 51.6
19	23 46.4	23 38.0	20 6.5	19 19.7	19 42.9
20	23 47.4	23 31.0	19 47.0	19 2.2	19 34.1
21	23 48.4	23 23.9	19 27.5	18 44.5	19 25.2
22	23 49.5	23 16.8	19 7.9	18 26.6	19 16.2
23	23 50.5	23 9.7	18 48.2	18 8.5	19 6.9
24	23 51.5	23 2.5	18 28.5	17 50.1	18 57.6
25	23 52.6	22 55.3	18 8.6	17 31.5	18 48.0
26	23 53.6	22 48.0	17 48.6	17 12.6	18 38.2
27	23 54.7	22 40.7	17 28.5	16 53.4	18 28.3
28	23 55.8	22 33.4	17 8.3	16 33.9	18 18.1
29	23 56.9	22 25.9	16 47.8	16 14.0	18 7.7
30	23 58.0	22 18.4	16 27.3	15 53.8	17 57.0
31	23 59.1	22 10.9	16 6.5	15 33.2	17 46.1
32	24 0.3	22 3.2	15 45.5	15 12.2	17 34.9
33	24 1.4	21 55.5	15 24.4	14 50.8	17 23.4
34	24 2.6	21 47.7	15 3.0	14 28.9	17 11.6
35	24 3.8	21 39.7	14 41.4	14 6.6	16 59.4
36	24 5.1	21 31.7	14 19.5	13 43.8	16 46.9
37	24 6.3	21 23.6	13 57.4	13 20.4	16 33.9
38	24 7.6	21 15.4	13 34.9	12 56.4	16 20.6
39	24 8.9	21 7.0	13 12.2	12 31.9	16 6.8
40	24 10.2	20 58.5	12 49.1	12 6.7	15 52.5
41	24 11.6	20 49.8	12 25.7	11 40.8	15 37.6
42	24 12.9	20 41.0	12 2.0	11 14.2	15 22.2
43	24 14.4	20 32.1	11 37.8	10 46.8	15 6.2
44	24 15.8	20 23.0	11 13.3	10 18.6	14 49.5
45	24 17.3	20 13.7	10 48.3	9 49.5	14 32.0
46	24 18.8	20 4.2	10 22.9	9 19.5	14 13.8
47	24 20.4	19 54.5	9 56.9	8 48.5	13 54.7
48	24 22.0	19 44.6	9 30.5	8 16.5	13 34.6
49	24 23.7	19 34.4	9 3.6	7 43.3	13 13.5
50	24 25.4	19 24.0	8 36.1	7 8.8	12 51.3
51	24 27.2	19 13.4	8 8.0	6 33.1	12 27.7
52	24 29.0	19 2.5	7 39.3	5 56.0	12 2.7
53	24 30.9	18 51.3	7 9.9	5 17.4	11 36.0
54	24 32.8	18 39.7	6 39.8	4 37.2	11 7.6
55	24 34.9	18 27.9	6 9.0	3 55.2	10 37.1
56	24 37.0	18 15.7	5 37.4	3 11.4	10 4.2
57	24 39.2	18 3.1	5 4.9	2 25.6	9 28.6
58	24 41.4	17 50.1	4 31.6	1 37.6	8 49.7
59	24 43.8	17 36.7	3 57.4	0 47.3	8 7.1
60	24 46.3	17 22.9	3 22.2	♏29 54.4	7 19.8
61	24 48.9	17 8.5	2 46.0	28 58.8	6 26.9
62	24 51.6	16 53.7	2 8.7	28 0.2	5 26.6
63	24 54.4	16 38.2	1 30.2	26 58.4	4 16.6
64	24 57.4	16 22.2	0 50.5	25 53.1	2 52.7
65	25 0.6	16 5.5	0 9.4	24 44.1	1 6.3
66	25 3.9	15 48.1	♎29 27.0	23 30.8	♐28 32.1
S LAT	5	6	Descendant	8	9

♒ 21° 37′ 27″ — 21h 36m 0s — MC 324° 0′ 0″

9h 40m 0s — MC 145° 0′ 0″ — ♌ 22° 38′ 57″

N LAT	11	12	Ascendant	2	3
°	° ′	° ′	° ′	° ′	° ′
0	♍24 33.2	♎26 56.5	♏27 17.0	♐25 24.7	♑23 9.7
5	24 37.2	26 18.6	25 35.6	24 5.2	22 32.4
10	24 41.3	25 41.8	23 56.6	22 44.9	21 54.0
15	24 45.3	25 5.4	22 18.7	21 22.5	21 14.0
16	24 46.1	24 58.2	21 59.2	21 5.6	21 5.7
17	24 47.0	24 50.9	21 39.6	20 48.6	20 57.4
18	24 47.8	24 43.6	21 20.0	20 31.4	20 48.9
19	24 48.6	24 36.3	21 0.3	20 14.0	20 40.3
20	24 49.5	24 29.0	20 40.6	19 56.5	20 31.5
21	24 50.3	24 21.7	20 20.8	19 38.7	20 22.7
22	24 51.2	24 14.3	20 0.9	19 20.7	20 13.6
23	24 52.0	24 6.9	19 41.0	19 2.5	20 4.5
24	24 52.9	23 59.4	19 21.0	18 44.1	19 55.1
25	24 53.8	23 51.9	19 0.8	18 25.3	19 45.6
26	24 54.6	23 44.4	18 40.5	18 6.3	19 35.9
27	24 55.5	23 36.8	18 20.1	17 47.0	19 25.9
28	24 56.4	23 29.1	17 59.5	17 27.4	19 15.8
29	24 57.4	23 21.4	17 38.8	17 7.4	19 5.4
30	24 58.3	23 13.6	17 17.9	16 47.1	18 54.8
31	24 59.2	23 5.8	16 56.9	16 26.3	18 43.9
32	25 0.2	22 57.8	16 35.6	16 5.2	18 32.7
33	25 1.2	22 49.8	16 14.1	15 43.7	18 21.3
34	25 2.1	22 41.7	15 52.4	15 21.7	18 9.5
35	25 3.2	22 33.4	15 30.4	14 59.2	17 57.4
36	25 4.2	22 25.1	15 8.2	14 36.2	17 44.9
37	25 5.2	22 16.6	14 45.7	14 12.6	17 32.0
38	25 6.3	22 8.1	14 22.8	13 48.5	17 18.7
39	25 7.4	21 59.4	13 59.7	13 23.7	17 4.9
40	25 8.5	21 50.5	13 36.3	12 58.3	16 50.6
41	25 9.6	21 41.5	13 12.5	12 32.2	16 35.8
42	25 10.7	21 32.4	12 48.3	12 5.4	16 20.4
43	25 11.9	21 23.1	12 23.8	11 37.8	16 4.5
44	25 13.1	21 13.6	11 58.8	11 9.3	15 47.8
45	25 14.4	21 3.9	11 33.4	10 40.0	15 30.4
46	25 15.7	20 54.0	11 7.5	10 9.7	15 12.2
47	25 17.0	20 43.9	10 41.1	9 38.4	14 53.1
48	25 18.3	20 33.6	10 14.2	9 6.0	14 33.1
49	25 19.7	20 23.1	9 46.8	8 32.4	14 12.0
50	25 21.1	20 12.2	9 18.8	7 57.6	13 49.8
51	25 22.6	20 1.2	8 50.2	7 21.5	13 26.2
52	25 24.1	19 49.8	8 20.9	6 43.9	13 1.2
53	25 25.7	19 38.1	7 51.0	6 4.8	12 34.6
54	25 27.3	19 26.1	7 20.3	5 24.1	12 6.2
55	25 29.0	19 13.8	6 48.9	4 41.6	11 35.7
56	25 30.8	19 1.1	6 16.7	3 57.2	11 2.8
57	25 32.6	18 48.0	5 43.6	3 10.7	10 27.2
58	25 34.5	18 34.5	5 9.7	2 22.0	9 48.3
59	25 36.5	18 20.5	4 34.8	1 30.9	9 5.6
60	25 38.5	18 6.1	3 58.9	0 37.3	8 18.3
61	25 40.7	17 51.1	3 21.9	♏29 40.8	7 25.2
62	25 43.0	17 35.6	2 43.9	28 41.2	6 24.8
63	25 45.3	17 19.5	2 4.6	27 38.4	5 14.6
64	25 47.8	17 2.8	1 24.1	26 31.9	3 50.2
65	25 50.4	16 45.4	0 42.3	25 21.6	2 2.8
66	25 53.2	16 27.3	♎29 59.0	24 7.0	♐29 25.6
S LAT	5	6	Descendant	8	9

♒ 22° 38′ 57″ — 21h 40m 0s — MC 325° 0′ 0″

9ʰ 44ᵐ 0ˢ MC 146° 0′ 0″

♌ 23° 40′ 37″

N LAT	11	12	Ascendant	2	3
0°	♍25° 38.5′	♎27° 59.7′	♏28° 15.0′	♐26° 19.8′	♑24° 6.4′
5	25 41.7	27 20.5	26 32.6	25 0.2	23 29.3
10	25 45.0	26 42.4	24 52.5	23 39.8	22 51.2
15	25 48.2	26 4.7	23 13.5	22 17.1	22 11.4
16	25 48.9	25 57.2	22 53.7	22 0.1	22 3.1
17	25 49.5	25 49.6	22 33.9	21 43.1	21 54.8
18	25 50.2	25 42.1	22 14.0	21 25.8	21 46.4
19	25 50.9	25 34.5	21 54.1	21 8.4	21 37.8
20	25 51.5	25 26.9	21 34.1	20 50.8	21 29.1
21	25 52.2	25 19.3	21 14.1	20 32.9	21 20.3
22	25 52.9	25 11.7	20 53.9	20 14.9	21 11.3
23	25 53.6	25 4.0	20 33.7	19 56.6	21 2.1
24	25 54.3	24 56.3	20 13.4	19 38.0	20 52.8
25	25 55.0	24 48.5	19 52.9	19 19.2	20 43.3
26	25 55.7	24 40.7	19 32.4	19 0.1	20 33.7
27	25 56.4	24 32.8	19 11.7	18 40.7	20 23.8
28	25 57.1	24 24.9	18 50.8	18 20.9	20 13.7
29	25 57.9	24 16.9	18 29.8	18 0.9	20 3.3
30	25 58.6	24 8.8	18 8.6	17 40.4	19 52.8
31	25 59.4	24 0.6	17 47.2	17 19.6	19 41.9
32	26 0.1	23 52.4	17 25.6	16 58.3	19 30.8
33	26 0.9	23 44.1	17 3.8	16 36.6	19 19.4
34	26 1.7	23 35.6	16 41.7	16 14.5	19 7.6
35	26 2.5	23 27.1	16 19.4	15 51.8	18 55.5
36	26 3.3	23 18.4	15 56.8	15 28.7	18 43.1
37	26 4.1	23 9.7	15 33.9	15 4.9	18 30.2
38	26 5.0	23 0.7	15 10.8	14 40.6	18 17.0
39	26 5.9	22 51.7	14 47.3	14 15.7	18 3.2
40	26 6.7	22 42.5	14 23.5	13 50.1	17 49.0
41	26 7.6	22 33.2	13 59.3	13 23.8	17 34.3
42	26 8.6	22 23.7	13 34.7	12 56.7	17 18.9
43	26 9.5	22 14.0	13 9.7	12 28.9	17 3.0
44	26 10.5	22 4.2	12 44.3	12 0.1	16 46.4
45	26 11.5	21 54.1	12 18.4	11 30.5	16 29.0
46	26 12.5	21 43.8	11 52.1	10 59.9	16 10.8
47	26 13.6	21 33.4	11 25.3	10 28.3	15 51.8
48	26 14.6	21 22.6	10 57.9	9 55.6	15 31.8
49	26 15.7	21 11.7	10 30.0	9 21.6	15 10.8
50	26 16.9	21 0.4	10 1.5	8 46.5	14 48.6
51	26 18.1	20 48.9	9 32.3	8 9.9	14 25.1
52	26 19.3	20 37.1	9 2.5	7 31.9	14 0.1
53	26 20.5	20 25.0	8 32.0	6 52.4	13 33.5
54	26 21.8	20 12.5	8 0.8	6 11.1	13 5.1
55	26 23.2	19 59.7	7 28.8	5 28.1	12 34.7
56	26 24.6	19 46.5	6 56.0	4 43.1	12 1.8
57	26 26.1	19 32.9	6 22.3	3 56.0	11 26.2
58	26 27.6	19 18.8	5 47.7	3 6.6	10 47.3
59	26 29.2	19 4.3	5 12.1	2 14.7	10 4.6
60	26 30.8	18 49.3	4 35.5	1 20.2	9 17.3
61	26 32.5	18 33.7	3 57.9	0 22.8	8 24.1
62	26 34.4	18 17.6	3 19.1	♏29 22.3	7 23.6
63	26 36.2	18 0.8	2 39.0	28 18.4	6 13.1
64	26 38.2	17 43.4	1 57.7	27 10.7	4 48.3
65	26 40.3	17 25.3	1 15.0	25 59.1	3 0.1
66	26 42.6	17 6.5	0 30.9	24 43.0	0 20.5
S LAT	5	6	Descendant	8	9

♒ 23° 40′ 37″

21ʰ 44ᵐ 0ˢ MC 326° 0′ 0″

9ʰ 48ᵐ 0ˢ MC 147° 0′ 0″

♌ 24° 42′ 28″

N LAT	11	12	Ascendant	2	3
0°	♍26° 43.8′	♎29° 2.8′	♏29° 12.8′	♐27° 14.8′	♑25° 3.3′
5	26 46.3	28 22.2	27 29.4	25 55.2	24 26.4
10	26 48.7	27 42.8	25 48.4	24 34.6	23 48.4
15	26 51.1	27 3.9	24 8.2	23 11.7	23 8.9
16	26 51.6	26 56.1	23 48.1	22 54.7	23 0.7
17	26 52.1	26 48.3	23 28.1	22 37.6	22 52.4
18	26 52.6	26 40.5	23 8.0	22 20.3	22 44.0
19	26 53.1	26 32.6	22 47.8	22 2.8	22 35.5
20	26 53.6	26 24.8	22 27.6	21 45.1	22 26.8
21	26 54.1	26 16.9	22 7.3	21 27.2	22 18.0
22	26 54.7	26 9.0	21 46.9	21 9.1	22 9.1
23	26 55.2	26 1.0	21 26.4	20 50.7	22 0.0
24	26 55.7	25 53.1	21 5.8	20 32.1	21 50.7
25	26 56.2	25 45.0	20 45.1	20 13.1	21 41.3
26	26 56.7	25 36.9	20 24.2	19 53.9	21 31.7
27	26 57.3	25 28.8	20 3.2	19 34.4	21 21.8
28	26 57.8	25 20.5	19 42.1	19 14.6	21 11.8
29	26 58.4	25 12.2	19 20.7	18 54.4	21 1.5
30	26 58.9	25 3.9	18 59.2	18 33.8	20 50.9
31	26 59.5	24 55.4	18 37.5	18 12.9	20 40.1
32	27 0.1	24 46.9	18 15.6	17 51.5	20 29.1
33	27 0.7	24 38.3	17 53.5	17 29.7	20 17.7
34	27 1.3	24 29.5	17 31.1	17 7.4	20 6.0
35	27 1.9	24 20.7	17 8.4	16 44.6	19 54.0
36	27 2.5	24 11.7	16 45.5	16 21.3	19 41.5
37	27 3.1	24 2.6	16 22.2	15 57.4	19 28.7
38	27 3.7	23 53.4	15 58.7	15 32.9	19 15.5
39	27 4.4	23 44.0	15 34.8	15 7.8	19 1.8
40	27 5.0	23 34.5	15 10.6	14 42.0	18 47.6
41	27 5.7	23 24.8	14 46.0	14 15.4	18 32.9
42	27 6.4	23 15.0	14 21.1	13 48.1	18 17.7
43	27 7.1	23 4.9	13 55.7	13 20.0	18 1.8
44	27 7.9	22 54.7	13 29.8	12 51.1	17 45.2
45	27 8.6	22 44.3	13 3.5	12 21.2	17 27.9
46	27 9.4	22 33.6	12 36.7	11 50.3	17 9.8
47	27 10.2	22 22.8	12 9.5	11 18.4	16 50.8
48	27 11.0	22 11.6	11 41.6	10 45.3	16 30.9
49	27 11.8	22 0.3	11 13.2	10 11.0	16 9.9
50	27 12.7	21 48.6	10 44.2	9 35.5	15 47.7
51	27 13.5	21 36.7	10 14.5	8 58.5	15 24.2
52	27 14.5	21 24.4	9 44.2	8 20.1	14 59.3
53	27 15.4	21 11.8	9 13.1	7 40.0	14 32.8
54	27 16.4	20 58.9	8 41.3	6 58.3	14 4.4
55	27 17.4	20 45.5	8 8.7	6 14.7	13 34.0
56	27 18.4	20 31.8	7 35.3	5 29.1	13 1.2
57	27 19.5	20 17.7	7 1.0	4 41.3	12 25.6
58	27 20.7	20 3.1	6 25.7	3 51.2	11 46.7
59	27 21.9	19 48.0	5 49.5	2 58.6	11 4.0
60	27 23.1	19 32.4	5 12.2	2 3.3	10 16.7
61	27 24.4	19 16.2	4 33.8	1 5.0	9 23.5
62	27 25.8	18 59.5	3 54.2	0 3.4	8 22.9
63	27 27.2	18 42.1	3 13.4	♏28 58.4	7 12.3
64	27 28.7	18 24.0	2 31.3	27 49.5	5 47.2
65	27 30.2	18 5.2	1 47.8	26 36.6	3 58.4
66	27 31.9	17 45.6	1 2.8	25 19.0	1 16.8
S LAT	5	6	Descendant	8	9

♒ 24° 42′ 28″

21ʰ 48ᵐ 0ˢ MC 327° 0′ 0″

9ʰ 52ᵐ 0ˢ — MC — 148° 0′ 0″

♌ 25° 44′ 30″

N LAT	11	12	Ascendant	2	3
°	° ′	° ′	° ′	° ′	° ′
0	♍27 49.2	♏ 0 5.7	♐ 0 10.5	♐28 9.9	♑26 0.2
5	27 50.8	♎29 23.8	♏28 26.2	26 50.2	25 23.6
10	27 52.5	28 43.1	26 44.1	25 29.5	24 45.9
15	27 54.1	28 2.9	25 2.8	24 6.4	24 6.5
16	27 54.4	27 54.9	24 42.5	23 49.3	23 58.4
17	27 54.7	27 46.8	24 22.2	23 32.1	23 50.2
18	27 55.1	27 38.7	24 1.9	23 14.8	23 41.8
19	27 55.4	27 30.7	23 41.4	22 57.2	23 33.3
20	27 55.7	27 22.5	23 21.0	22 39.5	23 24.7
21	27 56.1	27 14.4	23 0.4	22 21.5	23 16.0
22	27 56.4	27 6.2	22 39.8	22 3.3	23 7.1
23	27 56.8	26 58.0	22 19.0	21 44.9	22 58.0
24	27 57.1	26 49.7	21 58.1	21 26.1	22 48.8
25	27 57.5	26 41.4	21 37.2	21 7.2	22 39.4
26	27 57.8	26 33.1	21 16.0	20 47.9	22 29.8
27	27 58.2	26 24.6	20 54.7	20 28.3	22 20.1
28	27 58.5	26 16.1	20 33.3	20 8.3	22 10.0
29	27 58.9	26 7.6	20 11.7	19 48.0	21 59.8
30	27 59.3	25 58.9	19 49.9	19 27.4	21 49.3
31	27 59.7	25 50.2	19 27.8	19 6.3	21 38.6
32	28 0.0	25 41.3	19 5.6	18 44.8	21 27.5
33	28 0.4	25 32.4	18 43.1	18 22.8	21 16.2
34	28 0.8	25 23.3	18 20.4	18 0.4	21 4.6
35	28 1.2	25 14.2	17 57.4	17 37.5	20 52.6
36	28 1.6	25 4.9	17 34.1	17 14.0	20 40.2
37	28 2.1	24 55.5	17 10.6	16 49.9	20 27.5
38	28 2.5	24 46.0	16 46.6	16 25.3	20 14.3
39	28 2.9	24 36.3	16 22.4	15 59.9	20 0.6
40	28 3.4	24 26.4	15 57.8	15 33.9	19 46.5
41	28 3.8	24 16.4	15 32.8	15 7.2	19 31.9
42	28 4.3	24 6.2	15 7.4	14 39.7	19 16.6
43	28 4.7	23 55.8	14 41.6	14 11.4	19 0.8
44	28 5.2	23 45.2	14 15.4	13 42.1	18 44.3
45	28 5.7	23 34.4	13 48.6	13 12.0	18 27.0
46	28 6.2	23 23.4	13 21.4	12 40.8	18 9.0
47	28 6.8	23 12.1	12 53.6	12 8.6	17 50.1
48	28 7.3	23 0.6	12 25.3	11 35.2	17 30.2
49	28 7.9	22 48.8	11 56.4	11 0.5	17 9.2
50	28 8.4	22 36.7	11 26.9	10 24.6	16 47.1
51	28 9.0	22 24.4	10 56.7	9 47.2	16 23.7
52	28 9.6	22 11.6	10 25.8	9 8.4	15 58.9
53	28 10.3	21 58.6	9 54.2	8 27.9	15 32.4
54	28 10.9	21 45.2	9 21.9	7 45.6	15 4.1
55	28 11.6	21 31.4	8 48.7	7 1.4	14 33.7
56	28 12.3	21 17.2	8 14.6	6 15.2	14 0.9
57	28 13.0	21 2.5	7 39.7	5 26.8	13 25.4
58	28 13.8	20 47.3	7 3.8	4 36.0	12 46.6
59	28 14.6	20 31.7	6 26.8	3 42.6	12 3.9
60	28 15.4	20 15.5	5 48.8	2 46.4	11 16.6
61	28 16.3	19 58.7	5 9.7	1 47.2	10 23.4
62	28 17.2	19 41.4	4 29.4	0 44.6	9 22.8
63	28 18.1	19 23.3	3 47.8	♏29 38.5	8 12.1
64	28 19.1	19 4.5	3 4.9	28 28.4	6 46.9
65	28 20.2	18 45.0	2 20.5	27 14.0	4 57.7
66	28 21.3	18 24.7	1 34.6	25 55.0	2 14.6
S LAT	5	6	Descendant	8	9

♒ 25° 44′ 30″

21ʰ 52ᵐ 0ˢ — MC — 328° 0′ 0″

9ʰ 56ᵐ 0ˢ — MC — 149° 0′ 0″

♌ 26° 46′ 43″

N LAT	11	12	Ascendant	2	3
°	° ′	° ′	° ′	° ′	° ′
0	♍28 54.6	♏ 1 8.4	♐ 1 8.0	♐29 5.0	♑26 57.4
5	28 55.4	0 25.3	♏29 22.8	27 45.2	26 20.9
10	28 56.2	♎29 43.3	27 39.7	26 24.4	25 43.5
15	28 57.0	29 1.8	25 57.4	25 1.1	25 4.4
16	28 57.2	28 53.6	25 36.9	24 44.0	24 56.3
17	28 57.4	28 45.2	25 16.3	24 26.7	24 48.1
18	28 57.5	28 36.9	24 55.7	24 9.3	24 39.8
19	28 57.7	28 28.6	24 35.1	23 51.7	24 31.4
20	28 57.9	28 20.2	24 14.3	23 33.9	24 22.8
21	28 58.0	28 11.8	23 53.5	23 15.9	24 14.1
22	28 58.2	28 3.4	23 32.6	22 57.6	24 5.3
23	28 58.4	27 54.9	23 11.6	22 39.1	23 56.3
24	28 58.6	27 46.4	22 50.5	22 20.3	23 47.1
25	28 58.7	27 37.8	22 29.2	22 1.2	23 37.8
26	28 58.9	27 29.1	22 7.8	21 41.9	23 28.2
27	28 59.1	27 20.4	21 46.3	21 22.2	23 18.5
28	28 59.3	27 11.6	21 24.5	21 2.1	23 8.5
29	28 59.5	27 2.8	21 2.6	20 41.7	22 58.3
30	28 59.6	26 53.9	20 40.5	20 21.0	22 47.9
31	28 59.8	26 44.8	20 18.2	19 59.8	22 37.2
32	29 0.0	26 35.7	19 55.6	19 38.2	22 26.2
33	29 0.2	26 26.5	19 32.8	19 16.1	22 15.0
34	29 0.4	26 17.1	19 9.8	18 53.5	22 3.4
35	29 0.6	26 7.7	18 46.4	18 30.4	21 51.4
36	29 0.8	25 58.1	18 22.8	18 6.8	21 39.1
37	29 1.0	25 48.4	17 58.9	17 42.6	21 26.4
38	29 1.2	25 38.5	17 34.6	17 17.7	21 13.3
39	29 1.5	25 28.5	17 10.0	16 52.3	20 59.7
40	29 1.7	25 18.3	16 45.0	16 26.1	20 45.7
41	29 1.9	25 7.9	16 19.6	15 59.1	20 31.1
42	29 2.1	24 57.4	15 53.8	15 31.4	20 15.9
43	29 2.4	24 46.6	15 27.6	15 2.8	20 0.1
44	29 2.6	24 35.7	15 0.9	14 33.3	19 43.7
45	29 2.9	24 24.5	14 33.8	14 2.9	19 26.5
46	29 3.1	24 13.1	14 6.1	13 31.5	19 8.5
47	29 3.4	24 1.4	13 37.8	12 58.9	18 49.6
48	29 3.6	23 49.5	13 9.1	12 25.2	18 29.8
49	29 3.9	23 37.3	12 39.7	11 50.2	18 8.9
50	29 4.2	23 24.8	12 9.6	11 13.9	17 46.9
51	29 4.5	23 12.0	11 38.9	10 36.1	17 23.5
52	29 4.8	22 58.9	11 7.5	9 56.8	16 58.8
53	29 5.1	22 45.4	10 35.3	9 15.8	16 32.4
54	29 5.5	22 31.5	10 2.4	8 33.0	16 4.1
55	29 5.8	22 17.2	9 28.6	7 48.3	15 33.8
56	29 6.1	22 2.4	8 53.9	7 1.5	15 1.1
57	29 6.5	21 47.3	8 18.4	6 12.4	14 25.6
58	29 6.9	21 31.6	7 41.8	5 20.9	13 46.9
59	29 7.3	21 15.4	7 4.2	4 26.7	13 4.3
60	29 7.7	20 58.6	6 25.5	3 29.7	12 17.0
61	29 8.1	20 41.2	5 45.6	2 29.5	11 23.9
62	29 8.6	20 23.2	5 4.5	1 25.9	10 23.3
63	29 9.1	20 4.5	4 22.1	0 18.6	9 12.6
64	29 9.6	19 45.1	3 38.4	♏29 7.3	7 47.3
65	29 10.1	19 24.8	2 53.2	27 51.5	5 57.9
66	29 10.6	19 3.7	2 6.4	26 31.0	3 14.0
S LAT	5	6	Descendant	8	9

♒ 26° 46′ 43″

21ʰ 56ᵐ 0ˢ — MC — 329° 0′ 0″

10ʰ 0ᵐ 0ˢ MC 150° 0′ 0″

♌ 27° 49′ 5″

N LAT	11	12	Ascendant	2	3
°	° ′	° ′	° ′	° ′	° ′
0	♎ 0 0.0	♏ 2 10.9	♐ 2 5.4	♑ 0 0.0	♑27 54.6
5	0 0.0	1 26.6	0 19.3	♐28 40.2	27 18.4
10	0 0.0	0 43.4	♏28 35.3	27 19.3	26 41.3
15	0 0.0	0 0.7	26 51.8	25 55.8	26 2.4
16	0 0.0	♎29 52.1	26 31.1	25 38.7	25 54.4
17	0 0.0	29 43.6	26 10.3	25 21.4	25 46.2
18	0 0.0	29 35.0	25 49.5	25 3.9	25 38.0
19	0 0.0	29 26.4	25 28.6	24 46.3	25 29.6
20	0 0.0	29 17.8	25 7.7	24 28.4	25 21.1
21	0 0.0	29 9.1	24 46.6	24 10.3	25 12.4
22	0 0.0	29 0.4	24 25.4	23 52.0	25 3.7
23	0 0.0	28 51.7	24 4.2	23 33.4	24 54.7
24	0 0.0	28 42.9	23 42.8	23 14.6	24 45.6
25	0 0.0	28 34.0	23 21.3	22 55.4	24 36.3
26	0 0.0	28 25.1	22 59.6	22 36.0	24 26.8
27	0 0.0	28 16.1	22 37.7	22 16.2	24 17.1
28	0 0.0	28 7.1	22 15.7	21 56.0	24 7.2
29	0 0.0	27 58.0	21 53.5	21 35.6	23 57.1
30	0 0.0	27 48.7	21 31.1	21 14.7	23 46.7
31	0 0.0	27 39.4	21 8.5	20 53.4	23 36.1
32	0 0.0	27 30.0	20 45.6	20 31.6	23 25.2
33	0 0.0	27 20.5	20 22.5	20 9.4	23 13.9
34	0 0.0	27 10.9	19 59.1	19 46.7	23 2.4
35	0 0.0	27 1.1	19 35.4	19 23.5	22 50.5
36	0 0.0	26 51.2	19 11.5	18 59.7	22 38.3
37	0 0.0	26 41.2	18 47.2	18 35.4	22 25.6
38	0 0.0	26 31.0	18 22.5	18 10.4	22 12.6
39	0 0.0	26 20.6	17 57.6	17 44.7	21 59.1
40	0 0.0	26 10.1	17 32.2	17 18.3	21 45.1
41	0 0.0	25 59.4	17 6.4	16 51.2	21 30.5
42	0 0.0	25 48.5	16 40.2	16 23.2	21 15.4
43	0 0.0	25 37.4	16 13.6	15 54.4	20 59.7
44	0 0.0	25 26.1	15 46.5	15 24.7	20 43.3
45	0 0.0	25 14.6	15 18.9	14 54.0	20 26.2
46	0 0.0	25 2.8	14 50.8	14 22.3	20 8.3
47	0 0.0	24 50.7	14 22.1	13 49.4	19 49.5
48	0 0.0	24 38.4	13 52.8	13 15.4	19 29.7
49	0 0.0	24 25.8	13 22.9	12 40.0	19 8.9
50	0 0.0	24 12.9	12 52.4	12 3.3	18 47.0
51	0 0.0	23 59.6	12 21.1	11 25.2	18 23.7
52	0 0.0	23 46.1	11 49.2	10 45.4	17 59.0
53	0 0.0	23 32.1	11 16.5	10 3.9	17 32.7
54	0 0.0	23 17.7	10 42.9	9 20.6	17 4.5
55	0 0.0	23 2.9	10 8.6	8 35.4	16 34.3
56	0 0.0	22 47.7	9 33.3	7 47.9	16 1.6
57	0 0.0	22 32.0	8 57.0	6 58.2	15 26.2
58	0 0.0	22 15.8	8 19.8	6 6.0	14 47.6
59	0 0.0	21 59.0	7 41.5	5 11.0	14 5.1
60	0 0.0	21 41.6	7 2.1	4 13.1	13 17.9
61	0 0.0	21 23.7	6 21.5	3 11.9	12 24.9
62	0 0.0	21 5.0	5 39.7	2 7.3	11 24.3
63	0 0.0	20 45.7	4 56.5	0 58.8	10 13.7
64	0 0.0	20 25.5	4 11.9	♏29 46.2	8 48.5
65	0 0.0	20 4.6	3 25.8	28 29.0	6 59.1
66	0 0.0	19 42.7	2 38.2	27 6.9	4 15.0
S LAT	5	6	Descendant	8	9

♒ 27° 49′ 5″

22ʰ 0ᵐ 0ˢ MC 330° 0′ 0″

10ʰ 4ᵐ 0ˢ MC 151° 0′ 0″

♌ 28° 51′ 38″

N LAT	11	12	Ascendant	2	3
°	° ′	° ′	° ′	° ′	° ′
0	♎ 1 5.4	♏ 3 13.3	♐ 3 2.6	♑ 0 55.0	♑28 52.0
5	1 4.6	2 27.7	1 15.7	♐29 35.3	28 16.1
10	1 3.8	1 43.3	♏29 30.7	28 14.2	27 39.2
15	1 3.0	0 59.4	27 46.3	26 50.6	27 0.6
16	1 2.8	0 50.6	27 25.3	26 33.4	26 52.6
17	1 2.6	0 41.8	27 4.3	26 16.1	26 44.5
18	1 2.5	0 33.0	26 43.3	25 58.6	26 36.3
19	1 2.3	0 24.1	26 22.1	25 40.9	26 28.0
20	1 2.1	0 15.3	26 0.9	25 23.0	26 19.6
21	1 2.0	0 6.3	25 39.6	25 4.8	26 11.0
22	1 1.8	♎29 57.4	25 18.2	24 46.4	26 2.2
23	1 1.6	29 48.4	24 56.7	24 27.8	25 53.3
24	1 1.4	29 39.3	24 35.1	24 8.9	25 44.3
25	1 1.3	29 30.2	24 13.3	23 49.7	25 35.0
26	1 1.1	29 21.0	23 51.3	23 30.1	25 25.6
27	1 0.9	29 11.8	23 29.2	23 10.3	25 16.0
28	1 0.7	29 2.5	23 6.9	22 50.0	25 6.1
29	1 0.5	28 53.1	22 44.4	22 29.5	24 56.1
30	1 0.4	28 43.6	22 21.7	22 8.5	24 45.7
31	1 0.2	28 34.0	21 58.8	21 47.1	24 35.2
32	1 0.0	28 24.3	21 35.6	21 25.2	24 24.3
33	0 59.8	28 14.5	21 12.2	21 2.9	24 13.1
34	0 59.6	28 4.5	20 48.4	20 40.1	24 1.7
35	0 59.4	27 54.4	20 24.4	20 16.7	23 49.8
36	0 59.2	27 44.2	20 0.1	19 52.8	23 37.7
37	0 59.0	27 33.9	19 35.5	19 28.3	23 25.1
38	0 58.8	27 23.4	19 10.5	19 3.1	23 12.1
39	0 58.5	27 12.7	18 45.2	18 37.3	22 58.6
40	0 58.3	27 1.9	18 19.4	18 10.7	22 44.7
41	0 58.1	26 50.8	17 53.3	17 43.3	22 30.2
42	0 57.9	26 39.6	17 26.7	17 15.2	22 15.2
43	0 57.6	26 28.2	16 59.6	16 46.2	21 59.6
44	0 57.4	26 16.5	16 32.1	16 16.2	21 43.2
45	0 57.1	26 4.6	16 4.0	15 45.2	21 26.2
46	0 56.9	25 52.4	15 35.5	15 13.2	21 8.3
47	0 56.6	25 40.0	15 6.3	14 40.1	20 49.6
48	0 56.4	25 27.3	14 36.6	14 5.7	20 30.0
49	0 56.1	25 14.3	14 6.2	13 30.0	20 9.2
50	0 55.8	25 0.9	13 35.1	12 52.9	19 47.4
51	0 55.5	24 47.2	13 3.4	12 14.4	19 24.2
52	0 55.2	24 33.2	12 30.9	11 34.2	18 59.6
53	0 54.9	24 18.8	11 57.6	10 52.2	18 33.3
54	0 54.5	24 3.9	11 23.5	10 8.4	18 5.3
55	0 54.2	23 48.7	10 48.5	9 22.6	17 35.1
56	0 53.9	23 32.9	10 12.6	8 34.5	17 2.6
57	0 53.5	23 16.7	9 35.7	7 44.1	16 27.3
58	0 53.1	22 59.9	8 57.8	6 51.2	15 48.8
59	0 52.7	22 42.6	8 18.9	5 55.4	15 6.4
60	0 52.3	22 24.7	7 38.7	4 56.6	14 19.3
61	0 51.9	22 6.1	6 57.4	3 54.5	13 26.4
62	0 51.4	21 46.8	6 14.8	2 48.8	12 26.0
63	0 50.9	21 26.8	5 30.8	1 39.1	11 15.6
64	0 50.4	21 6.0	4 45.4	0 25.2	9 50.5
65	0 49.9	20 44.3	3 58.4	♏29 6.6	8 1.4
66	0 49.4	20 21.7	3 9.9	27 42.8	5 17.7
S LAT	5	6	Descendant	8	9

♒ 28° 51′ 38″

22ʰ 4ᵐ 0ˢ MC 331° 0′ 0″

10ʰ 8ᵐ 0ˢ MC 152° 0′ 0″

♌ 29° 54′ 21″

11	12	Ascendant	2	3	N LAT
° ′	° ′	° ′	° ′	° ′	°
♎ 2 10.8	♏ 4 15.5	♐ 3 59.8	♑ 1 50.1	♑29 49.5	0
2 9.2	3 28.7	2 12.0	0 30.3	29 13.9	5
2 7.5	2 43.2	0 26.1	♐29 9.2	28 37.3	10
2 5.9	1 58.0	♏28 40.6	27 45.4	27 59.0	15
2 5.6	1 49.0	28 19.5	27 28.3	27 51.1	16
2 5.3	1 39.9	27 58.3	27 10.9	27 43.0	17
2 4.9	1 30.9	27 37.0	26 53.3	27 34.9	18
2 4.6	1 21.8	27 15.6	26 35.6	27 26.6	19
2 4.3	1 12.6	26 54.2	26 17.6	27 18.2	20
2 3.9	1 3.5	26 32.6	25 59.4	27 9.7	21
2 3.6	0 54.3	26 11.0	25 41.0	27 1.0	22
2 3.2	0 45.0	25 49.2	25 22.3	26 52.2	23
2 2.9	0 35.7	25 27.3	25 3.3	26 43.2	24
2 2.5	0 26.3	25 5.3	24 44.0	26 34.0	25
2 2.2	0 16.9	24 43.1	24 24.4	26 24.6	26
2 1.8	0 7.4	24 20.7	24 4.4	26 15.0	27
2 1.5	♎29 57.8	23 58.1	23 44.2	26 5.3	28
2 1.1	29 48.1	23 35.3	23 23.5	25 55.3	29
2 0.7	29 38.3	23 12.3	23 2.4	25 45.0	30
2 0.3	29 28.4	22 49.1	22 40.9	25 34.5	31
2 0.0	29 18.5	22 25.6	22 18.9	25 23.7	32
1 59.6	29 8.3	22 1.8	21 56.5	25 12.6	33
1 59.2	28 58.1	21 37.8	21 33.5	25 1.2	34
1 58.8	28 47.8	21 13.5	21 10.1	24 49.4	35
1 58.4	28 37.2	20 48.8	20 46.0	24 37.3	36
1 57.9	28 26.6	20 23.8	20 21.3	24 24.8	37
1 57.5	28 15.8	19 58.5	19 56.0	24 11.9	38
1 57.1	28 4.8	19 32.8	19 30.0	23 58.5	39
1 56.6	27 53.6	19 6.7	19 3.2	23 44.6	40
1 56.2	27 42.2	18 40.1	18 35.7	23 30.2	41
1 55.7	27 30.7	18 13.1	18 7.3	23 15.3	42
1 55.3	27 18.9	17 45.7	17 38.1	22 59.7	43
1 54.8	27 6.8	17 17.7	17 7.9	22 43.5	44
1 54.3	26 54.6	16 49.2	16 36.6	22 26.5	45
1 53.8	26 42.0	16 20.2	16 4.3	22 8.7	46
1 53.2	26 29.2	15 50.6	15 30.9	21 50.1	47
1 52.7	26 16.1	15 20.4	14 56.2	21 30.5	48
1 52.1	26 2.7	14 49.5	14 20.2	21 9.9	49
1 51.6	25 48.9	14 17.9	13 42.7	20 48.1	50
1 51.0	25 34.8	13 45.6	13 3.7	20 25.0	51
1 50.4	25 20.3	13 12.6	12 23.1	20 0.5	52
1 49.7	25 5.5	12 38.7	11 40.7	19 34.4	53
1 49.1	24 50.1	12 4.1	10 56.4	19 6.4	54
1 48.4	24 34.4	11 28.5	10 10.0	18 36.4	55
1 47.7	24 18.1	10 52.0	9 21.3	18 4.0	56
1 47.0	24 1.4	10 14.4	8 30.2	17 28.8	57
1 46.2	23 44.1	9 35.9	7 36.5	16 50.4	58
1 45.4	23 26.2	8 56.2	6 39.9	16 8.2	59
1 44.6	23 7.7	8 15.3	5 40.2	15 21.2	60
1 43.7	22 48.5	7 33.3	4 37.1	14 28.5	61
1 42.8	22 28.5	6 49.9	3 30.3	13 28.3	62
1 41.9	22 7.9	6 5.1	2 19.5	12 18.1	63
1 40.9	21 46.3	5 18.8	1 4.2	10 53.3	64
1 39.8	21 23.9	4 31.0	♏29 44.1	9 4.6	65
1 38.7	21 0.6	3 41.6	28 18.7	6 22.0	66
5	6	Descendant	8	9	S LAT

♒ 29° 54′ 21″

22ʰ 8ᵐ 0ˢ MC 332° 0′ 0″

10ʰ 12ᵐ 0ˢ MC 153° 0′ 0″

♍ 0° 57′ 13″

11	12	Ascendant	2	3	N LAT
° ′	° ′	° ′	° ′	° ′	°
♎ 3 16.2	♏ 5 17.5	♐ 4 56.7	♑ 2 45.2	♒ 0 47.2	0
3 13.7	4 29.6	3 8.2	1 25.4	0 11.9	5
3 11.3	3 42.8	1 21.4	0 4.2	♑29 35.6	10
3 8.9	2 56.5	♏29 35.0	♐28 40.4	28 57.5	15
3 8.4	2 47.2	29 13.6	28 23.1	28 49.7	16
3 7.9	2 37.9	28 52.1	28 5.7	28 41.7	17
3 7.4	2 28.6	28 30.6	27 48.2	28 33.6	18
3 6.9	2 19.3	28 9.1	27 30.4	28 25.4	19
3 6.4	2 9.9	27 47.4	27 12.3	28 17.1	20
3 5.9	2 0.5	27 25.6	26 54.1	28 8.6	21
3 5.3	1 51.0	27 3.7	26 35.6	28 0.0	22
3 4.8	1 41.5	26 41.7	26 16.8	27 51.2	23
3 4.3	1 32.0	26 19.6	25 57.8	27 42.3	24
3 3.8	1 22.3	25 57.3	25 38.4	27 33.1	25
3 3.3	1 12.6	25 34.8	25 18.8	27 23.8	26
3 2.7	1 2.9	25 12.1	24 58.7	27 14.3	27
3 2.2	0 53.0	24 49.3	24 38.4	27 4.6	28
3 1.6	0 43.0	24 26.2	24 17.6	26 54.7	29
3 1.1	0 33.0	24 2.9	23 56.4	26 44.5	30
3 0.5	0 22.8	23 39.4	23 34.8	26 34.0	31
2 59.9	0 12.6	23 15.6	23 12.8	26 23.3	32
2 59.3	0 2.2	22 51.5	22 50.2	26 12.3	33
2 58.7	♎29 51.7	22 27.2	22 27.1	26 0.9	34
2 58.1	29 41.0	22 2.5	22 3.5	25 49.2	35
2 57.5	29 30.2	21 37.5	21 39.3	25 37.2	36
2 56.9	29 19.2	21 12.2	21 14.5	25 24.8	37
2 56.3	29 8.1	20 46.5	20 49.0	25 11.9	38
2 55.6	28 56.8	20 20.4	20 22.8	24 58.6	39
2 55.0	28 45.3	19 53.9	19 55.9	24 44.8	40
2 54.3	28 33.6	19 27.0	19 28.2	24 30.5	41
2 53.6	28 21.7	18 59.6	18 59.6	24 15.6	42
2 52.9	28 9.5	18 31.7	18 30.1	24 0.2	43
2 52.1	27 57.1	18 3.3	17 59.7	23 44.0	44
2 51.4	27 44.5	17 34.4	17 28.2	23 27.1	45
2 50.6	27 31.6	17 4.9	16 55.6	23 9.4	46
2 49.8	27 18.4	16 34.9	16 21.9	22 50.9	47
2 49.0	27 4.9	16 4.2	15 46.9	22 31.4	48
2 48.2	26 51.1	15 32.8	15 10.5	22 10.9	49
2 47.3	26 36.9	15 0.7	14 32.7	21 49.2	50
2 46.5	26 22.4	14 27.9	13 53.3	21 26.2	51
2 45.5	26 7.4	13 54.3	13 12.2	21 1.8	52
2 44.6	25 52.1	13 19.9	12 29.3	20 35.8	53
2 43.6	25 36.3	12 44.6	11 44.5	20 8.0	54
2 42.6	25 20.1	12 8.5	10 57.5	19 38.1	55
2 41.6	25 3.3	11 31.3	10 8.2	19 5.8	56
2 40.5	24 46.0	10 53.1	9 16.5	18 30.8	57
2 39.3	24 28.2	10 13.9	8 22.0	17 52.5	58
2 38.1	24 9.7	9 33.5	7 24.6	17 10.5	59
2 36.9	23 50.6	8 52.0	6 24.0	16 23.7	60
2 35.6	23 30.8	8 9.1	5 19.9	15 31.2	61
2 34.2	23 10.3	7 24.9	4 12.0	14 31.2	62
2 32.8	22 48.9	6 39.4	2 59.9	13 21.3	63
2 31.3	22 26.7	5 52.3	1 43.3	11 57.0	64
2 29.8	22 3.6	5 3.6	0 21.6	10 8.9	65
2 28.1	21 39.4	4 13.2	♏28 54.5	7 28.0	66
5	6	Descendant	8	9	S LAT

♓ 0° 57′ 13″

22ʰ 12ᵐ 0ˢ MC 333° 0′ 0″

10h 16m 0s		MC		154° 0′ 0″	
		♍ 2° 0′ 16″			
N LAT	11	12	Ascendant	2	3
°	° ′	° ′	° ′	° ′	° ′
0	♎ 4 21.5	♏ 6 19.4	♐ 5 53.6	♑ 3 40.2	♒ 1 45.0
5	4 18.3	5 30.3	4 4.3	2 20.5	1 10.0
10	4 15.0	4 42.4	2 16.6	0 59.3	0 34.0
15	4 11.8	3 54.9	0 29.2	♐ 29 35.3	♑ 29 56.3
16	4 11.1	3 45.4	0 7.6	29 18.1	29 48.5
17	4 10.5	3 35.9	♏ 29 46.0	29 0.7	29 40.6
18	4 9.8	3 26.3	29 24.3	28 43.0	29 32.5
19	4 9.1	3 16.7	29 2.5	28 25.2	29 24.4
20	4 8.5	3 7.1	28 40.6	28 7.2	29 16.1
21	4 7.8	2 57.4	28 18.6	27 48.9	29 7.7
22	4 7.1	2 47.7	27 56.4	27 30.3	28 59.2
23	4 6.4	2 38.0	27 34.2	27 11.5	28 50.5
24	4 5.7	2 28.2	27 11.8	26 52.4	28 41.6
25	4 5.0	2 18.3	26 49.2	26 33.0	28 32.5
26	4 4.3	2 8.3	26 26.5	26 13.2	28 23.3
27	4 3.6	1 58.3	26 3.6	25 53.1	28 13.8
28	4 2.9	1 48.1	25 40.4	25 32.7	28 4.2
29	4 2.1	1 37.9	25 17.1	25 11.8	27 54.3
30	4 1.4	1 27.6	24 53.5	24 50.6	27 44.2
31	4 0.6	1 17.2	24 29.7	24 28.9	27 33.8
32	3 59.9	1 6.6	24 5.6	24 6.7	27 23.2
33	3 59.1	0 56.0	23 41.2	23 44.1	27 12.2
34	3 58.3	0 45.1	23 16.6	23 20.9	27 0.9
35	3 57.5	0 34.2	22 51.6	22 57.1	26 49.3
36	3 56.7	0 23.1	22 26.3	22 32.8	26 37.4
37	3 55.9	0 11.8	22 0.6	22 7.8	26 25.0
38	3 55.0	0 0.4	21 34.5	21 42.2	26 12.2
39	3 54.1	♎ 29 48.7	21 8.1	21 15.8	25 59.0
40	3 53.3	29 36.9	20 41.2	20 48.7	25 45.3
41	3 52.4	29 24.9	20 13.9	20 20.8	25 31.1
42	3 51.4	29 12.6	19 46.1	19 52.1	25 16.3
43	3 50.5	29 0.1	19 17.8	19 22.4	25 0.9
44	3 49.5	28 47.4	18 49.0	18 51.7	24 44.8
45	3 48.5	28 34.4	18 19.7	18 20.0	24 28.0
46	3 47.5	28 21.1	17 49.7	17 47.1	24 10.5
47	3 46.4	28 7.5	17 19.2	17 13.1	23 52.0
48	3 45.4	27 53.6	16 48.0	16 37.7	23 32.7
49	3 44.3	27 39.4	16 16.1	16 1.0	23 12.2
50	3 43.1	27 24.8	15 43.6	15 22.8	22 50.7
51	3 41.9	27 9.9	15 10.2	14 43.0	22 27.8
52	3 40.7	26 54.5	14 36.1	14 1.5	22 3.5
53	3 39.5	26 38.7	14 1.1	13 18.2	21 37.6
54	3 38.2	26 22.4	13 25.3	12 32.8	21 9.9
55	3 36.8	26 5.7	12 48.5	11 45.3	20 40.2
56	3 35.4	25 48.4	12 10.7	10 55.4	20 8.1
57	3 33.9	25 30.6	11 31.9	10 2.9	19 33.2
58	3 32.4	25 12.2	10 51.9	9 7.7	18 55.1
59	3 30.8	24 53.2	10 10.9	8 9.5	18 13.2
60	3 29.2	24 33.5	9 28.6	7 7.9	17 26.7
61	3 27.5	24 13.1	8 45.0	6 2.8	16 34.4
62	3 25.6	23 51.9	8 0.0	4 53.8	15 34.8
63	3 23.8	23 29.9	7 13.6	3 40.5	14 25.2
64	3 21.8	23 7.0	6 25.7	2 22.4	13 1.4
65	3 19.7	22 43.2	5 36.1	0 59.2	11 14.2
66	3 17.4	22 18.3	4 44.9	♏ 29 30.3	8 35.6
S LAT	5	6	Descendant	8	9
		♓ 2° 0′ 16″			
22h 16m 0s		MC		334° 0′ 0″	

10h 20m 0s		MC		155° 0′ 0″	
		♍ 3° 3′ 27″			
N LAT	11	12	Ascendant	2	3
°	° ′	° ′	° ′	° ′	° ′
0	♎ 5 26.8	♏ 7 21.1	♐ 6 50.3	♑ 4 35.3	♒ 2 43.0
5	5 22.8	6 30.8	5 0.2	3 15.6	2 8.3
10	5 18.7	5 41.8	3 11.8	1 54.4	1 32.6
15	5 14.7	4 53.2	1 23.4	0 30.4	0 55.2
16	5 13.9	4 43.4	1 1.6	0 13.1	0 47.5
17	5 13.0	4 33.7	0 39.8	♐ 29 55.7	0 39.6
18	5 12.2	4 23.9	0 17.9	29 38.0	0 31.7
19	5 11.4	4 14.0	♏ 29 55.8	29 20.1	0 23.6
20	5 10.5	4 4.2	29 33.7	29 2.0	0 15.4
21	5 9.7	3 54.3	29 11.5	28 43.7	0 7.0
22	5 8.8	3 44.3	28 49.1	28 25.1	♑ 29 58.6
23	5 8.0	3 34.3	28 26.6	28 6.2	29 49.9
24	5 7.1	3 24.3	28 4.0	27 47.1	29 41.1
25	5 6.2	3 14.1	27 41.2	27 27.6	29 32.1
26	5 5.4	3 3.9	27 18.2	27 7.8	29 23.0
27	5 4.5	2 53.6	26 55.0	26 47.7	29 13.6
28	5 3.6	2 43.2	26 31.6	26 27.1	29 4.0
29	5 2.6	2 32.7	26 8.0	26 6.2	28 54.2
30	5 1.7	2 22.1	25 44.1	25 44.9	28 44.1
31	5 0.8	2 11.4	25 20.0	25 23.1	28 33.8
32	4 59.8	2 0.6	24 55.6	25 0.8	28 23.3
33	4 58.8	1 49.7	24 30.9	24 38.0	28 12.4
34	4 57.8	1 38.6	24 6.0	24 14.8	28 1.2
35	4 56.8	1 27.3	23 40.6	23 50.9	27 49.7
36	4 55.8	1 15.9	23 15.0	23 26.4	27 37.8
37	4 54.8	1 4.3	22 49.0	23 1.3	27 25.5
38	4 53.7	0 52.6	22 22.6	22 35.5	27 12.8
39	4 52.6	0 40.7	21 55.8	22 9.0	26 59.7
40	4 51.5	0 28.5	21 28.5	21 41.7	26 46.1
41	4 50.4	0 16.1	21 0.8	21 13.7	26 31.9
42	4 49.3	0 3.6	20 32.6	20 44.7	26 17.2
43	4 48.1	♎ 29 50.7	20 3.9	20 14.8	26 1.9
44	4 46.9	29 37.6	19 34.7	19 43.9	25 46.0
45	4 45.6	29 24.3	19 4.9	19 11.9	25 29.3
46	4 44.3	29 10.6	18 34.5	18 38.8	25 11.8
47	4 43.0	28 56.7	18 3.5	18 4.5	24 53.5
48	4 41.7	28 42.4	17 31.9	17 28.8	24 34.2
49	4 40.3	28 27.7	16 59.5	16 51.8	24 13.9
50	4 38.9	28 12.7	16 26.4	16 13.2	23 52.5
51	4 37.4	27 57.3	15 52.6	15 33.0	23 29.7
52	4 35.9	27 41.5	15 17.9	14 51.0	23 5.6
53	4 34.3	27 25.3	14 42.3	14 7.2	22 39.8
54	4 32.7	27 8.5	14 5.9	13 21.3	22 12.3
55	4 31.0	26 51.3	13 28.5	12 33.2	21 42.7
56	4 29.2	26 33.6	12 50.1	11 42.7	21 10.8
57	4 27.4	26 15.2	12 10.6	10 49.5	20 36.1
58	4 25.5	25 56.3	11 30.0	9 53.6	19 58.2
59	4 23.5	25 36.7	10 48.2	8 54.5	19 16.5
60	4 21.5	25 16.4	10 5.2	7 52.0	18 30.3
61	4 19.3	24 55.4	9 20.8	6 45.9	17 38.3
62	4 17.0	24 33.6	8 35.1	5 35.7	16 39.0
63	4 14.7	24 10.9	7 47.8	4 21.1	15 29.9
64	4 12.2	23 47.3	6 59.0	3 1.6	14 6.6
65	4 9.6	23 22.7	6 8.6	1 36.8	12 20.5
66	4 6.8	22 57.1	5 16.4	0 6.1	9 44.6
S LAT	5	6	Descendant	8	9
		♓ 3° 3′ 27″			
22h 20m 0s		MC		335° 0′ 0″	

10^h 24^m 0^s MC 156° 0′ 0″

♍ 4° 6′ 48″

N LAT	11	12	Ascendant	2	3
°	° ′	° ′	° ′	° ′	° ′
0	♎ 6 32.1	♏ 8 22.6	♐ 7 46.9	♑ 5 30.5	♒ 3 41.2
5	6 27.2	7 31.2	5 56.1	4 10.8	3 6.8
10	6 22.4	6 41.1	4 6.9	2 49.6	2 31.4
15	6 17.5	5 51.3	2 17.6	1 25.5	1 54.4
16	6 16.6	5 41.3	1 55.6	1 8.2	1 46.7
17	6 15.6	5 31.3	1 33.6	0 50.7	1 38.9
18	6 14.6	5 21.3	1 11.4	0 33.1	1 31.0
19	6 13.6	5 11.3	0 49.2	0 15.2	1 23.0
20	6 12.6	5 1.2	0 26.9	♐29 57.0	1 14.9
21	6 11.6	4 51.0	0 4.4	29 38.7	1 6.6
22	6 10.5	4 40.8	♏29 41.8	29 20.0	0 58.2
23	6 9.5	4 30.6	29 19.1	29 1.1	0 49.6
24	6 8.5	4 20.3	28 56.2	28 41.9	0 40.9
25	6 7.4	4 9.9	28 33.1	28 22.4	0 31.9
26	6 6.4	3 59.4	28 9.9	28 2.5	0 22.8
27	6 5.3	3 48.9	27 46.4	27 42.3	0 13.5
28	6 4.2	3 38.2	27 22.8	27 21.7	0 4.0
29	6 3.1	3 27.5	26 58.9	27 0.7	♑29 54.3
30	6 2.0	3 16.6	26 34.7	26 39.3	29 44.3
31	6 0.9	3 5.6	26 10.3	26 17.4	29 34.1
32	5 59.7	2 54.5	25 45.7	25 55.0	29 23.6
33	5 58.6	2 43.3	25 20.7	25 32.2	29 12.8
34	5 57.4	2 31.9	24 55.4	25 8.8	29 1.7
35	5 56.2	2 20.4	24 29.7	24 44.8	28 50.3
36	5 54.9	2 8.7	24 3.8	24 20.2	28 38.5
37	5 53.7	1 56.8	23 37.4	23 55.0	28 26.3
38	5 52.4	1 44.8	23 10.6	23 29.0	28 13.7
39	5 51.1	1 32.5	22 43.5	23 2.4	28 0.6
40	5 49.8	1 20.0	22 15.8	22 34.9	27 47.1
41	5 48.4	1 7.4	21 47.8	22 6.7	27 33.1
42	5 47.1	0 54.4	21 19.2	21 37.5	27 18.5
43	5 45.6	0 41.3	20 50.1	21 7.4	27 3.3
44	5 44.2	0 27.8	20 20.4	20 36.2	26 47.4
45	5 42.7	0 14.1	19 50.2	20 4.0	26 30.8
46	5 41.2	0 0.1	19 19.4	19 30.6	26 13.5
47	5 39.6	♎29 45.7	18 47.9	18 56.0	25 55.3
48	5 38.0	29 31.0	18 15.8	18 20.1	25 36.1
49	5 36.3	29 16.0	17 42.9	17 42.7	25 16.0
50	5 34.6	29 0.6	17 9.3	17 3.8	24 54.6
51	5 32.8	28 44.8	16 34.9	16 23.2	24 32.0
52	5 31.0	28 28.5	15 59.7	15 40.8	24 8.0
53	5 29.1	28 11.8	15 23.6	14 56.5	23 42.5
54	5 27.2	27 54.6	14 46.5	14 10.0	23 15.1
55	5 25.1	27 36.9	14 8.5	13 21.3	22 45.6
56	5 23.0	27 18.6	13 29.5	12 30.2	22 13.9
57	5 20.8	26 59.8	12 49.3	11 36.4	21 39.4
58	5 18.6	26 40.3	12 8.1	10 39.6	21 1.8
59	5 16.2	26 20.1	11 25.6	9 39.7	20 20.3
60	5 13.7	25 59.3	10 41.8	8 36.3	19 34.3
61	5 11.1	25 37.6	9 56.7	7 29.1	18 42.7
62	5 8.4	25 15.2	9 10.1	6 17.8	17 43.8
63	5 5.6	24 51.8	8 22.1	5 1.8	16 35.2
64	5 2.6	24 27.5	7 32.4	3 40.9	15 12.7
65	4 59.4	24 2.2	6 41.1	2 14.4	13 27.7
66	4 56.1	23 35.8	5 47.9	0 41.9	10 55.1
S LAT	5	6	Descendant	8	9

♓ 4° 6′ 48″

22^h 24^m 0^s MC 336° 0′ 0″

10^h 28^m 0^s MC 157° 0′ 0″

♍ 5° 10′ 18″

N LAT	11	12	Ascendant	2	3
°	° ′	° ′	° ′	° ′	° ′
0	♎ 7 37.4	♏ 9 23.9	♐ 8 43.3	♑ 6 25.6	♒ 4 39.5
5	7 31.7	8 31.5	6 51.9	5 6.1	4 5.5
10	7 26.0	7 40.2	5 1.9	3 44.9	3 30.4
15	7 20.4	6 49.3	3 11.7	2 20.7	2 53.7
16	7 19.2	6 39.1	2 49.5	2 3.4	2 46.1
17	7 18.1	6 28.9	2 27.3	1 45.9	2 38.4
18	7 16.9	6 18.7	2 4.9	1 28.2	2 30.6
19	7 15.8	6 8.4	1 42.5	1 10.3	2 22.6
20	7 14.6	5 58.0	1 20.0	0 52.1	2 14.5
21	7 13.4	5 47.7	0 57.3	0 33.7	2 6.3
22	7 12.2	5 37.2	0 34.5	0 15.1	1 58.0
23	7 11.0	5 26.8	0 11.5	♐29 56.1	1 49.5
24	7 9.8	5 16.2	♏29 48.4	29 36.8	1 40.8
25	7 8.6	5 5.6	29 25.1	29 17.3	1 32.0
26	7 7.3	4 54.8	29 1.6	28 57.3	1 23.0
27	7 6.1	4 44.0	28 37.9	28 37.1	1 13.8
28	7 4.8	4 33.1	28 14.0	28 16.4	1 4.3
29	7 3.6	4 22.1	27 49.8	27 55.3	0 54.7
30	7 2.3	4 11.0	27 25.4	27 33.8	0 44.8
31	7 0.9	3 59.8	27 0.7	27 11.9	0 34.6
32	6 59.6	3 48.4	26 35.7	26 49.4	0 24.2
33	6 58.2	3 36.9	26 10.4	26 26.5	0 13.5
34	6 56.9	3 25.2	25 44.8	26 3.0	0 2.5
35	6 55.5	3 13.4	25 18.9	25 38.9	♑29 51.1
36	6 54.0	3 1.4	24 52.6	25 14.2	29 39.4
37	6 52.6	2 49.2	24 25.9	24 48.8	29 27.3
38	6 51.1	2 36.9	23 58.7	24 22.7	29 14.8
39	6 49.6	2 24.3	23 31.2	23 55.9	29 1.9
40	6 48.0	2 11.5	23 3.2	23 28.3	28 48.4
41	6 46.5	1 58.5	22 34.8	22 59.8	28 34.5
42	6 44.8	1 45.3	22 5.8	22 30.5	28 20.0
43	6 43.2	1 31.7	21 36.3	22 0.2	28 4.9
44	6 41.5	1 18.0	21 6.2	21 28.8	27 49.2
45	6 39.8	1 3.9	20 35.6	20 56.4	27 32.7
46	6 38.0	0 49.5	20 4.3	20 22.7	27 15.5
47	6 36.2	0 34.8	19 32.3	19 47.8	26 57.4
48	6 34.3	0 19.7	18 59.7	19 11.5	26 38.4
49	6 32.3	0 4.3	18 26.4	18 33.8	26 18.3
50	6 30.3	♎29 48.4	17 52.2	17 54.5	25 57.2
51	6 28.3	29 32.2	17 17.3	17 13.5	25 34.7
52	6 26.1	29 15.5	16 41.5	16 30.7	25 10.9
53	6 23.9	28 58.3	16 4.8	15 45.9	24 45.5
54	6 21.6	28 40.7	15 27.2	14 59.0	24 18.3
55	6 19.3	28 22.5	14 48.6	14 9.7	23 49.0
56	6 16.8	28 3.7	14 8.9	13 17.9	23 17.5
57	6 14.3	27 44.3	13 28.1	12 23.4	22 43.2
58	6 11.6	27 24.3	12 46.1	11 25.9	22 5.8
59	6 8.8	27 3.6	12 2.9	10 25.1	21 24.7
60	6 6.0	26 42.1	11 18.4	9 20.8	20 39.0
61	6 2.9	26 19.8	10 32.5	8 12.5	19 47.7
62	5 59.8	25 56.7	9 45.2	7 0.0	18 49.3
63	5 56.5	25 32.7	8 56.3	5 42.7	17 41.2
64	5 53.0	25 7.7	8 5.7	4 20.2	16 19.6
65	5 49.3	24 41.7	7 13.5	2 52.1	14 36.0
66	5 45.4	24 14.5	6 19.4	1 17.6	12 6.8
S LAT	5	6	Descendant	8	9

♓ 5° 10′ 18″

22^h 28^m 0^s MC 337° 0′ 0″

10h 32m 0s	MC		158° 0′ 0″		
	♍ 6° 13′ 57″				
N LAT	11	12	Ascendant	2	3
°	° ′	° ′	° ′	° ′	° ′
0	♎ 8 42.5	♏10 25.0	♐ 9 39.7	♑ 7 20.8	♒ 5 38.0
5	8 36.1	9 31.5	7 47.6	6 1.4	5 4.3
10	8 29.6	8 39.2	5 56.8	4 40.2	4 29.6
15	8 23.2	7 47.3	4 5.8	3 16.0	3 53.2
16	8 21.9	7 36.8	3 43.4	2 58.7	3 45.7
17	8 20.5	7 26.4	3 21.0	2 41.2	3 38.1
18	8 19.2	7 15.9	2 58.5	2 23.5	3 30.3
19	8 17.9	7 5.4	2 35.8	2 5.5	3 22.4
20	8 16.6	6 54.8	2 13.1	1 47.3	3 14.4
21	8 15.2	6 44.2	1 50.2	1 28.9	3 6.3
22	8 13.9	6 33.6	1 27.1	1 10.2	2 58.0
23	8 12.5	6 22.8	1 3.9	0 51.2	2 49.6
24	8 11.1	6 12.0	0 40.6	0 31.9	2 41.0
25	8 9.7	6 1.2	0 17.0	0 12.3	2 32.3
26	8 8.3	5 50.2	♏29 53.3	♐29 52.3	2 23.3
27	8 6.9	5 39.1	29 29.3	29 32.0	2 14.2
28	8 5.4	5 28.0	29 5.2	29 11.2	2 4.8
29	8 4.0	5 16.7	28 40.7	28 50.1	1 55.3
30	8 2.5	5 5.3	28 16.0	28 28.5	1 45.5
31	8 1.0	4 53.8	27 51.1	28 6.5	1 35.4
32	7 59.5	4 42.2	27 25.8	27 44.0	1 25.1
33	7 57.9	4 30.4	27 0.2	27 20.9	1 14.5
34	7 56.3	4 18.5	26 34.3	26 57.3	1 3.5
35	7 54.7	4 6.4	26 8.0	26 33.1	0 52.3
36	7 53.1	3 54.1	25 41.4	26 8.3	0 40.6
37	7 51.4	3 41.6	25 14.3	25 42.8	0 28.6
38	7 49.7	3 28.9	24 46.9	25 16.6	0 16.2
39	7 48.0	3 16.1	24 19.0	24 49.6	0 3.4
40	7 46.3	3 3.0	23 50.6	24 21.9	♑29 50.1
41	7 44.5	2 49.6	23 21.8	23 53.2	29 36.3
42	7 42.6	2 36.1	22 52.4	23 23.7	29 21.9
43	7 40.7	2 22.2	22 22.5	22 53.2	29 6.9
44	7 38.8	2 8.1	21 52.0	22 21.6	28 51.3
45	7 36.8	1 53.6	21 20.9	21 48.9	28 34.9
46	7 34.8	1 38.9	20 49.2	21 15.0	28 17.8
47	7 32.7	1 23.8	20 16.8	20 39.8	27 59.9
48	7 30.5	1 8.3	19 43.7	20 3.2	27 41.0
49	7 28.3	0 52.5	19 9.8	19 25.2	27 21.1
50	7 26.0	0 36.2	18 35.2	18 45.5	27 0.1
51	7 23.7	0 19.6	17 59.7	18 4.1	26 37.8
52	7 21.2	0 2.4	17 23.4	17 20.9	26 14.1
53	7 18.7	♎29 44.8	16 46.1	16 35.6	25 48.9
54	7 16.1	29 26.7	16 7.9	15 48.2	25 21.9
55	7 13.4	29 8.0	15 28.7	14 58.3	24 52.8
56	7 10.6	28 48.7	14 48.3	14 5.9	24 21.5
57	7 7.7	28 28.8	14 6.9	13 10.7	23 47.5
58	7 4.6	28 8.2	13 24.2	12 12.4	23 10.4
59	7 1.5	27 46.9	12 40.3	11 10.7	22 29.5
60	6 58.2	27 24.9	11 55.0	10 5.4	21 44.2
61	6 54.7	27 2.0	11 8.3	8 56.1	20 53.3
62	6 51.1	26 38.3	10 20.2	7 42.3	19 55.4
63	6 47.3	26 13.6	9 30.4	6 23.7	18 48.0
64	6 43.3	25 47.9	8 39.0	4 59.7	17 27.2
65	6 39.1	25 21.1	7 45.9	3 29.8	15 45.2
66	6 34.7	24 53.1	6 50.9	1 53.4	13 19.8
S LAT	5	6	Descendant	8	9
	♓ 6° 13′ 57″				
22h 32m 0s	MC		338° 0′ 0″		

10h 36m 0s	MC		159° 0′ 0″		
	♍ 7° 17′ 44″				
N LAT	11	12	Ascendant	2	3
°	° ′	° ′	° ′	° ′	° ′
0	♎ 9 47.7	♏11 26.0	♐10 35.9	♑ 8 16.1	♒ 6 36.6
5	9 40.4	10 31.5	8 43.3	6 56.7	6 3.3
10	9 33.2	9 38.1	6 51.7	5 35.6	5 29.0
15	9 25.9	8 45.0	4 59.8	4 11.4	4 53.0
16	9 24.4	8 34.4	4 37.3	3 54.1	4 45.5
17	9 23.0	8 23.7	4 14.7	3 36.6	4 37.9
18	9 21.5	8 13.0	3 51.9	3 18.8	4 30.3
19	9 20.0	8 2.3	3 29.1	3 0.9	4 22.5
20	9 18.5	7 51.5	3 6.1	2 42.7	4 14.6
21	9 17.0	7 40.7	2 43.0	2 24.2	4 6.5
22	9 15.5	7 29.8	2 19.8	2 5.4	3 58.3
23	9 13.9	7 18.8	1 56.4	1 46.4	3 50.0
24	9 12.4	7 7.8	1 32.8	1 27.1	3 41.5
25	9 10.8	6 56.7	1 9.0	1 7.4	3 32.8
26	9 9.2	6 45.5	0 45.0	0 47.4	3 23.9
27	9 7.6	6 34.2	0 20.8	0 27.0	3 14.9
28	9 6.0	6 22.8	♏29 56.4	0 6.2	3 5.6
29	9 4.3	6 11.2	29 31.7	♐29 45.0	2 56.1
30	9 2.7	5 59.6	29 6.7	29 23.4	2 46.4
31	9 1.0	5 47.8	28 41.5	29 1.3	2 36.4
32	8 59.3	5 35.9	28 15.9	28 38.7	2 26.2
33	8 57.5	5 23.9	27 50.0	28 15.5	2 15.7
34	8 55.8	5 11.6	27 23.8	27 51.8	2 4.8
35	8 54.0	4 59.3	26 57.2	27 27.5	1 53.7
36	8 52.1	4 46.7	26 30.2	27 2.6	1 42.2
37	8 50.3	4 33.9	26 2.9	26 37.0	1 30.3
38	8 48.4	4 21.0	25 35.1	26 10.6	1 18.0
39	8 46.4	4 7.8	25 6.8	25 43.5	1 5.2
40	8 44.4	3 54.4	24 38.1	25 15.6	0 52.0
41	8 42.4	3 40.7	24 8.9	24 46.8	0 38.3
42	8 40.3	3 26.8	23 39.1	24 17.1	0 24.0
43	8 38.2	3 12.6	23 8.8	23 46.4	0 9.2
44	8 36.0	2 58.1	22 37.9	23 14.6	♑29 53.7
45	8 33.8	2 43.3	22 6.4	22 41.7	29 37.5
46	8 31.5	2 28.2	21 34.2	22 7.5	29 20.5
47	8 29.2	2 12.7	21 1.3	21 32.0	29 2.7
48	8 26.8	1 56.9	20 27.7	20 55.2	28 44.0
49	8 24.3	1 40.7	19 53.4	20 16.8	28 24.2
50	8 21.7	1 24.0	19 18.2	19 36.8	28 3.4
51	8 19.0	1 6.9	18 42.2	18 55.0	27 41.3
52	8 16.3	0 49.3	18 5.3	18 11.3	27 17.8
53	8 13.5	0 31.3	17 27.5	17 25.6	26 52.7
54	8 10.5	0 12.6	16 48.7	16 37.6	26 25.9
55	8 7.5	♎29 53.5	16 8.8	15 47.2	25 57.1
56	8 4.3	29 33.7	15 27.8	14 54.1	25 26.0
57	8 1.1	29 13.3	14 45.7	13 58.2	24 52.3
58	7 57.7	28 52.1	14 2.3	12 59.1	24 15.4
59	7 54.1	28 30.3	13 17.7	11 56.6	23 34.9
60	7 50.4	28 7.6	12 31.6	10 50.3	22 50.0
61	7 46.5	27 44.2	11 44.2	9 39.8	21 59.5
62	7 42.4	27 19.8	10 55.2	8 24.8	21 2.1
63	7 38.2	26 54.4	10 4.6	7 4.8	19 55.5
64	7 33.7	26 28.0	9 12.3	5 39.2	18 35.7
65	7 28.9	26 0.4	8 18.3	4 7.5	16 55.3
66	7 24.0	25 31.7	7 22.3	2 29.1	14 33.8
S LAT	5	6	Descendant	8	9
	♓ 7° 17′ 44″				
22h 36m 0s	MC		339° 0′ 0″		

10h 40m 0s — MC — 160° 0′ 0″

♍ 8° 21′ 40″

N LAT	11	12	Ascendant	2	3
0	♎10 52.7	♏12 26.7	♐11 32.1	♑ 9 11.4	♒ 7 35.4
5	10 44.7	11 31.2	9 38.8	7 52.1	7 2.5
10	10 36.7	10 36.8	7 46.6	6 31.1	6 28.5
15	10 28.6	9 42.7	5 53.9	5 6.9	5 52.9
16	10 27.0	9 31.9	5 31.1	4 49.6	5 45.5
17	10 25.3	9 21.0	5 8.3	4 32.1	5 38.0
18	10 23.7	9 10.1	4 45.4	4 14.3	5 30.4
19	10 22.0	8 59.1	4 22.4	3 56.3	5 22.7
20	10 20.4	8 48.1	3 59.2	3 38.1	5 14.9
21	10 18.7	8 37.0	3 35.9	3 19.6	5 6.9
22	10 17.0	8 25.9	3 12.4	3 0.8	4 58.8
23	10 15.3	8 14.7	2 48.8	2 41.8	4 50.6
24	10 13.6	8 3.4	2 25.0	2 22.4	4 42.1
25	10 11.8	7 52.1	2 1.0	2 2.7	4 33.5
26	10 10.1	7 40.6	1 36.8	1 42.6	4 24.8
27	10 8.3	7 29.1	1 12.3	1 22.2	4 15.8
28	10 6.5	7 17.4	0 47.6	1 1.4	4 6.6
29	10 4.7	7 5.7	0 22.6	0 40.1	3 57.2
30	10 2.8	6 53.8	♏29 57.4	0 18.4	3 47.6
31	10 0.9	6 41.7	29 31.9	♐29 56.2	3 37.7
32	9 59.0	6 29.6	29 6.0	29 33.6	3 27.6
33	9 57.1	6 17.3	28 39.9	29 10.3	3 17.2
34	9 55.2	6 4.8	28 13.3	28 46.5	3 6.4
35	9 53.2	5 52.1	27 46.4	28 22.1	2 55.4
36	9 51.1	5 39.2	27 19.1	27 57.1	2 43.9
37	9 49.0	5 26.2	26 51.4	27 31.4	2 32.2
38	9 46.9	5 12.9	26 23.3	27 4.9	2 20.0
39	9 44.8	4 59.4	25 54.7	26 37.6	2 7.3
40	9 42.6	4 45.7	25 25.6	26 9.6	1 54.3
41	9 40.3	4 31.7	24 56.0	25 40.6	1 40.7
42	9 38.0	4 17.5	24 25.8	25 10.7	1 26.5
43	9 35.7	4 3.0	23 55.1	24 39.8	1 11.8
44	9 33.3	3 48.1	23 23.8	24 7.8	0 56.4
45	9 30.8	3 33.0	22 51.8	23 34.6	0 40.4
46	9 28.2	3 17.5	22 19.2	23 0.2	0 23.5
47	9 25.6	3 1.7	21 45.9	22 24.5	0 5.9
48	9 23.0	2 45.4	21 11.8	21 47.3	♑29 47.3
49	9 20.2	2 28.8	20 36.9	21 8.6	29 27.7
50	9 17.3	2 11.7	20 1.3	20 28.3	29 7.0
51	9 14.4	1 54.2	19 24.7	19 46.1	28 45.1
52	9 11.4	1 36.2	18 47.3	19 2.0	28 21.8
53	9 8.2	1 17.7	18 8.9	18 15.8	27 57.0
54	9 5.0	0 58.6	17 29.4	17 27.2	27 30.4
55	9 1.6	0 38.9	16 48.9	16 36.3	27 1.8
56	8 58.1	0 18.7	16 7.3	15 42.6	26 31.0
57	8 54.4	♎29 57.7	15 24.5	14 45.9	25 57.5
58	8 50.6	29 36.0	14 40.4	13 46.0	25 21.0
59	8 46.7	29 13.6	13 55.0	12 42.6	24 40.8
60	8 42.6	28 50.4	13 8.3	11 35.4	23 56.3
61	8 38.2	28 26.3	12 20.0	10 23.8	23 6.3
62	8 33.7	28 1.2	11 30.2	9 7.5	22 9.5
63	8 29.0	27 35.2	10 38.8	7 46.1	21 3.6
64	8 24.0	27 8.0	9 45.6	6 18.9	19 45.0
65	8 18.7	26 39.8	8 50.6	4 45.4	18 6.3
66	8 13.2	26 10.2	7 53.7	3 4.8	15 48.9
S LAT	5	6	Descendant	8	9

♓ 8° 21′ 40″

22h 40m 0s — MC — 340° 0′ 0″

10h 44m 0s — MC — 161° 0′ 0″

♍ 9° 25′ 43″

N LAT	11	12	Ascendant	2	3
0	♎11 57.7	♏13 27.3	♐12 28.1	♑10 6.7	♒ 8 34.4
5	11 48.9	12 30.8	10 34.3	8 47.6	8 1.9
10	11 40.1	11 35.4	8 41.4	7 26.7	7 28.3
15	11 31.2	10 40.3	6 47.9	6 2.5	6 53.1
16	11 29.4	10 29.2	6 25.0	5 45.2	6 45.8
17	11 27.6	10 18.1	6 2.0	5 27.6	6 38.4
18	11 25.8	10 7.0	5 38.9	5 9.9	6 30.8
19	11 24.0	9 55.8	5 15.6	4 51.9	6 23.2
20	11 22.2	9 44.6	4 52.3	4 33.6	6 15.5
21	11 20.4	9 33.3	4 28.8	4 15.1	6 7.6
22	11 18.5	9 21.9	4 5.1	3 56.4	5 59.6
23	11 16.6	9 10.5	3 41.2	3 37.3	5 51.4
24	11 14.7	8 59.0	3 17.2	3 17.9	5 43.0
25	11 12.8	8 47.4	2 53.0	2 58.1	5 34.5
26	11 10.9	8 35.7	2 28.5	2 38.0	5 25.9
27	11 8.9	8 24.0	2 3.8	2 17.6	5 17.0
28	11 7.0	8 12.1	1 38.9	1 56.7	5 7.9
29	11 5.0	8 0.0	1 13.6	1 35.4	4 58.6
30	11 2.9	7 47.9	0 48.1	1 13.6	4 49.1
31	11 0.9	7 35.6	0 22.3	0 51.4	4 39.3
32	10 58.8	7 23.2	♏29 56.2	0 28.6	4 29.2
33	10 56.7	7 10.6	29 29.7	0 5.3	4 18.9
34	10 54.5	6 57.8	29 2.9	♐29 41.4	4 8.3
35	10 52.3	6 44.9	28 35.7	29 16.9	3 57.3
36	10 50.1	6 31.7	28 8.1	28 51.8	3 46.0
37	10 47.8	6 18.4	27 40.1	28 25.9	3 34.4
38	10 45.5	6 4.8	27 11.6	27 59.3	3 22.3
39	10 43.1	5 51.1	26 42.6	27 32.0	3 9.8
40	10 40.7	5 37.0	26 13.1	27 3.7	2 56.8
41	10 38.2	5 22.7	25 43.2	26 34.6	2 43.3
42	10 35.7	5 8.2	25 12.6	26 4.6	2 29.3
43	10 33.1	4 53.3	24 41.5	25 33.4	2 14.7
44	10 30.5	4 38.1	24 9.7	25 1.2	1 59.5
45	10 27.7	4 22.6	23 37.3	24 27.9	1 43.6
46	10 24.9	4 6.8	23 4.3	23 53.2	1 26.9
47	10 22.1	3 50.6	22 30.5	23 17.2	1 9.4
48	10 19.1	3 33.9	21 55.9	22 39.8	0 51.0
49	10 16.1	3 16.9	21 20.6	22 0.7	0 31.6
50	10 12.9	2 59.4	20 44.4	21 20.0	0 11.1
51	10 9.7	2 41.5	20 7.3	20 37.4	♑29 49.4
52	10 6.4	2 23.0	19 29.3	19 52.9	29 26.3
53	10 2.9	2 4.1	18 50.3	19 6.2	29 1.6
54	9 59.3	1 44.5	18 10.2	18 17.2	28 35.3
55	9 55.6	1 24.4	17 29.1	17 25.6	28 7.0
56	9 51.8	1 3.6	16 46.8	16 31.3	27 36.4
57	9 47.8	0 42.1	16 3.3	15 33.9	27 3.2
58	9 43.6	0 19.9	15 18.6	14 33.2	26 27.1
59	9 39.2	♎29 56.9	14 32.5	13 28.9	25 47.3
60	9 34.7	29 33.1	13 44.9	12 20.7	25 3.2
61	9 29.9	29 8.3	12 55.9	11 8.0	24 13.7
62	9 25.0	28 42.6	12 5.2	9 50.4	23 17.6
63	9 19.8	28 15.9	11 12.9	8 27.5	22 12.5
64	9 14.3	27 48.1	10 18.8	6 58.7	20 55.0
65	9 8.5	27 19.0	9 22.9	5 23.2	19 18.2
66	9 2.4	26 48.7	8 25.0	3 40.5	17 5.0
S LAT	5	6	Descendant	8	9

♓ 9° 25′ 43″

22h 44m 0s — MC — 341° 0′ 0″

10ʰ 48ᵐ 0ˢ MC 162° 0′ 0″

♍ 10° 29′ 54″

N LAT	11	12	Ascendant	2	3
°	° ′	° ′	° ′	° ′	° ′
0	♎13 2.6	♏14 27.7	♐13 24.0	♑11 2.1	♒9 33.6
5	12 53.0	13 30.3	11 29.7	9 43.2	9 1.4
10	12 43.4	12 33.9	9 36.1	8 22.3	8 28.2
15	12 33.8	11 37.7	7 41.8	6 58.3	7 53.4
16	12 31.9	11 26.4	7 18.8	6 40.9	7 46.2
17	12 29.9	11 15.1	6 55.6	6 23.4	7 38.9
18	12 27.9	11 3.8	6 32.3	6 5.6	7 31.5
19	12 26.0	10 52.4	6 8.9	5 47.6	7 23.9
20	12 24.0	10 40.9	5 45.3	5 29.3	7 16.3
21	12 22.0	10 29.4	5 21.6	5 10.8	7 8.5
22	12 19.9	10 17.9	4 57.7	4 52.0	7 0.5
23	12 17.9	10 6.2	4 33.7	4 32.9	6 52.4
24	12 15.8	9 54.5	4 9.4	4 13.5	6 44.2
25	12 13.8	9 42.7	3 45.0	3 53.7	6 35.8
26	12 11.6	9 30.8	3 20.3	3 33.6	6 27.2
27	12 9.5	9 18.7	2 55.3	3 13.1	6 18.4
28	12 7.4	9 6.6	2 30.1	2 52.1	6 9.4
29	12 5.2	8 54.3	2 4.7	2 30.8	6 0.2
30	12 3.0	8 41.9	1 38.9	2 9.0	5 50.8
31	12 0.7	8 29.4	1 12.8	1 46.7	5 41.1
32	11 58.5	8 16.7	0 46.4	1 23.8	5 31.2
33	11 56.1	8 3.8	0 19.7	1 0.5	5 21.0
34	11 53.8	7 50.8	♏29 52.5	0 36.5	5 10.4
35	11 51.4	7 37.6	29 25.0	0 11.9	4 59.6
36	11 49.0	7 24.2	28 57.1	♐29 46.7	4 48.4
37	11 46.5	7 10.6	28 28.7	29 20.7	4 36.8
38	11 44.0	6 56.7	27 59.9	28 54.0	4 24.9
39	11 41.4	6 42.6	27 30.6	28 26.5	4 12.5
40	11 38.8	6 28.3	27 0.7	27 58.1	3 59.7
41	11 36.1	6 13.7	26 30.4	27 28.9	3 46.3
42	11 33.3	5 58.8	25 59.4	26 58.6	3 32.5
43	11 30.5	5 43.6	25 27.9	26 27.3	3 18.0
44	11 27.6	5 28.1	24 55.8	25 54.9	3 2.9
45	11 24.6	5 12.2	24 22.9	25 21.3	2 47.2
46	11 21.6	4 56.0	23 49.4	24 46.5	2 30.6
47	11 18.5	4 39.4	23 15.1	24 10.2	2 13.3
48	11 15.2	4 22.4	22 40.1	23 32.4	1 55.1
49	11 11.9	4 5.0	22 4.2	22 53.1	1 35.8
50	11 8.5	3 47.1	21 27.5	22 12.0	1 15.5
51	11 5.0	3 28.8	20 49.9	21 29.1	0 54.0
52	11 1.3	3 9.9	20 11.3	20 44.1	0 31.1
53	10 57.6	2 50.4	19 31.7	19 56.9	0 6.7
54	10 53.7	2 30.4	18 51.1	19 7.4	♑29 40.6
55	10 49.6	2 9.8	18 9.3	18 15.3	29 12.6
56	10 45.4	1 48.5	17 26.4	17 20.3	28 42.3
57	10 41.1	1 26.5	16 42.2	16 22.2	28 9.5
58	10 36.5	1 3.7	15 56.7	15 20.7	27 33.6
59	10 31.8	0 40.1	15 9.9	14 15.5	26 54.3
60	10 26.8	0 15.7	14 21.6	13 6.2	26 10.6
61	10 21.6	♎29 50.4	13 31.7	11 52.4	25 21.8
62	10 16.2	29 24.0	12 40.2	10 33.5	24 26.3
63	10 10.5	28 56.6	11 47.0	9 9.1	23 22.1
64	10 4.5	28 28.0	10 52.1	7 38.6	22 5.8
65	9 58.2	27 58.3	9 55.2	6 1.2	20 31.0
66	9 51.6	27 27.1	8 56.3	4 16.2	18 21.9
S LAT	5	6	Descendant	8	9

♓ 10° 29′ 54″

22ʰ 48ᵐ 0ˢ MC 342° 0′ 0″

10ʰ 52ᵐ 0ˢ MC 163° 0′ 0″

♍ 11° 34′ 12″

N LAT	11	12	Ascendant	2	3
°	° ′	° ′	° ′	° ′	° ′
0	♎14 7.5	♏15 28.0	♐14 19.9	♑11 57.6	♒10 32.9
5	13 57.0	14 29.5	12 25.0	10 38.8	10 1.2
10	13 46.7	13 32.2	10 30.8	9 18.1	9 28.4
15	13 36.3	12 35.0	8 35.8	7 54.1	8 54.0
16	13 34.2	12 23.5	8 12.6	7 36.8	8 46.9
17	13 32.1	12 12.0	7 49.2	7 19.2	8 39.6
18	13 30.0	12 0.5	7 25.7	7 1.4	8 32.3
19	13 27.8	11 48.9	7 2.1	6 43.4	8 24.8
20	13 25.7	11 37.2	6 38.4	6 25.2	8 17.3
21	13 23.5	11 25.5	6 14.5	6 6.6	8 9.6
22	13 21.3	11 13.7	5 50.4	5 47.8	8 1.7
23	13 19.1	11 1.8	5 26.1	5 28.7	7 53.7
24	13 16.9	10 49.9	5 1.7	5 9.3	7 45.6
25	13 14.6	10 37.8	4 37.0	4 49.5	7 37.3
26	13 12.4	10 25.7	4 12.1	4 29.3	7 28.8
27	13 10.1	10 13.4	3 46.9	4 8.8	7 20.1
28	13 7.7	10 1.1	3 21.5	3 47.8	7 11.2
29	13 5.4	9 48.5	2 55.7	3 26.4	7 2.1
30	13 3.0	9 35.9	2 29.7	3 4.5	6 52.8
31	13 0.5	9 23.1	2 3.4	2 42.2	6 43.2
32	12 58.1	9 10.2	1 36.7	2 19.3	6 33.4
33	12 55.6	8 57.1	1 9.6	1 55.8	6 23.3
34	12 53.0	8 43.8	0 42.2	1 31.8	6 12.9
35	12 50.5	8 30.3	0 14.4	1 7.1	6 2.1
36	12 47.8	8 16.6	♏29 46.1	0 41.8	5 51.1
37	12 45.2	8 2.7	29 17.4	0 15.7	5 39.6
38	12 42.4	7 48.5	28 48.3	♐29 48.9	5 27.8
39	12 39.6	7 34.1	28 18.6	29 21.3	5 15.6
40	12 36.8	7 19.5	27 48.4	28 52.8	5 2.8
41	12 33.9	7 4.6	27 17.7	28 23.4	4 49.6
42	12 30.9	6 49.4	26 46.3	27 52.9	4 35.9
43	12 27.8	6 33.8	26 14.4	27 21.5	4 21.6
44	12 24.7	6 18.0	25 41.8	26 48.9	4 6.7
45	12 21.5	6 1.8	25 8.6	26 15.1	3 51.1
46	12 18.2	5 45.2	24 34.6	25 40.0	3 34.7
47	12 14.8	5 28.3	23 59.9	25 3.4	3 17.6
48	12 11.3	5 10.9	23 24.3	24 25.4	2 59.5
49	12 7.7	4 53.1	22 48.0	23 45.7	2 40.5
50	12 4.0	4 34.8	22 10.7	23 4.3	2 20.4
51	12 0.2	4 16.0	21 32.6	22 21.0	1 59.1
52	11 56.3	3 56.7	20 53.4	21 35.6	1 36.4
53	11 52.2	3 36.8	20 13.2	20 48.0	1 12.3
54	11 48.0	3 16.3	19 32.0	19 57.9	0 46.4
55	11 43.6	2 55.2	18 49.6	19 5.2	0 18.7
56	11 39.0	2 33.4	18 6.0	18 9.5	♑29 48.7
57	11 34.3	2 10.8	17 21.1	17 10.7	29 16.2
58	11 29.4	1 47.5	16 34.9	16 8.5	28 40.8
59	11 24.3	1 23.4	15 47.3	15 2.3	28 1.8
60	11 18.9	0 58.3	14 58.2	13 52.0	27 18.7
61	11 13.3	0 32.3	14 7.5	12 37.0	26 30.4
62	11 7.4	0 5.3	13 15.2	11 16.8	25 35.6
63	11 1.2	♎29 37.2	12 21.2	9 50.9	24 32.3
64	10 54.7	29 8.0	11 25.3	8 18.6	23 17.3
65	10 47.9	28 37.4	10 27.4	6 39.2	21 44.5
66	10 40.7	28 5.5	9 27.5	4 51.9	19 39.7
S LAT	5	6	Descendant	8	9

♓ 11° 34′ 12″

22ʰ 52ᵐ 0ˢ MC 343° 0′ 0″

10h 56m 0s — MC 164° 0′ 0″ — ♍ 12° 38′ 37″

11	12	Ascendant	2	3	N LAT
° ′	° ′	° ′	° ′	° ′	°
♎15 12.2	♏16 28.0	♐15 15.6	♑12 53.1	♒11 32.4	0
15 1.0	15 28.7	13 20.3	11 34.6	11 1.1	5
14 49.9	14 30.4	11 25.5	10 14.0	10 28.8	10
14 38.7	13 32.2	9 29.7	8 50.0	9 54.8	15
14 36.5	13 20.5	9 6.3	8 32.7	9 47.8	16
14 34.2	13 8.8	8 42.8	8 15.2	9 40.6	17
14 31.9	12 57.1	8 19.2	7 57.4	9 33.4	18
14 29.6	12 45.3	7 55.4	7 39.4	9 26.0	19
14 27.3	12 33.4	7 31.5	7 21.1	9 18.5	20
14 25.0	12 21.5	7 7.4	7 2.6	9 10.9	21
14 22.6	12 9.4	6 43.1	6 43.8	9 3.2	22
14 20.2	11 57.4	6 18.6	6 24.6	8 55.3	23
14 17.9	11 45.2	5 53.9	6 5.2	8 47.2	24
14 15.4	11 32.9	5 29.0	5 45.4	8 39.0	25
14 13.0	11 20.5	5 3.9	5 25.2	8 30.6	26
14 10.5	11 8.1	4 38.5	5 4.6	8 22.0	27
14 8.0	10 55.4	4 12.8	4 43.6	8 13.2	28
14 5.5	10 42.7	3 46.8	4 22.2	8 4.3	29
14 2.9	10 29.8	3 20.5	4 0.3	7 55.0	30
14 0.3	10 16.8	2 53.9	3 37.8	7 45.6	31
13 57.7	10 3.6	2 27.0	3 14.9	7 35.9	32
13 55.0	9 50.2	1 59.6	2 51.4	7 25.9	33
13 52.2	9 36.6	1 31.9	2 27.3	7 15.6	34
13 49.5	9 22.9	1 3.8	2 2.5	7 5.0	35
13 46.6	9 8.9	0 35.2	1 37.1	6 54.0	36
13 43.8	8 54.7	0 6.2	1 10.9	6 42.7	37
13 40.8	8 40.3	♏29 36.7	0 44.0	6 31.0	38
13 37.8	8 25.6	29 6.7	0 16.3	6 18.9	39
13 34.7	8 10.7	28 36.1	♐29 47.6	6 6.3	40
13 31.6	7 55.4	28 5.0	29 18.1	5 53.3	41
13 28.4	7 39.9	27 33.3	28 47.5	5 39.7	42
13 25.1	7 24.0	27 1.0	28 15.9	5 25.6	43
13 21.8	7 7.9	26 28.0	27 43.1	5 10.8	44
13 18.3	6 51.3	25 54.3	27 9.1	4 55.4	45
13 14.8	6 34.4	25 19.8	26 33.7	4 39.2	46
13 11.1	6 17.1	24 44.6	25 56.9	4 22.2	47
13 7.4	5 59.3	24 8.6	25 18.6	4 4.3	48
13 3.5	5 41.1	23 31.8	24 38.7	3 45.5	49
12 59.5	5 22.4	22 54.0	23 56.9	3 25.6	50
12 55.4	5 3.2	22 15.3	23 13.2	3 4.5	51
12 51.2	4 43.4	21 35.6	22 27.4	2 42.1	52
12 46.8	4 23.1	20 54.8	21 39.3	2 18.2	53
12 42.3	4 2.1	20 12.9	20 48.7	1 52.7	54
12 37.5	3 40.5	19 29.9	19 55.4	1 25.2	55
12 32.6	3 18.2	18 45.6	18 59.1	0 55.6	56
12 27.5	2 55.1	18 0.1	17 59.6	0 23.4	57
12 22.2	2 31.3	17 13.1	16 56.5	♑29 48.4	58
12 16.7	2 6.6	16 24.8	15 49.5	29 9.9	59
12 10.9	1 40.9	15 34.9	14 38.1	28 27.3	60
12 4.9	1 14.3	14 43.4	13 21.9	27 39.6	61
11 58.6	0 46.7	13 50.2	12 0.3	26 45.6	62
11 51.9	0 17.9	12 55.3	10 32.9	25 43.3	63
11 44.9	♎29 47.9	11 58.5	8 58.8	24 29.6	64
11 37.6	29 16.6	10 59.6	7 17.4	22 58.9	65
11 29.8	28 43.9	9 58.8	5 27.7	20 58.2	66
5	6	Descendant	8	9	S LAT

♓ 12° 38′ 37″ — 22h 56m 0s — MC 344° 0′ 0″

11h 0m 0s — MC 165° 0′ 0″ — ♍ 13° 43′ 9″

11	12	Ascendant	2	3	N LAT
° ′	° ′	° ′	° ′	° ′	°
♎16 16.8	♏17 27.9	♐16 11.3	♑13 48.7	♒12 32.1	0
16 4.9	16 27.6	14 15.5	12 30.4	12 1.2	5
15 53.0	15 28.4	12 20.1	11 10.0	11 29.3	10
15 41.1	14 29.3	10 23.7	9 46.1	10 55.8	15
15 38.7	14 17.4	10 0.1	9 28.8	10 48.9	16
15 36.2	14 5.5	9 36.4	9 11.3	10 41.8	17
15 33.8	13 53.5	9 12.6	8 53.5	10 34.7	18
15 31.3	13 41.5	8 48.7	8 35.5	10 27.4	19
15 28.9	13 29.5	8 24.6	8 17.3	10 20.0	20
15 26.4	13 17.3	8 0.3	7 58.7	10 12.5	21
15 23.9	13 5.1	7 35.8	7 39.9	10 4.8	22
15 21.3	12 52.8	7 11.1	7 20.8	9 57.0	23
15 18.8	12 40.4	6 46.2	7 1.3	9 49.1	24
15 16.2	12 27.9	6 21.1	6 41.4	9 41.0	25
15 13.6	12 15.3	5 55.7	6 21.2	9 32.7	26
15 10.9	12 2.6	5 30.1	6 0.6	9 24.2	27
15 8.2	11 49.8	5 4.2	5 39.6	9 15.5	28
15 5.5	11 36.8	4 38.0	5 18.1	9 6.7	29
15 2.8	11 23.6	4 11.4	4 56.2	8 57.6	30
15 0.0	11 10.4	3 44.6	4 33.7	8 48.2	31
14 57.2	10 56.9	3 17.3	4 10.7	8 38.6	32
14 54.3	10 43.3	2 49.7	3 47.2	8 28.7	33
14 51.4	10 29.5	2 21.7	3 23.0	8 18.6	34
14 48.4	10 15.4	1 53.3	2 58.2	8 8.1	35
14 45.4	10 1.2	1 24.4	2 32.6	7 57.3	36
14 42.3	9 46.7	0 55.0	2 6.4	7 46.1	37
14 39.2	9 32.0	0 25.2	1 39.4	7 34.5	38
14 35.9	9 17.0	♏29 54.8	1 11.5	7 22.6	39
14 32.7	9 1.8	29 23.9	0 42.8	7 10.2	40
14 29.3	8 46.2	28 52.4	0 13.1	6 57.3	41
14 25.9	8 30.4	28 20.3	♐29 42.3	6 43.8	42
14 22.4	8 14.2	27 47.6	29 10.5	6 29.8	43
14 18.8	7 57.7	27 14.2	28 37.6	6 15.2	44
14 15.1	7 40.8	26 40.0	28 3.3	6 0.0	45
14 11.3	7 23.5	26 5.2	27 27.8	5 44.0	46
14 7.4	7 5.8	25 29.5	26 50.7	5 27.2	47
14 3.4	6 47.7	24 53.0	26 12.2	5 9.5	48
13 59.3	6 29.1	24 15.6	25 31.9	4 50.9	49
13 55.0	6 10.0	23 37.3	24 49.8	4 31.3	50
13 50.6	5 50.3	22 58.1	24 5.7	4 10.4	51
13 46.1	5 30.2	22 17.8	23 19.5	3 48.3	52
13 41.4	5 9.4	21 36.4	22 30.9	3 24.6	53
13 36.5	4 47.9	20 53.9	21 39.8	2 59.4	54
13 31.4	4 25.8	20 10.2	20 45.9	2 32.2	55
13 26.2	4 3.0	19 25.3	19 49.0	2 2.9	56
13 20.7	3 39.4	18 39.0	18 48.8	1 31.2	57
13 15.0	3 15.0	17 51.4	17 44.8	0 56.5	58
13 9.1	2 49.7	17 2.3	16 36.9	0 18.5	59
13 2.9	2 23.5	16 11.6	15 24.4	♑29 36.5	60
12 56.5	1 56.2	15 19.3	14 7.0	28 49.4	61
12 49.7	1 27.9	14 25.2	12 44.1	27 56.2	62
12 42.5	0 58.4	13 29.4	11 15.1	26 54.9	63
12 35.1	0 27.7	12 31.6	9 39.2	25 42.6	64
12 27.2	♎29 55.6	11 31.8	7 55.6	24 14.0	65
12 18.8	29 22.1	10 29.9	6 3.4	22 17.4	66
5	6	Descendant	8	9	S LAT

♓ 13° 43′ 9″ — 23h 0m 0s — MC 345° 0′ 0″

11^h 4^m 0^s MC 166° 0′ 0″

♍ 14° 47′ 47″

11	12	Ascendant	2	3	N LAT
° ′	° ′	° ′	° ′	° ′	°
♎17 21.4	♏18 27.6	♐17 6.9	♑14 44.4	♒13 32.0	0
17 8.7	17 26.4	15 10.7	13 26.3	13 1.5	5
16 56.0	16 26.3	13 14.8	12 6.1	12 30.1	10
16 43.3	15 26.3	11 17.6	10 42.4	11 57.1	15
16 40.8	15 14.2	10 53.9	10 25.1	11 50.2	16
16 38.2	15 2.1	10 30.1	10 7.5	11 43.2	17
16 35.6	14 49.9	10 6.1	9 49.8	11 36.2	18
16 33.0	14 37.7	9 42.0	9 31.8	11 29.0	19
16 30.3	14 25.4	9 17.7	9 13.6	11 21.7	20
16 27.7	14 13.1	8 53.2	8 55.0	11 14.3	21
16 25.0	14 0.7	8 28.5	8 36.2	11 6.8	22
16 22.3	13 48.2	8 3.7	8 17.0	10 59.1	23
16 19.6	13 35.5	7 38.6	7 57.6	10 51.2	24
16 16.8	13 22.8	7 13.2	7 37.7	10 43.2	25
16 14.1	13 10.0	6 47.6	7 17.5	10 35.0	26
16 11.3	12 57.1	6 21.8	6 56.9	10 26.7	27
16 8.4	12 44.0	5 55.6	6 35.8	10 18.1	28
16 5.5	12 30.8	5 29.2	6 14.3	10 9.3	29
16 2.6	12 17.4	5 2.4	5 52.3	10 0.4	30
15 59.6	12 3.9	4 35.2	5 29.8	9 51.1	31
15 56.6	11 50.2	4 7.7	5 6.8	9 41.6	32
15 53.6	11 36.3	3 39.8	4 43.1	9 31.9	33
15 50.5	11 22.2	3 11.5	4 18.9	9 21.9	34
15 47.3	11 7.9	2 42.8	3 54.0	9 11.5	35
15 44.1	10 53.4	2 13.6	3 28.4	9 0.8	36
15 40.8	10 38.7	1 44.0	3 2.1	8 49.8	37
15 37.4	10 23.7	1 13.8	2 35.0	8 38.4	38
15 34.0	10 8.4	0 43.1	2 7.0	8 26.6	39
15 30.5	9 52.9	0 11.8	1 38.1	8 14.3	40
15 27.0	9 37.0	♏29 39.9	1 8.3	8 1.5	41
15 23.3	9 20.9	29 7.4	0 37.4	7 48.3	42
15 19.6	9 4.4	28 34.3	0 5.5	7 34.5	43
15 15.7	8 47.5	28 0.4	♐29 32.3	7 20.0	44
15 11.8	8 30.3	27 25.9	28 57.9	7 5.0	45
15 7.8	8 12.6	26 50.5	28 22.1	6 49.2	46
15 3.6	7 54.6	26 14.4	27 44.8	6 32.6	47
14 59.3	7 36.0	25 37.4	27 6.0	6 15.1	48
14 54.9	7 17.0	24 59.5	26 25.4	5 56.7	49
14 50.4	6 57.5	24 20.7	25 43.0	5 37.3	50
14 45.7	6 37.5	23 40.9	24 58.5	5 16.7	51
14 40.9	6 16.9	23 0.0	24 11.9	4 54.8	52
14 35.9	5 55.6	22 18.1	23 22.9	4 31.5	53
14 30.7	5 33.7	21 35.0	22 31.3	4 6.5	54
14 25.3	5 11.1	20 50.6	21 36.8	3 39.7	55
14 19.7	4 47.8	20 5.0	20 39.3	3 10.8	56
14 13.9	4 23.7	19 18.1	19 38.3	2 39.4	57
14 7.8	3 58.7	18 29.7	18 33.5	2 5.2	58
14 1.5	3 32.9	17 39.8	17 24.6	1 27.7	59
13 54.9	3 6.0	16 48.3	16 11.1	0 46.2	60
13 48.0	2 38.1	15 55.1	14 52.5	♑29 59.9	61
13 40.7	2 9.1	15 0.2	13 28.2	29 7.5	62
13 33.1	1 39.0	14 3.5	11 57.5	28 7.2	63
13 25.2	1 7.5	13 4.8	10 19.7	26 56.4	64
13 16.7	0 34.7	12 4.0	8 33.9	25 29.9	65
13 7.8	0 0.3	11 1.1	6 39.1	23 37.3	66
5	6	Descendant	8	9	S LAT

♓ 14° 47′ 47″

23^h 4^m 0^s MC 346° 0′ 0″

11^h 8^m 0^s MC 167° 0′ 0″

♍ 15° 52′ 32″

11	12	Ascendant	2	3	N LAT
° ′	° ′	° ′	° ′	° ′	°
♎18 25.8	♏19 27.1	♐18 2.4	♑15 40.1	♒14 32.0	0
18 12.3	18 25.1	16 5.8	14 22.3	14 2.0	5
17 59.0	17 24.1	14 9.3	13 2.3	13 31.1	10
17 45.5	16 23.1	12 11.5	11 38.7	12 58.5	15
17 42.8	16 10.9	11 47.7	11 21.4	12 51.7	16
17 40.0	15 58.6	11 23.7	11 3.9	12 44.9	17
17 37.3	15 46.2	10 59.5	10 46.2	12 37.9	18
17 34.5	15 33.8	10 35.3	10 28.2	12 30.9	19
17 31.7	15 21.3	10 10.8	10 10.0	12 23.7	20
17 28.9	15 8.8	9 46.1	9 51.5	12 16.4	21
17 26.1	14 56.1	9 21.3	9 32.6	12 8.9	22
17 23.2	14 43.4	8 56.2	9 13.5	12 1.3	23
17 20.4	14 30.6	8 30.9	8 54.0	11 53.6	24
17 17.4	14 17.7	8 5.4	8 34.2	11 45.7	25
17 14.5	14 4.6	7 39.6	8 13.9	11 37.6	26
17 11.5	13 51.5	7 13.5	7 53.3	11 29.4	27
17 8.5	13 38.2	6 47.1	7 32.2	11 20.9	28
17 5.4	13 24.7	6 20.4	7 10.7	11 12.3	29
17 2.3	13 11.1	5 53.4	6 48.7	11 3.4	30
16 59.2	12 57.4	5 26.0	6 26.1	10 54.3	31
16 56.0	12 43.4	4 58.2	6 3.0	10 45.0	32
16 52.8	12 29.3	4 30.0	5 39.4	10 35.3	33
16 49.5	12 14.9	4 1.5	5 15.1	10 25.4	34
16 46.1	12 0.4	3 32.4	4 50.1	10 15.2	35
16 42.7	11 45.6	3 2.9	4 24.4	10 4.7	36
16 39.2	11 30.6	2 32.9	3 58.0	9 53.8	37
16 35.7	11 15.3	2 2.4	3 30.8	9 42.5	38
16 32.0	10 59.8	1 31.4	3 2.8	9 30.9	39
16 28.3	10 43.9	0 59.7	2 33.8	9 18.8	40
16 24.5	10 27.8	0 27.5	2 3.8	9 6.2	41
16 20.7	10 11.3	♏29 54.6	1 32.8	8 53.1	42
16 16.7	9 54.5	29 21.1	1 0.7	8 39.4	43
16 12.6	9 37.3	28 46.8	0 27.4	8 25.2	44
16 8.5	9 19.7	28 11.8	♐29 52.8	8 10.3	45
16 4.2	9 1.7	27 36.0	29 16.7	7 54.7	46
15 59.8	8 43.3	26 59.4	28 39.2	7 38.3	47
15 55.2	8 24.4	26 21.9	28 0.1	7 21.1	48
15 50.6	8 5.0	25 43.5	27 19.3	7 3.0	49
15 45.8	7 45.1	25 4.2	26 36.5	6 43.8	50
15 40.8	7 24.6	24 23.8	25 51.7	6 23.4	51
15 35.6	7 3.6	23 42.4	25 4.7	6 1.8	52
15 30.3	6 41.9	22 59.8	24 15.2	5 38.8	53
15 24.8	6 19.5	22 16.1	23 23.1	5 14.1	54
15 19.1	5 56.4	21 31.1	22 28.0	4 47.6	55
15 13.1	5 32.6	20 44.8	21 29.8	4 19.1	56
15 7.0	5 7.9	19 57.1	20 28.1	3 48.2	57
15 0.5	4 42.4	19 8.0	19 22.5	3 14.4	58
14 53.8	4 16.0	18 17.3	18 12.7	2 37.5	59
14 46.8	3 48.5	17 25.0	16 58.1	1 56.6	60
14 39.5	3 20.0	16 31.0	15 38.2	1 10.9	61
14 31.8	2 50.3	15 35.2	14 12.5	0 19.4	62
14 23.7	2 19.5	14 37.6	12 40.1	♑29 20.2	63
14 15.2	1 47.3	13 37.9	11 0.4	28 10.8	64
14 6.2	1 13.6	12 36.1	9 12.4	26 46.6	65
13 56.8	0 38.5	11 32.2	7 14.9	24 57.9	66
5	6	Descendant	8	9	S LAT

♓ 15° 52′ 32″

23^h 8^m 0^s MC 347° 0′ 0″

11h 12m 0s — MC — 168° 0′ 0″

♍ 16° 57′ 21″

11	12	Ascendant	2	3	N LAT
° ′	° ′	° ′	° ′	° ′	°
♎19 30.1	♏20 26.4	♐18 57.9	♑16 36.0	♒15 32.3	0
19 15.9	19 23.6	17 0.9	15 18.4	15 2.8	5
19 1.8	18 21.7	15 3.9	13 58.6	14 32.2	10
18 47.6	17 19.9	13 5.5	12 35.2	14 0.2	15
18 44.7	17 7.4	12 41.5	12 18.0	13 53.5	16
18 41.8	16 54.9	12 17.3	12 0.5	13 46.8	17
18 38.9	16 42.4	11 53.0	11 42.8	13 39.9	18
18 36.0	16 29.8	11 28.6	11 24.8	13 32.9	19
18 33.1	16 17.1	11 3.9	11 6.6	13 25.9	20
18 30.1	16 4.3	10 39.1	10 48.1	13 18.7	21
18 27.1	15 51.5	10 14.1	10 29.3	13 11.3	22
18 24.1	15 38.6	9 48.8	10 10.1	13 3.8	23
18 21.0	15 25.6	9 23.3	9 50.6	12 56.2	24
18 18.0	15 12.4	8 57.6	9 30.8	12 48.4	25
18 14.8	14 59.2	8 31.6	9 10.5	12 40.5	26
18 11.7	14 45.8	8 5.3	8 49.9	12 32.4	27
18 8.5	14 32.3	7 38.7	8 28.8	12 24.0	28
18 5.3	14 18.6	7 11.7	8 7.3	12 15.5	29
18 2.0	14 4.8	6 44.4	7 45.2	12 6.8	30
17 58.7	13 50.8	6 16.8	7 22.6	11 57.8	31
17 55.3	13 36.6	5 48.7	6 59.5	11 48.6	32
17 51.9	13 22.2	5 20.3	6 35.8	11 39.1	33
17 48.4	13 7.6	4 51.4	6 11.5	11 29.3	34
17 44.9	12 52.8	4 22.1	5 46.4	11 19.2	35
17 41.2	12 37.7	3 52.3	5 20.7	11 8.9	36
17 37.6	12 22.4	3 22.0	4 54.2	10 58.1	37
17 33.8	12 6.9	2 51.1	4 26.9	10 47.0	38
17 30.0	11 51.0	2 19.7	3 58.8	10 35.5	39
17 26.1	11 34.9	1 47.8	3 29.7	10 23.5	40
17 22.1	11 18.5	1 15.1	2 59.6	10 11.1	41
17 18.0	11 1.7	0 41.9	2 28.5	9 58.2	42
17 13.8	10 44.5	0 7.9	1 56.2	9 44.8	43
17 9.5	10 27.0	♏29 33.2	1 22.7	9 30.7	44
17 5.1	10 9.1	28 57.8	0 47.9	9 16.0	45
17 0.5	9 50.7	28 21.6	0 11.7	9 0.6	46
16 55.9	9 32.0	27 44.5	♐29 34.0	8 44.5	47
16 51.1	9 12.7	27 6.5	28 54.6	8 27.5	48
16 46.1	8 52.9	26 27.6	28 13.4	8 9.6	49
16 41.1	8 32.6	25 47.7	27 30.4	7 50.6	50
16 35.8	8 11.7	25 6.8	26 45.2	7 30.6	51
16 30.4	7 50.2	24 24.8	25 57.8	7 9.2	52
16 24.7	7 28.1	23 41.6	25 7.8	6 46.5	53
16 18.9	7 5.2	22 57.2	24 15.2	6 22.2	54
16 12.8	6 41.7	22 11.6	23 19.6	5 56.1	55
16 6.6	6 17.3	21 24.6	22 20.8	5 27.9	56
16 0.0	5 52.1	20 36.2	21 18.3	4 57.4	57
15 53.2	5 26.1	19 46.3	20 11.9	4 24.2	58
15 46.1	4 59.0	18 54.9	19 1.1	3 47.7	59
15 38.7	4 31.0	18 1.8	17 45.4	3 7.5	60
15 30.9	4 1.8	17 6.9	16 24.3	2 22.6	61
15 22.8	3 31.5	16 10.3	14 57.1	1 31.9	62
15 14.2	2 59.9	15 11.7	13 23.0	0 33.9	63
15 5.2	2 27.0	14 11.0	11 41.3	♑29 26.0	64
14 55.7	1 52.6	13 8.3	9 50.9	28 3.9	65
14 45.7	1 16.6	12 3.2	7 50.7	26 19.0	66
5	6	Descendant	8	9	S LAT

♓ 16° 57′ 21″

23h 12m 0s — MC — 348° 0′ 0″

11h 16m 0s — MC — 169° 0′ 0″

♍ 18° 2′ 16″

11	12	Ascendant	2	3	N LAT
° ′	° ′	° ′	° ′	° ′	°
♎20 34.3	♏21 25.6	♐19 53.3	♑17 31.9	♒16 32.7	0
20 19.3	20 21.9	17 55.9	16 14.7	16 3.6	5
20 4.5	19 19.2	15 58.5	14 55.1	15 33.6	10
19 49.6	18 16.5	13 59.4	13 31.9	15 2.1	15
19 46.5	18 3.9	13 35.3	13 14.7	14 55.5	16
19 43.5	17 51.2	13 11.0	12 57.2	14 48.9	17
19 40.4	17 38.4	12 46.5	12 39.5	14 42.1	18
19 37.4	17 25.6	12 21.9	12 21.6	14 35.3	19
19 34.3	17 12.8	11 57.1	12 3.4	14 28.3	20
19 31.2	16 59.8	11 32.1	11 44.9	14 21.2	21
19 28.0	16 46.8	11 6.9	11 26.1	14 14.0	22
19 24.8	16 33.7	10 41.5	11 6.9	14 6.6	23
19 21.6	16 20.4	10 15.8	10 47.5	13 59.1	24
19 18.4	16 7.1	9 49.8	10 27.6	13 51.4	25
19 15.1	15 53.6	9 23.6	10 7.4	13 43.6	26
19 11.8	15 40.0	8 57.1	9 46.7	13 35.6	27
19 8.4	15 26.3	8 30.3	9 25.6	13 27.4	28
19 5.0	15 12.4	8 3.1	9 4.1	13 19.0	29
19 1.6	14 58.3	7 35.6	8 42.0	13 10.4	30
18 58.1	14 44.1	7 7.7	8 19.4	13 1.5	31
18 54.5	14 29.7	6 39.4	7 56.2	12 52.5	32
18 50.9	14 15.0	6 10.6	7 32.5	12 43.1	33
18 47.3	14 0.2	5 41.5	7 8.1	12 33.5	34
18 43.5	13 45.1	5 11.9	6 43.0	12 23.6	35
18 39.7	13 29.8	4 41.8	6 17.3	12 13.3	36
18 35.9	13 14.3	4 11.1	5 50.7	12 2.7	37
18 31.9	12 58.4	3 40.0	5 23.3	11 51.8	38
18 27.9	12 42.3	3 8.2	4 55.1	11 40.4	39
18 23.7	12 25.9	2 35.9	4 25.9	11 28.7	40
18 19.5	12 9.1	2 2.9	3 55.7	11 16.4	41
18 15.2	11 52.0	1 29.2	3 24.4	11 3.7	42
18 10.8	11 34.6	0 54.9	2 52.0	10 50.4	43
18 6.3	11 16.7	0 19.8	2 18.4	10 36.6	44
18 1.6	10 58.5	♏29 43.9	1 43.4	10 22.1	45
17 56.8	10 39.8	29 7.2	1 7.0	10 6.9	46
17 51.9	10 20.6	28 29.7	0 29.0	9 51.0	47
17 46.9	10 1.0	27 51.2	♐29 49.4	9 34.2	48
17 41.7	9 40.8	27 11.8	29 8.0	9 16.6	49
17 36.3	9 20.1	26 31.3	28 24.6	8 57.9	50
17 30.8	8 58.8	25 49.9	27 39.1	8 38.1	51
17 25.0	8 36.9	25 7.3	26 51.2	8 17.1	52
17 19.1	8 14.3	24 23.5	26 0.9	7 54.7	53
17 12.9	7 51.0	23 38.5	25 7.7	7 30.7	54
17 6.6	7 26.9	22 52.2	24 11.6	7 5.0	55
16 59.9	7 2.1	22 4.5	23 12.1	6 37.2	56
16 53.0	6 36.3	21 15.4	22 8.9	6 7.2	57
16 45.8	6 9.7	20 24.7	21 1.6	5 34.4	58
16 38.3	5 42.1	19 32.5	19 49.9	4 58.6	59
16 30.5	5 13.4	18 38.6	18 33.1	4 19.0	60
16 22.3	4 43.6	17 42.9	17 10.7	3 34.8	61
16 13.7	4 12.6	16 45.3	15 42.0	2 45.1	62
16 4.7	3 40.3	15 45.7	14 6.2	1 48.2	63
15 55.2	3 6.6	14 44.1	12 22.5	0 41.8	64
15 45.1	2 31.5	13 40.3	10 29.6	♑29 21.9	65
15 34.5	1 54.7	12 34.3	8 26.5	27 40.8	66
5	6	Descendant	8	9	S LAT

♓ 18° 2′ 16″

23h 16m 0s — MC — 349° 0′ 0″

11ʰ 20ᵐ 0ˢ		MC		170° 0′ 0″	
		♍ 19° 7′ 15″			**N**
11	**12**	**Ascendant**	**2**	**3**	**LAT**
° ′	° ′	° ′	° ′	° ′	°
♎21 38.3	♏22 24.6	♐20 48.6	♑18 27.9	♒17 33.3	**0**
21 22.7	21 20.1	18 50.9	17 11.0	17 4.7	**5**
21 7.1	20 16.6	16 53.0	15 51.7	16 35.2	**10**
20 51.4	19 13.0	14 53.4	14 28.7	16 4.2	**15**
20 48.3	19 0.2	14 29.1	14 11.5	15 57.7	**16**
20 45.1	18 47.3	14 4.7	13 54.1	15 51.2	**17**
20 41.9	18 34.4	13 40.1	13 36.4	15 44.6	**18**
20 38.6	18 21.4	13 15.3	13 18.5	15 37.8	**19**
20 35.4	18 8.4	12 50.3	13 0.4	15 31.0	**20**
20 32.1	17 55.2	12 25.2	12 41.9	15 24.0	**21**
20 28.8	17 42.0	11 59.8	12 23.1	15 16.9	**22**
20 25.5	17 28.7	11 34.2	12 4.0	15 9.6	**23**
20 22.1	17 15.3	11 8.3	11 44.5	15 2.2	**24**
20 18.7	17 1.7	10 42.2	11 24.7	14 54.7	**25**
20 15.3	16 48.0	10 15.7	11 4.4	14 47.0	**26**
20 11.8	16 34.2	9 49.0	10 43.8	14 39.1	**27**
20 8.3	16 20.3	9 21.9	10 22.7	14 31.0	**28**
20 4.7	16 6.1	8 54.5	10 1.1	14 22.8	**29**
20 1.1	15 51.8	8 26.8	9 39.0	14 14.3	**30**
19 57.4	15 37.4	7 58.6	9 16.4	14 5.6	**31**
19 53.7	15 22.7	7 30.0	8 53.2	13 56.6	**32**
19 49.9	15 7.8	7 1.1	8 29.4	13 47.4	**33**
19 46.1	14 52.8	6 31.6	8 5.0	13 37.9	**34**
19 42.1	14 37.4	6 1.7	7 39.9	13 28.2	**35**
19 38.1	14 21.9	5 31.3	7 14.1	13 18.1	**36**
19 34.1	14 6.0	5 0.4	6 47.4	13 7.7	**37**
19 29.9	13 49.9	4 28.9	6 20.0	12 56.9	**38**
19 25.7	13 33.5	3 56.8	5 51.7	12 45.7	**39**
19 21.4	13 16.8	3 24.1	5 22.4	12 34.1	**40**
19 16.9	12 59.8	2 50.7	4 52.1	12 22.1	**41**
19 12.4	12 42.4	2 16.7	4 20.7	12 9.5	**42**
19 7.8	12 24.6	1 41.9	3 48.2	11 56.4	**43**
19 3.0	12 6.4	1 6.4	3 14.4	11 42.8	**44**
18 58.1	11 47.8	0 30.1	2 39.2	11 28.5	**45**
18 53.1	11 28.8	♏29 52.9	2 2.6	11 13.6	**46**
18 47.9	11 9.3	29 14.9	1 24.4	10 57.9	**47**
18 42.6	10 49.2	28 35.9	0 44.6	10 41.4	**48**
18 37.1	10 28.7	27 56.0	0 2.9	10 24.0	**49**
18 31.5	10 7.6	27 15.0	♐29 19.2	10 5.6	**50**
18 25.7	9 45.8	26 33.0	28 33.3	9 46.1	**51**
18 19.6	9 23.5	25 49.8	27 45.1	9 25.4	**52**
18 13.4	9 0.4	25 5.4	26 54.3	9 3.3	**53**
18 6.9	8 36.7	24 19.8	26 0.6	8 39.7	**54**
18 0.2	8 12.1	23 32.8	25 3.9	8 14.3	**55**
17 53.2	7 46.8	22 44.4	24 3.8	7 47.0	**56**
17 46.0	7 20.5	21 54.6	22 59.8	7 17.5	**57**
17 38.4	6 53.3	21 3.2	21 51.8	6 45.3	**58**
17 30.5	6 25.1	20 10.1	20 39.0	6 10.0	**59**
17 22.3	5 55.8	19 15.4	19 21.1	5 31.0	**60**
17 13.6	5 25.4	18 18.8	17 57.4	4 47.7	**61**
17 4.6	4 53.7	17 20.3	16 27.2	3 58.9	**62**
16 55.1	4 20.7	16 19.8	14 49.7	3 3.2	**63**
16 45.1	3 46.3	15 17.2	13 3.8	1 58.3	**64**
16 34.5	3 10.3	14 12.4	11 8.5	0 40.6	**65**
16 23.3	2 32.7	13 5.2	9 2.4	♑29 3.1	**66**
5	**6**	**Descendant**	**8**	**9**	**S**
		♓ 19° 7′ 15″			**LAT**
23ʰ 20ᵐ 0ˢ		MC		350° 0′ 0″	

11ʰ 24ᵐ 0ˢ		MC		171° 0′ 0″	
		♍ 20° 12′ 19″			**N**
11	**12**	**Ascendant**	**2**	**3**	**LAT**
° ′	° ′	° ′	° ′	° ′	°
♎22 42.3	♏23 23.4	♐21 43.9	♑19 24.1	♒18 34.0	**0**
22 25.9	22 18.2	19 45.9	18 7.5	18 6.0	**5**
22 9.6	21 13.8	17 47.6	16 48.5	17 37.0	**10**
21 53.2	20 9.4	15 47.4	15 25.7	17 6.5	**15**
21 49.9	19 56.4	15 22.9	15 8.5	17 0.2	**16**
21 46.6	19 43.4	14 58.4	14 51.2	16 53.8	**17**
21 43.2	19 30.3	14 33.6	14 33.5	16 47.2	**18**
21 39.8	19 17.1	14 8.7	14 15.7	16 40.6	**19**
21 36.4	19 3.9	13 43.6	13 57.5	16 33.9	**20**
21 33.0	18 50.6	13 18.3	13 39.0	16 27.0	**21**
21 29.5	18 37.1	12 52.7	13 20.3	16 20.0	**22**
21 26.1	18 23.6	12 26.9	13 1.2	16 12.9	**23**
21 22.5	18 10.0	12 0.9	12 41.7	16 5.6	**24**
21 19.0	17 56.2	11 34.5	12 21.9	15 58.2	**25**
21 15.4	17 42.3	11 7.9	12 1.7	15 50.6	**26**
21 11.7	17 28.3	10 40.9	11 41.0	15 42.9	**27**
21 8.0	17 14.1	10 13.7	11 19.9	15 34.9	**28**
21 4.3	16 59.8	9 46.0	10 58.3	15 26.8	**29**
21 0.5	16 45.3	9 18.0	10 36.3	15 18.5	**30**
20 56.7	16 30.6	8 49.6	10 13.6	15 9.9	**31**
20 52.8	16 15.7	8 20.8	9 50.4	15 1.1	**32**
20 48.8	16 0.6	7 51.6	9 26.6	14 52.0	**33**
20 44.8	15 45.3	7 21.8	9 2.2	14 42.7	**34**
20 40.7	15 29.7	6 51.6	8 37.0	14 33.1	**35**
20 36.5	15 13.9	6 20.9	8 11.1	14 23.2	**36**
20 32.2	14 57.8	5 49.7	7 44.5	14 12.9	**37**
20 27.9	14 41.4	5 17.8	7 16.9	14 2.3	**38**
20 23.4	14 24.7	4 45.4	6 48.5	13 51.3	**39**
20 18.9	14 7.7	4 12.4	6 19.2	13 39.9	**40**
20 14.3	13 50.4	3 38.6	5 48.8	13 28.0	**41**
20 9.5	13 32.6	3 4.2	5 17.3	13 15.7	**42**
20 4.6	13 14.6	2 29.1	4 44.6	13 2.8	**43**
19 59.7	12 56.1	1 53.1	4 10.7	12 49.4	**44**
19 54.5	12 37.1	1 16.4	3 35.3	12 35.3	**45**
19 49.3	12 17.7	0 38.8	2 58.6	12 20.6	**46**
19 43.9	11 57.9	0 0.3	2 20.2	12 5.2	**47**
19 38.3	11 37.5	♏29 20.8	1 40.1	11 48.9	**48**
19 32.6	11 16.5	28 40.4	0 58.1	11 31.8	**49**
19 26.6	10 55.0	27 58.8	0 14.1	11 13.7	**50**
19 20.5	10 32.9	27 16.2	♐29 27.9	10 54.5	**51**
19 14.2	10 10.1	26 32.5	28 39.3	10 34.1	**52**
19 7.6	9 46.6	25 47.5	27 48.1	10 12.4	**53**
19 0.9	9 22.4	25 1.2	26 53.9	9 49.1	**54**
18 53.8	8 57.3	24 13.5	25 56.7	9 24.2	**55**
18 46.5	8 31.4	23 24.4	24 55.9	8 57.3	**56**
18 38.9	8 4.7	22 33.8	23 51.2	8 28.2	**57**
18 30.9	7 36.9	21 41.7	22 42.3	7 56.6	**58**
18 22.7	7 8.1	20 47.8	21 28.6	7 21.9	**59**
18 14.0	6 38.2	19 52.2	20 9.6	6 43.7	**60**
18 4.9	6 7.1	18 54.8	18 44.6	6 1.1	**61**
17 55.4	5 34.8	17 55.4	17 12.8	5 13.3	**62**
17 45.4	5 1.0	16 53.9	15 33.5	4 18.8	**63**
17 34.9	4 25.8	15 50.3	13 45.4	3 15.5	**64**
17 23.8	3 49.1	14 44.4	11 47.5	2 0.0	**65**
17 12.1	3 10.6	13 36.2	9 38.3	0 26.0	**66**
5	**6**	**Descendant**	**8**	**9**	**S**
		♓ 20° 12′ 19″			**LAT**
23ʰ 24ᵐ 0ˢ		MC		351° 0′ 0″	

$11^{h}\ 28^{m}\ 0^{s}$ MC 172° 0′ 0″

♍ 21° 17′ 27″

N LAT	11	12	Ascendant	2	3
0°	♎23° 46.0′	♏24° 22.0′	♐22° 39.2′	♑20° 20.3′	♒19° 35.0′
5	23 29.0	23 16.1	20 40.9	19 4.1	19 7.5
10	23 12.0	22 11.0	18 42.1	17 45.4	18 39.0
15	22 54.9	21 5.7	16 41.4	16 22.8	18 9.1
16	22 51.4	20 52.5	16 16.8	16 5.7	18 2.9
17	22 47.9	20 39.3	15 52.1	15 48.4	17 56.5
18	22 44.4	20 26.1	15 27.2	15 30.8	17 50.1
19	22 40.9	20 12.7	15 2.2	15 13.0	17 43.6
20	22 37.3	19 59.3	14 36.9	14 54.8	17 37.0
21	22 33.8	19 45.8	14 11.4	14 36.4	17 30.3
22	22 30.2	19 32.2	13 45.7	14 17.7	17 23.4
23	22 26.5	19 18.5	13 19.7	13 58.6	17 16.4
24	22 22.8	19 4.6	12 53.5	13 39.2	17 9.3
25	22 19.1	18 50.7	12 27.0	13 19.4	17 2.0
26	22 15.4	18 36.6	12 0.1	12 59.2	16 54.5
27	22 11.6	18 22.4	11 33.0	12 38.5	16 46.9
28	22 7.7	18 8.0	11 5.5	12 17.4	16 39.1
29	22 3.8	17 53.4	10 37.6	11 55.8	16 31.1
30	21 59.8	17 38.7	10 9.4	11 33.8	16 22.9
31	21 55.8	17 23.8	9 40.7	11 11.1	16 14.5
32	21 51.8	17 8.6	9 11.7	10 47.9	16 5.8
33	21 47.6	16 53.3	8 42.2	10 24.1	15 56.9
34	21 43.4	16 37.7	8 12.2	9 59.6	15 47.8
35	21 39.1	16 21.9	7 41.7	9 34.4	15 38.3
36	21 34.7	16 5.8	7 10.7	9 8.5	15 28.5
37	21 30.3	15 49.5	6 39.1	8 41.8	15 18.5
38	21 25.7	15 32.8	6 6.9	8 14.2	15 8.0
39	21 21.1	15 15.9	5 34.2	7 45.7	14 57.2
40	21 16.4	14 58.6	5 0.8	7 16.3	14 46.0
41	21 11.5	14 40.9	4 26.7	6 45.8	14 34.3
42	21 6.5	14 22.9	3 51.9	6 14.2	14 22.2
43	21 1.5	14 4.5	3 16.3	5 41.4	14 9.5
44	20 56.2	13 45.7	2 39.9	5 7.3	13 56.3
45	20 50.9	13 26.4	2 2.8	4 31.8	13 42.5
46	20 45.4	13 6.7	1 24.7	3 54.9	13 28.0
47	20 39.7	12 46.5	0 45.7	3 16.3	13 12.8
48	20 33.9	12 25.7	0 5.8	2 36.0	12 56.9
49	20 27.9	12 4.4	♏29 24.8	1 53.7	12 40.0
50	20 21.7	11 42.5	28 42.7	1 9.5	12 22.2
51	20 15.3	11 19.9	27 59.6	0 22.9	12 3.3
52	20 8.7	10 56.7	27 15.2	♐29 33.9	11 43.3
53	20 1.8	10 32.7	26 29.6	28 42.3	11 21.9
54	19 54.7	10 8.0	25 42.6	27 47.7	10 59.1
55	19 47.4	9 42.5	24 54.3	26 49.8	10 34.5
56	19 39.7	9 16.1	24 4.5	25 48.4	10 8.1
57	19 31.7	8 48.8	23 13.2	24 43.0	9 39.5
58	19 23.4	8 20.5	22 20.2	23 33.3	9 8.4
59	19 14.7	7 51.1	21 25.6	22 18.6	8 34.4
60	19 5.7	7 20.6	20 29.1	20 58.4	7 56.9
61	18 56.2	6 48.8	19 30.8	19 32.1	7 15.2
62	18 46.2	6 15.8	18 30.4	17 58.8	6 28.3
63	18 35.7	5 41.3	17 28.0	16 17.6	5 35.0
64	18 24.7	5 5.4	16 23.4	14 27.3	4 33.3
65	18 13.1	4 27.8	15 16.5	12 26.8	3 20.0
66	18 0.8	3 48.5	14 7.1	10 14.3	1 49.4
S LAT	5	6	Descendant	8	9

♓ 21° 17′ 27″

$23^{h}\ 28^{m}\ 0^{s}$ MC 352° 0′ 0″

$11^{h}\ 32^{m}\ 0^{s}$ MC 173° 0′ 0″

♍ 22° 22′ 38″

N LAT	11	12	Ascendant	2	3
0°	♎24° 49.7′	♏25° 20.5′	♐23° 34.4′	♑21° 16.7′	♒20° 36.1′
5	24 31.9	24 13.8	21 35.8	20 0.8	20 9.2
10	24 14.2	23 8.0	19 36.7	18 42.4	19 41.3
15	23 56.4	22 1.9	17 35.4	17 20.1	19 11.9
16	23 52.8	21 48.6	17 10.7	17 3.1	19 5.8
17	23 49.2	21 35.2	16 45.9	16 45.8	18 59.6
18	23 45.5	21 21.7	16 20.9	16 28.3	18 53.3
19	23 41.9	21 8.2	15 55.7	16 10.5	18 46.9
20	23 38.2	20 54.6	15 30.3	15 52.4	18 40.4
21	23 34.4	20 40.9	15 4.6	15 34.0	18 33.8
22	23 30.7	20 27.1	14 38.7	15 15.3	18 27.0
23	23 26.9	20 13.2	14 12.6	14 56.2	18 20.1
24	23 23.0	19 59.2	13 46.2	14 36.8	18 13.1
25	23 19.2	19 45.1	13 19.5	14 17.1	18 6.0
26	23 15.3	19 30.8	12 52.4	13 56.9	17 58.7
27	23 11.3	19 16.3	12 25.1	13 36.2	17 51.2
28	23 7.3	19 1.7	11 57.4	13 15.2	17 43.5
29	23 3.2	18 47.0	11 29.3	12 53.6	17 35.7
30	22 59.1	18 32.0	11 0.8	12 31.5	17 27.6
31	22 54.9	18 16.9	10 31.9	12 8.9	17 19.4
32	22 50.7	18 1.5	10 2.6	11 45.7	17 10.9
33	22 46.3	17 46.0	9 32.8	11 21.8	17 2.1
34	22 41.9	17 30.1	9 2.6	10 57.3	16 53.1
35	22 37.5	17 14.1	8 31.8	10 32.1	16 43.8
36	22 32.9	16 57.7	8 0.5	10 6.1	16 34.2
37	22 28.3	16 41.1	7 28.6	9 39.4	16 24.3
38	22 23.5	16 24.2	6 56.1	9 11.8	16 14.1
39	22 18.7	16 7.0	6 23.0	8 43.2	16 3.4
40	22 13.7	15 49.4	5 49.3	8 13.7	15 52.4
41	22 8.7	15 31.5	5 14.8	7 43.1	15 41.0
42	22 3.5	15 13.2	4 39.6	7 11.4	15 29.0
43	21 58.2	14 54.4	4 3.7	6 38.5	15 16.6
44	21 52.8	14 35.3	3 26.9	6 4.3	15 3.6
45	21 47.2	14 15.7	2 49.3	5 28.7	14 50.0
46	21 41.4	13 55.6	2 10.7	4 51.6	14 35.8
47	21 35.5	13 35.0	1 31.3	4 12.8	14 20.9
48	21 29.4	13 13.9	0 50.8	3 32.3	14 5.2
49	21 23.2	12 52.2	0 9.3	2 49.8	13 48.6
50	21 16.7	12 29.9	♏29 26.7	2 5.2	13 31.1
51	21 10.0	12 6.9	28 43.0	1 18.4	13 12.6
52	21 3.1	11 43.3	27 58.0	0 29.0	12 52.9
53	20 56.0	11 18.9	27 11.8	♐29 36.9	12 31.9
54	20 48.6	10 53.7	26 24.2	28 41.8	12 9.4
55	20 40.9	10 27.7	25 35.2	27 43.4	11 45.4
56	20 32.8	10 0.8	24 44.6	26 41.4	11 19.4
57	20 24.5	9 32.9	23 52.6	25 35.3	10 51.4
58	20 15.8	9 4.0	22 58.8	24 24.7	10 20.8
59	20 6.8	8 34.0	22 3.3	23 9.0	9 47.4
60	19 57.3	8 2.9	21 6.0	21 47.7	9 10.7
61	19 47.3	7 30.5	20 6.8	20 20.0	8 29.8
62	19 36.9	6 56.8	19 5.5	18 45.1	7 44.0
63	19 26.0	6 21.6	18 2.1	17 2.0	6 51.9
64	19 14.4	5 44.9	16 56.5	15 9.5	5 51.8
65	19 2.3	5 6.5	15 48.4	13 6.2	4 40.6
66	18 49.4	4 26.3	14 38.0	10 50.3	3 13.3
S LAT	5	6	Descendant	8	9

♓ 22° 22′ 38″

$23^{h}\ 32^{m}\ 0^{s}$ MC 353° 0′ 0″

11h 36m 0s		MC		174° 0′ 0″	
		♍ 23° 27′ 53″			
11	12	Ascendant	2	3	N LAT
° ′	° ′	° ′	° ′	° ′	°
♎25 53.2	♏26 18.8	♐24 29.5	♑22 13.1	♒21 37.4	0
25 34.7	25 11.4	22 30.8	20 57.6	21 11.0	5
25 16.4	24 4.8	20 31.3	19 39.6	20 43.7	10
24 57.8	22 58.0	18 29.5	18 17.6	20 14.9	15
24 54.1	22 44.5	18 4.7	18 0.6	20 8.9	16
24 50.3	22 30.9	17 39.7	17 43.4	20 2.8	17
24 46.5	22 17.3	17 14.6	17 25.9	19 56.6	18
24 42.7	22 3.6	16 49.2	17 8.2	19 50.4	19
24 38.9	21 49.8	16 23.7	16 50.1	19 44.0	20
24 35.0	21 36.0	15 57.9	16 31.8	19 37.5	21
24 31.1	21 22.0	15 31.8	16 13.1	19 30.9	22
24 27.1	21 7.9	15 5.5	15 54.1	19 24.2	23
24 23.2	20 53.7	14 38.9	15 34.7	19 17.3	24
24 19.1	20 39.4	14 12.0	15 15.0	19 10.3	25
24 15.1	20 24.9	13 44.8	14 54.8	19 3.1	26
24 10.9	20 10.3	13 17.3	14 34.2	18 55.7	27
24 6.8	19 55.5	12 49.3	14 13.1	18 48.2	28
24 2.5	19 40.5	12 21.0	13 51.6	18 40.5	29
23 58.2	19 25.3	11 52.4	13 29.5	18 32.6	30
23 53.9	19 9.9	11 23.2	13 6.9	18 24.5	31
23 49.5	18 54.4	10 53.7	12 43.7	18 16.2	32
23 45.0	18 38.6	10 23.6	12 19.8	18 7.6	33
23 40.4	18 22.5	9 53.1	11 55.3	17 58.8	34
23 35.7	18 6.2	9 22.0	11 30.1	17 49.6	35
23 31.0	17 49.6	8 50.4	11 4.1	17 40.2	36
23 26.2	17 32.7	8 18.2	10 37.3	17 30.5	37
23 21.2	17 15.6	7 45.4	10 9.6	17 20.4	38
23 16.2	16 58.1	7 12.0	9 41.0	17 10.0	39
23 11.1	16 40.2	6 37.9	9 11.5	16 59.2	40
23 5.8	16 22.0	6 3.1	8 40.8	16 47.9	41
23 0.4	16 3.4	5 27.5	8 9.0	16 36.2	42
22 54.9	15 44.3	4 51.1	7 36.0	16 24.0	43
22 49.2	15 24.9	4 13.9	7 1.7	16 11.3	44
22 43.4	15 5.0	3 35.9	6 25.9	15 57.9	45
22 37.4	14 44.5	2 56.9	5 48.6	15 44.0	46
22 31.3	14 23.6	2 17.0	5 9.7	15 29.3	47
22 24.9	14 2.1	1 36.0	4 28.9	15 13.9	48
22 18.4	13 40.0	0 54.0	3 46.2	14 57.6	49
22 11.7	13 17.3	0 10.8	3 1.4	14 40.5	50
22 4.7	12 53.9	♏29 26.5	2 14.2	14 22.2	51
21 57.5	12 29.8	28 40.9	1 24.5	14 2.9	52
21 50.0	12 5.0	27 54.1	0 32.0	13 42.3	53
21 42.3	11 39.3	27 5.8	♐29 36.4	13 20.3	54
21 34.3	11 12.8	26 16.1	28 37.5	12 56.6	55
21 25.9	10 45.4	25 24.9	27 34.8	12 31.2	56
21 17.2	10 17.0	24 32.0	26 28.0	12 3.7	57
21 8.2	9 47.5	23 37.5	25 16.6	11 33.7	58
20 58.7	9 17.0	22 41.2	23 59.9	11 1.0	59
20 48.8	8 45.2	21 43.0	22 37.5	10 25.0	60
20 38.5	8 12.2	20 42.8	21 8.4	9 45.0	61
20 27.6	7 37.7	19 40.6	19 31.9	9 0.2	62
20 16.1	7 1.8	18 36.2	17 46.9	8 9.4	63
20 4.1	6 24.3	17 29.5	15 52.0	7 10.9	64
19 51.4	5 45.1	16 20.4	13 45.8	6 1.9	65
19 37.9	5 4.1	15 8.8	11 26.4	4 37.8	66
5	6	Descendant	8	9	S LAT
		♓ 23° 27′ 53″			
23h 36m 0s		MC		354° 0′ 0″	

11h 40m 0s		MC		175° 0′ 0″	
		♍ 24° 33′ 10″			
11	12	Ascendant	2	3	N LAT
° ′	° ′	° ′	° ′	° ′	°
♎26 56.5	♏27 17.0	♐25 24.7	♑23 9.7	♒22 38.9	0
26 37.4	26 8.9	23 25.7	21 54.6	22 13.1	5
26 18.3	25 1.6	21 25.9	20 37.0	21 46.3	10
25 59.1	23 54.0	19 23.6	19 15.3	21 18.1	15
25 55.3	23 40.3	18 58.7	18 58.4	21 12.2	16
25 51.3	23 26.6	18 33.6	18 41.2	21 6.3	17
25 47.4	23 12.8	18 8.3	18 23.8	21 0.2	18
25 43.5	22 58.9	17 42.9	18 6.1	20 54.1	19
25 39.5	22 45.0	17 17.2	17 48.1	20 47.8	20
25 35.4	22 31.0	16 51.2	17 29.8	20 41.5	21
25 31.4	22 16.8	16 25.0	17 11.1	20 35.0	22
25 27.3	22 2.5	15 58.5	16 52.2	20 28.4	23
25 23.2	21 48.1	15 31.8	16 32.8	20 21.7	24
25 19.0	21 33.6	15 4.7	16 13.1	20 14.8	25
25 14.8	21 18.9	14 37.3	15 53.0	20 7.8	26
25 10.5	21 4.1	14 9.5	15 32.4	20 0.6	27
25 6.1	20 49.1	13 41.4	15 11.4	19 53.2	28
25 1.7	20 33.9	13 12.9	14 49.8	19 45.6	29
24 57.3	20 18.5	12 44.0	14 27.8	19 37.9	30
24 52.8	20 3.0	12 14.6	14 5.2	19 29.9	31
24 48.2	19 47.2	11 44.8	13 42.0	19 21.8	32
24 43.5	19 31.1	11 14.5	13 18.1	19 13.4	33
24 38.8	19 14.8	10 43.7	12 53.6	19 4.7	34
24 33.9	18 58.3	10 12.4	12 28.4	18 55.8	35
24 29.0	18 41.5	9 40.5	12 2.4	18 46.5	36
24 24.0	18 24.3	9 8.0	11 35.5	18 37.0	37
24 18.9	18 6.9	8 34.9	11 7.8	18 27.1	38
24 13.6	17 49.1	8 1.1	10 39.2	18 16.9	39
24 8.3	17 31.0	7 26.6	10 9.6	18 6.3	40
24 2.8	17 12.5	6 51.4	9 38.8	17 55.2	41
23 57.2	16 53.6	6 15.5	9 7.0	17 43.8	42
23 51.5	16 34.2	5 38.7	8 33.9	17 31.8	43
23 45.6	16 14.5	5 1.1	7 59.4	17 19.3	44
23 39.5	15 54.2	4 22.6	7 23.5	17 6.2	45
23 33.3	15 33.4	3 43.2	6 46.1	16 52.5	46
23 26.9	15 12.1	3 2.8	6 7.0	16 38.1	47
23 20.3	14 50.3	2 21.3	5 26.0	16 23.0	48
23 13.5	14 27.8	1 38.7	4 43.1	16 7.1	49
23 6.5	14 4.7	0 55.1	3 58.0	15 50.2	50
22 59.3	13 40.9	0 10.1	3 10.5	15 32.3	51
22 51.8	13 16.4	♏29 24.0	2 20.4	15 13.4	52
22 44.0	12 51.1	28 36.5	1 27.5	14 53.2	53
22 36.0	12 25.0	27 47.5	0 31.5	14 31.6	54
22 27.6	11 58.0	26 57.1	♐29 32.1	14 8.4	55
22 19.0	11 30.0	26 5.2	28 28.8	13 43.5	56
22 9.9	11 1.1	25 11.6	27 21.2	13 16.5	57
22 0.5	10 31.0	24 16.2	26 8.9	12 47.2	58
21 50.6	9 59.9	23 19.1	24 51.3	12 15.1	59
21 40.3	9 27.5	22 20.0	23 27.7	11 39.9	60
21 29.5	8 53.8	21 18.9	21 57.3	11 0.8	61
21 18.2	8 18.7	20 15.7	20 19.1	10 17.1	62
21 6.3	7 42.0	19 10.3	18 32.1	9 27.6	63
20 53.7	7 3.7	18 2.6	16 34.9	8 30.6	64
20 40.5	6 23.7	16 52.4	14 25.7	7 23.8	65
20 26.4	5 41.8	15 39.6	12 2.6	6 2.8	66
5	6	Descendant	8	9	S LAT
		♓ 24° 33′ 10″			
23h 40m 0s		MC		355° 0′ 0″	

11h 44m 0s		MC		176° 0′ 0″
		♍ 25° 38′ 29″		

N LAT	11	12	Ascendant	2	3
	° ′	° ′	° ′	° ′	° ′
0	♎27 59.7	♏28 15.0	♐26 19.8	♑24 6.4	♒23 40.6
5	27 39.9	27 6.3	24 20.6	22 51.8	23 15.3
10	27 20.2	25 58.2	22 20.5	21 34.5	22 49.1
15	27 0.3	24 49.8	20 17.7	20 13.2	22 21.5
16	26 56.3	24 36.0	19 52.7	19 56.3	22 15.8
17	26 52.2	24 22.2	19 27.5	19 39.2	22 10.0
18	26 48.2	24 8.2	19 2.1	19 21.8	22 4.1
19	26 44.1	23 54.2	18 36.5	19 4.2	21 58.0
20	26 39.9	23 40.1	18 10.7	18 46.2	21 51.9
21	26 35.8	23 25.9	17 44.6	18 28.0	21 45.7
22	26 31.6	23 11.5	17 18.3	18 9.4	21 39.4
23	26 27.3	22 57.1	16 51.6	17 50.5	21 32.9
24	26 23.1	22 42.5	16 24.7	17 31.2	21 26.3
25	26 18.7	22 27.8	15 57.4	17 11.5	21 19.6
26	26 14.3	22 12.9	15 29.9	16 51.4	21 12.7
27	26 9.9	21 57.9	15 1.9	16 30.9	21 5.6
28	26 5.4	21 42.7	14 33.6	16 9.9	20 58.4
29	26 0.9	21 27.3	14 4.9	15 48.4	20 51.0
30	25 56.2	21 11.7	13 35.7	15 26.3	20 43.4
31	25 51.6	20 55.9	13 6.1	15 3.7	20 35.7
32	25 46.8	20 39.9	12 36.1	14 40.5	20 27.7
33	25 42.0	20 23.6	12 5.5	14 16.7	20 19.4
34	25 37.0	20 7.1	11 34.5	13 52.2	20 10.9
35	25 32.0	19 50.3	11 2.9	13 26.9	20 2.2
36	25 26.9	19 33.3	10 30.7	13 0.9	19 53.1
37	25 21.7	19 15.9	9 57.9	12 34.1	19 43.8
38	25 16.4	18 58.2	9 24.4	12 6.4	19 34.1
39	25 11.0	18 40.1	8 50.3	11 37.7	19 24.1
40	25 5.4	18 21.7	8 15.5	11 8.0	19 13.7
41	24 59.8	18 2.9	7 39.9	10 37.2	19 2.9
42	24 53.9	17 43.7	7 3.6	10 5.3	18 51.6
43	24 48.0	17 24.1	6 26.4	9 32.1	18 39.9
44	24 41.9	17 4.0	5 48.4	8 57.6	18 27.6
45	24 35.6	16 43.4	5 9.5	8 21.6	18 14.8
46	24 29.2	16 22.3	4 29.6	7 44.0	18 1.4
47	24 22.5	16 0.7	3 48.7	7 4.7	17 47.3
48	24 15.7	15 38.4	3 6.7	6 23.5	17 32.5
49	24 8.6	15 15.6	2 23.6	5 40.4	17 16.9
50	24 1.4	14 52.1	1 39.4	4 55.0	17 0.4
51	23 53.8	14 27.9	0 53.9	4 7.3	16 42.9
52	23 46.1	14 2.9	0 7.1	3 16.9	16 24.3
53	23 38.0	13 37.2	♏29 19.0	2 23.6	16 4.5
54	23 29.6	13 10.6	28 29.4	1 27.1	15 43.3
55	23 20.9	12 43.1	27 38.3	0 27.1	15 20.7
56	23 11.9	12 14.6	26 45.5	♐29 23.2	14 56.2
57	23 2.5	11 45.1	25 51.2	28 14.9	14 29.8
58	22 52.7	11 14.5	24 55.0	27 1.8	14 1.1
59	22 42.5	10 42.8	23 57.0	25 43.2	13 29.8
60	22 31.7	10 9.8	22 57.1	24 18.4	12 55.4
61	22 20.5	9 35.4	21 55.0	22 46.6	12 17.2
62	22 8.7	8 59.6	20 50.9	21 6.8	11 34.5
63	21 56.3	8 22.2	19 44.4	19 17.8	10 46.3
64	21 43.2	7 43.1	18 35.6	17 18.1	9 51.0
65	21 29.4	7 2.2	17 24.3	15 5.8	8 46.2
66	21 14.8	6 19.4	16 10.4	12 38.9	7 28.2
S LAT	5	6	Descendant	8	9

		♓ 25° 38′ 29″		
23h 44m 0s		MC		356° 0′ 0″

11h 48m 0s		MC		177° 0′ 0″
		♍ 26° 43′ 50″		

N LAT	11	12	Ascendant	2	3
	° ′	° ′	° ′	° ′	° ′
0	♎29 2.8	♏29 12.8	♐27 14.8	♑25 3.3	♒24 42.5
5	28 42.3	28 3.4	25 15.5	23 49.1	24 17.8
10	28 21.9	26 54.7	23 15.1	22 32.2	23 52.2
15	28 1.4	25 45.6	21 11.9	21 11.3	23 25.2
16	27 57.2	25 31.7	20 46.8	20 54.4	23 19.6
17	27 53.0	25 17.6	20 21.5	20 37.4	23 13.9
18	27 48.8	25 3.5	19 56.0	20 20.1	23 8.1
19	27 44.6	24 49.3	19 30.3	20 2.5	23 2.2
20	27 40.3	24 35.1	19 4.3	19 44.6	22 56.2
21	27 36.0	24 20.7	18 38.1	19 26.4	22 50.2
22	27 31.7	24 6.2	18 11.6	19 7.9	22 44.0
23	27 27.3	23 51.6	17 44.8	18 49.0	22 37.6
24	27 22.8	23 36.8	17 17.7	18 29.8	22 31.2
25	27 18.4	23 21.9	16 50.3	18 10.2	22 24.6
26	27 13.8	23 6.9	16 22.5	17 50.1	22 17.9
27	27 9.2	22 51.6	15 54.4	17 29.6	22 11.0
28	27 4.6	22 36.2	15 25.9	17 8.7	22 3.9
29	26 59.9	22 20.7	14 56.9	16 47.2	21 56.7
30	26 55.1	22 4.9	14 27.6	16 25.2	21 49.3
31	26 50.3	21 48.9	13 57.8	16 2.6	21 41.6
32	26 45.3	21 32.6	13 27.5	15 39.4	21 33.8
33	26 40.3	21 16.1	12 56.7	15 15.6	21 25.8
34	26 35.2	20 59.4	12 25.3	14 51.1	21 17.4
35	26 30.0	20 42.4	11 53.5	14 25.9	21 8.9
36	26 24.8	20 25.0	11 21.0	13 59.8	21 0.0
37	26 19.4	20 7.4	10 47.9	13 33.0	20 50.9
38	26 13.9	19 49.5	10 14.1	13 5.2	20 41.4
39	26 8.2	19 31.1	9 39.7	12 36.5	20 31.6
40	26 2.5	19 12.5	9 4.5	12 6.8	20 21.4
41	25 56.6	18 53.4	8 28.6	11 36.0	20 10.9
42	25 50.6	18 33.9	7 51.9	11 4.0	19 59.8
43	25 44.4	18 14.0	7 14.3	10 30.7	19 48.4
44	25 38.1	17 53.5	6 35.9	9 56.1	19 36.4
45	25 31.6	17 32.6	5 56.5	9 20.0	19 23.8
46	25 24.9	17 11.2	5 16.1	8 42.3	19 10.7
47	25 18.0	16 49.2	4 34.7	8 2.8	18 56.9
48	25 10.9	16 26.6	3 52.3	7 21.5	18 42.4
49	25 3.6	16 3.4	3 8.6	6 38.1	18 27.1
50	24 56.1	15 39.5	2 23.8	5 52.5	18 10.9
51	24 48.3	15 14.8	1 37.8	5 4.5	17 53.8
52	24 40.2	14 49.5	0 50.4	4 13.8	17 35.6
53	24 31.9	14 23.3	0 1.6	3 20.1	17 16.2
54	24 23.2	13 56.2	♏29 11.3	2 23.2	16 55.6
55	24 14.2	13 28.2	28 19.5	1 22.7	16 33.4
56	24 4.8	12 59.2	27 26.0	0 18.2	16 9.5
57	23 55.1	12 29.2	26 30.9	♐29 9.2	15 43.7
58	23 44.9	11 58.0	25 33.9	27 55.2	15 15.6
59	23 34.2	11 25.7	24 35.0	26 35.7	14 45.0
60	23 23.1	10 52.0	23 34.2	25 9.7	14 11.4
61	23 11.4	10 17.0	22 31.2	23 36.5	13 34.1
62	22 59.2	9 40.4	21 26.0	21 55.0	12 52.6
63	22 46.3	9 2.3	20 18.6	20 3.9	12 5.6
64	22 32.7	8 22.4	19 8.7	18 1.6	11 11.9
65	22 18.4	7 40.7	17 56.2	15 46.3	10 9.3
66	22 3.2	6 57.0	16 41.1	13 15.3	8 54.2
S LAT	5	6	Descendant	8	9

		♓ 26° 43′ 50″		
23h 48m 0s		MC		357° 0′ 0″

11h 52m 0s — MC 178° 0′ 0″ — ♍ 27° 49′ 13″

N LAT	11	12	Ascendant	2	3
°	° ′	° ′	° ′	° ′	° ′
0	♏ 0 5.7	♐ 0 10.5	♐28 9.9	♑26 0.2	♒25 44.5
5	♎29 44.5	♏29 0.5	26 10.5	24 46.5	25 20.4
10	29 23.5	27 51.1	24 9.8	23 30.1	24 55.4
15	29 2.3	26 41.3	22 6.2	22 9.5	24 29.1
16	28 58.0	26 27.2	21 41.0	21 52.8	24 23.6
17	28 53.7	26 13.0	21 15.6	21 35.8	24 18.0
18	28 49.3	25 58.8	20 50.0	21 18.6	24 12.4
19	28 45.0	25 44.4	20 24.1	21 1.0	24 6.6
20	28 40.5	25 30.0	19 58.0	20 43.2	24 0.8
21	28 36.1	25 15.4	19 31.6	20 25.1	23 54.9
22	28 31.6	25 0.8	19 5.0	20 6.6	23 48.8
23	28 27.1	24 46.0	18 38.1	19 47.8	23 42.6
24	28 22.5	24 31.0	18 10.8	19 28.7	23 36.3
25	28 17.9	24 16.0	17 43.2	19 9.1	23 29.9
26	28 13.2	24 0.7	17 15.3	18 49.1	23 23.3
27	28 8.5	23 45.3	16 47.0	18 28.6	23 16.6
28	28 3.7	23 29.7	16 18.2	18 7.7	23 9.7
29	27 58.8	23 13.9	15 49.1	17 46.3	23 2.6
30	27 53.9	22 58.0	15 19.5	17 24.3	22 55.4
31	27 48.8	22 41.7	14 49.5	17 1.7	22 47.9
32	27 43.8	22 25.3	14 19.0	16 38.6	22 40.3
33	27 38.6	22 8.6	13 47.9	16 14.8	22 32.4
34	27 33.3	21 51.6	13 16.3	15 50.3	22 24.3
35	27 28.0	21 34.4	12 44.2	15 25.1	22 15.9
36	27 22.5	21 16.8	12 11.4	14 59.1	22 7.2
37	27 16.9	20 58.9	11 38.0	14 32.2	21 58.3
38	27 11.2	20 40.7	11 3.9	14 4.5	21 49.0
39	27 5.4	20 22.1	10 29.2	13 35.7	21 39.5
40	26 59.5	20 3.2	9 53.7	13 6.0	21 29.5
41	26 53.4	19 43.8	9 17.4	12 35.1	21 19.2
42	26 47.2	19 24.0	8 40.3	12 3.1	21 8.4
43	26 40.8	19 3.8	8 2.3	11 29.7	20 57.2
44	26 34.2	18 43.1	7 23.4	10 55.0	20 45.4
45	26 27.5	18 21.8	6 43.6	10 18.8	20 33.2
46	26 20.6	18 0.1	6 2.8	9 41.0	20 20.3
47	26 13.5	17 37.7	5 20.9	9 1.4	20 6.8
48	26 6.1	17 14.8	4 37.9	8 19.9	19 52.7
49	25 58.6	16 51.1	3 53.8	7 36.3	19 37.7
50	25 50.8	16 26.8	3 8.4	6 50.5	19 21.9
51	25 42.7	16 1.8	2 21.8	6 2.2	19 5.2
52	25 34.3	15 36.0	1 33.7	5 11.1	18 47.4
53	25 25.7	15 9.3	0 44.3	4 17.1	18 28.4
54	25 16.7	14 41.8	♏29 53.3	3 19.8	18 8.2
55	25 7.4	14 13.3	29 0.8	2 18.8	17 46.5
56	24 57.7	13 43.8	28 6.6	1 13.7	17 23.2
57	24 47.5	13 13.2	27 10.6	0 4.0	16 58.0
58	24 37.0	12 41.5	26 12.8	♐28 49.2	16 30.6
59	24 26.0	12 8.5	25 13.1	27 28.7	16 0.7
60	24 14.4	11 34.2	24 11.3	26 1.5	15 27.9
61	24 2.3	10 58.5	23 7.4	24 26.9	14 51.6
62	23 49.6	10 21.3	22 1.2	22 43.7	14 11.2
63	23 36.2	9 42.4	20 52.7	20 50.5	13 25.5
64	23 22.1	9 1.7	19 41.7	18 45.6	12 33.4
65	23 7.2	8 19.2	18 28.1	16 27.0	11 32.9
66	22 51.5	7 34.5	17 11.8	13 51.8	10 20.7
S LAT	5	6	Descendant	8	9

♓ 27° 49′ 13″ — 23h 52m 0s — MC 358° 0′ 0″

11h 56m 0s — MC 179° 0′ 0″ — ♍ 28° 54′ 36″

N LAT	11	12	Ascendant	2	3
°	° ′	° ′	° ′	° ′	° ′
0	♏ 1 8.4	♐ 1 8.0	♐29 5.0	♑26 57.4	♒26 46.7
5	0 46.6	♏29 57.5	27 5.4	25 44.1	26 23.2
10	0 24.9	28 47.4	25 4.5	24 28.1	25 58.9
15	0 3.1	27 36.9	23 0.5	23 8.0	25 33.2
16	♎29 58.7	27 22.7	22 35.2	22 51.3	25 27.8
17	29 54.2	27 8.3	22 9.7	22 34.4	25 22.4
18	29 49.7	26 53.9	21 44.0	22 17.3	25 16.9
19	29 45.2	26 39.4	21 18.0	21 59.8	25 11.3
20	29 40.7	26 24.8	20 51.8	21 42.1	25 5.6
21	29 36.1	26 10.1	20 25.3	21 24.0	24 59.8
22	29 31.4	25 55.3	19 58.5	21 5.6	24 53.9
23	29 26.8	25 40.3	19 31.4	20 46.9	24 47.8
24	29 22.1	25 25.2	19 4.0	20 27.8	24 41.7
25	29 17.3	25 10.0	18 36.3	20 8.2	24 35.4
26	29 12.5	24 54.5	18 8.2	19 48.3	24 29.0
27	29 7.6	24 39.0	17 39.7	19 27.9	24 22.4
28	29 2.6	24 23.2	17 10.8	19 7.0	24 15.7
29	28 57.6	24 7.2	16 41.4	18 45.6	24 8.8
30	28 52.5	23 51.0	16 11.6	18 23.7	24 1.7
31	28 47.3	23 34.6	15 41.4	18 1.2	23 54.4
32	28 42.1	23 17.9	15 10.6	17 38.1	23 47.0
33	28 36.7	23 1.0	14 39.3	17 14.3	23 39.3
34	28 31.3	22 43.8	14 7.5	16 49.9	23 31.4
35	28 25.8	22 26.3	13 35.0	16 24.6	23 23.2
36	28 20.1	22 8.5	13 2.0	15 58.6	23 14.7
37	28 14.4	21 50.4	12 28.3	15 31.8	23 6.0
38	28 8.5	21 31.9	11 53.9	15 4.0	22 57.0
39	28 2.5	21 13.1	11 18.8	14 35.3	22 47.6
40	27 56.4	20 53.9	10 43.0	14 5.5	22 37.9
41	27 50.1	20 34.2	10 6.3	13 34.7	22 27.8
42	27 43.7	20 14.2	9 28.8	13 2.6	22 17.3
43	27 37.1	19 53.6	8 50.5	12 29.2	22 6.3
44	27 30.3	19 32.6	8 11.2	11 54.4	21 54.9
45	27 23.3	19 11.0	7 30.9	11 18.1	21 42.9
46	27 16.2	18 48.9	6 49.6	10 40.1	21 30.3
47	27 8.8	18 26.2	6 7.3	10 0.4	21 17.2
48	27 1.3	18 2.9	5 23.8	9 18.8	21 3.3
49	26 53.4	17 38.9	4 39.1	8 35.0	20 48.7
50	26 45.4	17 14.2	3 53.2	7 49.0	20 33.3
51	26 37.0	16 48.8	3 5.9	7 0.4	20 16.9
52	26 28.4	16 22.5	2 17.3	6 9.1	19 59.6
53	26 19.4	15 55.4	1 27.1	5 14.7	19 41.1
54	26 10.1	15 27.4	0 35.5	4 16.9	19 21.3
55	26 0.5	14 58.4	♏29 42.2	3 15.4	19 0.2
56	25 50.4	14 28.4	28 47.3	2 9.7	18 37.4
57	25 40.0	13 57.3	27 50.5	0 59.4	18 12.8
58	25 29.0	13 25.0	26 51.9	♐29 43.8	17 46.1
59	25 17.6	12 51.4	25 51.3	28 22.2	17 17.0
60	25 5.7	12 16.4	24 48.6	26 54.0	16 45.0
61	24 53.1	11 40.0	23 43.7	25 17.9	16 9.7
62	24 39.9	11 2.1	22 36.5	23 33.0	15 30.3
63	24 26.1	10 22.4	21 26.9	21 37.7	14 46.1
64	24 11.5	9 41.0	20 14.8	19 30.1	13 55.6
65	23 56.0	8 57.5	19 0.0	17 8.0	12 57.1
66	23 39.7	8 12.0	17 42.5	14 28.5	11 47.7
S LAT	5	6	Descendant	8	9

♓ 28° 54′ 36″ — 23h 56m 0s — MC 359° 0′ 0″

12h 0m 0s — MC 180° 0′ 0″ — ♎ 0° 0′ 0″

11	12	Ascendant	2	3	N LAT
° ′	° ′	° ′	° ′	° ′	°
♏ 2 10.9	♐ 2 5.4	♑ 0 0.0	♑27 54.6	♒27 49.1	0
1 48.5	0 54.3	♐28 0.4	26 41.9	27 26.2	5
1 26.2	♏29 43.6	25 59.2	25 26.3	27 2.5	10
1 3.7	28 32.4	23 54.9	24 6.7	26 37.5	15
0 59.2	28 18.0	23 29.5	23 50.1	26 32.3	16
0 54.6	28 3.6	23 3.9	23 33.3	26 27.0	17
0 50.0	27 49.0	22 38.1	23 16.2	26 21.6	18
0 45.3	27 34.3	22 12.0	22 58.8	26 16.1	19
0 40.6	27 19.6	21 45.7	22 41.1	26 10.6	20
0 35.9	27 4.7	21 19.0	22 23.2	26 4.9	21
0 31.2	26 49.7	20 52.1	22 4.8	25 59.2	22
0 26.3	26 34.6	20 24.9	21 46.2	25 53.3	23
0 21.5	26 19.3	19 57.3	21 27.1	25 47.3	24
0 16.6	26 3.9	19 29.4	21 7.7	25 41.2	25
0 11.6	25 48.3	19 1.1	20 47.8	25 34.9	26
0 6.6	25 32.5	18 32.5	20 27.4	25 28.5	27
0 1.5	25 16.6	18 3.4	20 6.6	25 22.0	28
♎29 56.3	25 0.4	17 33.8	19 45.3	25 15.2	29
29 51.0	24 44.0	17 3.9	19 23.4	25 8.3	30
29 45.7	24 27.4	16 33.4	19 1.0	25 1.3	31
29 40.3	24 10.5	16 2.4	18 37.9	24 54.0	32
29 34.8	23 53.4	15 30.9	18 14.2	24 46.5	33
29 29.2	23 36.0	14 58.7	17 49.7	24 38.7	34
29 23.5	23 18.3	14 26.0	17 24.6	24 30.8	35
29 17.7	23 0.2	13 52.7	16 58.6	24 22.5	36
29 11.7	22 41.9	13 18.7	16 31.7	24 14.0	37
29 5.7	22 23.1	12 44.0	16 4.0	24 5.2	38
28 59.5	22 4.1	12 8.6	15 35.3	23 56.1	39
28 53.2	21 44.6	11 32.4	15 5.5	23 46.6	40
28 46.7	21 24.6	10 55.4	14 34.6	23 36.7	41
28 40.1	21 4.3	10 17.5	14 2.4	23 26.5	42
28 33.3	20 43.5	9 38.8	13 29.0	23 15.8	43
28 26.3	20 22.1	8 59.1	12 54.2	23 4.6	44
28 19.1	20 0.2	8 18.4	12 17.8	22 52.9	45
28 11.7	19 37.8	7 36.6	11 39.8	22 40.7	46
28 4.1	19 14.7	6 53.8	10 59.9	22 27.9	47
27 56.3	18 51.1	6 9.8	10 18.1	22 14.3	48
27 48.2	18 26.7	5 24.5	9 34.2	22 0.1	49
27 39.9	18 1.6	4 38.0	8 47.9	21 45.0	50
27 31.3	17 35.7	3 50.2	7 59.1	21 29.1	51
27 22.3	17 9.1	3 0.9	7 7.5	21 12.2	52
27 13.1	16 41.5	2 10.1	6 12.8	20 54.2	53
27 3.5	16 13.0	1 17.8	5 14.6	20 34.9	54
26 53.5	15 43.5	0 23.8	4 12.6	20 14.3	55
26 43.1	15 13.0	♏29 28.1	3 6.4	19 52.1	56
26 32.3	14 41.3	28 30.5	1 55.4	19 28.1	57
26 21.0	14 8.4	27 31.0	0 39.0	19 2.1	58
26 9.2	13 34.2	26 29.5	♐29 16.5	18 33.8	59
25 56.8	12 58.6	25 25.8	27 47.0	18 2.7	60
25 43.9	12 21.5	24 20.0	26 9.6	17 28.3	61
25 30.2	11 42.9	23 11.8	24 22.9	16 50.1	62
25 15.9	11 2.4	22 1.1	22 25.3	16 7.1	63
25 0.7	10 20.2	20 47.8	20 15.1	15 18.3	64
24 44.7	9 35.9	19 31.9	17 49.5	14 21.8	65
24 27.8	8 49.4	18 13.1	15 5.2	13 15.2	66
5	6	Descendant	8	9	S LAT

♈ 0° 0′ 0″ — 0h 0m 0s — MC 0° 0′ 0″

12h 4m 0s — MC 181° 0′ 0″ — ♎ 1° 5′ 24″

11	12	Ascendant	2	3	N LAT
° ′	° ′	° ′	° ′	° ′	°
♏ 3 13.3	♐ 3 2.6	♑ 0 55.0	♑28 52.0	♒28 51.6	0
2 50.3	1 51.0	♐28 55.4	27 39.8	28 29.4	5
2 27.4	0 39.7	26 54.0	26 24.8	28 6.3	10
2 4.3	♏29 27.9	24 49.4	25 5.6	27 42.0	15
1 59.6	29 13.3	24 23.9	24 49.1	27 36.9	16
1 54.9	28 58.7	23 58.2	24 32.3	27 31.8	17
1 50.1	28 44.0	23 32.3	24 15.3	27 26.5	18
1 45.3	28 29.2	23 6.1	23 58.0	27 21.2	19
1 40.5	28 14.3	22 39.6	23 40.5	27 15.8	20
1 35.7	27 59.3	22 12.9	23 22.6	27 10.3	21
1 30.8	27 44.1	21 45.8	23 4.3	27 4.7	22
1 25.8	27 28.8	21 18.5	22 45.7	26 59.0	23
1 20.8	27 13.4	20 50.8	22 26.7	26 53.2	24
1 15.7	26 57.8	20 22.7	22 7.4	26 47.2	25
1 10.6	26 42.0	19 54.3	21 47.6	26 41.1	26
1 5.4	26 26.1	19 25.4	21 27.3	26 34.9	27
1 0.2	26 9.9	18 56.1	21 6.5	26 28.5	28
0 54.9	25 53.6	18 26.4	20 45.3	26 21.9	29
0 49.5	25 37.0	17 56.2	20 23.4	26 15.2	30
0 44.0	25 20.2	17 25.5	20 1.0	26 8.3	31
0 38.4	25 3.1	16 54.3	19 38.0	26 1.2	32
0 32.7	24 45.7	16 22.5	19 14.3	25 53.9	33
0 27.0	24 28.1	15 50.2	18 50.0	25 46.4	34
0 21.1	24 10.2	15 17.2	18 24.8	25 38.6	35
0 15.1	23 51.9	14 43.6	17 58.9	25 30.6	36
0 9.0	23 33.3	14 9.3	17 32.1	25 22.3	37
0 2.8	23 14.3	13 34.3	17 4.3	25 13.7	38
♎29 56.4	22 55.0	12 58.6	16 35.6	25 4.9	39
29 49.9	22 35.2	12 22.0	16 5.8	24 55.6	40
29 43.2	22 15.1	11 44.6	15 34.9	24 46.0	41
29 36.4	21 54.4	11 6.4	15 2.8	24 36.0	42
29 29.4	21 33.3	10 27.2	14 29.3	24 25.6	43
29 22.2	21 11.6	9 47.1	13 54.4	24 14.7	44
29 14.8	20 49.4	9 6.0	13 17.9	24 3.4	45
29 7.2	20 26.7	8 23.7	12 39.8	23 51.4	46
28 59.3	20 3.3	7 40.4	11 59.9	23 38.9	47
28 51.3	19 39.2	6 55.9	11 18.0	23 25.8	48
28 42.9	19 14.5	6 10.1	10 33.9	23 11.9	49
28 34.3	18 49.0	5 23.0	9 47.4	22 57.2	50
28 25.4	18 22.7	4 34.6	8 58.4	22 41.7	51
28 16.2	17 55.6	3 44.7	8 6.5	22 25.2	52
28 6.7	17 27.6	2 53.2	7 11.4	22 7.6	53
27 56.8	16 58.6	2 0.2	6 12.9	21 48.9	54
27 46.5	16 28.6	1 5.5	5 10.4	21 28.8	55
27 35.8	15 57.6	0 9.0	4 3.6	21 7.2	56
27 24.6	15 25.3	♏29 10.6	2 52.0	20 43.9	57
27 12.9	14 51.9	28 10.2	1 34.8	20 18.6	58
27 0.7	14 17.1	27 7.8	0 11.3	19 51.0	59
26 47.9	13 40.8	26 3.2	♐28 40.7	19 20.8	60
26 34.5	13 3.0	24 56.3	27 1.8	18 47.5	61
26 20.4	12 23.6	23 47.1	25 13.4	18 10.4	62
26 5.6	11 42.4	22 35.3	23 13.6	17 28.8	63
25 49.9	10 59.3	21 20.9	21 0.5	16 41.5	64
25 33.4	10 14.2	20 3.7	18 31.3	15 47.1	65
25 15.8	9 26.8	18 43.7	15 42.2	14 43.1	66
5	6	Descendant	8	9	S LAT

♈ 1° 5′ 24″ — 0h 4m 0s — MC 1° 0′ 0″

$12^h\ 8^m\ 0^s$ MC 182° 0′ 0″

♎ 2° 10′ 47″

N LAT	11	12	Ascendant	2	3
°	° ′	° ′	° ′	° ′	° ′
0	♏ 4 15.5	♐ 3 59.8	♑ 1 50.1	♑29 49.5	♒29 54.3
5	3 51.9	2 47.5	♐29 50.4	28 37.9	29 32.8
10	3 28.4	1 35.7	27 48.9	27 23.4	29 10.3
15	3 4.6	0 23.2	25 43.9	26 4.7	28 46.7
16	2 59.8	0 8.5	25 18.4	25 48.3	28 41.8
17	2 55.0	♏29 53.8	24 52.6	25 31.6	28 36.8
18	2 50.1	29 38.9	24 26.5	25 14.7	28 31.7
19	2 45.2	29 24.0	24 0.3	24 57.5	28 26.5
20	2 40.2	29 8.9	23 33.7	24 40.0	28 21.3
21	2 35.3	28 53.8	23 6.8	24 22.2	28 15.9
22	2 30.2	28 38.5	22 39.7	24 4.1	28 10.5
23	2 25.1	28 23.0	22 12.2	23 45.5	28 4.9
24	2 20.0	28 7.4	21 44.3	23 26.6	27 59.2
25	2 14.8	27 51.6	21 16.1	23 7.3	27 53.5
26	2 9.5	27 35.7	20 47.5	22 47.6	27 47.5
27	2 4.2	27 19.6	20 18.5	22 27.4	27 41.5
28	1 58.8	27 3.2	19 49.0	22 6.7	27 35.3
29	1 53.3	26 46.7	19 19.1	21 45.5	27 28.9
30	1 47.8	26 29.9	18 48.7	21 23.8	27 22.4
31	1 42.2	26 12.9	18 17.8	21 1.5	27 15.7
32	1 36.4	25 55.6	17 46.4	20 38.5	27 8.8
33	1 30.6	25 38.1	17 14.4	20 14.9	27 1.7
34	1 24.7	25 20.2	16 41.8	19 50.5	26 54.3
35	1 18.6	25 2.1	16 8.5	19 25.4	26 46.8
36	1 12.5	24 43.6	15 34.6	18 59.5	26 39.0
37	1 6.2	24 24.7	15 0.1	18 32.7	26 30.9
38	0 59.8	24 5.5	14 24.8	18 5.0	26 22.6
39	0 53.2	23 45.9	13 48.7	17 36.3	26 13.9
40	0 46.5	23 25.9	13 11.8	17 6.6	26 4.9
41	0 39.6	23 5.4	12 34.1	16 35.7	25 55.6
42	0 32.6	22 44.5	11 55.4	16 3.5	25 45.9
43	0 25.4	22 23.1	11 15.9	15 30.0	25 35.8
44	0 18.0	22 1.1	10 35.3	14 55.1	25 25.2
45	0 10.4	21 38.6	9 53.7	14 18.6	25 14.1
46	0 2.5	21 15.5	9 11.1	13 40.4	25 2.5
47	♎29 54.5	20 51.8	8 27.2	13 0.4	24 50.3
48	29 46.2	20 27.4	7 42.2	12 18.3	24 37.5
49	29 37.6	20 2.3	6 55.9	11 34.1	24 24.0
50	29 28.7	19 36.4	6 8.2	10 47.5	24 9.8
51	29 19.5	19 9.7	5 19.2	9 58.2	23 54.7
52	29 10.1	18 42.1	4 28.6	9 6.1	23 38.6
53	29 0.2	18 13.7	3 36.5	8 10.7	23 21.6
54	28 50.0	17 44.2	2 42.8	7 11.8	23 3.3
55	28 39.4	17 13.8	1 47.3	6 8.9	22 43.8
56	28 28.3	16 42.2	0 50.0	5 1.5	22 22.8
57	28 16.8	16 9.4	♏29 50.8	3 49.2	22 0.2
58	28 4.7	15 35.3	28 49.5	2 31.2	21 35.6
59	27 52.2	14 59.9	27 46.2	1 6.8	21 8.8
60	27 39.0	14 23.0	26 40.6	♐29 35.1	20 39.5
61	27 25.1	13 44.5	25 32.8	27 54.8	20 7.2
62	27 10.6	13 4.4	24 22.4	26 4.5	19 31.3
63	26 55.2	12 22.4	23 9.5	24 2.5	18 51.0
64	26 39.1	11 38.5	21 53.9	21 46.6	18 5.4
65	26 21.9	10 52.4	20 35.5	19 13.5	17 12.9
66	26 3.8	10 4.1	19 14.2	16 19.4	16 11.6
S LAT	5	6	Descendant	8	9

♈ 2° 10′ 47″

$0^h\ 8^m\ 0^s$ MC 2° 0′ 0″

$12^h\ 12^m\ 0^s$ MC 183° 0′ 0″

♎ 3° 16′ 10″

N LAT	11	12	Ascendant	2	3
°	° ′	° ′	° ′	° ′	° ′
0	♏ 5 17.5	♐ 4 56.7	♑ 2 45.2	♒ 0 47.2	♓ 0 57.2
5	4 53.3	3 44.0	0 45.5	♑29 36.1	0 36.3
10	4 29.2	2 31.6	♐28 43.8	28 22.2	0 14.5
15	4 4.9	1 18.5	26 38.6	27 4.0	♒29 51.6
16	3 59.9	1 3.7	26 12.9	26 47.7	29 46.8
17	3 55.0	0 48.8	25 47.0	26 31.2	29 42.0
18	3 50.0	0 33.8	25 20.9	26 14.3	29 37.0
19	3 44.9	0 18.7	24 54.5	25 57.2	29 32.0
20	3 39.8	0 3.5	24 27.9	25 39.8	29 26.9
21	3 34.7	♏29 48.2	24 0.9	25 22.1	29 21.7
22	3 29.6	29 32.7	23 33.6	25 4.1	29 16.5
23	3 24.3	29 17.1	23 6.0	24 45.6	29 11.1
24	3 19.1	29 1.4	22 38.0	24 26.8	29 5.6
25	3 13.7	28 45.4	22 9.6	24 7.6	28 59.9
26	3 8.3	28 29.3	21 40.9	23 48.0	28 54.2
27	3 2.9	28 13.0	21 11.7	23 27.8	28 48.3
28	2 57.3	27 56.5	20 42.1	23 7.2	28 42.3
29	2 51.7	27 39.8	20 12.0	22 46.1	28 36.1
30	2 46.0	27 22.8	19 41.4	22 24.4	28 29.8
31	2 40.2	27 5.6	19 10.3	22 2.2	28 23.2
32	2 34.3	26 48.1	18 38.6	21 39.3	28 16.5
33	2 28.3	26 30.4	18 6.4	21 15.7	28 9.6
34	2 22.2	26 12.3	17 33.5	20 51.4	28 2.5
35	2 16.0	25 53.9	17 0.0	20 26.4	27 55.2
36	2 9.7	25 35.2	16 25.9	20 0.5	27 47.6
37	2 3.3	25 16.2	15 51.0	19 33.8	27 39.8
38	1 56.7	24 56.7	15 15.4	19 6.2	27 31.7
39	1 49.9	24 36.8	14 39.0	18 37.5	27 23.3
40	1 43.0	24 16.6	14 1.8	18 7.7	27 14.6
41	1 36.0	23 55.8	13 23.7	17 36.8	27 5.5
42	1 28.7	23 34.6	12 44.7	17 4.7	26 56.1
43	1 21.3	23 12.9	12 4.7	16 31.2	26 46.2
44	1 13.7	22 50.7	11 23.7	15 56.2	26 36.0
45	1 5.9	22 27.8	10 41.7	15 19.7	26 25.2
46	0 57.8	22 4.4	9 58.6	14 41.4	26 13.9
47	0 49.5	21 40.3	9 14.2	14 1.3	26 2.1
48	0 41.0	21 15.5	8 28.7	13 19.2	25 49.7
49	0 32.1	20 50.0	7 41.8	12 34.9	25 36.6
50	0 23.0	20 23.8	6 53.6	11 48.1	25 22.7
51	0 13.6	19 56.7	6 3.9	10 58.6	25 8.0
52	0 3.8	19 28.7	5 12.7	10 6.2	24 52.5
53	♎29 53.7	18 59.8	4 19.9	9 10.6	24 35.9
54	29 43.1	18 29.9	3 25.5	8 11.3	24 18.2
55	29 32.2	17 58.9	2 29.2	7 8.0	23 59.2
56	29 20.8	17 26.8	1 31.1	6 0.1	23 38.9
57	29 8.9	16 53.4	0 31.1	4 47.2	23 16.9
58	28 56.5	16 18.8	♏29 29.0	3 28.4	22 53.1
59	28 43.5	15 42.7	28 24.7	2 3.1	22 27.1
60	28 29.9	15 5.2	27 18.2	0 30.1	21 58.7
61	28 15.6	14 26.0	26 9.2	♐28 48.4	21 27.4
62	28 0.6	13 45.1	24 57.8	26 56.4	20 52.6
63	27 44.8	13 2.3	23 43.8	24 52.1	20 13.7
64	27 28.1	12 17.5	22 27.0	22 33.2	19 29.7
65	27 10.4	11 30.6	21 7.4	19 56.2	18 39.3
66	26 51.7	10 41.3	19 44.8	16 56.7	17 40.5
S LAT	5	6	Descendant	8	9

♈ 3° 16′ 10″

$0^h\ 12^m\ 0^s$ MC 3° 0′ 0″

12h 16m 0s — MC — 184° 0′ 0″

♎ 4° 21′ 31″

11	12	Ascendant	2	3	N LAT
° ′	° ′	° ′	° ′	° ′	°
♏ 6 19.4	♐ 5 53.6	♑ 3 40.2	♒ 1 45.0	♓ 2 0.3	0
5 54.6	4 40.3	1 40.6	0 34.6	1 40.0	5
5 29.9	3 27.4	♐29 38.8	♑29 21.2	1 18.9	10
5 5.0	2 13.7	27 33.3	28 3.5	0 56.7	15
4 59.9	1 58.7	27 7.6	27 47.4	0 52.1	16
4 54.8	1 43.7	26 41.6	27 30.9	0 47.4	17
4 49.7	1 28.6	26 15.4	27 14.2	0 42.6	18
4 44.5	1 13.4	25 48.9	26 57.2	0 37.8	19
4 39.3	0 58.0	25 22.2	26 39.9	0 32.8	20
4 34.1	0 42.6	24 55.1	26 22.3	0 27.8	21
4 28.8	0 27.0	24 27.7	26 4.3	0 22.7	22
4 23.4	0 11.2	23 59.9	25 46.0	0 17.4	23
4 18.0	♏29 55.3	23 31.8	25 27.3	0 12.1	24
4 12.5	29 39.2	23 3.3	25 8.2	0 6.7	25
4 7.0	29 22.9	22 34.4	24 48.6	0 1.1	26
4 1.4	29 6.4	22 5.0	24 28.6	♒29 55.4	27
3 55.7	28 49.8	21 35.2	24 8.1	29 49.5	28
3 49.9	28 32.9	21 5.0	23 47.0	29 43.5	29
3 44.1	28 15.7	20 34.2	23 25.4	29 37.4	30
3 38.2	27 58.3	20 2.9	23 3.3	29 31.1	31
3 32.1	27 40.6	19 31.0	22 40.4	29 24.6	32
3 26.0	27 22.7	18 58.5	22 16.9	29 17.9	33
3 19.7	27 4.4	18 25.4	21 52.7	29 11.0	34
3 13.4	26 45.8	17 51.7	21 27.8	29 3.9	35
3 6.9	26 26.9	17 17.2	21 2.0	28 56.6	36
3 0.2	26 7.6	16 42.1	20 35.3	28 49.0	37
2 53.5	25 47.9	16 6.2	20 7.7	28 41.1	38
2 46.5	25 27.8	15 29.5	19 39.0	28 33.0	39
2 39.5	25 7.2	14 51.9	19 9.3	28 24.5	40
2 32.2	24 46.2	14 13.4	18 38.4	28 15.7	41
2 24.8	24 24.7	13 34.1	18 6.3	28 6.6	42
2 17.2	24 2.7	12 53.7	17 32.8	27 57.0	43
2 9.3	23 40.2	12 12.3	16 57.8	27 47.1	44
2 1.3	23 17.0	11 29.8	16 21.3	27 36.6	45
1 53.0	22 53.3	10 46.2	15 43.0	27 25.7	46
1 44.5	22 28.9	10 1.4	15 2.8	27 14.2	47
1 35.7	22 3.7	9 15.3	14 20.6	27 2.2	48
1 26.6	21 37.9	8 27.9	13 36.2	26 49.5	49
1 17.2	21 11.2	7 39.1	12 49.2	26 36.0	50
1 7.5	20 43.7	6 48.8	11 59.6	26 21.8	51
0 57.5	20 15.3	5 57.0	11 7.0	26 6.7	52
0 47.1	19 45.9	5 3.5	10 11.0	25 50.6	53
0 36.2	19 15.5	4 8.3	9 11.4	25 33.5	54
0 25.0	18 44.0	3 11.3	8 7.7	25 15.1	55
0 13.2	18 11.4	2 12.4	6 59.3	24 55.4	56
0 1.0	17 37.5	1 11.5	5 45.8	24 34.1	57
♎29 48.2	17 2.2	0 8.5	4 26.3	24 11.0	58
29 34.8	16 25.5	♏29 3.3	3 0.0	23 45.9	59
29 20.8	15 47.3	27 55.8	1 26.0	23 18.4	60
29 6.1	15 7.5	26 45.8	♐29 42.8	22 48.1	61
28 50.6	14 25.8	25 33.3	27 49.0	22 14.6	62
28 34.3	13 42.2	24 18.1	25 42.4	21 37.0	63
28 17.1	12 56.6	23 0.1	23 20.5	20 54.7	64
27 58.8	12 8.8	21 39.2	20 39.5	20 6.2	65
27 39.5	11 18.5	20 15.3	17 34.3	19 9.8	66
5	6	Descendant	8	9	S LAT

♈ 4° 21′ 31″

0h 16m 0s — MC — 4° 0′ 0″

12h 20m 0s — MC — 185° 0′ 0″

♎ 5° 26′ 50″

11	12	Ascendant	2	3	N LAT
° ′	° ′	° ′	° ′	° ′	°
♏ 7 21.1	♐ 6 50.3	♑ 4 35.3	♒ 2 43.0	♓ 3 3.5	0
6 55.7	5 36.6	2 35.7	1 33.2	2 43.9	5
6 30.5	4 23.1	0 33.9	0 20.4	2 23.5	10
6 4.9	3 8.8	♐28 28.1	♑29 3.3	2 2.0	15
5 59.7	2 53.7	28 2.3	28 47.3	1 57.5	16
5 54.5	2 38.6	27 36.3	28 30.9	1 53.0	17
5 49.3	2 23.3	27 10.0	28 14.3	1 48.4	18
5 44.0	2 8.0	26 43.4	27 57.4	1 43.7	19
5 38.7	1 52.5	26 16.6	27 40.2	1 38.9	20
5 33.3	1 36.9	25 49.4	27 22.7	1 34.0	21
5 27.8	1 21.1	25 21.9	27 4.9	1 29.1	22
5 22.4	1 5.2	24 54.0	26 46.7	1 24.0	23
5 16.8	0 49.2	24 25.8	26 28.0	1 18.9	24
5 11.2	0 32.9	23 57.1	26 9.0	1 13.6	25
5 5.6	0 16.5	23 28.1	25 49.6	1 8.2	26
4 59.8	♏29 59.8	22 58.6	25 29.6	1 2.7	27
4 54.0	29 43.0	22 28.6	25 9.2	0 57.0	28
4 48.1	29 25.9	21 58.1	24 48.3	0 51.2	29
4 42.1	29 8.6	21 27.2	24 26.8	0 45.3	30
4 36.0	28 51.0	20 55.6	24 4.7	0 39.2	31
4 29.8	28 33.1	20 23.5	23 41.9	0 32.9	32
4 23.5	28 14.9	19 50.8	23 18.5	0 26.4	33
4 17.1	27 56.5	19 17.5	22 54.4	0 19.8	34
4 10.6	27 37.7	18 43.5	22 29.5	0 12.9	35
4 3.9	27 18.5	18 8.8	22 3.8	0 5.8	36
3 57.1	26 59.0	17 33.4	21 37.2	♒29 58.4	37
3 50.2	26 39.0	16 57.2	21 9.6	29 50.8	38
3 43.1	26 18.7	16 20.1	20 41.0	29 42.9	39
3 35.8	25 57.9	15 42.2	20 11.4	29 34.7	40
3 28.4	25 36.6	15 3.4	19 40.5	29 26.2	41
3 20.7	25 14.9	14 23.7	19 8.4	29 17.4	42
3 12.9	24 52.6	13 42.9	18 34.9	29 8.1	43
3 4.9	24 29.7	13 1.1	18 0.0	28 58.5	44
2 56.6	24 6.3	12 18.2	17 23.4	28 48.4	45
2 48.1	23 42.2	11 34.1	16 45.1	28 37.8	46
2 39.4	23 17.4	10 48.8	16 4.9	28 26.7	47
2 30.3	22 51.9	10 2.2	15 22.6	28 15.0	48
2 21.0	22 25.7	9 14.2	14 38.0	28 2.7	49
2 11.4	21 58.6	8 24.8	13 51.0	27 49.7	50
2 1.4	21 30.7	7 33.9	13 1.2	27 36.0	51
1 51.1	21 1.9	6 41.4	12 8.4	27 21.3	52
1 40.4	20 32.0	5 47.2	11 12.2	27 5.8	53
1 29.2	20 1.2	4 51.3	10 12.2	26 49.2	54
1 17.6	19 29.2	3 53.5	9 8.1	26 31.4	55
1 5.6	18 56.0	2 53.8	7 59.3	26 12.3	56
0 53.0	18 21.5	1 52.1	6 45.1	25 51.7	57
0 39.8	17 45.7	0 48.1	5 24.9	25 29.4	58
0 26.1	17 8.4	♏29 42.0	3 57.8	25 5.1	59
0 11.6	16 29.5	28 33.4	2 22.6	24 38.6	60
♎29 56.5	15 48.9	27 22.4	0 38.0	24 9.4	61
29 40.5	15 6.5	26 8.8	♐28 42.4	23 37.0	62
29 23.7	14 22.1	24 52.4	26 33.5	23 0.8	63
29 6.0	13 35.6	23 33.2	24 8.5	22 20.1	64
28 47.1	12 46.9	22 11.0	21 23.3	21 33.6	65
28 27.2	11 55.7	20 45.7	18 12.2	20 39.7	66
5	6	Descendant	8	9	S LAT

♈ 5° 26′ 50″

0h 20m 0s — MC — 5° 0′ 0″

12h 24m 0s — MC 186° 0′ 0″ — ♎ 6° 32′ 7″

N LAT	11	12	Ascendant	2	3
°	° ′	° ′	° ′	° ′	° ′
0	♏ 8 22.6	♐ 7 46.9	♑ 5 30.5	♒ 3 41.2	♓ 4 6.8
5	7 56.7	6 32.7	3 30.9	2 32.0	3 47.9
10	7 30.8	5 18.7	1 29.0	1 19.8	3 28.3
15	7 4.7	4 3.9	♐29 23.0	0 3.4	3 7.5
16	6 59.4	3 48.7	28 57.2	♑29 47.4	3 3.2
17	6 54.1	3 33.4	28 31.1	29 31.2	2 58.8
18	6 48.7	3 18.0	28 4.7	29 14.7	2 54.3
19	6 43.3	3 2.5	27 38.1	28 57.9	2 49.8
20	6 37.9	2 46.9	27 11.1	28 40.8	2 45.2
21	6 32.4	2 31.2	26 43.8	28 23.4	2 40.5
22	6 26.8	2 15.3	26 16.2	28 5.7	2 35.7
23	6 21.2	1 59.2	25 48.2	27 47.6	2 30.8
24	6 15.5	1 43.0	25 19.9	27 29.1	2 25.9
25	6 9.8	1 26.6	24 51.1	27 10.2	2 20.8
26	6 4.0	1 10.0	24 21.9	26 50.8	2 15.6
27	5 58.1	0 53.2	23 52.2	26 31.0	2 10.2
28	5 52.1	0 36.2	23 22.1	26 10.7	2 4.8
29	5 46.1	0 18.9	22 51.5	25 49.8	1 59.2
30	5 40.0	0 1.4	22 20.3	25 28.4	1 53.4
31	5 33.7	♏29 43.6	21 48.6	25 6.4	1 47.5
32	5 27.4	29 25.6	21 16.3	24 43.8	1 41.4
33	5 20.9	29 7.2	20 43.4	24 20.5	1 35.2
34	5 14.4	28 48.5	20 9.8	23 56.4	1 28.8
35	5 7.7	28 29.5	19 35.6	23 31.6	1 22.1
36	5 0.8	28 10.1	19 0.6	23 6.0	1 15.2
37	4 53.9	27 50.4	18 24.9	22 39.4	1 8.1
38	4 46.8	27 30.2	17 48.3	22 11.9	1 0.8
39	4 39.5	27 9.6	17 11.0	21 43.4	0 53.2
40	4 32.0	26 48.6	16 32.8	21 13.8	0 45.3
41	4 24.4	26 27.0	15 53.6	20 43.0	0 37.0
42	4 16.6	26 5.0	15 13.5	20 11.0	0 28.5
43	4 8.6	25 42.4	14 32.3	19 37.5	0 19.5
44	4 0.3	25 19.3	13 50.1	19 2.6	0 10.2
45	3 51.9	24 55.5	13 6.7	18 26.0	0 0.5
46	3 43.1	24 31.1	12 22.2	17 47.7	♒29 50.2
47	3 34.2	24 6.0	11 36.4	17 7.4	29 39.5
48	3 24.9	23 40.1	10 49.2	16 25.1	29 28.2
49	3 15.3	23 13.5	10 0.7	15 40.5	29 16.3
50	3 5.4	22 46.1	9 10.7	14 53.3	29 3.8
51	2 55.2	22 17.8	8 19.2	14 3.4	28 50.5
52	2 44.6	21 48.5	7 26.0	13 10.4	28 36.4
53	2 33.6	21 18.2	6 31.1	12 14.0	28 21.3
54	2 22.2	20 46.9	5 34.5	11 13.7	28 5.3
55	2 10.3	20 14.4	4 35.9	10 9.3	27 48.1
56	1 57.9	19 40.6	3 35.4	9 0.0	27 29.7
57	1 44.9	19 5.6	2 32.8	7 45.3	27 9.8
58	1 31.4	18 29.2	1 27.9	6 24.3	26 48.3
59	1 17.2	17 51.2	0 20.8	4 56.3	26 24.9
60	1 2.4	17 11.6	♏29 11.2	3 20.0	25 59.2
61	0 46.8	16 30.3	27 59.1	1 34.1	25 31.1
62	0 30.4	15 47.2	26 44.3	♐29 36.6	24 59.9
63	0 13.1	15 2.0	25 26.8	27 25.4	24 25.2
64	♎29 54.8	14 14.6	24 6.3	24 57.2	23 46.0
65	29 35.4	13 24.9	22 42.8	22 7.7	23 1.4
66	29 14.8	12 32.7	21 16.2	18 50.3	22 9.9
S LAT	5	6	Descendant	8	9

♈ 6° 32′ 7″ — 0h 24m 0s — MC 6° 0′ 0″

12h 28m 0s — MC 187° 0′ 0″ — ♎ 7° 37′ 22″

N LAT	11	12	Ascendant	2	3
°	° ′	° ′	° ′	° ′	° ′
0	♏ 9 23.9	♐ 8 43.3	♑ 6 25.6	♒ 4 39.5	♓ 5 10.3
5	8 57.4	7 28.7	4 26.2	3 31.0	4 52.1
10	8 31.1	6 14.3	2 24.2	2 19.5	4 33.2
15	8 4.4	4 58.8	0 18.1	1 3.6	4 13.2
16	7 59.0	4 43.5	♐29 52.2	0 47.8	4 9.0
17	7 53.5	4 28.1	29 26.0	0 31.7	4 4.8
18	7 48.0	4 12.6	28 59.6	0 15.3	4 0.5
19	7 42.5	3 57.0	28 32.8	♑29 58.7	3 56.1
20	7 36.9	3 41.3	28 5.8	29 41.7	3 51.7
21	7 31.3	3 25.4	27 38.4	29 24.4	3 47.2
22	7 25.6	3 9.3	27 10.7	29 6.8	3 42.6
23	7 19.9	2 53.2	26 42.6	28 48.8	3 37.9
24	7 14.1	2 36.8	26 14.1	28 30.4	3 33.0
25	7 8.2	2 20.2	25 45.2	28 11.6	3 28.1
26	7 2.3	2 3.5	25 15.9	27 52.4	3 23.1
27	6 56.3	1 46.5	24 46.1	27 32.7	3 18.0
28	6 50.2	1 29.3	24 15.8	27 12.5	3 12.7
29	6 44.0	1 11.9	23 45.0	26 51.7	3 7.3
30	6 37.7	0 54.2	23 13.6	26 30.5	3 1.8
31	6 31.3	0 36.3	22 41.7	26 8.6	2 56.1
32	6 24.8	0 18.0	22 9.2	25 46.0	2 50.2
33	6 18.2	♏29 59.5	21 36.1	25 22.8	2 44.2
34	6 11.5	29 40.6	21 2.3	24 58.9	2 38.0
35	6 4.7	29 21.4	20 27.8	24 34.1	2 31.6
36	5 57.7	29 1.8	19 52.6	24 8.6	2 25.0
37	5 50.5	28 41.8	19 16.6	23 42.1	2 18.1
38	5 43.3	28 21.4	18 39.7	23 14.7	2 11.0
39	5 35.8	28 0.6	18 2.1	22 46.3	2 3.7
40	5 28.2	27 39.3	17 23.5	22 16.8	1 56.1
41	5 20.4	27 17.5	16 44.0	21 46.0	1 48.1
42	5 12.4	26 55.2	16 3.5	21 14.0	1 39.9
43	5 4.2	26 32.3	15 21.9	20 40.6	1 31.3
44	4 55.7	26 8.8	14 39.3	20 5.7	1 22.3
45	4 47.0	25 44.8	13 55.5	19 29.2	1 12.9
46	4 38.1	25 20.0	13 10.5	18 50.8	1 3.0
47	4 28.9	24 54.6	12 24.1	18 10.6	0 52.6
48	4 19.4	24 28.4	11 36.5	17 28.2	0 41.8
49	4 9.6	24 1.4	10 47.4	16 43.5	0 30.3
50	3 59.4	23 33.6	9 56.8	15 56.3	0 18.2
51	3 48.9	23 4.8	9 4.6	15 6.2	0 5.4
52	3 38.1	22 35.1	8 10.8	14 13.1	♒29 51.7
53	3 26.8	22 4.4	7 15.2	13 16.4	29 37.3
54	3 15.0	21 32.6	6 17.8	12 15.9	29 21.8
55	3 2.8	20 59.5	5 18.5	11 11.1	29 5.3
56	2 50.1	20 25.3	4 17.1	10 1.4	28 47.5
57	2 36.8	19 49.7	3 13.6	8 46.2	28 28.3
58	2 22.9	19 12.6	2 7.8	7 24.6	28 7.6
59	2 8.3	18 34.1	0 59.7	5 55.7	27 45.0
60	1 53.1	17 53.8	♏29 49.1	4 18.3	27 20.4
61	1 37.0	17 11.8	28 35.9	2 31.0	26 53.3
62	1 20.2	16 27.8	27 20.0	0 31.7	26 23.4
63	1 2.3	15 41.8	26 1.2	♐28 18.1	25 50.0
64	0 43.5	14 53.6	24 39.5	25 46.8	25 12.5
65	0 23.5	14 3.0	23 14.6	22 52.7	24 29.8
66	0 2.4	13 9.7	21 46.6	19 28.8	23 40.7
S LAT	5	6	Descendant	8	9

♈ 7° 37′ 22″ — 0h 28m 0s — MC 7° 0′ 0″

12h 32m 0s — MC — 188° 0′ 0″

♎ 8° 42′ 33″

11	12	Ascendant	2	3	N LAT
° ′	° ′	° ′	° ′	° ′	°
♏10 25.0	♐ 9 39.7	♑ 7 20.8	♒ 5 38.0	♓ 6 14.0	0
9 58.0	8 24.7	5 21.5	4 30.2	5 56.5	5
9 31.1	7 9.7	3 19.5	3 19.3	5 38.3	10
9 3.9	5 53.8	1 13.2	2 4.1	5 19.1	15
8 58.4	5 38.4	0 47.2	1 48.4	5 15.0	16
8 52.8	5 22.8	0 21.0	1 32.5	5 11.0	17
8 47.2	5 7.2	♐29 54.5	1 16.2	5 6.9	18
8 41.5	4 51.5	29 27.7	0 59.7	5 2.7	19
8 35.8	4 35.6	29 0.6	0 42.9	4 58.4	20
8 30.1	4 19.6	28 33.1	0 25.7	4 54.0	21
8 24.3	4 3.4	28 5.3	0 8.2	4 49.6	22
8 18.4	3 47.1	27 37.1	♑29 50.3	4 45.1	23
8 12.5	3 30.6	27 8.5	29 32.1	4 40.4	24
8 6.5	3 13.9	26 39.5	29 13.4	4 35.7	25
8 0.5	2 57.0	26 10.0	28 54.3	4 30.9	26
7 54.3	2 39.8	25 40.1	28 34.7	4 26.0	27
7 48.1	2 22.5	25 9.6	28 14.6	4 20.9	28
7 41.8	2 4.9	24 38.7	27 54.0	4 15.7	29
7 35.3	1 47.0	24 7.1	27 32.8	4 10.4	30
7 28.8	1 28.9	23 35.0	27 11.0	4 4.9	31
7 22.2	1 10.5	23 2.3	26 48.6	3 59.3	32
7 15.4	0 51.7	22 29.0	26 25.5	3 53.5	33
7 8.6	0 32.6	21 55.0	26 1.7	3 47.5	34
7 1.6	0 13.2	21 20.2	25 37.1	3 41.3	35
6 54.4	♏29 53.4	20 44.7	25 11.6	3 34.9	36
6 47.1	29 33.2	20 8.5	24 45.3	3 28.4	37
6 39.7	29 12.6	19 31.4	24 17.9	3 21.5	38
6 32.0	28 51.5	18 53.4	23 49.6	3 14.5	39
6 24.2	28 30.0	18 14.5	23 20.1	3 7.1	40
6 16.3	28 7.9	17 34.6	22 49.5	2 59.5	41
6 8.1	27 45.3	16 53.7	22 17.5	2 51.5	42
5 59.6	27 22.2	16 11.8	21 44.2	2 43.3	43
5 51.0	26 58.4	15 28.7	21 9.3	2 34.6	44
5 42.1	26 34.0	14 44.5	20 32.8	2 25.5	45
5 33.0	26 9.0	13 59.0	19 54.5	2 16.1	46
5 23.5	25 43.2	13 12.1	19 14.3	2 6.1	47
5 13.8	25 16.7	12 23.9	18 31.9	1 55.6	48
5 3.7	24 49.3	11 34.3	17 47.2	1 44.6	49
4 53.3	24 21.1	10 43.1	16 59.9	1 32.9	50
4 42.6	23 51.9	9 50.3	16 9.7	1 20.6	51
4 31.4	23 21.8	8 55.8	15 16.4	1 7.5	52
4 19.8	22 50.6	7 59.5	14 19.6	0 53.6	53
4 7.8	22 18.3	7 1.4	13 18.9	0 38.7	54
3 55.3	21 44.8	6 1.2	12 13.7	0 22.8	55
3 42.2	21 10.0	4 59.0	11 3.6	0 5.7	56
3 28.6	20 33.8	3 54.6	9 47.9	♒29 47.3	57
3 14.3	19 56.1	2 47.9	8 25.6	29 27.3	58
2 59.4	19 16.9	1 38.7	6 55.9	29 5.7	59
2 43.7	18 36.0	0 27.1	5 17.5	28 42.0	60
2 27.2	17 53.2	♏29 12.8	3 28.8	28 16.0	61
2 9.9	17 8.5	27 55.7	1 27.8	27 47.3	62
1 51.5	16 21.6	26 35.7	♐29 11.8	27 15.3	63
1 32.2	15 32.5	25 12.6	26 37.2	26 39.4	64
1 11.6	14 40.9	23 46.4	23 38.5	25 58.7	65
0 49.8	13 46.7	22 16.9	20 7.7	25 11.8	66
5	6	Descendant	8	9	S LAT

♈ 8° 42′ 33″

0h 32m 0s — MC — 8° 0′ 0″

12h 36m 0s — MC — 189° 0′ 0″

♎ 9° 47′ 41″

11	12	Ascendant	2	3	N LAT
° ′	° ′	° ′	° ′	° ′	°
♏11 26.0	♐10 35.9	♑ 8 16.1	♒ 6 36.6	♓ 7 17.7	0
10 58.5	9 20.5	6 16.9	5 29.5	7 1.0	5
10 31.0	8 5.1	4 14.9	4 19.4	6 43.5	10
10 3.2	6 48.6	2 8.5	3 4.9	6 25.1	15
9 57.6	6 33.1	1 42.5	2 49.3	6 21.2	16
9 51.9	6 17.5	1 16.2	2 33.5	6 17.3	17
9 46.2	6 1.7	0 49.6	2 17.4	6 13.4	18
9 40.4	5 45.9	0 22.8	2 1.0	6 9.4	19
9 34.6	5 29.9	♐29 55.6	1 44.3	6 5.3	20
9 28.8	5 13.7	29 28.0	1 27.3	6 1.1	21
9 22.9	4 57.4	29 0.1	1 9.9	5 56.8	22
9 16.9	4 41.0	28 31.8	0 52.2	5 52.5	23
9 10.8	4 24.3	28 3.1	0 34.0	5 48.0	24
9 4.7	4 7.5	27 33.9	0 15.5	5 43.5	25
8 58.5	3 50.4	27 4.3	♑29 56.5	5 38.9	26
8 52.2	3 33.1	26 34.3	29 37.0	5 34.1	27
8 45.9	3 15.6	26 3.7	29 17.1	5 29.3	28
8 39.4	2 57.9	25 32.5	28 56.6	5 24.3	29
8 32.9	2 39.8	25 0.9	28 35.5	5 19.2	30
8 26.2	2 21.5	24 28.6	28 13.9	5 13.9	31
8 19.4	2 2.9	23 55.7	27 51.6	5 8.5	32
8 12.5	1 44.0	23 22.1	27 28.6	5 3.0	33
8 5.5	1 24.7	22 47.9	27 4.9	4 57.2	34
7 58.4	1 5.1	22 12.9	26 40.4	4 51.3	35
7 51.0	0 45.0	21 37.1	26 15.1	4 45.2	36
7 43.6	0 24.6	21 0.6	25 48.8	4 38.9	37
7 36.0	0 3.8	20 23.2	25 21.6	4 32.3	38
7 28.2	♏29 42.5	19 44.9	24 53.4	4 25.5	39
7 20.2	29 20.7	19 5.7	24 24.0	4 18.5	40
7 12.0	28 58.4	18 25.5	23 53.4	4 11.1	41
7 3.6	28 35.5	17 44.2	23 21.6	4 3.5	42
6 55.0	28 12.1	17 1.9	22 48.3	3 55.5	43
6 46.2	27 48.0	16 18.4	22 13.5	3 47.2	44
6 37.1	27 23.4	15 33.7	21 37.1	3 38.5	45
6 27.7	26 58.0	14 47.7	20 58.8	3 29.4	46
6 18.1	26 31.9	14 0.4	20 18.6	3 19.9	47
6 8.1	26 5.0	13 11.6	19 36.2	3 9.8	48
5 57.8	25 37.2	12 21.4	18 51.5	2 59.2	49
5 47.2	25 8.6	11 29.6	18 4.1	2 48.0	50
5 36.2	24 39.0	10 36.2	17 13.9	2 36.2	51
5 24.7	24 8.5	9 41.0	16 20.5	2 23.6	52
5 12.9	23 36.8	8 44.0	15 23.5	2 10.2	53
5 0.5	23 4.0	7 45.1	14 22.6	1 56.0	54
4 47.7	22 30.0	6 44.1	13 17.1	1 40.7	55
4 34.3	21 54.7	5 41.1	12 6.7	1 24.3	56
4 20.3	21 17.9	4 35.7	10 50.4	1 6.6	57
4 5.6	20 39.7	3 28.1	9 27.6	0 47.5	58
3 50.3	19 59.8	2 17.9	7 57.1	0 26.7	59
3 34.2	19 18.1	1 5.2	6 17.6	0 4.0	60
3 17.3	18 34.7	♏29 49.7	4 27.6	♒29 39.2	61
2 59.5	17 49.1	28 31.4	2 24.8	29 11.7	62
2 40.6	17 1.5	27 10.2	0 6.5	28 41.1	63
2 20.7	16 11.4	25 45.8	♐27 28.5	28 6.8	64
1 59.6	15 18.9	24 18.2	24 25.1	27 28.0	65
1 37.2	14 23.6	22 47.3	20 47.0	26 43.4	66
5	6	Descendant	8	9	S LAT

♈ 9° 47′ 41″

0h 36m 0s — MC — 9° 0′ 0″

12h 40m 0s		MC ♎ 10° 52′ 45″		190° 0′ 0″	N
11	12	Ascendant	2	3	LAT
° ′	° ′	° ′	° ′	° ′	°
♏12 26.7	♐11 32.1	♑ 9 11.4	♒ 7 35.4	♓ 8 21.7	0
11 58.7	10 16.3	7 12.3	6 29.1	8 5.6	5
11 30.8	9 0.4	5 10.4	5 19.7	7 48.9	10
11 2.5	7 43.5	3 3.8	4 5.9	7 31.3	15
10 56.7	7 27.8	2 37.8	3 50.4	7 27.6	16
10 50.9	7 12.1	2 11.5	3 34.8	7 23.9	17
10 45.1	6 56.2	1 44.9	3 18.8	7 20.1	18
10 39.2	6 40.2	1 17.9	3 2.5	7 16.2	19
10 33.3	6 24.1	0 50.7	2 46.0	7 12.3	20
10 27.3	6 7.8	0 23.0	2 29.1	7 8.3	21
10 21.3	5 51.4	♐29 55.0	2 11.9	7 4.2	22
10 15.2	5 34.8	29 26.7	1 54.3	7 0.1	23
10 9.0	5 18.0	28 57.8	1 36.3	6 55.8	24
10 2.8	5 1.0	28 28.6	1 17.9	6 51.5	25
9 56.4	4 43.8	27 58.9	0 59.0	6 47.1	26
9 50.0	4 26.4	27 28.6	0 39.7	6 42.5	27
9 43.6	4 8.7	26 57.9	0 19.9	6 37.9	28
9 37.0	3 50.8	26 26.6	♑29 59.6	6 33.1	29
9 30.3	3 32.6	25 54.8	29 38.6	6 28.2	30
9 23.5	3 14.1	25 22.3	29 17.1	6 23.2	31
9 16.6	2 55.4	24 49.2	28 55.0	6 18.0	32
9 9.5	2 36.2	24 15.4	28 32.1	6 12.7	33
9 2.3	2 16.8	23 41.0	28 8.5	6 7.2	34
8 55.0	1 56.9	23 5.7	27 44.1	6 1.5	35
8 47.6	1 36.7	22 29.8	27 18.9	5 55.6	36
8 40.0	1 16.1	21 52.9	26 52.8	5 49.6	37
8 32.2	0 55.0	21 15.3	26 25.7	5 43.3	38
8 24.2	0 33.4	20 36.7	25 57.6	5 36.8	39
8 16.1	0 11.4	19 57.1	25 28.4	5 30.1	40
8 7.7	♏29 48.8	19 16.5	24 57.9	5 23.0	41
7 59.1	29 25.7	18 34.9	24 26.1	5 15.7	42
7 50.3	29 2.0	17 52.2	23 53.0	5 8.1	43
7 41.3	28 37.7	17 8.3	23 18.2	5 0.2	44
7 32.0	28 12.7	16 23.1	22 41.9	4 51.8	45
7 22.4	27 47.0	15 36.7	22 3.7	4 43.1	46
7 12.6	27 20.5	14 48.9	21 23.5	4 33.9	47
7 2.4	26 53.3	13 59.6	20 41.1	4 24.3	48
6 51.8	26 25.2	13 8.8	19 56.4	4 14.2	49
6 40.9	25 56.2	12 16.4	19 9.0	4 3.4	50
6 29.7	25 26.2	11 22.3	18 18.8	3 52.1	51
6 17.9	24 55.2	10 26.5	17 25.3	3 40.1	52
6 5.8	24 23.1	9 28.7	16 28.2	3 27.2	53
5 53.2	23 49.8	8 29.0	15 27.0	3 13.6	54
5 40.0	23 15.3	7 27.3	14 21.3	2 59.0	55
5 26.3	22 39.4	6 23.3	13 10.5	2 43.3	56
5 11.9	22 2.1	5 17.1	11 53.8	2 26.4	57
4 56.9	21 23.2	4 8.4	10 30.4	2 8.1	58
4 41.2	20 42.7	2 57.2	8 59.2	1 48.2	59
4 24.7	20 0.3	1 43.4	7 18.7	1 26.5	60
4 7.3	19 16.1	0 26.8	5 27.4	1 2.7	61
3 49.0	18 29.8	♏29 7.3	3 22.9	0 36.5	62
3 29.7	17 41.2	27 44.7	1 2.2	0 7.3	63
3 9.2	16 50.3	26 19.0	♐28 20.9	♒29 34.7	64
2 47.5	15 56.8	24 50.0	25 12.6	28 57.7	65
2 24.4	15 0.4	23 17.6	21 26.8	28 15.5	66
5	6	Descendant	8	9	S LAT
0h 40m 0s		MC ♈ 10° 52′ 45″		10° 0′ 0″	

12h 44m 0s		MC ♎ 11° 57′ 44″		191° 0′ 0″	N
11	12	Ascendant	2	3	LAT
° ′	° ′	° ′	° ′	° ′	°
♏13 27.3	♐12 28.1	♑10 6.7	♒ 8 34.4	♓ 9 25.7	0
12 58.8	11 11.9	8 7.9	7 28.8	9 10.4	5
12 30.4	9 55.7	6 6.0	6 20.2	8 54.5	10
12 1.5	8 38.2	3 59.4	5 7.1	8 37.7	15
11 55.7	8 22.5	3 33.3	4 51.8	8 34.2	16
11 49.8	8 6.6	3 6.9	4 36.3	8 30.6	17
11 43.8	7 50.7	2 40.3	4 20.5	8 27.0	18
11 37.8	7 34.6	2 13.3	4 4.4	8 23.3	19
11 31.8	7 18.3	1 45.9	3 48.0	8 19.5	20
11 25.7	7 1.9	1 18.2	3 31.3	8 15.7	21
11 19.6	6 45.4	0 50.2	3 14.2	8 11.8	22
11 13.4	6 28.6	0 21.7	2 56.7	8 7.9	23
11 7.1	6 11.7	♐29 52.8	2 38.9	8 3.8	24
11 0.7	5 54.6	29 23.4	2 20.6	7 59.7	25
10 54.3	5 37.3	28 53.6	2 1.9	7 55.4	26
10 47.7	5 19.7	28 23.2	1 42.7	7 51.1	27
10 41.1	5 1.9	27 52.3	1 23.1	7 46.7	28
10 34.4	4 43.8	27 20.9	1 2.9	7 42.1	29
10 27.6	4 25.4	26 48.9	0 42.1	7 37.4	30
10 20.6	4 6.8	26 16.3	0 20.7	7 32.6	31
10 13.6	3 47.8	25 43.0	♑29 58.7	7 27.7	32
10 6.4	3 28.5	25 9.0	29 36.0	7 22.6	33
9 59.1	3 8.8	24 34.3	29 12.6	7 17.3	34
9 51.6	2 48.8	23 58.9	28 48.3	7 11.9	35
9 44.0	2 28.4	23 22.6	28 23.2	7 6.3	36
9 36.2	2 7.5	22 45.5	27 57.3	7 0.6	37
9 28.3	1 46.2	22 7.6	27 30.3	6 54.6	38
9 20.2	1 24.4	21 28.7	27 2.3	6 48.4	39
9 11.8	1 2.2	20 48.8	26 33.2	6 41.9	40
9 3.3	0 39.4	20 7.9	26 2.9	6 35.2	41
8 54.6	0 16.0	19 25.9	25 31.2	6 28.2	42
8 45.6	♏29 52.0	18 42.8	24 58.1	6 20.9	43
8 36.3	29 27.4	17 58.4	24 23.5	6 13.3	44
8 26.8	29 2.1	17 12.8	23 47.2	6 5.4	45
8 17.0	28 36.1	16 25.9	23 9.1	5 57.1	46
8 6.9	28 9.3	15 37.6	22 29.0	5 48.3	47
7 56.5	27 41.7	14 47.8	21 46.7	5 39.1	48
7 45.8	27 13.2	13 56.4	21 2.0	5 29.4	49
7 34.6	26 43.8	13 3.4	20 14.6	5 19.2	50
7 23.1	26 13.4	12 8.7	19 24.3	5 8.3	51
7 11.1	25 42.0	11 12.1	18 30.8	4 56.8	52
6 58.7	25 9.4	10 13.6	17 33.6	4 44.6	53
6 45.7	24 35.7	9 13.2	16 32.3	4 31.6	54
6 32.2	24 0.6	8 10.6	15 26.4	4 17.6	55
6 18.2	23 24.2	7 5.7	14 15.3	4 2.6	56
6 3.5	22 46.3	5 58.6	12 58.2	3 46.5	57
5 48.1	22 6.8	4 48.9	11 34.2	3 29.0	58
5 32.0	21 25.6	3 36.7	10 2.2	3 10.1	59
5 15.1	20 42.5	2 21.7	8 20.9	2 49.4	60
4 57.3	19 57.5	1 4.0	6 28.3	2 26.8	61
4 38.5	19 10.4	♏29 43.2	4 22.1	2 1.8	62
4 18.7	18 21.0	28 19.4	1 59.0	1 34.0	63
3 57.6	17 29.2	26 52.3	♐29 14.4	1 3.0	64
3 35.3	16 34.6	25 21.8	26 1.1	0 27.9	65
3 11.6	15 37.2	23 47.9	22 7.1	♒29 47.9	66
5	6	Descendant	8	9	S LAT
0h 44m 0s		MC ♈ 11° 57′ 44″		11° 0′ 0″	

12ʰ 48ᵐ 0ˢ		MC ♎ 13° 2′ 39″		192° 0′ 0″

N LAT	11	12	Ascendant	2	3
°	° ′	° ′	° ′	° ′	° ′
0	♏14 27.7	♐13 24.0	♑11 2.1	♒ 9 33.6	♓10 29.9
5	13 58.8	12 7.5	9 3.5	8 28.8	10 15.3
10	13 29.8	10 50.9	7 1.7	7 20.9	10 0.2
15	13 0.4	9 33.0	4 55.0	6 8.6	9 44.2
16	12 54.5	9 17.1	4 28.9	5 53.5	9 40.8
17	12 48.5	9 1.2	4 2.5	5 38.1	9 37.5
18	12 42.4	8 45.1	3 35.8	5 22.5	9 34.0
19	12 36.3	8 28.9	3 8.8	5 6.5	9 30.5
20	12 30.2	8 12.5	2 41.4	4 50.3	9 26.9
21	12 24.0	7 56.0	2 13.6	4 33.7	9 23.3
22	12 17.7	7 39.3	1 45.5	4 16.8	9 19.6
23	12 11.4	7 22.5	1 16.9	3 59.5	9 15.8
24	12 5.0	7 5.4	0 47.9	3 41.8	9 12.0
25	11 58.5	6 48.1	0 18.4	3 23.7	9 8.0
26	11 51.9	6 30.7	♐29 48.5	3 5.1	9 4.0
27	11 45.3	6 13.0	29 18.0	2 46.1	8 59.9
28	11 38.5	5 55.0	28 47.0	2 26.6	8 55.7
29	11 31.7	5 36.7	28 15.4	2 6.6	8 51.3
30	11 24.7	5 18.2	27 43.2	1 45.9	8 46.9
31	11 17.7	4 59.4	27 10.4	1 24.7	8 42.3
32	11 10.5	4 40.3	26 37.0	1 2.9	8 37.6
33	11 3.2	4 20.8	26 2.8	0 40.3	8 32.7
34	10 55.7	4 0.9	25 27.9	0 17.0	8 27.7
35	10 48.1	3 40.7	24 52.2	♑29 52.9	8 22.6
36	10 40.3	3 20.1	24 15.7	29 28.0	8 17.3
37	10 32.4	2 59.0	23 38.4	29 2.2	8 11.8
38	10 24.3	2 37.5	23 0.1	28 35.4	8 6.1
39	10 16.0	2 15.5	22 20.9	28 7.5	8 0.1
40	10 7.5	1 53.0	21 40.7	27 38.5	7 54.0
41	9 58.8	1 29.9	20 59.5	27 8.3	7 47.6
42	9 49.9	1 6.3	20 17.1	26 36.8	7 41.0
43	9 40.7	0 42.0	19 33.6	26 3.9	7 34.0
44	9 31.3	0 17.1	18 48.8	25 29.4	7 26.8
45	9 21.6	♏29 51.5	18 2.8	24 53.2	7 19.2
46	9 11.6	29 25.2	17 15.4	24 15.2	7 11.3
47	9 1.3	28 58.0	16 26.6	23 35.2	7 3.0
48	8 50.6	28 30.1	15 36.2	22 52.9	6 54.2
49	8 39.6	28 1.3	14 44.3	22 8.3	6 45.0
50	8 28.2	27 31.5	13 50.6	21 20.9	6 35.2
51	8 16.4	27 0.7	12 55.3	20 30.6	6 24.9
52	8 4.2	26 28.8	11 58.0	19 37.0	6 14.0
53	7 51.5	25 55.8	10 58.8	18 39.8	6 2.3
54	7 38.2	25 21.5	9 57.5	17 38.3	5 49.9
55	7 24.4	24 46.0	8 54.1	16 32.3	5 36.6
56	7 10.0	24 9.0	7 48.4	15 20.9	5 22.3
57	6 55.0	23 30.5	6 40.2	14 3.4	5 7.0
58	6 39.3	22 50.4	5 29.6	12 39.0	4 50.4
59	6 22.8	22 8.5	4 16.3	11 6.3	4 32.4
60	6 5.4	21 24.8	3 0.2	9 24.0	4 12.7
61	5 47.2	20 39.0	1 41.3	7 30.2	3 51.2
62	5 27.9	19 51.1	0 19.2	5 22.4	3 27.5
63	5 7.5	19 0.8	♏28 54.1	2 57.0	3 1.1
64	4 46.0	18 8.0	27 25.6	0 9.0	2 31.7
65	4 23.1	17 12.5	25 53.7	♐26 50.6	1 58.5
66	3 58.7	16 14.0	24 18.2	22 48.1	1 20.7
S LAT	5	6	Descendant	8	9

0ʰ 48ᵐ 0ˢ		MC ♈ 13° 2′ 39″		12° 0′ 0″

12ʰ 52ᵐ 0ˢ		MC ♎ 14° 7′ 28″		193° 0′ 0″

N LAT	11	12	Ascendant	2	3
°	° ′	° ′	° ′	° ′	° ′
0	♏15 28.0	♐14 19.9	♑11 57.6	♒10 32.9	♓11 34.2
5	14 58.5	13 3.0	9 59.2	9 28.9	11 20.4
10	14 29.1	11 46.0	7 57.6	8 21.9	11 6.0
15	13 59.2	10 27.6	5 50.8	7 10.4	10 50.9
16	13 53.2	10 11.7	5 24.7	6 55.4	10 47.7
17	13 47.0	9 55.6	4 58.3	6 40.2	10 44.5
18	13 40.9	9 39.5	4 31.5	6 24.7	10 41.2
19	13 34.7	9 23.1	4 4.4	6 8.9	10 37.9
20	13 28.4	9 6.7	3 37.0	5 52.8	10 34.5
21	13 22.1	8 50.0	3 9.2	5 36.4	10 31.1
22	13 15.7	8 33.2	2 40.9	5 19.7	10 27.6
23	13 9.3	8 16.3	2 12.3	5 2.5	10 24.0
24	13 2.8	7 59.1	1 43.2	4 45.0	10 20.3
25	12 56.2	7 41.7	1 13.6	4 27.1	10 16.6
26	12 49.5	7 24.1	0 43.6	4 8.7	10 12.8
27	12 42.7	7 6.2	0 13.0	3 49.8	10 8.8
28	12 35.9	6 48.1	♐29 41.8	3 30.5	10 4.8
29	12 28.9	6 29.7	29 10.1	3 10.6	10 0.7
30	12 21.8	6 11.0	28 37.8	2 50.2	9 56.5
31	12 14.6	5 52.0	28 4.8	2 29.1	9 52.1
32	12 7.3	5 32.7	27 31.2	2 7.4	9 47.7
33	11 59.8	5 13.1	26 56.8	1 45.0	9 43.1
34	11 52.2	4 53.0	26 21.7	1 21.9	9 38.3
35	11 44.5	4 32.6	25 45.8	0 58.0	9 33.5
36	11 36.5	4 11.8	25 9.1	0 33.2	9 28.4
37	11 28.5	3 50.5	24 31.5	0 7.5	9 23.2
38	11 20.2	3 28.8	23 52.9	♑29 40.9	9 17.8
39	11 11.7	3 6.6	23 13.4	29 13.2	9 12.1
40	11 3.1	2 43.8	22 32.9	28 44.4	9 6.3
41	10 54.2	2 20.5	21 51.3	28 14.3	9 0.3
42	10 45.1	1 56.6	21 8.6	27 43.0	8 53.9
43	10 35.7	1 32.1	20 24.7	27 10.2	8 47.4
44	10 26.1	1 6.9	19 39.5	26 35.8	8 40.5
45	10 16.2	0 41.0	18 53.0	25 59.7	8 33.3
46	10 6.0	0 14.3	18 5.2	25 21.9	8 25.8
47	9 55.5	♏29 46.9	17 15.8	24 41.9	8 17.9
48	9 44.6	29 18.6	16 24.9	23 59.8	8 9.6
49	9 33.4	28 49.4	15 32.4	23 15.2	8 0.8
50	9 21.7	28 19.2	14 38.1	22 27.9	7 51.6
51	9 9.7	27 48.0	13 42.1	21 37.7	7 41.8
52	8 57.2	27 15.7	12 44.1	20 44.1	7 31.4
53	8 44.2	26 42.2	11 44.2	19 46.8	7 20.3
54	8 30.6	26 7.4	10 42.1	18 45.3	7 8.5
55	8 16.5	25 31.4	9 37.8	17 39.0	6 55.9
56	8 1.8	24 53.8	8 31.2	16 27.4	6 42.4
57	7 46.4	24 14.7	7 22.1	15 9.6	6 27.8
58	7 30.3	23 34.0	6 10.4	13 44.7	6 12.1
59	7 13.4	22 51.5	4 56.1	12 11.5	5 55.0
60	6 55.7	22 7.0	3 38.8	10 28.3	5 36.4
61	6 37.0	21 20.5	2 18.7	8 33.4	5 16.0
62	6 17.2	20 31.7	0 55.4	6 23.9	4 53.6
63	5 56.3	19 40.6	♏29 28.8	3 56.2	4 28.7
64	5 34.2	18 46.8	27 58.9	1 4.9	4 0.9
65	5 10.7	17 50.2	26 25.5	♐27 41.3	3 29.6
66	4 45.7	16 50.7	24 48.4	23 29.7	2 53.9
S LAT	5	6	Descendant	8	9

0ʰ 52ᵐ 0ˢ		MC ♈ 14° 7′ 28″		13° 0′ 0″

12ʰ 56ᵐ 0ˢ — MC — 194° 0′ 0″

♎ 15° 12′ 13″

N LAT	11	12	Ascendant	2	3
°	° ′	° ′	° ′	° ′	° ′
0	♏16 28.0	♐15 15.6	♑12 53.1	♒11 32.4	♓12 38.6
5	15 58.1	13 58.5	10 55.0	10 29.3	12 25.6
10	15 28.2	12 41.1	8 53.5	9 23.1	12 12.0
15	14 57.8	11 22.3	6 46.8	8 12.4	11 57.7
16	14 51.7	11 6.3	6 20.6	7 57.6	11 54.7
17	14 45.5	10 50.1	5 54.2	7 42.6	11 51.7
18	14 39.2	10 33.8	5 27.4	7 27.2	11 48.6
19	14 32.9	10 17.4	5 0.3	7 11.6	11 45.4
20	14 26.5	10 0.8	4 32.8	6 55.7	11 42.2
21	14 20.1	9 44.1	4 4.9	6 39.5	11 39.0
22	14 13.6	9 27.2	3 36.6	6 22.9	11 35.7
23	14 7.1	9 10.1	3 7.9	6 5.9	11 32.3
24	14 0.5	8 52.8	2 38.7	5 48.6	11 28.8
25	13 53.7	8 35.2	2 9.1	5 30.8	11 25.3
26	13 46.9	8 17.5	1 38.9	5 12.6	11 21.7
27	13 40.0	7 59.5	1 8.2	4 53.9	11 18.0
28	13 33.1	7 41.2	0 36.9	4 34.7	11 14.2
29	13 26.0	7 22.7	0 5.1	4 15.0	11 10.3
30	13 18.7	7 3.9	♐29 32.6	3 54.8	11 6.3
31	13 11.4	6 44.7	28 59.5	3 33.9	11 2.2
32	13 3.9	6 25.2	28 25.6	3 12.4	10 58.0
33	12 56.4	6 5.4	27 51.1	2 50.1	10 53.6
34	12 48.6	5 45.2	27 15.8	2 27.2	10 49.1
35	12 40.7	5 24.6	26 39.7	2 3.5	10 44.5
36	12 32.7	5 3.5	26 2.7	1 38.9	10 39.7
37	12 24.4	4 42.1	25 24.8	1 13.4	10 34.8
38	12 16.0	4 20.1	24 46.0	0 46.9	10 29.7
39	12 7.4	3 57.7	24 6.2	0 19.4	10 24.4
40	11 58.6	3 34.7	23 25.4	♑29 50.7	10 18.9
41	11 49.5	3 11.1	22 43.5	29 20.9	10 13.1
42	11 40.2	2 47.0	22 0.4	28 49.7	10 7.2
43	11 30.7	2 22.2	21 16.1	28 17.0	10 0.9
44	11 20.9	1 56.7	20 30.5	27 42.8	9 54.4
45	11 10.8	1 30.5	19 43.6	27 6.9	9 47.7
46	11 0.4	1 3.5	18 55.2	26 29.2	9 40.5
47	10 49.6	0 35.7	18 5.3	25 49.4	9 33.1
48	10 38.5	0 7.1	17 13.9	25 7.3	9 25.2
49	10 27.1	♏29 37.5	16 20.8	24 22.9	9 16.9
50	10 15.2	29 7.0	15 25.9	23 35.7	9 8.2
51	10 2.9	28 35.3	14 29.2	22 45.5	8 58.9
52	9 50.1	28 2.6	13 30.6	21 51.9	8 49.1
53	9 36.8	27 28.6	12 29.8	20 54.6	8 38.6
54	9 23.0	26 53.4	11 27.0	19 53.0	8 27.5
55	9 8.6	26 16.8	10 21.8	18 46.7	8 15.6
56	8 53.5	25 38.7	9 14.3	17 34.9	8 2.8
57	8 37.8	24 59.0	8 4.2	16 16.8	7 49.1
58	8 21.3	24 17.7	6 51.5	14 51.5	7 34.2
59	8 4.0	23 34.5	5 36.0	13 17.7	7 18.1
60	7 45.9	22 49.3	4 17.6	11 33.8	7 0.5
61	7 26.7	22 2.0	2 56.2	9 37.7	6 41.3
62	7 6.5	21 12.4	1 31.6	7 26.7	6 20.1
63	6 45.1	20 20.3	0 3.6	4 56.8	5 56.6
64	6 22.4	19 25.6	♏28 32.3	2 2.2	5 30.4
65	5 58.3	18 28.0	26 57.3	♐28 33.2	5 1.0
66	5 32.6	17 27.3	25 18.7	24 12.2	4 27.5
S LAT	5	6	Descendant	8	9

♈ 15° 12′ 13″

0ʰ 56ᵐ 0ˢ — MC — 14° 0′ 0″

13ʰ 0ᵐ 0ˢ — MC — 195° 0′ 0″

♎ 16° 16′ 51″

N LAT	11	12	Ascendant	2	3
°	° ′	° ′	° ′	° ′	° ′
0	♏17 27.9	♐16 11.3	♑13 48.7	♒12 32.1	♓13 43.2
5	16 57.5	14 53.9	11 50.9	11 29.8	13 30.9
10	16 27.2	13 36.1	9 49.6	10 24.5	13 18.1
15	15 56.3	12 16.9	7 42.9	9 14.7	13 4.6
16	15 50.1	12 0.8	7 16.7	9 0.1	13 1.8
17	15 43.7	11 44.5	6 50.3	8 45.2	12 59.0
18	15 37.4	11 28.2	6 23.5	8 30.1	12 56.1
19	15 31.0	11 11.6	5 56.3	8 14.6	12 53.1
20	15 24.5	10 54.9	5 28.8	7 58.9	12 50.1
21	15 18.0	10 38.1	5 0.8	7 42.8	12 47.0
22	15 11.4	10 21.1	4 32.5	7 26.4	12 43.9
23	15 4.7	10 3.8	4 3.7	7 9.6	12 40.7
24	14 58.0	9 46.4	3 34.4	6 52.4	12 37.5
25	14 51.2	9 28.8	3 4.7	6 34.9	12 34.1
26	14 44.3	9 10.9	2 34.4	6 16.8	12 30.7
27	14 37.2	8 52.8	2 3.6	5 58.4	12 27.3
28	14 30.1	8 34.4	1 32.3	5 39.4	12 23.7
29	14 22.9	8 15.7	1 0.3	5 19.8	12 20.0
30	14 15.6	7 56.7	0 27.7	4 59.8	12 16.3
31	14 8.1	7 37.4	♐29 54.4	4 39.1	12 12.4
32	14 0.5	7 17.7	29 20.4	4 17.7	12 8.4
33	13 52.8	6 57.7	28 45.6	3 55.7	12 4.3
34	13 44.9	6 37.3	28 10.1	3 32.9	12 0.1
35	13 36.9	6 16.6	27 33.8	3 9.4	11 55.8
36	13 28.7	5 55.3	26 56.6	2 45.0	11 51.3
37	13 20.3	5 33.6	26 18.5	2 19.7	11 46.6
38	13 11.7	5 11.5	25 39.4	1 53.4	11 41.8
39	13 2.9	4 48.8	24 59.3	1 26.1	11 36.8
40	12 54.0	4 25.6	24 18.2	0 57.6	11 31.6
41	12 44.7	4 1.8	23 35.9	0 27.9	11 26.2
42	12 35.3	3 37.4	22 52.5	♑29 56.9	11 20.6
43	12 25.6	3 12.3	22 7.8	29 24.4	11 14.7
44	12 15.6	2 46.5	21 21.8	28 50.4	11 8.6
45	12 5.3	2 20.1	20 34.4	28 14.7	11 2.2
46	11 54.6	1 52.8	19 45.6	27 37.1	10 55.5
47	11 43.7	1 24.7	18 55.2	26 57.5	10 48.5
48	11 32.4	0 55.7	18 3.2	26 15.6	10 41.1
49	11 20.7	0 25.7	17 9.5	25 31.2	10 33.3
50	11 8.6	♏29 54.8	16 14.0	24 44.2	10 25.0
51	10 56.0	29 22.8	15 16.6	23 54.0	10 16.3
52	10 43.0	28 49.6	14 17.2	23 0.5	10 7.1
53	10 29.4	28 15.2	13 15.7	22 3.3	9 57.3
54	10 15.3	27 39.4	12 12.1	21 1.7	9 46.8
55	10 0.6	27 2.3	11 6.0	19 55.3	9 35.6
56	9 45.2	26 23.7	9 57.5	18 43.4	9 23.5
57	9 29.1	25 43.4	8 46.5	17 25.1	9 10.6
58	9 12.3	25 1.4	7 32.7	15 59.4	8 56.6
59	8 54.6	24 17.5	6 16.1	14 25.1	8 41.4
60	8 36.0	23 31.6	4 56.5	12 40.4	8 24.9
61	8 16.4	22 43.5	3 33.8	10 43.4	8 6.8
62	7 55.7	21 53.1	2 7.9	8 30.9	7 46.9
63	7 33.7	21 0.1	0 38.6	5 58.8	7 24.9
64	7 10.5	20 4.4	♏29 5.7	3 0.9	7 0.3
65	6 45.8	19 5.7	27 29.2	♐29 26.6	6 32.8
66	6 19.4	18 3.9	25 48.9	24 55.6	6 1.5
S LAT	5	6	Descendant	8	9

♈ 16° 16′ 51″

1ʰ 0ᵐ 0ˢ — MC — 15° 0′ 0″

13ʰ 4ᵐ 0ˢ — MC 196° 0′ 0″ — ♎ 17° 21′ 23″

N LAT	11	12	Ascendant	2	3
0°	♏18° 27.6′	♐17° 6.9′	♑14° 44.4′	♒13° 32.0′	♓14° 47.8′
5	17 56.8	15 49.2	12 46.9	12 30.6	14 36.3
10	17 26.0	14 31.1	10 45.8	11 26.1	14 24.4
15	16 54.7	13 11.5	8 39.2	10 17.2	14 11.7
16	16 48.3	12 55.3	8 13.0	10 2.8	14 9.1
17	16 41.9	12 38.9	7 46.5	9 48.1	14 6.4
18	16 35.4	12 22.5	7 19.7	9 33.2	14 3.7
19	16 28.9	12 5.8	6 52.5	9 17.9	14 0.9
20	16 22.4	11 49.1	6 24.9	9 2.3	13 58.1
21	16 15.7	11 32.1	5 57.0	8 46.5	13 55.2
22	16 9.0	11 15.0	5 28.6	8 30.2	13 52.3
23	16 2.3	10 57.6	4 59.7	8 13.6	13 49.3
24	15 55.4	10 40.1	4 30.4	7 56.7	13 46.3
25	15 48.5	10 22.3	4 0.6	7 39.3	13 43.2
26	15 41.4	10 4.3	3 30.2	7 21.4	13 40.0
27	15 34.3	9 46.0	2 59.3	7 3.1	13 36.7
28	15 27.1	9 27.5	2 27.8	6 44.4	13 33.4
29	15 19.7	9 8.7	1 55.7	6 25.0	13 29.9
30	15 12.3	8 49.5	1 23.0	6 5.1	13 26.4
31	15 4.7	8 30.1	0 49.5	5 44.6	13 22.8
32	14 57.0	8 10.3	0 15.4	5 23.5	13 19.1
33	14 49.1	7 50.1	♐29 40.4	5 1.7	13 15.2
34	14 41.1	7 29.5	29 4.7	4 39.1	13 11.3
35	14 32.9	7 8.6	28 28.2	4 15.8	13 7.2
36	14 24.6	6 47.2	27 50.8	3 51.6	13 3.0
37	14 16.1	6 25.3	27 12.4	3 26.5	12 58.6
38	14 7.3	6 2.9	26 33.1	3 0.4	12 54.1
39	13 58.4	5 40.0	25 52.7	2 33.3	12 49.4
40	13 49.3	5 16.5	25 11.2	2 5.0	12 44.6
41	13 39.9	4 52.5	24 28.6	1 35.5	12 39.5
42	13 30.2	4 27.8	23 44.8	1 4.7	12 34.2
43	13 20.3	4 2.5	22 59.8	0 32.5	12 28.8
44	13 10.2	3 36.5	22 13.3	♑29 58.6	12 23.0
45	12 59.7	3 9.7	21 25.5	29 23.1	12 17.0
46	12 48.9	2 42.1	20 36.2	28 45.7	12 10.7
47	12 37.7	2 13.7	19 45.3	28 6.2	12 4.2
48	12 26.2	1 44.3	18 52.8	27 24.5	11 57.2
49	12 14.2	1 14.0	17 58.5	26 40.3	11 49.9
50	12 1.9	0 42.7	17 2.4	25 53.4	11 42.2
51	11 49.1	0 10.2	16 4.3	25 3.4	11 34.0
52	11 35.8	♏29 36.6	15 4.2	24 10.0	11 25.4
53	11 21.9	29 1.8	14 1.9	23 12.8	11 16.1
54	11 7.5	28 25.5	12 57.4	22 11.3	11 6.3
55	10 52.5	27 47.9	11 50.5	21 4.8	10 55.8
56	10 36.8	27 8.7	10 41.1	19 52.8	10 44.5
57	10 20.4	26 27.8	9 29.0	18 34.3	10 32.4
58	10 3.1	25 45.1	8 14.2	17 8.4	10 19.3
59	9 45.0	25 0.6	6 56.4	15 33.6	10 5.1
60	9 26.0	24 13.9	5 35.6	13 48.3	9 49.7
61	9 6.0	23 25.0	4 11.6	11 50.3	9 32.8
62	8 44.8	22 33.7	2 44.3	9 36.4	9 14.1
63	8 22.3	21 39.8	1 13.5	7 2.3	8 53.6
64	7 58.5	20 43.1	♏29 39.2	4 1.1	8 30.6
65	7 33.2	19 43.4	28 1.0	0 21.4	8 4.9
66	7 6.1	18 40.4	26 19.1	♐25 40.1	7 35.8
S LAT	5	6	Descendant	8	9

♈ 17° 21′ 23″ — 1ʰ 4ᵐ 0ˢ — MC 16° 0′ 0″

13ʰ 8ᵐ 0ˢ — MC 197° 0′ 0″ — ♎ 18° 25′ 48″

N LAT	11	12	Ascendant	2	3
0°	♏19° 27.1′	♐18° 2.4′	♑15° 40.1′	♒14° 32.0′	♓15° 52.5′
5	18 55.8	16 44.5	13 43.1	13 31.6	15 41.8
10	18 24.6	15 26.1	11 42.2	12 28.0	15 30.7
15	17 52.8	14 6.1	9 35.6	11 20.1	15 18.9
16	17 46.4	13 49.8	9 9.5	11 5.8	15 16.5
17	17 39.9	13 33.3	8 43.0	10 51.3	15 14.0
18	17 33.3	13 16.8	8 16.1	10 36.5	15 11.4
19	17 26.7	13 0.1	7 48.9	10 21.5	15 8.9
20	17 20.1	12 43.2	7 21.3	10 6.1	15 6.2
21	17 13.3	12 26.1	6 53.3	9 50.4	15 3.6
22	17 6.5	12 8.9	6 24.8	9 34.4	15 0.8
23	16 59.7	11 51.4	5 55.9	9 18.0	14 58.1
24	16 52.7	11 33.8	5 26.6	9 1.2	14 55.2
25	16 45.7	11 15.9	4 56.7	8 44.0	14 52.3
26	16 38.5	10 57.7	4 26.2	8 26.4	14 49.3
27	16 31.3	10 39.4	3 55.2	8 8.3	14 46.3
28	16 23.9	10 20.7	3 23.6	7 49.7	14 43.2
29	16 16.5	10 1.7	2 51.4	7 30.6	14 40.0
30	16 8.9	9 42.4	2 18.5	7 10.9	14 36.7
31	16 1.2	9 22.8	1 44.9	6 50.6	14 33.3
32	15 53.3	9 2.9	1 10.6	6 29.7	14 29.9
33	15 45.4	8 42.5	0 35.5	6 8.1	14 26.3
34	15 37.2	8 21.8	♐29 59.6	5 45.8	14 22.6
35	15 28.9	8 0.6	29 22.9	5 22.6	14 18.8
36	15 20.4	7 39.0	28 45.2	4 58.7	14 14.9
37	15 11.7	7 16.9	28 6.6	4 33.8	14 10.8
38	15 2.9	6 54.4	27 27.0	4 7.9	14 6.6
39	14 53.8	6 31.2	26 46.4	3 41.0	14 2.2
40	14 44.5	6 7.6	26 4.6	3 13.0	13 57.7
41	14 34.9	5 43.3	25 21.7	2 43.7	13 53.0
42	14 25.1	5 18.4	24 37.5	2 13.1	13 48.1
43	14 15.0	4 52.8	23 52.1	1 41.1	13 43.0
44	14 4.7	4 26.5	23 5.2	1 7.5	13 37.6
45	13 54.0	3 59.4	22 17.0	0 32.1	13 32.0
46	13 43.0	3 31.5	21 27.2	♑29 54.9	13 26.2
47	13 31.6	3 2.7	20 35.8	29 15.7	13 20.0
48	13 19.9	2 33.0	19 42.7	28 34.2	13 13.6
49	13 7.7	2 2.4	18 47.8	27 50.2	13 6.8
50	12 55.1	1 30.6	17 51.0	27 3.4	12 59.6
51	12 42.1	0 57.8	16 52.3	26 13.6	12 51.9
52	12 28.5	0 23.7	15 51.5	25 20.3	12 43.9
53	12 14.4	♏29 48.4	14 48.4	24 23.2	12 35.3
54	11 59.7	29 11.7	13 43.1	23 21.8	12 26.1
55	11 44.3	28 33.5	12 35.2	22 15.3	12 16.3
56	11 28.3	27 53.7	11 24.9	21 3.3	12 5.8
57	11 11.5	27 12.3	10 11.8	19 44.7	11 54.6
58	10 53.9	26 29.0	8 55.8	18 18.5	11 42.4
59	10 35.4	25 43.7	7 36.9	16 43.4	11 29.1
60	10 16.0	24 56.3	6 14.9	14 57.5	11 14.7
61	9 55.5	24 6.6	4 49.6	12 58.6	10 59.0
62	9 33.8	23 14.4	3 20.9	10 43.5	10 41.7
63	9 10.8	22 19.6	1 48.6	8 7.4	10 22.6
64	8 46.4	21 21.9	0 12.7	5 3.0	10 1.3
65	8 20.5	20 21.1	♏28 32.9	1 18.0	9 37.4
66	7 52.7	19 16.9	26 49.2	♐26 25.9	9 10.4
S LAT	5	6	Descendant	8	9

♈ 18° 25′ 48″ — 1ʰ 8ᵐ 0ˢ — MC 17° 0′ 0″

13ʰ 12ᵐ 0ˢ		MC ♎ 19° 30′ 6″		198° 0′ 0″	N
11	**12**	**Ascendant**	**2**	**3**	**LAT**
° ′	° ′	° ′	° ′	° ′	°
♏20 26.4	♐18 57.9	♑16 36.0	♒15 32.3	♓16 57.4	**0**
19 54.8	17 39.7	14 39.3	14 32.7	16 47.5	**5**
19 23.1	16 21.0	12 38.7	13 30.1	16 37.2	**10**
18 50.9	15 0.6	10 32.3	12 23.1	16 26.3	**15**
18 44.3	14 44.3	10 6.1	12 9.1	16 24.0	**16**
18 37.7	14 27.7	9 39.6	11 54.8	16 21.7	**17**
18 31.1	14 11.1	9 12.7	11 40.2	16 19.3	**18**
18 24.4	13 54.3	8 45.5	11 25.3	16 16.9	**19**
18 17.6	13 37.3	8 17.9	11 10.2	16 14.5	**20**
18 10.8	13 20.1	7 49.8	10 54.7	16 12.0	**21**
18 3.9	13 2.8	7 21.4	10 38.9	16 9.5	**22**
17 56.9	12 45.2	6 52.4	10 22.7	16 6.9	**23**
17 49.9	12 27.5	6 23.0	10 6.1	16 4.3	**24**
17 42.7	12 9.5	5 53.0	9 49.1	16 1.6	**25**
17 35.5	11 51.2	5 22.5	9 31.7	15 58.9	**26**
17 28.1	11 32.7	4 51.4	9 13.8	15 56.0	**27**
17 20.7	11 13.9	4 19.7	8 55.5	15 53.2	**28**
17 13.1	10 54.8	3 47.4	8 36.6	15 50.2	**29**
17 5.4	10 35.3	3 14.3	8 17.1	15 47.2	**30**
16 57.6	10 15.6	2 40.6	7 57.0	15 44.0	**31**
16 49.6	9 55.5	2 6.2	7 36.3	15 40.8	**32**
16 41.5	9 35.0	1 30.9	7 15.0	15 37.5	**33**
16 33.2	9 14.1	0 54.8	6 52.8	15 34.1	**34**
16 24.8	8 52.7	0 17.9	6 29.9	15 30.6	**35**
16 16.1	8 30.9	♐29 40.0	6 6.2	15 26.9	**36**
16 7.3	8 8.7	29 1.2	5 41.6	15 23.2	**37**
15 58.3	7 45.9	28 21.3	5 15.9	15 19.3	**38**
15 49.1	7 22.5	27 40.4	4 49.3	15 15.2	**39**
15 39.6	6 58.6	26 58.3	4 21.5	15 11.0	**40**
15 29.9	6 34.1	26 15.1	3 52.4	15 6.7	**41**
15 19.9	6 9.0	25 30.6	3 22.1	15 2.1	**42**
15 9.6	5 43.1	24 44.7	2 50.3	14 57.4	**43**
14 59.1	5 16.5	23 57.5	2 16.9	14 52.4	**44**
14 48.2	4 49.1	23 8.8	1 41.8	14 47.3	**45**
14 37.0	4 20.9	22 18.5	1 4.9	14 41.8	**46**
14 25.4	3 51.9	21 26.6	0 25.9	14 36.1	**47**
14 13.5	3 21.8	20 32.9	♑29 44.6	14 30.1	**48**
14 1.1	2 50.8	19 37.4	29 0.8	14 23.8	**49**
13 48.3	2 18.7	18 40.0	28 14.2	14 17.2	**50**
13 35.0	1 45.4	17 40.6	27 24.6	14 10.1	**51**
13 21.1	1 10.9	16 39.0	26 31.5	14 2.6	**52**
13 6.8	0 35.1	15 35.2	25 34.5	13 54.7	**53**
12 51.8	♏29 57.9	14 29.0	24 33.2	13 46.2	**54**
12 36.1	29 19.2	13 20.3	23 26.9	13 37.1	**55**
12 19.8	28 38.9	12 8.9	22 14.8	13 27.4	**56**
12 2.6	27 56.8	10 54.8	20 56.2	13 17.0	**57**
11 44.7	27 12.8	9 37.7	19 29.9	13 5.7	**58**
11 25.8	26 26.9	8 17.6	17 54.4	12 53.4	**59**
11 5.9	25 38.7	6 54.3	16 8.1	12 40.1	**60**
10 44.9	24 48.2	5 27.7	14 8.4	12 25.6	**61**
10 22.8	23 55.2	3 57.6	11 52.0	12 9.6	**62**
9 59.3	22 59.4	2 23.8	9 14.1	11 51.9	**63**
9 34.3	22 0.6	0 46.3	6 6.7	11 32.2	**64**
9 7.7	20 58.7	♏29 4.8	2 16.4	11 10.2	**65**
8 39.3	19 53.3	27 19.4	♐27 13.1	10 45.3	**66**
5	**6**	**Descendant**	**8**	**9**	**S LAT**
1ʰ 12ᵐ 0ˢ		♈ 19° 30′ 6″ MC		18° 0′ 0″	

13ʰ 16ᵐ 0ˢ		MC ♎ 20° 34′ 17″		199° 0′ 0″	N
11	**12**	**Ascendant**	**2**	**3**	**LAT**
° ′	° ′	° ′	° ′	° ′	°
♏21 25.6	♐19 53.3	♑17 31.9	♒16 32.7	♓18 2.3	**0**
20 53.5	18 34.9	15 35.7	15 34.1	17 53.2	**5**
20 21.4	17 15.9	13 35.3	14 32.5	17 43.7	**10**
19 48.8	15 55.2	11 29.1	13 26.5	17 33.7	**15**
19 42.2	15 38.7	11 2.9	13 12.6	17 31.6	**16**
19 35.5	15 22.1	10 36.4	12 58.5	17 29.5	**17**
19 28.7	15 5.4	10 9.6	12 44.2	17 27.3	**18**
19 21.9	14 48.5	9 42.3	12 29.5	17 25.1	**19**
19 15.1	14 31.4	9 14.7	12 14.6	17 22.9	**20**
19 8.2	14 14.2	8 46.6	11 59.3	17 20.6	**21**
19 1.2	13 56.7	8 18.1	11 43.7	17 18.3	**22**
18 54.1	13 39.0	7 49.1	11 27.7	17 15.9	**23**
18 46.9	13 21.2	7 19.6	11 11.3	17 13.5	**24**
18 39.7	13 3.0	6 49.6	10 54.6	17 11.0	**25**
18 32.3	12 44.7	6 19.0	10 37.4	17 8.5	**26**
18 24.8	12 26.0	5 47.8	10 19.7	17 5.9	**27**
18 17.3	12 7.1	5 16.0	10 1.6	17 3.3	**28**
18 9.6	11 47.9	4 43.6	9 42.9	17 0.5	**29**
18 1.8	11 28.3	4 10.5	9 23.7	16 57.7	**30**
17 53.8	11 8.4	3 36.6	9 3.9	16 54.9	**31**
17 45.7	10 48.1	3 2.0	8 43.4	16 51.9	**32**
17 37.5	10 27.5	2 26.6	8 22.2	16 48.9	**33**
17 29.1	10 6.4	1 50.3	8 0.4	16 45.7	**34**
17 20.5	9 44.9	1 13.2	7 37.7	16 42.5	**35**
17 11.7	9 22.9	0 35.1	7 14.2	16 39.1	**36**
17 2.8	9 0.4	♐29 56.0	6 49.8	16 35.7	**37**
16 53.6	8 37.5	29 15.9	6 24.5	16 32.1	**38**
16 44.2	8 13.9	28 34.7	5 58.0	16 28.4	**39**
16 34.6	7 49.8	27 52.4	5 30.5	16 24.5	**40**
16 24.7	7 25.0	27 8.8	5 1.7	16 20.5	**41**
16 14.6	6 59.6	26 23.9	4 31.6	16 16.3	**42**
16 4.2	6 33.5	25 37.7	4 0.1	16 12.0	**43**
15 53.4	6 6.7	24 50.0	3 27.0	16 7.4	**44**
15 42.4	5 39.0	24 0.9	2 52.2	16 2.6	**45**
15 31.0	5 10.5	23 10.1	2 15.5	15 57.7	**46**
15 19.2	4 41.1	22 17.7	1 36.7	15 52.4	**47**
15 7.0	4 10.7	21 23.4	0 55.7	15 46.9	**48**
14 54.4	3 39.3	20 27.4	0 12.2	15 41.1	**49**
14 41.4	3 6.8	19 29.4	♑29 25.9	15 35.0	**50**
14 27.8	2 33.1	18 29.3	28 36.4	15 28.5	**51**
14 13.7	1 58.2	17 26.9	27 43.6	15 21.6	**52**
13 59.1	1 21.9	16 22.3	26 46.8	15 14.3	**53**
13 43.8	0 44.2	15 15.2	25 45.6	15 6.5	**54**
13 27.8	0 5.0	14 5.6	24 39.4	14 58.2	**55**
13 11.1	♏29 24.1	12 53.2	23 27.4	14 49.2	**56**
12 53.7	28 41.4	11 38.0	22 8.8	14 39.6	**57**
12 35.3	27 56.8	10 19.9	20 42.4	14 29.2	**58**
12 16.1	27 10.1	8 58.6	19 6.7	14 18.0	**59**
11 55.8	26 21.2	7 34.0	17 20.0	14 5.8	**60**
11 34.3	25 29.8	6 6.0	15 19.6	13 52.4	**61**
11 11.7	24 35.9	4 34.4	13 2.2	13 37.7	**62**
10 47.6	23 39.2	2 59.1	10 22.5	13 21.5	**63**
10 22.1	22 39.4	1 19.9	7 12.3	13 3.4	**64**
9 54.8	21 36.3	♏29 36.8	3 16.8	12 43.2	**65**
9 25.7	20 29.7	27 49.5	♐28 2.0	12 20.4	**66**
5	**6**	**Descendant**	**8**	**9**	**S LAT**
1ʰ 16ᵐ 0ˢ		♈ 20° 34′ 17″ MC		19° 0′ 0″	

13h 20m 0s MC 200° 0′ 0″

♎ 21° 38′ 20″

11	12	Ascendant	2	3	N LAT
° ′	° ′	° ′	° ′	° ′	°
♏22 24.6	♐20 48.6	♑18 27.9	♒17 33.3	♓19 7.3	0
21 52.1	19 30.0	16 32.1	16 35.7	18 59.0	5
21 19.6	18 10.8	14 32.1	15 35.1	18 50.3	10
20 46.6	16 49.7	12 26.0	14 30.1	18 41.2	15
20 39.8	16 33.2	11 59.9	14 16.5	18 39.3	16
20 33.1	16 16.5	11 33.4	14 2.6	18 37.4	17
20 26.2	15 59.7	11 6.6	13 48.4	18 35.4	18
20 19.3	15 42.7	10 39.3	13 34.0	18 33.4	19
20 12.4	15 25.5	10 11.7	13 19.3	18 31.4	20
20 5.4	15 8.2	9 43.6	13 4.2	18 29.3	21
19 58.3	14 50.6	9 15.1	12 48.8	18 27.2	22
19 51.1	14 32.9	8 46.0	12 33.1	18 25.0	23
19 43.8	14 14.9	8 16.5	12 16.9	18 22.8	24
19 36.5	13 56.7	7 46.4	12 0.4	18 20.6	25
19 29.0	13 38.2	7 15.8	11 43.4	18 18.3	26
19 21.5	13 19.4	6 44.5	11 26.0	18 15.9	27
19 13.8	13 0.4	6 12.6	11 8.1	18 13.5	28
19 6.0	12 41.0	5 40.1	10 49.7	18 11.0	29
18 58.1	12 21.3	5 6.9	10 30.7	18 8.4	30
18 50.0	12 1.2	4 32.9	10 11.1	18 5.8	31
18 41.8	11 40.8	3 58.1	9 50.9	18 3.1	32
18 33.4	11 20.0	3 22.6	9 30.0	18 0.4	33
18 24.9	10 58.8	2 46.1	9 8.4	17 57.5	34
18 16.2	10 37.1	2 8.8	8 46.0	17 54.5	35
18 7.3	10 14.9	1 30.5	8 22.7	17 51.5	36
17 58.2	9 52.3	0 51.2	7 58.6	17 48.3	37
17 48.9	9 29.1	0 10.9	7 33.5	17 45.1	38
17 39.3	9 5.3	♐29 29.4	7 7.4	17 41.7	39
17 29.6	8 41.0	28 46.7	6 40.1	17 38.2	40
17 19.5	8 16.0	28 2.8	6 11.6	17 34.5	41
17 9.2	7 50.4	27 17.6	5 41.8	17 30.7	42
16 58.6	7 24.0	26 31.0	5 10.5	17 26.7	43
16 47.7	6 56.9	25 43.0	4 37.7	17 22.6	44
16 36.5	6 28.9	24 53.4	4 3.2	17 18.2	45
16 24.9	6 0.1	24 2.1	3 26.8	17 13.7	46
16 12.9	5 30.4	23 9.2	2 48.3	17 8.9	47
16 0.5	4 59.7	22 14.4	2 7.6	17 3.9	48
15 47.7	4 27.9	21 17.8	1 24.3	16 58.6	49
15 34.4	3 55.0	20 19.1	0 38.3	16 53.0	50
15 20.6	3 20.9	19 18.2	♑29 49.1	16 47.1	51
15 6.3	2 45.6	18 15.2	28 56.5	16 40.8	52
14 51.3	2 8.8	17 9.7	28 0.0	16 34.1	53
14 35.7	1 30.6	16 1.8	26 59.0	16 27.0	54
14 19.5	0 50.8	14 51.2	25 53.0	16 19.4	55
14 2.5	0 9.4	13 37.9	24 41.2	16 11.3	56
13 44.7	♏29 26.0	12 21.6	23 22.6	16 2.5	57
13 26.0	28 40.8	11 2.3	21 56.2	15 53.1	58
13 6.3	27 53.4	9 39.7	20 20.4	15 42.8	59
12 45.6	27 3.7	8 13.8	18 33.3	15 31.7	60
12 23.7	26 11.5	6 44.4	16 32.4	15 19.5	61
12 0.5	25 16.7	5 11.3	14 14.1	15 6.1	62
11 35.9	24 19.0	3 34.5	11 32.8	14 51.4	63
11 9.8	23 18.1	1 53.6	8 20.0	14 34.9	64
10 41.9	22 13.9	0 8.7	4 19.3	14 16.6	65
10 12.0	21 6.0	♏28 19.6	♐28 52.8	13 55.9	66
5	6	Descendant	8	9	S LAT

♈ 21° 38′ 20″

1h 20m 0s MC 20° 0′ 0″

13h 24m 0s MC 201° 0′ 0″

♎ 22° 42′ 16″

N LAT	11	12	Ascendant	2	3
°	° ′	° ′	° ′	° ′	° ′
0	♏23 23.4	♐21 43.9	♑19 24.1	♒18 34.0	♓20 12.3
5	22 50.5	20 25.1	17 28.7	17 37.5	20 4.9
10	22 17.7	19 5.6	15 29.1	16 37.9	19 57.1
15	21 44.2	17 44.3	13 23.2	15 34.0	19 48.8
16	21 37.4	17 27.6	12 57.1	15 20.6	19 47.1
17	21 30.5	17 10.9	12 30.7	15 6.9	19 45.4
18	21 23.6	16 54.0	12 3.8	14 53.0	19 43.6
19	21 16.6	16 36.9	11 36.6	14 38.8	19 41.8
20	21 9.6	16 19.7	11 8.9	14 24.2	19 39.9
21	21 2.5	16 2.2	10 40.8	14 9.4	19 38.1
22	20 55.3	15 44.6	10 12.3	13 54.3	19 36.2
23	20 48.0	15 26.7	9 43.2	13 38.7	19 34.2
24	20 40.6	15 8.7	9 13.6	13 22.8	19 32.2
25	20 33.2	14 50.3	8 43.5	13 6.5	19 30.2
26	20 25.6	14 31.7	8 12.8	12 49.8	19 28.1
27	20 18.0	14 12.8	7 41.5	12 32.6	19 26.0
28	20 10.2	13 53.7	7 9.5	12 15.0	19 23.8
29	20 2.3	13 34.2	6 36.9	11 56.8	19 21.6
30	19 54.2	13 14.3	6 3.6	11 38.1	19 19.3
31	19 46.0	12 54.1	5 29.5	11 18.7	19 16.9
32	19 37.7	12 33.6	4 54.6	10 58.8	19 14.5
33	19 29.2	12 12.6	4 18.9	10 38.2	19 12.0
34	19 20.6	11 51.2	3 42.3	10 16.8	19 9.4
35	19 11.7	11 29.4	3 4.8	9 54.7	19 6.7
36	19 2.7	11 7.0	2 26.3	9 31.7	19 4.0
37	18 53.5	10 44.2	1 46.8	9 7.9	19 1.1
38	18 44.0	10 20.8	1 6.2	8 43.1	18 58.2
39	18 34.3	9 56.9	0 24.4	8 17.2	18 55.1
40	18 24.4	9 32.3	♐29 41.5	7 50.2	18 51.9
41	18 14.2	9 7.1	28 57.3	7 22.1	18 48.6
42	18 3.8	8 41.2	28 11.7	6 52.5	18 45.2
43	17 53.0	8 14.6	27 24.7	6 21.6	18 41.6
44	17 41.9	7 47.2	26 36.3	5 49.1	18 37.9
45	17 30.5	7 18.9	25 46.2	5 14.9	18 33.9
46	17 18.7	6 49.8	24 54.5	4 38.8	18 29.8
47	17 6.5	6 19.8	24 1.1	4 0.6	18 25.5
48	16 53.9	5 48.7	23 5.7	3 20.2	18 21.0
49	16 40.9	5 16.6	22 8.5	2 37.3	18 16.2
50	16 27.4	4 43.3	21 9.1	1 51.6	18 11.2
51	16 13.3	4 8.8	20 7.6	1 2.7	18 5.8
52	15 58.7	3 33.0	19 3.8	0 10.4	18 0.2
53	15 43.5	2 55.8	17 57.5	♑29 14.1	17 54.2
54	15 27.7	2 17.1	16 48.7	28 13.4	17 47.7
55	15 11.1	1 36.8	15 37.1	27 7.6	17 40.9
56	14 53.8	0 54.7	14 22.8	25 56.0	17 33.5
57	14 35.6	0 10.8	13 5.4	24 37.6	17 25.6
58	14 16.5	♏29 24.8	11 44.9	23 11.2	17 17.1
59	13 56.4	28 36.7	10 21.1	21 35.4	17 7.9
60	13 35.3	27 46.3	8 53.9	19 48.1	16 57.8
61	13 12.9	26 53.3	7 23.1	17 46.8	16 46.8
62	12 49.2	25 57.5	5 48.5	15 27.7	16 34.8
63	12 24.1	24 58.8	4 10.0	12 45.0	16 21.5
64	11 57.4	23 56.9	2 27.4	9 29.7	16 6.7
65	11 28.8	22 51.5	0 40.7	5 24.3	15 50.2
66	10 58.3	21 42.3	♏28 49.7	♐29 45.9	15 31.6
S LAT	5	6	Descendant	8	9

♈ 22° 42′ 16″

1h 24m 0s MC 21° 0′ 0″

13ʰ 28ᵐ 0ˢ — MC — 202° 0′ 0″

♎ 23° 46′ 3″

N LAT	11	12	Ascendant	2	3
°	° ′	° ′	° ′	° ′	° ′
0	♏24 22.0	♐22 39.2	♑20 20.3	♒19 35.0	♓21 17.5
5	23 48.8	21 20.2	18 25.5	18 39.5	21 10.8
10	23 15.5	20 0.5	16 26.2	17 41.0	21 3.9
15	22 41.7	18 38.8	14 20.6	16 38.2	20 56.5
16	22 34.8	18 22.1	13 54.5	16 25.0	20 55.0
17	22 27.8	18 5.3	13 28.1	16 11.5	20 53.4
18	22 20.8	17 48.3	13 1.3	15 57.8	20 51.9
19	22 13.8	17 31.2	12 34.1	15 43.8	20 50.3
20	22 6.6	17 13.8	12 6.4	15 29.5	20 48.6
21	21 59.4	16 56.3	11 38.3	15 15.0	20 47.0
22	21 52.2	16 38.6	11 9.7	15 0.0	20 45.3
23	21 44.8	16 20.6	10 40.7	14 44.8	20 43.5
24	21 37.3	16 2.4	10 11.0	14 29.1	20 41.7
25	21 29.8	15 44.0	9 40.9	14 13.1	20 39.9
26	21 22.1	15 25.3	9 10.1	13 56.6	20 38.1
27	21 14.3	15 6.3	8 38.8	13 39.7	20 36.2
28	21 6.5	14 47.0	8 6.7	13 22.2	20 34.2
29	20 58.4	14 27.4	7 34.0	13 4.3	20 32.2
30	20 50.3	14 7.4	7 0.6	12 45.9	20 30.2
31	20 42.0	13 47.1	6 26.4	12 26.8	20 28.1
32	20 33.6	13 26.4	5 51.4	12 7.1	20 25.9
33	20 25.0	13 5.3	5 15.5	11 46.8	20 23.7
34	20 16.2	12 43.7	4 38.8	11 25.7	20 21.4
35	20 7.2	12 21.7	4 1.1	11 3.9	20 19.0
36	19 58.1	11 59.2	3 22.4	10 41.2	20 16.6
37	19 48.7	11 36.2	2 42.7	10 17.7	20 14.0
38	19 39.1	11 12.6	2 1.8	9 53.2	20 11.4
39	19 29.3	10 48.4	1 19.8	9 27.6	20 8.7
40	19 19.2	10 23.7	0 36.6	9 1.0	20 5.8
41	19 8.9	9 58.2	♐29 52.1	8 33.1	20 2.9
42	18 58.2	9 32.1	29 6.2	8 3.9	19 59.8
43	18 47.3	9 5.2	28 18.8	7 33.3	19 56.6
44	18 36.0	8 37.5	27 30.0	7 1.1	19 53.3
45	18 24.4	8 9.0	26 39.5	6 27.2	19 49.8
46	18 12.5	7 39.6	25 47.3	5 51.5	19 46.1
47	18 0.1	7 9.2	24 53.3	5 13.7	19 42.3
48	17 47.3	6 37.9	23 57.4	4 33.6	19 38.3
49	17 34.0	6 5.4	22 59.6	3 51.0	19 34.0
50	17 20.3	5 31.7	21 59.6	3 5.7	19 29.5
51	17 6.0	4 56.8	20 57.3	2 17.2	19 24.8
52	16 51.1	4 20.6	19 52.7	1 25.2	19 19.7
53	16 35.6	3 42.9	18 45.6	0 29.3	19 14.4
54	16 19.5	3 3.7	17 35.9	♑29 28.9	19 8.6
55	16 2.6	2 22.8	16 23.4	28 23.4	19 2.5
56	15 45.0	1 40.2	15 8.0	27 12.0	18 56.0
57	15 26.5	0 55.6	13 49.6	25 53.8	18 49.0
58	15 7.0	0 9.0	12 27.9	24 27.6	18 41.4
59	14 46.5	♏29 20.2	11 2.8	22 51.8	18 33.1
60	14 25.0	28 28.9	9 34.2	21 4.5	18 24.2
61	14 2.1	27 35.0	8 1.9	19 2.9	18 14.4
62	13 37.9	26 38.4	6 25.7	16 43.1	18 3.7
63	13 12.3	25 38.6	4 45.6	13 59.3	17 51.9
64	12 44.9	24 35.6	3 1.3	10 41.7	17 38.7
65	12 15.7	23 29.0	1 12.7	6 31.9	17 24.0
66	11 44.4	22 18.5	♏29 19.8	0 41.6	17 7.5
S LAT	5	6	Descendant	8	9

♈ 23° 46′ 3″

1ʰ 28ᵐ 0ˢ — MC — 22° 0′ 0″

13ʰ 32ᵐ 0ˢ — MC — 203° 0′ 0″

♎ 24° 49′ 42″

N LAT	11	12	Ascendant	2	3
°	° ′	° ′	° ′	° ′	° ′
0	♏25 20.5	♐23 34.4	♑21 16.7	♒20 36.1	♓22 22.6
5	24 46.9	22 15.2	19 22.4	19 41.7	22 16.8
10	24 13.3	20 55.3	17 23.5	18 44.3	22 10.7
15	23 39.0	19 33.3	15 18.2	17 42.6	22 4.3
16	23 32.1	19 16.6	14 52.2	17 29.6	22 3.0
17	23 25.0	18 59.7	14 25.8	17 16.4	22 1.6
18	23 17.9	18 42.6	13 59.0	17 2.9	22 0.2
19	23 10.8	18 25.4	13 31.8	16 49.2	21 58.8
20	23 3.6	18 8.0	13 4.1	16 35.2	21 57.4
21	22 56.3	17 50.4	12 36.0	16 20.8	21 55.9
22	22 48.9	17 32.6	12 7.5	16 6.1	21 54.4
23	22 41.5	17 14.6	11 38.4	15 51.1	21 52.9
24	22 33.9	16 56.3	11 8.7	15 35.7	21 51.3
25	22 26.2	16 37.7	10 38.5	15 19.9	21 49.8
26	22 18.5	16 18.9	10 7.7	15 3.7	21 48.1
27	22 10.6	15 59.8	9 36.3	14 47.0	21 46.5
28	22 2.6	15 40.4	9 4.2	14 29.9	21 44.8
29	21 54.5	15 20.7	8 31.4	14 12.3	21 43.0
30	21 46.3	15 0.6	7 57.9	13 54.1	21 41.2
31	21 37.9	14 40.1	7 23.6	13 35.3	21 39.4
32	21 29.3	14 19.3	6 48.5	13 15.9	21 37.5
33	21 20.6	13 58.0	6 12.5	12 55.9	21 35.5
34	21 11.7	13 36.3	5 35.6	12 35.1	21 33.5
35	21 2.6	13 14.1	4 57.7	12 13.6	21 31.4
36	20 53.3	12 51.4	4 18.8	11 51.2	21 29.3
37	20 43.8	12 28.2	3 38.9	11 28.0	21 27.0
38	20 34.1	12 4.5	2 57.8	11 3.8	21 24.7
39	20 24.1	11 40.1	2 15.6	10 38.6	21 22.3
40	20 13.9	11 15.1	1 32.1	10 12.2	21 19.9
41	20 3.4	10 49.5	0 47.2	9 44.7	21 17.3
42	19 52.6	10 23.1	0 1.0	9 15.9	21 14.6
43	19 41.5	9 56.0	♐29 13.3	8 45.6	21 11.8
44	19 30.1	9 28.0	28 24.0	8 13.8	21 8.9
45	19 18.3	8 59.2	27 33.1	7 40.3	21 5.8
46	19 6.1	8 29.5	26 40.5	7 4.9	21 2.6
47	18 53.6	7 58.8	25 46.0	6 27.5	20 59.2
48	18 40.6	7 27.1	24 49.6	5 47.8	20 55.7
49	18 27.1	6 54.3	23 51.1	5 5.6	20 51.9
50	18 13.1	6 20.2	22 50.4	4 20.6	20 48.0
51	17 58.6	5 44.9	21 47.4	3 32.5	20 43.8
52	17 43.5	5 8.2	20 42.1	2 40.9	20 39.4
53	17 27.7	4 30.1	19 34.1	1 45.4	20 34.7
54	17 11.3	3 50.4	18 23.5	0 45.4	20 29.7
55	16 54.1	3 9.0	17 10.1	♑29 40.2	20 24.4
56	16 36.1	2 25.7	15 53.6	28 29.2	20 18.6
57	16 17.3	1 40.6	14 34.0	27 11.3	20 12.4
58	15 57.5	0 53.3	13 11.1	25 45.3	20 5.8
59	15 36.6	0 3.7	11 44.7	24 9.7	19 58.6
60	15 14.6	♏29 11.6	10 14.7	22 22.4	19 50.7
61	14 51.3	28 16.9	8 40.9	20 20.6	19 42.2
62	14 26.6	27 19.3	7 3.2	18 0.4	19 32.8
63	14 0.3	26 18.5	5 21.3	15 15.6	19 22.4
64	13 32.4	25 14.4	3 35.3	11 56.2	19 10.9
65	13 2.5	24 6.5	1 44.8	7 42.2	18 58.1
66	12 30.5	22 54.7	♏29 49.9	1 40.4	18 43.6
S LAT	5	6	Descendant	8	9

♈ 24° 49′ 42″

1ʰ 32ᵐ 0ˢ — MC — 23° 0′ 0″

13ʰ 36ᵐ 0ˢ	MC	♎ 25° 53′ 12″	204° 0′ 0″		N LAT	13ʰ 40ᵐ 0ˢ	MC	♎ 26° 56′ 33″	205° 0′ 0″	
11	**12**	**Ascendant**	**2**	**3**		**11**	**12**	**Ascendant**	**2**	**3**
° ′	° ′	° ′	° ′	° ′	°	° ′	° ′	° ′	° ′	° ′
♏26 18.8	♐24 29.5	♑22 13.1	♒21 37.4	♓23 27.9	**0**	♏27 17.0	♐25 24.7	♑23 9.7	♒22 38.9	♓24 33.2
25 44.9	23 10.2	20 19.4	20 44.1	23 22.9	**5**	26 42.7	24 5.2	21 16.6	21 46.7	24 29.0
25 10.9	21 50.1	18 21.0	19 47.8	23 17.7	**10**	26 8.4	22 44.9	19 18.7	20 51.6	24 24.6
24 36.3	20 27.9	16 16.0	18 47.3	23 12.1	**15**	25 33.4	21 22.5	17 14.0	19 52.2	24 20.0
24 29.2	20 11.1	15 50.0	18 34.6	23 11.0	**16**	25 26.2	21 5.6	16 48.1	19 39.8	24 19.1
24 22.1	19 54.1	15 23.7	18 21.6	23 9.8	**17**	25 19.0	20 48.6	16 21.8	19 27.0	24 18.1
24 14.9	19 37.0	14 56.9	18 8.4	23 8.6	**18**	25 11.8	20 31.4	15 55.1	19 14.1	24 17.1
24 7.7	19 19.7	14 29.7	17 54.9	23 7.4	**19**	25 4.5	20 14.0	15 27.9	19 0.8	24 16.1
24 0.4	19 2.2	14 2.1	17 41.1	23 6.2	**20**	24 57.1	19 56.5	15 0.3	18 47.3	24 15.1
23 53.0	18 44.5	13 34.0	17 27.0	23 4.9	**21**	24 49.6	19 38.7	14 32.3	18 33.5	24 14.0
23 45.5	18 26.6	13 5.4	17 12.6	23 3.6	**22**	24 42.1	19 20.7	14 3.7	18 19.3	24 12.9
23 38.0	18 8.5	12 36.3	16 57.8	23 2.3	**23**	24 34.4	19 2.5	13 34.6	18 4.8	24 11.9
23 30.4	17 50.1	12 6.7	16 42.7	23 1.0	**24**	24 26.7	18 44.1	13 4.9	17 49.9	24 10.7
23 22.6	17 31.5	11 36.5	16 27.1	22 59.6	**25**	24 18.9	18 25.3	12 34.7	17 34.7	24 9.6
23 14.8	17 12.6	11 5.6	16 11.2	22 58.3	**26**	24 10.9	18 6.3	12 3.8	17 19.0	24 8.4
23 6.8	16 53.4	10 34.2	15 54.8	22 56.8	**27**	24 2.9	17 47.0	11 32.3	17 2.9	24 7.2
22 58.7	16 33.9	10 2.0	15 37.9	22 55.4	**28**	23 54.7	17 27.4	11 0.1	16 46.4	24 6.0
22 50.5	16 14.0	9 29.2	15 20.6	22 53.9	**29**	23 46.4	17 7.4	10 27.2	16 29.3	24 4.8
22 42.1	15 53.8	8 55.5	15 2.7	22 52.3	**30**	23 37.9	16 47.1	9 53.5	16 11.7	24 3.5
22 33.6	15 33.2	8 21.1	14 44.2	22 50.7	**31**	23 29.3	16 26.3	9 19.0	15 53.6	24 2.2
22 24.9	15 12.2	7 45.9	14 25.1	22 49.1	**32**	23 20.5	16 5.2	8 43.7	15 34.8	24 0.8
22 16.1	14 50.8	7 9.8	14 5.4	22 47.4	**33**	23 11.5	15 43.7	8 7.5	15 15.4	23 59.4
22 7.1	14 28.9	6 32.7	13 44.9	22 45.7	**34**	23 2.4	15 21.7	7 30.3	14 55.2	23 58.0
21 57.9	14 6.6	5 54.7	13 23.7	22 43.9	**35**	22 53.1	14 59.2	6 52.1	14 34.3	23 56.5
21 48.5	13 43.8	5 15.7	13 1.7	22 42.1	**36**	22 43.6	14 36.2	6 12.9	14 12.7	23 54.9
21 38.8	13 20.4	4 35.5	12 38.8	22 40.1	**37**	22 33.8	14 12.6	5 32.6	13 50.1	23 53.3
21 29.0	12 56.4	3 54.3	12 14.9	22 38.2	**38**	22 23.8	13 48.5	4 51.1	13 26.6	23 51.7
21 18.9	12 31.9	3 11.8	11 50.1	22 36.1	**39**	22 13.6	13 23.7	4 8.3	13 2.1	23 50.0
21 8.5	12 6.7	2 28.0	11 24.1	22 34.0	**40**	22 3.1	12 58.3	3 24.3	12 36.5	23 48.2
20 57.9	11 40.8	1 42.8	10 56.9	22 31.8	**41**	21 52.3	12 32.2	2 38.8	12 9.7	23 46.3
20 46.9	11 14.2	0 56.3	10 28.5	22 29.5	**42**	21 41.2	12 5.4	1 52.0	11 41.7	23 44.4
20 35.7	10 46.8	0 8.2	9 58.6	22 27.0	**43**	21 29.8	11 37.8	1 3.5	11 12.2	23 42.4
20 24.1	10 18.6	♐29 18.6	9 27.2	22 24.5	**44**	21 18.0	11 9.3	0 13.5	10 41.2	23 40.3
20 12.1	9 49.5	28 27.2	8 54.0	22 21.9	**45**	21 5.9	10 40.0	♐29 21.7	10 8.5	23 38.1
19 59.8	9 19.5	27 34.1	8 19.1	22 19.1	**46**	20 53.3	10 9.7	28 28.1	9 33.9	23 35.8
19 47.0	8 48.5	26 39.1	7 42.0	22 16.3	**47**	20 40.4	9 38.4	27 32.6	8 57.3	23 33.4
19 33.8	8 16.5	25 42.1	7 2.8	22 13.2	**48**	20 26.9	9 6.0	26 35.0	8 18.5	23 30.8
19 20.1	7 43.3	24 43.0	6 21.0	22 10.0	**49**	20 13.0	8 32.4	25 35.3	7 37.2	23 28.2
19 5.9	7 8.8	23 41.7	5 36.4	22 6.6	**50**	19 58.6	7 57.6	24 33.3	6 53.1	23 25.3
18 51.1	6 33.1	22 38.0	4 48.8	22 3.0	**51**	19 43.6	7 21.5	23 28.9	6 5.9	23 22.4
18 35.8	5 56.0	21 31.8	3 57.6	21 59.2	**52**	19 28.0	6 43.9	22 22.0	5 15.3	23 19.2
18 19.7	5 17.4	20 23.1	3 2.5	21 55.2	**53**	19 11.7	6 4.8	21 12.4	4 20.6	23 15.8
18 3.0	4 37.2	19 11.5	2 2.9	21 50.9	**54**	18 54.7	5 24.1	19 59.9	3 21.6	23 12.2
17 45.6	3 55.2	17 57.1	0 58.2	21 46.3	**55**	18 36.9	4 41.6	18 44.5	2 17.3	23 8.4
17 27.3	3 11.4	16 39.5	♑29 47.6	21 41.4	**56**	18 18.3	3 57.2	17 25.9	1 7.2	23 4.3
17 8.0	2 25.6	15 18.8	28 30.0	21 36.1	**57**	17 58.8	3 10.7	16 3.9	♑29 50.1	22 59.9
16 47.9	1 37.6	13 54.6	27 4.4	21 30.4	**58**	17 38.2	2 22.0	14 38.5	28 24.9	22 55.1
16 26.6	0 47.3	12 26.9	25 29.0	21 24.2	**59**	17 16.5	1 30.9	13 9.5	26 49.9	22 49.9
16 4.1	♏29 54.4	10 55.5	23 41.9	21 17.5	**60**	16 53.6	0 37.3	11 36.6	25 3.0	22 44.3
15 40.4	28 58.8	9 20.2	21 40.1	21 10.1	**61**	16 29.4	♏29 40.8	9 59.7	23 1.4	22 38.2
15 15.2	28 0.2	7 40.8	19 19.6	21 2.1	**62**	16 3.7	28 41.2	8 18.6	20 40.8	22 31.5
14 48.4	26 58.4	5 57.2	16 34.1	20 53.2	**63**	15 36.3	27 38.4	6 33.3	17 54.9	22 24.1
14 19.8	25 53.1	4 9.3	13 13.1	20 43.3	**64**	15 7.1	26 31.9	4 43.5	14 32.6	22 15.8
13 49.2	24 44.1	2 16.9	8 55.5	20 32.3	**65**	14 35.8	25 21.6	2 49.1	10 12.0	22 6.7
13 16.4	23 30.8	0 20.0	2 42.8	20 19.9	**66**	14 2.3	24 7.0	0 50.1	3 49.3	21 56.4
5	**6**	**Descendant**	**8**	**9**	**S LAT**	**5**	**6**	**Descendant**	**8**	**9**
		♈ 25° 53′ 12″						♈ 26° 56′ 33″		
1ʰ 36ᵐ 0ˢ		MC		24° 0′ 0″		1ʰ 40ᵐ 0ˢ		MC		25° 0′ 0″

13ʰ 44ᵐ 0ˢ — MC — 206° 0′ 0″

♎ 27° 59′ 44″

11	12	Ascendant	2	3	N LAT
° ′	° ′	° ′	° ′	° ′	°
♏28 15.0	♐26 19.8	♑24 6.4	♒23 40.6	♓25 38.5	**0**
27 40.4	25 0.2	22 13.9	22 49.6	25 35.1	**5**
27 5.7	23 39.8	20 16.5	21 55.6	25 31.7	**10**
26 30.3	22 17.1	18 12.3	20 57.5	25 28.0	**15**
26 23.1	22 0.1	17 46.4	20 45.2	25 27.2	**16**
26 15.8	21 43.1	17 20.1	20 32.8	25 26.4	**17**
26 8.5	21 25.8	16 53.5	20 20.1	25 25.6	**18**
26 1.1	21 8.4	16 26.4	20 7.1	25 24.8	**19**
25 53.7	20 50.8	15 58.8	19 53.8	25 24.0	**20**
25 46.1	20 32.9	15 30.8	19 40.3	25 23.2	**21**
25 38.5	20 14.9	15 2.2	19 26.4	25 22.3	**22**
25 30.8	19 56.6	14 33.1	19 12.2	25 21.4	**23**
25 22.9	19 38.0	14 3.5	18 57.6	25 20.5	**24**
25 15.0	19 19.2	13 33.2	18 42.6	25 19.6	**25**
25 7.0	19 0.1	13 2.4	18 27.3	25 18.7	**26**
24 58.8	18 40.7	12 30.8	18 11.5	25 17.7	**27**
24 50.5	18 20.9	11 58.6	17 55.2	25 16.8	**28**
24 42.1	18 0.9	11 25.6	17 38.4	25 15.7	**29**
24 33.6	17 40.4	10 51.9	17 21.2	25 14.7	**30**
24 24.8	17 19.6	10 17.3	17 3.3	25 13.7	**31**
24 16.0	16 58.3	9 41.9	16 44.9	25 12.6	**32**
24 6.9	16 36.6	9 5.5	16 25.8	25 11.4	**33**
23 57.7	16 14.5	8 28.2	16 6.0	25 10.3	**34**
23 48.2	15 51.8	7 49.9	15 45.5	25 9.1	**35**
23 38.6	15 28.7	7 10.5	15 24.1	25 7.9	**36**
23 28.7	15 4.9	6 30.0	15 2.0	25 6.6	**37**
23 18.6	14 40.6	5 48.3	14 38.8	25 5.3	**38**
23 8.2	14 15.7	5 5.3	14 14.7	25 3.9	**39**
22 57.5	13 50.1	4 21.0	13 49.5	25 2.5	**40**
22 46.6	13 23.8	3 35.3	13 23.1	25 1.0	**41**
22 35.3	12 56.7	2 48.1	12 55.5	24 59.4	**42**
22 23.8	12 28.9	1 59.3	12 26.4	24 57.8	**43**
22 11.8	12 0.1	1 8.9	11 55.9	24 56.1	**44**
21 59.5	11 30.5	0 16.7	11 23.6	24 54.4	**45**
21 46.8	10 59.9	♐29 22.6	10 49.5	24 52.5	**46**
21 33.7	10 28.3	28 26.6	10 13.4	24 50.6	**47**
21 20.1	9 55.6	27 28.5	9 35.1	24 48.6	**48**
21 5.9	9 21.6	26 28.1	8 54.3	24 46.4	**49**
20 51.3	8 46.5	25 25.5	8 10.7	24 44.2	**50**
20 36.0	8 9.9	24 20.4	7 24.0	24 41.8	**51**
20 20.2	7 31.9	23 12.6	6 33.9	24 39.2	**52**
20 3.6	6 52.4	22 2.1	5 39.8	24 36.5	**53**
19 46.4	6 11.1	20 48.7	4 41.3	24 33.7	**54**
19 28.3	5 28.1	19 32.3	3 37.6	24 30.6	**55**
19 9.3	4 43.1	18 12.5	2 28.0	24 27.3	**56**
18 49.4	3 56.0	16 49.4	1 11.5	24 23.8	**57**
18 28.5	3 6.6	15 22.7	♑29 46.8	24 19.9	**58**
18 6.4	2 14.7	13 52.3	28 12.3	24 15.8	**59**
17 43.1	1 20.2	12 17.9	26 25.8	24 11.3	**60**
17 18.4	0 22.8	10 39.4	24 24.5	24 6.4	**61**
16 52.1	♏29 22.3	8 56.7	22 4.1	24 1.0	**62**
16 24.2	28 18.4	7 9.5	19 18.0	23 55.1	**63**
15 54.4	27 10.7	5 17.7	15 54.9	23 48.5	**64**
15 22.4	25 59.1	3 21.3	11 31.9	23 41.2	**65**
14 48.0	24 43.0	1 20.1	5 0.5	23 32.9	**66**
5	6	Descendant	8	9	S LAT

♈ 27° 59′ 44″

1ʰ 44ᵐ 0ˢ — MC — 26° 0′ 0″

13ʰ 48ᵐ 0ˢ — MC — 207° 0′ 0″

♎ 29° 2′ 47″

N LAT	11	12	Ascendant	2	3
°	° ′	° ′	° ′	° ′	° ′
0	♏29 12.8	♐27 14.8	♑25 3.3	♒24 42.5	♓26 43.8
5	28 37.9	25 55.2	23 11.4	23 52.6	26 41.3
10	28 2.9	24 34.6	21 14.6	22 59.8	26 38.7
15	27 27.2	23 11.7	19 10.8	22 3.0	26 35.9
16	27 19.9	22 54.7	18 45.0	21 51.0	26 35.4
17	27 12.5	22 37.6	18 18.8	21 38.8	26 34.8
18	27 5.1	22 20.3	17 52.1	21 26.4	26 34.2
19	26 57.7	22 2.8	17 25.1	21 13.6	26 33.6
20	26 50.1	21 45.1	16 57.6	21 0.7	26 33.0
21	26 42.5	21 27.2	16 29.6	20 47.4	26 32.3
22	26 34.8	21 9.1	16 1.0	20 33.8	26 31.7
23	26 27.0	20 50.7	15 32.0	20 19.8	26 31.0
24	26 19.1	20 32.1	15 2.3	20 5.6	26 30.4
25	26 11.1	20 13.1	14 32.1	19 50.9	26 29.7
26	26 2.9	19 53.9	14 1.2	19 35.8	26 29.0
27	25 54.7	19 34.4	13 29.6	19 20.3	26 28.3
28	25 46.3	19 14.6	12 57.4	19 4.4	26 27.5
29	25 37.8	18 54.4	12 24.4	18 48.0	26 26.8
30	25 29.1	18 33.8	11 50.6	18 31.0	26 26.0
31	25 20.3	18 12.9	11 15.9	18 13.5	26 25.2
32	25 11.3	17 51.5	10 40.4	17 55.4	26 24.4
33	25 2.2	17 29.7	10 4.0	17 36.6	26 23.5
34	24 52.8	17 7.4	9 26.5	17 17.2	26 22.7
35	24 43.2	16 44.6	8 48.1	16 57.1	26 21.8
36	24 33.5	16 21.3	8 8.5	16 36.1	26 20.8
37	24 23.5	15 57.4	7 27.8	16 14.3	26 19.9
38	24 13.2	15 32.9	6 45.9	15 51.6	26 18.9
39	24 2.7	15 7.8	6 2.7	15 27.9	26 17.9
40	23 51.9	14 42.0	5 18.2	15 3.1	26 16.8
41	23 40.8	14 15.4	4 32.2	14 37.2	26 15.7
42	23 29.5	13 48.1	3 44.7	14 9.9	26 14.5
43	23 17.7	13 20.0	2 55.5	13 41.3	26 13.3
44	23 5.6	12 51.1	2 4.7	13 11.2	26 12.1
45	22 53.1	12 21.2	1 12.1	12 39.4	26 10.7
46	22 40.3	11 50.3	0 17.6	12 5.9	26 9.3
47	22 26.9	11 18.4	♐29 21.0	11 30.3	26 7.9
48	22 13.1	10 45.3	28 22.4	10 52.5	26 6.4
49	21 58.8	10 11.0	27 21.4	10 12.2	26 4.8
50	21 43.9	9 35.5	26 18.1	9 29.2	26 3.1
51	21 28.4	8 58.5	25 12.2	8 43.1	26 1.3
52	21 12.3	8 20.1	24 3.7	7 53.5	25 59.4
53	20 55.5	7 40.0	22 52.3	7 0.0	25 57.3
54	20 38.0	6 58.3	21 38.0	6 2.1	25 55.2
55	20 19.6	6 14.7	20 20.5	4 59.1	25 52.9
56	20 0.3	5 29.1	18 59.7	3 50.1	25 50.4
57	19 40.1	4 41.3	17 35.3	2 34.2	25 47.7
58	19 18.7	3 51.2	16 7.3	1 10.1	25 44.9
59	18 56.3	2 58.6	14 35.4	♑29 36.2	25 41.8
60	18 32.5	2 3.3	12 59.5	27 50.4	25 38.4
61	18 7.3	1 5.0	11 19.4	25 49.6	25 34.7
62	17 40.5	0 3.4	9 34.9	23 29.5	25 30.7
63	17 12.0	♏28 58.4	7 45.8	20 43.5	25 26.2
64	16 41.5	27 49.5	5 52.1	17 20.0	25 21.3
65	16 8.9	26 36.6	3 53.6	12 55.4	25 15.8
66	15 33.7	25 19.0	1 50.2	6 16.9	25 9.6
S LAT	5	6	Descendant	8	9

♈ 29° 2′ 47″

1ʰ 48ᵐ 0ˢ — MC — 27° 0′ 0″

13ʰ 52ᵐ 0ˢ MC 208° 0′ 0″

♏ 0° 5′ 39″

11	12	Ascendant	2	3	N LAT
° ′	° ′	° ′	° ′	° ′	°
♐ 0 10.5	♐28 9.9	♑26 0.2	♒25 44.5	♓27 49.2	**0**
♏29 35.3	26 50.2	24 9.1	24 55.8	27 47.5	**5**
28 59.9	25 29.5	22 12.8	24 4.3	27 45.8	**10**
28 23.9	24 6.4	20 9.5	23 8.7	27 43.9	**15**
28 16.5	23 49.3	19 43.7	22 57.0	27 43.6	**16**
28 9.1	23 32.1	19 17.6	22 45.1	27 43.2	**17**
28 1.6	23 14.8	18 51.1	22 32.9	27 42.8	**18**
27 54.1	22 57.2	18 24.1	22 20.5	27 42.4	**19**
27 46.5	22 39.5	17 56.6	22 7.8	27 42.0	**20**
27 38.8	22 21.5	17 28.6	21 54.8	27 41.5	**21**
27 31.0	22 3.3	17 0.2	21 41.5	27 41.1	**22**
27 23.1	21 44.9	16 31.1	21 27.9	27 40.7	**23**
27 15.1	21 26.1	16 1.5	21 13.9	27 40.2	**24**
27 7.0	21 7.2	15 31.3	20 59.5	27 39.8	**25**
26 58.8	20 47.9	15 0.4	20 44.8	27 39.3	**26**
26 50.5	20 28.3	14 28.8	20 29.6	27 38.8	**27**
26 42.0	20 8.3	13 56.5	20 14.0	27 38.3	**28**
26 33.4	19 48.0	13 23.5	19 57.9	27 37.8	**29**
26 24.6	19 27.4	12 49.6	19 41.3	27 37.3	**30**
26 15.7	19 6.3	12 14.9	19 24.1	27 36.8	**31**
26 6.6	18 44.8	11 39.3	19 6.3	27 36.2	**32**
25 57.3	18 22.8	11 2.8	18 48.0	27 35.7	**33**
25 47.9	18 0.4	10 25.2	18 28.9	27 35.1	**34**
25 38.2	17 37.5	9 46.7	18 9.1	27 34.5	**35**
25 28.3	17 14.0	9 7.0	17 48.6	27 33.9	**36**
25 18.2	16 49.9	8 26.1	17 27.2	27 33.2	**37**
25 7.8	16 25.3	7 44.0	17 4.9	27 32.6	**38**
24 57.2	15 59.9	7 0.6	16 41.6	27 31.9	**39**
24 46.3	15 33.9	6 15.8	16 17.3	27 31.2	**40**
24 35.0	15 7.2	5 29.5	15 51.8	27 30.4	**41**
24 23.5	14 39.7	4 41.7	15 25.0	27 29.7	**42**
24 11.6	14 11.4	3 52.2	14 56.9	27 28.8	**43**
23 59.4	13 42.1	3 1.0	14 27.3	27 28.0	**44**
23 46.7	13 12.0	2 8.0	13 56.0	27 27.1	**45**
23 33.6	12 40.8	1 13.0	13 22.9	27 26.2	**46**
23 20.1	12 8.6	0 16.0	12 47.9	27 25.2	**47**
23 6.1	11 35.2	♐29 16.7	12 10.6	27 24.2	**48**
22 51.6	11 0.5	28 15.2	11 30.9	27 23.1	**49**
22 36.5	10 24.6	27 11.2	10 48.5	27 22.0	**50**
22 20.8	9 47.2	26 4.6	10 3.1	27 20.8	**51**
22 4.4	9 8.4	24 55.3	9 14.1	27 19.5	**52**
21 47.4	8 27.9	23 43.0	8 21.3	27 18.2	**53**
21 29.5	7 45.6	22 27.7	7 24.1	27 16.7	**54**
21 10.8	7 1.4	21 9.1	6 21.7	27 15.2	**55**
20 51.2	6 15.2	19 47.2	5 13.5	27 13.6	**56**
20 30.6	5 26.8	18 21.6	3 58.3	27 11.8	**57**
20 9.0	4 36.0	16 52.3	2 34.9	27 9.9	**58**
19 46.1	3 42.6	15 18.9	1 1.8	27 7.8	**59**
19 21.9	2 46.4	13 41.5	♑29 16.7	27 5.6	**60**
18 56.2	1 47.2	11 59.7	27 16.5	27 3.1	**61**
18 28.9	0 44.6	10 13.4	24 57.1	27 0.4	**62**
17 59.8	♏29 38.5	8 22.4	22 11.5	26 57.4	**63**
17 28.7	28 28.4	6 26.6	18 48.1	26 54.2	**64**
16 55.3	27 14.0	4 25.9	14 22.6	26 50.5	**65**
16 19.3	25 55.0	2 20.3	7 39.1	26 46.4	**66**
5	**6**	**Descendant**	**8**	**9**	**S LAT**

♉ 0° 5′ 39″

1ʰ 52ᵐ 0ˢ MC 28° 0′ 0″

13ʰ 56ᵐ 0ˢ MC 209° 0′ 0″

♏ 1° 8′ 22″

11	12	Ascendant	2	3	N LAT
° ′	° ′	° ′	° ′	° ′	°
♐ 1 8.0	♐29 5.0	♑26 57.4	♒26 46.7	♓28 54.6	**0**
0 32.5	27 45.2	25 6.9	25 59.2	28 53.8	**5**
♏29 56.9	26 24.4	23 11.3	25 9.0	28 52.9	**10**
29 20.5	25 1.1	21 8.4	24 14.8	28 52.0	**15**
29 13.1	24 44.0	20 42.8	24 3.3	28 51.8	**16**
29 5.6	24 26.7	20 16.7	23 51.7	28 51.6	**17**
28 58.0	24 9.3	19 50.3	23 39.8	28 51.4	**18**
28 50.4	23 51.7	19 23.3	23 27.6	28 51.2	**19**
28 42.7	23 33.9	18 55.9	23 15.2	28 51.0	**20**
28 34.9	23 15.9	18 28.0	23 2.5	28 50.8	**21**
28 27.1	22 57.6	17 59.6	22 49.5	28 50.5	**22**
28 19.1	22 39.1	17 30.6	22 36.2	28 50.3	**23**
28 11.0	22 20.3	17 1.0	22 22.5	28 50.1	**24**
28 2.9	22 1.2	16 30.8	22 8.5	28 49.9	**25**
27 54.6	21 41.9	15 59.9	21 54.1	28 49.6	**26**
27 46.1	21 22.2	15 28.3	21 39.2	28 49.4	**27**
27 37.6	21 2.1	14 56.0	21 23.9	28 49.2	**28**
27 28.9	20 41.7	14 22.9	21 8.2	28 48.9	**29**
27 20.0	20 21.0	13 49.0	20 51.9	28 48.6	**30**
27 11.0	19 59.8	13 14.3	20 35.1	28 48.4	**31**
27 1.8	19 38.2	12 38.6	20 17.7	28 48.1	**32**
26 52.4	19 16.1	12 2.0	19 59.7	28 47.8	**33**
26 42.9	18 53.5	11 24.4	19 41.1	28 47.5	**34**
26 33.1	18 30.4	10 45.7	19 21.7	28 47.2	**35**
26 23.1	18 6.8	10 5.8	19 1.6	28 46.9	**36**
26 12.8	17 42.6	9 24.8	18 40.6	28 46.6	**37**
26 2.3	17 17.7	8 42.5	18 18.7	28 46.3	**38**
25 51.6	16 52.3	7 58.9	17 55.9	28 45.9	**39**
25 40.5	16 26.1	7 13.9	17 32.0	28 45.6	**40**
25 29.2	15 59.1	6 27.3	17 7.0	28 45.2	**41**
25 17.5	15 31.4	5 39.2	16 40.7	28 44.8	**42**
25 5.4	15 2.8	4 49.4	16 13.1	28 44.4	**43**
24 53.0	14 33.3	3 57.9	15 44.0	28 44.0	**44**
24 40.2	14 2.9	3 4.4	15 13.3	28 43.6	**45**
24 27.0	13 31.5	2 9.0	14 40.7	28 43.1	**46**
24 13.3	12 58.9	1 11.4	14 6.3	28 42.6	**47**
23 59.1	12 25.2	0 11.6	13 29.6	28 42.1	**48**
23 44.3	11 50.2	♐29 9.5	12 50.6	28 41.6	**49**
23 29.0	11 13.9	28 4.8	12 8.8	28 41.0	**50**
23 13.1	10 36.1	26 57.5	11 24.0	28 40.4	**51**
22 56.5	9 56.8	25 47.3	10 35.8	28 39.8	**52**
22 39.1	9 15.8	24 34.2	9 43.7	28 39.1	**53**
22 21.0	8 33.0	23 17.9	8 47.2	28 38.4	**54**
22 2.0	7 48.3	21 58.3	7 45.6	28 37.6	**55**
21 42.1	7 1.5	20 35.2	6 38.1	28 36.8	**56**
21 21.2	6 12.4	19 8.3	5 23.8	28 35.9	**57**
20 59.1	5 20.9	17 37.6	4 1.3	28 34.9	**58**
20 35.8	4 26.7	16 2.8	2 29.0	28 33.9	**59**
20 11.2	3 29.7	14 23.7	0 44.7	28 32.8	**60**
19 45.0	2 29.5	12 40.2	♑28 45.5	28 31.5	**61**
19 17.2	1 25.9	10 52.1	26 26.8	28 30.2	**62**
18 47.5	0 18.6	8 59.1	23 42.0	28 28.7	**63**
18 15.7	♏29 7.3	7 1.3	20 19.1	28 27.1	**64**
17 41.6	27 51.5	4 58.3	15 53.6	28 25.2	**65**
17 4.8	26 31.0	2 50.3	9 7.5	28 23.2	**66**
5	**6**	**Descendant**	**8**	**9**	**S LAT**

♉ 1° 8′ 22″

1ʰ 56ᵐ 0ˢ MC 29° 0′ 0″

14^h 0^m 0^s MC 210° 0′ 0″

♏ 2° 10′ 55″

N LAT	11	12	Ascendant	2	3
	° ′	° ′	° ′	° ′	° ′
0	♐ 2 5.4	♑ 0 0.0	♑27 54.6	♒27 49.1	♈ 0 0.0
5	1 29.6	♐28 40.2	26 4.9	27 2.9	0 0.0
10	0 53.7	27 19.3	24 9.9	26 13.9	0 0.0
15	0 17.0	25 55.8	22 7.6	25 21.0	0 0.0
16	0 9.5	25 38.7	21 42.1	25 9.9	0 0.0
17	0 1.9	25 21.4	21 16.1	24 58.5	0 0.0
18	♏29 54.3	25 3.9	20 49.7	24 46.9	0 0.0
19	29 46.6	24 46.3	20 22.9	24 35.1	0 0.0
20	29 38.9	24 28.4	19 55.5	24 23.0	0 0.0
21	29 31.0	24 10.3	19 27.7	24 10.6	0 0.0
22	29 23.1	23 52.0	18 59.3	23 57.9	0 0.0
23	29 15.0	23 33.4	18 30.3	23 44.9	0 0.0
24	29 6.9	23 14.6	18 0.8	23 31.5	0 0.0
25	28 58.6	22 55.4	17 30.6	23 17.8	0 0.0
26	28 50.2	22 36.0	16 59.7	23 3.7	0 0.0
27	28 41.7	22 16.2	16 28.2	22 49.2	0 0.0
28	28 33.1	21 56.0	15 55.9	22 34.3	0 0.0
29	28 24.3	21 35.6	15 22.8	22 18.9	0 0.0
30	28 15.4	21 14.7	14 48.9	22 3.0	0 0.0
31	28 6.2	20 53.4	14 14.1	21 46.6	0 0.0
32	27 56.9	20 31.6	13 38.3	21 29.6	0 0.0
33	27 47.5	20 9.4	13 1.7	21 12.0	0 0.0
34	27 37.8	19 46.7	12 23.9	20 53.7	0 0.0
35	27 27.9	19 23.5	11 45.1	20 34.7	0 0.0
36	27 17.8	18 59.7	11 5.2	20 15.0	0 0.0
37	27 7.4	18 35.4	10 24.0	19 54.5	0 0.0
38	26 56.8	18 10.4	9 41.5	19 33.1	0 0.0
39	26 45.9	17 44.7	8 57.7	19 10.7	0 0.0
40	26 34.7	17 18.3	8 12.4	18 47.3	0 0.0
41	26 23.2	16 51.2	7 25.7	18 22.7	0 0.0
42	26 11.4	16 23.2	6 37.3	17 57.0	0 0.0
43	25 59.2	15 54.4	5 47.1	17 29.9	0 0.0
44	25 46.6	15 24.7	4 55.2	17 1.4	0 0.0
45	25 33.7	14 54.0	4 1.4	16 31.2	0 0.0
46	25 20.2	14 22.3	3 5.5	15 59.3	0 0.0
47	25 6.4	13 49.4	2 7.4	15 25.4	0 0.0
48	24 52.0	13 15.4	1 7.1	14 49.4	0 0.0
49	24 37.0	12 40.0	0 4.3	14 11.0	0 0.0
50	24 21.5	12 3.3	♐28 58.9	13 29.9	0 0.0
51	24 5.3	11 25.2	27 50.9	12 45.9	0 0.0
52	23 48.5	10 45.4	26 39.9	11 58.4	0 0.0
53	23 30.9	10 3.9	25 25.9	11 7.1	0 0.0
54	23 12.5	9 20.6	24 8.6	10 11.4	0 0.0
55	22 53.2	8 35.4	22 47.9	9 10.6	0 0.0
56	22 33.0	7 47.9	21 23.6	8 4.1	0 0.0
57	22 11.7	6 58.2	19 55.5	6 50.6	0 0.0
58	21 49.3	6 6.0	18 23.4	5 29.1	0 0.0
59	21 25.6	5 11.0	16 47.1	3 57.8	0 0.0
60	21 0.5	4 13.1	15 6.4	2 14.6	0 0.0
61	20 33.8	3 11.9	13 21.1	0 16.4	0 0.0
62	20 5.5	2 7.3	11 31.0	♑27 58.8	0 0.0
63	19 35.2	0 58.8	9 36.1	25 15.1	0 0.0
64	19 2.7	♏29 46.2	7 36.0	21 53.3	0 0.0
65	18 27.8	28 29.0	5 30.8	17 28.6	0 0.0
66	17 50.2	27 6.9	3 20.4	10 42.5	0 0.0
S LAT	5	6	Descendant	8	9

♉ 2° 10′ 55″

2^h 0^m 0^s MC 30° 0′ 0″

14^h 4^m 0^s MC 211° 0′ 0″

♏ 3° 13′ 17″

N LAT	11	12	Ascendant	2	3
	° ′	° ′	° ′	° ′	° ′
0	♐ 3 2.6	♑ 0 55.0	♑28 52.0	♒28 51.6	♈ 1 5.4
5	2 26.6	♐29 35.3	27 3.1	28 6.7	1 6.2
10	1 50.4	28 14.2	25 8.8	27 19.0	1 7.1
15	1 13.4	26 50.6	23 7.1	26 27.6	1 8.0
16	1 5.8	26 33.4	22 41.6	26 16.7	1 8.2
17	0 58.2	26 16.1	22 15.8	26 5.6	1 8.4
18	0 50.5	25 58.6	21 49.5	25 54.3	1 8.6
19	0 42.7	25 40.9	21 22.7	25 42.8	1 8.8
20	0 34.9	25 23.0	20 55.4	25 31.0	1 9.0
21	0 27.0	25 4.8	20 27.7	25 18.9	1 9.2
22	0 19.0	24 46.4	19 59.3	25 6.5	1 9.5
23	0 10.8	24 27.8	19 30.4	24 53.8	1 9.7
24	0 2.6	24 8.9	19 0.9	24 40.8	1 9.9
25	♏29 54.3	23 49.7	18 30.8	24 27.5	1 10.1
26	29 45.8	23 30.1	17 59.9	24 13.7	1 10.4
27	29 37.2	23 10.3	17 28.4	23 59.6	1 10.6
28	29 28.5	22 50.0	16 56.1	23 45.0	1 10.8
29	29 19.6	22 29.5	16 23.0	23 30.0	1 11.1
30	29 10.6	22 8.5	15 49.1	23 14.4	1 11.4
31	29 1.4	21 47.1	15 14.2	22 58.4	1 11.6
32	28 52.0	21 25.2	14 38.5	22 41.8	1 11.9
33	28 42.4	21 2.9	14 1.7	22 24.6	1 12.2
34	28 32.6	20 40.1	13 23.9	22 6.8	1 12.5
35	28 22.6	20 16.7	12 45.0	21 48.2	1 12.8
36	28 12.4	19 52.8	12 4.9	21 29.0	1 13.1
37	28 1.9	19 28.3	11 23.6	21 8.9	1 13.4
38	27 51.2	19 3.1	10 41.0	20 47.9	1 13.7
39	27 40.2	18 37.3	9 57.0	20 26.0	1 14.1
40	27 28.9	18 10.7	9 11.5	20 3.1	1 14.4
41	27 17.2	17 43.3	8 24.5	19 39.1	1 14.8
42	27 5.3	17 15.2	7 35.8	19 13.9	1 15.2
43	26 52.9	16 46.2	6 45.4	18 47.4	1 15.6
44	26 40.2	16 16.2	5 53.1	18 19.4	1 16.0
45	26 27.1	15 45.2	4 58.8	17 49.9	1 16.4
46	26 13.5	15 13.2	4 2.5	17 18.6	1 16.9
47	25 59.4	14 40.1	3 3.9	16 45.4	1 17.4
48	25 44.8	14 5.7	2 3.0	16 10.1	1 17.9
49	25 29.7	13 30.0	0 59.7	15 32.3	1 18.4
50	25 13.9	12 52.9	♐29 53.6	14 52.0	1 19.0
51	24 57.6	12 14.4	28 44.8	14 8.7	1 19.6
52	24 40.5	11 34.2	27 33.0	13 22.0	1 20.2
53	24 22.6	10 52.2	26 18.1	12 31.6	1 20.9
54	24 4.0	10 8.4	24 59.8	11 36.7	1 21.6
55	23 44.4	9 22.6	23 38.0	10 36.9	1 22.4
56	23 23.8	8 34.5	22 12.5	9 31.3	1 23.2
57	23 2.2	7 44.1	20 43.1	8 18.9	1 24.1
58	22 39.4	6 51.2	19 9.6	6 58.5	1 25.1
59	22 15.3	5 55.4	17 31.7	5 28.3	1 26.1
60	21 49.7	4 56.6	15 49.4	3 46.3	1 27.2
61	21 22.6	3 54.5	14 2.3	1 49.4	1 28.5
62	20 53.7	2 48.8	12 10.3	♑29 33.1	1 29.8
63	20 22.8	1 39.1	10 13.2	26 50.8	1 31.3
64	19 49.7	0 25.2	8 11.0	23 30.5	1 32.9
65	19 14.0	♏29 6.6	6 3.4	19 7.6	1 34.8
66	18 35.5	27 42.8	3 50.5	12 24.2	1 36.8
S LAT	5	6	Descendant	8	9

♉ 3° 13′ 17″

2^h 4^m 0^s MC 31° 0′ 0″

14ʰ 8ᵐ 0ˢ — MC 212° 0′ 0″ — ♏ 4° 15′ 30″

N LAT	11	12	Ascendant	2	3
0	♐ 3 59.8	♑ 1 50.1	♑29 49.5	♒29 54.3	♈ 2 10.8
5	3 23.4	0 30.3	28 1.4	29 10.7	2 12.5
10	2 46.9	♐29 9.2	26 7.9	28 24.4	2 14.2
15	2 9.6	27 45.4	24 6.8	27 34.3	2 16.1
16	2 2.0	27 28.3	23 41.4	27 23.8	2 16.4
17	1 54.3	27 10.9	23 15.7	27 13.0	2 16.8
18	1 46.6	26 53.3	22 49.5	27 2.0	2 17.2
19	1 38.8	26 35.6	22 22.8	26 50.8	2 17.6
20	1 30.8	26 17.6	21 55.6	26 39.3	2 18.0
21	1 22.9	25 59.4	21 27.9	26 27.5	2 18.5
22	1 14.8	25 41.0	20 59.7	26 15.5	2 18.9
23	1 6.6	25 22.3	20 30.8	26 3.1	2 19.3
24	0 58.3	25 3.3	20 1.4	25 50.5	2 19.8
25	0 49.9	24 44.0	19 31.3	25 37.4	2 20.2
26	0 41.3	24 24.4	19 0.5	25 24.0	2 20.7
27	0 32.7	24 4.4	18 29.0	25 10.3	2 21.2
28	0 23.8	23 44.2	17 56.7	24 56.1	2 21.7
29	0 14.9	23 23.5	17 23.6	24 41.4	2 22.2
30	0 5.7	23 2.4	16 49.7	24 26.3	2 22.7
31	♏29 56.4	22 40.9	16 14.8	24 10.7	2 23.2
32	29 47.0	22 18.9	15 39.0	23 54.5	2 23.8
33	29 37.3	21 56.5	15 2.2	23 37.7	2 24.3
34	29 27.4	21 33.5	14 24.4	23 20.3	2 24.9
35	29 17.3	21 10.1	13 45.4	23 2.2	2 25.5
36	29 7.0	20 46.0	13 5.2	22 43.4	2 26.1
37	28 56.4	20 21.3	12 23.8	22 23.8	2 26.8
38	28 45.5	19 56.0	11 41.0	22 3.3	2 27.4
39	28 34.4	19 30.0	10 56.8	21 42.0	2 28.1
40	28 23.0	19 3.2	10 11.1	21 19.6	2 28.8
41	28 11.2	18 35.7	9 23.8	20 56.1	2 29.6
42	27 59.1	18 7.3	8 34.9	20 31.5	2 30.3
43	27 46.6	17 38.1	7 44.1	20 5.5	2 31.2
44	27 33.7	17 7.9	6 51.5	19 38.2	2 32.0
45	27 20.4	16 36.6	5 56.9	19 9.2	2 32.9
46	27 6.7	16 4.3	5 0.1	18 38.6	2 33.8
47	26 52.4	15 30.9	4 1.0	18 6.1	2 34.8
48	26 37.7	14 56.2	2 59.6	17 31.5	2 35.8
49	26 22.3	14 20.2	1 55.6	16 54.5	2 36.9
50	26 6.4	13 42.7	0 48.9	16 15.0	2 38.0
51	25 49.8	13 3.7	♐29 39.3	15 32.5	2 39.2
52	25 32.4	12 23.1	28 26.7	14 46.7	2 40.5
53	25 14.3	11 40.7	27 10.9	13 57.1	2 41.8
54	24 55.4	10 56.4	25 51.6	13 3.3	2 43.3
55	24 35.5	10 10.0	24 28.7	12 4.5	2 44.8
56	24 14.6	9 21.3	23 2.0	10 59.9	2 46.4
57	23 52.7	8 30.2	21 31.3	9 48.7	2 48.2
58	23 29.5	7 36.5	19 56.3	8 29.4	2 50.1
59	23 4.9	6 39.9	18 16.8	7 0.5	2 52.2
60	22 38.9	5 40.2	16 32.8	5 19.8	2 54.4
61	22 11.3	4 37.1	14 43.8	3 24.4	2 56.9
62	21 41.8	3 30.3	12 49.8	1 9.7	2 59.6
63	21 10.3	2 19.5	10 50.6	♑28 29.2	3 2.6
64	20 36.5	1 4.2	8 46.0	25 10.9	3 5.8
65	20 0.1	♏29 44.1	6 36.1	20 50.8	3 9.5
66	19 20.7	28 18.7	4 20.6	14 12.6	3 13.6
S LAT	5	6	Descendant	8	9

♉ 4° 15′ 30″ — 2ʰ 8ᵐ 0ˢ — MC 32° 0′ 0″

14ʰ 12ᵐ 0ˢ — MC 213° 0′ 0″ — ♏ 5° 17′ 32″

N LAT	11	12	Ascendant	2	3
0	♐ 4 56.7	♑ 2 45.2	♒ 0 47.2	♓ 0 57.2	♈ 3 16.2
5	4 20.2	1 25.4	♑29 0.0	0 14.9	3 18.7
10	3 43.4	0 4.2	27 7.2	♒29 29.9	3 21.3
15	3 5.8	♐28 40.4	25 6.8	28 41.4	3 24.1
16	2 58.1	28 23.1	24 41.5	28 31.1	3 24.6
17	2 50.4	28 5.7	24 15.9	28 20.7	3 25.2
18	2 42.5	27 48.2	23 49.8	28 10.0	3 25.8
19	2 34.7	27 30.4	23 23.2	27 59.1	3 26.4
20	2 26.7	27 12.3	22 56.1	27 47.9	3 27.0
21	2 18.6	26 54.1	22 28.5	27 36.5	3 27.7
22	2 10.5	26 35.6	22 0.4	27 24.8	3 28.3
23	2 2.2	26 16.8	21 31.6	27 12.7	3 29.0
24	1 53.8	25 57.8	21 2.2	27 0.4	3 29.6
25	1 45.4	25 38.4	20 32.2	26 47.8	3 30.3
26	1 36.7	25 18.8	20 1.4	26 34.7	3 31.0
27	1 28.0	24 58.7	19 30.0	26 21.3	3 31.7
28	1 19.1	24 38.4	18 57.7	26 7.5	3 32.5
29	1 10.0	24 17.6	18 24.6	25 53.3	3 33.2
30	1 0.8	23 56.4	17 50.7	25 38.5	3 34.0
31	0 51.4	23 34.8	17 15.9	25 23.3	3 34.8
32	0 41.9	23 12.8	16 40.0	25 7.5	3 35.6
33	0 32.1	22 50.2	16 3.2	24 51.2	3 36.5
34	0 22.1	22 27.1	15 25.3	24 34.3	3 37.3
35	0 11.9	22 3.5	14 46.2	24 16.6	3 38.2
36	0 1.5	21 39.3	14 5.9	23 58.3	3 39.2
37	♏29 50.8	21 14.5	13 24.4	23 39.2	3 40.1
38	29 39.8	20 49.0	12 41.5	23 19.3	3 41.1
39	29 28.6	20 22.8	11 57.1	22 58.4	3 42.1
40	29 17.0	19 55.9	11 11.2	22 36.6	3 43.2
41	29 5.1	19 28.2	10 23.7	22 13.7	3 44.3
42	28 52.8	18 59.6	9 34.5	21 49.6	3 45.5
43	28 40.2	18 30.1	8 43.5	21 24.3	3 46.7
44	28 27.2	17 59.7	7 50.5	20 57.6	3 47.9
45	28 13.7	17 28.2	6 55.5	20 29.3	3 49.3
46	27 59.8	16 55.6	5 58.3	19 59.4	3 50.7
47	27 45.4	16 21.9	4 58.7	19 27.6	3 52.1
48	27 30.4	15 46.9	3 56.8	18 53.7	3 53.6
49	27 14.9	15 10.5	2 52.2	18 17.6	3 55.2
50	26 58.7	14 32.7	1 44.8	17 38.8	3 56.9
51	26 41.9	13 53.3	0 34.5	16 57.2	3 58.7
52	26 24.4	13 12.2	♐29 21.0	16 12.3	4 0.6
53	26 6.0	12 29.3	28 4.2	15 23.7	4 2.7
54	25 46.8	11 44.5	26 43.9	14 30.9	4 4.8
55	25 26.6	10 57.5	25 19.9	13 33.2	4 7.1
56	25 5.4	10 8.2	23 52.0	12 29.8	4 9.6
57	24 43.1	9 16.5	22 19.9	11 19.8	4 12.3
58	24 19.5	8 22.0	20 43.4	10 1.8	4 15.1
59	23 54.6	7 24.6	19 2.4	8 34.4	4 18.2
60	23 28.1	6 24.0	17 16.5	6 55.2	4 21.6
61	23 0.0	5 19.9	15 25.7	5 1.4	4 25.3
62	22 30.0	4 12.0	13 29.6	2 48.7	4 29.3
63	21 57.9	2 59.9	11 28.2	0 10.2	4 33.8
64	21 23.4	1 43.3	9 21.3	♑26 54.6	4 38.7
65	20 46.2	0 21.6	7 8.8	22 37.9	4 44.2
66	20 5.9	♏28 54.5	4 50.7	16 7.5	4 50.4
S LAT	5	6	Descendant	8	9

♉ 5° 17′ 32″ — 2ʰ 12ᵐ 0ˢ — MC 33° 0′ 0″

14ʰ 16ᵐ 0ˢ MC 214° 0′ 0″

♏ 6° 19′ 23″

N LAT	11	12	Ascendant	2	3
°	° ′	° ′	° ′	° ′	° ′
0	♐ 5 53.6	♑ 3 40.2	♒ 1 45.0	♓ 2 0.3	♈ 4 21.5
5	5 16.8	2 20.5	♑29 58.7	1 19.3	4 24.9
10	4 39.7	0 59.3	28 6.7	0 35.7	4 28.3
15	4 1.8	♐29 35.3	26 7.0	♒29 48.6	4 32.0
16	3 54.1	29 18.1	25 41.9	29 38.7	4 32.8
17	3 46.3	29 0.7	25 16.4	29 28.6	4 33.6
18	3 38.4	28 43.0	24 50.4	29 18.2	4 34.4
19	3 30.5	28 25.2	24 23.9	29 7.6	4 35.2
20	3 22.4	28 7.2	23 57.0	28 56.8	4 36.0
21	3 14.3	27 48.9	23 29.5	28 45.7	4 36.8
22	3 6.1	27 30.3	23 1.4	28 34.3	4 37.7
23	2 57.8	27 11.5	22 32.7	28 22.7	4 38.6
24	2 49.3	26 52.4	22 3.4	28 10.7	4 39.5
25	2 40.8	26 33.0	21 33.4	27 58.4	4 40.4
26	2 32.1	26 13.2	21 2.8	27 45.8	4 41.3
27	2 23.2	25 53.1	20 31.3	27 32.7	4 42.3
28	2 14.3	25 32.7	19 59.1	27 19.3	4 43.2
29	2 5.1	25 11.8	19 26.1	27 5.5	4 44.3
30	1 55.8	24 50.6	18 52.2	26 51.2	4 45.3
31	1 46.4	24 28.9	18 17.3	26 36.4	4 46.3
32	1 36.7	24 6.7	17 41.5	26 21.0	4 47.4
33	1 26.8	23 44.1	17 4.6	26 5.2	4 48.6
34	1 16.8	23 20.9	16 26.7	25 48.7	4 49.7
35	1 6.5	22 57.1	15 47.5	25 31.5	4 50.9
36	0 55.9	22 32.8	15 7.2	25 13.7	4 52.1
37	0 45.1	22 7.8	14 25.5	24 55.1	4 53.4
38	0 34.0	21 42.2	13 42.5	24 35.7	4 54.7
39	0 22.7	21 15.8	12 58.0	24 15.4	4 56.1
40	0 11.0	20 48.7	12 11.9	23 54.1	4 57.5
41	♏29 59.0	20 20.8	11 24.2	23 31.8	4 59.0
42	29 46.6	19 52.1	10 34.7	23 8.4	5 0.6
43	29 33.8	19 22.4	9 43.4	22 43.7	5 2.2
44	29 20.6	18 51.7	8 50.1	22 17.6	5 3.9
45	29 7.0	18 20.0	7 54.7	21 50.1	5 5.6
46	28 52.9	17 47.1	6 57.1	21 20.9	5 7.5
47	28 38.3	17 13.1	5 57.1	20 49.8	5 9.4
48	28 23.2	16 37.7	4 54.5	20 16.8	5 11.4
49	28 7.5	16 1.0	3 49.3	19 41.5	5 13.6
50	27 51.1	15 22.8	2 41.3	19 3.6	5 15.8
51	27 34.1	14 43.0	1 30.2	18 22.9	5 18.2
52	27 16.3	14 1.5	0 15.9	17 39.0	5 20.8
53	26 57.6	13 18.2	♐28 58.2	16 51.5	5 23.5
54	26 38.1	12 32.8	27 36.9	15 59.7	5 26.3
55	26 17.7	11 45.3	26 11.8	15 3.2	5 29.4
56	25 56.2	10 55.4	24 42.6	14 1.1	5 32.7
57	25 33.5	10 2.9	23 9.1	12 52.4	5 36.2
58	25 9.6	9 7.7	21 31.1	11 35.9	5 40.1
59	24 44.2	8 9.5	19 48.4	10 9.9	5 44.2
60	24 17.3	7 7.9	18 0.8	8 32.5	5 48.7
61	23 48.7	6 2.8	16 8.0	6 40.6	5 53.6
62	23 18.1	4 53.8	14 9.8	4 29.9	5 59.0
63	22 45.3	3 40.5	12 6.1	1 53.9	6 4.9
64	22 10.1	2 22.4	9 56.7	♑28 41.4	6 11.5
65	21 32.1	0 59.2	7 41.7	24 29.1	6 18.8
66	20 50.9	♏29 30.3	5 20.8	18 8.6	6 27.1
S LAT	5	6	Descendant	8	9

♉ 6° 19′ 23″

2ʰ 16ᵐ 0ˢ MC 34° 0′ 0″

14ʰ 20ᵐ 0ˢ MC 215° 0′ 0″

♏ 7° 21′ 3″

N LAT	11	12	Ascendant	2	3
°	° ′	° ′	° ′	° ′	° ′
0	♐ 6 50.3	♑ 4 35.3	♒ 2 43.0	♓ 3 3.5	♈ 5 26.8
5	6 13.2	3 15.6	0 57.6	2 23.8	5 31.0
10	5 36.0	1 54.4	♑29 6.5	1 41.7	5 35.4
15	4 57.8	0 30.4	27 7.5	0 56.2	5 40.0
16	4 50.0	0 13.1	26 42.6	0 46.5	5 40.9
17	4 42.1	♐29 55.7	26 17.2	0 36.7	5 41.9
18	4 34.2	29 38.0	25 51.3	0 26.7	5 42.9
19	4 26.2	29 20.1	25 25.0	0 16.4	5 43.9
20	4 18.1	29 2.0	24 58.1	0 5.9	5 44.9
21	4 9.9	28 43.7	24 30.7	♒29 55.2	5 46.0
22	4 1.6	28 25.1	24 2.8	29 44.2	5 47.1
23	3 53.2	28 6.2	23 34.2	29 32.9	5 48.1
24	3 44.7	27 47.1	23 5.0	29 21.3	5 49.3
25	3 36.1	27 27.6	22 35.1	29 9.4	5 50.4
26	3 27.3	27 7.8	22 4.5	28 57.1	5 51.6
27	3 18.4	26 47.7	21 33.1	28 44.5	5 52.8
28	3 9.4	26 27.1	21 1.0	28 31.5	5 54.0
29	3 0.2	26 6.2	20 28.0	28 18.0	5 55.2
30	2 50.8	25 44.9	19 54.1	28 4.2	5 56.5
31	2 41.2	25 23.1	19 19.2	27 49.8	5 57.8
32	2 31.5	25 0.8	18 43.4	27 34.9	5 59.2
33	2 21.5	24 38.0	18 6.5	27 19.5	6 0.6
34	2 11.4	24 14.8	17 28.5	27 3.5	6 2.0
35	2 1.0	23 50.9	16 49.4	26 46.9	6 3.5
36	1 50.3	23 26.4	16 8.9	26 29.5	6 5.1
37	1 39.4	23 1.3	15 27.2	26 11.5	6 6.7
38	1 28.2	22 35.5	14 44.0	25 52.6	6 8.3
39	1 16.7	22 9.0	13 59.4	25 32.9	6 10.0
40	1 4.9	21 41.7	13 13.1	25 12.2	6 11.8
41	0 52.8	21 13.7	12 25.2	24 50.5	6 13.7
42	0 40.3	20 44.7	11 35.5	24 27.7	6 15.6
43	0 27.4	20 14.8	10 43.9	24 3.7	6 17.6
44	0 14.0	19 43.9	9 50.3	23 38.4	6 19.7
45	0 0.3	19 11.9	8 54.5	23 11.5	6 21.9
46	♏29 46.0	18 38.8	7 56.5	22 43.1	6 24.2
47	29 31.3	18 4.5	6 56.0	22 12.8	6 26.6
48	29 15.9	17 28.8	5 53.0	21 40.6	6 29.2
49	29 0.0	16 51.8	4 47.2	21 6.2	6 31.8
50	28 43.4	16 13.2	3 38.4	20 29.3	6 34.7
51	28 26.2	15 33.0	2 26.6	19 49.6	6 37.6
52	28 8.1	14 51.0	1 11.5	19 6.7	6 40.8
53	27 49.3	14 7.2	♐29 52.8	18 20.3	6 44.2
54	27 29.5	13 21.3	28 30.5	17 29.7	6 47.8
55	27 8.8	12 33.2	27 4.2	16 34.4	6 51.6
56	26 46.9	11 42.7	25 33.7	15 33.6	6 55.7
57	26 23.9	10 49.5	23 58.9	14 26.4	7 0.1
58	25 59.6	9 53.6	22 19.3	13 11.4	7 4.9
59	25 33.8	8 54.5	20 35.0	11 47.2	7 10.1
60	25 6.5	7 52.0	18 45.5	10 11.7	7 15.7
61	24 37.3	6 45.9	16 50.6	8 21.8	7 21.8
62	24 6.2	5 35.7	14 50.3	6 13.5	7 28.5
63	23 32.8	4 21.1	12 44.2	3 40.4	7 35.9
64	22 56.9	3 1.6	10 32.4	0 31.4	7 44.2
65	22 18.1	1 36.8	8 14.6	♑26 24.3	7 53.3
66	21 35.9	0 6.1	5 50.9	20 15.4	8 3.6
S LAT	5	6	Descendant	8	9

♉ 7° 21′ 3″

2ʰ 20ᵐ 0ˢ MC 35° 0′ 0″

$14^h\ 24^m\ 0^s$		MC		216° 0′ 0″	N	$14^h\ 28^m\ 0^s$		MC		217° 0′ 0″
		♏ 8° 22′ 33″						♏ 9° 23′ 53″		
11	12	Ascendant	2	3	LAT	11	12	Ascendant	2	3
° ′	° ′	° ′	° ′	° ′	°	° ′	° ′	° ′	° ′	° ′
♐ 7 46.9	♑ 5 30.5	♒ 3 41.2	♓ 4 6.8	♈ 6 32.1	0	♐ 8 43.3	♑ 6 25.6	♒ 4 39.5	♓ 5 10.3	♈ 7 37.4
7 9.6	4 10.8	1 56.7	3 28.6	6 37.1	5	8 5.9	5 6.1	2 56.0	4 33.5	7 43.2
6 32.1	2 49.6	0 6.5	2 47.9	6 42.3	10	7 28.1	3 44.9	1 6.7	3 54.3	7 49.3
5 53.7	1 25.5	♑28 8.3	2 3.9	6 47.9	15	6 49.4	2 20.7	♑29 9.4	3 11.9	7 55.7
5 45.8	1 8.2	27 43.5	1 54.6	6 49.0	16	6 41.5	2 3.4	28 44.7	3 2.9	7 57.0
5 37.9	0 50.7	27 18.3	1 45.1	6 50.2	17	6 33.6	1 45.9	28 19.6	2 53.7	7 58.4
5 29.9	0 33.1	26 52.5	1 35.4	6 51.4	18	6 25.5	1 28.2	27 54.1	2 44.4	7 59.8
5 21.8	0 15.2	26 26.3	1 25.5	6 52.6	19	6 17.4	1 10.3	27 28.0	2 34.8	8 1.2
5 13.7	♐29 57.0	25 59.6	1 15.4	6 53.8	20	6 9.2	0 52.1	27 1.4	2 25.0	8 2.6
5 5.4	29 38.7	25 32.3	1 5.0	6 55.1	21	6 0.9	0 33.7	26 34.3	2 15.0	8 4.1
4 57.1	29 20.0	25 4.5	0 54.3	6 56.4	22	5 52.5	0 15.1	26 6.6	2 4.7	8 5.6
4 48.6	29 1.1	24 36.0	0 43.4	6 57.7	23	5 43.9	♐29 56.1	25 38.2	1 54.2	8 7.1
4 40.0	28 41.9	24 6.9	0 32.2	6 59.0	24	5 35.3	29 36.8	25 9.2	1 43.4	8 8.7
4 31.3	28 22.4	23 37.1	0 20.7	7 0.4	25	5 26.5	29 17.3	24 39.5	1 32.2	8 10.2
4 22.5	28 2.5	23 6.6	0 8.8	7 1.7	26	5 17.6	28 57.3	24 9.1	1 20.8	8 11.9
4 13.5	27 42.3	22 35.3	♒29 56.6	7 3.2	27	5 8.6	28 37.1	23 37.9	1 9.0	8 13.5
4 4.4	27 21.7	22 3.2	29 44.0	7 4.6	28	4 59.4	28 16.4	23 5.9	0 56.8	8 15.2
3 55.1	27 0.7	21 30.3	29 31.0	7 6.1	29	4 50.0	27 55.3	22 33.0	0 44.3	8 17.0
3 45.7	26 39.3	20 56.4	29 17.5	7 7.7	30	4 40.5	27 33.8	21 59.2	0 31.3	8 18.8
3 36.0	26 17.4	20 21.6	29 3.6	7 9.3	31	4 30.8	27 11.9	21 24.4	0 17.8	8 20.6
3 26.2	25 55.0	19 45.8	28 49.2	7 10.9	32	4 20.8	26 49.4	20 48.7	0 3.9	8 22.5
3 16.2	25 32.2	19 8.9	28 34.3	7 12.6	33	4 10.7	26 26.5	20 11.8	♒29 49.5	8 24.5
3 5.9	25 8.8	18 30.9	28 18.8	7 14.3	34	4 0.4	26 3.0	19 33.8	29 34.5	8 26.5
2 55.4	24 44.8	17 51.7	28 2.7	7 16.1	35	3 49.8	25 38.9	18 54.6	29 18.9	8 28.6
2 44.6	24 20.2	17 11.2	27 45.9	7 17.9	36	3 38.9	25 14.2	18 14.0	29 2.6	8 30.7
2 33.6	23 55.0	16 29.4	27 28.4	7 19.9	37	3 27.8	24 48.8	17 32.1	28 45.7	8 33.0
2 22.3	23 29.0	15 46.1	27 10.1	7 21.8	38	3 16.4	24 22.7	16 48.8	28 28.0	8 35.3
2 10.7	23 2.4	15 1.3	26 50.9	7 23.9	39	3 4.7	23 55.9	16 3.9	28 9.5	8 37.7
1 58.8	22 34.9	14 14.9	26 30.9	7 26.0	40	2 52.7	23 28.3	15 17.4	27 50.0	8 40.1
1 46.5	22 6.7	13 26.8	26 9.8	7 28.2	41	2 40.3	22 59.8	14 29.1	27 29.7	8 42.7
1 33.9	21 37.5	12 36.9	25 47.7	7 30.5	42	2 27.5	22 30.5	13 38.9	27 8.2	8 45.4
1 20.9	21 7.4	11 45.0	25 24.4	7 33.0	43	2 14.3	22 0.2	12 46.8	26 45.6	8 48.2
1 7.4	20 36.2	10 51.1	24 59.7	7 35.5	44	2 0.7	21 28.8	11 52.6	26 21.7	8 51.1
0 53.5	20 4.0	9 55.0	24 33.7	7 38.1	45	1 46.7	20 56.4	10 56.2	25 56.4	8 54.2
0 39.1	19 30.6	8 56.6	24 6.0	7 40.9	46	1 32.1	20 22.7	9 57.3	25 29.6	8 57.4
0 24.1	18 56.0	7 55.6	23 36.6	7 43.7	47	1 17.0	19 47.8	8 55.9	25 1.1	9 0.8
0 8.6	18 20.1	6 52.0	23 5.3	7 46.8	48	1 1.3	19 11.5	7 51.8	24 30.7	9 4.3
♏29 52.5	17 42.7	5 45.7	22 31.8	7 50.0	49	0 45.0	18 33.8	6 44.9	23 58.1	9 8.1
29 35.8	17 3.8	4 36.3	21 55.8	7 53.4	50	0 28.1	17 54.5	5 34.8	23 23.2	9 12.0
29 18.3	16 23.2	3 23.7	21 17.1	7 57.0	51	0 10.3	17 13.5	4 21.5	22 45.7	9 16.2
29 0.0	15 40.8	2 7.7	20 35.4	8 0.8	52	♏29 51.8	16 30.7	3 4.7	22 5.1	9 20.6
28 40.9	14 56.5	0 48.2	19 50.1	8 4.8	53	29 32.5	15 45.9	1 44.2	21 21.0	9 25.3
28 20.8	14 10.0	♐29 24.8	19 0.8	8 9.1	54	29 12.2	14 59.0	0 19.7	20 33.0	9 30.3
27 59.8	13 21.3	27 57.3	18 6.9	8 13.7	55	28 50.9	14 9.7	♐28 51.1	19 40.5	9 35.6
27 37.7	12 30.2	26 25.5	17 7.5	8 18.6	56	28 28.4	13 17.9	27 18.0	18 42.7	9 41.4
27 14.3	11 36.4	24 49.2	16 1.8	8 23.9	57	28 4.7	12 23.4	25 40.2	17 38.7	9 47.6
26 49.6	10 39.6	23 8.1	14 48.6	8 29.6	58	27 39.6	11 25.9	23 57.5	16 27.2	9 54.2
26 23.4	9 39.7	21 22.0	13 26.2	8 35.8	59	27 13.0	10 25.1	22 9.6	15 6.9	10 1.4
25 55.6	8 36.3	19 30.6	11 52.7	8 42.5	60	26 44.7	9 20.8	20 16.3	13 35.6	10 9.3
25 26.0	7 29.1	17 33.7	10 5.1	8 49.9	61	26 14.6	8 12.5	18 17.3	11 50.6	10 17.8
24 54.2	6 17.8	15 31.1	7 59.5	8 57.9	62	25 42.3	7 0.0	16 12.4	9 47.8	10 27.2
24 20.2	5 1.8	13 22.7	5 29.5	9 6.8	63	25 7.6	5 42.7	14 1.4	7 21.3	10 37.6
23 43.6	3 40.9	11 8.2	2 24.5	9 16.7	64	24 30.2	4 20.2	11 44.3	4 20.8	10 49.1
23 3.9	2 14.4	8 47.7	♑28 23.5	9 27.7	65	23 49.7	2 52.1	9 20.9	0 26.4	11 1.9
22 20.8	0 41.9	6 21.1	22 27.6	9 40.1	66	23 5.6	1 17.6	6 51.3	♑24 44.6	11 16.4
5	6	Descendant	8	9	S LAT	5	6	Descendant	8	9
		♉ 8° 22′ 33″						♉ 9° 23′ 53″		
$2^h\ 24^m\ 0^s$		MC		36° 0′ 0″		$2^h\ 28^m\ 0^s$		MC		37° 0′ 0″

$14^h\ 32^m\ 0^s$ MC 218° 0′ 0″

♏ 10° 25′ 1″

11	12	Ascendant	2	3	N LAT
° ′	° ′	° ′	° ′	° ′	°
♐ 9 39.7	♑ 7 20.8	♒ 5 38.0	♓ 6 14.0	♈ 8 42.5	0
9 2.0	6 1.4	3 55.6	5 38.6	8 49.2	5
8 24.0	4 40.2	2 7.2	5 0.9	8 56.1	10
7 45.1	3 16.0	0 10.7	4 20.1	9 3.5	15
7 37.2	2 58.7	♑29 46.3	4 11.4	9 5.0	16
7 29.2	2 41.2	29 21.3	4 2.6	9 6.6	17
7 21.1	2 23.5	28 55.9	3 53.6	9 8.1	18
7 12.9	2 5.5	28 30.0	3 44.4	9 9.7	19
7 4.6	1 47.3	28 3.6	3 35.0	9 11.4	20
6 56.2	1 28.9	27 36.6	3 25.3	9 13.0	21
6 47.8	1 10.2	27 9.0	3 15.4	9 14.7	22
6 39.2	0 51.2	26 40.8	3 5.3	9 16.5	23
6 30.5	0 31.9	26 11.9	2 54.9	9 18.3	24
6 21.6	0 12.3	25 42.3	2 44.1	9 20.1	25
6 12.7	♐29 52.3	25 12.0	2 33.1	9 21.9	26
6 3.6	29 32.0	24 40.9	2 21.7	9 23.8	27
5 54.3	29 11.2	24 9.0	2 10.0	9 25.8	28
5 44.9	28 50.1	23 36.2	1 57.9	9 27.8	29
5 35.3	28 28.5	23 2.5	1 45.4	9 29.8	30
5 25.5	28 6.5	22 27.8	1 32.4	9 31.9	31
5 15.5	27 44.0	21 52.0	1 19.0	9 34.1	32
5 5.2	27 20.9	21 15.2	1 5.1	9 36.3	33
4 54.8	26 57.3	20 37.2	0 50.6	9 38.6	34
4 44.1	26 33.1	19 58.0	0 35.5	9 41.0	35
4 33.2	26 8.3	19 17.4	0 19.8	9 43.4	36
4 22.0	25 42.8	18 35.5	0 3.5	9 46.0	37
4 10.5	25 16.6	17 52.0	♒29 46.4	9 48.6	38
3 58.6	24 49.6	17 7.1	29 28.5	9 51.3	39
3 46.5	24 21.9	16 20.4	29 9.7	9 54.2	40
3 34.0	23 53.2	15 32.0	28 50.0	9 57.1	41
3 21.1	23 23.7	14 41.6	28 29.3	10 0.2	42
3 7.8	22 53.2	13 49.3	28 7.5	10 3.4	43
2 54.0	22 21.6	12 54.8	27 44.4	10 6.7	44
2 39.8	21 48.9	11 58.0	27 19.9	10 10.2	45
2 25.1	21 15.0	10 58.8	26 54.0	10 13.9	46
2 9.9	20 39.8	9 56.9	26 26.3	10 17.7	47
1 54.0	20 3.2	8 52.3	25 56.9	10 21.7	48
1 37.5	19 25.2	7 44.8	25 25.4	10 26.0	49
1 20.3	18 45.5	6 34.1	24 51.5	10 30.5	50
1 2.4	18 4.1	5 20.0	24 15.1	10 35.2	51
0 43.7	17 20.9	4 2.4	23 35.7	10 40.3	52
0 24.1	16 35.6	2 40.9	22 53.0	10 45.6	53
0 3.5	15 48.2	1 15.4	22 6.4	10 51.4	54
♏29 41.9	14 58.3	♐29 45.5	21 15.4	10 57.5	55
29 19.1	14 5.9	28 11.1	20 19.2	11 4.0	56
28 55.1	13 10.7	26 31.9	19 16.9	11 11.0	57
28 29.6	12 12.4	24 47.6	18 7.4	11 18.6	58
28 2.6	11 10.7	22 57.9	16 49.2	11 26.9	59
27 33.9	10 5.4	21 2.5	15 20.3	11 35.8	60
27 3.2	8 56.1	19 1.3	13 38.0	11 45.6	61
26 30.3	7 42.3	16 54.1	11 38.5	11 56.3	62
25 55.0	6 23.7	14 40.5	9 15.8	12 8.1	63
25 16.8	4 59.7	12 20.6	6 20.2	12 21.3	64
24 35.5	3 29.8	9 54.3	2 33.0	12 36.0	65
23 50.4	1 53.4	7 21.5	♑27 6.0	12 52.5	66

| 5 | 6 | Descendant | 8 | 9 | S LAT |

♉ 10° 25′ 1″

$2^h\ 32^m\ 0^s$ MC 38° 0′ 0″

$14^h\ 36^m\ 0^s$ MC 219° 0′ 0″

♏ 11° 25′ 58″

11	12	Ascendant	2	3	N LAT
° ′	° ′	° ′	° ′	° ′	°
♐10 35.9	♑ 8 16.1	♒ 6 36.6	♓ 7 17.7	♈ 9 47.7	0
9 58.1	6 56.7	4 55.3	6 43.8	9 55.1	5
9 19.9	5 35.6	3 7.9	6 7.7	10 2.9	10
8 40.7	4 11.4	1 12.4	5 28.5	10 11.2	15
8 32.7	3 54.1	0 48.1	5 20.2	10 12.9	16
8 24.7	3 36.6	0 23.3	5 11.7	10 14.6	17
8 16.5	3 18.8	♑29 58.1	5 3.1	10 16.4	18
8 8.3	3 0.9	29 32.4	4 54.2	10 18.2	19
8 0.0	2 42.7	29 6.1	4 45.2	10 20.1	20
7 51.5	2 24.2	28 39.2	4 35.9	10 21.9	21
7 43.0	2 5.4	28 11.8	4 26.4	10 23.8	22
7 34.4	1 46.4	27 43.7	4 16.6	10 25.8	23
7 25.6	1 27.1	27 15.0	4 6.6	10 27.8	24
7 16.7	1 7.4	26 45.5	3 56.3	10 29.8	25
7 7.7	0 47.4	26 15.3	3 45.7	10 31.9	26
6 58.5	0 27.0	25 44.4	3 34.8	10 34.0	27
6 49.2	0 6.2	25 12.5	3 23.5	10 36.2	28
6 39.6	♐29 45.0	24 39.8	3 11.8	10 38.4	29
6 30.0	29 23.4	24 6.2	2 59.8	10 40.7	30
6 20.1	29 1.3	23 31.6	2 47.3	10 43.1	31
6 10.0	28 38.7	22 55.9	2 34.4	10 45.5	32
5 59.7	28 15.5	22 19.1	2 21.0	10 48.0	33
5 49.2	27 51.8	21 41.1	2 7.1	10 50.6	34
5 38.4	27 27.5	21 1.9	1 52.6	10 53.3	35
5 27.4	27 2.6	20 21.3	1 37.5	10 56.0	36
5 16.1	26 37.0	19 39.4	1 21.7	10 58.9	37
5 4.5	26 10.6	18 55.9	1 5.3	11 1.8	38
4 52.6	25 43.5	18 10.8	0 48.0	11 4.9	39
4 40.3	25 15.6	17 24.1	0 29.9	11 8.1	40
4 27.7	24 46.8	16 35.5	0 11.0	11 11.4	41
4 14.6	24 17.1	15 45.0	♒29 51.0	11 14.8	42
4 1.2	23 46.4	14 52.4	29 29.9	11 18.4	43
3 47.3	23 14.6	13 57.6	29 7.6	11 22.1	44
3 33.0	22 41.7	13 0.5	28 44.0	11 26.1	45
3 18.1	22 7.5	12 0.9	28 19.0	11 30.2	46
3 2.7	21 32.0	10 58.7	27 52.3	11 34.5	47
2 46.7	20 55.2	9 53.6	27 23.8	11 39.0	48
2 30.0	20 16.8	8 45.5	26 53.4	11 43.8	49
2 12.6	19 36.8	7 34.1	26 20.7	11 48.8	50
1 54.5	18 55.0	6 19.3	25 45.4	11 54.2	51
1 35.5	18 11.3	5 0.8	25 7.3	11 59.8	52
1 15.7	17 25.6	3 38.4	24 26.0	12 5.8	53
0 54.9	16 37.6	2 11.8	23 40.9	12 12.3	54
0 32.9	15 47.2	0 40.8	22 51.4	12 19.1	55
0 9.9	14 54.1	♐29 5.0	21 57.0	12 26.5	56
♏29 45.5	13 58.2	27 24.3	20 56.6	12 34.4	57
29 19.6	12 59.1	25 38.3	19 49.2	12 42.9	58
28 52.2	11 56.6	23 46.7	18 33.3	12 52.1	59
28 23.0	10 50.3	21 49.3	17 6.9	13 2.2	60
27 51.8	9 39.8	19 45.9	15 27.6	13 13.2	61
27 18.3	8 24.8	17 36.1	13 31.4	13 25.2	62
26 42.3	7 4.8	15 20.0	11 12.9	13 38.5	63
26 3.4	5 39.2	12 57.2	8 22.6	13 53.3	64
25 21.1	4 7.5	10 27.8	4 43.2	14 9.8	65
24 35.0	2 29.1	7 51.7	♑29 31.5	14 28.4	66

| 5 | 6 | Descendant | 8 | 9 | S LAT |

♉ 11° 25′ 58″

$2^h\ 36^m\ 0^s$ MC 39° 0′ 0″

$14^{h}\ 40^{m}\ 0^{s}$		MC		220° 0′ 0″	
		♏ 12° 26′ 45″			
11	**12**	**Ascendant**	**2**	**3**	**N LAT**
° ′	° ′	° ′	° ′	° ′	°
♐11 32.1	♑ 9 11.4	♒ 7 35.4	♓ 8 21.7	♈10 52.7	**0**
10 54.0	7 52.1	5 55.2	7 49.2	11 1.0	**5**
10 15.6	6 31.1	4 8.9	7 14.6	11 9.7	**10**
9 36.3	5 6.9	2 14.3	6 37.1	11 18.8	**15**
9 28.2	4 49.6	1 50.2	6 29.2	11 20.7	**16**
9 20.1	4 32.1	1 25.7	6 21.1	11 22.6	**17**
9 11.9	4 14.3	1 0.6	6 12.8	11 24.6	**18**
9 3.6	3 56.3	0 35.0	6 4.3	11 26.6	**19**
8 55.2	3 38.1	0 8.9	5 55.6	11 28.6	**20**
8 46.8	3 19.6	♑29 42.3	5 46.7	11 30.7	**21**
8 38.2	3 0.8	29 15.0	5 37.6	11 32.8	**22**
8 29.5	2 41.8	28 47.1	5 28.3	11 35.0	**23**
8 20.7	2 22.4	28 18.5	5 18.7	11 37.2	**24**
8 11.7	2 2.7	27 49.2	5 8.8	11 39.4	**25**
8 2.6	1 42.6	27 19.1	4 58.6	11 41.7	**26**
7 53.4	1 22.2	26 48.3	4 48.1	11 44.1	**27**
7 44.0	1 1.4	26 16.6	4 37.3	11 46.5	**28**
7 34.4	0 40.1	25 44.0	4 26.1	11 49.0	**29**
7 24.6	0 18.4	25 10.4	4 14.6	11 51.6	**30**
7 14.7	♐29 56.2	24 35.9	4 2.6	11 54.2	**31**
7 4.5	29 33.6	24 0.3	3 50.2	11 56.9	**32**
6 54.2	29 10.3	23 23.6	3 37.4	11 59.6	**33**
6 43.5	28 46.5	22 45.6	3 24.0	12 2.5	**34**
6 32.7	28 22.1	22 6.4	3 10.1	12 5.5	**35**
6 21.6	27 57.1	21 25.9	2 55.6	12 8.5	**36**
6 10.2	27 31.4	20 43.9	2 40.4	12 11.7	**37**
5 58.5	27 4.9	20 0.4	2 24.6	12 14.9	**38**
5 46.4	26 37.6	19 15.2	2 8.0	12 18.3	**39**
5 34.1	26 9.6	18 28.4	1 50.7	12 21.8	**40**
5 21.3	25 40.6	17 39.7	1 32.4	12 25.5	**41**
5 8.2	25 10.7	16 49.0	1 13.2	12 29.3	**42**
4 54.6	24 39.8	15 56.2	0 52.9	12 33.3	**43**
4 40.6	24 7.8	15 1.2	0 31.5	12 37.4	**44**
4 26.1	23 34.6	14 3.8	0 8.7	12 41.8	**45**
4 11.1	23 0.2	13 3.9	♒29 44.6	12 46.3	**46**
3 55.5	22 24.5	12 1.2	29 18.9	12 51.1	**47**
3 39.3	21 47.3	10 55.6	28 51.5	12 56.1	**48**
3 22.5	21 8.6	9 47.0	28 22.2	13 1.4	**49**
3 4.9	20 28.3	8 35.0	27 50.6	13 7.0	**50**
2 46.6	19 46.1	7 19.4	27 16.7	13 12.9	**51**
2 27.4	19 2.0	6 0.1	26 39.9	13 19.2	**52**
2 7.3	18 15.8	4 36.7	26 0.0	13 25.9	**53**
1 46.2	17 27.2	3 9.0	25 16.4	13 33.0	**54**
1 24.0	16 36.3	1 36.8	24 28.7	13 40.6	**55**
1 0.6	15 42.6	♐29 59.7	23 36.1	13 48.7	**56**
0 35.9	14 45.9	28 17.4	22 37.7	13 57.5	**57**
0 9.6	13 46.0	26 29.7	21 32.4	14 6.9	**58**
♏29 41.8	12 42.6	24 36.2	20 19.0	14 17.2	**59**
29 12.1	11 35.4	22 36.7	18 55.4	14 28.3	**60**
28 40.4	10 23.8	20 31.0	17 19.2	14 40.5	**61**
28 6.3	9 7.5	18 18.7	15 26.7	14 53.9	**62**
27 29.6	7 46.1	15 59.8	13 12.6	15 8.6	**63**
26 50.0	6 18.9	13 34.0	10 28.0	15 25.1	**64**
26 6.8	4 45.4	11 1.4	6 56.9	15 43.4	**65**
25 19.6	3 4.8	8 22.0	2 0.6	16 4.1	**66**
5	**6**	**Descendant**	**8**	**9**	**S LAT**
		♉ 12° 26′ 45″			
$2^{h}\ 40^{m}\ 0^{s}$		MC		40° 0′ 0″	

$14^{h}\ 44^{m}\ 0^{s}$		MC		221° 0′ 0″	
		♏ 13° 27′ 20″			
N LAT	**11**	**12**	**Ascendant**	**2**	**3**
°	° ′	° ′	° ′	° ′	° ′
0	♐12 28.1	♑10 6.7	♒ 8 34.4	♓ 9 25.7	♈11 57.7
5	11 49.9	8 47.6	6 55.3	8 54.7	12 6.8
10	11 11.3	7 26.7	5 10.1	8 21.7	12 16.3
15	10 31.7	6 2.5	3 16.6	7 45.9	12 26.3
16	10 23.6	5 45.2	2 52.7	7 38.3	12 28.4
17	10 15.5	5 27.6	2 28.3	7 30.6	12 30.5
18	10 7.2	5 9.9	2 3.4	7 22.7	12 32.7
19	9 58.9	4 51.9	1 38.1	7 14.6	12 34.9
20	9 50.5	4 33.6	1 12.1	7 6.3	12 37.1
21	9 41.9	4 15.1	0 45.6	6 57.8	12 39.4
22	9 33.3	3 56.4	0 18.5	6 49.1	12 41.7
23	9 24.5	3 37.3	♑29 50.8	6 40.2	12 44.1
24	9 15.7	3 17.9	29 22.4	6 31.0	12 46.5
25	9 6.6	2 58.1	28 53.2	6 21.5	12 49.0
26	8 57.5	2 38.0	28 23.3	6 11.8	12 51.5
27	8 48.2	2 17.6	27 52.6	6 1.8	12 54.1
28	8 38.7	1 56.7	27 21.0	5 51.4	12 56.7
29	8 29.1	1 35.4	26 48.6	5 40.7	12 59.5
30	8 19.2	1 13.6	26 15.1	5 29.7	13 2.3
31	8 9.2	0 51.4	25 40.7	5 18.2	13 5.1
32	7 59.0	0 28.6	25 5.2	5 6.4	13 8.1
33	7 48.5	0 5.3	24 28.5	4 54.1	13 11.1
34	7 37.9	♐29 41.4	23 50.7	4 41.3	13 14.3
35	7 26.9	29 16.9	23 11.5	4 27.9	13 17.5
36	7 15.7	28 51.8	22 31.0	4 14.0	13 20.9
37	7 4.2	28 25.9	21 49.0	3 59.5	13 24.3
38	6 52.4	27 59.3	21 5.5	3 44.4	13 27.9
39	6 40.3	27 32.0	20 20.3	3 28.5	13 31.6
40	6 27.8	27 3.7	19 33.4	3 11.8	13 35.5
41	6 15.0	26 34.6	18 44.6	2 54.3	13 39.5
42	6 1.7	26 4.6	17 53.7	2 35.9	13 43.7
43	5 48.0	25 33.4	17 0.8	2 16.5	13 48.0
44	5 33.9	25 1.2	16 5.5	1 55.9	13 52.6
45	5 19.2	24 27.9	15 7.9	1 34.1	13 57.4
46	5 4.1	23 53.2	14 7.6	1 10.9	14 2.3
47	4 48.3	23 17.2	13 4.5	0 46.3	14 7.6
48	4 32.0	22 39.8	11 58.5	0 19.9	14 13.1
49	4 14.9	22 0.7	10 49.3	♒29 51.7	14 18.9
50	3 57.2	21 20.0	9 36.6	29 21.4	14 25.0
51	3 38.6	20 37.4	8 20.4	28 48.8	14 31.5
52	3 19.2	19 52.9	7 0.2	28 13.4	14 38.4
53	2 58.9	19 6.2	5 35.9	27 35.0	14 45.7
54	2 37.5	18 17.2	4 7.1	26 53.1	14 53.5
55	2 15.1	17 25.6	2 33.6	26 7.1	15 1.8
56	1 51.3	16 31.3	0 55.1	25 16.4	15 10.8
57	1 26.3	15 33.9	♐29 11.3	24 20.1	15 20.4
58	0 59.7	14 33.2	27 21.8	23 17.2	15 30.8
59	0 31.4	13 28.9	25 26.4	22 6.3	15 42.0
60	0 1.3	12 20.7	23 24.8	20 45.6	15 54.2
61	♏29 29.0	11 8.0	21 16.6	19 12.7	16 7.6
62	28 54.3	9 50.4	19 1.8	17 24.2	16 22.3
63	28 17.0	8 27.5	16 40.0	15 14.8	16 38.5
64	27 36.5	6 58.7	14 11.1	12 36.3	16 56.6
65	26 52.4	5 23.2	11 35.3	9 13.8	17 16.8
66	26 4.1	3 40.5	8 52.4	4 33.2	17 39.6
S LAT	**5**	**6**	**Descendant**	**8**	**9**
			♉ 13° 27′ 20″		
	$2^{h}\ 44^{m}\ 0^{s}$		MC		41° 0′ 0″

14h 48m 0s — MC 222° 0′ 0″ — ♏ 14° 27′ 45″

N LAT	11	12	Ascendant	2	3
°	° ′	° ′	° ′	° ′	° ′
0	♐13 24.0	♑11 2.1	♒ 9 33.6	♓10 29.9	♈13 2.6
5	12 45.6	9 43.2	7 55.6	10 0.4	13 12.5
10	12 6.9	8 22.3	6 11.6	9 29.0	13 22.8
15	11 27.1	6 58.3	4 19.1	8 54.9	13 33.7
16	11 19.0	6 40.9	3 55.4	8 47.7	13 36.0
17	11 10.8	6 23.4	3 31.3	8 40.3	13 38.3
18	11 2.5	6 5.6	3 6.6	8 32.8	13 40.7
19	10 54.1	5 47.6	2 41.4	8 25.1	13 43.1
20	10 45.6	5 29.3	2 15.7	8 17.2	13 45.5
21	10 37.0	5 10.8	1 49.4	8 9.1	13 48.0
22	10 28.3	4 52.0	1 22.5	8 0.8	13 50.5
23	10 19.5	4 32.9	0 54.9	7 52.3	13 53.1
24	10 10.6	4 13.5	0 26.7	7 43.5	13 55.7
25	10 1.5	3 53.7	♑29 57.7	7 34.5	13 58.4
26	9 52.3	3 33.6	29 27.9	7 25.3	14 1.1
27	9 43.0	3 13.1	28 57.4	7 15.7	14 4.0
28	9 33.4	2 52.1	28 26.0	7 5.8	14 6.8
29	9 23.7	2 30.8	27 53.7	6 55.6	14 9.8
30	9 13.8	2 9.0	27 20.4	6 45.1	14 12.8
31	9 3.7	1 46.7	26 46.0	6 34.2	14 16.0
32	8 53.4	1 23.8	26 10.6	6 22.9	14 19.2
33	8 42.9	1 0.5	25 34.1	6 11.1	14 22.5
34	8 32.1	0 36.5	24 56.3	5 58.9	14 25.9
35	8 21.1	0 11.9	24 17.2	5 46.2	14 29.4
36	8 9.8	♐29 46.7	23 36.7	5 32.9	14 33.1
37	7 58.2	29 20.7	22 54.8	5 19.1	14 36.8
38	7 46.3	28 54.0	22 11.2	5 4.6	14 40.7
39	7 34.1	28 26.5	21 26.0	4 49.4	14 44.8
40	7 21.5	27 58.1	20 39.0	4 33.5	14 49.0
41	7 8.6	27 28.9	19 50.1	4 16.8	14 53.3
42	6 55.2	26 58.6	18 59.2	3 59.2	14 57.9
43	6 41.4	26 27.3	18 6.1	3 40.6	15 2.6
44	6 27.1	25 54.9	17 10.6	3 20.9	15 7.6
45	6 12.4	25 21.3	16 12.7	3 0.0	15 12.7
46	5 57.1	24 46.5	15 12.1	2 37.9	15 18.2
47	5 41.2	24 10.2	14 8.7	2 14.3	15 23.9
48	5 24.6	23 32.4	13 2.2	1 49.1	15 29.9
49	5 7.4	22 53.1	11 52.4	1 22.1	15 36.2
50	4 49.5	22 12.0	10 39.2	0 53.0	15 42.8
51	4 30.7	21 29.1	9 22.2	0 21.7	15 49.9
52	4 11.1	20 44.1	8 1.2	♒29 47.8	15 57.4
53	3 50.5	19 56.9	6 35.9	29 11.0	16 5.3
54	3 28.9	19 7.4	5 6.1	28 30.7	16 13.8
55	3 6.1	18 15.3	3 31.3	27 46.6	16 22.9
56	2 42.1	17 20.3	1 51.4	26 57.9	16 32.6
57	2 16.7	16 22.2	0 6.0	26 3.8	16 43.0
58	1 49.7	15 20.7	♐28 14.7	25 3.4	16 54.3
59	1 21.0	14 15.5	26 17.3	23 55.2	17 6.6
60	0 50.4	13 6.2	24 13.5	22 37.6	17 19.9
61	0 17.6	11 52.4	22 2.9	21 8.3	17 34.4
62	♏29 42.4	10 33.5	19 45.3	19 23.9	17 50.4
63	29 4.3	9 9.1	17 20.6	17 19.5	18 8.1
64	28 23.0	7 38.6	14 48.6	14 47.5	18 27.8
65	27 38.0	6 1.2	12 9.3	11 34.0	18 49.8
66	26 48.6	4 16.2	9 22.7	7 9.0	19 14.7
S LAT	5	6	Descendant	8	9

♉ 14° 27′ 45″ — 2h 48m 0s — MC 42° 0′ 0″

14h 52m 0s — MC 223° 0′ 0″ — ♏ 15° 27′ 59″

N LAT	11	12	Ascendant	2	3
°	° ′	° ′	° ′	° ′	° ′
0	♐14 19.9	♑11 57.6	♒10 32.9	♓11 34.2	♈14 7.5
5	13 41.3	10 38.8	8 56.2	11 6.3	14 18.2
10	13 2.4	9 18.1	7 13.3	10 36.5	14 29.3
15	12 22.4	7 54.1	5 22.0	10 4.1	14 41.1
16	12 14.3	7 36.8	4 58.5	9 57.3	14 43.5
17	12 6.0	7 19.2	4 34.6	9 50.3	14 46.0
18	11 57.7	7 1.4	4 10.1	9 43.1	14 48.6
19	11 49.2	6 43.4	3 45.2	9 35.8	14 51.1
20	11 40.7	6 25.2	3 19.6	9 28.3	14 53.8
21	11 32.1	6 6.6	2 53.5	9 20.6	14 56.4
22	11 23.3	5 47.8	2 26.8	9 12.8	14 59.2
23	11 14.5	5 28.7	1 59.4	9 4.7	15 1.9
24	11 5.5	5 9.3	1 31.4	8 56.3	15 4.8
25	10 56.4	4 49.5	1 2.6	8 47.8	15 7.7
26	10 47.1	4 29.3	0 33.0	8 39.0	15 10.7
27	10 37.7	4 8.8	0 2.7	8 29.9	15 13.7
28	10 28.1	3 47.8	♑29 31.4	8 20.5	15 16.8
29	10 18.3	3 26.4	28 59.2	8 10.8	15 20.0
30	10 8.4	3 4.5	28 26.1	8 0.8	15 23.3
31	9 58.2	2 42.2	27 51.9	7 50.4	15 26.7
32	9 47.8	2 19.3	27 16.6	7 39.7	15 30.1
33	9 37.2	1 55.8	26 40.2	7 28.5	15 33.7
34	9 26.4	1 31.8	26 2.5	7 16.9	15 37.4
35	9 15.3	1 7.1	25 23.5	7 4.8	15 41.2
36	9 3.9	0 41.8	24 43.1	6 52.2	15 45.1
37	8 52.2	0 15.7	24 1.2	6 39.0	15 49.2
38	8 40.3	♐29 48.9	23 17.7	6 25.2	15 53.4
39	8 27.9	29 21.3	22 32.5	6 10.8	15 57.8
40	8 15.3	28 52.8	21 45.4	5 55.6	16 2.3
41	8 2.2	28 23.4	20 56.5	5 39.7	16 7.0
42	7 48.7	27 52.9	20 5.4	5 22.9	16 11.9
43	7 34.8	27 21.5	19 12.2	5 5.2	16 17.0
44	7 20.4	26 48.9	18 16.5	4 46.4	16 22.4
45	7 5.5	26 15.1	17 18.4	4 26.6	16 28.0
46	6 50.0	25 40.0	16 17.5	4 5.4	16 33.8
47	6 34.0	25 3.4	15 13.7	3 42.9	16 40.0
48	6 17.3	24 25.4	14 6.7	3 18.9	16 46.4
49	5 59.9	23 45.7	12 56.5	2 53.1	16 53.2
50	5 41.8	23 4.3	11 42.6	2 25.4	17 0.4
51	5 22.8	22 21.0	10 24.9	1 55.5	17 8.1
52	5 3.0	21 35.6	9 3.1	1 23.1	17 16.1
53	4 42.2	20 48.0	7 36.9	0 47.9	17 24.7
54	4 20.3	19 57.9	6 5.9	0 9.4	17 33.9
55	3 57.2	19 5.2	4 29.9	♒29 27.2	17 43.7
56	3 32.9	18 9.5	2 48.6	28 40.6	17 54.2
57	3 7.2	17 10.7	1 1.6	27 48.9	18 5.4
58	2 39.8	16 8.5	♐29 8.5	26 51.0	18 17.6
59	2 10.7	15 2.3	27 9.1	25 45.7	18 30.9
60	1 39.6	13 52.0	25 2.9	24 31.4	18 45.3
61	1 6.3	12 37.0	22 49.8	23 5.7	19 1.0
62	0 30.4	11 16.8	20 29.5	21 25.7	19 18.3
63	♏29 51.6	9 50.9	18 1.7	19 26.7	19 37.4
64	29 9.5	8 18.6	15 26.4	17 1.4	19 58.7
65	28 23.5	6 39.2	12 43.5	13 57.2	20 22.6
66	27 32.9	4 51.9	9 53.2	9 47.8	20 49.6
S LAT	5	6	Descendant	8	9

♉ 15° 27′ 59″ — 2h 52m 0s — MC 43° 0′ 0″

14ʰ 56ᵐ 0ˢ MC 224° 0′ 0″

♏ 16° 28′ 1″

N LAT	11	12	Ascendant	2	3
°	° ′	° ′	° ′	° ′	° ′
0	♐15 15.6	♑12 53.1	♒11 32.4	♓12 38.6	♈15 12.2
5	14 36.9	11 34.6	9 57.0	12 12.2	15 23.7
10	13 57.9	10 14.0	8 15.3	11 44.1	15 35.6
15	13 17.7	8 50.0	6 25.2	11 13.5	15 48.3
16	13 9.5	8 32.7	6 1.9	11 7.0	15 50.9
17	13 1.2	8 15.2	5 38.2	11 0.4	15 53.6
18	12 52.8	7 57.4	5 14.0	10 53.6	15 56.3
19	12 44.3	7 39.4	4 49.2	10 46.7	15 59.1
20	12 35.8	7 21.1	4 23.9	10 39.6	16 1.9
21	12 27.1	7 2.6	3 58.1	10 32.4	16 4.8
22	12 18.3	6 43.8	3 31.6	10 24.9	16 7.7
23	12 9.4	6 24.6	3 4.4	10 17.3	16 10.7
24	12 0.4	6 5.2	2 36.5	10 9.4	16 13.7
25	11 51.2	5 45.4	2 8.0	10 1.3	16 16.8
26	11 41.9	5 25.2	1 38.6	9 53.0	16 20.0
27	11 32.4	5 4.6	1 8.4	9 44.4	16 23.3
28	11 22.7	4 43.6	0 37.3	9 35.5	16 26.6
29	11 12.9	4 22.2	0 5.3	9 26.3	16 30.1
30	11 2.9	4 0.3	♑29 32.4	9 16.8	16 33.6
31	10 52.7	3 37.8	28 58.3	9 7.0	16 37.2
32	10 42.2	3 14.9	28 23.2	8 56.8	16 40.9
33	10 31.6	2 51.4	27 46.9	8 46.2	16 44.8
34	10 20.6	2 27.3	27 9.3	8 35.2	16 48.7
35	10 9.5	2 2.5	26 30.4	8 23.7	16 52.8
36	9 58.0	1 37.1	25 50.1	8 11.8	16 57.0
37	9 46.2	1 10.9	25 8.2	7 59.3	17 1.4
38	9 34.2	0 44.0	24 24.8	7 46.2	17 5.9
39	9 21.7	0 16.3	23 39.6	7 32.5	17 10.6
40	9 9.0	♐29 47.6	22 52.5	7 18.2	17 15.4
41	8 55.8	29 18.1	22 3.5	7 3.1	17 20.5
42	8 42.2	28 47.5	21 12.4	6 47.1	17 25.8
43	8 28.2	28 15.9	20 19.0	6 30.3	17 31.2
44	8 13.6	27 43.1	19 23.2	6 12.5	17 37.0
45	7 58.6	27 9.1	18 24.9	5 53.6	17 43.0
46	7 43.0	26 33.7	17 23.7	5 33.6	17 49.3
47	7 26.8	25 56.9	16 19.5	5 12.2	17 55.8
48	7 10.0	25 18.6	15 12.2	4 49.3	18 2.8
49	6 52.4	24 38.7	14 1.4	4 24.8	18 10.1
50	6 34.1	23 56.9	12 47.0	3 58.5	18 17.8
51	6 14.9	23 13.2	11 28.6	3 30.0	18 26.0
52	5 54.9	22 27.4	10 5.9	2 59.2	18 34.6
53	5 33.8	21 39.3	8 38.8	2 25.7	18 43.9
54	5 11.7	20 48.7	7 6.7	1 49.1	18 53.7
55	4 48.4	19 55.4	5 29.5	1 8.9	19 4.2
56	4 23.7	18 59.1	3 46.7	0 24.5	19 15.5
57	3 57.6	17 59.6	1 58.1	♒29 35.2	19 27.6
58	3 29.9	16 56.5	0 3.2	28 40.0	19 40.7
59	3 0.4	15 49.5	♐28 1.6	27 37.7	19 54.9
60	2 28.8	14 38.1	25 53.2	26 26.8	20 10.3
61	1 54.9	13 21.9	23 37.4	25 5.1	20 27.2
62	1 18.4	12 0.3	21 14.2	23 29.7	20 45.9
63	0 38.9	10 32.9	18 43.3	21 36.2	21 6.4
64	♏29 56.0	8 58.8	16 4.5	19 17.9	21 29.4
65	29 9.0	7 17.4	13 17.9	16 23.4	21 55.1
66	28 17.3	5 27.7	10 23.7	12 29.4	22 24.2
S LAT	5	6	Descendant	8	9

♉ 16° 28′ 1″

2ʰ 56ᵐ 0ˢ MC 44° 0′ 0″

15ʰ 0ᵐ 0ˢ MC 225° 0′ 0″

♏ 17° 27′ 54″

N LAT	11	12	Ascendant	2	3
°	° ′	° ′	° ′	° ′	° ′
0	♐16 11.3	♑13 48.7	♒12 32.1	♓13 43.2	♈16 16.8
5	15 32.5	12 30.4	10 57.9	13 18.3	16 29.1
10	14 53.2	11 10.0	9 17.5	12 51.8	16 41.9
15	14 12.9	9 46.1	7 28.7	12 23.1	16 55.4
16	14 4.7	9 28.8	7 5.6	12 17.0	16 58.2
17	13 56.3	9 11.3	6 42.2	12 10.7	17 1.0
18	13 47.9	8 53.5	6 18.2	12 4.4	17 3.9
19	13 39.4	8 35.5	5 53.7	11 57.8	17 6.9
20	13 30.8	8 17.3	5 28.6	11 51.2	17 9.9
21	13 22.1	7 58.7	5 3.0	11 44.3	17 13.0
22	13 13.2	7 39.9	4 36.7	11 37.3	17 16.1
23	13 4.3	7 20.8	4 9.8	11 30.1	17 19.3
24	12 55.2	7 1.3	3 42.1	11 22.7	17 22.5
25	12 46.0	6 41.4	3 13.7	11 15.0	17 25.9
26	12 36.6	6 21.2	2 44.6	11 7.2	17 29.3
27	12 27.0	6 0.6	2 14.6	10 59.1	17 32.7
28	12 17.3	5 39.6	1 43.7	10 50.7	17 36.3
29	12 7.4	5 18.1	1 11.9	10 42.1	17 40.0
30	11 57.4	4 56.2	0 39.1	10 33.1	17 43.7
31	11 47.1	4 33.7	0 5.3	10 23.8	17 47.6
32	11 36.6	4 10.7	♑29 30.3	10 14.2	17 51.6
33	11 25.8	3 47.2	28 54.2	10 4.2	17 55.7
34	11 14.9	3 23.0	28 16.7	9 53.9	17 59.9
35	11 3.6	2 58.2	27 38.0	9 43.0	18 4.2
36	10 52.1	2 32.6	26 57.7	9 31.7	18 8.7
37	10 40.2	2 6.4	26 16.0	9 20.0	18 13.4
38	10 28.1	1 39.4	25 32.6	9 7.6	18 18.2
39	10 15.5	1 11.5	24 47.4	8 54.7	18 23.2
40	10 2.7	0 42.8	24 0.4	8 41.1	18 28.4
41	9 49.4	0 13.1	23 11.4	8 26.9	18 33.8
42	9 35.7	♐29 42.3	22 20.2	8 11.8	18 39.4
43	9 21.5	29 10.5	21 26.7	7 55.9	18 45.3
44	9 6.9	28 37.6	20 30.8	7 39.1	18 51.4
45	8 51.7	28 3.3	19 32.2	7 21.3	18 57.8
46	8 36.0	27 27.8	18 30.8	7 2.3	19 4.5
47	8 19.7	26 50.7	17 26.3	6 42.1	19 11.5
48	8 2.7	26 12.2	16 18.6	6 20.5	19 18.9
49	7 44.9	25 31.9	15 7.4	5 57.3	19 26.7
50	7 26.4	24 49.8	13 52.3	5 32.3	19 35.0
51	7 7.1	24 5.7	12 33.3	5 5.4	19 43.7
52	6 46.8	23 19.5	11 9.8	4 36.2	19 52.9
53	6 25.5	22 30.9	9 41.7	4 4.5	20 2.7
54	6 3.2	21 39.8	8 8.6	3 29.7	20 13.2
55	5 39.6	20 45.9	6 30.1	2 51.6	20 24.4
56	5 14.6	19 49.0	4 45.9	2 9.5	20 36.5
57	4 48.2	18 48.8	2 55.5	1 22.7	20 49.4
58	4 20.1	17 44.8	0 58.7	0 30.3	21 3.4
59	3 50.1	16 36.9	♐28 55.1	♒29 31.2	21 18.6
60	3 18.0	15 24.4	26 44.2	28 23.8	21 35.1
61	2 43.6	14 7.0	24 25.8	27 6.2	21 53.2
62	2 6.5	12 44.1	21 59.6	25 35.6	22 13.1
63	1 26.3	11 15.1	19 25.4	23 48.0	22 35.1
64	0 42.5	9 39.2	16 43.0	21 37.0	22 59.7
65	♏29 54.5	7 55.6	13 52.6	18 52.3	23 27.2
66	29 1.5	6 3.4	10 54.3	15 13.6	23 58.5
S LAT	5	6	Descendant	8	9

♉ 17° 27′ 54″

3ʰ 0ᵐ 0ˢ MC 45° 0′ 0″

15h 4m 0s		MC ♏ 18° 27′ 35″		226° 0′ 0″	
11	**12**	**Ascendant**	**2**	**3**	**N LAT**
° ′	° ′	° ′	° ′	° ′	°
♐17 6.9	♑14 44.4	♒13 32.0	♓14 47.8	♈17 21.4	0
16 28.0	13 26.3	11 59.1	14 24.6	17 34.4	5
15 48.6	12 6.1	10 20.0	13 59.7	17 48.0	10
15 8.1	10 42.4	8 32.5	13 32.8	18 2.3	15
14 59.8	10 25.1	8 9.7	13 27.1	18 5.3	16
14 51.4	10 7.5	7 46.5	13 21.2	18 8.3	17
14 42.9	9 49.8	7 22.7	13 15.2	18 11.4	18
14 34.4	9 31.8	6 58.5	13 9.1	18 14.6	19
14 25.7	9 13.6	6 33.7	13 2.9	18 17.8	20
14 17.0	8 55.0	6 8.3	12 56.5	18 21.0	21
14 8.1	8 36.2	5 42.2	12 49.9	18 24.3	22
13 59.1	8 17.0	5 15.5	12 43.1	18 27.7	23
13 50.0	7 57.6	4 48.1	12 36.2	18 31.2	24
13 40.7	7 37.7	4 20.0	12 29.0	18 34.7	25
13 31.3	7 17.5	3 51.1	12 21.6	18 38.3	26
13 21.7	6 56.9	3 21.3	12 14.0	18 42.0	27
13 11.9	6 35.8	2 50.7	12 6.2	18 45.8	28
13 2.0	6 14.3	2 19.1	11 58.1	18 49.7	29
12 51.8	5 52.3	1 46.5	11 49.7	18 53.7	30
12 41.5	5 29.8	1 12.8	11 41.0	18 57.8	31
12 30.9	5 6.8	0 38.0	11 31.9	19 2.0	32
12 20.1	4 43.1	0 2.0	11 22.6	19 6.4	33
12 9.1	4 18.9	♑29 24.8	11 12.8	19 10.9	34
11 57.7	3 54.0	28 46.1	11 2.6	19 15.5	35
11 46.1	3 28.4	28 6.0	10 52.1	19 20.3	36
11 34.2	3 2.1	27 24.4	10 41.0	19 25.2	37
11 21.9	2 35.0	26 41.1	10 29.4	19 30.3	38
11 9.3	2 7.0	25 56.0	10 17.2	19 35.6	39
10 56.4	1 38.1	25 9.0	10 4.5	19 41.1	40
10 43.0	1 8.3	24 20.0	9 51.1	19 46.9	41
10 29.2	0 37.4	23 28.8	9 36.9	19 52.8	42
10 14.9	0 5.5	22 35.3	9 22.0	19 59.1	43
10 0.2	♐29 32.3	21 39.2	9 6.2	20 5.6	44
9 44.9	28 57.9	20 40.5	8 49.4	20 12.3	45
9 29.0	28 22.1	19 38.8	8 31.6	20 19.5	46
9 12.5	27 44.8	18 34.1	8 12.5	20 26.9	47
8 55.4	27 6.0	17 26.0	7 52.2	20 34.8	48
8 37.5	26 25.4	16 14.3	7 30.3	20 43.1	49
8 18.8	25 43.0	14 58.7	7 6.8	20 51.8	50
7 59.3	24 58.5	13 39.0	6 41.5	21 1.1	51
7 38.8	24 11.9	12 14.7	6 14.0	21 10.9	52
7 17.3	23 22.9	10 45.7	5 44.0	21 21.4	53
6 54.6	22 31.3	9 11.5	5 11.3	21 32.5	54
6 30.8	21 36.8	7 31.7	4 35.3	21 44.4	55
6 5.5	20 39.3	5 46.0	3 55.6	21 57.2	56
5 38.7	19 38.3	3 54.0	3 11.4	22 10.9	57
5 10.2	18 33.5	1 55.3	2 21.9	22 25.8	58
4 39.9	17 24.6	♐29 49.4	1 26.1	22 41.9	59
4 7.3	16 11.1	27 36.1	0 22.5	22 59.5	60
3 32.3	14 52.5	25 14.9	♒29 9.2	23 18.7	61
2 54.6	13 28.2	22 45.6	27 43.6	23 39.9	62
2 13.7	11 57.5	20 8.0	26 1.9	24 3.4	63
1 29.0	10 19.7	17 22.0	23 58.5	24 29.6	64
0 40.0	8 33.9	14 27.5	21 23.9	24 59.0	65
♏29 45.7	6 39.1	11 24.9	18 0.4	25 32.5	66
5	**6**	**Descendant**	**8**	**9**	**S LAT**
3h 4m 0s		MC ♉ 18° 27′ 35″		46° 0′ 0″	

15h 8m 0s		MC ♏ 19° 27′ 5″		227° 0′ 0″	
N LAT	**11**	**12**	**Ascendant**	**2**	**3**
°	° ′	° ′	° ′	° ′	° ′
0	♐18 2.4	♑15 40.1	♒14 32.0	♓15 52.5	♈18 25.8
5	17 23.4	14 22.3	13 0.6	15 30.9	18 39.6
10	16 43.8	13 2.3	11 22.8	15 7.8	18 54.0
15	16 3.2	11 38.7	9 36.6	14 42.6	19 9.1
16	15 54.8	11 21.4	9 14.1	14 37.3	19 12.3
17	15 46.4	11 3.9	8 51.1	14 31.9	19 15.5
18	15 37.9	10 46.2	8 27.6	14 26.3	19 18.8
19	15 29.4	10 28.2	8 3.7	14 20.6	19 22.1
20	15 20.7	10 10.0	7 39.1	14 14.8	19 25.5
21	15 11.9	9 51.5	7 14.0	14 8.8	19 28.9
22	15 2.9	9 32.6	6 48.2	14 2.6	19 32.4
23	14 53.9	9 13.5	6 21.7	13 56.3	19 36.0
24	14 44.7	8 54.0	5 54.6	13 49.8	19 39.7
25	14 35.4	8 34.2	5 26.7	13 43.2	19 43.4
26	14 25.9	8 13.9	4 58.0	13 36.3	19 47.2
27	14 16.3	7 53.3	4 28.5	13 29.2	19 51.2
28	14 6.5	7 32.2	3 58.1	13 21.9	19 55.2
29	13 56.5	7 10.7	3 26.7	13 14.3	19 59.3
30	13 46.3	6 48.7	2 54.3	8 43.0	20 3.5
31	13 35.9	6 26.1	2 20.9	12 58.3	20 7.9
32	13 25.2	6 3.0	1 46.3	12 49.9	20 12.3
33	13 14.4	5 39.4	1 10.5	12 41.2	20 16.9
34	13 3.3	5 15.1	0 33.4	12 32.1	20 21.7
35	12 51.9	4 50.1	♑29 55.0	12 22.6	20 26.5
36	12 40.2	4 24.4	29 15.0	12 12.7	20 31.6
37	12 28.2	3 58.0	28 33.5	12 2.3	20 36.8
38	12 15.8	3 30.8	27 50.3	11 51.5	20 42.2
39	12 3.2	3 2.8	27 5.3	11 40.2	20 47.9
40	11 50.1	2 33.8	26 18.4	11 28.2	20 53.7
41	11 36.6	2 3.8	25 29.4	11 15.7	20 59.7
42	11 22.7	1 32.8	24 38.2	11 2.5	21 6.1
43	11 8.3	1 0.7	23 44.7	10 48.5	21 12.6
44	10 53.5	0 27.4	22 48.5	10 33.7	21 19.5
45	10 38.1	♐29 52.8	21 49.6	10 18.0	21 26.7
46	10 22.1	29 16.7	20 47.8	10 1.3	21 34.2
47	10 5.4	28 39.2	19 42.8	9 43.5	21 42.1
48	9 48.1	28 0.1	18 34.3	9 24.4	21 50.4
49	9 30.1	27 19.3	17 22.2	9 4.0	21 59.2
50	9 11.2	26 36.5	16 6.1	8 42.0	22 8.4
51	8 51.5	25 51.7	14 45.7	8 18.2	22 18.2
52	8 30.8	25 4.7	13 20.7	7 52.5	22 28.6
53	8 9.1	24 15.2	11 50.8	7 24.4	22 39.7
54	7 46.2	23 23.1	10 15.5	6 53.7	22 51.5
55	7 22.0	22 28.0	8 34.5	6 20.0	23 4.1
56	6 56.5	21 29.8	6 47.3	5 42.7	23 17.6
57	6 29.3	20 28.1	4 53.6	5 1.2	23 32.2
58	6 0.5	19 22.5	2 52.9	4 14.8	23 47.9
59	5 29.7	18 12.7	0 44.8	3 22.3	24 5.0
60	4 56.6	16 58.1	♐28 28.9	2 22.6	24 23.6
61	4 21.1	15 38.2	26 4.9	1 13.7	24 44.0
62	3 42.7	14 12.5	23 32.4	♒29 53.4	25 6.4
63	3 1.0	12 40.1	20 51.3	28 18.0	25 31.3
64	2 15.5	11 0.4	18 1.4	26 22.4	25 59.1
65	1 25.4	9 12.4	15 2.7	23 58.0	26 30.4
66	0 29.8	7 14.9	11 55.6	20 49.5	27 6.1
S LAT	**5**	**6**	**Descendant**	**8**	**9**
	3h 8m 0s		MC ♉ 19° 27′ 5″		47° 0′ 0″

15h 12m 0s — MC — 228° 0′ 0″

♏ 20° 26′ 26″

11	12	Ascendant	2	3	N LAT
° ′	° ′	° ′	° ′	° ′	°
♐18 57.9	♑16 36.0	♒15 32.3	♓16 57.4	♈19 30.1	0
18 18.7	15 18.4	14 2.2	16 37.3	19 44.7	5
17 39.0	13 58.6	12 25.8	16 15.9	19 59.8	10
16 58.2	12 35.2	10 41.0	15 52.6	20 15.8	15
16 49.9	12 18.0	10 18.8	15 47.7	20 19.2	16
16 41.4	12 0.5	9 56.1	15 42.7	20 22.5	17
16 32.9	11 42.8	9 32.9	15 37.5	20 26.0	18
16 24.3	11 24.8	9 9.2	15 32.2	20 29.5	19
16 15.6	11 6.6	8 44.9	15 26.8	20 33.1	20
16 6.7	10 48.1	8 20.0	15 21.3	20 36.7	21
15 57.8	10 29.3	7 54.5	15 15.6	20 40.4	22
15 48.7	10 10.1	7 28.4	15 9.7	20 44.2	23
15 39.5	9 50.6	7 1.5	15 3.7	20 48.0	24
15 30.1	9 30.8	6 33.9	14 57.5	20 52.0	25
15 20.6	9 10.5	6 5.4	14 51.2	20 56.0	26
15 10.9	8 49.9	5 36.2	14 44.6	21 0.1	27
15 1.0	8 28.8	5 6.0	14 37.8	21 4.3	28
14 51.0	8 7.3	4 34.9	14 30.8	21 8.7	29
14 40.7	7 45.2	4 2.8	14 23.5	21 13.1	30
14 30.3	7 22.6	3 29.6	14 16.0	21 17.7	31
14 19.6	6 59.5	2 55.2	14 8.1	21 22.4	32
14 8.6	6 35.8	2 19.6	14 0.0	21 27.3	33
13 57.4	6 11.5	1 42.8	13 51.6	21 32.3	34
13 46.0	5 46.4	1 4.5	13 42.8	21 37.4	35
13 34.2	5 20.7	0 24.7	13 33.6	21 42.7	36
13 22.1	4 54.2	♑29 43.4	13 24.0	21 48.2	37
13 9.7	4 26.9	29 0.3	13 13.9	21 53.9	38
12 57.0	3 58.8	28 15.4	13 3.4	21 59.9	39
12 43.8	3 29.7	27 28.6	12 52.3	22 6.0	40
12 30.3	2 59.6	26 39.7	12 40.7	22 12.4	41
12 16.2	2 28.5	25 48.5	12 28.4	22 19.0	42
12 1.8	1 56.2	24 54.9	12 15.4	22 26.0	43
11 46.8	1 22.7	23 58.8	12 1.7	22 33.2	44
11 31.2	0 47.9	22 59.8	11 47.1	22 40.8	45
11 15.1	0 11.7	21 57.8	11 31.6	22 48.7	46
10 58.4	♐29 34.0	20 52.5	11 15.0	22 57.0	47
10 40.9	28 54.6	19 43.8	10 57.3	23 5.8	48
10 22.7	28 13.4	18 31.2	10 38.3	23 15.0	49
10 3.6	27 30.4	17 14.6	10 17.8	23 24.8	50
9 43.7	26 45.2	15 53.6	9 55.7	23 35.1	51
9 22.8	25 57.8	14 27.9	9 31.7	23 46.0	52
9 0.9	25 7.8	12 57.0	9 5.6	23 57.7	53
8 37.7	24 15.2	11 20.6	8 37.0	24 10.1	54
8 13.3	23 19.6	9 38.4	8 5.5	24 23.4	55
7 47.5	22 20.8	7 49.7	7 30.8	24 37.7	56
7 20.0	21 18.3	5 54.3	6 52.1	24 53.0	57
6 50.8	20 11.9	3 51.6	6 8.8	25 9.6	58
6 19.5	19 1.1	1 41.2	5 19.9	25 27.6	59
5 46.0	17 45.4	♐29 22.7	4 24.1	25 47.3	60
5 10.0	16 24.3	26 55.7	3 19.9	26 8.8	61
4 30.9	14 57.1	24 20.0	2 5.0	26 32.5	62
3 48.5	13 23.0	21 35.1	0 36.1	26 58.9	63
3 2.0	11 41.3	18 41.2	♒28 48.5	27 28.3	64
2 10.8	9 50.9	15 38.2	26 34.5	28 1.5	65
1 13.9	7 50.7	12 26.5	23 40.8	28 39.3	66
5	6	Descendant	8	9	S LAT

♉ 20° 26′ 26″

3h 12m 0s — MC — 48° 0′ 0″

15h 16m 0s — MC — 229° 0′ 0″

♏ 21° 25′ 35″

11	12	Ascendant	2	3	N LAT
° ′	° ′	° ′	° ′	° ′	°
♐19 53.3	♑17 31.9	♒16 32.7	♓18 2.3	♈20 34.3	0
19 14.0	16 14.7	15 4.1	17 43.9	20 49.6	5
18 34.2	14 55.1	13 29.1	17 24.2	21 5.5	10
17 53.3	13 31.9	11 45.8	17 2.8	21 22.3	15
17 44.9	13 14.7	11 23.8	16 58.2	21 25.8	16
17 36.4	12 57.2	11 1.4	16 53.6	21 29.4	17
17 27.8	12 39.5	10 38.5	16 48.8	21 33.0	18
17 19.2	12 21.6	10 15.1	16 44.0	21 36.7	19
17 10.4	12 3.4	9 51.1	16 39.0	21 40.5	20
17 1.5	11 44.9	9 26.5	16 33.9	21 44.3	21
16 52.5	11 26.1	9 1.3	16 28.7	21 48.2	22
16 43.4	11 6.9	8 35.4	16 23.3	21 52.1	23
16 34.2	10 47.5	8 8.8	16 17.8	21 56.2	24
16 24.8	10 27.6	7 41.5	16 12.1	22 0.3	25
16 15.2	10 7.4	7 13.3	16 6.2	22 4.6	26
16 5.5	9 46.7	6 44.4	16 0.2	22 8.9	27
15 55.5	9 25.6	6 14.5	15 53.9	22 13.3	28
15 45.4	9 4.1	5 43.6	15 47.4	22 17.9	29
15 35.1	8 42.0	5 11.8	15 40.7	22 22.6	30
15 24.6	8 19.4	4 38.8	15 33.8	22 27.4	31
15 13.9	7 56.2	4 4.7	15 26.6	22 32.3	32
15 2.9	7 32.5	3 29.4	15 19.1	22 37.4	33
14 51.6	7 8.1	2 52.7	15 11.4	22 42.7	34
14 40.1	6 43.0	2 14.7	15 3.2	22 48.1	35
14 28.3	6 17.3	1 35.1	14 54.8	22 53.7	36
14 16.1	5 50.7	0 53.9	14 45.9	22 59.4	37
14 3.6	5 23.3	0 11.1	14 36.7	23 5.4	38
13 50.8	4 55.1	♑29 26.3	14 27.0	23 11.6	39
13 37.6	4 25.9	28 39.6	14 16.8	23 18.1	40
13 23.9	3 55.7	27 50.8	14 6.0	23 24.8	41
13 9.8	3 24.4	26 59.7	13 54.7	23 31.8	42
12 55.2	2 52.0	26 6.1	13 42.7	23 39.1	43
12 40.1	2 18.4	25 9.9	13 30.1	23 46.7	44
12 24.5	1 43.4	24 10.9	13 16.6	23 54.6	45
12 8.2	1 7.0	23 8.7	13 2.3	24 2.9	46
11 51.1	0 29.0	22 3.3	12 47.0	24 11.7	47
11 33.7	♐29 49.4	20 54.3	12 30.6	24 20.9	48
11 15.3	29 8.0	19 41.4	12 13.1	24 30.6	49
10 56.1	28 24.6	18 24.3	11 54.2	24 40.8	50
10 36.0	27 39.1	17 2.7	11 33.7	24 51.7	51
10 14.9	26 51.2	15 36.2	11 11.6	25 3.2	52
9 52.8	26 0.9	14 4.5	10 47.4	25 15.4	53
9 29.4	25 7.7	12 27.0	10 21.0	25 28.4	54
9 4.7	24 11.6	10 43.5	9 51.9	25 42.4	55
8 38.5	23 12.1	8 53.4	9 19.8	25 57.4	56
8 10.7	22 8.9	6 56.2	8 44.0	26 13.5	57
7 41.1	21 1.6	4 51.5	8 3.9	26 31.0	58
7 9.5	19 49.9	2 38.7	7 18.7	26 49.9	59
6 35.5	18 33.1	0 17.6	6 27.1	27 10.6	60
5 58.8	17 10.7	♐27 47.6	5 27.6	27 33.2	61
5 19.2	15 42.0	25 8.4	4 18.2	27 58.2	62
4 35.9	14 6.2	22 19.7	2 56.0	28 26.0	63
3 48.6	12 22.5	19 21.6	1 16.6	28 57.0	64
2 56.3	10 29.6	16 14.0	♒29 13.3	29 32.1	65
1 58.0	8 26.5	12 57.4	26 34.3	♉0 12.1	66
5	6	Descendant	8	9	S LAT

♉ 21° 25′ 35″

3h 16m 0s — MC — 49° 0′ 0″

15ʰ 20ᵐ 0ˢ — MC 230° 0′ 0″

♏ 22° 24′ 34″

N LAT	11	12	Ascendant	2	3
°	° ′	° ′	° ′	° ′	° ′
0	♐20 48.6	♑18 27.9	♒17 33.3	♓19 7.3	♈21 38.3
5	20 9.2	17 11.0	16 6.1	18 50.5	21 54.4
10	19 29.3	15 51.7	14 32.7	18 32.5	22 11.1
15	18 48.3	14 28.7	12 50.8	18 13.0	22 28.7
16	18 39.9	14 11.5	12 29.2	18 8.9	22 32.4
17	18 31.3	13 54.1	12 7.1	18 4.7	22 36.1
18	18 22.8	13 36.4	11 44.5	18 0.3	22 39.9
19	18 14.1	13 18.5	11 21.4	17 55.9	22 43.8
20	18 5.3	13 0.4	10 57.7	17 51.4	22 47.7
21	17 56.4	12 41.9	10 33.4	17 46.7	22 51.7
22	17 47.3	12 23.1	10 8.5	17 41.9	22 55.8
23	17 38.2	12 4.0	9 42.9	17 37.0	22 59.9
24	17 28.9	11 44.5	9 16.6	17 32.0	23 4.2
25	17 19.4	11 24.7	8 49.6	17 26.8	23 8.5
26	17 9.8	11 4.4	8 21.7	17 21.4	23 12.9
27	17 0.0	10 43.8	7 53.0	17 15.9	23 17.5
28	16 50.1	10 22.7	7 23.4	17 10.2	23 22.1
29	16 39.9	10 1.1	6 52.9	17 4.3	23 26.9
30	16 29.6	9 39.0	6 21.3	16 58.2	23 31.8
31	16 19.0	9 16.4	5 48.6	16 51.9	23 36.8
32	16 8.2	8 53.2	5 14.8	16 45.3	23 42.0
33	15 57.2	8 29.4	4 39.7	16 38.5	23 47.3
34	15 45.8	8 5.0	4 3.3	16 31.4	23 52.8
35	15 34.2	7 39.9	3 25.5	16 24.0	23 58.5
36	15 22.3	7 14.1	2 46.2	16 16.2	24 4.4
37	15 10.1	6 47.4	2 5.3	16 8.2	24 10.4
38	14 57.6	6 20.0	1 22.6	15 59.7	24 16.7
39	14 44.6	5 51.7	0 38.0	15 50.8	24 23.2
40	14 31.3	5 22.4	♑29 51.5	15 41.5	24 29.9
41	14 17.6	4 52.1	29 2.8	15 31.7	24 37.0
42	14 3.4	4 20.7	28 11.8	15 21.3	24 44.3
43	13 48.7	3 48.2	27 18.3	15 10.4	24 51.9
44	13 33.5	3 14.4	26 22.1	14 58.8	24 59.8
45	13 17.7	2 39.2	25 23.0	14 46.5	25 8.2
46	13 1.3	2 2.6	24 20.8	14 33.4	25 16.9
47	12 44.3	1 24.4	23 15.1	14 19.4	25 26.1
48	12 26.6	0 44.6	22 5.9	14 4.5	25 35.7
49	12 8.0	0 2.9	20 52.6	13 48.4	25 45.8
50	11 48.7	♐29 19.2	19 35.1	13 31.1	25 56.6
51	11 28.4	28 33.3	18 13.0	13 12.4	26 7.9
52	11 7.1	27 45.1	16 45.8	12 52.1	26 19.9
53	10 44.7	26 54.3	15 13.2	12 30.0	26 32.8
54	10 21.1	26 0.6	13 34.7	12 5.8	26 46.4
55	9 56.1	25 3.9	11 49.9	11 39.1	27 1.0
56	9 29.7	24 3.8	9 58.3	11 9.7	27 16.7
57	9 1.5	22 59.8	7 59.3	10 36.9	27 33.6
58	8 31.5	21 51.8	5 52.6	10 0.1	27 51.9
59	7 59.5	20 39.0	3 37.5	9 18.6	28 11.8
60	7 25.0	19 21.1	1 13.6	8 31.2	28 33.5
61	6 47.8	17 57.4	♐28 40.4	7 36.7	28 57.3
62	6 7.5	16 27.2	25 57.6	6 33.1	29 23.5
63	5 23.5	14 49.7	23 5.0	5 17.7	29 52.7
64	4 35.2	13 3.8	20 2.5	3 46.7	♉ 0 25.3
65	3 41.7	11 8.5	16 50.2	1 54.1	1 2.3
66	2 42.0	9 2.4	13 28.4	♒29 29.8	1 44.5
S LAT	5	6	Descendant	8	9

♉ 22° 24′ 34″

3ʰ 20ᵐ 0ˢ — MC 50° 0′ 0″

15ʰ 24ᵐ 0ˢ — MC 231° 0′ 0″

♏ 23° 23′ 23″

N LAT	11	12	Ascendant	2	3
°	° ′	° ′	° ′	° ′	° ′
0	♐21 43.9	♑19 24.1	♒18 34.0	♓20 12.3	♈22 42.3
5	21 4.4	18 7.5	17 8.4	19 57.2	22 59.0
10	20 24.4	16 48.5	15 36.6	19 41.0	23 16.5
15	19 43.2	15 25.7	13 56.2	19 23.4	23 34.9
16	19 34.8	15 8.5	13 34.9	19 19.7	23 38.8
17	19 26.3	14 51.2	13 13.2	19 15.8	23 42.7
18	19 17.6	14 33.5	12 50.9	19 11.9	23 46.6
19	19 8.9	14 15.7	12 28.1	19 7.9	23 50.6
20	19 0.1	13 57.5	12 4.7	19 3.9	23 54.7
21	18 51.1	13 39.0	11 40.7	18 59.6	23 58.9
22	18 42.1	13 20.3	11 16.1	18 55.3	24 3.2
23	18 32.9	13 1.2	10 50.9	18 50.9	24 7.5
24	18 23.5	12 41.7	10 24.9	18 46.4	24 12.0
25	18 14.1	12 21.9	9 58.2	18 41.7	24 16.5
26	18 4.4	12 1.7	9 30.6	18 36.8	24 21.1
27	17 54.6	11 41.0	9 2.2	18 31.8	24 25.9
28	17 44.6	11 19.9	8 33.0	18 26.7	24 30.7
29	17 34.4	10 58.3	8 2.7	18 21.4	24 35.7
30	17 24.0	10 36.3	7 31.4	18 15.8	24 40.8
31	17 13.4	10 13.6	6 59.1	18 10.1	24 46.1
32	17 2.5	9 50.4	6 25.5	18 4.2	24 51.5
33	16 51.4	9 26.6	5 50.8	17 58.0	24 57.0
34	16 40.1	9 2.2	5 14.6	17 51.6	25 2.8
35	16 28.4	8 37.0	4 37.1	17 44.9	25 8.7
36	16 16.4	8 11.1	3 58.0	17 37.9	25 14.8
37	16 4.1	7 44.5	3 17.3	17 30.6	25 21.1
38	15 51.5	7 16.9	2 34.9	17 23.0	25 27.7
39	15 38.5	6 48.5	1 50.6	17 14.9	25 34.5
40	15 25.1	6 19.2	1 4.2	17 6.5	25 41.5
41	15 11.3	5 48.8	0 15.7	16 57.6	25 48.9
42	14 57.0	5 17.3	♑29 24.8	16 48.3	25 56.5
43	14 42.2	4 44.6	28 31.4	16 38.4	26 4.5
44	14 26.9	4 10.7	27 35.2	16 27.9	26 12.8
45	14 11.0	3 35.3	26 36.1	16 16.8	26 21.5
46	13 54.5	2 58.6	25 33.8	16 4.9	26 30.6
47	13 37.4	2 20.2	24 28.1	15 52.3	26 40.1
48	13 19.5	1 40.1	23 18.6	15 38.7	26 50.2
49	13 0.8	0 58.1	22 5.1	15 24.2	27 0.8
50	12 41.3	0 14.1	20 47.2	15 8.5	27 12.0
51	12 20.8	♐29 27.9	19 24.5	14 51.6	27 23.8
52	11 59.3	28 39.3	17 56.7	14 33.2	27 36.4
53	11 36.7	27 48.1	16 23.2	14 13.1	27 49.8
54	11 12.8	26 53.9	14 43.7	13 51.2	28 4.0
55	10 47.6	25 56.7	12 57.7	13 27.0	28 19.3
56	10 20.9	24 55.9	11 4.6	13 0.3	28 35.7
57	9 52.4	23 51.2	9 3.8	12 30.6	28 53.4
58	9 22.0	22 42.3	6 55.0	11 57.2	29 12.5
59	8 49.5	21 28.6	4 37.5	11 19.6	29 33.3
60	8 14.6	20 9.6	2 10.7	10 36.6	29 56.0
61	7 36.8	18 44.6	♐29 34.3	9 47.1	♉ 0 20.8
62	6 55.8	17 12.8	26 47.9	8 49.4	0 48.3
63	6 11.0	15 33.5	23 51.1	7 41.1	1 18.9
64	5 21.8	13 45.4	20 44.0	6 18.7	1 53.2
65	4 27.2	11 47.5	17 26.7	4 36.9	2 32.0
66	3 25.9	9 38.3	13 59.6	2 27.1	3 16.6
S LAT	5	6	Descendant	8	9

♉ 23° 23′ 23″

3ʰ 24ᵐ 0ˢ — MC 51° 0′ 0″

15ʰ 28ᵐ 0ˢ		MC		232° 0′ 0″
		♏ 24° 22′ 2″		

N LAT	11	12	Ascendant	2	3
°	° ′	° ′	° ′	° ′	° ′
0	♐22 39.2	♑20 20.3	♒19 35.0	♓21 17.5	♈23 46.0
5	21 59.6	19 4.1	18 11.0	21 4.0	24 3.5
10	21 19.5	17 45.4	16 40.7	20 49.6	24 21.7
15	20 38.2	16 22.8	15 2.0	20 33.9	24 40.9
16	20 29.7	16 5.7	14 41.0	20 30.5	24 45.0
17	20 21.2	15 48.4	14 19.5	20 27.1	24 49.0
18	20 12.5	15 30.8	13 57.6	20 23.7	24 53.1
19	20 3.8	15 13.0	13 35.1	20 20.1	24 57.3
20	19 54.9	14 54.8	13 12.1	20 16.5	25 1.6
21	19 45.9	14 36.4	12 48.4	20 12.7	25 6.0
22	19 36.8	14 17.7	12 24.2	20 8.9	25 10.4
23	19 27.6	13 58.6	11 59.2	20 4.9	25 14.9
24	19 18.2	13 39.2	11 33.6	20 0.9	25 19.6
25	19 8.7	13 19.4	11 7.2	19 56.7	25 24.3
26	18 59.0	12 59.2	10 40.0	19 52.4	25 29.1
27	18 49.2	12 38.5	10 12.0	19 47.9	25 34.0
28	18 39.1	12 17.4	9 43.0	19 43.3	25 39.1
29	18 28.9	11 55.8	9 13.1	19 38.6	25 44.3
30	18 18.4	11 33.8	8 42.1	19 33.6	25 49.6
31	18 7.8	11 11.1	8 10.1	19 28.5	25 55.1
32	17 56.9	10 47.9	7 36.9	19 23.3	26 0.7
33	17 45.7	10 24.1	7 2.4	19 17.7	26 6.5
34	17 34.3	9 59.6	6 26.6	19 12.0	26 12.5
35	17 22.6	9 34.4	5 49.4	19 6.1	26 18.7
36	17 10.5	9 8.5	5 10.6	18 59.8	26 25.1
37	16 58.2	8 41.8	4 30.2	18 53.3	26 31.6
38	16 45.5	8 14.2	3 48.0	18 46.5	26 38.5
39	16 32.4	7 45.7	3 3.9	18 39.3	26 45.5
40	16 18.9	7 16.3	2 17.8	18 31.8	26 52.9
41	16 5.0	6 45.8	1 29.4	18 23.9	27 0.5
42	15 50.6	6 14.2	0 38.7	18 15.5	27 8.5
43	15 35.8	5 41.4	♑29 45.4	18 6.7	27 16.7
44	15 20.4	5 7.3	28 49.4	17 57.3	27 25.4
45	15 4.4	4 31.8	27 50.3	17 47.4	27 34.5
46	14 47.7	3 54.9	26 48.0	17 36.8	27 43.9
47	14 30.5	3 16.3	25 42.2	17 25.5	27 53.9
48	14 12.4	2 36.0	24 32.6	17 13.4	28 4.4
49	13 53.6	1 53.7	23 18.8	17 0.4	28 15.4
50	13 33.9	1 9.5	22 0.5	16 46.4	28 27.1
51	13 13.3	0 22.9	20 37.4	16 31.2	28 39.4
52	12 51.6	♐29 33.9	19 8.9	16 14.8	28 52.5
53	12 28.8	28 42.3	17 34.6	15 56.9	29 6.4
54	12 4.7	27 47.7	15 54.1	15 37.2	29 21.3
55	11 39.2	26 49.8	14 6.9	15 15.6	29 37.2
56	11 12.1	25 48.4	12 12.3	14 51.7	29 54.3
57	10 43.4	24 43.0	10 9.8	14 25.1	♉ 0 12.7
58	10 12.6	23 33.3	7 58.8	13 55.3	0 32.7
59	9 39.7	22 18.6	5 38.8	13 21.5	0 54.3
60	9 4.3	20 58.4	3 9.1	12 43.1	1 18.0
61	8 25.9	19 32.1	0 29.4	11 58.7	1 44.0
62	7 44.3	17 58.8	♐27 39.1	11 7.1	2 12.7
63	6 58.7	16 17.6	24 38.1	10 5.9	2 44.7
64	6 8.5	14 27.3	21 26.2	8 52.3	3 20.6
65	5 12.6	12 26.8	18 3.7	7 21.5	4 1.3
66	4 9.9	10 14.3	14 30.9	5 26.2	4 48.2
S LAT	5	6	Descendant	8	9

		♉ 24° 22′ 2″		
3ʰ 28ᵐ 0ˢ		MC		52° 0′ 0″

15ʰ 32ᵐ 0ˢ		MC		233° 0′ 0″
		♏ 25° 20′ 30″		

N LAT	11	12	Ascendant	2	3
°	° ′	° ′	° ′	° ′	° ′
0	♐23 34.4	♑21 16.7	♒20 36.1	♓22 22.6	♈24 49.7
5	22 54.7	20 0.8	19 13.7	22 10.8	25 7.9
10	22 14.5	18 42.4	17 45.1	21 58.2	25 26.8
15	21 33.1	17 20.1	16 8.0	21 44.4	25 46.8
16	21 24.6	17 3.1	15 47.4	21 41.5	25 51.0
17	21 16.1	16 45.8	15 26.3	21 38.5	25 55.2
18	21 7.4	16 28.3	15 4.7	21 35.5	25 59.5
19	20 58.6	16 10.5	14 42.5	21 32.4	26 3.9
20	20 49.7	15 52.4	14 19.8	21 29.2	26 8.3
21	20 40.7	15 34.0	13 56.6	21 25.9	26 12.8
22	20 31.6	15 15.3	13 32.6	21 22.5	26 17.4
23	20 22.3	14 56.2	13 8.1	21 19.0	26 22.1
24	20 12.9	14 36.8	12 42.8	21 15.5	26 27.0
25	20 3.4	14 17.1	12 16.7	21 11.8	26 31.9
26	19 53.6	13 56.9	11 49.9	21 8.0	26 36.9
27	19 43.7	13 36.2	11 22.2	21 4.1	26 42.0
28	19 33.7	13 15.2	10 53.6	21 0.1	26 47.3
29	19 23.4	12 53.6	10 24.0	20 55.9	26 52.7
30	19 12.9	12 31.5	9 53.4	20 51.6	26 58.2
31	19 2.2	12 8.9	9 21.7	20 47.1	27 3.9
32	18 51.2	11 45.7	8 48.9	20 42.5	27 9.8
33	18 40.0	11 21.8	8 14.8	20 37.7	27 15.8
34	18 28.5	10 57.3	7 39.3	20 32.6	27 22.0
35	18 16.8	10 32.1	7 2.4	20 27.4	27 28.4
36	18 4.7	10 6.1	6 23.9	20 21.9	27 35.0
37	17 52.3	9 39.4	5 43.8	20 16.2	27 41.9
38	17 39.5	9 11.8	5 1.9	20 10.2	27 49.0
39	17 26.3	8 43.2	4 18.1	20 3.9	27 56.3
40	17 12.8	8 13.7	3 32.2	19 57.3	28 3.9
41	16 58.8	7 43.1	2 44.1	19 50.4	28 11.9
42	16 44.3	7 11.4	1 53.6	19 43.0	28 20.1
43	16 29.4	6 38.5	1 0.5	19 35.2	28 28.7
44	16 13.9	6 4.3	0 4.6	19 27.0	28 37.7
45	15 57.8	5 28.7	♑29 5.7	19 18.3	28 47.1
46	15 41.0	4 51.6	28 3.4	19 9.0	28 57.0
47	15 23.6	4 12.8	26 57.5	18 59.0	29 7.4
48	15 5.5	3 32.3	25 47.8	18 48.4	29 18.2
49	14 46.5	2 49.8	24 33.8	18 37.0	29 29.7
50	14 26.6	2 5.2	23 15.2	18 24.7	29 41.8
51	14 5.8	1 18.4	21 51.6	18 11.3	29 54.6
52	13 43.9	0 29.0	20 22.5	17 56.9	♉ 0 8.3
53	13 20.9	♐29 36.9	18 47.5	17 41.1	0 22.7
54	12 56.6	28 41.8	17 6.0	17 23.8	0 38.2
55	12 30.8	27 43.4	15 17.6	17 4.8	0 54.7
56	12 3.5	26 41.4	13 21.5	16 43.8	1 12.5
57	11 34.4	25 35.3	11 17.2	16 20.3	1 31.7
58	11 3.3	24 24.7	9 4.1	15 54.1	1 52.4
59	10 30.0	23 9.0	6 41.5	15 24.4	2 15.0
60	9 54.0	21 47.7	4 8.9	14 50.5	2 39.6
61	9 15.1	20 20.0	1 25.7	14 11.4	3 6.7
62	8 32.8	18 45.1	♐28 31.5	13 25.9	3 36.6
63	7 46.4	17 2.0	25 26.0	12 32.1	4 10.0
64	6 55.2	15 9.5	22 9.1	11 27.3	4 47.5
65	5 58.1	13 6.2	18 41.0	10 7.7	5 30.2
66	4 53.8	10 50.3	15 2.3	8 26.8	6 19.3
S LAT	5	6	Descendant	8	9

		♉ 25° 20′ 30″		
3ʰ 32ᵐ 0ˢ		MC		53° 0′ 0″

$15^h\ 36^m\ 0^s$ MC 234° 0′ 0″

♏ 26° 18′ 49″

N LAT	11	12	Ascendant	2	3
°	° ′	° ′	° ′	° ′	° ′
0	♐24 29.5	♑22 13.1	♒21 37.4	♓23 27.9	♈25 53.2
5	23 49.8	20 57.6	20 16.7	23 17.7	26 12.1
10	23 9.6	19 39.6	18 49.7	23 6.9	26 31.7
15	22 28.1	18 17.6	17 14.4	22 55.1	26 52.5
16	22 19.5	18 0.6	16 54.1	22 52.6	26 56.8
17	22 10.9	17 43.4	16 33.4	22 50.0	27 1.2
18	22 2.2	17 25.9	16 12.1	22 47.4	27 5.7
19	21 53.4	17 8.2	15 50.3	22 44.7	27 10.2
20	21 44.5	16 50.1	15 28.0	22 42.0	27 14.8
21	21 35.5	16 31.8	15 5.1	22 39.1	27 19.5
22	21 26.3	16 13.1	14 41.5	22 36.2	27 24.3
23	21 17.0	15 54.1	14 17.3	22 33.3	27 29.2
24	21 7.6	15 34.7	13 52.4	22 30.2	27 34.1
25	20 58.0	15 15.0	13 26.7	22 27.1	27 39.2
26	20 48.3	14 54.8	13 0.3	22 23.8	27 44.4
27	20 38.3	14 34.2	12 33.0	22 20.5	27 49.8
28	20 28.2	14 13.1	12 4.7	22 17.0	27 55.2
29	20 17.9	13 51.6	11 35.5	22 13.4	28 0.8
30	20 7.4	13 29.5	11 5.3	22 9.7	28 6.6
31	19 56.6	13 6.9	10 34.0	22 5.9	28 12.5
32	19 45.6	12 43.7	10 1.5	22 1.9	28 18.6
33	19 34.3	12 19.8	9 27.8	21 57.7	28 24.8
34	19 22.8	11 55.3	8 52.6	21 53.4	28 31.2
35	19 11.0	11 30.1	8 16.1	21 48.9	28 37.9
36	18 58.8	11 4.1	7 38.0	21 44.2	28 44.8
37	18 46.4	10 37.3	6 58.2	21 39.3	28 51.9
38	18 33.5	10 9.6	6 16.7	21 34.1	28 59.2
39	18 20.3	9 41.0	5 33.2	21 28.7	29 6.8
40	18 6.7	9 11.5	4 47.6	21 23.0	29 14.7
41	17 52.6	8 40.8	3 59.8	21 17.0	29 23.0
42	17 38.1	8 9.0	3 9.5	21 10.7	29 31.5
43	17 23.0	7 36.0	2 16.6	21 4.1	29 40.5
44	17 7.4	7 1.7	1 20.9	20 57.0	29 49.8
45	16 51.2	6 25.9	0 22.1	20 49.5	29 59.5
46	16 34.4	5 48.6	♑29 19.9	20 41.5	♉ 0 9.8
47	16 16.8	5 9.7	28 14.1	20 32.9	0 20.5
48	15 58.5	4 28.9	27 4.2	20 23.7	0 31.8
49	15 39.4	3 46.2	25 50.1	20 13.9	0 43.7
50	15 19.4	3 1.4	24 31.2	20 3.3	0 56.2
51	14 58.4	2 14.2	23 7.2	19 51.8	1 9.5
52	14 36.4	1 24.5	21 37.6	19 39.4	1 23.6
53	14 13.1	0 32.0	20 1.9	19 25.8	1 38.7
54	13 48.6	♐29 36.4	18 19.5	19 10.9	1 54.7
55	13 22.6	28 37.5	16 29.8	18 54.6	2 11.9
56	12 55.0	27 34.8	14 32.3	18 36.4	2 30.3
57	12 25.6	26 28.0	12 26.3	18 16.2	2 50.2
58	11 54.1	25 16.6	10 11.0	17 53.6	3 11.7
59	11 20.3	23 59.9	7 45.8	17 28.0	3 35.1
60	10 43.9	22 37.5	5 10.2	16 58.8	4 0.8
61	10 4.4	21 8.4	2 23.4	16 25.1	4 28.9
62	9 21.4	19 31.9	♐29 25.1	15 45.9	5 0.1
63	8 34.2	17 46.9	26 14.9	14 59.5	5 34.8
64	7 42.0	15 52.0	22 52.7	14 3.7	6 14.0
65	6 43.7	13 45.8	19 18.9	12 55.3	6 58.6
66	5 37.7	11 26.4	15 33.9	11 28.9	7 50.1
S LAT	**5**	**6**	**Descendant**	**8**	**9**

♉ 26° 18′ 49″

$3^h\ 36^m\ 0^s$ MC 54° 0′ 0″

$15^h\ 40^m\ 0^s$ MC 235° 0′ 0″

♏ 27° 16′ 58″

N LAT	11	12	Ascendant	2	3
°	° ′	° ′	° ′	° ′	° ′
0	♐25 24.7	♑23 9.7	♒22 38.9	♓24 33.2	♈26 56.5
5	24 44.9	21 54.6	21 19.9	24 24.7	27 16.1
10	24 4.6	20 37.0	19 54.6	24 15.7	27 36.5
15	23 23.0	19 15.3	18 21.1	24 5.8	27 58.0
16	23 14.4	18 58.4	18 1.2	24 3.7	28 2.5
17	23 5.8	18 41.2	17 40.8	24 1.6	28 7.0
18	22 57.1	18 23.8	17 19.9	23 59.4	28 11.6
19	22 48.3	18 6.1	16 58.5	23 57.1	28 16.3
20	22 39.3	17 48.1	16 36.6	23 54.8	28 21.1
21	22 30.3	17 29.8	16 14.0	23 52.5	28 26.0
22	22 21.1	17 11.1	15 50.9	23 50.1	28 30.9
23	22 11.8	16 52.2	15 27.0	23 47.6	28 36.0
24	22 2.3	16 32.8	15 2.5	23 45.0	28 41.1
25	21 52.7	16 13.1	14 37.2	23 42.4	28 46.4
26	21 42.9	15 53.0	14 11.2	23 39.7	28 51.8
27	21 32.9	15 32.4	13 44.2	23 36.9	28 57.3
28	21 22.8	15 11.4	13 16.4	23 34.0	29 3.0
29	21 12.4	14 49.8	12 47.6	23 31.0	29 8.8
30	21 1.8	14 27.8	12 17.8	23 27.9	29 14.7
31	20 51.0	14 5.2	11 46.9	23 24.7	29 20.8
32	20 40.0	13 42.0	11 14.8	23 21.4	29 27.1
33	20 28.7	13 18.1	10 41.4	23 17.9	29 33.6
34	20 17.1	12 53.6	10 6.7	23 14.3	29 40.2
35	20 5.2	12 28.4	9 30.6	23 10.5	29 47.1
36	19 53.0	12 2.4	8 52.8	23 6.6	29 54.2
37	19 40.5	11 35.5	8 13.5	23 2.5	♉ 0 1.6
38	19 27.6	11 7.8	7 32.3	22 58.2	0 9.2
39	19 14.3	10 39.2	6 49.1	22 53.7	0 17.1
40	19 0.6	10 9.6	6 3.9	22 48.9	0 25.3
41	18 46.5	9 38.8	5 16.4	22 43.9	0 33.8
42	18 31.9	9 7.0	4 26.4	22 38.6	0 42.6
43	18 16.7	8 33.9	3 33.8	22 33.1	0 51.9
44	18 1.0	7 59.4	2 38.3	22 27.1	1 1.5
45	17 44.7	7 23.5	1 39.7	22 20.9	1 11.6
46	17 27.8	6 46.1	0 37.6	22 14.2	1 22.2
47	17 10.1	6 7.0	♑29 31.8	22 7.0	1 33.3
48	16 51.7	5 26.0	28 22.0	21 59.4	1 45.0
49	16 32.5	4 43.1	27 7.8	21 51.1	1 57.3
50	16 12.3	3 58.0	25 48.7	21 42.3	2 10.3
51	15 51.2	3 10.5	24 24.4	21 32.7	2 24.0
52	15 28.9	2 20.4	22 54.3	21 22.3	2 38.7
53	15 5.5	1 27.5	21 17.9	21 10.9	2 54.2
54	14 40.7	0 31.5	19 34.6	20 58.4	3 10.8
55	14 14.4	♐29 32.1	17 43.8	20 44.7	3 28.6
56	13 46.6	28 28.8	15 44.8	20 29.6	3 47.7
57	13 16.8	27 21.2	13 37.0	20 12.7	4 8.3
58	12 45.0	26 8.9	11 19.5	19 53.7	4 30.6
59	12 10.8	24 51.3	8 51.8	19 32.2	4 54.9
60	11 33.9	23 27.7	6 13.0	19 7.8	5 21.4
61	10 53.8	21 57.3	3 22.6	18 39.6	5 50.6
62	10 10.2	20 19.1	0 20.0	18 6.8	6 23.0
63	9 22.1	18 32.1	♐27 4.9	17 28.0	6 59.2
64	8 28.9	16 34.9	23 37.2	16 41.3	7 39.9
65	7 29.3	14 25.7	19 57.3	15 44.1	8 26.4
66	6 21.5	12 2.6	16 5.7	14 32.1	9 20.3
S LAT	**5**	**6**	**Descendant**	**8**	**9**

♉ 27° 16′ 58″

$3^h\ 40^m\ 0^s$ MC 55° 0′ 0″

$15^h\ 44^m\ 0^s$ MC 236° 0′ 0″

♏ 28° 14′ 58″

11	12	Ascendant	2	3	N LAT
° ′	° ′	° ′	° ′	° ′	°
♐26 19.8	♑24 6.4	♒23 40.6	♓25 38.5	♈27 59.7	**0**
25 40.0	22 51.8	22 23.3	25 31.7	28 20.0	**5**
24 59.6	21 34.5	20 59.8	25 24.5	28 41.1	**10**
24 17.9	20 13.2	19 28.1	25 16.6	29 3.3	**15**
24 9.3	19 56.3	19 8.6	25 14.9	29 7.9	**16**
24 0.7	19 39.2	18 48.6	25 13.2	29 12.6	**17**
23 51.9	19 21.8	18 28.1	25 11.4	29 17.4	**18**
23 43.1	19 4.2	18 7.1	25 9.6	29 22.2	**19**
23 34.1	18 46.2	17 45.5	25 7.8	29 27.2	**20**
23 25.1	18 28.0	17 23.4	25 5.9	29 32.2	**21**
23 15.8	18 9.4	17 0.6	25 4.0	29 37.3	**22**
23 6.5	17 50.5	16 37.2	25 2.0	29 42.6	**23**
22 57.0	17 31.2	16 13.1	24 59.9	29 47.9	**24**
22 47.4	17 11.5	15 48.2	24 57.8	29 53.3	**25**
22 37.5	16 51.4	15 22.5	24 55.6	29 58.9	**26**
22 27.5	16 30.9	14 56.0	24 53.4	♉ 0 4.6	**27**
22 17.4	16 9.9	14 28.6	24 51.1	0 10.5	**28**
22 7.0	15 48.4	14 0.3	24 48.7	0 16.5	**29**
21 56.4	15 26.3	13 30.9	24 46.2	0 22.6	**30**
21 45.5	15 3.7	13 0.4	24 43.6	0 28.9	**31**
21 34.4	14 40.5	12 28.7	24 41.0	0 35.4	**32**
21 23.1	14 16.7	11 55.8	24 38.2	0 42.1	**33**
21 11.5	13 52.2	11 21.5	24 35.3	0 49.0	**34**
20 59.5	13 26.9	10 45.8	24 32.3	0 56.1	**35**
20 47.3	13 0.9	10 8.5	24 29.1	1 3.4	**36**
20 34.7	12 34.1	9 29.5	24 25.8	1 11.0	**37**
20 21.7	12 6.4	8 48.7	24 22.4	1 18.9	**38**
20 8.4	11 37.7	8 6.0	24 18.8	1 27.0	**39**
19 54.6	11 8.0	7 21.1	24 15.0	1 35.5	**40**
19 40.4	10 37.2	6 34.0	24 10.9	1 44.3	**41**
19 25.7	10 5.3	5 44.4	24 6.7	1 53.4	**42**
19 10.5	9 32.1	4 52.0	24 2.2	2 3.0	**43**
18 54.7	8 57.6	3 56.8	23 57.5	2 12.9	**44**
18 38.3	8 21.6	2 58.4	23 52.5	2 23.4	**45**
18 21.3	7 44.0	1 56.6	23 47.1	2 34.3	**46**
18 3.5	7 4.7	0 50.9	23 41.4	2 45.8	**47**
17 44.9	6 23.5	♑29 41.1	23 35.2	2 57.8	**48**
17 25.6	5 40.4	28 26.9	23 28.6	3 10.5	**49**
17 5.3	4 55.0	27 7.7	23 21.5	3 24.0	**50**
16 44.0	4 7.3	25 43.0	23 13.8	3 38.2	**51**
16 21.5	3 16.9	24 12.5	23 5.4	3 53.3	**52**
15 57.9	2 23.6	22 35.5	22 56.3	4 9.4	**53**
15 32.9	1 27.1	20 51.4	22 46.3	4 26.5	**54**
15 6.4	0 27.1	18 59.5	22 35.3	4 44.9	**55**
14 38.2	♐29 23.2	16 59.1	22 23.1	5 4.6	**56**
14 8.2	28 14.9	14 49.5	22 9.6	5 25.9	**57**
13 36.0	27 1.8	12 29.9	21 54.3	5 49.0	**58**
13 1.4	25 43.2	9 59.5	21 37.1	6 14.1	**59**
12 24.0	24 18.4	7 17.5	21 17.5	6 41.6	**60**
11 43.4	22 46.6	4 23.3	20 54.8	7 11.9	**61**
10 59.0	21 6.8	1 16.2	20 28.5	7 45.4	**62**
10 10.1	19 17.8	♐27 56.0	19 57.3	8 23.0	**63**
9 15.9	17 18.1	24 22.6	19 19.9	9 5.3	**64**
8 14.9	15 5.8	20 36.2	18 34.0	9 53.8	**65**
7 5.4	12 38.9	16 37.7	17 36.4	10 50.2	**66**
5	6	Descendant	8	9	S LAT

♉ 28° 14′ 58″

$3^h\ 44^m\ 0^s$ MC 56° 0′ 0″

$15^h\ 48^m\ 0^s$ MC 237° 0′ 0″

♏ 29° 12′ 48″

11	12	Ascendant	2	3	N LAT
° ′	° ′	° ′	° ′	° ′	°
♐27 14.8	♑25 3.3	♒24 42.5	♓26 43.8	♈29 2.8	**0**
26 35.0	23 49.1	23 26.9	26 38.8	29 23.7	**5**
25 54.6	22 32.2	22 5.2	26 33.3	29 45.5	**10**
25 12.8	21 11.3	20 35.5	26 27.4	♉ 0 8.4	**15**
25 4.2	20 54.4	20 16.3	26 26.1	0 13.2	**16**
24 55.6	20 37.4	19 56.7	26 24.8	0 18.0	**17**
24 46.8	20 20.1	19 36.6	26 23.5	0 23.0	**18**
24 37.9	20 2.5	19 16.0	26 22.2	0 28.0	**19**
24 29.0	19 44.6	18 54.9	26 20.8	0 33.1	**20**
24 19.9	19 26.4	18 33.1	26 19.4	0 38.3	**21**
24 10.6	19 7.9	18 10.8	26 17.9	0 43.5	**22**
24 1.3	18 49.0	17 47.8	26 16.4	0 48.9	**23**
23 51.7	18 29.8	17 24.1	26 14.9	0 54.4	**24**
23 42.1	18 10.2	16 59.7	26 13.3	1 0.1	**25**
23 32.2	17 50.1	16 34.4	26 11.7	1 5.8	**26**
23 22.2	17 29.6	16 8.4	26 10.0	1 11.7	**27**
23 12.0	17 8.7	15 41.4	26 8.3	1 17.7	**28**
23 1.6	16 47.2	15 13.5	26 6.5	1 23.9	**29**
22 50.9	16 25.2	14 44.5	26 4.6	1 30.2	**30**
22 40.0	16 2.6	14 14.5	26 2.7	1 36.8	**31**
22 28.9	15 39.4	13 43.3	26 0.6	1 43.5	**32**
22 17.5	15 15.6	13 10.8	25 58.6	1 50.4	**33**
22 5.9	14 51.1	12 37.0	25 56.4	1 57.5	**34**
21 53.9	14 25.9	12 1.7	25 54.1	2 4.8	**35**
21 41.6	13 59.8	11 24.9	25 51.8	2 12.4	**36**
21 28.9	13 33.0	10 46.3	25 49.3	2 20.2	**37**
21 15.9	13 5.2	10 6.0	25 46.7	2 28.3	**38**
21 2.5	12 36.5	9 23.7	25 44.0	2 36.7	**39**
20 48.7	12 6.8	8 39.3	25 41.1	2 45.4	**40**
20 34.4	11 36.0	7 52.5	25 38.1	2 54.5	**41**
20 19.6	11 4.0	7 3.3	25 34.9	3 3.9	**42**
20 4.3	10 30.7	6 11.4	25 31.6	3 13.8	**43**
19 48.4	9 56.1	5 16.5	25 28.0	3 24.0	**44**
19 32.0	9 20.0	4 18.4	25 24.2	3 34.8	**45**
19 14.8	8 42.3	3 16.8	25 20.2	3 46.1	**46**
18 56.9	8 2.8	2 11.3	25 15.9	3 57.9	**47**
18 38.3	7 21.5	1 1.7	25 11.2	4 10.3	**48**
18 18.8	6 38.1	♑29 47.4	25 6.3	4 23.4	**49**
17 58.3	5 52.5	28 28.1	25 0.9	4 37.3	**50**
17 36.8	5 4.5	27 3.3	24 55.1	4 52.0	**51**
17 14.3	4 13.8	25 32.4	24 48.9	5 7.5	**52**
16 50.4	3 20.1	23 54.8	24 42.0	5 24.1	**53**
16 25.2	2 23.2	22 9.9	24 34.5	5 41.8	**54**
15 58.5	1 22.7	20 17.0	24 26.2	6 0.8	**55**
15 30.0	0 18.2	18 15.3	24 17.0	6 21.1	**56**
14 59.7	♐29 9.2	16 3.9	24 6.8	6 43.1	**57**
14 27.1	27 55.2	13 42.1	23 55.4	7 6.9	**58**
13 52.1	26 35.7	11 9.0	23 42.4	7 32.9	**59**
13 14.2	25 9.7	8 23.8	23 27.6	8 1.3	**60**
12 33.0	23 36.5	5 25.6	23 10.6	8 32.6	**61**
11 48.0	21 55.0	2 14.0	22 50.8	9 7.4	**62**
10 58.3	20 3.9	♐28 48.4	22 27.3	9 46.3	**63**
10 2.9	18 1.6	25 8.9	21 59.2	10 30.3	**64**
9 0.6	15 46.3	21 15.8	21 24.8	11 20.7	**65**
7 49.2	13 15.3	17 9.9	20 41.6	12 19.5	**66**
5	6	Descendant	8	9	S LAT

♉ 29° 12′ 48″

$3^h\ 48^m\ 0^s$ MC 57° 0′ 0″

15h 52m 0s — MC 238° 0′ 0″ — ♐ 0° 10′ 29″

11	12	Ascendant	2	3	N LAT
° ′	° ′	° ′	° ′	° ′	°
♐28 9.9	♑26 0.2	♒25 44.5	♓27 49.2	♉ 0 5.7	0
27 30.0	24 46.5	24 30.7	27 45.8	0 27.2	5
26 49.6	23 30.1	23 10.9	27 42.2	0 49.7	10
26 7.7	22 9.5	21 43.1	27 38.2	1 13.3	15
25 59.2	21 52.8	21 24.4	27 37.4	1 18.2	16
25 50.5	21 35.8	21 5.2	27 36.5	1 23.2	17
25 41.7	21 18.6	20 45.5	27 35.7	1 28.3	18
25 32.8	21 1.0	20 25.3	27 34.8	1 33.5	19
25 23.8	20 43.2	20 4.6	27 33.8	1 38.7	20
25 14.7	20 25.1	19 43.3	27 32.9	1 44.1	21
25 5.4	20 6.6	19 21.4	27 31.9	1 49.5	22
24 56.1	19 47.8	18 58.8	27 30.9	1 55.1	23
24 46.5	19 28.7	18 35.6	27 29.9	2 0.8	24
24 36.8	19 9.1	18 11.6	27 28.8	2 6.5	25
24 26.9	18 49.1	17 46.8	27 27.8	2 12.5	26
24 16.9	18 28.6	17 21.2	27 26.6	2 18.5	27
24 6.6	18 7.7	16 54.7	27 25.5	2 24.7	28
23 56.2	17 46.3	16 27.3	27 24.3	2 31.1	29
23 45.5	17 24.3	15 58.8	27 23.0	2 37.6	30
23 34.6	17 1.7	15 29.2	27 21.7	2 44.3	31
23 23.4	16 38.6	14 58.5	27 20.4	2 51.2	32
23 12.0	16 14.8	14 26.5	27 19.0	2 58.3	33
23 0.3	15 50.3	13 53.2	27 17.6	3 5.7	34
22 48.3	15 25.1	13 18.4	27 16.0	3 13.2	35
22 35.9	14 59.1	12 42.1	27 14.5	3 21.0	36
22 23.2	14 32.2	12 4.0	27 12.8	3 29.1	37
22 10.2	14 4.5	11 24.2	27 11.1	3 37.4	38
21 56.7	13 35.7	10 42.3	27 9.3	3 46.1	39
21 42.8	13 6.0	9 58.4	27 7.4	3 55.1	40
21 28.5	12 35.1	9 12.1	27 5.3	4 4.4	41
21 13.6	12 3.1	8 23.3	27 3.2	4 14.1	42
20 58.2	11 29.7	7 31.8	27 1.0	4 24.2	43
20 42.3	10 55.0	6 37.3	26 58.6	4 34.8	44
20 25.7	10 18.8	5 39.6	26 56.1	4 45.9	45
20 8.4	9 41.0	4 38.3	26 53.4	4 57.5	46
19 50.5	9 1.4	3 33.1	26 50.5	5 9.7	47
19 31.7	8 19.9	2 23.6	26 47.4	5 22.5	48
19 12.1	7 36.3	1 9.5	26 44.1	5 36.0	49
18 51.5	6 50.5	♑29 50.2	26 40.5	5 50.2	50
18 29.8	6 2.2	28 25.2	26 36.7	6 5.3	51
18 7.1	5 11.1	26 54.0	26 32.5	6 21.4	52
17 43.1	4 17.1	25 16.0	26 27.9	6 38.4	53
17 17.6	3 19.8	23 30.4	26 22.9	6 56.7	54
16 50.7	2 18.8	21 36.5	26 17.3	7 16.2	55
16 22.0	1 13.7	19 33.4	26 11.2	7 37.2	56
15 51.3	0 4.0	17 20.3	26 4.4	7 59.8	57
15 18.4	♐28 49.2	14 56.4	25 56.7	8 24.4	58
14 42.9	27 28.7	12 20.5	25 48.1	8 51.2	59
14 4.6	26 1.5	9 32.0	25 38.2	9 20.5	60
13 22.8	24 26.9	6 29.8	25 26.8	9 52.8	61
12 37.1	22 43.7	3 13.4	25 13.6	10 28.7	62
11 46.5	20 50.5	♐29 42.3	24 57.9	11 9.0	63
10 50.1	18 45.6	25 56.4	24 39.1	11 54.6	64
9 46.4	16 27.0	21 56.0	24 16.2	12 47.1	65
8 33.1	13 51.8	17 42.3	23 47.4	13 48.4	66
5	6	Descendant	8	9	S LAT

♊ 0° 10′ 29″ — 3h 52m 0s — MC 58° 0′ 0″

15h 56m 0s — MC 239° 0′ 0″ — ♐ 1° 8′ 1″

11	12	Ascendant	2	3	N LAT
° ′	° ′	° ′	° ′	° ′	°
♐29 5.0	♑26 57.4	♒26 46.7	♓28 54.6	♉ 1 8.4	0
28 25.1	25 44.1	25 34.8	28 52.9	1 30.6	5
27 44.6	24 28.1	24 16.9	28 51.1	1 53.7	10
27 2.7	23 8.0	22 51.1	28 49.1	2 18.0	15
26 54.1	22 51.3	22 32.8	28 48.7	2 23.1	16
26 45.4	22 34.4	22 14.0	28 48.3	2 28.2	17
26 36.6	22 17.3	21 54.8	28 47.8	2 33.5	18
26 27.7	21 59.8	21 35.0	28 47.4	2 38.8	19
26 18.7	21 42.1	21 14.7	28 46.9	2 44.2	20
26 9.5	21 24.0	20 53.9	28 46.4	2 49.7	21
26 0.3	21 5.6	20 32.4	28 46.0	2 55.3	22
25 50.9	20 46.9	20 10.3	28 45.5	3 1.0	23
25 41.3	20 27.8	19 47.5	28 44.9	3 6.8	24
25 31.6	20 8.2	19 24.0	28 44.4	3 12.8	25
25 21.7	19 48.3	18 59.7	28 43.9	3 18.9	26
25 11.6	19 27.9	18 34.6	28 43.3	3 25.1	27
25 1.3	19 7.0	18 8.6	28 42.7	3 31.5	28
24 50.8	18 45.6	17 41.6	28 42.1	3 38.1	29
24 40.1	18 23.7	17 13.7	28 41.5	3 44.8	30
24 29.2	18 1.2	16 44.6	28 40.9	3 51.7	31
24 18.0	17 38.1	16 14.4	28 40.2	3 58.8	32
24 6.5	17 14.3	15 42.9	28 39.5	4 6.1	33
23 54.8	16 49.9	15 10.1	28 38.8	4 13.6	34
23 42.7	16 24.6	14 35.9	28 38.0	4 21.4	35
23 30.3	15 58.6	14 0.0	28 37.2	4 29.4	36
23 17.6	15 31.8	13 22.5	28 36.4	4 37.7	37
23 4.5	15 4.0	12 43.2	28 35.5	4 46.3	38
22 51.0	14 35.3	12 1.9	28 34.6	4 55.1	39
22 37.0	14 5.5	11 18.5	28 33.7	5 4.4	40
22 22.6	13 34.7	10 32.7	28 32.7	5 14.0	41
22 7.7	13 2.6	9 44.4	28 31.6	5 24.0	42
21 52.2	12 29.2	8 53.4	28 30.5	5 34.4	43
21 36.2	11 54.4	7 59.4	28 29.3	5 45.3	44
21 19.5	11 18.1	7 2.0	28 28.0	5 56.6	45
21 2.2	10 40.1	6 1.1	28 26.7	6 8.6	46
20 44.1	10 0.4	4 56.2	28 25.2	6 21.1	47
20 25.2	9 18.8	3 47.1	28 23.7	6 34.2	48
20 5.4	8 35.0	2 33.1	28 22.0	6 48.1	49
19 44.7	7 49.0	1 13.9	28 20.2	7 2.8	50
19 23.0	7 0.4	♑29 48.9	28 18.3	7 18.3	51
19 0.0	6 9.1	28 17.5	28 16.2	7 34.8	52
18 35.8	5 14.7	26 39.0	28 13.9	7 52.4	53
18 10.2	4 16.9	24 52.8	28 11.4	8 11.1	54
17 43.0	3 15.4	22 57.9	28 8.6	8 31.2	55
17 14.0	2 9.7	20 53.6	28 5.6	8 52.8	56
16 43.0	0 59.4	18 38.9	28 2.2	9 16.1	57
16 9.8	♐29 43.8	16 12.8	27 58.3	9 41.4	58
15 33.9	28 22.2	13 34.2	27 54.0	10 9.0	59
14 55.1	26 54.0	10 42.2	27 49.0	10 39.2	60
14 12.8	25 17.9	7 36.0	27 43.3	11 12.5	61
13 26.3	23 33.0	4 14.6	27 36.7	11 49.6	62
12 34.9	21 37.7	0 37.6	27 28.9	12 31.2	63
11 37.4	19 30.1	♐26 44.9	27 19.5	13 18.5	64
10 32.2	17 8.0	22 37.1	27 8.0	14 12.9	65
9 16.9	14 28.5	18 15.0	26 53.6	15 16.9	66
5	6	Descendant	8	9	S LAT

♊ 1° 8′ 1″ — 3h 56m 0s — MC 59° 0′ 0″

16h 0m 0s	MC	240° 0′ 0″			
	♐ 2° 5′ 24″				
N LAT	11	12	Ascendant	2	3
0	♑ 0 0.0	♑27 54.6	♒27 49.1	♈ 0 0.0	♉ 2 10.9
5	♐29 20.1	26 41.9	26 39.0	0 0.0	2 33.8
10	28 39.6	25 26.3	25 23.1	0 0.0	2 57.5
15	27 57.7	24 6.7	23 59.4	0 0.0	3 22.5
16	27 49.0	23 50.1	23 41.5	0 0.0	3 27.7
17	27 40.3	23 33.3	23 23.2	0 0.0	3 33.0
18	27 31.5	23 16.2	23 4.4	0 0.0	3 38.4
19	27 22.6	22 58.8	22 45.1	0 0.0	3 43.9
20	27 13.6	22 41.1	22 25.2	0 0.0	3 49.4
21	27 4.4	22 23.2	22 4.8	0 0.0	3 55.1
22	26 55.1	22 4.8	21 43.8	0 0.0	4 0.8
23	26 45.7	21 46.2	21 22.2	0 0.0	4 6.7
24	26 36.1	21 27.1	20 59.9	0 0.0	4 12.7
25	26 26.4	21 7.7	20 36.9	0 0.0	4 18.8
26	26 16.5	20 47.8	20 13.1	0 0.0	4 25.1
27	26 6.4	20 27.4	19 48.5	0 0.0	4 31.5
28	25 56.1	20 6.6	19 23.0	0 0.0	4 38.0
29	25 45.6	19 45.3	18 56.6	0 0.0	4 44.8
30	25 34.8	19 23.4	18 29.1	0 0.0	4 51.7
31	25 23.9	19 1.0	18 0.6	0 0.0	4 58.7
32	25 12.6	18 37.9	17 30.9	0 0.0	5 6.0
33	25 1.1	18 14.2	17 0.0	0 0.0	5 13.5
34	24 49.3	17 49.7	16 27.8	0 0.0	5 21.3
35	24 37.2	17 24.6	15 54.1	0 0.0	5 29.2
36	24 24.8	16 58.6	15 18.8	0 0.0	5 37.5
37	24 12.0	16 31.7	14 41.9	0 0.0	5 46.0
38	23 58.9	16 4.0	14 3.1	0 0.0	5 54.8
39	23 45.3	15 35.3	13 22.4	0 0.0	6 3.9
40	23 31.3	15 5.5	12 39.5	0 0.0	6 13.4
41	23 16.8	14 34.6	11 54.3	0 0.0	6 23.3
42	23 1.8	14 2.4	11 6.6	0 0.0	6 33.5
43	22 46.3	13 29.0	10 16.1	0 0.0	6 44.2
44	22 30.2	12 54.2	9 22.6	0 0.0	6 55.4
45	22 13.4	12 17.8	8 25.8	0 0.0	7 7.1
46	21 56.0	11 39.8	7 25.3	0 0.0	7 19.3
47	21 37.8	10 59.9	6 20.8	0 0.0	7 32.1
48	21 18.8	10 18.1	5 12.0	0 0.0	7 45.7
49	20 59.0	9 34.2	3 58.3	0 0.0	7 59.9
50	20 38.1	8 47.9	2 39.2	0 0.0	8 15.0
51	20 16.2	7 59.1	1 14.3	0 0.0	8 30.9
52	19 53.1	7 7.5	♑29 42.8	0 0.0	8 47.8
53	19 28.7	6 12.8	28 4.0	0 0.0	9 5.8
54	19 2.9	5 14.6	26 17.2	0 0.0	9 25.1
55	18 35.4	4 12.6	24 21.6	0 0.0	9 45.7
56	18 6.2	3 6.4	22 16.1	0 0.0	10 7.9
57	17 34.9	1 55.4	19 59.7	0 0.0	10 31.9
58	17 1.3	0 39.0	17 31.5	0 0.0	10 57.9
59	16 25.1	♐29 16.5	14 50.2	0 0.0	11 26.2
60	15 45.8	27 47.0	11 54.8	0 0.0	11 57.3
61	15 2.9	26 9.6	8 44.2	0 0.0	12 31.7
62	14 15.8	24 22.9	5 17.7	0 0.0	13 9.9
63	13 23.5	22 25.3	1 34.6	0 0.0	13 52.9
64	12 24.8	20 15.1	♐27 34.8	0 0.0	14 41.7
65	11 18.2	17 49.5	23 18.9	0 0.0	15 38.2
66	10 0.8	15 5.2	18 48.0	0 0.0	16 44.8
S LAT	5	6	Descendant	8	9
	♊ 2° 5′ 24″				
4h 0m 0s	MC	60° 0′ 0″			

16h 4m 0s	MC	241° 0′ 0″			
	♐ 3° 2′ 39″				
N LAT	11	12	Ascendant	2	3
0	♑ 0 55.0	♑28 52.0	♒28 51.6	♈ 1 5.4	♉ 3 13.3
5	0 15.2	27 39.8	27 43.5	1 7.1	3 36.8
10	♐29 34.6	26 24.8	26 29.6	1 8.9	4 1.1
15	28 52.6	25 5.6	25 8.0	1 10.9	4 26.8
16	28 44.0	24 49.1	24 50.6	1 11.3	4 32.2
17	28 35.3	24 32.3	24 32.7	1 11.7	4 37.6
18	28 26.5	24 15.3	24 14.3	1 12.2	4 43.1
19	28 17.6	23 58.0	23 55.5	1 12.6	4 48.7
20	28 8.5	23 40.5	23 36.1	1 13.1	4 54.4
21	27 59.4	23 22.6	23 16.2	1 13.6	5 0.2
22	27 50.1	23 4.3	22 55.7	1 14.0	5 6.1
23	27 40.6	22 45.7	22 34.6	1 14.5	5 12.2
24	27 31.0	22 26.7	22 12.8	1 15.1	5 18.3
25	27 21.2	22 7.4	21 50.3	1 15.6	5 24.6
26	27 11.3	21 47.6	21 27.0	1 16.1	5 31.0
27	27 1.2	21 27.3	21 2.9	1 16.7	5 37.6
28	26 50.9	21 6.5	20 37.9	1 17.3	5 44.3
29	26 40.3	20 45.3	20 12.1	1 17.9	5 51.2
30	26 29.6	20 23.4	19 45.2	1 18.5	5 58.3
31	26 18.6	20 1.0	19 17.2	1 19.1	6 5.6
32	26 7.3	19 38.0	18 48.1	1 19.8	6 13.0
33	25 55.8	19 14.3	18 17.8	1 20.5	6 20.7
34	25 44.0	18 50.0	17 46.1	1 21.2	6 28.6
35	25 31.8	18 24.8	17 13.0	1 22.0	6 36.8
36	25 19.4	17 58.9	16 38.4	1 22.8	6 45.3
37	25 6.5	17 32.1	16 2.1	1 23.6	6 54.0
38	24 53.3	17 4.3	15 23.9	1 24.5	7 3.0
39	24 39.7	16 35.6	14 43.8	1 25.4	7 12.4
40	24 25.6	16 5.8	14 1.6	1 26.3	7 22.1
41	24 11.1	15 34.9	13 17.0	1 27.3	7 32.2
42	23 56.1	15 2.8	12 29.9	1 28.4	7 42.7
43	23 40.5	14 29.3	11 40.0	1 29.5	7 53.7
44	23 24.3	13 54.4	10 47.1	1 30.7	8 5.1
45	23 7.5	13 17.9	9 50.8	1 32.0	8 17.1
46	22 49.9	12 39.8	8 50.9	1 33.3	8 29.7
47	22 31.7	11 59.9	7 46.9	1 34.8	8 42.8
48	22 12.6	11 18.0	6 38.5	1 36.3	8 56.7
49	21 52.6	10 33.9	5 25.1	1 38.0	9 11.3
50	21 31.6	9 47.4	4 6.4	1 39.8	9 26.7
51	21 9.5	8 58.4	2 41.5	1 41.7	9 43.1
52	20 46.3	8 6.5	1 10.0	1 43.8	10 0.4
53	20 21.7	7 11.4	♑29 31.1	1 46.1	10 18.9
54	19 55.7	6 12.9	27 43.8	1 48.6	10 38.7
55	19 28.0	5 10.4	25 47.4	1 51.4	10 59.8
56	18 58.6	4 3.6	23 40.8	1 54.4	11 22.6
57	18 27.0	2 52.0	21 23.0	1 57.8	11 47.2
58	17 53.1	1 34.8	18 52.6	2 1.7	12 13.9
59	17 16.4	0 11.3	16 8.6	2 6.0	12 43.0
60	16 36.6	♐28 40.7	13 9.7	2 11.0	13 15.0
61	15 53.2	27 1.8	9 54.8	2 16.7	13 50.3
62	15 5.3	25 13.4	6 22.9	2 23.3	14 29.7
63	14 12.2	23 13.6	2 33.4	2 31.1	15 13.9
64	13 12.4	21 0.5	♐28 26.1	2 40.5	16 4.4
65	12 4.2	18 31.3	24 1.7	2 52.0	17 2.9
66	10 44.7	15 42.2	19 21.4	3 6.4	18 12.3
S LAT	5	6	Descendant	8	9
	♊ 3° 2′ 39″				
4h 4m 0s	MC	61° 0′ 0″			

16ʰ 8ᵐ 0ˢ		MC		242° 0′ 0″
		♐ 3° 59′ 45″			

N LAT	11	12	Ascendant	2	3
	° ′	° ′	° ′	° ′	° ′
0	♑ 1 50.1	♑29 49.5	♒29 54.3	♈ 2 10.8	♉ 4 15.5
5	1 10.2	28 37.9	28 48.1	2 14.2	4 39.6
10	0 29.6	27 23.4	27 36.3	2 17.8	5 4.6
15	♐29 47.7	26 4.7	26 16.9	2 21.8	5 30.9
16	29 39.0	25 48.3	25 59.9	2 22.6	5 36.4
17	29 30.3	25 31.6	25 42.5	2 23.5	5 42.0
18	29 21.5	25 14.7	25 24.6	2 24.3	5 47.6
19	29 12.6	24 57.5	25 6.3	2 25.2	5 53.4
20	29 3.5	24 40.0	24 47.4	2 26.2	5 59.2
21	28 54.3	24 22.2	24 28.0	2 27.1	6 5.1
22	28 45.0	24 4.1	24 8.0	2 28.1	6 11.2
23	28 35.5	23 45.5	23 47.4	2 29.1	6 17.4
24	28 25.9	23 26.6	23 26.1	2 30.1	6 23.7
25	28 16.2	23 7.3	23 4.1	2 31.2	6 30.1
26	28 6.2	22 47.6	22 41.4	2 32.2	6 36.7
27	27 56.1	22 27.4	22 17.8	2 33.4	6 43.4
28	27 45.7	22 6.7	21 53.4	2 34.5	6 50.3
29	27 35.2	21 45.5	21 28.1	2 35.7	6 57.4
30	27 24.4	21 23.8	21 1.8	2 37.0	7 4.6
31	27 13.3	21 1.5	20 34.5	2 38.3	7 12.1
32	27 2.1	20 38.5	20 6.0	2 39.6	7 19.7
33	26 50.5	20 14.9	19 36.3	2 41.0	7 27.6
34	26 38.7	19 50.5	19 5.2	2 42.4	7 35.7
35	26 26.5	19 25.4	18 32.8	2 44.0	7 44.1
36	26 14.0	18 59.5	17 58.8	2 45.5	7 52.8
37	26 1.1	18 32.7	17 23.1	2 47.2	8 1.7
38	25 47.9	18 5.0	16 45.6	2 48.9	8 11.0
39	25 34.2	17 36.3	16 6.2	2 50.7	8 20.5
40	25 20.1	17 6.6	15 24.6	2 52.6	8 30.5
41	25 5.5	16 35.7	14 40.7	2 54.7	8 40.8
42	24 50.4	16 3.5	13 54.3	2 56.8	8 51.6
43	24 34.7	15 30.0	13 5.1	2 59.0	9 2.8
44	24 18.5	14 55.1	12 12.8	3 1.4	9 14.6
45	24 1.6	14 18.6	11 17.2	3 3.9	9 26.8
46	23 44.0	13 40.4	10 17.8	3 6.6	9 39.7
47	23 25.6	13 0.4	9 14.5	3 9.5	9 53.2
48	23 6.4	12 18.3	8 6.6	3 12.6	10 7.3
49	22 46.3	11 34.1	6 53.7	3 15.9	10 22.3
50	22 25.2	10 47.5	5 35.3	3 19.5	10 38.1
51	22 3.0	9 58.2	4 10.7	3 23.3	10 54.8
52	21 39.6	9 6.1	2 39.3	3 27.5	11 12.6
53	21 14.9	8 10.7	1 0.2	3 32.1	11 31.6
54	20 48.7	7 11.8	♑29 12.7	3 37.1	11 51.8
55	20 20.8	6 8.9	27 15.7	3 42.7	12 13.5
56	19 51.1	5 1.5	25 8.1	3 48.8	12 36.8
57	19 19.2	3 49.2	22 48.7	3 55.6	13 2.0
58	18 44.9	2 31.2	20 16.4	4 3.3	13 29.4
59	18 7.9	1 6.8	17 29.6	4 11.9	13 59.3
60	17 27.7	♐29 35.1	14 27.2	4 21.8	14 32.1
61	16 43.7	27 54.8	11 7.8	4 33.2	15 8.4
62	15 55.1	26 4.5	7 30.4	4 46.4	15 48.8
63	15 1.1	24 2.5	3 34.2	5 2.1	16 34.5
64	14 0.2	21 46.6	♐29 19.0	5 20.9	17 26.6
65	12 50.4	19 13.5	24 45.4	5 43.8	18 27.1
66	11 28.6	16 19.4	19 55.1	6 12.6	19 39.3
S LAT	5	6	Descendant	8	9

		♊ 3° 59′ 45″		
4ʰ 8ᵐ 0ˢ		MC		62° 0′ 0″

16ʰ 12ᵐ 0ˢ		MC		243° 0′ 0″
		♐ 4° 56′ 43″			

N LAT	11	12	Ascendant	2	3
	° ′	° ′	° ′	° ′	° ′
0	♑ 2 45.2	♒ 0 47.2	♓ 0 57.2	♈ 3 16.2	♉ 5 17.5
5	2 5.3	♑29 36.1	♒29 53.0	3 21.2	5 42.2
10	1 24.7	28 22.2	28 43.3	3 26.7	6 7.8
15	0 42.7	27 4.0	27 26.2	3 32.6	6 34.8
16	0 34.1	26 47.7	27 9.6	3 33.9	6 40.4
17	0 25.4	26 31.2	26 52.7	3 35.2	6 46.1
18	0 16.5	26 14.3	26 35.3	3 36.5	6 51.9
19	0 7.6	25 57.2	26 17.4	3 37.8	6 57.8
20	♐29 58.5	25 39.8	25 59.0	3 39.2	7 3.7
21	29 49.3	25 22.1	25 40.1	3 40.6	7 9.8
22	29 40.0	25 4.1	25 20.7	3 42.1	7 16.0
23	29 30.5	24 45.6	25 0.6	3 43.6	7 22.4
24	29 20.9	24 26.8	24 39.8	3 45.1	7 28.8
25	29 11.1	24 7.6	24 18.4	3 46.7	7 35.4
26	29 1.1	23 48.0	23 56.2	3 48.3	7 42.1
27	28 51.0	23 27.8	23 33.3	3 50.0	7 49.0
28	28 40.6	23 7.2	23 9.5	3 51.7	7 56.1
29	28 30.0	22 46.1	22 44.8	3 53.5	8 3.3
30	28 19.2	22 24.4	22 19.1	3 55.4	8 10.7
31	28 8.2	22 2.2	21 52.3	3 57.3	8 18.4
32	27 56.9	21 39.3	21 24.5	3 59.4	8 26.2
33	27 45.3	21 15.7	20 55.4	4 1.4	8 34.2
34	27 33.4	20 51.4	20 25.1	4 3.6	8 42.6
35	27 21.2	20 26.4	19 53.3	4 5.9	8 51.1
36	27 8.7	20 0.5	19 20.0	4 8.2	9 0.0
37	26 55.8	19 33.8	18 45.0	4 10.7	9 9.1
38	26 42.5	19 6.1	18 8.2	4 13.3	9 18.6
39	26 28.8	18 37.5	17 29.5	4 16.0	9 28.4
40	26 14.6	18 7.7	16 48.7	4 18.9	9 38.6
41	26 0.0	17 36.8	16 5.5	4 21.9	9 49.1
42	25 44.8	17 4.7	15 19.8	4 25.1	10 0.2
43	25 29.1	16 31.2	14 31.3	4 28.4	10 11.6
44	25 12.8	15 56.2	13 39.8	4 32.0	10 23.6
45	24 55.8	15 19.7	12 44.9	4 35.8	10 36.2
46	24 38.1	14 41.4	11 46.2	4 39.8	10 49.3
47	24 19.7	14 1.3	10 43.5	4 44.1	11 3.1
48	24 0.4	13 19.2	9 36.2	4 48.8	11 17.6
49	23 40.2	12 34.9	8 23.9	4 53.7	11 32.9
50	23 19.0	11 48.1	7 6.0	4 59.1	11 49.1
51	22 56.6	10 58.6	5 41.8	5 4.9	12 6.2
52	22 33.1	10 6.2	4 10.6	5 11.1	12 24.4
53	22 8.2	9 10.6	2 31.6	5 18.0	12 43.8
54	21 41.8	8 11.3	0 43.9	5 25.5	13 4.4
55	21 13.7	7 8.0	♑28 46.4	5 33.8	13 26.6
56	20 43.7	6 0.1	26 37.9	5 43.0	13 50.5
57	20 11.6	4 47.2	24 17.2	5 53.2	14 16.3
58	19 37.0	3 28.4	21 42.9	6 4.6	14 44.4
59	18 59.6	2 3.1	18 53.5	6 17.6	15 15.0
60	18 18.9	0 30.1	15 47.6	6 32.4	15 48.6
61	17 34.3	♐28 48.4	12 23.6	6 49.4	16 25.9
62	16 45.1	26 56.4	8 40.4	7 9.2	17 7.4
63	15 50.2	24 52.1	4 37.1	7 32.7	17 54.4
64	14 48.1	22 33.2	0 13.6	8 0.8	18 48.1
65	13 36.8	19 56.2	♐25 30.3	8 35.2	19 50.7
66	12 12.6	16 56.7	20 29.2	9 18.4	21 5.8
S LAT	5	6	Descendant	8	9

		♊ 4° 56′ 43″		
4ʰ 12ᵐ 0ˢ		MC		63° 0′ 0″

16ʰ 16ᵐ 0ˢ — MC 244° 0′ 0″ — ♐ 5° 53′ 34″

N LAT	11	12	Ascendant	2	3
°	° ′	° ′	° ′	° ′	° ′
0	♑ 3 40.2	♒ 1 45.0	♓ 2 0.3	♈ 4 21.5	♉ 6 19.4
5	3 0.4	0 34.6	0 58.1	4 28.3	6 44.7
10	2 19.8	♑29 21.2	♒29 50.5	4 35.5	7 10.9
15	1 37.8	28 3.5	28 35.7	4 43.4	7 38.5
16	1 29.2	27 47.4	28 19.6	4 45.1	7 44.2
17	1 20.5	27 30.9	28 3.2	4 46.8	7 50.0
18	1 11.6	27 14.2	27 46.3	4 48.6	7 55.9
19	1 2.7	26 57.2	27 28.9	4 50.4	8 2.0
20	0 53.6	26 39.9	27 11.1	4 52.2	8 8.1
21	0 44.4	26 22.3	26 52.7	4 54.1	8 14.3
22	0 35.1	26 4.3	26 33.7	4 56.0	8 20.6
23	0 25.6	25 46.0	26 14.2	4 58.0	8 27.1
24	0 15.9	25 27.3	25 54.0	5 0.1	8 33.7
25	0 6.1	25 8.2	25 33.2	5 2.2	8 40.4
26	♐29 56.2	24 48.6	25 11.6	5 4.4	8 47.3
27	29 46.0	24 28.6	24 49.2	5 6.6	8 54.4
28	29 35.6	24 8.1	24 26.0	5 8.9	9 1.6
29	29 25.0	23 47.0	24 2.0	5 11.3	9 9.0
30	29 14.2	23 25.4	23 36.9	5 13.8	9 16.6
31	29 3.1	23 3.3	23 10.8	5 16.4	9 24.3
32	28 51.8	22 40.4	22 43.6	5 19.0	9 32.3
33	28 40.2	22 16.9	22 15.3	5 21.8	9 40.6
34	28 28.3	21 52.7	21 45.6	5 24.7	9 49.1
35	28 16.0	21 27.8	21 14.5	5 27.7	9 57.8
36	28 3.5	21 2.0	20 41.9	5 30.9	10 6.9
37	27 50.5	20 35.3	20 7.7	5 34.2	10 16.2
38	27 37.2	20 7.7	19 31.7	5 37.6	10 25.9
39	27 23.5	19 39.0	18 53.8	5 41.2	10 35.9
40	27 9.3	19 9.3	18 13.7	5 45.0	10 46.3
41	26 54.6	18 38.4	17 31.4	5 49.1	10 57.1
42	26 39.4	18 6.3	16 46.5	5 53.3	11 8.4
43	26 23.6	17 32.8	15 58.8	5 57.8	11 20.1
44	26 7.2	16 57.8	15 8.0	6 2.5	11 32.4
45	25 50.2	16 21.3	14 13.9	6 7.5	11 45.2
46	25 32.4	15 43.0	13 16.1	6 12.9	11 58.6
47	25 13.9	15 2.8	12 14.1	6 18.6	12 12.7
48	24 54.5	14 20.6	11 7.6	6 24.8	12 27.5
49	24 34.2	13 36.2	9 55.9	6 31.4	12 43.1
50	24 12.9	12 49.2	8 38.6	6 38.5	12 59.6
51	23 50.4	11 59.6	7 14.9	6 46.2	13 17.1
52	23 26.7	11 7.0	5 44.1	6 54.6	13 35.7
53	23 1.7	10 11.0	4 5.3	7 3.7	13 55.5
54	22 35.1	9 11.4	2 17.5	7 13.7	14 16.7
55	22 6.8	8 7.7	0 19.7	7 24.7	14 39.3
56	21 36.6	6 59.3	♑28 10.5	7 36.9	15 3.8
57	21 4.2	5 45.8	25 48.6	7 50.4	15 30.2
58	20 29.3	4 26.3	23 12.4	8 5.7	15 58.9
59	19 51.5	3 0.0	20 20.4	8 22.9	16 30.2
60	19 10.3	1 26.0	17 10.9	8 42.5	17 4.6
61	18 25.2	♐29 42.8	13 42.3	9 5.2	17 42.8
62	17 35.3	27 49.0	9 53.2	9 31.5	18 25.5
63	16 39.5	25 42.4	5 42.5	10 2.7	19 13.7
64	15 36.3	23 20.5	1 10.1	10 40.1	20 9.0
65	14 23.3	20 39.5	♐26 16.5	11 26.0	21 13.8
66	12 56.7	17 34.3	21 3.8	12 23.6	22 31.8
S LAT	5	6	Descendant	8	9

♊ 5° 53′ 34″ — 4ʰ 16ᵐ 0ˢ — MC 64° 0′ 0″

16ʰ 20ᵐ 0ˢ — MC 245° 0′ 0″ — ♐ 6° 50′ 16″

N LAT	11	12	Ascendant	2	3
°	° ′	° ′	° ′	° ′	° ′
0	♑ 4 35.3	♒ 2 43.0	♓ 3 3.5	♈ 5 26.8	♉ 7 21.1
5	3 55.5	1 33.2	2 3.3	5 35.3	7 46.9
10	3 15.0	0 20.4	0 58.0	5 44.3	8 13.7
15	2 33.0	♑29 3.3	♒29 45.5	5 54.2	8 41.9
16	2 24.3	28 47.3	29 29.9	5 56.3	8 47.8
17	2 15.6	28 30.9	29 14.0	5 58.4	8 53.7
18	2 6.8	28 14.3	28 57.6	6 0.6	8 59.8
19	1 57.8	27 57.4	28 40.8	6 2.9	9 5.9
20	1 48.7	27 40.2	28 23.4	6 5.2	9 12.2
21	1 39.5	27 22.7	28 5.6	6 7.5	9 18.5
22	1 30.2	27 4.9	27 47.2	6 9.9	9 25.0
23	1 20.7	26 46.7	27 28.2	6 12.4	9 31.6
24	1 11.0	26 28.0	27 8.7	6 15.0	9 38.3
25	1 1.2	26 9.0	26 48.4	6 17.6	9 45.2
26	0 51.2	25 49.6	26 27.4	6 20.3	9 52.2
27	0 41.0	25 29.6	26 5.7	6 23.1	9 59.4
28	0 30.7	25 9.2	25 43.1	6 26.0	10 6.8
29	0 20.0	24 48.3	25 19.7	6 29.0	10 14.4
30	0 9.2	24 26.8	24 55.3	6 32.1	10 22.1
31	♐29 58.1	24 4.7	24 29.9	6 35.3	10 30.1
32	29 46.8	23 41.9	24 3.4	6 38.6	10 38.2
33	29 35.1	23 18.5	23 35.8	6 42.1	10 46.6
34	29 23.2	22 54.4	23 6.8	6 45.7	10 55.3
35	29 10.9	22 29.5	22 36.5	6 49.5	11 4.2
36	28 58.3	22 3.8	22 4.7	6 53.4	11 13.5
37	28 45.4	21 37.2	21 31.3	6 57.5	11 23.0
38	28 32.0	21 9.6	20 56.1	7 1.8	11 32.9
39	28 18.2	20 41.0	20 19.0	7 6.3	11 43.1
40	28 4.0	20 11.4	19 39.8	7 11.1	11 53.7
41	27 49.3	19 40.5	18 58.3	7 16.1	12 4.8
42	27 34.0	19 8.4	18 14.2	7 21.4	12 16.2
43	27 18.2	18 34.9	17 27.4	7 26.9	12 28.2
44	27 1.7	18 0.0	16 37.6	7 32.9	12 40.7
45	26 44.6	17 23.4	15 44.4	7 39.1	12 53.8
46	26 26.8	16 45.1	14 47.4	7 45.8	13 7.5
47	26 8.2	16 4.9	13 46.3	7 53.0	13 21.9
48	25 48.7	15 22.6	12 40.6	8 0.6	13 37.0
49	25 28.3	14 38.0	11 29.8	8 8.9	13 52.9
50	25 6.9	13 51.0	10 13.2	8 17.7	14 9.8
51	24 44.3	13 1.2	8 50.1	8 27.3	14 27.7
52	24 20.5	12 8.4	7 19.8	8 37.7	14 46.6
53	23 55.3	11 12.2	5 41.4	8 49.1	15 6.8
54	23 28.5	10 12.2	3 53.7	9 1.6	15 28.4
55	23 0.1	9 8.1	1 55.7	9 15.3	15 51.6
56	22 29.6	7 59.3	♑29 45.9	9 30.4	16 16.5
57	21 56.9	6 45.1	27 22.9	9 47.3	16 43.5
58	21 21.7	5 24.9	24 45.1	10 6.3	17 12.8
59	20 43.6	3 57.8	21 50.6	10 27.8	17 44.9
60	20 2.0	2 22.6	18 37.6	10 52.2	18 20.1
61	19 16.3	0 38.0	15 4.2	11 20.4	18 59.2
62	18 25.7	♐28 42.4	11 9.0	11 53.2	19 42.9
63	17 29.0	26 33.5	6 50.6	12 32.0	20 32.4
64	16 24.6	24 8.5	2 8.7	13 18.7	21 29.4
65	15 10.0	21 23.3	♐27 4.1	14 15.9	22 36.2
66	13 40.8	18 12.2	21 38.9	15 27.9	23 57.2
S LAT	5	6	Descendant	8	9

♊ 6° 50′ 16″ — 4ʰ 20ᵐ 0ˢ — MC 65° 0′ 0″

16h 24m 0s MC 246° 0′ 0″

♐ 7° 46′ 52″

11	12	Ascendant	2	3	N LAT
° ′	° ′	° ′	° ′	° ′	°
♑ 5 30.5	♒ 3 41.2	♓ 4 6.8	♈ 6 32.1	♉ 8 22.6	0
4 50.7	2 32.0	3 8.8	6 42.3	8 49.0	5
4 10.2	1 19.8	2 5.6	6 53.1	9 16.3	10
3 28.2	0 3.4	0 55.6	7 4.9	9 45.1	15
3 19.5	♑29 47.4	0 40.5	7 7.4	9 51.1	16
3 10.8	29 31.2	0 25.1	7 10.0	9 57.2	17
3 1.9	29 14.7	0 9.2	7 12.6	10 3.4	18
2 53.0	28 57.9	♒29 52.9	7 15.3	10 9.6	19
2 43.9	28 40.8	29 36.2	7 18.0	10 16.0	20
2 34.7	28 23.4	29 18.9	7 20.9	10 22.5	21
2 25.4	28 5.7	29 1.1	7 23.8	10 29.1	22
2 15.9	27 47.6	28 42.7	7 26.7	10 35.8	23
2 6.2	27 29.1	28 23.7	7 29.8	10 42.7	24
1 56.4	27 10.2	28 4.1	7 32.9	10 49.7	25
1 46.4	26 50.8	27 43.7	7 36.2	10 56.9	26
1 36.2	26 31.0	27 22.6	7 39.5	11 4.3	27
1 25.8	26 10.7	27 0.7	7 43.0	11 11.8	28
1 15.2	25 49.8	26 38.0	7 46.6	11 19.5	29
1 4.3	25 28.4	26 14.3	7 50.3	11 27.4	30
0 53.2	25 6.4	25 49.6	7 54.1	11 35.5	31
0 41.8	24 43.8	25 23.9	7 58.1	11 43.8	32
0 30.2	24 20.5	24 57.0	8 2.3	11 52.4	33
0 18.2	23 56.4	24 28.8	8 6.6	12 1.2	34
0 6.0	23 31.6	23 59.3	8 11.1	12 10.4	35
♐29 53.3	23 6.0	23 28.3	8 15.8	12 19.8	36
29 40.3	22 39.4	22 55.7	8 20.7	12 29.5	37
29 26.9	22 11.9	22 21.4	8 25.9	12 39.6	38
29 13.1	21 43.4	21 45.1	8 31.3	12 50.0	39
28 58.8	21 13.8	21 6.8	8 37.0	13 0.8	40
28 44.1	20 43.0	20 26.2	8 43.0	13 12.1	41
28 28.8	20 11.0	19 43.2	8 49.3	13 23.8	42
28 12.9	19 37.5	18 57.3	8 55.9	13 36.0	43
27 56.4	19 2.6	18 8.4	9 3.0	13 48.7	44
27 39.2	18 26.0	17 16.2	9 10.5	14 2.1	45
27 21.4	17 47.7	16 20.2	9 18.5	14 16.0	46
27 2.7	17 7.4	15 20.1	9 27.1	14 30.7	47
26 43.1	16 25.1	14 15.4	9 36.3	14 46.1	48
26 22.6	15 40.5	13 5.4	9 46.1	15 2.4	49
26 1.1	14 53.3	11 49.7	9 56.7	15 19.5	50
25 38.4	14 3.4	10 27.4	10 8.2	15 37.8	51
25 14.5	13 10.4	8 57.8	10 20.6	15 57.1	52
24 49.1	12 14.0	7 19.9	10 34.2	16 17.7	53
24 22.2	11 13.7	5 32.5	10 49.1	16 39.7	54
23 53.5	10 9.3	3 34.5	11 5.4	17 3.4	55
23 22.8	9 0.0	1 24.4	11 23.6	17 28.8	56
22 49.9	7 45.3	♑29 0.5	11 43.8	17 56.3	57
22 14.4	6 24.3	26 21.1	12 6.4	18 26.3	58
21 35.9	4 56.3	23 24.2	12 32.0	18 59.0	59
20 53.8	3 20.0	20 7.8	13 1.2	19 35.0	60
20 7.6	1 34.1	16 29.7	13 34.9	20 15.0	61
19 16.3	♐29 36.6	12 28.1	14 14.1	20 59.8	62
18 18.8	27 25.4	8 1.6	15 0.5	21 50.6	63
17 13.2	24 57.2	3 9.7	15 56.3	22 49.1	64
15 56.8	22 7.7	♐27 53.2	17 4.7	23 58.1	65
14 25.1	18 50.3	22 14.6	18 31.1	25 22.2	66
5	6	Descendant	8	9	S LAT

♊ 7° 46′ 52″

4h 24m 0s MC 66° 0′ 0″

16h 28m 0s MC 247° 0′ 0″

♐ 8° 43′ 20″

11	12	Ascendant	2	3	N LAT
° ′	° ′	° ′	° ′	° ′	°
♑ 6 25.6	♒ 4 39.5	♓ 5 10.3	♈ 7 37.4	♉ 9 23.9	0
5 45.9	3 31.0	4 14.4	7 49.2	9 50.8	5
5 5.4	2 19.5	3 13.5	8 1.8	10 18.7	10
4 23.4	1 3.6	2 5.9	8 15.6	10 48.1	15
4 14.8	0 47.8	1 51.4	8 18.5	10 54.2	16
4 6.1	0 31.7	1 36.5	8 21.5	11 0.4	17
3 57.2	0 15.3	1 21.2	8 24.5	11 6.7	18
3 48.3	♑29 58.7	1 5.5	8 27.6	11 13.1	19
3 39.2	29 41.7	0 49.2	8 30.8	11 19.6	20
3 30.0	29 24.4	0 32.6	8 34.1	11 26.2	21
3 20.6	29 6.8	0 15.3	8 37.5	11 33.0	22
3 11.1	28 48.8	♒29 57.6	8 41.0	11 39.9	23
3 1.5	28 30.4	29 39.2	8 44.5	11 46.9	24
2 51.6	28 11.6	29 20.2	8 48.2	11 54.0	25
2 41.6	27 52.4	29 0.5	8 52.0	12 1.3	26
2 31.4	27 32.7	28 40.1	8 55.9	12 8.8	27
2 21.0	27 12.5	28 18.8	8 59.9	12 16.5	28
2 10.4	26 51.7	27 56.8	9 4.1	12 24.3	29
1 59.5	26 30.5	27 33.8	9 8.4	12 32.4	30
1 48.4	26 8.6	27 9.9	9 12.9	12 40.6	31
1 37.0	25 46.0	26 44.9	9 17.5	12 49.1	32
1 25.3	25 22.8	26 18.8	9 22.3	12 57.9	33
1 13.4	24 58.9	25 51.5	9 27.4	13 6.9	34
1 1.1	24 34.1	25 22.8	9 32.6	13 16.2	35
0 48.4	24 8.6	24 52.6	9 38.1	13 25.8	36
0 35.4	23 42.1	24 20.9	9 43.8	13 35.7	37
0 22.0	23 14.7	23 47.5	9 49.8	13 45.9	38
0 8.1	22 46.3	23 12.2	9 56.1	13 56.6	39
♐29 53.8	22 16.8	22 34.9	10 2.7	14 7.6	40
29 39.0	21 46.0	21 55.3	10 9.6	14 19.0	41
29 23.7	21 14.0	21 13.2	10 17.0	14 31.0	42
29 7.8	20 40.6	20 28.4	10 24.8	14 43.4	43
28 51.2	20 5.7	19 40.6	10 33.0	14 56.4	44
28 34.0	19 29.2	18 49.4	10 41.7	15 10.0	45
28 16.0	18 50.8	17 54.5	10 51.0	15 24.2	46
27 57.3	18 10.6	16 55.5	11 1.0	15 39.1	47
27 37.7	17 28.2	15 51.8	11 11.6	15 54.8	48
27 17.1	16 43.5	14 43.0	11 23.0	16 11.4	49
26 55.4	15 56.3	13 28.2	11 35.3	16 28.9	50
26 32.6	15 6.2	12 6.9	11 48.7	16 47.4	51
26 8.6	14 13.1	10 38.2	12 3.1	17 7.1	52
25 43.1	13 16.4	9 1.0	12 18.9	17 28.1	53
25 16.0	12 15.9	7 14.1	12 36.2	17 50.6	54
24 47.1	11 11.1	5 16.3	12 55.2	18 14.6	55
24 16.3	10 1.4	3 6.0	13 16.2	18 40.6	56
23 43.1	8 46.2	0 41.5	13 39.7	19 8.6	57
23 7.3	7 24.6	♑28 0.8	14 5.9	19 39.2	58
22 28.4	5 55.7	25 1.6	14 35.6	20 12.6	59
21 45.9	4 18.3	21 41.8	15 9.5	20 49.3	60
20 59.2	2 31.0	17 58.9	15 48.6	21 30.2	61
20 7.2	0 31.7	13 50.9	16 34.1	22 16.0	62
19 8.8	♐28 18.1	9 15.9	17 27.9	23 8.1	63
18 2.0	25 46.8	4 13.4	18 32.7	24 8.2	64
16 43.9	22 52.7	♐28 44.2	19 52.3	25 19.4	65
15 9.4	19 28.8	22 51.0	21 33.2	26 46.7	66
5	6	Descendant	8	9	S LAT

♊ 8° 43′ 20″

4h 28m 0s MC 67° 0′ 0″

16ʰ 32ᵐ 0ˢ — MC 248° 0′ 0″ — ♐ 9° 39′ 41″

N LAT	11	12	Ascendant	2	3
0	♑ 7 20.8	♒ 5 38.0	♓ 6 14.0	♈ 8 42.5	♉10 25.0
5	6 41.1	4 30.2	5 20.2	8 56.0	10 52.5
10	6 0.7	3 19.3	4 21.7	9 10.4	11 21.0
15	5 18.7	2 4.1	3 16.6	9 26.1	11 50.9
16	5 10.1	1 48.4	3 2.6	9 29.5	11 57.1
17	5 1.4	1 32.5	2 48.2	9 32.9	12 3.5
18	4 52.5	1 16.2	2 33.5	9 36.3	12 9.9
19	4 43.6	0 59.7	2 18.3	9 39.9	12 16.4
20	4 34.5	0 42.9	2 2.7	9 43.5	12 23.0
21	4 25.3	0 25.7	1 46.6	9 47.3	12 29.7
22	4 15.9	0 8.2	1 30.0	9 51.1	12 36.6
23	4 6.4	♑29 50.3	1 12.8	9 55.1	12 43.6
24	3 56.8	29 32.1	0 55.1	9 59.1	12 50.7
25	3 46.9	29 13.4	0 36.7	10 3.3	12 58.0
26	3 36.9	28 54.3	0 17.7	10 7.6	13 5.5
27	3 26.7	28 34.7	♒29 58.0	10 12.1	13 13.1
28	3 16.3	28 14.6	29 37.5	10 16.7	13 20.9
29	3 5.7	27 54.0	29 16.1	10 21.4	13 28.9
30	2 54.8	27 32.8	28 53.9	10 26.4	13 37.1
31	2 43.7	27 11.0	28 30.8	10 31.5	13 45.5
32	2 32.3	26 48.6	28 6.6	10 36.7	13 54.2
33	2 20.6	26 25.5	27 41.3	10 42.3	14 3.1
34	2 8.6	26 1.7	27 14.8	10 48.0	14 12.2
35	1 56.3	25 37.1	26 47.0	10 53.9	14 21.7
36	1 43.6	25 11.6	26 17.8	11 0.2	14 31.5
37	1 30.6	24 45.3	25 47.0	11 6.7	14 41.5
38	1 17.1	24 17.9	25 14.5	11 13.5	14 52.0
39	1 3.3	23 49.6	24 40.2	11 20.7	15 2.8
40	0 48.9	23 20.1	24 3.9	11 28.2	15 14.0
41	0 34.1	22 49.5	23 25.4	11 36.1	15 25.7
42	0 18.7	22 17.5	22 44.4	11 44.5	15 37.8
43	0 2.7	21 44.2	22 0.7	11 53.3	15 50.5
44	♐29 46.1	21 9.3	21 14.0	12 2.7	16 3.7
45	29 28.9	20 32.8	20 24.0	12 12.6	16 17.5
46	29 10.9	19 54.5	19 30.3	12 23.2	16 32.0
47	28 52.0	19 14.3	18 32.5	12 34.5	16 47.2
48	28 32.3	18 31.9	17 30.0	12 46.6	17 3.1
49	28 11.7	17 47.2	16 22.4	12 59.6	17 20.0
50	27 50.0	16 59.9	15 8.8	13 13.6	17 37.8
51	27 27.1	16 9.7	13 48.7	13 28.8	17 56.7
52	27 2.9	15 16.4	12 21.0	13 45.2	18 16.7
53	26 37.3	14 19.6	10 44.7	14 3.1	18 38.1
54	26 10.0	13 18.9	8 58.5	14 22.8	19 0.9
55	25 41.0	12 13.7	7 1.2	14 44.4	19 25.5
56	25 9.9	11 3.6	4 51.0	15 8.3	19 51.9
57	24 36.5	9 47.9	2 26.1	15 34.9	20 20.5
58	24 0.4	8 25.6	♑29 44.2	16 4.7	20 51.6
59	23 21.2	6 55.9	26 43.1	16 38.5	21 25.6
60	22 38.3	5 17.5	23 20.0	17 16.9	22 3.1
61	21 51.0	3 28.8	19 32.4	18 1.3	22 44.8
62	20 58.4	1 27.8	15 17.7	18 52.9	23 31.7
63	19 59.1	♐29 11.8	10 33.9	19 54.1	24 25.0
64	18 51.1	26 37.2	5 20.1	21 7.7	25 26.7
65	17 31.3	23 38.5	♐29 37.1	22 38.5	26 40.0
66	15 53.8	20 7.7	23 28.2	24 33.8	28 10.6
S LAT	5	6	Descendant	8	9

♊ 9° 39′ 41″ — 4ʰ 32ᵐ 0ˢ — MC 68° 0′ 0″

16ʰ 36ᵐ 0ˢ — MC 249° 0′ 0″ — ♐ 10° 35′ 56″

N LAT	11	12	Ascendant	2	3
0	♑ 8 16.1	♒ 6 36.6	♓ 7 17.7	♈ 9 47.7	♉11 26.0
5	7 36.5	5 29.5	6 26.2	10 2.8	11 54.0
10	6 56.0	4 19.4	5 30.0	10 19.0	12 23.0
15	6 14.1	3 4.9	4 27.5	10 36.6	12 53.5
16	6 5.5	2 49.3	4 14.0	10 40.3	12 59.8
17	5 56.8	2 33.5	4 0.2	10 44.2	13 6.2
18	5 47.9	2 17.4	3 46.1	10 48.1	13 12.8
19	5 39.0	2 1.0	3 31.5	10 52.1	13 19.4
20	5 29.9	1 44.3	3 16.4	10 56.1	13 26.1
21	5 20.7	1 27.3	3 0.9	11 0.4	13 33.0
22	5 11.4	1 9.9	2 44.9	11 4.7	13 40.0
23	5 1.9	0 52.2	2 28.4	11 9.1	13 47.1
24	4 52.2	0 34.0	2 11.4	11 13.6	13 54.4
25	4 42.4	0 15.5	1 53.7	11 18.3	14 1.8
26	4 32.3	♑29 56.5	1 35.3	11 23.2	14 9.4
27	4 22.1	29 37.0	1 16.3	11 28.2	14 17.1
28	4 11.7	29 17.1	0 56.6	11 33.3	14 25.1
29	4 1.1	28 56.6	0 36.0	11 38.6	14 33.2
30	3 50.2	28 35.5	0 14.6	11 44.2	14 41.5
31	3 39.0	28 13.9	♒29 52.2	11 49.9	14 50.1
32	3 27.6	27 51.6	29 28.9	11 55.8	14 58.9
33	3 16.0	27 28.6	29 4.4	12 2.0	15 8.0
34	3 4.0	27 4.9	28 38.8	12 8.4	15 17.3
35	2 51.6	26 40.4	28 11.9	12 15.1	15 26.9
36	2 38.9	26 15.1	27 43.6	12 22.1	15 36.8
37	2 25.9	25 48.8	27 13.9	12 29.4	15 47.1
38	2 12.4	25 21.6	26 42.4	12 37.0	15 57.7
39	1 58.5	24 53.4	26 9.2	12 45.1	16 8.7
40	1 44.1	24 24.0	25 33.9	12 53.5	16 20.1
41	1 29.3	23 53.4	24 56.5	13 2.4	16 32.0
42	1 13.9	23 21.6	24 16.7	13 11.7	16 44.3
43	0 57.9	22 48.3	23 34.2	13 21.6	16 57.2
44	0 41.2	22 13.5	22 48.7	13 32.1	17 10.6
45	0 23.9	21 37.1	22 0.0	13 43.2	17 24.7
46	0 5.8	20 58.8	21 7.6	13 55.1	17 39.4
47	♐29 47.0	20 18.6	20 11.1	14 7.7	17 54.8
48	29 27.2	19 36.2	19 10.0	14 21.3	18 11.1
49	29 6.5	18 51.5	18 3.7	14 35.8	18 28.2
50	28 44.7	18 4.1	16 51.5	14 51.5	18 46.3
51	28 21.7	17 13.9	15 32.7	15 8.4	19 5.5
52	27 57.4	16 20.5	14 6.2	15 26.8	19 25.9
53	27 31.6	15 23.5	12 31.1	15 46.9	19 47.6
54	27 4.2	14 22.6	10 45.9	16 8.8	20 10.9
55	26 35.0	13 17.1	8 49.3	16 33.0	20 35.8
56	26 3.7	12 6.7	6 39.5	16 59.7	21 2.7
57	25 30.1	10 50.4	4 14.4	17 29.4	21 31.8
58	24 53.7	9 27.6	1 31.8	18 2.8	22 3.4
59	24 14.2	7 57.1	♑28 28.9	18 40.4	22 38.1
60	23 30.9	6 17.6	25 2.8	19 23.4	23 16.3
61	22 43.1	4 27.6	21 10.5	20 12.9	23 58.9
62	21 49.8	2 24.8	16 49.0	21 10.6	24 46.7
63	20 49.7	0 6.5	11 56.0	22 18.9	25 41.2
64	19 40.5	♐27 28.5	6 30.2	23 41.3	26 44.5
65	18 18.9	24 25.1	0 32.4	25 23.1	28 0.0
66	16 38.5	20 47.0	♐24 6.2	27 32.9	29 34.0
S LAT	5	6	Descendant	8	9

♊ 10° 35′ 56″ — 4ʰ 36ᵐ 0ˢ — MC 69° 0′ 0″

16h 40m 0s		MC	250° 0′ 0″		N
		♐ 11° 32′ 4″			
11	**12**	**Ascendant**	**2**	**3**	**LAT**
° ′	° ′	° ′	° ′	° ′	°
♑ 9 11.4	♒ 7 35.4	♓ 8 21.7	♈10 52.7	♉12 26.7	**0**
8 31.8	6 29.1	7 32.4	11 9.5	12 55.3	**5**
7 51.5	5 19.7	6 38.6	11 27.5	13 24.8	**10**
7 9.6	4 5.9	5 38.7	11 47.0	13 55.8	**15**
7 1.0	3 50.4	5 25.8	11 51.1	14 2.3	**16**
6 52.3	3 34.8	5 12.5	11 55.3	14 8.8	**17**
6 43.4	3 18.8	4 58.9	11 59.7	14 15.4	**18**
6 34.5	3 2.5	4 44.9	12 4.1	14 22.2	**19**
6 25.4	2 46.0	4 30.5	12 8.6	14 29.0	**20**
6 16.2	2 29.1	4 15.6	12 13.3	14 36.0	**21**
6 6.9	2 11.9	4 0.3	12 18.1	14 43.1	**22**
5 57.4	1 54.3	3 44.4	12 23.0	14 50.4	**23**
5 47.7	1 36.3	3 28.0	12 28.0	14 57.8	**24**
5 37.9	1 17.9	3 11.0	12 33.2	15 5.3	**25**
5 27.8	0 59.0	2 53.4	12 38.6	15 13.0	**26**
5 17.6	0 39.7	2 35.2	12 44.1	15 20.9	**27**
5 7.2	0 19.9	2 16.1	12 49.8	15 29.0	**28**
4 56.6	♑29 59.6	1 56.4	12 55.7	15 37.2	**29**
4 45.7	29 38.6	1 35.8	13 1.8	15 45.7	**30**
4 34.5	29 17.1	1 14.2	13 8.1	15 54.4	**31**
4 23.1	28 55.0	0 51.7	13 14.7	16 3.4	**32**
4 11.4	28 32.1	0 28.2	13 21.5	16 12.6	**33**
3 59.4	28 8.5	0 3.5	13 28.6	16 22.1	**34**
3 47.1	27 44.1	♒29 37.6	13 36.0	16 31.8	**35**
3 34.4	27 18.9	29 10.3	13 43.8	16 41.9	**36**
3 21.3	26 52.8	28 41.5	13 51.8	16 52.3	**37**
3 7.8	26 25.7	28 11.1	14 0.3	17 3.1	**38**
2 53.9	25 57.6	27 39.0	14 9.2	17 14.3	**39**
2 39.5	25 28.4	27 4.9	14 18.5	17 25.9	**40**
2 24.6	24 57.9	26 28.7	14 28.3	17 37.9	**41**
2 9.2	24 26.1	25 50.1	14 38.7	17 50.5	**42**
1 53.1	23 53.0	25 8.9	14 49.6	18 3.6	**43**
1 36.5	23 18.2	24 24.8	15 1.2	18 17.2	**44**
1 19.1	22 41.9	23 37.4	15 13.5	18 31.5	**45**
1 1.0	22 3.7	22 46.4	15 26.6	18 46.4	**46**
0 42.1	21 23.5	21 51.4	15 40.6	19 2.1	**47**
0 22.2	20 41.1	20 51.8	15 55.5	19 18.6	**48**
0 1.4	19 56.4	19 47.0	16 11.6	19 36.0	**49**
♐29 39.5	19 9.0	18 36.3	16 28.9	19 54.4	**50**
29 16.5	18 18.8	17 19.0	16 47.6	20 13.9	**51**
28 52.1	17 25.3	15 54.0	17 7.9	20 34.6	**52**
28 26.2	16 28.2	14 20.2	17 30.0	20 56.7	**53**
27 58.7	15 27.0	12 36.3	17 54.2	21 20.3	**54**
27 29.3	14 21.3	10 40.8	18 20.9	21 45.7	**55**
26 57.8	13 10.5	8 31.7	18 50.3	22 13.0	**56**
26 24.0	11 53.8	6 6.8	19 23.1	22 42.5	**57**
25 47.3	10 30.4	3 23.6	19 59.9	23 14.7	**58**
25 7.5	8 59.2	0 19.3	20 41.4	23 50.0	**59**
24 23.8	7 18.7	♑26 50.4	21 28.8	24 29.0	**60**
23 35.5	5 27.4	22 53.6	22 23.3	25 12.3	**61**
22 41.6	3 22.9	18 25.3	23 26.9	26 1.1	**62**
21 40.6	1 2.2	13 22.7	24 42.3	26 56.8	**63**
20 30.2	♐28 20.9	7 44.1	26 13.3	28 1.7	**64**
19 6.7	25 12.6	1 30.3	28 5.9	29 19.4	**65**
17 23.2	21 26.8	♐24 45.2	♉ 0 30.2	♊ 0 56.9	**66**
5	**6**	**Descendant**	**8**	**9**	**S**
		♊ 11° 32′ 4″			**LAT**
4h 40m 0s		MC	70° 0′ 0″		

16h 44m 0s		MC	251° 0′ 0″		N
		♐ 12° 28′ 6″			
11	**12**	**Ascendant**	**2**	**3**	**LAT**
° ′	° ′	° ′	° ′	° ′	°
♑10 6.7	♒ 8 34.4	♓ 9 25.7	♈11 57.7	♉13 27.3	**0**
9 27.2	7 28.8	8 38.7	12 16.1	13 56.4	**5**
8 47.0	6 20.2	7 47.3	12 35.8	14 26.4	**10**
8 5.1	5 7.1	6 50.1	12 57.2	14 57.9	**15**
7 56.5	4 51.8	6 37.8	13 1.8	15 4.5	**16**
7 47.8	4 36.3	6 25.1	13 6.4	15 11.1	**17**
7 39.0	4 20.5	6 12.1	13 11.2	15 17.9	**18**
7 30.1	4 4.4	5 58.7	13 16.0	15 24.7	**19**
7 21.0	3 48.0	5 44.9	13 21.0	15 31.7	**20**
7 11.8	3 31.3	5 30.7	13 26.1	15 38.8	**21**
7 2.5	3 14.2	5 16.0	13 31.3	15 46.0	**22**
6 53.0	2 56.7	5 0.8	13 36.7	15 53.4	**23**
6 43.3	2 38.9	4 45.1	13 42.2	16 0.9	**24**
6 33.5	2 20.6	4 28.8	13 47.9	16 8.6	**25**
6 23.5	2 1.9	4 11.9	13 53.8	16 16.4	**26**
6 13.2	1 42.7	3 54.4	13 59.8	16 24.4	**27**
6 2.8	1 23.1	3 36.2	14 6.1	16 32.6	**28**
5 52.2	1 2.9	3 17.2	14 12.6	16 41.0	**29**
5 41.3	0 42.1	2 57.4	14 19.3	16 49.6	**30**
5 30.1	0 20.7	2 36.8	14 26.2	16 58.5	**31**
5 18.7	♑29 58.7	2 15.2	14 33.4	17 7.5	**32**
5 7.0	29 36.0	1 52.6	14 40.9	17 16.9	**33**
4 55.0	29 12.6	1 28.8	14 48.6	17 26.5	**34**
4 42.7	28 48.3	1 3.9	14 56.8	17 36.4	**35**
4 29.9	28 23.2	0 37.6	15 5.2	17 46.7	**36**
4 16.9	27 57.3	0 10.0	15 14.1	17 57.3	**37**
4 3.4	27 30.3	♒29 40.7	15 23.3	18 8.2	**38**
3 49.4	27 2.3	29 9.7	15 33.0	18 19.6	**39**
3 35.0	26 33.2	28 36.8	15 43.2	18 31.3	**40**
3 20.1	26 2.9	28 1.9	15 54.0	18 43.6	**41**
3 4.6	25 31.2	27 24.6	16 5.3	18 56.3	**42**
2 48.6	24 58.1	26 44.7	16 17.3	19 9.6	**43**
2 31.9	24 23.5	26 2.0	16 29.9	19 23.4	**44**
2 14.5	23 47.2	25 16.2	16 43.4	19 37.9	**45**
1 56.3	23 9.1	24 26.7	16 57.7	19 53.1	**46**
1 37.3	22 29.0	23 33.3	17 13.0	20 9.0	**47**
1 17.4	21 46.7	22 35.3	17 29.4	20 25.8	**48**
0 56.6	21 2.0	21 32.2	17 46.9	20 43.4	**49**
0 34.6	20 14.6	20 23.2	18 5.8	21 2.1	**50**
0 11.4	19 24.3	19 7.6	18 26.3	21 21.9	**51**
♐29 47.0	18 30.8	17 44.3	18 48.4	21 42.9	**52**
29 21.0	17 33.6	16 12.2	19 12.6	22 5.3	**53**
28 53.3	16 32.3	14 29.9	19 39.0	22 29.3	**54**
28 23.8	15 26.4	12 35.7	20 8.1	22 55.0	**55**
27 52.2	14 15.3	10 27.6	20 40.2	23 22.8	**56**
27 18.1	12 58.2	8 3.3	21 16.0	23 52.8	**57**
26 41.2	11 34.2	5 20.1	21 56.1	24 25.6	**58**
26 1.0	10 2.2	2 14.7	22 41.3	25 1.4	**59**
25 16.9	8 20.9	♑28 43.4	23 32.9	25 41.0	**60**
24 28.2	6 28.3	24 42.2	24 32.4	26 25.2	**61**
23 33.6	4 22.1	20 7.2	25 41.8	27 14.9	**62**
22 31.8	1 59.0	14 54.6	27 4.0	28 11.8	**63**
21 20.2	♐29 14.4	9 2.5	28 43.4	29 18.2	**64**
19 55.0	26 1.1	2 31.2	♉ 0 46.7	♊ 0 38.1	**65**
18 8.2	22 7.1	♐25 25.4	3 25.7	2 19.2	**66**
5	**6**	**Descendant**	**8**	**9**	**S**
		♊ 12° 28′ 6″			**LAT**
4h 44m 0s		MC	71° 0′ 0″		

16ʰ 48ᵐ 0ˢ — MC 252° 0′ 0″ — ♐ 13° 24′ 2″

11	12	Ascendant	2	3	N LAT
° ′	° ′	° ′	° ′	° ′	°
♑11 2.1	♒ 9 33.6	♓10 29.9	♈13 2.6	♉14 27.7	**0**
10 22.7	8 28.8	9 45.2	13 22.7	14 57.2	**5**
9 42.5	7 20.9	8 56.3	13 44.1	15 27.8	**10**
9 0.8	6 8.6	8 1.7	14 7.4	15 59.8	**15**
8 52.2	5 53.5	7 50.0	14 12.3	16 6.5	**16**
8 43.5	5 38.1	7 37.9	14 17.3	16 13.2	**17**
8 34.7	5 22.5	7 25.5	14 22.5	16 20.1	**18**
8 25.7	5 6.5	7 12.8	14 27.8	16 27.1	**19**
8 16.7	4 50.3	6 59.6	14 33.2	16 34.1	**20**
8 7.5	4 33.7	6 46.0	14 38.7	16 41.3	**21**
7 58.1	4 16.8	6 32.0	14 44.4	16 48.7	**22**
7 48.7	3 59.5	6 17.5	14 50.3	16 56.2	**23**
7 39.0	3 41.8	6 2.5	14 56.3	17 3.8	**24**
7 29.2	3 23.7	5 47.0	15 2.5	17 11.6	**25**
7 19.2	3 5.1	5 30.9	15 8.8	17 19.5	**26**
7 9.0	2 46.1	5 14.1	15 15.4	17 27.6	**27**
6 58.5	2 26.6	4 56.7	15 22.2	17 36.0	**28**
6 47.9	2 6.6	4 38.5	15 29.2	17 44.5	**29**
6 37.0	1 45.9	4 19.6	15 36.5	17 53.2	**30**
6 25.9	1 24.7	3 59.9	15 44.0	18 2.2	**31**
6 14.5	1 2.9	3 39.2	15 51.9	18 11.4	**32**
6 2.7	0 40.3	3 17.5	16 0.0	18 20.9	**33**
5 50.7	0 17.0	2 54.8	16 8.4	18 30.7	**34**
5 38.4	♑29 52.9	2 30.9	16 17.2	18 40.8	**35**
5 25.7	29 28.0	2 5.7	16 26.4	18 51.1	**36**
5 12.6	29 2.2	1 39.2	16 36.0	19 1.9	**37**
4 59.0	28 35.4	1 11.1	16 46.1	19 13.0	**38**
4 45.1	28 7.5	0 41.3	16 56.6	19 24.5	**39**
4 30.7	27 38.5	0 9.7	17 7.7	19 36.5	**40**
4 15.7	27 8.3	♒29 36.1	17 19.3	19 48.9	**41**
4 0.2	26 36.8	29 0.2	17 31.6	20 1.8	**42**
3 44.2	26 3.9	28 21.8	17 44.6	20 15.2	**43**
3 27.4	25 29.4	27 40.6	17 58.3	20 29.3	**44**
3 10.0	24 53.2	26 56.3	18 12.9	20 44.0	**45**
2 51.8	24 15.2	26 8.5	18 28.4	20 59.4	**46**
2 32.8	23 35.2	25 16.8	18 45.0	21 15.5	**47**
2 12.8	22 52.9	24 20.6	19 2.7	21 32.5	**48**
1 51.9	22 8.3	23 19.3	19 21.7	21 50.4	**49**
1 29.9	21 20.9	22 12.2	19 42.2	22 9.4	**50**
1 6.6	20 30.6	20 58.6	20 4.3	22 29.4	**51**
0 42.1	19 37.0	19 37.2	20 28.3	22 50.8	**52**
0 16.0	18 39.8	18 7.1	20 54.4	23 13.5	**53**
♐29 48.2	17 38.3	16 26.6	21 23.0	23 37.8	**54**
29 18.5	16 32.3	14 34.1	21 54.5	24 3.9	**55**
28 46.7	15 20.9	12 27.5	22 29.2	24 32.1	**56**
28 12.4	14 3.4	10 4.3	23 7.9	25 2.6	**57**
27 35.3	12 39.0	7 21.4	23 51.2	25 35.8	**58**
26 54.8	11 6.3	4 15.4	24 40.1	26 12.3	**59**
26 10.4	9 24.0	0 42.1	25 35.9	26 52.5	**60**
25 21.2	7 30.2	♑26 37.0	26 40.1	27 37.4	**61**
24 26.0	5 22.4	21 55.2	27 55.0	28 28.1	**62**
23 23.3	2 57.0	16 32.4	29 23.9	29 26.1	**63**
22 10.6	0 9.0	10 25.8	♉ 1 11.5	♊ 0 34.0	**64**
20 43.5	♐26 50.6	3 35.6	3 25.5	1 56.1	**65**
18 53.4	22 48.1	♐26 6.9	6 19.2	3 41.0	**66**
5	6	Descendant	8	9	S LAT

♊ 13° 24′ 2″ — 4ʰ 48ᵐ 0ˢ — MC 72° 0′ 0″

16ʰ 52ᵐ 0ˢ — MC 253° 0′ 0″ — ♐ 14° 19′ 53″

N LAT	11	12	Ascendant	2	3
°	° ′	° ′	° ′	° ′	° ′
0	♑11 57.6	♒10 32.9	♓11 34.2	♈14 7.5	♉15 28.0
5	11 18.3	9 28.9	10 51.8	14 29.1	15 58.0
10	10 38.2	8 21.9	10 5.4	14 52.2	16 28.9
15	9 56.5	7 10.4	9 13.6	15 17.4	17 1.5
16	9 47.9	6 55.4	9 2.5	15 22.7	17 8.3
17	9 39.2	6 40.2	8 51.0	15 28.1	17 15.1
18	9 30.4	6 24.7	8 39.2	15 33.7	17 22.1
19	9 21.5	6 8.9	8 27.1	15 39.4	17 29.1
20	9 12.4	5 52.8	8 14.6	15 45.2	17 36.3
21	9 3.3	5 36.4	8 1.7	15 51.2	17 43.6
22	8 53.9	5 19.7	7 48.3	15 57.4	17 51.1
23	8 44.5	5 2.5	7 34.6	16 3.7	17 58.7
24	8 34.8	4 45.0	7 20.3	16 10.2	18 6.4
25	8 25.0	4 27.1	7 5.5	16 16.8	18 14.3
26	8 15.0	4 8.7	6 50.2	16 23.7	18 22.4
27	8 4.8	3 49.8	6 34.2	16 30.8	18 30.6
28	7 54.4	3 30.5	6 17.6	16 38.1	18 39.1
29	7 43.7	3 10.6	6 0.3	16 45.7	18 47.7
30	7 32.9	2 50.2	5 42.3	16 53.5	18 56.6
31	7 21.7	2 29.1	5 23.5	17 1.7	19 5.7
32	7 10.3	2 7.4	5 3.7	17 10.1	19 15.0
33	6 58.6	1 45.0	4 43.1	17 18.8	19 24.7
34	6 46.6	1 21.9	4 21.4	17 27.9	19 34.6
35	6 34.2	0 58.0	3 58.5	17 37.4	19 44.8
36	6 21.5	0 33.2	3 34.5	17 47.3	19 55.3
37	6 8.4	0 7.5	3 9.1	17 57.7	20 6.2
38	5 54.9	♑29 40.9	2 42.2	18 8.5	20 17.5
39	5 40.9	29 13.2	2 13.7	18 19.8	20 29.1
40	5 26.5	28 44.4	1 43.5	18 31.8	20 41.2
41	5 11.5	28 14.3	1 11.2	18 44.3	20 53.8
42	4 56.0	27 43.0	0 36.8	18 57.5	21 6.9
43	4 39.9	27 10.2	0 0.0	19 11.5	21 20.6
44	4 23.2	26 35.8	♒29 20.4	19 26.3	21 34.8
45	4 5.7	25 59.7	28 37.8	19 42.0	21 49.7
46	3 47.5	25 21.9	27 51.8	19 58.7	22 5.3
47	3 28.4	24 41.9	27 1.9	20 16.5	22 21.7
48	3 8.4	23 59.8	26 7.6	20 35.6	22 38.9
49	2 47.4	23 15.2	25 8.4	20 56.0	22 57.0
50	2 25.4	22 27.9	24 3.4	21 18.0	23 16.2
51	2 2.0	21 37.7	22 51.9	21 41.8	23 36.6
52	1 37.4	20 44.1	21 32.8	22 7.5	23 58.2
53	1 11.2	19 46.8	20 4.8	22 35.6	24 21.2
54	0 43.3	18 45.3	18 26.6	23 6.3	24 45.9
55	0 13.5	17 39.0	16 36.2	23 40.0	25 12.4
56	♐29 41.6	16 27.4	14 31.5	24 17.3	25 40.9
57	29 7.1	15 9.6	12 9.8	24 58.8	26 11.8
58	28 29.7	13 44.7	9 27.9	25 45.2	26 45.6
59	27 49.0	12 11.5	6 21.8	26 37.7	27 22.5
60	27 4.2	10 28.3	2 47.1	27 37.4	28 3.4
61	26 14.5	8 33.4	♑28 38.4	28 46.3	28 49.1
62	25 18.8	6 23.9	23 50.1	♉ 0 6.6	29 40.6
63	24 15.3	3 56.2	18 16.9	1 42.0	♊ 0 39.8
64	23 1.4	1 4.9	11 55.0	3 37.6	1 49.2
65	21 32.4	♐27 41.3	4 44.2	6 2.0	3 13.4
66	19 38.9	23 29.7	♐26 50.1	9 10.5	5 2.1
S LAT	5	6	Descendant	8	9

♊ 14° 19′ 53″ — 4ʰ 52ᵐ 0ˢ — MC 73° 0′ 0″

16ʰ 56ᵐ 0ˢ MC 254° 0′ 0″

♐ 15° 15′ 39″

N LAT	11	12	Ascendant	2	3
°	° ′	° ′	° ′	° ′	° ′
0	♑12 53.1	♒11 32.4	♓12 38.6	♈15 12.2	♉16 28.0
5	12 13.9	10 29.3	11 58.6	15 35.4	16 58.5
10	11 33.9	9 23.1	11 14.7	16 0.3	17 29.9
15	10 52.3	8 12.4	10 25.7	16 27.2	18 2.9
16	10 43.7	7 57.6	10 15.2	16 32.9	18 9.8
17	10 35.1	7 42.6	10 4.3	16 38.8	18 16.8
18	10 26.3	7 27.2	9 53.2	16 44.8	18 23.8
19	10 17.4	7 11.6	9 41.7	16 50.9	18 31.0
20	10 8.3	6 55.7	9 29.8	16 57.1	18 38.3
21	9 59.2	6 39.5	9 17.6	17 3.5	18 45.7
22	9 49.9	6 22.9	9 5.0	17 10.1	18 53.2
23	9 40.4	6 5.9	8 51.9	17 16.9	19 0.9
24	9 30.8	5 48.6	8 38.4	17 23.8	19 8.8
25	9 21.0	5 30.8	8 24.4	17 31.0	19 16.8
26	9 11.0	5 12.6	8 9.8	17 38.4	19 25.0
27	9 0.8	4 53.9	7 54.7	17 46.0	19 33.3
28	8 50.4	4 34.7	7 39.0	17 53.8	19 41.9
29	8 39.7	4 15.0	7 22.6	18 1.9	19 50.7
30	8 28.8	3 54.8	7 5.4	18 10.3	19 59.6
31	8 17.7	3 33.9	6 47.6	18 19.0	20 8.9
32	8 6.3	3 12.4	6 28.8	18 28.1	20 18.4
33	7 54.6	2 50.1	6 9.2	18 37.4	20 28.1
34	7 42.6	2 27.2	5 48.5	18 47.2	20 38.1
35	7 30.2	2 3.5	5 26.8	18 57.4	20 48.5
36	7 17.5	1 38.9	5 3.9	19 7.9	20 59.2
37	7 4.4	1 13.4	4 39.8	19 19.0	21 10.2
38	6 50.9	0 46.9	4 14.2	19 30.6	21 21.6
39	6 36.9	0 19.4	3 47.0	19 42.8	21 33.4
40	6 22.5	♑29 50.7	3 18.1	19 55.5	21 45.7
41	6 7.5	29 20.9	2 47.4	20 8.9	21 58.5
42	5 52.0	28 49.7	2 14.5	20 23.1	22 11.7
43	5 35.9	28 17.0	1 39.3	20 38.0	22 25.5
44	5 19.1	27 42.8	1 1.4	20 53.8	22 40.0
45	5 1.6	27 6.9	0 20.6	21 10.6	22 55.0
46	4 43.3	26 29.2	♒29 36.5	21 28.4	23 10.8
47	4 24.2	25 49.4	28 48.6	21 47.5	23 27.4
48	4 4.2	25 7.3	27 56.4	22 7.8	23 44.9
49	3 43.2	24 22.9	26 59.4	22 29.7	24 3.3
50	3 21.0	23 35.7	25 56.7	22 53.2	24 22.7
51	2 57.7	22 45.5	24 47.6	23 18.5	24 43.3
52	2 32.9	21 51.9	23 31.0	23 46.0	25 5.2
53	2 6.7	20 54.6	22 5.6	24 16.0	25 28.5
54	1 38.7	19 53.0	20 29.9	24 48.7	25 53.5
55	1 8.8	18 46.7	18 42.0	25 24.7	26 20.3
56	0 36.6	17 34.9	16 39.7	26 4.4	26 49.2
57	0 2.0	16 16.8	14 20.1	26 48.6	27 20.6
58	♐29 24.4	14 51.5	11 39.7	27 38.1	27 54.8
59	28 43.4	13 17.7	8 34.3	28 33.9	28 32.3
60	27 58.3	11 33.8	4 58.8	29 37.5	29 13.8
61	27 8.2	9 37.7	0 47.1	♉ 0 50.8	♊ 0 0.1
62	26 11.9	7 26.7	♑25 52.7	2 16.4	0 52.5
63	25 7.6	4 56.8	20 8.9	3 58.1	1 52.8
64	23 52.6	2 2.2	13 30.9	6 1.5	3 3.6
65	22 21.8	♐28 33.2	5 57.5	8 36.1	4 30.1
66	20 24.7	24 12.2	♐27 35.2	11 59.6	6 22.7
S LAT	5	6	Descendant	8	9

♊ 15° 15′ 39″

4ʰ 56ᵐ 0ˢ MC 74° 0′ 0″

17ʰ 0ᵐ 0ˢ MC 255° 0′ 0″

♐ 16° 11′ 19″

N LAT	11	12	Ascendant	2	3
°	° ′	° ′	° ′	° ′	° ′
0	♑13 48.7	♒12 32.1	♓13 43.2	♈16 16.8	♉17 27.9
5	13 9.6	11 29.8	13 5.5	16 41.7	17 58.8
10	12 29.7	10 24.5	12 24.2	17 8.2	18 30.7
15	11 48.2	9 14.7	11 38.1	17 36.9	19 4.2
16	11 39.7	9 0.1	11 28.1	17 43.0	19 11.1
17	11 31.0	8 45.2	11 17.9	17 49.3	19 18.2
18	11 22.2	8 30.1	11 7.4	17 55.6	19 25.3
19	11 13.3	8 14.6	10 56.5	18 2.2	19 32.6
20	11 4.3	7 58.9	10 45.4	18 8.8	19 40.0
21	10 55.2	7 42.8	10 33.8	18 15.7	19 47.5
22	10 45.9	7 26.4	10 21.9	18 22.7	19 55.2
23	10 36.4	7 9.6	10 9.6	18 29.9	20 3.0
24	10 26.8	6 52.4	9 56.8	18 37.3	20 10.9
25	10 17.0	6 34.9	9 43.6	18 45.0	20 19.0
26	10 7.0	6 16.8	9 29.9	18 52.8	20 27.3
27	9 56.9	5 58.4	9 15.6	19 0.9	20 35.8
28	9 46.5	5 39.4	9 0.7	19 9.3	20 44.5
29	9 35.8	5 19.8	8 45.2	19 17.9	20 53.3
30	9 25.0	4 59.8	8 29.0	19 26.9	21 2.4
31	9 13.8	4 39.1	8 12.1	19 36.2	21 11.8
32	9 2.4	4 17.7	7 54.4	19 45.8	21 21.4
33	8 50.7	3 55.7	7 35.8	19 55.8	21 31.3
34	8 38.7	3 32.9	7 16.3	20 6.1	21 41.4
35	8 26.4	3 9.4	6 55.7	20 17.0	21 51.9
36	8 13.6	2 45.0	6 34.0	20 28.3	22 2.7
37	8 0.5	2 19.7	6 11.1	20 40.0	22 13.9
38	7 47.0	1 53.4	5 46.8	20 52.4	22 25.5
39	7 33.0	1 26.1	5 21.1	21 5.3	22 37.4
40	7 18.6	0 57.6	4 53.7	21 18.9	22 49.8
41	7 3.6	0 27.9	4 24.4	21 33.1	23 2.7
42	6 48.1	♑29 56.9	3 53.2	21 48.2	23 16.2
43	6 32.0	29 24.4	3 19.6	22 4.1	23 30.2
44	6 15.2	28 50.4	2 43.6	22 20.9	23 44.8
45	5 57.7	28 14.7	2 4.7	22 38.7	24 0.0
46	5 39.4	27 37.1	1 22.6	22 57.7	24 16.0
47	5 20.3	26 57.5	0 36.8	23 17.9	24 32.8
48	5 0.2	26 15.6	♒29 46.9	23 39.5	24 50.5
49	4 39.1	25 31.2	28 52.2	24 2.7	25 9.1
50	4 17.0	24 44.2	27 52.1	24 27.7	25 28.7
51	3 53.5	23 54.0	26 45.6	24 54.6	25 49.6
52	3 28.7	23 0.5	25 31.8	25 23.8	26 11.7
53	3 2.4	22 3.3	24 9.3	25 55.5	26 35.4
54	2 34.3	21 1.7	22 36.5	26 30.3	27 0.6
55	2 4.3	19 55.3	20 51.7	27 8.4	27 27.8
56	1 32.0	18 43.4	18 52.3	27 50.5	27 57.1
57	0 57.2	17 25.1	16 35.4	28 37.3	28 28.8
58	0 19.4	15 59.4	13 57.2	29 29.7	29 3.5
59	♐29 38.2	14 25.1	10 53.3	♉ 0 28.8	29 41.5
60	28 52.7	12 40.4	7 17.8	1 36.2	♊ 0 23.5
61	28 2.2	10 43.4	3 3.9	2 53.8	1 10.6
62	27 5.4	8 30.9	♑28 3.9	4 24.4	2 3.8
63	26 0.3	5 58.8	22 9.6	6 12.0	3 5.1
64	24 44.2	3 0.9	15 14.6	8 23.0	4 17.4
65	23 11.6	♐29 26.6	7 16.6	11 7.7	5 46.0
66	21 10.8	24 55.6	♐28 22.5	14 46.4	7 42.6
S LAT	5	6	Descendant	8	9

♊ 16° 11′ 19″

5ʰ 0ᵐ 0ˢ MC 75° 0′ 0″

17ʰ 4ᵐ 0ˢ | MC | 256° 0′ 0″

♐ 17° 6′ 55″

N LAT	11	12	Ascendant	2	3
0°	♑14° 44.4′	♒13° 32.0′	♓14° 47.8′	♈17° 21.4′	♉18° 27.6′
5	14 5.4	12 30.6	14 12.5	17 47.8	18 58.9
10	13 25.6	11 26.1	13 33.8	18 15.9	19 31.2
15	12 44.2	10 17.2	12 50.6	18 46.5	20 5.2
16	12 35.7	10 2.8	12 41.3	18 53.0	20 12.2
17	12 27.1	9 48.1	12 31.7	18 59.6	20 19.4
18	12 18.3	9 33.2	12 21.8	19 6.4	20 26.6
19	12 9.4	9 17.9	12 11.6	19 13.3	20 34.0
20	12 0.4	9 2.3	12 1.2	19 20.4	20 41.5
21	11 51.3	8 46.5	11 50.3	19 27.6	20 49.1
22	11 42.0	8 30.2	11 39.1	19 35.1	20 56.8
23	11 32.6	8 13.6	11 27.6	19 42.7	21 4.7
24	11 23.0	7 56.7	11 15.6	19 50.6	21 12.8
25	11 13.2	7 39.3	11 3.2	19 58.7	21 21.0
26	11 3.2	7 21.4	10 50.3	20 7.0	21 29.4
27	10 53.1	7 3.1	10 36.8	20 15.6	21 38.0
28	10 42.7	6 44.4	10 22.9	20 24.5	21 46.8
29	10 32.1	6 25.0	10 8.3	20 33.7	21 55.7
30	10 21.2	6 5.1	9 53.1	20 43.2	22 5.0
31	10 10.1	5 44.6	9 37.1	20 53.0	22 14.4
32	9 58.7	5 23.5	9 20.5	21 3.2	22 24.1
33	9 47.0	5 1.7	9 3.0	21 13.8	22 34.1
34	9 35.0	4 39.1	8 44.6	21 24.8	22 44.4
35	9 22.7	4 15.8	8 25.2	21 36.3	22 55.0
36	9 10.0	3 51.6	8 4.7	21 48.2	23 6.0
37	8 56.9	3 26.5	7 43.1	22 0.7	23 17.3
38	8 43.3	3 0.4	7 20.2	22 13.8	23 29.0
39	8 29.4	2 33.3	6 55.9	22 27.5	23 41.1
40	8 14.9	2 5.0	6 30.0	22 41.8	23 53.7
41	7 59.9	1 35.5	6 2.4	22 56.9	24 6.7
42	7 44.4	1 4.7	5 32.8	23 12.9	24 20.3
43	7 28.3	0 32.5	5 1.1	23 29.7	24 34.4
44	7 11.5	♑29 58.6	4 26.9	23 47.5	24 49.2
45	6 53.9	29 23.1	3 50.0	24 6.4	25 4.6
46	6 35.6	28 45.7	3 10.0	24 26.4	25 20.8
47	6 16.5	28 6.2	2 26.5	24 47.8	25 37.8
48	5 56.4	27 24.5	1 39.0	25 10.7	25 55.7
49	5 35.3	26 40.3	0 46.9	25 35.2	26 14.5
50	5 13.1	25 53.4	♒29 49.5	26 1.5	26 34.4
51	4 49.6	25 3.4	28 46.0	26 30.0	26 55.5
52	4 24.7	24 10.0	27 35.2	27 0.8	27 17.9
53	3 58.3	23 12.8	26 15.9	27 34.3	27 41.8
54	3 30.2	22 11.3	24 46.6	28 10.9	28 7.3
55	3 0.0	21 4.8	23 5.2	28 51.1	28 34.8
56	2 27.7	19 52.8	21 9.3	29 35.5	29 4.4
57	1 52.7	18 34.3	18 55.7	♉ 0 24.8	29 36.6
58	1 14.7	17 8.4	16 20.6	1 20.0	♊ 0 11.6
59	0 33.2	15 33.6	13 19.0	2 22.3	0 50.1
60	♐29 47.5	13 48.3	9 44.5	3 33.2	1 32.7
61	28 56.6	11 50.3	5 29.5	4 54.9	2 20.4
62	27 59.3	9 36.4	0 24.7	6 30.3	3 14.4
63	26 53.5	7 2.3	♑24 20.3	8 23.8	4 16.7
64	25 36.3	4 1.1	17 7.4	10 42.1	5 30.4
65	24 1.8	0 21.4	8 42.6	13 36.6	7 1.1
66	21 57.3	♐25 40.1	♐29 12.7	17 30.6	9 1.8
S LAT	5	6	Descendant	8	9

♊ 17° 6′ 55″

5ʰ 4ᵐ 0ˢ | MC | 76° 0′ 0″

17ʰ 8ᵐ 0ˢ | MC | 257° 0′ 0″

♐ 18° 2′ 27″

N LAT	11	12	Ascendant	2	3
0°	♑15° 40.1′	♒14° 32.0′	♓15° 52.5′	♈18° 25.8′	♉19° 27.1′
5	15 1.3	13 31.6	15 19.6	18 53.7	19 58.8
10	14 21.6	12 28.0	14 43.6	19 23.5	20 31.6
15	13 40.4	11 20.1	14 3.3	19 55.9	21 6.0
16	13 31.8	11 5.8	13 54.6	20 2.7	21 13.1
17	13 23.2	10 51.3	13 45.7	20 9.7	21 20.4
18	13 14.5	10 36.5	13 36.4	20 16.9	21 27.7
19	13 5.6	10 21.5	13 27.0	20 24.2	21 35.2
20	12 56.7	10 6.1	13 17.2	20 31.7	21 42.7
21	12 47.6	9 50.4	13 7.1	20 39.4	21 50.4
22	12 38.3	9 34.4	12 56.6	20 47.2	21 58.3
23	12 28.9	9 18.0	12 45.8	20 55.3	22 6.3
24	12 19.3	9 1.2	12 34.6	21 3.7	22 14.4
25	12 9.5	8 44.0	12 23.0	21 12.2	22 22.7
26	11 59.6	8 26.4	12 11.0	21 21.0	22 31.2
27	11 49.4	8 8.3	11 58.4	21 30.1	22 39.9
28	11 39.1	7 49.7	11 45.3	21 39.5	22 48.8
29	11 28.5	7 30.6	11 31.7	21 49.2	22 57.9
30	11 17.6	7 10.9	11 17.5	21 59.2	23 7.2
31	11 6.5	6 50.6	11 2.6	22 9.6	23 16.8
32	10 55.1	6 29.7	10 47.0	22 20.3	23 26.6
33	10 43.5	6 8.1	10 30.6	22 31.5	23 36.7
34	10 31.5	5 45.8	10 13.4	22 43.1	23 47.1
35	10 19.1	5 22.6	9 55.2	22 55.2	23 57.9
36	10 6.4	4 58.7	9 36.1	23 7.8	24 8.9
37	9 53.3	4 33.8	9 15.8	23 21.0	24 20.4
38	9 39.8	4 7.9	8 54.3	23 34.8	24 32.2
39	9 25.9	3 41.0	8 31.5	23 49.2	24 44.4
40	9 11.4	3 13.0	8 7.2	24 4.4	24 57.2
41	8 56.4	2 43.7	7 41.2	24 20.3	25 10.4
42	8 40.9	2 13.1	7 13.4	24 37.1	25 24.1
43	8 24.8	1 41.1	6 43.5	24 54.8	25 38.4
44	8 8.0	1 7.5	6 11.3	25 13.6	25 53.3
45	7 50.4	0 32.1	5 36.6	25 33.4	26 8.9
46	7 32.1	♑29 54.9	4 58.8	25 54.6	26 25.3
47	7 13.0	29 15.7	4 17.7	26 17.1	26 42.4
48	6 52.9	28 34.2	3 32.8	26 41.1	27 0.5
49	6 31.7	27 50.2	2 43.5	27 6.9	27 19.5
50	6 9.5	27 3.4	1 49.0	27 34.6	27 39.6
51	5 45.9	26 13.6	0 48.6	28 4.5	28 0.9
52	5 21.0	25 20.3	♒29 41.2	28 36.9	28 23.6
53	4 54.6	24 23.2	28 25.5	29 12.1	28 47.7
54	4 26.3	23 21.8	26 59.9	29 50.6	29 13.6
55	3 56.1	22 15.3	25 22.5	♉ 0 32.8	29 41.3
56	3 23.6	21 3.3	23 30.7	1 19.4	♊ 0 11.3
57	2 48.5	19 44.7	21 21.3	2 11.1	0 43.8
58	2 10.4	18 18.5	18 50.2	3 9.0	1 19.2
59	1 28.7	16 43.4	15 51.9	4 14.3	1 58.2
60	0 42.7	14 57.5	12 19.6	5 28.6	2 41.3
61	♐29 51.4	12 58.6	8 4.7	6 54.3	3 29.6
62	28 53.6	10 43.5	2 56.3	8 34.3	4 24.4
63	27 47.1	8 7.4	♑26 42.3	10 33.3	5 27.7
64	26 28.8	5 3.0	19 11.2	12 58.6	6 42.7
65	24 52.6	1 18.0	10 16.8	16 2.8	8 15.5
66	22 44.3	♐26 25.9	0 6.2	20 12.2	10 20.3
S LAT	5	6	Descendant	8	9

♊ 18° 2′ 27″

5ʰ 8ᵐ 0ˢ | MC | 77° 0′ 0″

17ʰ 12ᵐ 0ˢ		MC		258° 0′ 0″	
		♐ 18° 57′ 54″			
11	**12**	**Ascendant**	**2**	**3**	**N LAT**
° ′	° ′	° ′	° ′	° ′	°
♑16 36.0	♒15 32.3	♓16 57.4	♈19 30.1	♉20 26.4	**0**
15 57.3	14 32.7	16 26.9	19 59.6	20 58.6	**5**
15 17.8	13 30.1	15 53.5	20 31.0	21 31.8	**10**
14 36.6	12 23.1	15 16.2	21 5.1	22 6.6	**15**
14 28.1	12 9.1	15 8.1	21 12.3	22 13.8	**16**
14 19.5	11 54.8	14 59.8	21 19.7	22 21.1	**17**
14 10.8	11 40.2	14 51.3	21 27.2	22 28.5	**18**
14 2.0	11 25.3	14 42.5	21 34.9	22 36.1	**19**
13 53.0	11 10.2	14 33.4	21 42.8	22 43.7	**20**
13 43.9	10 54.7	14 24.1	21 50.9	22 51.5	**21**
13 34.7	10 38.9	14 14.4	21 59.2	22 59.5	**22**
13 25.3	10 22.7	14 4.3	22 7.7	23 7.6	**23**
13 15.7	10 6.1	13 53.9	22 16.5	23 15.8	**24**
13 6.0	9 49.1	13 43.2	22 25.5	23 24.2	**25**
12 56.1	9 31.7	13 32.0	22 34.7	23 32.8	**26**
12 45.9	9 13.8	13 20.3	22 44.3	23 41.6	**27**
12 35.6	8 55.5	13 8.2	22 54.2	23 50.6	**28**
12 25.0	8 36.6	12 55.5	23 4.4	23 59.8	**29**
12 14.2	8 17.1	12 42.3	23 14.9	24 9.2	**30**
12 3.1	7 57.0	12 28.4	23 25.8	24 18.9	**31**
11 51.7	7 36.3	12 13.9	23 37.1	24 28.8	**32**
11 40.1	7 15.0	11 58.7	23 48.9	24 39.0	**33**
11 28.1	6 52.8	11 42.7	24 1.1	24 49.6	**34**
11 15.8	6 29.9	11 25.8	24 13.8	25 0.4	**35**
11 3.1	6 6.2	11 7.9	24 27.1	25 11.6	**36**
10 50.0	5 41.6	10 49.0	24 40.9	25 23.2	**37**
10 36.5	5 15.9	10 29.0	24 55.4	25 35.1	**38**
10 22.5	4 49.3	10 7.7	25 10.6	25 47.5	**39**
10 8.1	4 21.5	9 45.0	25 26.5	26 0.3	**40**
9 53.1	3 52.4	9 20.8	25 43.2	26 13.7	**41**
9 37.6	3 22.1	8 54.8	26 0.8	26 27.5	**42**
9 21.5	2 50.3	8 26.9	26 19.4	26 42.0	**43**
9 4.6	2 16.9	7 56.8	26 39.1	26 57.1	**44**
8 47.1	1 41.8	7 24.3	27 0.0	27 12.8	**45**
8 28.8	1 4.9	6 48.9	27 22.1	27 29.4	**46**
8 9.6	0 25.9	6 10.4	27 45.7	27 46.7	**47**
7 49.5	♑29 44.6	5 28.2	28 10.9	28 4.9	**48**
7 28.4	29 0.8	4 41.8	28 37.9	28 24.2	**49**
7 6.1	28 14.2	3 50.5	29 7.0	28 44.5	**50**
6 42.5	27 24.6	2 53.4	29 38.3	29 6.0	**51**
6 17.6	26 31.5	1 49.7	♉ 0 12.2	29 28.9	**52**
5 51.0	25 34.5	0 38.0	0 49.0	29 53.3	**53**
5 22.7	24 33.2	♒29 16.6	1 29.3	♊ 0 19.4	**54**
4 52.4	23 26.9	27 43.7	2 13.4	0 47.4	**55**
4 19.9	22 14.8	25 56.7	3 2.1	1 17.7	**56**
3 44.6	20 56.2	23 52.2	3 56.2	1 50.5	**57**
3 6.3	19 29.9	21 26.0	4 56.6	2 26.4	**58**
2 24.4	17 54.4	18 32.3	6 4.8	3 5.7	**59**
1 38.2	16 8.1	15 3.6	7 22.4	3 49.4	**60**
0 46.6	14 8.4	10 50.3	8 51.7	4 38.2	**61**
♐29 48.3	11 52.0	5 39.9	10 36.1	5 33.7	**62**
28 41.2	9 14.1	♑29 17.4	12 40.5	6 37.9	**63**
27 21.9	6 6.7	21 27.7	15 12.5	7 54.2	**64**
25 44.0	2 16.4	12 1.1	18 26.0	9 29.0	**65**
23 31.7	♐27 13.1	1 4.0	22 51.0	11 38.1	**66**
5	**6**	**Descendant**	**8**	**9**	**S LAT**
		♊ 18° 57′ 54″			
5ʰ 12ᵐ 0ˢ		MC		78° 0′ 0″	

17ʰ 16ᵐ 0ˢ		MC		259° 0′ 0″	
		♐ 19° 53′ 18″			
N LAT	**11**	**12**	**Ascendant**	**2**	**3**
°	° ′	° ′	° ′	° ′	° ′
0	♑17 31.9	♒16 32.7	♓18 2.3	♈20 34.3	♉21 25.6
5	16 53.4	15 34.1	17 34.3	21 5.3	21 58.1
10	16 14.0	14 32.5	17 3.6	21 38.3	22 31.7
15	15 33.0	13 26.5	16 29.2	22 14.1	23 6.9
16	15 24.5	13 12.6	16 21.8	22 21.7	23 14.2
17	15 15.9	12 58.5	16 14.2	22 29.4	23 21.6
18	15 7.3	12 44.2	16 6.3	22 37.3	23 29.2
19	14 58.5	12 29.5	15 58.2	22 45.4	23 36.8
20	14 49.5	12 14.6	15 49.9	22 53.7	23 44.5
21	14 40.4	11 59.3	15 41.2	23 2.2	23 52.4
22	14 31.2	11 43.7	15 32.3	23 10.9	24 0.4
23	14 21.9	11 27.7	15 23.1	23 19.8	24 8.6
24	14 12.3	11 11.3	15 13.5	23 29.0	24 17.0
25	14 2.6	10 54.6	15 3.6	23 38.5	24 25.5
26	13 52.7	10 37.4	14 53.3	23 48.2	24 34.1
27	13 42.6	10 19.7	14 42.5	23 58.2	24 43.0
28	13 32.3	10 1.6	14 31.3	24 8.6	24 52.1
29	13 21.7	9 42.9	14 19.7	24 19.3	25 1.4
30	13 10.9	9 23.7	14 7.5	24 30.3	25 10.9
31	12 59.8	9 3.9	13 54.7	24 41.8	25 20.7
32	12 48.5	8 43.4	13 41.3	24 53.6	25 30.8
33	12 36.8	8 22.2	13 27.2	25 5.9	25 41.1
34	12 24.9	8 0.4	13 12.4	25 18.7	25 51.7
35	12 12.6	7 37.7	12 56.8	25 32.1	26 2.7
36	11 59.9	7 14.2	12 40.3	25 46.0	26 14.0
37	11 46.8	6 49.8	12 22.9	26 0.5	26 25.6
38	11 33.3	6 24.5	12 4.3	26 15.6	26 37.7
39	11 19.4	5 58.0	11 44.6	26 31.5	26 50.2
40	11 5.0	5 30.5	11 23.6	26 48.2	27 3.2
41	10 50.0	5 1.7	11 1.2	27 5.7	27 16.7
42	10 34.5	4 31.6	10 37.1	27 24.1	27 30.7
43	10 18.4	4 0.1	10 11.2	27 43.5	27 45.3
44	10 1.6	3 27.0	9 43.3	28 4.1	28 0.5
45	9 44.0	2 52.2	9 13.1	28 25.9	28 16.4
46	9 25.7	2 15.5	8 40.2	28 49.1	28 33.1
47	9 6.5	1 36.7	8 4.3	29 13.7	28 50.6
48	8 46.4	0 55.7	7 25.0	29 40.1	29 9.0
49	8 25.2	0 12.2	6 41.7	♉ 0 8.3	29 28.4
50	8 2.9	♑29 25.9	5 53.8	0 38.6	29 48.9
51	7 39.4	28 36.4	5 0.4	1 11.2	♊ 0 10.6
52	7 14.4	27 43.6	4 0.7	1 46.6	0 33.7
53	6 47.8	26 46.8	2 53.2	2 25.0	0 58.4
54	6 19.5	25 45.6	1 36.6	3 6.9	1 24.7
55	5 49.1	24 39.4	0 8.8	3 52.9	1 53.0
56	5 16.4	23 27.4	♒28 27.2	4 43.6	2 23.6
57	4 41.1	22 8.8	26 28.5	5 39.9	2 56.8
58	4 2.7	20 42.4	24 8.2	6 42.8	3 32.9
59	3 20.6	19 6.7	21 20.4	7 53.7	4 12.7
60	2 34.1	17 20.0	17 57.0	9 14.4	4 56.8
61	1 42.2	15 19.6	13 47.3	10 47.3	5 46.3
62	0 43.5	13 2.2	8 36.8	12 35.8	6 42.4
63	♐29 35.8	10 22.5	2 7.6	14 45.2	7 47.5
64	28 15.5	7 12.3	♑23 59.7	17 23.7	9 5.0
65	26 35.9	3 16.8	13 58.0	20 46.2	10 41.8
66	24 19.8	♐28 2.0	2 7.2	25 26.8	12 55.0
S LAT	**5**	**6**	**Descendant**	**8**	**9**
			♊ 19° 53′ 18″		
	5ʰ 16ᵐ 0ˢ		MC		79° 0′ 0″

17h 20m 0s	MC ♐ 20° 48′ 39″	260° 0′ 0″			N LAT
11	**12**	**Ascendant**	**2**	**3**	
° ′	° ′	° ′	° ′	° ′	°
♑18 27.9	♒17 33.3	♓19 7.3	♈21 38.3	♉22 24.6	**0**
17 49.6	16 35.7	18 41.8	22 10.8	22 57.5	**5**
17 10.3	15 35.1	18 13.8	22 45.4	23 31.5	**10**
16 29.5	14 30.1	17 42.4	23 22.9	24 7.1	**15**
16 21.0	14 16.5	17 35.7	23 30.8	24 14.5	**16**
16 12.5	14 2.6	17 28.7	23 38.9	24 22.0	**17**
16 3.8	13 48.4	17 21.5	23 47.2	24 29.6	**18**
15 55.1	13 34.0	17 14.2	23 55.7	24 37.3	**19**
15 46.1	13 19.3	17 6.5	24 4.4	24 45.1	**20**
15 37.1	13 4.2	16 58.7	24 13.3	24 53.1	**21**
15 27.9	12 48.8	16 50.5	24 22.4	25 1.2	**22**
15 18.6	12 33.1	16 42.1	24 31.7	25 9.4	**23**
15 9.1	12 16.9	16 33.3	24 41.3	25 17.9	**24**
14 59.4	12 0.4	16 24.3	24 51.2	25 26.5	**25**
14 49.5	11 43.4	16 14.8	25 1.4	25 35.2	**26**
14 39.4	11 26.0	16 5.0	25 11.9	25 44.2	**27**
14 29.1	11 8.1	15 54.8	25 22.7	25 53.4	**28**
14 18.5	10 49.7	15 44.1	25 33.9	26 2.8	**29**
14 7.8	10 30.7	15 33.0	25 45.4	26 12.4	**30**
13 56.7	10 11.1	15 21.3	25 57.4	26 22.3	**31**
13 45.4	9 50.9	15 9.0	26 9.8	26 32.4	**32**
13 33.8	9 30.0	14 56.1	26 22.6	26 42.8	**33**
13 21.8	9 8.4	14 42.6	26 36.0	26 53.6	**34**
13 9.5	8 46.0	14 28.3	26 49.9	27 4.6	**35**
12 56.9	8 22.7	14 13.2	27 4.4	27 16.1	**36**
12 43.8	7 58.6	13 57.2	27 19.6	27 27.8	**37**
12 30.4	7 33.5	13 40.2	27 35.4	27 40.0	**38**
12 16.5	7 7.4	13 22.2	27 52.0	27 52.7	**39**
12 2.0	6 40.1	13 2.9	28 9.3	28 5.7	**40**
11 47.1	6 11.6	12 42.3	28 27.6	28 19.3	**41**
11 31.6	5 41.8	12 20.2	28 46.8	28 33.5	**42**
11 15.5	5 10.5	11 56.4	29 7.1	28 48.2	**43**
10 58.7	4 37.7	11 30.7	29 28.5	29 3.6	**44**
10 41.2	4 3.2	11 2.9	29 51.3	29 19.6	**45**
10 22.8	3 26.8	10 32.6	♉ 0 15.4	29 36.5	**46**
10 3.7	2 48.3	9 59.6	0 41.1	29 54.1	**47**
9 43.5	2 7.6	9 23.3	1 8.5	♊ 0 12.7	**48**
9 22.4	1 24.3	8 43.3	1 37.8	0 32.3	**49**
9 0.1	0 38.3	7 58.9	2 9.4	0 53.0	**50**
8 36.5	♑29 49.1	7 9.5	2 43.3	1 14.9	**51**
8 11.5	28 56.5	6 14.0	3 20.1	1 38.2	**52**
7 44.9	28 0.0	5 11.2	4 0.0	2 3.0	**53**
7 16.5	26 59.0	3 59.7	4 43.6	2 29.6	**54**
6 46.0	25 53.0	2 37.5	5 31.3	2 58.2	**55**
6 13.3	24 41.2	1 2.1	6 23.9	3 29.0	**56**
5 37.9	23 22.6	♒29 10.0	7 22.3	4 2.5	**57**
4 59.3	21 56.2	26 56.9	8 27.6	4 39.0	**58**
4 17.1	20 20.4	24 16.4	9 41.0	5 19.2	**59**
3 30.4	18 33.3	21 0.2	11 4.6	6 3.7	**60**
2 38.2	16 32.4	16 56.3	12 40.8	6 53.7	**61**
1 39.1	14 14.1	11 48.6	14 33.3	7 50.5	**62**
0 30.8	11 32.8	5 15.2	16 47.4	8 56.4	**63**
♐29 9.7	8 20.0	♑26 50.1	19 32.0	10 15.0	**64**
27 28.5	4 19.3	16 10.6	23 3.1	11 53.7	**65**
25 8.5	♐28 52.8	3 17.4	27 59.4	14 11.1	**66**
5	**6**	**Descendant**	**8**	**9**	**S LAT**
5h 20m 0s	MC ♊ 20° 48′ 39″	80° 0′ 0″			

17h 24m 0s	MC ♐ 21° 43′ 56″	261° 0′ 0″			N LAT
11	**12**	**Ascendant**	**2**	**3**	
° ′	° ′	° ′	° ′	° ′	°
♑19 24.1	♒18 34.0	♓20 12.3	♈22 42.3	♉23 23.4	**0**
18 45.9	17 37.5	19 49.3	23 16.2	23 56.7	**5**
18 6.8	16 37.9	19 24.1	23 52.3	24 31.0	**10**
17 26.1	15 34.0	18 55.8	24 31.5	25 7.0	**15**
17 17.7	15 20.6	18 49.7	24 39.8	25 14.5	**16**
17 9.2	15 6.9	18 43.4	24 48.3	25 22.1	**17**
17 0.5	14 53.0	18 36.9	24 56.9	25 29.7	**18**
16 51.8	14 38.8	18 30.2	25 5.8	25 37.5	**19**
16 42.9	14 24.2	18 23.4	25 14.8	25 45.4	**20**
16 33.9	14 9.4	18 16.2	25 24.1	25 53.5	**21**
16 24.7	13 54.3	18 8.9	25 33.6	26 1.7	**22**
16 15.4	13 38.7	18 1.3	25 43.4	26 10.0	**23**
16 5.9	13 22.8	17 53.4	25 53.4	26 18.5	**24**
15 56.3	13 6.5	17 45.2	26 3.7	26 27.2	**25**
15 46.4	12 49.8	17 36.6	26 14.3	26 36.1	**26**
15 36.4	12 32.6	17 27.8	26 25.2	26 45.1	**27**
15 26.1	12 15.0	17 18.5	26 36.5	26 54.4	**28**
15 15.6	11 56.8	17 8.9	26 48.2	27 3.9	**29**
15 4.8	11 38.1	16 58.8	27 0.2	27 13.6	**30**
14 53.8	11 18.7	16 48.2	27 12.7	27 23.6	**31**
14 42.5	10 58.8	16 37.1	27 25.6	27 33.8	**32**
14 30.9	10 38.2	16 25.4	27 39.0	27 44.3	**33**
14 19.0	10 16.8	16 13.2	27 52.9	27 55.2	**34**
14 6.7	9 54.7	16 0.2	28 7.4	28 6.3	**35**
13 54.1	9 31.7	15 46.5	28 22.5	28 17.8	**36**
13 41.1	9 7.9	15 32.0	28 38.3	28 29.7	**37**
13 27.6	8 43.1	15 16.7	28 54.7	28 42.0	**38**
13 13.7	8 17.2	15 0.3	29 12.0	28 54.8	**39**
12 59.3	7 50.2	14 42.8	29 30.1	29 8.0	**40**
12 44.4	7 22.1	14 24.1	29 49.0	29 21.7	**41**
12 28.9	6 52.5	14 4.0	♉ 0 9.0	29 36.0	**42**
12 12.8	6 21.6	13 42.4	0 30.1	29 50.8	**43**
11 56.0	5 49.1	13 19.0	0 52.4	♊ 0 6.3	**44**
11 38.5	5 14.9	12 53.7	1 16.0	0 22.5	**45**
11 20.2	4 38.8	12 26.1	1 41.0	0 39.5	**46**
11 1.0	4 0.6	11 56.0	2 7.7	0 57.3	**47**
10 40.9	3 20.2	11 22.9	2 36.2	1 16.0	**48**
10 19.8	2 37.3	10 46.3	3 6.6	1 35.8	**49**
9 57.4	1 51.6	10 5.8	3 39.3	1 56.6	**50**
9 33.8	1 2.7	9 20.5	4 14.6	2 18.7	**51**
9 8.8	0 10.4	8 29.5	4 52.7	2 42.2	**52**
8 42.2	♑29 14.1	7 31.8	5 34.0	3 7.3	**53**
8 13.8	28 13.4	6 25.9	6 19.1	3 34.1	**54**
7 43.3	27 7.6	5 9.9	7 8.6	4 2.9	**55**
7 10.5	25 56.0	3 41.3	8 3.0	4 34.0	**56**
6 35.0	24 37.6	1 56.8	9 3.4	5 7.7	**57**
5 56.4	23 11.2	♒29 52.0	10 10.8	5 44.6	**58**
5 14.0	21 35.4	27 20.5	11 26.7	6 25.1	**59**
4 27.1	19 48.1	24 13.5	12 53.1	7 10.0	**60**
3 34.7	17 46.8	20 18.3	14 32.4	8 0.5	**61**
2 35.2	15 27.7	15 16.6	16 28.6	8 57.9	**62**
1 26.5	12 45.0	8 42.7	18 47.1	10 4.5	**63**
0 4.5	9 29.7	0 2.9	21 37.4	11 24.3	**64**
♐28 21.8	5 24.3	♑18 43.3	25 16.8	13 4.7	**65**
25 58.0	♐29 45.9	4 36.8	♊ 0 28.5	15 26.2	**66**
5	**6**	**Descendant**	**8**	**9**	**S LAT**
5h 24m 0s	MC ♊ 21° 43′ 56″	81° 0′ 0″			

17h 28m 0s — MC 262° 0′ 0″ — ♐ 22° 39′ 10″

N LAT	11	12	Ascendant	2	3
	° ′	° ′	° ′	° ′	° ′
0	♑20 20.3	♒19 35.0	♓21 17.5	♈23 46.0	♉24 22.0
5	19 42.3	18 39.5	20 57.0	24 21.4	24 55.7
10	19 3.4	17 41.0	20 34.5	24 59.1	25 30.4
15	18 22.9	16 38.2	20 9.3	25 39.9	26 6.8
16	18 14.5	16 25.0	20 3.8	25 48.6	26 14.3
17	18 6.0	16 11.5	19 58.2	25 57.4	26 21.9
18	17 57.4	15 57.8	19 52.4	26 6.4	26 29.7
19	17 48.7	15 43.8	19 46.5	26 15.6	26 37.6
20	17 39.8	15 29.5	19 40.3	26 25.0	26 45.6
21	17 30.9	15 15.0	19 34.0	26 34.7	26 53.7
22	17 21.7	15 0.0	19 27.4	26 44.6	27 2.0
23	17 12.4	14 44.8	19 20.6	26 54.7	27 10.4
24	17 3.0	14 29.1	19 13.6	27 5.1	27 19.0
25	16 53.3	14 13.1	19 6.3	27 15.9	27 27.7
26	16 43.5	13 56.6	18 58.7	27 26.9	27 36.7
27	16 33.5	13 39.7	18 50.7	27 38.3	27 45.8
28	16 23.2	13 22.2	18 42.5	27 50.0	27 55.2
29	16 12.8	13 4.3	18 33.9	28 2.1	28 4.7
30	16 2.0	12 45.9	18 24.8	28 14.6	28 14.5
31	15 51.0	12 26.8	18 15.4	28 27.6	28 24.6
32	15 39.8	12 7.1	18 5.5	28 41.0	28 34.9
33	15 28.2	11 46.8	17 55.1	28 54.9	28 45.5
34	15 16.3	11 25.7	17 44.1	29 9.4	28 56.5
35	15 4.1	11 3.9	17 32.5	29 24.5	29 7.7
36	14 51.5	10 41.2	17 20.3	29 40.2	29 19.4
37	14 38.5	10 17.7	17 7.3	29 56.5	29 31.4
38	14 25.1	9 53.2	16 53.6	♉ 0 13.6	29 43.8
39	14 11.2	9 27.6	16 38.9	0 31.5	29 56.6
40	13 56.8	9 1.0	16 23.2	0 50.3	♊ 0 9.9
41	13 41.9	8 33.1	16 6.5	1 10.0	0 23.7
42	13 26.4	8 3.9	15 48.5	1 30.7	0 38.1
43	13 10.4	7 33.3	15 29.1	1 52.5	0 53.1
44	12 53.6	7 1.1	15 8.1	2 15.6	1 8.7
45	12 36.1	6 27.2	14 45.4	2 40.1	1 25.1
46	12 17.8	5 51.5	14 20.6	3 6.0	1 42.2
47	11 58.7	5 13.7	13 53.5	3 33.7	2 0.1
48	11 38.6	4 33.6	13 23.7	4 3.1	2 19.0
49	11 17.4	3 51.0	12 50.8	4 34.6	2 38.9
50	10 55.1	3 5.7	12 14.2	5 8.5	2 59.9
51	10 31.5	2 17.2	11 33.2	5 44.9	3 22.2
52	10 6.5	1 25.2	10 47.2	6 24.3	3 45.9
53	9 39.8	0 29.3	9 54.8	7 7.0	4 11.1
54	9 11.4	♑29 28.9	8 54.9	7 53.6	4 38.1
55	8 40.9	28 23.4	7 45.7	8 44.6	5 7.2
56	8 8.1	27 12.0	6 24.7	9 40.8	5 38.5
57	7 32.5	25 53.8	4 48.8	10 43.1	6 12.5
58	6 53.8	24 27.6	2 53.5	11 52.6	6 49.6
59	6 11.3	22 51.8	0 32.7	13 10.8	7 30.5
60	5 24.2	21 4.5	♒27 37.2	14 39.7	8 15.8
61	4 31.6	19 2.9	23 53.8	16 22.0	9 6.7
62	3 31.8	16 43.1	19 2.4	18 21.5	10 4.6
63	2 22.6	13 59.3	12 32.9	20 44.2	11 12.0
64	0 59.9	10 41.7	3 42.7	23 39.8	12 32.8
65	♐29 15.9	6 31.9	♑21 42.0	27 27.0	14 14.8
66	26 48.4	0 41.6	6 8.9	♊ 2 54.0	16 40.2
S LAT	5	6	Descendant	8	9

♊ 22° 39′ 10″ — 5h 28m 0s — MC 82° 0′ 0″

17h 32m 0s — MC 263° 0′ 0″ — ♐ 23° 34′ 22″

N LAT	11	12	Ascendant	2	3
	° ′	° ′	° ′	° ′	° ′
0	♑21 16.7	♒20 36.1	♓22 22.6	♈24 49.7	♉25 20.5
5	20 38.9	19 41.7	22 4.7	25 26.5	25 54.5
10	20 0.1	18 44.3	21 45.0	26 5.7	26 29.6
15	19 19.8	17 42.6	21 22.9	26 48.1	27 6.3
16	19 11.4	17 29.6	21 18.1	26 57.1	27 13.9
17	19 3.0	17 16.4	21 13.2	27 6.3	27 21.6
18	18 54.4	17 2.9	21 8.1	27 15.6	27 29.4
19	18 45.7	16 49.2	21 2.9	27 25.2	27 37.4
20	18 36.9	16 35.2	20 57.5	27 35.0	27 45.5
21	18 28.0	16 20.8	20 51.9	27 45.0	27 53.7
22	18 18.9	16 6.1	20 46.2	27 55.3	28 2.0
23	18 9.6	15 51.1	20 40.2	28 5.8	28 10.5
24	18 0.2	15 35.7	20 34.0	28 16.6	28 19.2
25	17 50.6	15 19.9	20 27.6	28 27.8	28 28.0
26	17 40.8	15 3.7	20 20.9	28 39.2	28 37.0
27	17 30.8	14 47.0	20 13.9	28 51.0	28 46.2
28	17 20.6	14 29.9	20 6.7	29 3.2	28 55.7
29	17 10.1	14 12.3	19 59.1	29 15.7	29 5.3
30	16 59.4	13 54.1	19 51.2	29 28.7	29 15.2
31	16 48.5	13 35.3	19 42.9	29 42.2	29 25.4
32	16 37.2	13 15.9	19 34.2	29 56.1	29 35.8
33	16 25.7	12 55.9	19 25.0	♉ 0 10.5	29 46.5
34	16 13.8	12 35.1	19 15.4	0 25.5	29 57.5
35	16 1.6	12 13.6	19 5.2	0 41.1	♊ 0 8.9
36	15 49.1	11 51.2	18 54.4	0 57.4	0 20.6
37	15 36.1	11 28.0	18 43.0	1 14.3	0 32.7
38	15 22.7	11 3.8	18 30.9	1 32.0	0 45.2
39	15 8.9	10 38.6	18 18.0	1 50.5	0 58.1
40	14 54.5	10 12.2	18 4.2	2 10.0	1 11.6
41	14 39.7	9 44.7	17 49.4	2 30.3	1 25.5
42	14 24.2	9 15.9	17 33.5	2 51.8	1 40.0
43	14 8.1	8 45.6	17 16.4	3 14.4	1 55.1
44	13 51.4	8 13.8	16 58.0	3 38.3	2 10.8
45	13 34.0	7 40.3	16 37.9	4 3.6	2 27.3
46	13 15.7	7 4.9	16 16.0	4 30.4	2 44.5
47	12 56.6	6 27.5	15 52.0	4 58.9	3 2.6
48	12 36.5	5 47.8	15 25.7	5 29.3	3 21.6
49	12 15.3	5 5.6	14 56.5	6 1.9	3 41.7
50	11 53.0	4 20.6	14 24.0	6 36.8	4 2.8
51	11 29.4	3 32.5	13 47.7	7 14.3	4 25.3
52	11 4.4	2 40.9	13 6.7	7 54.9	4 49.1
53	10 37.8	1 45.4	12 20.1	8 39.0	5 14.5
54	10 9.3	0 45.4	11 26.7	9 27.0	5 41.7
55	9 38.8	♑29 40.2	10 24.7	10 19.5	6 11.0
56	9 5.9	28 29.2	9 12.0	11 17.3	6 42.5
57	8 30.3	27 11.3	7 45.5	12 21.3	7 16.8
58	7 51.6	25 45.3	6 1.1	13 32.8	7 54.2
59	7 9.0	24 9.7	3 52.7	14 53.1	8 35.3
60	6 21.8	22 22.4	1 11.2	16 24.4	9 21.0
61	5 29.0	20 20.6	♒27 43.1	18 9.4	10 12.3
62	4 29.0	18 0.4	23 7.0	20 12.2	11 10.7
63	3 19.3	15 15.6	16 48.8	22 38.7	12 18.8
64	1 56.0	11 56.2	7 55.4	25 39.2	13 40.4
65	0 10.7	7 42.2	♑25 14.9	29 33.6	15 24.0
66	♐27 39.7	1 40.4	7 58.7	♊ 5 15.4	17 53.2
S LAT	5	6	Descendant	8	9

♊ 23° 34′ 22″ — 5h 32m 0s — MC 83° 0′ 0″

17ʰ 36ᵐ 0ˢ MC 264° 0′ 0″

♐ 24° 29′ 31″

N LAT	11	12	Ascendant	2	3
°	° ′	° ′	° ′	° ′	° ′
0	♑22 13.1	♒21 37.4	♓23 27.9	♈25 53.2	♉26 18.8
5	21 35.6	20 44.1	23 12.5	26 31.4	26 53.2
10	20 57.0	19 47.8	22 55.5	27 12.1	27 28.6
15	20 16.8	18 47.3	22 36.6	27 56.1	28 5.6
16	20 8.5	18 34.6	22 32.4	28 5.4	28 13.3
17	20 0.1	18 21.6	22 28.2	28 14.9	28 21.1
18	19 51.6	18 8.4	22 23.9	28 24.6	28 29.0
19	19 42.9	17 54.9	22 19.4	28 34.5	28 37.0
20	19 34.1	17 41.1	22 14.7	28 44.6	28 45.1
21	19 25.2	17 27.0	22 10.0	28 55.0	28 53.4
22	19 16.2	17 12.6	22 5.0	29 5.7	29 1.8
23	19 6.9	16 57.8	21 59.9	29 16.6	29 10.4
24	18 57.6	16 42.7	21 54.6	29 27.8	29 19.1
25	18 48.0	16 27.1	21 49.0	29 39.3	29 28.1
26	18 38.2	16 11.2	21 43.3	29 51.2	29 37.2
27	18 28.3	15 54.8	21 37.3	♉ 0 3.4	29 46.5
28	18 18.1	15 37.9	21 31.1	0 16.0	29 56.0
29	18 7.7	15 20.6	21 24.6	0 29.0	♊ 0 5.7
30	17 57.0	15 2.7	21 17.7	0 42.5	0 15.7
31	17 46.1	14 44.2	21 10.6	0 56.4	0 25.9
32	17 34.9	14 25.1	21 3.1	1 10.8	0 36.4
33	17 23.4	14 5.4	20 55.2	1 25.7	0 47.2
34	17 11.6	13 44.9	20 46.9	1 41.2	0 58.3
35	16 59.4	13 23.7	20 38.2	1 57.3	1 9.7
36	16 46.9	13 1.7	20 28.9	2 14.1	1 21.5
37	16 33.9	12 38.8	20 19.1	2 31.6	1 33.7
38	16 20.6	12 14.9	20 8.6	2 49.9	1 46.3
39	16 6.8	11 50.1	19 57.5	3 9.1	1 59.4
40	15 52.5	11 24.1	19 45.6	3 29.1	2 12.9
41	15 37.6	10 56.9	19 32.8	3 50.2	2 26.9
42	15 22.2	10 28.5	19 19.2	4 12.3	2 41.5
43	15 6.2	9 58.6	19 4.4	4 35.6	2 56.7
44	14 49.5	9 27.2	18 48.4	5 0.3	3 12.6
45	14 32.0	8 54.0	18 31.1	5 26.3	3 29.2
46	14 13.8	8 19.1	18 12.2	5 54.0	3 46.5
47	13 54.7	7 42.0	17 51.4	6 23.4	4 4.7
48	13 34.6	7 2.8	17 28.6	6 54.7	4 23.9
49	13 13.5	6 21.0	17 3.3	7 28.2	4 44.0
50	12 51.3	5 36.4	16 35.2	8 4.2	5 5.4
51	12 27.7	4 48.8	16 3.7	8 42.9	5 28.0
52	12 2.7	3 57.6	15 28.0	9 24.6	5 52.0
53	11 36.0	3 2.5	14 47.5	10 9.9	6 17.5
54	11 7.6	2 2.9	14 0.8	10 59.2	6 44.9
55	10 37.1	0 58.2	13 6.6	11 53.1	7 14.4
56	10 4.2	♑29 47.6	12 2.9	12 52.5	7 46.1
57	9 28.6	28 30.0	10 46.8	13 58.2	8 20.6
58	8 49.7	27 4.4	9 14.5	15 11.4	8 58.2
59	8 7.1	25 29.0	7 20.2	16 33.8	9 39.7
60	7 19.8	23 41.9	4 55.4	18 7.3	10 25.7
61	6 26.9	21 40.1	1 46.4	19 54.9	11 17.3
62	5 26.6	19 19.6	♒27 31.4	22 0.5	12 16.2
63	4 16.6	16 34.1	21 33.1	24 30.5	13 24.8
64	2 52.7	13 13.1	12 47.4	27 35.5	14 47.3
65	1 6.3	8 55.5	♑29 33.7	♊ 1 36.5	16 32.3
66	♐28 32.1	2 42.8	10 14.8	7 32.4	19 4.9
S LAT	5	6	Descendant	8	9

♊ 24° 29′ 31″

5ʰ 36ᵐ 0ˢ MC 84° 0′ 0″

17ʰ 40ᵐ 0ˢ MC 265° 0′ 0″

♐ 25° 24′ 39″

N LAT	11	12	Ascendant	2	3
°	° ′	° ′	° ′	° ′	° ′
0	♑23 9.7	♒22 38.9	♓24 33.2	♈26 56.5	♉27 17.0
5	22 32.4	21 46.7	24 20.3	27 36.2	27 51.7
10	21 54.0	20 51.6	24 6.2	28 18.3	28 27.4
15	21 14.0	19 52.2	23 50.3	29 3.8	29 4.8
16	21 5.7	19 39.8	23 46.9	29 13.5	29 12.5
17	20 57.4	19 27.0	23 43.4	29 23.3	29 20.4
18	20 48.9	19 14.1	23 39.7	29 33.3	29 28.3
19	20 40.3	19 0.8	23 36.0	29 43.6	29 36.4
20	20 31.5	18 47.3	23 32.1	29 54.1	29 44.6
21	20 22.7	18 33.5	23 28.1	♉ 0 4.8	29 53.0
22	20 13.6	18 19.3	23 24.0	0 15.8	♊ 0 1.4
23	20 4.5	18 4.8	23 19.7	0 27.1	0 10.1
24	19 55.1	17 49.9	23 15.3	0 38.7	0 18.9
25	19 45.6	17 34.7	23 10.6	0 50.6	0 27.9
26	19 35.9	17 19.0	23 5.8	1 2.9	0 37.0
27	19 25.9	17 2.9	23 0.8	1 15.5	0 46.4
28	19 15.8	16 46.4	22 55.6	1 28.5	0 56.0
29	19 5.4	16 29.3	22 50.2	1 42.0	1 5.8
30	18 54.8	16 11.7	22 44.5	1 55.8	1 15.9
31	18 43.9	15 53.6	22 38.5	2 10.2	1 26.2
32	18 32.7	15 34.8	22 32.2	2 25.1	1 36.7
33	18 21.3	15 15.4	22 25.7	2 40.5	1 47.6
34	18 9.5	14 55.2	22 18.7	2 56.5	1 58.8
35	17 57.4	14 34.3	22 11.4	3 13.1	2 10.3
36	17 44.9	14 12.7	22 3.6	3 30.5	2 22.2
37	17 32.0	13 50.1	21 55.4	3 48.5	2 34.5
38	17 18.7	13 26.6	21 46.7	4 7.4	2 47.2
39	17 4.9	13 2.1	21 37.3	4 27.1	3 0.3
40	16 50.6	12 36.5	21 27.4	4 47.8	3 13.9
41	16 35.8	12 9.7	21 16.7	5 9.5	3 28.1
42	16 20.4	11 41.7	21 5.2	5 32.3	3 42.8
43	16 4.5	11 12.2	20 52.9	5 56.3	3 58.1
44	15 47.8	10 41.2	20 39.5	6 21.6	4 14.0
45	15 30.4	10 8.5	20 24.9	6 48.5	4 30.7
46	15 12.2	9 33.9	20 9.0	7 16.9	4 48.2
47	14 53.1	8 57.3	19 51.6	7 47.2	5 6.5
48	14 33.1	8 18.5	19 32.5	8 19.4	5 25.8
49	14 12.0	7 37.2	19 11.2	8 53.8	5 46.1
50	13 49.8	6 53.1	18 47.5	9 30.7	6 7.5
51	13 26.2	6 5.9	18 21.0	10 10.4	6 30.3
52	13 1.2	5 15.3	17 50.9	10 53.3	6 54.4
53	12 34.6	4 20.6	17 16.6	11 39.7	7 20.2
54	12 6.2	3 21.6	16 37.2	12 30.3	7 47.7
55	11 35.7	2 17.3	15 51.2	13 25.6	8 17.3
56	11 2.8	1 7.2	14 57.1	14 26.4	8 49.2
57	10 27.2	♑29 50.1	13 52.2	15 33.6	9 23.9
58	9 48.3	28 24.9	12 33.2	16 48.6	10 1.8
59	9 5.6	26 49.9	10 54.8	18 12.8	10 43.5
60	8 18.3	25 3.0	8 49.1	19 48.3	11 29.7
61	7 25.2	23 1.4	6 3.3	21 38.2	12 21.7
62	6 24.8	20 40.8	2 15.9	23 46.5	13 21.0
63	5 14.6	17 54.9	♒26 48.0	26 19.6	14 30.1
64	3 50.2	14 32.6	18 26.3	29 28.6	15 53.4
65	2 2.8	10 12.0	4 54.5	♊ 3 35.7	17 39.5
66	♐29 25.6	3 49.3	♑13 11.5	9 44.6	20 15.4
S LAT	5	6	Descendant	8	9

♊ 25° 24′ 39″

5ʰ 40ᵐ 0ˢ MC 85° 0′ 0″

17ʰ 44ᵐ 0ˢ MC 266° 0′ 0″

♐ 26° 19′ 45″

N LAT	11	12	Ascendant	2	3
°	° ′	° ′	° ′	° ′	° ′
0	♑24 6.4	♒23 40.6	♓25 38.5	♈27 59.7	♉28 15.0
5	23 29.3	22 49.6	25 28.2	28 40.7	28 50.0
10	22 51.2	21 55.6	25 16.9	29 24.3	29 26.0
15	22 11.4	20 57.5	25 4.2	♉ 0 11.4	♊ 0 3.7
16	22 3.1	20 45.2	25 1.4	0 21.3	0 11.5
17	21 54.8	20 32.8	24 58.6	0 31.4	0 19.4
18	21 46.4	20 20.1	24 55.7	0 41.8	0 27.5
19	21 37.8	20 7.1	24 52.7	0 52.4	0 35.6
20	21 29.1	19 53.8	24 49.6	1 3.2	0 43.9
21	21 20.3	19 40.3	24 46.4	1 14.3	0 52.3
22	21 11.3	19 26.4	24 43.1	1 25.7	1 0.8
23	21 2.1	19 12.2	24 39.6	1 37.3	1 9.5
24	20 52.8	18 57.6	24 36.1	1 49.3	1 18.4
25	20 43.3	18 42.6	24 32.4	2 1.6	1 27.5
26	20 33.7	18 27.3	24 28.5	2 14.2	1 36.7
27	20 23.8	18 11.5	24 24.5	2 27.3	1 46.2
28	20 13.7	17 55.2	24 20.3	2 40.7	1 55.8
29	20 3.3	17 38.4	24 16.0	2 54.5	2 5.7
30	19 52.8	17 21.2	24 11.4	3 8.8	2 15.8
31	19 41.9	17 3.3	24 6.6	3 23.6	2 26.2
32	19 30.8	16 44.9	24 1.6	3 39.0	2 36.8
33	19 19.4	16 25.8	23 56.3	3 54.8	2 47.8
34	19 7.6	16 6.0	23 50.7	4 11.3	2 59.1
35	18 55.5	15 45.5	23 44.8	4 28.5	3 10.7
36	18 43.1	15 24.1	23 38.6	4 46.3	3 22.6
37	18 30.2	15 2.0	23 32.0	5 4.9	3 35.0
38	18 17.0	14 38.8	23 25.0	5 24.3	3 47.8
39	18 3.2	14 14.7	23 17.5	5 44.6	4 1.0
40	17 49.0	13 49.5	23 9.5	6 5.9	4 14.7
41	17 34.3	13 23.1	23 0.9	6 28.2	4 28.9
42	17 18.9	12 55.5	22 51.7	6 51.6	4 43.7
43	17 3.0	12 26.4	22 41.8	7 16.3	4 59.1
44	16 46.4	11 55.9	22 31.0	7 42.4	5 15.2
45	16 29.0	11 23.6	22 19.3	8 9.9	5 32.0
46	16 10.8	10 49.5	22 6.5	8 39.1	5 49.5
47	15 51.8	10 13.4	21 52.5	9 10.2	6 8.0
48	15 31.8	9 35.1	21 37.1	9 43.2	6 27.3
49	15 10.8	8 54.3	21 19.9	10 18.5	6 47.8
50	14 48.6	8 10.7	21 0.8	10 56.4	7 9.3
51	14 25.1	7 24.0	20 39.4	11 37.1	7 32.2
52	14 0.1	6 33.9	20 15.1	12 21.0	7 56.5
53	13 33.5	5 39.8	19 47.4	13 8.5	8 22.4
54	13 5.1	4 41.3	19 15.4	14 0.3	8 50.1
55	12 34.7	3 37.6	18 38.1	14 56.8	9 19.8
56	12 1.8	2 28.0	17 54.1	15 58.9	9 51.9
57	11 26.2	1 11.5	17 1.2	17 7.6	10 26.8
58	10 47.3	♑29 46.8	15 56.5	18 24.1	11 4.9
59	10 4.6	28 12.3	14 35.6	19 50.1	11 46.8
60	9 17.3	26 25.8	12 51.5	21 27.5	12 33.3
61	8 24.1	24 24.5	10 32.9	23 19.4	13 25.6
62	7 23.6	22 4.1	7 19.9	25 30.1	14 25.2
63	6 13.1	19 18.0	2 34.6	28 6.1	15 34.8
64	4 48.3	15 54.9	♒24 58.9	♊ 1 18.6	16 58.6
65	3 0.1	11 31.9	11 40.0	5 30.9	18 45.8
66	0 20.5	5 0.5	♑17 16.2	11 51.4	21 24.4
S LAT	5	6	Descendant	8	9

♊ 26° 19′ 45″

5ʰ 44ᵐ 0ˢ MC 86° 0′ 0″

17ʰ 48ᵐ 0ˢ MC 267° 0′ 0″

♐ 27° 14′ 50″

N LAT	11	12	Ascendant	2	3
°	° ′	° ′	° ′	° ′	° ′
0	♑25 3.3	♒24 42.5	♓26 43.8	♈29 2.8	♉29 12.8
5	24 26.4	23 52.6	26 36.1	29 45.1	29 48.1
10	23 48.4	22 59.8	26 27.6	♉ 0 30.1	♊ 0 24.4
15	23 8.9	22 3.0	26 18.1	1 18.6	1 2.5
16	23 0.7	21 51.0	26 16.0	1 28.9	1 10.3
17	22 52.4	21 38.8	26 13.9	1 39.3	1 18.3
18	22 44.0	21 26.4	26 11.7	1 50.0	1 26.4
19	22 35.5	21 13.6	26 9.5	2 0.9	1 34.6
20	22 26.8	21 0.7	26 7.1	2 12.1	1 42.9
21	22 18.0	20 47.4	26 4.7	2 23.5	1 51.4
22	22 9.1	20 33.8	26 2.2	2 35.2	2 0.0
23	22 0.0	20 19.8	25 59.6	2 47.3	2 8.8
24	21 50.7	20 5.6	25 57.0	2 59.6	2 17.7
25	21 41.3	19 50.9	25 54.2	3 12.2	2 26.9
26	21 31.7	19 35.8	25 51.3	3 25.3	2 36.2
27	21 21.8	19 20.3	25 48.3	3 38.7	2 45.7
28	21 11.8	19 4.4	25 45.1	3 52.5	2 55.4
29	21 1.5	18 48.0	25 41.9	4 6.7	3 5.3
30	20 50.9	18 31.0	25 38.4	4 21.5	3 15.5
31	20 40.1	18 13.5	25 34.8	4 36.7	3 26.0
32	20 29.1	17 55.4	25 31.1	4 52.5	3 36.7
33	20 17.7	17 36.6	25 27.1	5 8.8	3 47.7
34	20 6.0	17 17.2	25 22.9	5 25.7	3 59.1
35	19 54.0	16 57.1	25 18.5	5 43.4	4 10.8
36	19 41.5	16 36.1	25 13.8	6 1.7	4 22.8
37	19 28.7	16 14.3	25 8.8	6 20.8	4 35.2
38	19 15.5	15 51.6	25 3.5	6 40.7	4 48.1
39	19 1.8	15 27.9	24 57.9	7 1.6	5 1.4
40	18 47.6	15 3.1	24 51.9	7 23.4	5 15.2
41	18 32.9	14 37.2	24 45.4	7 46.3	5 29.5
42	18 17.7	14 9.9	24 38.5	8 10.4	5 44.4
43	18 1.8	13 41.3	24 31.0	8 35.7	5 59.8
44	17 45.2	13 11.2	24 22.9	9 2.4	6 16.0
45	17 27.9	12 39.4	24 14.1	9 30.7	6 32.9
46	17 9.8	12 5.9	24 4.5	10 0.6	6 50.6
47	16 50.8	11 30.3	23 53.9	10 32.4	7 9.1
48	16 30.9	10 52.5	23 42.2	11 6.3	7 28.6
49	16 9.9	10 12.2	23 29.3	11 42.4	7 49.1
50	15 47.7	9 29.2	23 14.9	12 21.2	8 10.8
51	15 24.2	8 43.1	22 58.7	13 2.8	8 33.8
52	14 59.3	7 53.5	22 40.4	13 47.7	8 58.2
53	14 32.8	7 0.0	22 19.4	14 36.3	9 24.2
54	14 4.4	6 2.1	21 55.2	15 29.1	9 52.0
55	13 34.0	4 59.1	21 26.9	16 26.8	10 21.9
56	13 1.2	3 50.1	20 53.4	17 30.2	10 54.2
57	12 25.6	2 34.2	20 13.2	18 40.2	11 29.2
58	11 46.7	1 10.1	19 23.8	19 58.2	12 7.5
59	11 4.0	♑29 36.2	18 21.7	21 25.6	12 49.5
60	10 16.7	27 50.4	17 1.5	23 4.8	13 36.3
61	9 23.5	25 49.6	15 13.8	24 58.6	14 28.8
62	8 22.9	23 29.5	12 41.8	27 11.3	15 28.8
63	7 12.3	20 43.5	8 52.2	29 49.8	16 38.7
64	5 47.2	17 20.0	2 29.9	♊ 3 5.4	18 3.0
65	3 58.4	12 55.4	♒20 18.4	7 22.1	19 51.1
66	1 16.8	6 16.9	♑23 24.9	13 52.5	22 32.0
S LAT	5	6	Descendant	8	9

♊ 27° 14′ 50″

5ʰ 48ᵐ 0ˢ MC 87° 0′ 0″

17h 52m 0s MC 268° 0′ 0″

♐ 28° 9′ 54″

N LAT	11	12	Ascendant	2	3
0°	♑26 0.2	♒25 44.5	♓27 49.2	♉ 0 5.7	♊ 0 10.5
5	25 23.6	24 55.8	27 44.1	0 49.3	0 46.1
10	24 45.9	24 4.3	27 38.4	1 35.6	1 22.7
15	24 6.5	23 8.7	27 32.0	2 25.7	2 1.0
16	23 58.4	22 57.0	27 30.7	2 36.2	2 8.9
17	23 50.2	22 45.1	27 29.2	2 47.0	2 17.0
18	23 41.8	22 32.9	27 27.8	2 58.0	2 25.1
19	23 33.3	22 20.5	27 26.3	3 9.2	2 33.4
20	23 24.7	22 7.8	27 24.7	3 20.7	2 41.8
21	23 16.0	21 54.8	27 23.1	3 32.5	2 50.3
22	23 7.1	21 41.5	27 21.4	3 44.5	2 59.0
23	22 58.0	21 27.9	27 19.7	3 56.9	3 7.8
24	22 48.8	21 13.9	27 17.9	4 9.5	3 16.8
25	22 39.4	20 59.5	27 16.1	4 22.6	3 26.0
26	22 29.8	20 44.8	27 14.2	4 36.0	3 35.4
27	22 20.1	20 29.6	27 12.1	4 49.7	3 45.0
28	22 10.0	20 14.0	27 10.1	5 3.9	3 54.7
29	21 59.8	19 57.9	27 7.9	5 18.6	4 4.7
30	21 49.3	19 41.3	27 5.6	5 33.7	4 15.0
31	21 38.6	19 24.1	27 3.2	5 49.3	4 25.5
32	21 27.5	19 6.3	27 0.6	6 5.5	4 36.3
33	21 16.2	18 48.0	26 58.0	6 22.3	4 47.4
34	21 4.6	18 28.9	26 55.2	6 39.7	4 58.8
35	20 52.6	18 9.1	26 52.2	6 57.8	5 10.6
36	20 40.2	17 48.6	26 49.1	7 16.6	5 22.7
37	20 27.5	17 27.2	26 45.8	7 36.2	5 35.2
38	20 14.3	17 4.9	26 42.3	7 56.7	5 48.1
39	20 0.6	16 41.6	26 38.5	8 18.0	6 1.5
40	19 46.5	16 17.3	26 34.5	8 40.4	6 15.4
41	19 31.9	15 51.8	26 30.2	9 3.9	6 29.8
42	19 16.6	15 25.0	26 25.5	9 28.5	6 44.7
43	19 0.8	14 56.9	26 20.5	9 54.5	7 0.3
44	18 44.3	14 27.3	26 15.1	10 21.8	7 16.5
45	18 27.0	13 56.0	26 9.2	10 50.8	7 33.5
46	18 9.0	13 22.9	26 2.8	11 21.4	7 51.3
47	17 50.1	12 47.9	25 55.7	11 53.9	8 9.9
48	17 30.2	12 10.6	25 47.9	12 28.5	8 29.5
49	17 9.2	11 30.9	25 39.3	13 5.5	8 50.1
50	16 47.1	10 48.5	25 29.6	13 45.0	9 11.9
51	16 23.7	10 3.1	25 18.7	14 27.5	9 35.0
52	15 58.9	9 14.1	25 6.4	15 13.3	9 59.5
53	15 32.4	8 21.3	24 52.4	16 2.9	10 25.6
54	15 4.1	7 24.1	24 36.1	16 56.7	10 53.6
55	14 33.7	6 21.7	24 17.1	17 55.5	11 23.6
56	14 0.9	5 13.5	23 54.6	19 0.1	11 56.0
57	13 25.4	3 58.3	23 27.4	20 11.3	12 31.2
58	12 46.6	2 34.9	22 54.1	21 30.6	13 9.6
59	12 3.9	1 1.8	22 12.1	22 59.5	13 51.8
60	11 16.6	♑29 16.7	21 17.5	24 40.2	14 38.8
61	10 23.4	27 16.5	20 3.8	26 35.6	15 31.5
62	9 22.8	24 57.1	18 18.7	28 50.3	16 31.7
63	8 12.1	22 11.5	15 37.2	♊ 1 30.8	17 41.9
64	6 46.9	18 48.1	10 58.7	4 49.1	19 6.7
65	4 57.7	14 22.6	1 18.2	9 9.2	20 55.4
66	2 14.6	7 39.1	♒ 3 47.9	15 47.4	23 38.0
S LAT	5	6	Descendant	8	9

♊ 28° 9′ 54″

5h 52m 0s MC 88° 0′ 0″

17h 56m 0s MC 269° 0′ 0″

♐ 29° 4′ 57″

N LAT	11	12	Ascendant	2	3
0°	♑26 57.4	♒26 46.7	♓28 54.6	♉ 1 8.4	♊ 1 8.0
5	26 20.9	25 59.2	28 52.0	1 53.3	1 43.9
10	25 43.5	25 9.0	28 49.2	2 41.0	2 20.8
15	25 4.4	24 14.8	28 46.0	3 32.4	2 59.4
16	24 56.3	24 3.3	28 45.3	3 43.3	3 7.4
17	24 48.1	23 51.7	28 44.6	3 54.4	3 15.5
18	24 39.8	23 39.8	28 43.9	4 5.7	3 23.7
19	24 31.4	23 27.6	28 43.1	4 17.2	3 32.0
20	24 22.8	23 15.2	28 42.3	4 29.0	3 40.4
21	24 14.1	23 2.5	28 41.5	4 41.1	3 49.0
22	24 5.3	22 49.5	28 40.7	4 53.5	3 57.8
23	23 56.3	22 36.2	28 39.9	5 6.2	4 6.7
24	23 47.1	22 22.5	28 39.0	5 19.2	4 15.7
25	23 37.8	22 8.5	28 38.0	5 32.5	4 25.0
26	23 28.2	21 54.1	28 37.1	5 46.3	4 34.4
27	23 18.5	21 39.2	28 36.1	6 0.4	4 44.0
28	23 8.5	21 23.9	28 35.0	6 15.0	4 53.9
29	22 58.3	21 8.2	28 33.9	6 30.0	5 3.9
30	22 47.9	20 51.9	28 32.8	6 45.6	5 14.3
31	22 37.2	20 35.1	28 31.6	7 1.6	5 24.8
32	22 26.2	20 17.7	28 30.3	7 18.2	5 35.7
33	22 15.0	19 59.7	28 29.0	7 35.4	5 46.9
34	22 3.4	19 41.1	28 27.6	7 53.2	5 58.3
35	21 51.4	19 21.7	28 26.1	8 11.8	6 10.2
36	21 39.1	19 1.6	28 24.5	8 31.0	6 22.3
37	21 26.4	18 40.6	28 22.9	8 51.1	6 34.9
38	21 13.3	18 18.7	28 21.1	9 12.1	6 47.9
39	20 59.7	17 55.9	28 19.2	9 34.0	7 1.4
40	20 45.7	17 32.0	28 17.2	9 56.9	7 15.3
41	20 31.1	17 7.0	28 15.0	10 20.9	7 29.8
42	20 15.9	16 40.7	28 12.7	10 46.1	7 44.8
43	20 0.1	16 13.1	28 10.2	11 12.6	8 0.4
44	19 43.7	15 44.0	28 7.5	11 40.6	8 16.8
45	19 26.5	15 13.3	28 4.6	12 10.1	8 33.8
46	19 8.5	14 40.7	28 1.3	12 41.4	8 51.7
47	18 49.6	14 6.3	27 57.8	13 14.6	9 10.4
48	18 29.8	13 29.6	27 53.9	13 49.9	9 30.0
49	18 8.9	12 50.6	27 49.5	14 27.7	9 50.8
50	17 46.9	12 8.8	27 44.7	15 8.0	10 12.6
51	17 23.5	11 24.0	27 39.3	15 51.3	10 35.8
52	16 58.8	10 35.8	27 33.1	16 38.0	11 0.4
53	16 32.4	9 43.7	27 26.0	17 28.4	11 26.7
54	16 4.1	8 47.2	27 17.9	18 23.3	11 54.7
55	15 33.8	7 45.6	27 8.3	19 23.1	12 24.9
56	15 1.1	6 38.1	26 57.0	20 28.7	12 57.4
57	14 25.6	5 23.8	26 43.3	21 41.1	13 32.7
58	13 46.9	4 1.3	26 26.5	23 1.5	14 11.2
59	13 4.3	2 29.0	26 5.3	24 31.7	14 53.6
60	12 17.0	0 44.7	25 37.7	26 13.7	15 40.7
61	11 23.9	♑28 45.5	25 0.2	28 10.6	16 33.6
62	10 23.3	26 26.8	24 6.5	♊ 0 26.9	17 34.0
63	9 12.6	23 42.0	22 42.9	3 9.2	18 44.4
64	7 47.3	20 19.1	20 15.5	6 29.5	20 9.5
65	5 57.9	15 53.6	14 47.1	10 52.4	21 58.6
66	3 14.0	9 7.5	♒23 43.5	17 35.8	24 42.3
S LAT	5	6	Descendant	8	9

♊ 29° 4′ 57″

5h 56m 0s MC 89° 0′ 0″

18ʰ 0ᵐ 0ˢ		MC		270° 0′ 0″	
		♑ 0° 0′ 0″			N
11	12	Ascendant	2	3	LAT
° ′	° ′	° ′	° ′	° ′	°
♑27 54.6	♒27 49.1	♈ 0 0.0	♉ 2 10.9	♊ 2 5.4	0
27 18.4	27 2.9	0 0.0	2 57.1	2 41.6	5
26 41.3	26 13.9	0 0.0	3 46.1	3 18.7	10
26 2.4	25 21.0	0 0.0	4 39.0	3 57.6	15
25 54.4	25 9.9	0 0.0	4 50.1	4 5.6	16
25 46.2	24 58.5	0 0.0	5 1.5	4 13.8	17
25 38.0	24 46.9	0 0.0	5 13.1	4 22.0	18
25 29.6	24 35.1	0 0.0	5 24.9	4 30.4	19
25 21.1	24 23.0	0 0.0	5 37.0	4 38.9	20
25 12.4	24 10.6	0 0.0	5 49.4	4 47.6	21
25 3.7	23 57.9	0 0.0	6 2.1	4 56.3	22
24 54.7	23 44.9	0 0.0	6 15.1	5 5.3	23
24 45.6	23 31.5	0 0.0	6 28.5	5 14.4	24
24 36.3	23 17.8	0 0.0	6 42.2	5 23.7	25
24 26.8	23 3.7	0 0.0	6 56.3	5 33.2	26
24 17.1	22 49.2	0 0.0	7 10.8	5 42.9	27
24 7.2	22 34.3	0 0.0	7 25.7	5 52.8	28
23 57.1	22 18.9	0 0.0	7 41.1	6 2.9	29
23 46.7	22 3.0	0 0.0	7 57.0	6 13.3	30
23 36.1	21 46.6	0 0.0	8 13.4	6 23.9	31
23 25.2	21 29.6	0 0.0	8 30.4	6 34.8	32
23 13.9	21 12.0	0 0.0	8 48.0	6 46.1	33
23 2.4	20 53.7	0 0.0	9 6.3	6 57.6	34
22 50.5	20 34.7	0 0.0	9 25.3	7 9.5	35
22 38.3	20 15.0	0 0.0	9 45.0	7 21.7	36
22 25.6	19 54.5	0 0.0	10 5.5	7 34.4	37
22 12.6	19 33.1	0 0.0	10 26.9	7 47.4	38
21 59.1	19 10.7	0 0.0	10 49.3	8 0.9	39
21 45.1	18 47.3	0 0.0	11 12.7	8 14.9	40
21 30.5	18 22.7	0 0.0	11 37.3	8 29.5	41
21 15.4	17 57.0	0 0.0	12 3.0	8 44.6	42
20 59.7	17 29.9	0 0.0	12 30.1	9 0.3	43
20 43.3	17 1.4	0 0.0	12 58.6	9 16.7	44
20 26.2	16 31.2	0 0.0	13 28.8	9 33.8	45
20 8.3	15 59.3	0 0.0	14 0.7	9 51.7	46
19 49.5	15 25.4	0 0.0	14 34.6	10 10.5	47
19 29.7	14 49.4	0 0.0	15 10.6	10 30.3	48
19 8.9	14 11.0	0 0.0	15 49.0	10 51.1	49
18 47.0	13 29.9	0 0.0	16 30.1	11 13.0	50
18 23.7	12 45.9	0 0.0	17 14.1	11 36.3	51
17 59.0	11 58.4	0 0.0	18 1.6	12 1.0	52
17 32.7	11 7.1	0 0.0	18 52.9	12 27.3	53
17 4.5	10 11.4	0 0.0	19 48.6	12 55.5	54
16 34.3	9 10.6	0 0.0	20 49.4	13 25.7	55
16 1.6	8 4.1	0 0.0	21 55.9	13 58.4	56
15 26.2	6 50.6	0 0.0	23 9.4	14 33.8	57
14 47.6	5 29.1	0 0.0	24 30.9	15 12.4	58
14 5.1	3 57.8	0 0.0	26 2.2	15 54.9	59
13 17.9	2 14.6	0 0.0	27 45.4	16 42.1	60
12 24.9	0 16.4	0 0.0	29 43.6	17 35.1	61
11 24.3	♑27 58.8	0 0.0	♊ 2 1.2	18 35.7	62
10 13.7	25 15.1	0 0.0	4 44.9	19 46.3	63
8 48.5	21 53.3	0 0.0	8 6.7	21 11.5	64
6 59.1	17 28.6	0 0.0	12 31.4	23 0.9	65
4 15.0	10 42.5	0 0.0	19 17.5	25 45.0	66
5	6	Descendant	8	9	S LAT
		♋ 0° 0′ 0″			
6ʰ 0ᵐ 0ˢ		MC		90° 0′ 0″	

18ʰ 4ᵐ 0ˢ		MC		271° 0′ 0″	
		♑ 0° 55′ 3″			N
11	12	Ascendant	2	3	LAT
° ′	° ′	° ′	° ′	° ′	°
♑28 52.0	♒28 51.6	♈ 1 5.4	♉ 3 13.3	♊ 3 2.6	0
28 16.1	28 6.7	1 8.0	4 0.8	3 39.1	5
27 39.2	27 19.0	1 10.8	4 51.0	4 16.5	10
27 0.6	26 27.6	1 14.0	5 45.2	4 55.6	15
26 52.6	26 16.7	1 14.7	5 56.7	5 3.7	16
26 44.5	26 5.6	1 15.4	6 8.3	5 11.9	17
26 36.3	25 54.3	1 16.1	6 20.2	5 20.2	18
26 28.0	25 42.8	1 16.9	6 32.4	5 28.6	19
26 19.6	25 31.0	1 17.7	6 44.8	5 37.2	20
26 11.0	25 18.9	1 18.5	6 57.5	5 45.9	21
26 2.2	25 6.5	1 19.3	7 10.5	5 54.7	22
25 53.3	24 53.8	1 20.1	7 23.8	6 3.7	23
25 44.3	24 40.8	1 21.0	7 37.5	6 12.9	24
25 35.0	24 27.5	1 22.0	7 51.5	6 22.2	25
25 25.6	24 13.7	1 22.9	8 5.9	6 31.8	26
25 16.0	23 59.6	1 23.9	8 20.8	6 41.5	27
25 6.1	23 45.0	1 25.0	8 36.1	6 51.5	28
24 56.1	23 30.0	1 26.1	8 51.8	7 1.7	29
24 45.7	23 14.4	1 27.2	9 8.1	7 12.1	30
24 35.2	22 58.4	1 28.4	9 24.9	7 22.8	31
24 24.3	22 41.8	1 29.7	9 42.3	7 33.8	32
24 13.1	22 24.6	1 31.0	10 0.3	7 45.0	33
24 1.7	22 6.8	1 32.4	10 18.9	7 56.6	34
23 49.8	21 48.2	1 33.9	10 38.3	8 8.6	35
23 37.7	21 29.0	1 35.5	10 58.4	8 20.9	36
23 25.1	21 8.9	1 37.1	11 19.4	8 33.6	37
23 12.1	20 47.9	1 38.9	11 41.3	8 46.7	38
22 58.6	20 26.0	1 40.8	12 4.1	9 0.3	39
22 44.7	20 3.1	1 42.8	12 28.0	9 14.3	40
22 30.2	19 39.1	1 45.0	12 53.0	9 28.9	41
22 15.2	19 13.9	1 47.3	13 19.3	9 44.1	42
21 59.6	18 47.4	1 49.8	13 46.9	9 59.9	43
21 43.2	18 19.4	1 52.5	14 16.0	10 16.3	44
21 26.2	17 49.9	1 55.4	14 46.7	10 33.5	45
21 8.3	17 18.6	1 58.7	15 19.3	10 51.5	46
20 49.6	16 45.4	2 2.2	15 53.7	11 10.4	47
20 30.0	16 10.1	2 6.1	16 30.4	11 30.2	48
20 9.2	15 32.3	2 10.5	17 9.4	11 51.1	49
19 47.4	14 52.0	2 15.3	17 51.2	12 13.1	50
19 24.2	14 8.7	2 20.7	18 36.0	12 36.5	51
18 59.6	13 22.0	2 26.9	19 24.2	13 1.2	52
18 33.3	12 31.6	2 34.0	20 16.3	13 27.6	53
18 5.3	11 36.7	2 42.1	21 12.8	13 55.9	54
17 35.1	10 36.9	2 51.7	22 14.4	14 26.2	55
17 2.6	9 31.3	3 3.0	23 21.9	14 58.9	56
16 27.3	8 18.9	3 16.7	24 36.2	15 34.4	57
15 48.8	6 58.5	3 33.5	25 58.7	16 13.1	58
15 6.4	5 28.3	3 54.7	27 31.0	16 55.7	59
14 19.3	3 46.3	4 22.3	29 15.3	17 43.0	60
13 26.4	1 49.4	4 59.8	♊ 1 14.5	18 36.1	61
12 26.0	♑29 33.1	5 53.5	3 33.2	19 36.7	62
11 15.6	26 50.8	7 17.1	6 18.0	20 47.4	63
9 50.5	23 30.5	9 44.5	9 40.9	22 12.7	64
8 1.4	19 7.6	15 12.9	14 6.4	24 2.1	65
5 17.7	12 24.2	♉ 6 16.5	20 52.5	26 46.0	66
5	6	Descendant	8	9	S LAT
		♋ 0° 55′ 3″			
6ʰ 4ᵐ 0ˢ		MC		91° 0′ 0″	

18^h 8^m 0^s		MC		272° 0′ 0″	
		♑ 1° 50′ 6″			
11	**12**	**Ascendant**	**2**	**3**	**N LAT**
♑29 49.5	♒29 54.3	♈ 2 10.8	♉ 4 15.5	♊ 3 59.8	0°
29 13.9	29 10.7	2 15.9	5 4.2	4 36.4	5
28 37.3	28 24.4	2 21.6	5 55.7	5 14.1	10
27 59.0	27 34.3	2 28.0	6 51.3	5 53.5	15
27 51.1	27 23.8	2 29.3	7 3.0	6 1.6	16
27 43.0	27 13.0	2 30.8	7 14.9	6 9.8	17
27 34.9	27 2.0	2 32.2	7 27.1	6 18.2	18
27 26.6	26 50.8	2 33.7	7 39.5	6 26.7	19
27 18.2	26 39.3	2 35.3	7 52.2	6 35.3	20
27 9.7	26 27.5	2 36.9	8 5.2	6 44.0	21
27 1.0	26 15.5	2 38.6	8 18.5	6 52.9	22
26 52.2	26 3.1	2 40.3	8 32.1	7 2.0	23
26 43.2	25 50.5	2 42.1	8 46.1	7 11.2	24
26 34.0	25 37.4	2 43.9	9 0.5	7 20.6	25
26 24.6	25 24.0	2 45.8	9 15.2	7 30.2	26
26 15.0	25 10.3	2 47.9	9 30.4	7 39.9	27
26 5.3	24 56.1	2 49.9	9 46.0	7 50.0	28
25 55.3	24 41.4	2 52.1	10 2.1	8 0.2	29
25 45.0	24 26.3	2 54.4	10 18.7	8 10.7	30
25 34.5	24 10.7	2 56.8	10 35.9	8 21.4	31
25 23.7	23 54.5	2 59.4	10 53.7	8 32.5	32
25 12.6	23 37.7	3 2.0	11 12.0	8 43.8	33
25 1.2	23 20.3	3 4.8	11 31.1	8 55.4	34
24 49.4	23 2.2	3 7.8	11 50.9	9 7.4	35
24 37.3	22 43.4	3 10.9	12 11.4	9 19.8	36
24 24.8	22 23.8	3 14.2	12 32.8	9 32.5	37
24 11.9	22 3.3	3 17.7	12 55.1	9 45.7	38
23 58.5	21 42.0	3 21.5	13 18.4	9 59.4	39
23 44.6	21 19.6	3 25.5	13 42.7	10 13.5	40
23 30.2	20 56.1	3 29.8	14 8.2	10 28.1	41
23 15.3	20 31.5	3 34.5	14 35.0	10 43.4	42
22 59.7	20 5.5	3 39.5	15 3.1	10 59.2	43
22 43.5	19 38.2	3 44.9	15 32.7	11 15.7	44
22 26.5	19 9.2	3 50.8	16 4.0	11 33.0	45
22 8.7	18 38.6	3 57.2	16 37.1	11 51.0	46
21 50.1	18 6.1	4 4.3	17 12.1	12 9.9	47
21 30.5	17 31.5	4 12.1	17 49.4	12 29.8	48
21 9.9	16 54.5	4 20.7	18 29.1	12 50.8	49
20 48.1	16 15.0	4 30.4	19 11.5	13 12.9	50
20 25.0	15 32.5	4 41.3	19 56.9	13 36.3	51
20 0.5	14 46.7	4 53.6	20 45.9	14 1.1	52
19 34.4	13 57.1	5 7.6	21 38.7	14 27.6	53
19 6.4	13 3.3	5 23.9	22 35.9	14 55.9	54
18 36.4	12 4.5	5 42.9	23 38.3	15 26.3	55
18 4.0	10 59.9	6 5.4	24 46.5	15 59.1	56
17 28.8	9 48.7	6 32.6	26 1.7	16 34.6	57
16 50.4	8 29.4	7 5.9	27 25.1	17 13.4	58
16 8.2	7 0.5	7 47.9	28 58.2	17 56.1	59
15 21.2	5 19.8	8 42.5	♊ 0 43.3	18 43.4	60
14 28.5	3 24.4	9 56.2	2 43.5	19 36.6	61
13 28.3	1 9.7	11 41.3	5 2.9	20 37.2	62
12 18.1	♑28 29.2	14 22.8	7 48.5	21 47.9	63
10 53.3	25 10.9	19 1.3	11 11.9	23 13.1	64
9 4.6	20 50.8	28 41.8	15 37.4	25 2.3	65
6 22.0	14 12.6	♉26 12.1	22 20.9	27 45.4	66
5	**6**	**Descendant**	**8**	**9**	**S LAT**
		♋ 1° 50′ 6″			
6^h 8^m 0^s		MC		92° 0′ 0″	

18^h 12^m 0^s		MC		273° 0′ 0″	
		♑ 2° 45′ 10″			
11	**12**	**Ascendant**	**2**	**3**	**N LAT**
♒ 0 47.2	♓ 0 57.2	♈ 3 16.2	♉ 5 17.5	♊ 4 56.7	0°
0 11.9	0 14.9	3 23.9	6 7.4	5 33.6	5
♑29 35.6	♒29 29.9	3 32.4	7 0.2	6 11.6	10
28 57.5	28 41.4	3 41.9	7 57.0	6 51.1	15
28 49.7	28 31.1	3 44.0	8 9.0	6 59.3	16
28 41.7	28 20.7	3 46.1	8 21.2	7 7.6	17
28 33.6	28 10.0	3 48.3	8 33.6	7 16.0	18
28 25.4	27 59.1	3 50.5	8 46.4	7 24.5	19
28 17.1	27 47.9	3 52.9	8 59.3	7 33.2	20
28 8.6	27 36.5	3 55.3	9 12.6	7 42.0	21
28 0.0	27 24.8	3 57.8	9 26.2	7 50.9	22
27 51.2	27 12.7	4 0.4	9 40.2	8 0.0	23
27 42.3	27 0.4	4 3.0	9 54.4	8 9.3	24
27 33.1	26 47.8	4 5.8	10 9.1	8 18.7	25
27 23.8	26 34.7	4 8.7	10 24.2	8 28.3	26
27 14.3	26 21.3	4 11.7	10 39.7	8 38.2	27
27 4.6	26 7.5	4 14.9	10 55.6	8 48.2	28
26 54.7	25 53.3	4 18.1	11 12.0	8 58.5	29
26 44.5	25 38.5	4 21.6	11 29.0	9 9.1	30
26 34.0	25 23.3	4 25.2	11 46.5	9 19.9	31
26 23.3	25 7.5	4 28.9	12 4.6	9 30.9	32
26 12.3	24 51.2	4 32.9	12 23.4	9 42.3	33
26 0.9	24 34.3	4 37.1	12 42.8	9 54.0	34
25 49.2	24 16.6	4 41.5	13 2.9	10 6.0	35
25 37.2	23 58.3	4 46.2	13 23.9	10 18.5	36
25 24.8	23 39.2	4 51.2	13 45.7	10 31.3	37
25 11.9	23 19.3	4 56.5	14 8.4	10 44.5	38
24 58.6	22 58.4	5 2.1	14 32.1	10 58.2	39
24 44.8	22 36.6	5 8.1	14 56.9	11 12.4	40
24 30.5	22 13.7	5 14.6	15 22.8	11 27.1	41
24 15.6	21 49.6	5 21.5	15 50.1	11 42.3	42
24 0.2	21 24.3	5 29.0	16 18.7	11 58.2	43
23 44.0	20 57.6	5 37.1	16 48.8	12 14.8	44
23 27.1	20 29.3	5 45.9	17 20.6	12 32.1	45
23 9.4	19 59.4	5 55.5	17 54.1	12 50.2	46
22 50.9	19 27.6	6 6.1	18 29.7	13 9.2	47
22 31.4	18 53.7	6 17.8	19 7.5	13 29.1	48
22 10.9	18 17.6	6 30.7	19 47.8	13 50.1	49
21 49.2	17 38.8	6 45.1	20 30.8	14 12.3	50
21 26.2	16 57.2	7 1.3	21 16.9	14 35.8	51
21 1.8	16 12.3	7 19.6	22 6.5	15 0.7	52
20 35.8	15 23.7	7 40.6	23 0.0	15 27.2	53
20 8.0	14 30.9	8 4.8	23 57.9	15 55.6	54
19 38.1	13 33.2	8 33.1	25 0.9	16 26.0	55
19 5.8	12 29.8	9 6.6	26 9.9	16 58.8	56
18 30.8	11 19.8	9 46.8	27 25.8	17 34.4	57
17 52.5	10 1.8	10 36.2	28 49.9	18 13.3	58
17 10.5	8 34.4	11 38.3	♊ 0 23.8	18 56.0	59
16 23.7	6 55.2	12 58.5	2 9.6	19 43.3	60
15 31.2	5 1.4	14 46.2	4 10.4	20 36.5	61
14 31.2	2 48.7	17 18.2	6 30.5	21 37.1	62
13 21.3	0 10.2	21 7.8	9 16.5	22 47.7	63
11 57.0	♑26 54.6	27 30.1	12 40.0	24 12.8	64
10 8.9	22 37.9	♉ 9 41.6	17 4.6	26 1.6	65
7 28.0	16 7.5	♊ 6 35.1	23 43.1	28 43.2	66
5	**6**	**Descendant**	**8**	**9**	**S LAT**
		♋ 2° 45′ 10″			
6^h 12^m 0^s		MC		93° 0′ 0″	

18ʰ 16ᵐ 0ˢ MC 274° 0′ 0″

♑ 3° 40′ 15″

N LAT	11	12	Ascendant	2	3
°	° ′	° ′	° ′	° ′	° ′
0	♒ 1 45.0	♓ 2 0.3	♈ 4 21.5	♉ 6 19.4	♊ 5 53.6
5	1 10.0	1 19.3	4 31.8	7 10.4	6 30.7
10	0 34.0	0 35.7	4 43.1	8 4.4	7 8.8
15	♑29 56.3	♒29 48.6	4 55.8	9 2.5	7 48.6
16	29 48.5	29 38.7	4 58.6	9 14.8	7 56.9
17	29 40.6	29 28.6	5 1.4	9 27.2	8 5.2
18	29 32.5	29 18.2	5 4.3	9 39.9	8 13.6
19	29 24.4	29 7.6	5 7.3	9 52.9	8 22.2
20	29 16.1	28 56.8	5 10.4	10 6.2	8 30.9
21	29 7.7	28 45.7	5 13.6	10 19.7	8 39.7
22	28 59.2	28 34.3	5 16.9	10 33.6	8 48.7
23	28 50.5	28 22.7	5 20.4	10 47.8	8 57.9
24	28 41.6	28 10.7	5 23.9	11 2.4	9 7.2
25	28 32.5	27 58.4	5 27.6	11 17.4	9 16.7
26	28 23.3	27 45.8	5 31.5	11 32.7	9 26.3
27	28 13.8	27 32.7	5 35.5	11 48.5	9 36.2
28	28 4.2	27 19.3	5 39.7	12 4.8	9 46.3
29	27 54.3	27 5.5	5 44.0	12 21.6	9 56.7
30	27 44.2	26 51.2	5 48.6	12 38.8	10 7.2
31	27 33.8	26 36.4	5 53.4	12 56.7	10 18.1
32	27 23.2	26 21.0	5 58.4	13 15.1	10 29.2
33	27 12.2	26 5.2	6 3.7	13 34.2	10 40.6
34	27 0.9	25 48.7	6 9.3	13 54.0	10 52.4
35	26 49.3	25 31.5	6 15.2	14 14.5	11 4.5
36	26 37.4	25 13.7	6 21.4	14 35.9	11 16.9
37	26 25.0	24 55.1	6 28.0	14 58.0	11 29.8
38	26 12.2	24 35.7	6 35.0	15 21.2	11 43.0
39	25 59.0	24 15.4	6 42.5	15 45.3	11 56.8
40	25 45.3	23 54.1	6 50.5	16 10.5	12 11.0
41	25 31.1	23 31.8	6 59.1	16 36.9	12 25.7
42	25 16.3	23 8.4	7 8.3	17 4.5	12 41.1
43	25 0.9	22 43.7	7 18.2	17 33.6	12 57.0
44	24 44.8	22 17.6	7 29.0	18 4.1	13 13.6
45	24 28.0	21 50.1	7 40.7	18 36.4	13 31.0
46	24 10.5	21 20.9	7 53.5	19 10.5	13 49.2
47	23 52.0	20 49.8	8 7.5	19 46.6	14 8.2
48	23 32.7	20 16.8	8 22.9	20 24.9	14 28.2
49	23 12.2	19 41.5	8 40.1	21 5.7	14 49.2
50	22 50.7	19 3.6	8 59.2	21 49.3	15 11.4
51	22 27.8	18 22.9	9 20.6	22 36.0	15 34.9
52	22 3.5	17 39.0	9 44.9	23 26.1	15 59.9
53	21 37.6	16 51.5	10 12.6	24 20.2	16 26.5
54	21 9.9	15 59.7	10 44.6	25 18.7	16 54.9
55	20 40.2	15 3.2	11 21.9	26 22.4	17 25.3
56	20 8.1	14 1.1	12 5.9	27 32.0	17 58.2
57	19 33.2	12 52.4	12 58.8	28 48.5	18 33.8
58	18 55.1	11 35.9	14 3.5	♊ 0 13.2	19 12.7
59	18 13.2	10 9.9	15 24.4	1 47.7	19 55.4
60	17 26.7	8 32.5	17 8.5	3 34.2	20 42.7
61	16 34.4	6 40.6	19 27.1	5 35.5	21 35.9
62	15 34.8	4 29.9	22 40.1	7 55.9	22 36.4
63	14 25.2	1 53.9	27 25.4	10 42.0	23 46.9
64	13 1.4	♑28 41.4	♉ 5 1.1	14 5.1	25 11.7
65	11 14.2	24 29.1	18 20.0	18 28.1	26 59.9
66	8 35.6	18 8.6	♊12 43.8	24 59.5	29 39.5
S LAT	5	6	Descendant	8	9

♋ 3° 40′ 15″

6ʰ 16ᵐ 0ˢ MC 94° 0′ 0″

18ʰ 20ᵐ 0ˢ MC 275° 0′ 0″

♑ 4° 35′ 21″

N LAT	11	12	Ascendant	2	3
°	° ′	° ′	° ′	° ′	° ′
0	♒ 2 43.0	♓ 3 3.5	♈ 5 26.8	♉ 7 21.1	♊ 6 50.3
5	2 8.3	2 23.8	5 39.7	8 13.3	7 27.6
10	1 32.6	1 41.7	5 53.8	9 8.4	8 6.0
15	0 55.2	0 56.2	6 9.7	10 7.8	8 46.0
16	0 47.5	0 46.5	6 13.1	10 20.2	8 54.3
17	0 39.6	0 36.7	6 16.6	10 33.0	9 2.6
18	0 31.7	0 26.7	6 20.3	10 45.9	9 11.1
19	0 23.6	0 16.4	6 24.0	10 59.2	9 19.7
20	0 15.4	0 5.9	6 27.9	11 12.7	9 28.5
21	0 7.0	♒29 55.2	6 31.9	11 26.5	9 37.3
22	♑29 58.6	29 44.2	6 36.0	11 40.7	9 46.4
23	29 49.9	29 32.9	6 40.3	11 55.2	9 55.5
24	29 41.1	29 21.3	6 44.7	12 10.1	10 4.9
25	29 32.1	29 9.4	6 49.4	12 25.3	10 14.4
26	29 23.0	28 57.1	6 54.2	12 41.0	10 24.1
27	29 13.6	28 44.5	6 59.2	12 57.1	10 34.1
28	29 4.0	28 31.5	7 4.4	13 13.6	10 44.2
29	28 54.2	28 18.0	7 9.8	13 30.7	10 54.6
30	28 44.1	28 4.2	7 15.5	13 48.3	11 5.2
31	28 33.8	27 49.8	7 21.5	14 6.4	11 16.1
32	28 23.3	27 34.9	7 27.8	14 25.2	11 27.3
33	28 12.4	27 19.5	7 34.3	14 44.6	11 38.7
34	28 1.2	27 3.5	7 41.3	15 4.8	11 50.5
35	27 49.7	26 46.9	7 48.6	15 25.7	12 2.6
36	27 37.8	26 29.5	7 56.4	15 47.3	12 15.1
37	27 25.5	26 11.5	8 4.6	16 9.9	12 28.0
38	27 12.8	25 52.6	8 13.3	16 33.4	12 41.3
39	26 59.7	25 32.9	8 22.7	16 57.9	12 55.1
40	26 46.1	25 12.2	8 32.6	17 23.5	13 9.4
41	26 31.9	24 50.5	8 43.3	17 50.3	13 24.2
42	26 17.2	24 27.7	8 54.8	18 18.3	13 39.6
43	26 1.9	24 3.7	9 7.1	18 47.8	13 55.5
44	25 46.0	23 38.4	9 20.5	19 18.8	14 12.2
45	25 29.3	23 11.5	9 35.1	19 51.5	14 29.6
46	25 11.8	22 43.1	9 51.0	20 26.1	14 47.8
47	24 53.5	22 12.8	10 8.4	21 2.7	15 6.9
48	24 34.2	21 40.6	10 27.5	21 41.5	15 26.9
49	24 13.9	21 6.2	10 48.8	22 22.8	15 48.0
50	23 52.5	20 29.3	11 12.5	23 6.9	16 10.2
51	23 29.7	19 49.6	11 39.0	23 54.1	16 33.8
52	23 5.6	19 6.7	12 9.1	24 44.7	16 58.8
53	22 39.8	18 20.3	12 43.4	25 39.4	17 25.4
54	22 12.3	17 29.7	13 22.8	26 38.4	17 53.8
55	21 42.7	16 34.4	14 8.8	27 42.7	18 24.3
56	21 10.8	15 33.6	15 2.9	28 52.8	18 57.2
57	20 36.1	14 26.4	16 7.8	♊ 0 9.9	19 32.8
58	19 58.2	13 11.4	17 26.8	1 35.1	20 11.7
59	19 16.5	11 47.2	19 5.2	3 10.1	20 54.4
60	18 30.3	10 11.7	21 10.9	4 57.0	21 41.7
61	17 38.3	8 21.8	23 56.7	6 58.6	22 34.8
62	16 39.0	6 13.5	27 44.1	9 19.2	23 35.2
63	15 29.9	3 40.4	♉ 3 12.0	12 5.1	24 45.4
64	14 6.6	0 31.4	11 33.7	15 27.4	26 9.8
65	12 20.5	♑26 24.3	25 5.5	19 48.0	27 57.2
66	9 44.6	20 15.4	♊16 48.5	26 10.7	♋ 0 34.4
S LAT	5	6	Descendant	8	9

♋ 4° 35′ 21″

6ʰ 20ᵐ 0ˢ MC 95° 0′ 0″

$18^h\ 24^m\ 0^s$ MC 276° 0′ 0″

♑ 5° 30′ 29″

11	12	Ascendant	2	3	N LAT
° ′	° ′	° ′	° ′	° ′	°
♒ 3 41.2	♓ 4 6.8	♈ 6 32.1	♉ 8 22.6	♊ 7 46.9	0
3 6.8	3 28.6	6 47.5	9 15.9	8 24.4	5
2 31.4	2 47.9	7 4.5	10 12.2	9 3.0	10
1 54.4	2 3.9	7 23.4	11 12.7	9 43.2	15
1 46.7	1 54.6	7 27.6	11 25.4	9 51.5	16
1 38.9	1 45.1	7 31.8	11 38.4	9 59.9	17
1 31.0	1 35.4	7 36.1	11 51.6	10 8.4	18
1 23.0	1 25.5	7 40.6	12 5.1	10 17.1	19
1 14.9	1 15.4	7 45.3	12 18.9	10 25.9	20
1 6.6	1 5.0	7 50.0	12 33.0	10 34.8	21
0 58.2	0 54.3	7 55.0	12 47.4	10 43.8	22
0 49.6	0 43.4	8 0.1	13 2.2	10 53.1	23
0 40.9	0 32.2	8 5.4	13 17.3	11 2.4	24
0 31.9	0 20.7	8 11.0	13 32.9	11 12.0	25
0 22.8	0 8.8	8 16.7	13 48.8	11 21.8	26
0 13.5	♒29 56.6	8 22.7	14 5.2	11 31.7	27
0 4.0	29 44.0	8 28.9	14 22.1	11 41.9	28
♑29 54.3	29 31.0	8 35.4	14 39.4	11 52.3	29
29 44.3	29 17.5	8 42.3	14 57.3	12 3.0	30
29 34.1	29 3.6	8 49.4	15 15.8	12 13.9	31
29 23.6	28 49.2	8 56.9	15 34.9	12 25.1	32
29 12.8	28 34.3	9 4.8	15 54.6	12 36.6	33
29 1.7	28 18.8	9 13.1	16 15.1	12 48.4	34
28 50.3	28 2.7	9 21.8	16 36.3	13 0.6	35
28 38.5	27 45.9	9 31.1	16 58.3	13 13.1	36
28 26.3	27 28.4	9 40.9	17 21.2	13 26.1	37
28 13.7	27 10.1	9 51.4	17 45.1	13 39.4	38
28 0.6	26 50.9	10 2.5	18 9.9	13 53.2	39
27 47.1	26 30.9	10 14.4	18 35.9	14 7.5	40
27 33.1	26 9.8	10 27.2	19 3.1	14 22.4	41
27 18.5	25 47.7	10 40.8	19 31.5	14 37.8	42
27 3.3	25 24.4	10 55.6	20 1.4	14 53.8	43
26 47.4	24 59.7	11 11.6	20 32.8	15 10.5	44
26 30.8	24 33.7	11 28.9	21 6.0	15 28.0	45
26 13.5	24 6.0	11 47.8	21 40.9	15 46.2	46
25 55.3	23 36.6	12 8.6	22 18.0	16 5.3	47
25 36.1	23 5.3	12 31.4	22 57.2	16 25.4	48
25 16.0	22 31.8	12 56.7	23 39.0	16 46.5	49
24 54.6	21 55.8	13 24.8	24 23.6	17 8.7	50
24 32.0	21 17.1	13 56.3	25 11.2	17 32.3	51
24 8.0	20 35.4	14 32.0	26 2.4	17 57.3	52
23 42.5	19 50.1	15 12.5	26 57.5	18 24.0	53
23 15.1	19 0.8	15 59.2	27 57.1	18 52.4	54
22 45.6	18 6.9	16 53.4	29 1.8	19 22.9	55
22 13.9	17 7.5	17 57.1	♊ 0 12.4	19 55.8	56
21 39.4	16 1.8	19 13.2	1 30.0	20 31.4	57
21 1.8	14 48.6	20 45.5	2 55.6	21 10.3	58
20 20.3	13 26.2	22 39.8	4 31.0	21 52.9	59
19 34.3	11 52.7	25 4.6	6 18.1	22 40.2	60
18 42.7	10 5.1	28 13.6	8 19.9	23 33.1	61
17 43.8	7 59.5	♉ 2 28.6	10 40.4	24 33.4	62
16 35.2	5 29.5	8 26.9	13 25.9	25 43.4	63
15 12.7	2 24.5	17 12.6	16 46.9	27 7.3	64
13 27.7	♑28 23.5	♊ 0 26.3	21 4.5	28 53.7	65
10 55.1	22 27.6	19 45.2	27 17.2	♋ 1 27.9	66
5	6	Descendant	8	9	S LAT

♋ 5° 30′ 29″

$6^h\ 24^m\ 0^s$ MC 96° 0′ 0″

$18^h\ 28^m\ 0^s$ MC 277° 0′ 0″

♑ 6° 25′ 38″

11	12	Ascendant	2	3	N LAT
° ′	° ′	° ′	° ′	° ′	°
♒ 4 39.5	♓ 5 10.3	♈ 7 37.4	♉ 9 23.9	♊ 8 43.3	0
4 5.5	4 33.5	7 55.3	10 18.3	9 21.1	5
3 30.4	3 54.3	8 15.0	11 15.7	9 59.9	10
2 53.7	3 11.9	8 37.1	12 17.4	10 40.2	15
2 46.1	3 2.9	8 41.9	12 30.4	10 48.6	16
2 38.4	2 53.7	8 46.8	12 43.6	10 57.0	17
2 30.6	2 44.4	8 51.9	12 57.1	11 5.6	18
2 22.6	2 34.8	8 57.1	13 10.8	11 14.3	19
2 14.5	2 25.0	9 2.5	13 24.8	11 23.1	20
2 6.3	2 15.0	9 8.1	13 39.2	11 32.0	21
1 58.0	2 4.7	9 13.8	13 53.9	11 41.1	22
1 49.5	1 54.2	9 19.8	14 8.9	11 50.4	23
1 40.8	1 43.4	9 26.0	14 24.3	11 59.8	24
1 32.0	1 32.2	9 32.4	14 40.1	12 9.4	25
1 23.0	1 20.8	9 39.1	14 56.3	12 19.2	26
1 13.8	1 9.0	9 46.1	15 13.0	12 29.2	27
1 4.3	0 56.8	9 53.3	15 30.1	12 39.4	28
0 54.7	0 44.3	10 0.9	15 47.7	12 49.9	29
0 44.8	0 31.3	10 8.8	16 5.9	13 0.6	30
0 34.6	0 17.8	10 17.1	16 24.7	13 11.5	31
0 24.2	0 3.9	10 25.8	16 44.1	13 22.8	32
0 13.5	♒29 49.5	10 35.0	17 4.1	13 34.3	33
0 2.5	29 34.5	10 44.6	17 24.9	13 46.2	34
♑29 51.1	29 18.9	10 54.8	17 46.4	13 58.4	35
29 39.4	29 2.6	11 5.6	18 8.8	14 10.9	36
29 27.3	28 45.7	11 17.0	18 32.0	14 23.9	37
29 14.8	28 28.0	11 29.1	18 56.2	14 37.3	38
29 1.9	28 9.5	11 42.0	19 21.4	14 51.1	39
28 48.4	27 50.0	11 55.8	19 47.8	15 5.5	40
28 34.5	27 29.7	12 10.6	20 15.3	15 20.3	41
28 20.0	27 8.2	12 26.5	20 44.1	15 35.8	42
28 4.9	26 45.6	12 43.6	21 14.4	15 51.9	43
27 49.2	26 21.7	13 2.0	21 46.2	16 8.6	44
27 32.7	25 56.4	13 22.1	22 19.7	16 26.0	45
27 15.5	25 29.6	13 44.0	22 55.1	16 44.3	46
26 57.4	25 1.1	14 8.0	23 32.5	17 3.4	47
26 38.4	24 30.7	14 34.3	24 12.2	17 23.5	48
26 18.3	23 58.1	15 3.5	24 54.4	17 44.7	49
25 57.2	23 23.2	15 36.0	25 39.4	18 7.0	50
25 34.7	22 45.7	16 12.3	26 27.5	18 30.6	51
25 10.9	22 5.1	16 53.3	27 19.1	18 55.6	52
24 45.5	21 21.0	17 39.9	28 14.6	19 22.2	53
24 18.3	20 33.0	18 33.3	29 14.6	19 50.7	54
23 49.0	19 40.5	19 35.3	♊ 0 19.8	20 21.2	55
23 17.5	18 42.7	20 48.0	1 30.8	20 54.1	56
22 43.2	17 38.7	22 14.5	2 48.7	21 29.7	57
22 5.8	16 27.2	23 58.9	4 14.7	22 8.4	58
21 24.7	15 6.9	26 7.3	5 50.3	22 51.0	59
20 39.0	13 35.6	28 48.8	7 37.6	23 38.2	60
19 47.7	11 50.6	♉ 2 16.9	9 39.4	24 31.0	61
18 49.3	9 47.8	6 53.0	11 59.6	25 31.0	62
17 41.2	7 21.3	13 11.2	14 44.4	26 40.7	63
16 19.6	4 20.8	22 4.6	18 3.8	28 4.0	64
14 36.0	0 26.4	♊ 4 45.1	22 17.8	29 49.3	65
12 6.8	♑24 44.6	22 1.3	28 19.6	♋ 2 20.3	66
5	6	Descendant	8	9	S LAT

♋ 6° 25′ 38″

$6^h\ 28^m\ 0^s$ MC 97° 0′ 0″

18ʰ 32ᵐ 0ˢ — MC 278° 0′ 0″ — ♑ 7° 20′ 50″

11	12	Ascendant	2	3	N LAT
° ′	° ′	° ′	° ′	° ′	°
♒ 5 38.0	♓ 6 14.0	♈ 8 42.5	♉10 25.0	♊ 9 39.7	**0**
5 4.3	5 38.6	9 3.0	11 20.5	10 17.7	**5**
4 29.6	5 0.9	9 25.5	12 19.0	10 56.6	**10**
3 53.2	4 20.1	9 50.7	13 21.8	11 37.1	**15**
3 45.7	4 11.4	9 56.2	13 35.0	11 45.5	**16**
3 38.1	4 2.6	10 1.8	13 48.5	11 54.0	**17**
3 30.3	3 53.6	10 7.6	14 2.2	12 2.6	**18**
3 22.4	3 44.4	10 13.5	14 16.2	12 11.3	**19**
3 14.4	3 35.0	10 19.7	14 30.5	12 20.2	**20**
3 6.3	3 25.3	10 26.0	14 45.0	12 29.1	**21**
2 58.0	3 15.4	10 32.6	15 0.0	12 38.3	**22**
2 49.6	3 5.3	10 39.4	15 15.2	12 47.6	**23**
2 41.0	2 54.9	10 46.4	15 30.9	12 57.0	**24**
2 32.3	2 44.1	10 53.7	15 46.9	13 6.7	**25**
2 23.3	2 33.1	11 1.3	16 3.4	13 16.5	**26**
2 14.2	2 21.7	11 9.3	16 20.3	13 26.5	**27**
2 4.8	2 10.0	11 17.5	16 37.8	13 36.8	**28**
1 55.3	1 57.9	11 26.1	16 55.7	13 47.2	**29**
1 45.5	1 45.4	11 35.2	17 14.1	13 58.0	**30**
1 35.4	1 32.4	11 44.6	17 33.2	14 9.0	**31**
1 25.1	1 19.0	11 54.5	17 52.9	14 20.2	**32**
1 14.5	1 5.1	12 4.9	18 13.2	14 31.8	**33**
1 3.5	0 50.6	12 15.9	18 34.3	14 43.7	**34**
0 52.3	0 35.5	12 27.5	18 56.1	14 55.9	**35**
0 40.6	0 19.8	12 39.7	19 18.8	15 8.5	**36**
0 28.6	0 3.5	12 52.7	19 42.3	15 21.5	**37**
0 16.2	♒29 46.4	13 6.4	20 6.8	15 34.9	**38**
0 3.4	29 28.5	13 21.1	20 32.4	15 48.8	**39**
♑29 50.1	29 9.7	13 36.8	20 59.0	16 3.2	**40**
29 36.3	28 50.0	13 53.5	21 26.9	16 18.1	**41**
29 21.9	28 29.3	14 11.5	21 56.1	16 33.6	**42**
29 6.9	28 7.5	14 30.9	22 26.7	16 49.6	**43**
28 51.3	27 44.4	14 51.9	22 58.9	17 6.4	**44**
28 34.9	27 19.9	15 14.6	23 32.8	17 23.9	**45**
28 17.8	26 54.0	15 39.4	24 8.5	17 42.2	**46**
27 59.9	26 26.3	16 6.5	24 46.3	18 1.3	**47**
27 41.0	25 56.9	16 36.3	25 26.4	18 21.4	**48**
27 21.1	25 25.4	17 9.2	26 9.0	18 42.6	**49**
27 0.1	24 51.5	17 45.8	26 54.3	19 4.9	**50**
26 37.8	24 15.1	18 26.8	27 42.8	19 28.5	**51**
26 14.1	23 35.7	19 12.8	28 34.8	19 53.5	**52**
25 48.9	22 53.0	20 5.2	29 30.7	20 20.2	**53**
25 21.9	22 6.4	21 5.1	♊ 0 31.1	20 48.6	**54**
24 52.8	21 15.4	22 14.3	1 36.6	21 19.1	**55**
24 21.5	20 19.2	23 35.3	2 48.0	21 51.9	**56**
23 47.5	19 16.9	25 11.2	4 6.2	22 27.5	**57**
23 10.4	18 7.4	27 6.5	5 32.4	23 6.2	**58**
22 29.5	16 49.2	29 27.3	7 8.2	23 48.7	**59**
21 44.2	15 20.3	♉ 2 22.8	8 55.5	24 35.8	**60**
20 53.3	13 38.0	6 6.2	10 57.1	25 28.4	**61**
19 55.4	11 38.5	10 57.6	13 16.9	26 28.2	**62**
18 48.0	9 15.8	17 27.1	16 0.7	27 37.4	**63**
17 27.2	6 20.2	26 17.3	19 18.3	29 0.1	**64**
15 45.2	2 33.0	♊ 8 18.0	23 28.1	♋ 0 44.1	**65**
13 19.8	♑27 6.0	23 51.1	29 18.4	3 11.6	**66**
5	6	Descendant	8	9	S LAT

♋ 7° 20′ 50″ — 6ʰ 32ᵐ 0ˢ — MC 98° 0′ 0″

18ʰ 36ᵐ 0ˢ — MC 279° 0′ 0″ — ♑ 8° 16′ 4″

11	12	Ascendant	2	3	N LAT
° ′	° ′	° ′	° ′	° ′	°
♒ 6 36.6	♓ 7 17.7	♈ 9 47.7	♉11 26.0	♊10 35.9	**0**
6 3.3	6 43.8	10 10.7	12 22.5	11 14.1	**5**
5 29.0	6 7.7	10 35.9	13 22.1	11 53.2	**10**
4 53.0	5 28.5	11 4.2	14 26.0	12 33.9	**15**
4 45.5	5 20.2	11 10.3	14 39.4	12 42.3	**16**
4 37.9	5 11.7	11 16.6	14 53.1	12 50.8	**17**
4 30.3	5 3.1	11 23.1	15 7.0	12 59.5	**18**
4 22.5	4 54.2	11 29.8	15 21.2	13 8.2	**19**
4 14.6	4 45.2	11 36.6	15 35.8	13 17.1	**20**
4 6.5	4 35.9	11 43.8	15 50.6	13 26.1	**21**
3 58.3	4 26.4	11 51.1	16 5.7	13 35.3	**22**
3 50.0	4 16.6	11 58.7	16 21.3	13 44.6	**23**
3 41.5	4 6.6	12 6.6	16 37.2	13 54.1	**24**
3 32.8	3 56.3	12 14.8	16 53.5	14 3.7	**25**
3 23.9	3 45.7	12 23.4	17 10.2	14 13.6	**26**
3 14.9	3 34.8	12 32.2	17 27.4	14 23.6	**27**
3 5.6	3 23.5	12 41.5	17 45.0	14 33.9	**28**
2 56.1	3 11.8	12 51.1	18 3.2	14 44.4	**29**
2 46.4	2 59.8	13 1.2	18 21.9	14 55.2	**30**
2 36.4	2 47.3	13 11.8	18 41.3	15 6.2	**31**
2 26.2	2 34.4	13 22.9	19 1.2	15 17.5	**32**
2 15.7	2 21.0	13 34.6	19 21.8	15 29.1	**33**
2 4.8	2 7.1	13 46.8	19 43.2	15 41.0	**34**
1 53.7	1 52.6	13 59.8	20 5.3	15 53.3	**35**
1 42.2	1 37.5	14 13.5	20 28.3	16 5.9	**36**
1 30.3	1 21.7	14 28.0	20 52.1	16 18.9	**37**
1 18.0	1 5.3	14 43.3	21 16.9	16 32.4	**38**
1 5.2	0 48.0	14 59.7	21 42.8	16 46.3	**39**
0 52.0	0 29.9	15 17.2	22 9.8	17 0.7	**40**
0 38.3	0 11.0	15 35.9	22 37.9	17 15.6	**41**
0 24.0	♒29 51.0	15 56.0	23 7.5	17 31.1	**42**
0 9.2	29 29.9	16 17.6	23 38.4	17 47.2	**43**
♑29 53.7	29 7.6	16 41.0	24 10.9	18 4.0	**44**
29 37.5	28 44.0	17 6.3	24 45.1	18 21.5	**45**
29 20.5	28 19.0	17 33.9	25 21.2	18 39.8	**46**
29 2.7	27 52.3	18 4.0	25 59.4	18 59.0	**47**
28 44.0	27 23.8	18 37.1	26 39.8	19 19.1	**48**
28 24.2	26 53.4	19 13.7	27 22.7	19 40.2	**49**
28 3.4	26 20.7	19 54.2	28 8.4	20 2.6	**50**
27 41.3	25 45.4	20 39.5	28 57.3	20 26.2	**51**
27 17.8	25 7.3	21 30.5	29 49.6	20 51.2	**52**
26 52.7	24 26.0	22 28.2	♊ 0 45.9	21 17.8	**53**
26 25.9	23 40.9	23 34.1	1 46.6	21 46.2	**54**
25 57.1	22 51.4	24 50.1	2 52.4	22 16.7	**55**
25 26.0	21 57.0	26 18.7	4 4.0	22 49.5	**56**
24 52.3	20 56.6	28 3.2	5 22.4	23 25.0	**57**
24 15.4	19 49.2	♉ 0 8.0	6 48.8	24 3.6	**58**
23 34.9	18 33.3	2 39.5	8 24.6	24 46.0	**59**
22 50.0	17 6.9	5 46.5	10 11.9	25 32.9	**60**
21 59.5	15 27.6	9 41.7	12 13.2	26 25.3	**61**
21 2.1	13 31.4	14 43.4	14 32.3	27 24.8	**62**
19 55.5	11 12.9	21 17.3	17 15.0	28 33.5	**63**
18 35.7	8 22.6	29 57.1	20 30.3	29 55.5	**64**
16 55.3	4 43.2	♊11 16.7	24 35.7	♋ 1 38.2	**65**
14 33.8	♑29 31.5	25 23.2	♋ 0 14.1	4 2.0	**66**
5	6	Descendant	8	9	S LAT

♋ 8° 16′ 4″ — 6ʰ 36ᵐ 0ˢ — MC 99° 0′ 0″

18ʰ 40ᵐ 0ˢ MC 280° 0′ 0″

♑ 9° 11′ 21″

11	12	Ascendant	2	3	N LAT
° ′	° ′	° ′	° ′	° ′	°
♒ 7 35.4	♓ 8 21.7	♈10 52.7	♉12 26.7	♊11 32.1	0
7 2.5	7 49.2	11 18.2	13 24.3	12 10.4	5
6 28.5	7 14.6	11 46.2	14 24.9	12 49.7	10
5 52.9	6 37.1	12 17.6	15 29.9	13 30.5	15
5 45.5	6 29.2	12 24.3	15 43.5	13 39.0	16
5 38.0	6 21.1	12 31.3	15 57.4	13 47.5	17
5 30.4	6 12.8	12 38.5	16 11.6	13 56.2	18
5 22.7	6 4.3	12 45.8	16 26.0	14 4.9	19
5 14.9	5 55.6	12 53.5	16 40.7	14 13.9	20
5 6.9	5 46.7	13 1.3	16 55.8	14 22.9	21
4 58.8	5 37.6	13 9.5	17 11.2	14 32.1	22
4 50.6	5 28.3	13 17.9	17 26.9	14 41.4	23
4 42.1	5 18.7	13 26.7	17 43.1	14 50.9	24
4 33.5	5 8.8	13 35.7	17 59.6	15 0.6	25
4 24.8	4 58.6	13 45.2	18 16.6	15 10.5	26
4 15.8	4 48.1	13 55.0	18 34.0	15 20.6	27
4 6.6	4 37.3	14 5.2	18 51.9	15 30.9	28
3 57.2	4 26.1	14 15.9	19 10.3	15 41.5	29
3 47.6	4 14.6	14 27.0	19 29.3	15 52.2	30
3 37.7	4 2.6	14 38.7	19 48.9	16 3.3	31
3 27.6	3 50.2	14 51.0	20 9.1	16 14.6	32
3 17.2	3 37.4	15 3.9	20 30.0	16 26.2	33
3 6.4	3 24.0	15 17.4	20 51.6	16 38.2	34
2 55.4	3 10.1	15 31.7	21 14.0	16 50.5	35
2 43.9	2 55.6	15 46.8	21 37.3	17 3.1	36
2 32.2	2 40.4	16 2.8	22 1.4	17 16.2	37
2 20.0	2 24.6	16 19.8	22 26.5	17 29.6	38
2 7.3	2 8.0	16 37.8	22 52.6	17 43.5	39
1 54.3	1 50.7	16 57.1	23 19.9	17 58.0	40
1 40.7	1 32.4	17 17.7	23 48.4	18 12.9	41
1 26.5	1 13.2	17 39.8	24 18.2	18 28.4	42
1 11.8	0 52.9	18 3.6	24 49.5	18 44.5	43
0 56.4	0 31.5	18 29.3	25 22.3	19 1.3	44
0 40.4	0 8.7	18 57.1	25 56.8	19 18.8	45
0 23.5	♒29 44.6	19 27.4	26 33.2	19 37.2	46
0 5.9	29 18.9	20 0.4	27 11.7	19 56.3	47
♑29 47.3	28 51.5	20 36.7	27 52.4	20 16.5	48
29 27.7	28 22.2	21 16.7	28 35.7	20 37.6	49
29 7.0	27 50.6	22 1.1	29 21.7	20 59.9	50
28 45.1	27 16.7	22 50.5	♊ 0 10.9	21 23.5	51
28 21.8	26 39.9	23 46.0	1 3.5	21 48.5	52
27 57.0	26 0.0	24 48.8	2 0.0	22 15.1	53
27 30.4	25 16.4	26 0.3	3 1.0	22 43.5	54
27 1.8	24 28.7	27 22.5	4 7.0	23 14.0	55
26 31.0	23 36.1	28 57.9	5 18.8	23 46.7	56
25 57.5	22 37.7	♉ 0 50.0	6 37.4	24 22.1	57
25 21.0	21 32.4	3 3.1	8 3.8	25 0.7	58
24 40.8	20 19.0	5 43.6	9 39.6	25 42.9	59
23 56.3	18 55.4	8 59.8	11 26.7	26 29.6	60
23 6.3	17 19.2	13 3.7	13 27.6	27 21.8	61
22 9.5	15 26.7	18 11.4	15 45.9	28 20.9	62
21 3.6	13 12.6	24 44.8	18 27.2	29 29.2	63
19 45.0	10 28.0	♊ 3 9.9	21 40.0	♋ 0 50.3	64
18 6.3	6 56.9	13 49.4	25 40.7	2 31.5	65
15 48.9	2 0.6	26 42.6	♋ 1 7.2	4 51.5	66
5	6	Descendant	8	9	S LAT

♋ 9° 11′ 21″

6ʰ 40ᵐ 0ˢ MC 100° 0′ 0″

18ʰ 44ᵐ 0ˢ MC 281° 0′ 0″

♑ 10° 6′ 42″

11	12	Ascendant	2	3	N LAT
° ′	° ′	° ′	° ′	° ′	°
♒ 8 34.4	♓ 9 25.7	♈11 57.7	♉13 27.3	♊12 28.1	0
8 1.9	8 54.7	12 25.7	14 25.9	13 6.6	5
7 28.3	8 21.7	12 56.4	15 27.5	13 46.0	10
6 53.1	7 45.9	13 30.8	16 33.5	14 27.0	15
6 45.8	7 38.3	13 38.2	16 47.4	14 35.5	16
6 38.4	7 30.6	13 45.8	17 1.5	14 44.1	17
6 30.8	7 22.7	13 53.7	17 15.8	14 52.7	18
6 23.2	7 14.6	14 1.8	17 30.5	15 1.5	19
6 15.5	7 6.3	14 10.1	17 45.4	15 10.5	20
6 7.6	6 57.8	14 18.8	18 0.7	15 19.6	21
5 59.6	6 49.1	14 27.7	18 16.3	15 28.8	22
5 51.4	6 40.2	14 36.9	18 32.3	15 38.1	23
5 43.0	6 31.0	14 46.5	18 48.7	15 47.7	24
5 34.5	6 21.5	14 56.4	19 5.4	15 57.4	25
5 25.9	6 11.8	15 6.7	19 22.6	16 7.3	26
5 17.0	6 1.8	15 17.5	19 40.3	16 17.4	27
5 7.9	5 51.4	15 28.7	19 58.4	16 27.7	28
4 58.6	5 40.7	15 40.3	20 17.1	16 38.3	29
4 49.1	5 29.7	15 52.5	20 36.3	16 49.1	30
4 39.3	5 18.2	16 5.3	20 56.1	17 0.2	31
4 29.2	5 6.4	16 18.7	21 16.6	17 11.5	32
4 18.9	4 54.1	16 32.8	21 37.8	17 23.2	33
4 8.3	4 41.3	16 47.6	21 59.6	17 35.1	34
3 57.3	4 27.9	17 3.2	22 22.3	17 47.4	35
3 46.0	4 14.0	17 19.7	22 45.8	18 0.1	36
3 34.4	3 59.5	17 37.1	23 10.2	18 13.2	37
3 22.3	3 44.4	17 55.7	23 35.5	18 26.7	38
3 9.8	3 28.5	18 15.4	24 2.0	18 40.6	39
2 56.8	3 11.8	18 36.4	24 29.5	18 55.0	40
2 43.3	2 54.3	18 58.8	24 58.3	19 10.0	41
2 29.3	2 35.9	19 22.9	25 28.4	19 25.5	42
2 14.7	2 16.5	19 48.8	25 59.9	19 41.6	43
1 59.5	1 55.9	20 16.7	26 33.0	19 58.4	44
1 43.6	1 34.1	20 46.9	27 7.8	20 16.0	45
1 26.9	1 10.9	21 19.8	27 44.5	20 34.3	46
1 9.4	0 46.3	21 55.7	28 23.3	20 53.5	47
0 51.0	0 19.9	22 35.0	29 4.3	21 13.6	48
0 31.6	♒29 51.7	23 18.3	29 47.8	21 34.8	49
0 11.1	29 21.4	24 6.2	♊ 0 34.1	21 57.1	50
♑29 49.4	28 48.8	24 59.6	1 23.6	22 20.6	51
29 26.3	28 13.4	25 59.3	2 16.4	22 45.6	52
29 1.6	27 35.0	27 6.8	3 13.2	23 12.2	53
28 35.3	26 53.1	28 23.4	4 14.4	23 40.5	54
28 7.0	26 7.1	29 51.2	5 20.6	24 10.9	55
27 36.4	25 16.4	♉ 1 32.8	6 32.6	24 43.6	56
27 3.2	24 20.1	3 31.5	7 51.2	25 18.9	57
26 27.1	23 17.2	5 51.8	9 17.6	25 57.3	58
25 47.3	22 6.3	8 39.6	10 53.3	26 39.4	59
25 3.2	20 45.6	12 3.0	12 40.0	27 25.9	60
24 13.7	19 12.7	16 12.7	14 40.4	28 17.8	61
23 17.6	17 24.2	21 23.2	16 57.8	29 16.5	62
22 12.5	15 14.8	27 52.4	19 37.5	♋ 0 24.2	63
20 55.0	12 36.3	♊ 6 0.3	22 47.7	1 44.5	64
19 18.2	9 13.8	16 2.0	26 43.2	3 24.1	65
17 5.0	4 33.2	27 52.8	♋ 1 58.0	5 40.2	66
5	6	Descendant	8	9	S LAT

♋ 10° 6′ 42″

6ʰ 44ᵐ 0ˢ MC 101° 0′ 0″

18h 48m 0s — MC — 282° 0′ 0″

♑ 11° 2′ 6″

N LAT	11	12	Ascendant	2	3
°	° ′	° ′	° ′	° ′	° ′
0	♒ 9 33.6	♓10 29.9	♈13 2.6	♉14 27.7	♊13 24.0
5	9 1.4	10 0.4	13 33.1	15 27.3	14 2.7
10	8 28.2	9 29.0	14 6.5	16 29.9	14 42.2
15	7 53.4	8 54.9	14 43.8	17 36.9	15 23.4
16	7 46.2	8 47.7	14 51.9	17 50.9	15 31.9
17	7 38.9	8 40.3	15 0.2	18 5.2	15 40.5
18	7 31.5	8 32.8	15 8.7	18 19.8	15 49.2
19	7 23.9	8 25.1	15 17.5	18 34.7	15 58.0
20	7 16.3	8 17.2	15 26.6	18 49.8	16 7.0
21	7 8.5	8 9.1	15 35.9	19 5.3	16 16.1
22	7 0.5	8 0.8	15 45.6	19 21.1	16 25.3
23	6 52.4	7 52.3	15 55.7	19 37.3	16 34.7
24	6 44.2	7 43.5	16 6.1	19 53.9	16 44.3
25	6 35.8	7 34.5	16 16.8	20 10.9	16 54.0
26	6 27.2	7 25.3	16 28.0	20 28.3	17 3.9
27	6 18.4	7 15.7	16 39.7	20 46.2	17 14.1
28	6 9.4	7 5.8	16 51.8	21 4.5	17 24.4
29	6 0.2	6 55.6	17 4.5	21 23.4	17 35.0
30	5 50.8	6 45.1	17 17.7	21 42.9	17 45.8
31	5 41.1	6 34.2	17 31.6	22 3.0	17 56.9
32	5 31.2	6 22.9	17 46.1	22 23.7	18 8.3
33	5 21.0	6 11.1	18 1.3	22 45.0	18 19.9
34	5 10.4	5 58.9	18 17.3	23 7.2	18 31.9
35	4 59.6	5 46.2	18 34.2	23 30.1	18 44.2
36	4 48.4	5 32.9	18 52.1	23 53.8	18 56.9
37	4 36.8	5 19.1	19 11.0	24 18.4	19 10.0
38	4 24.9	5 4.6	19 31.0	24 44.1	19 23.5
39	4 12.5	4 49.4	19 52.3	25 10.7	19 37.5
40	3 59.7	4 33.5	20 15.0	25 38.5	19 51.9
41	3 46.3	4 16.8	20 39.2	26 7.6	20 6.9
42	3 32.5	3 59.2	21 5.2	26 37.9	20 22.4
43	3 18.0	3 40.6	21 33.1	27 9.7	20 38.5
44	3 2.9	3 20.9	22 3.2	27 43.1	20 55.4
45	2 47.2	3 0.0	22 35.7	28 18.2	21 12.9
46	2 30.6	2 37.9	23 11.1	28 55.1	21 31.2
47	2 13.3	2 14.3	23 49.6	29 34.1	21 50.4
48	1 55.1	1 49.1	24 31.8	♊ 0 15.4	22 10.5
49	1 35.8	1 22.1	25 18.2	0 59.2	22 31.6
50	1 15.5	0 53.0	26 9.5	1 45.8	22 53.9
51	0 54.0	0 21.7	27 6.6	2 35.4	23 17.5
52	0 31.1	♒29 47.8	28 10.3	3 28.5	23 42.4
53	0 6.7	29 11.0	29 22.0	4 25.5	24 9.0
54	♑29 40.6	28 30.7	♉ 0 43.4	5 26.8	24 37.3
55	29 12.6	27 46.6	2 16.3	6 33.1	25 7.6
56	28 42.3	26 57.9	4 3.3	7 45.2	25 40.1
57	28 9.5	26 3.8	6 7.8	9 3.8	26 15.4
58	27 33.6	25 3.4	8 34.0	10 30.1	26 53.7
59	26 54.3	23 55.2	11 27.7	12 5.6	27 35.6
60	26 10.6	22 37.6	14 56.4	13 51.9	28 21.8
61	25 21.8	21 8.3	19 9.7	15 51.6	29 13.4
62	24 26.3	19 23.9	24 20.1	18 8.0	♋ 0 11.7
63	23 22.1	17 19.5	♊ 0 42.6	20 45.9	1 18.8
64	22 5.8	14 47.5	8 32.3	23 53.3	2 38.1
65	20 31.0	11 34.0	17 58.9	27 43.6	4 16.0
66	18 21.9	7 9.0	28 56.0	♋ 2 46.9	6 28.3
S LAT	5	6	Descendant	8	9

♋ 11° 2′ 6″

6h 48m 0s — MC — 102° 0′ 0″

18h 52m 0s — MC — 283° 0′ 0″

♑ 11° 57′ 33″

N LAT	11	12	Ascendant	2	3
°	° ′	° ′	° ′	° ′	° ′
0	♒10 32.9	♓11 34.2	♈14 7.5	♉15 28.0	♊14 19.9
5	10 1.2	11 6.3	14 40.4	16 28.4	14 58.7
10	9 28.4	10 36.5	15 16.4	17 32.0	15 38.4
15	8 54.0	10 4.1	15 56.7	18 39.9	16 19.6
16	8 46.9	9 57.3	16 5.4	18 54.2	16 28.2
17	8 39.6	9 50.3	16 14.3	19 8.7	16 36.8
18	8 32.3	9 43.1	16 23.6	19 23.5	16 45.5
19	8 24.8	9 35.8	16 33.0	19 38.5	16 54.4
20	8 17.3	9 28.3	16 42.8	19 53.9	17 3.3
21	8 9.6	9 20.6	16 52.9	20 9.6	17 12.4
22	8 1.7	9 12.8	17 3.4	20 25.6	17 21.7
23	7 53.7	9 4.7	17 14.2	20 42.0	17 31.1
24	7 45.6	8 56.3	17 25.4	20 58.8	17 40.7
25	7 37.3	8 47.8	17 37.0	21 16.0	17 50.5
26	7 28.8	8 39.0	17 49.0	21 33.6	18 0.4
27	7 20.1	8 29.9	18 1.6	21 51.7	18 10.6
28	7 11.2	8 20.5	18 14.7	22 10.3	18 20.9
29	7 2.1	8 10.8	18 28.3	22 29.4	18 31.5
30	6 52.8	8 0.8	18 42.5	22 49.1	18 42.4
31	6 43.2	7 50.4	18 57.4	23 9.4	18 53.5
32	6 33.4	7 39.7	19 13.0	23 30.3	19 4.9
33	6 23.3	7 28.5	19 29.4	23 51.9	19 16.5
34	6 12.9	7 16.9	19 46.6	24 14.2	19 28.5
35	6 2.1	7 4.8	20 4.8	24 37.4	19 40.9
36	5 51.1	6 52.2	20 23.9	25 1.3	19 53.6
37	5 39.6	6 39.0	20 44.2	25 26.2	20 6.7
38	5 27.8	6 25.2	21 5.7	25 52.1	20 20.2
39	5 15.6	6 10.8	21 28.5	26 19.0	20 34.1
40	5 2.8	5 55.6	21 52.8	26 47.0	20 48.6
41	4 49.6	5 39.7	22 18.8	27 16.3	21 3.6
42	4 35.9	5 22.9	22 46.6	27 46.9	21 19.1
43	4 21.6	5 5.2	23 16.5	28 18.9	21 35.2
44	4 6.7	4 46.4	23 48.7	28 52.5	21 52.0
45	3 51.1	4 26.6	24 23.4	29 27.9	22 9.6
46	3 34.7	4 5.4	25 1.2	♊ 0 5.1	22 27.9
47	3 17.6	3 42.9	25 42.3	0 44.3	22 47.0
48	2 59.5	3 18.9	26 27.2	1 25.8	23 7.1
49	2 40.5	2 53.1	27 16.5	2 9.8	23 28.3
50	2 20.4	2 25.4	28 11.0	2 56.6	23 50.5
51	1 59.1	1 55.5	29 11.4	3 46.4	24 14.1
52	1 36.4	1 23.1	♉ 0 18.8	4 39.7	24 39.0
53	1 12.3	0 47.9	1 34.5	5 36.8	25 5.4
54	0 46.4	0 9.4	3 0.1	6 38.2	25 33.7
55	0 18.7	♒29 27.2	4 37.5	7 44.7	26 3.9
56	♑29 48.7	28 40.6	6 29.3	8 56.7	26 36.4
57	29 16.2	27 48.9	8 38.7	10 15.3	27 11.5
58	28 40.8	26 51.0	11 9.8	11 41.5	27 49.6
59	28 1.8	25 45.7	14 8.1	13 16.6	28 31.3
60	27 18.7	24 31.4	17 40.4	15 2.5	29 17.3
61	26 30.4	23 5.7	21 55.3	17 1.4	♋ 0 8.6
62	25 35.6	21 25.7	27 3.7	19 16.5	1 6.4
63	24 32.3	19 26.7	♊ 3 17.7	21 52.6	2 12.9
64	23 17.3	17 1.4	10 48.8	24 57.0	3 31.2
65	21 44.5	13 57.2	19 43.2	28 42.0	5 7.4
66	19 39.7	9 47.8	29 53.8	♋ 3 34.1	7 15.7
S LAT	5	6	Descendant	8	9

♋ 11° 57′ 33″

6h 52m 0s — MC — 103° 0′ 0″

18ʰ 56ᵐ 0ˢ		MC		284° 0′ 0″	
		♑ 12° 53′ 5″			N
11	12	Ascendant	2	3	LAT
° ′	° ′	° ′	° ′	° ′	°
♒11 32.4	♓12 38.6	♈15 12.2	♉16 28.0	♊15 15.6	0
11 1.1	12 12.2	15 47.5	17 29.4	15 54.6	5
10 28.8	11 44.1	16 26.2	18 33.9	16 34.4	10
9 54.8	11 13.5	17 9.4	19 42.8	17 15.8	15
9 47.8	11 7.0	17 18.7	19 57.2	17 24.3	16
9 40.6	11 0.4	17 28.3	20 11.9	17 32.9	17
9 33.4	10 53.6	17 38.2	20 26.8	17 41.7	18
9 26.0	10 46.7	17 48.4	20 42.1	17 50.6	19
9 18.5	10 39.6	17 58.8	20 57.7	17 59.6	20
9 10.9	10 32.4	18 9.7	21 13.5	18 8.7	21
9 3.2	10 24.9	18 20.9	21 29.8	18 18.0	22
8 55.3	10 17.3	18 32.4	21 46.4	18 27.4	23
8 47.2	10 9.4	18 44.4	22 3.3	18 37.0	24
8 39.0	10 1.3	18 56.8	22 20.7	18 46.8	25
8 30.6	9 53.0	19 9.7	22 38.6	18 56.8	26
8 22.0	9 44.4	19 23.2	22 56.9	19 6.9	27
8 13.2	9 35.5	19 37.1	23 15.6	19 17.3	28
8 4.3	9 26.3	19 51.7	23 35.0	19 27.9	29
7 55.0	9 16.8	20 6.9	23 54.9	19 38.8	30
7 45.6	9 7.0	20 22.9	24 15.4	19 49.9	31
7 35.9	8 56.8	20 39.5	24 36.5	20 1.3	32
7 25.9	8 46.2	20 57.0	24 58.3	20 13.0	33
7 15.6	8 35.2	21 15.4	25 20.9	20 25.0	34
7 5.0	8 23.7	21 34.8	25 44.2	20 37.3	35
6 54.0	8 11.8	21 55.3	26 8.4	20 50.0	36
6 42.7	7 59.3	22 16.9	26 33.5	21 3.1	37
6 31.0	7 46.2	22 39.8	26 59.6	21 16.7	38
6 18.9	7 32.5	23 4.1	27 26.7	21 30.6	39
6 6.3	7 18.2	23 30.0	27 55.0	21 45.1	40
5 53.3	7 3.1	23 57.6	28 24.5	22 0.1	41
5 39.7	6 47.1	24 27.2	28 55.3	22 15.6	42
5 25.6	6 30.3	24 58.9	29 27.5	22 31.7	43
5 10.8	6 12.5	25 33.1	♊ 0 1.4	22 48.5	44
4 55.4	5 53.6	26 10.0	0 36.9	23 6.1	45
4 39.2	5 33.6	26 50.0	1 14.3	23 24.4	46
4 22.2	5 12.2	27 33.5	1 53.8	23 43.5	47
4 4.3	4 49.3	28 21.0	2 35.5	24 3.6	48
3 45.5	4 24.8	29 13.1	3 19.7	24 24.7	49
3 25.6	3 58.5	♉ 0 10.5	4 6.6	24 46.9	50
3 4.5	3 30.0	1 14.0	4 56.6	25 10.4	51
2 42.1	2 59.2	2 24.8	5 50.0	25 35.3	52
2 18.2	2 25.7	3 44.1	6 47.2	26 1.7	53
1 52.7	1 49.1	5 13.4	7 48.7	26 29.8	54
1 25.2	1 8.9	6 54.8	8 55.2	27 0.0	55
0 55.6	0 24.5	8 50.7	10 7.2	27 32.3	56
0 23.4	♒29 35.2	11 4.3	11 25.7	28 7.3	57
♑29 48.4	28 40.0	13 39.4	12 51.6	28 45.3	58
29 9.9	27 37.7	16 41.0	14 26.4	29 26.8	59
28 27.3	26 26.8	20 15.5	16 11.7	♋ 0 12.5	60
27 39.6	25 5.1	24 30.5	18 9.7	1 3.4	61
26 45.6	23 29.7	29 35.3	20 23.6	2 0.7	62
25 43.3	21 36.2	♊ 5 39.7	22 57.7	3 6.5	63
24 29.6	19 17.9	12 52.6	25 58.9	4 23.7	64
22 58.9	16 23.4	21 17.4	29 38.6	5 58.2	65
20 58.2	12 29.4	♋ 0 47.3	♋ 4 19.9	8 2.7	66
5	6	Descendant	8	9	S
		♋ 12° 53′ 5″			LAT
6ʰ 56ᵐ 0ˢ		MC		104° 0′ 0″	

19ʰ 0ᵐ 0ˢ		MC		285° 0′ 0″	
		♑ 13° 48′ 41″			N
11	12	Ascendant	2	3	LAT
° ′	° ′	° ′	° ′	° ′	°
♒12 32.1	♓13 43.2	♈16 16.8	♉17 27.9	♊16 11.3	0
12 1.2	13 18.3	16 54.5	18 30.2	16 50.4	5
11 29.3	12 51.8	17 35.8	19 35.5	17 30.3	10
10 55.8	12 23.1	18 21.9	20 45.3	18 11.8	15
10 48.9	12 17.0	18 31.9	20 59.9	18 20.3	16
10 41.8	12 10.7	18 42.1	21 14.8	18 29.0	17
10 34.7	12 4.4	18 52.6	21 29.9	18 37.8	18
10 27.4	11 57.8	19 3.5	21 45.4	18 46.7	19
10 20.0	11 51.2	19 14.6	22 1.1	18 55.7	20
10 12.5	11 44.3	19 26.2	22 17.2	19 4.8	21
10 4.8	11 37.3	19 38.1	22 33.6	19 14.1	22
9 57.0	11 30.1	19 50.4	22 50.4	19 23.6	23
9 49.1	11 22.7	20 3.2	23 7.6	19 33.2	24
9 41.0	11 15.0	20 16.4	23 25.1	19 43.0	25
9 32.7	11 7.2	20 30.1	23 43.2	19 53.0	26
9 24.2	10 59.1	20 44.4	24 1.6	20 3.1	27
9 15.5	10 50.7	20 59.3	24 20.6	20 13.5	28
9 6.7	10 42.1	21 14.8	24 40.2	20 24.2	29
8 57.6	10 33.1	21 31.0	25 0.2	20 35.0	30
8 48.2	10 23.8	21 47.9	25 20.9	20 46.2	31
8 38.6	10 14.2	22 5.6	25 42.3	20 57.6	32
8 28.7	10 4.2	22 24.2	26 4.3	21 9.3	33
8 18.6	9 53.9	22 43.7	26 27.1	21 21.3	34
8 8.1	9 43.0	23 4.3	26 50.6	21 33.6	35
7 57.3	9 31.7	23 26.0	27 15.0	21 46.4	36
7 46.1	9 20.0	23 48.9	27 40.3	21 59.5	37
7 34.5	9 7.6	24 13.2	28 6.6	22 13.0	38
7 22.6	8 54.7	24 38.9	28 33.9	22 27.0	39
7 10.2	8 41.1	25 6.3	29 2.4	22 41.4	40
6 57.3	8 26.9	25 35.6	29 32.1	22 56.4	41
6 43.8	8 11.8	26 6.8	♊ 0 3.1	23 11.9	42
6 29.8	7 55.9	26 40.4	0 35.6	23 28.0	43
6 15.2	7 39.1	27 16.4	1 9.6	23 44.8	44
6 0.0	7 21.3	27 55.3	1 45.3	24 2.3	45
5 44.0	7 2.3	28 37.4	2 22.9	24 20.6	46
5 27.2	6 42.1	29 23.2	3 2.5	24 39.7	47
5 9.5	6 20.5	♉ 0 13.1	3 44.4	24 59.8	48
4 50.9	5 57.3	1 7.8	4 28.8	25 20.9	49
4 31.3	5 32.3	2 7.9	5 15.8	25 43.0	50
4 10.4	5 5.4	3 14.4	6 6.0	26 6.5	51
3 48.3	4 36.2	4 28.2	6 59.5	26 31.3	52
3 24.6	4 4.5	5 50.7	7 56.7	26 57.6	53
2 59.4	3 29.7	7 23.5	8 58.3	27 25.7	54
2 32.2	2 51.6	9 8.3	10 4.7	27 55.7	55
2 2.9	2 9.5	11 7.7	11 16.6	28 28.0	56
1 31.2	1 22.7	13 24.6	12 34.9	29 2.8	57
0 56.5	0 30.3	16 2.8	14 0.6	29 40.6	58
0 18.5	♒29 31.2	19 6.7	15 34.9	♋ 0 21.8	59
♑29 36.5	28 23.8	22 42.2	17 19.6	1 7.3	60
28 49.4	27 6.2	26 56.1	19 16.6	1 57.8	61
27 56.2	25 35.6	♊ 1 56.1	21 29.1	2 54.6	62
26 54.9	23 48.0	7 50.4	24 1.2	3 59.7	63
25 42.6	21 37.0	14 45.4	26 59.1	5 15.8	64
24 14.0	18 52.3	22 43.4	♋ 0 33.4	6 48.4	65
22 17.4	15 13.6	♋ 1 37.5	5 4.4	8 49.2	66
5	6	Descendant	8	9	S
		♋ 13° 48′ 41″			LAT
7ʰ 0ᵐ 0ˢ		MC		105° 0′ 0″	

19ʰ 4ᵐ 0ˢ		MC		286° 0′ 0″	
		♑ 14° 44′ 21″			
N LAT	11	12	Ascendant	2	3
°	° ′	° ′	° ′	° ′	° ′
0	♒13 32.0	♓14 47.8	♈17 21.4	♉18 27.6	♊17 6.9
5	13 1.5	14 24.6	18 1.4	19 30.7	17 46.1
10	12 30.1	13 59.7	18 45.3	20 36.9	18 26.1
15	11 57.1	13 32.8	19 34.3	21 47.6	19 7.7
16	11 50.2	13 27.1	19 44.8	22 2.4	19 16.3
17	11 43.2	13 21.2	19 55.7	22 17.4	19 24.9
18	11 36.2	13 15.2	20 6.8	22 32.8	19 33.7
19	11 29.0	13 9.1	20 18.3	22 48.4	19 42.6
20	11 21.7	13 2.9	20 30.2	23 4.3	19 51.7
21	11 14.3	12 56.5	20 42.4	23 20.5	20 0.8
22	11 6.8	12 49.9	20 55.0	23 37.1	20 10.1
23	10 59.1	12 43.1	21 8.1	23 54.1	20 19.6
24	10 51.2	12 36.2	21 21.6	24 11.4	20 29.2
25	10 43.2	12 29.0	21 35.6	24 29.2	20 39.0
26	10 35.0	12 21.6	21 50.2	24 47.4	20 49.0
27	10 26.7	12 14.0	22 5.3	25 6.1	20 59.2
28	10 18.1	12 6.2	22 21.0	25 25.3	21 9.6
29	10 9.3	11 58.1	22 37.4	25 45.0	21 20.3
30	10 0.4	11 49.7	22 54.6	26 5.2	21 31.2
31	9 51.1	11 41.0	23 12.4	26 26.1	21 42.3
32	9 41.6	11 31.9	23 31.2	26 47.6	21 53.7
33	9 31.9	11 22.6	23 50.8	27 9.9	22 5.4
34	9 21.9	11 12.8	24 11.5	27 32.8	22 17.4
35	9 11.5	11 2.6	24 33.2	27 56.5	22 29.8
36	9 0.8	10 52.1	24 56.1	28 21.1	22 42.5
37	8 49.8	10 41.0	25 20.2	28 46.6	22 55.6
38	8 38.4	10 29.4	25 45.8	29 13.1	23 9.1
39	8 26.6	10 17.2	26 13.0	29 40.6	23 23.1
40	8 14.3	10 4.5	26 41.9	♊ 0 9.3	23 37.5
41	8 1.5	9 51.1	27 12.6	0 39.1	23 52.5
42	7 48.3	9 36.9	27 45.5	1 10.3	24 8.0
43	7 34.5	9 22.0	28 20.7	1 43.0	24 24.1
44	7 20.0	9 6.2	28 58.6	2 17.2	24 40.9
45	7 5.0	8 49.4	29 39.4	2 53.1	24 58.4
46	6 49.2	8 31.6	♉ 0 23.5	3 30.8	25 16.7
47	6 32.6	8 12.5	1 11.4	4 10.6	25 35.8
48	6 15.1	7 52.2	2 3.6	4 52.7	25 55.8
49	5 56.7	7 30.3	3 0.6	5 37.1	26 16.8
50	5 37.3	7 6.8	4 3.3	6 24.3	26 39.0
51	5 16.7	6 41.5	5 12.4	7 14.5	27 2.3
52	4 54.8	6 14.0	6 29.0	8 8.1	27 27.1
53	4 31.5	5 44.0	7 54.4	9 5.4	27 53.3
54	4 6.5	5 11.3	9 30.1	10 7.0	28 21.3
55	3 39.7	4 35.3	11 18.0	11 13.3	28 51.2
56	3 10.8	3 55.6	13 20.3	12 25.1	29 23.4
57	2 39.4	3 11.4	15 39.9	13 43.2	29 58.0
58	2 5.2	2 21.9	18 20.3	15 8.5	♋ 0 35.6
59	1 27.7	1 26.1	21 25.7	16 42.3	1 16.6
60	0 46.2	0 22.5	25 1.2	18 26.2	2 1.7
61	♑29 59.9	♒29 9.2	29 12.9	20 22.3	2 51.8
62	29 7.5	27 43.6	♊ 4 7.3	22 33.3	3 48.1
63	28 7.2	26 1.9	9 51.1	25 3.2	4 52.4
64	26 56.4	23 58.5	16 29.1	27 57.8	6 7.4
65	25 29.9	21 23.9	24 2.5	♋ 1 26.8	7 38.2
66	23 37.3	18 0.4	♋ 2 24.8	5 47.8	9 35.3
S LAT	5	6	Descendant	8	9
		♋ 14° 44′ 21″			
7ʰ 4ᵐ 0ˢ		MC		106° 0′ 0″	

19ʰ 8ᵐ 0ˢ		MC		287° 0′ 0″	
		♑ 15° 40′ 7″			
N LAT	11	12	Ascendant	2	3
°	° ′	° ′	° ′	° ′	° ′
0	♒14 32.0	♓15 52.5	♈18 25.8	♉19 27.1	♊18 2.4
5	14 2.0	15 30.9	19 8.2	20 31.1	18 41.7
10	13 31.1	15 7.8	19 54.6	21 38.1	19 21.8
15	12 58.5	14 42.6	20 46.4	22 49.6	20 3.5
16	12 51.7	14 37.3	20 57.5	23 4.6	20 12.1
17	12 44.9	14 31.9	21 9.0	23 19.8	20 20.8
18	12 37.9	14 26.3	21 20.8	23 35.3	20 29.6
19	12 30.9	14 20.6	21 32.9	23 51.1	20 38.5
20	12 23.7	14 14.8	21 45.4	24 7.2	20 47.6
21	12 16.4	14 8.8	21 58.3	24 23.6	20 56.7
22	12 8.9	14 2.6	22 11.7	24 40.3	21 6.1
23	12 1.3	13 56.3	22 25.4	24 57.5	21 15.5
24	11 53.6	13 49.8	22 39.7	25 15.0	21 25.2
25	11 45.7	13 43.2	22 54.5	25 32.9	21 35.0
26	11 37.6	13 36.3	23 9.8	25 51.3	21 45.0
27	11 29.4	13 29.2	23 25.8	26 10.2	21 55.2
28	11 20.9	13 21.9	23 42.4	26 29.5	22 5.6
29	11 12.3	13 14.3	23 59.7	26 49.4	22 16.3
30	11 3.4	13 6.5	24 17.7	27 9.8	22 27.1
31	10 54.3	12 58.3	24 36.5	27 30.9	22 38.3
32	10 45.0	12 49.9	24 56.3	27 52.6	22 49.7
33	10 35.3	12 41.2	25 16.9	28 15.0	23 1.4
34	10 25.4	12 32.1	25 38.6	28 38.1	23 13.4
35	10 15.2	12 22.6	26 1.5	29 2.0	23 25.8
36	10 4.7	12 12.7	26 25.5	29 26.8	23 38.5
37	9 53.8	12 2.3	26 50.9	29 52.5	23 51.6
38	9 42.5	11 51.5	27 17.8	♊ 0 19.1	24 5.1
39	9 30.9	11 40.2	27 46.3	0 46.8	24 19.1
40	9 18.8	11 28.2	28 16.5	1 15.6	24 33.5
41	9 6.2	11 15.7	28 48.8	1 45.7	24 48.5
42	8 53.1	11 2.5	29 23.2	2 17.0	25 4.0
43	8 39.4	10 48.5	♉ 0 0.0	2 49.8	25 20.1
44	8 25.2	10 33.7	0 39.6	3 24.2	25 36.8
45	8 10.3	10 18.0	1 22.2	4 0.3	25 54.3
46	7 54.7	10 1.3	2 8.2	4 38.1	26 12.5
47	7 38.3	9 43.5	2 58.1	5 18.1	26 31.6
48	7 21.1	9 24.4	3 52.4	6 0.2	26 51.6
49	7 3.0	9 4.0	4 51.6	6 44.8	27 12.6
50	6 43.8	8 42.0	5 56.6	7 32.1	27 34.6
51	6 23.4	8 18.2	7 8.1	8 22.3	27 58.0
52	6 1.8	7 52.5	8 27.2	9 15.9	28 22.6
53	5 38.8	7 24.4	9 55.2	10 13.2	28 48.8
54	5 14.1	6 53.7	11 33.4	11 14.7	29 16.7
55	4 47.6	6 20.0	13 23.8	12 21.0	29 46.5
56	4 19.1	5 42.7	15 28.5	13 32.6	♋ 0 18.4
57	3 48.2	5 1.2	17 50.2	14 50.4	0 52.9
58	3 14.4	4 14.8	20 32.1	16 15.3	1 30.3
59	2 37.5	3 22.3	23 38.2	17 48.5	2 11.0
60	1 56.6	2 22.6	27 12.9	19 31.7	2 55.8
61	1 10.9	1 13.7	♊ 1 21.6	21 26.6	3 45.5
62	0 19.4	♒29 53.4	6 9.9	23 36.1	4 41.2
63	♑29 20.2	28 18.0	11 43.1	26 3.8	5 44.7
64	28 10.8	26 22.4	18 5.0	28 55.1	6 58.6
65	26 46.6	23 58.0	25 15.8	♋ 2 18.7	8 27.6
66	24 57.9	20 49.5	♋ 3 9.9	6 30.3	10 21.1
S LAT	5	6	Descendant	8	9
		♋ 15° 40′ 7″			
7ʰ 8ᵐ 0ˢ		MC		107° 0′ 0″	

19h 12m 0s		MC ♑ 16° 35′ 58″		288° 0′ 0″

N LAT	11	12	Ascendant	2	3
°	° ′	° ′	° ′	° ′	° ′
0	♒15 32.3	♓16 57.4	♈19 30.1	♉20 26.4	♊18 57.9
5	15 2.8	16 37.3	20 14.8	21 31.2	19 37.3
10	14 32.2	16 15.9	21 3.7	22 39.1	20 17.5
15	14 0.2	15 52.6	21 58.3	23 51.4	20 59.2
16	13 53.5	15 47.7	22 10.0	24 6.5	21 7.8
17	13 46.8	15 42.7	22 22.1	24 21.9	21 16.5
18	13 39.9	15 37.5	22 34.5	24 37.5	21 25.3
19	13 32.9	15 32.2	22 47.2	24 53.5	21 34.3
20	13 25.9	15 26.8	23 0.4	25 9.7	21 43.3
21	13 18.7	15 21.3	23 14.0	25 26.3	21 52.5
22	13 11.3	15 15.6	23 28.0	25 43.2	22 1.9
23	13 3.8	15 9.7	23 42.5	26 0.5	22 11.3
24	12 56.2	15 3.7	23 57.5	26 18.2	22 21.0
25	12 48.4	14 57.5	24 13.0	26 36.3	22 30.8
26	12 40.5	14 51.2	24 29.1	26 54.9	22 40.8
27	12 32.4	14 44.6	24 45.9	27 13.9	22 51.0
28	12 24.0	14 37.8	25 3.3	27 33.4	23 1.5
29	12 15.5	14 30.8	25 21.5	27 53.4	23 12.1
30	12 6.8	14 23.5	25 40.4	28 14.1	23 23.0
31	11 57.8	14 16.0	26 0.1	28 35.3	23 34.1
32	11 48.6	14 8.1	26 20.8	28 57.1	23 45.5
33	11 39.1	14 0.0	26 42.5	29 19.7	23 57.3
34	11 29.3	13 51.6	27 5.2	29 43.0	24 9.3
35	11 19.2	13 42.8	27 29.1	♊ 0 7.1	24 21.6
36	11 8.9	13 33.6	27 54.3	0 32.0	24 34.3
37	10 58.1	13 24.0	28 20.8	0 57.8	24 47.4
38	10 47.0	13 13.9	28 48.9	1 24.6	25 1.0
39	10 35.5	13 3.4	29 18.7	1 52.5	25 14.9
40	10 23.5	12 52.3	29 50.3	2 21.5	25 29.3
41	10 11.1	12 40.7	♉ 0 23.9	2 51.7	25 44.3
42	9 58.2	12 28.4	0 59.8	3 23.2	25 59.8
43	9 44.8	12 15.4	1 38.2	3 56.1	26 15.8
44	9 30.7	12 1.7	2 19.4	4 30.6	26 32.6
45	9 16.0	11 47.1	3 3.7	5 6.8	26 50.0
46	9 0.6	11 31.6	3 51.5	5 44.8	27 8.2
47	8 44.5	11 15.0	4 43.2	6 24.8	27 27.2
48	8 27.5	10 57.3	5 39.4	7 7.1	27 47.2
49	8 9.6	10 38.3	6 40.7	7 51.7	28 8.1
50	7 50.6	10 17.8	7 47.8	8 39.1	28 30.1
51	7 30.6	9 55.7	9 1.4	9 29.4	28 53.4
52	7 9.2	9 31.7	10 22.8	10 23.0	29 17.9
53	6 46.5	9 5.6	11 52.9	11 20.2	29 44.0
54	6 22.2	8 37.0	13 33.4	12 21.7	♋ 0 11.8
55	5 56.1	8 5.5	15 25.9	13 27.7	0 41.5
56	5 27.9	7 30.8	17 32.5	14 39.1	1 13.3
57	4 57.4	6 52.1	19 55.7	15 56.6	1 47.6
58	4 24.2	6 8.8	22 38.6	17 21.0	2 24.7
59	3 47.7	5 19.9	25 44.6	18 53.7	3 5.2
60	3 7.5	4 24.1	29 17.9	20 36.0	3 49.6
61	2 22.6	3 19.9	♊ 3 23.0	22 29.8	4 38.8
62	1 31.9	2 5.0	8 4.8	24 37.6	5 34.0
63	0 33.9	0 36.1	13 27.6	27 3.0	6 36.7
64	♑29 26.0	♒28 48.5	19 34.2	29 51.0	7 49.4
65	28 3.9	26 34.5	26 24.4	♋ 3 9.4	9 16.5
66	26 19.0	23 40.8	♋ 3 53.1	7 11.9	11 6.6
S LAT	5	6	Descendant	8	9

7h 12m 0s		MC ♋ 16° 35′ 58″		108° 0′ 0″

19h 16m 0s		MC ♑ 17° 31′ 54″		289° 0′ 0″

N LAT	11	12	Ascendant	2	3
°	° ′	° ′	° ′	° ′	° ′
0	♒16 32.7	♓18 2.3	♈20 34.3	♉21 25.6	♊19 53.3
5	16 3.6	17 43.9	21 21.3	22 31.2	20 32.8
10	15 33.6	17 24.2	22 12.7	23 39.8	21 13.0
15	15 2.1	17 2.8	23 9.9	24 52.9	21 54.9
16	14 55.5	16 58.2	23 22.2	25 8.2	22 3.5
17	14 48.9	16 53.6	23 34.9	25 23.7	22 12.2
18	14 42.1	16 48.8	23 47.9	25 39.5	22 21.0
19	14 35.3	16 44.0	24 1.3	25 55.6	22 29.9
20	14 28.3	16 39.0	24 15.1	26 12.0	22 39.0
21	14 21.2	16 33.9	24 29.3	26 28.7	22 48.2
22	14 14.0	16 28.7	24 44.0	26 45.8	22 57.5
23	14 6.6	16 23.3	24 59.2	27 3.3	23 7.0
24	13 59.1	16 17.8	25 14.9	27 21.1	23 16.7
25	13 51.4	16 12.1	25 31.2	27 39.4	23 26.5
26	13 43.6	16 6.2	25 48.1	27 58.1	23 36.5
27	13 35.6	16 0.2	26 5.6	28 17.3	23 46.8
28	13 27.4	15 53.9	26 23.8	28 36.9	23 57.2
29	13 19.0	15 47.4	26 42.8	28 57.1	24 7.8
30	13 10.4	15 40.7	27 2.6	29 17.9	24 18.7
31	13 1.5	15 33.8	27 23.2	29 39.3	24 29.9
32	12 52.5	15 26.6	27 44.8	♊ 0 1.3	24 41.3
33	12 43.1	15 19.1	28 7.4	0 24.0	24 53.0
34	12 33.5	15 11.4	28 31.2	0 47.4	25 5.0
35	12 23.6	15 3.2	28 56.1	1 11.7	25 17.3
36	12 13.3	14 54.8	29 22.4	1 36.8	25 30.1
37	12 2.7	14 45.9	29 50.0	2 2.7	25 43.1
38	11 51.8	14 36.7	♉ 0 19.3	2 29.7	25 56.6
39	11 40.4	14 27.0	0 50.3	2 57.7	26 10.6
40	11 28.7	14 16.8	1 23.2	3 26.8	26 25.0
41	11 16.4	14 6.0	1 58.1	3 57.1	26 39.9
42	11 3.7	13 54.7	2 35.4	4 28.8	26 55.4
43	10 50.4	13 42.7	3 15.3	5 1.9	27 11.4
44	10 36.6	13 30.1	3 58.0	5 36.5	27 28.1
45	10 22.1	13 16.6	4 43.8	6 12.8	27 45.5
46	10 6.9	13 2.3	5 33.3	6 50.9	28 3.7
47	9 51.0	12 47.0	6 26.7	7 31.0	28 22.7
48	9 34.2	12 30.6	7 24.7	8 13.3	28 42.6
49	9 16.6	12 13.1	8 27.8	8 58.0	29 3.4
50	8 57.9	11 54.2	9 36.8	9 45.4	29 25.4
51	8 38.1	11 33.7	10 52.4	10 35.7	29 48.5
52	8 17.1	11 11.6	12 15.7	11 29.2	♋ 0 13.0
53	7 54.7	10 47.4	13 47.8	12 26.4	0 39.0
54	7 30.7	10 21.0	15 30.1	13 27.7	1 6.7
55	7 5.0	9 51.9	17 24.3	14 33.6	1 36.2
56	6 37.2	9 19.8	19 32.4	15 44.7	2 7.8
57	6 7.2	8 44.0	21 56.7	17 1.8	2 41.9
58	5 34.4	8 3.9	24 39.9	18 25.8	3 18.8
59	4 58.6	7 18.7	27 45.3	19 57.8	3 59.0
60	4 19.0	6 27.1	♊ 1 16.6	21 39.1	4 43.1
61	3 34.8	5 27.6	5 17.8	23 31.7	5 31.8
62	2 45.1	4 18.2	9 52.8	25 37.9	6 26.4
63	1 48.2	2 56.0	15 5.4	28 1.0	7 28.2
64	0 41.8	1 16.6	20 57.5	♋ 0 45.6	8 39.8
65	♑29 21.9	♒29 13.3	27 28.8	3 58.9	10 5.0
66	27 40.8	26 34.3	♋ 4 34.6	7 52.9	11 51.8
S LAT	5	6	Descendant	8	9

7h 16m 0s		MC ♋ 17° 31′ 54″		109° 0′ 0″

19ʰ 20ᵐ 0ˢ		MC		290° 0′ 0″	
		♑ 18° 27′ 56″			
N LAT	11	12	Ascendant	2	3
°	° ′	° ′	° ′	° ′	° ′
0	♒17 33.3	♓19 7.3	♈21 38.3	♉22 24.6	♊20 48.6
5	17 4.7	18 50.5	22 27.6	23 30.9	21 28.2
10	16 35.2	18 32.5	23 21.4	24 40.3	22 8.5
15	16 4.2	18 13.0	24 21.3	25 54.1	22 50.4
16	15 57.7	18 8.9	24 34.2	26 9.6	22 59.0
17	15 51.2	18 4.7	24 47.5	26 25.2	23 7.7
18	15 44.6	18 0.3	25 1.1	26 41.2	23 16.6
19	15 37.8	17 55.9	25 15.1	26 57.5	23 25.5
20	15 31.0	17 51.4	25 29.5	27 14.0	23 34.6
21	15 24.0	17 46.7	25 44.4	27 30.9	23 43.8
22	15 16.9	17 41.9	25 59.7	27 48.1	23 53.1
23	15 9.6	17 37.0	26 15.6	28 5.7	24 2.6
24	15 2.2	17 32.0	26 32.0	28 23.7	24 12.3
25	14 54.7	17 26.8	26 49.0	28 42.1	24 22.1
26	14 47.0	17 21.4	27 6.6	29 1.0	24 32.2
27	14 39.1	17 15.9	27 24.8	29 20.3	24 42.4
28	14 31.0	17 10.2	27 43.9	29 40.1	24 52.8
29	14 22.8	17 4.3	28 3.6	♊ 0 0.4	25 3.4
30	14 14.3	16 58.2	28 24.2	0 21.4	25 14.3
31	14 5.6	16 51.9	28 45.8	0 42.9	25 25.5
32	13 56.6	16 45.3	29 8.3	1 5.0	25 36.9
33	13 47.4	16 38.5	29 31.8	1 27.9	25 48.6
34	13 37.9	16 31.4	29 56.5	1 51.5	26 0.6
35	13 28.2	16 24.0	♉ 0 22.4	2 15.9	26 12.9
36	13 18.1	16 16.2	0 49.7	2 41.1	26 25.6
37	13 7.7	16 8.2	1 18.5	3 7.2	26 38.7
38	12 56.9	15 59.7	1 48.9	3 34.3	26 52.2
39	12 45.7	15 50.8	2 21.0	4 2.4	27 6.1
40	12 34.1	15 41.5	2 55.1	4 31.6	27 20.5
41	12 22.1	15 31.7	3 31.3	5 2.1	27 35.4
42	12 9.5	15 21.3	4 9.9	5 33.9	27 50.8
43	11 56.4	15 10.4	4 51.1	6 7.0	28 6.9
44	11 42.8	14 58.8	5 35.2	6 41.8	28 23.5
45	11 28.5	14 46.5	6 22.6	7 18.1	28 40.9
46	11 13.6	14 33.4	7 13.6	7 56.3	28 59.0
47	10 57.9	14 19.4	8 8.6	8 36.5	29 17.9
48	10 41.4	14 4.5	9 8.2	9 18.9	29 37.8
49	10 24.0	13 48.4	10 13.0	10 3.6	29 58.6
50	10 5.6	13 31.1	11 23.7	10 51.0	♋ 0 20.5
51	9 46.1	13 12.4	12 41.0	11 41.2	0 43.5
52	9 25.4	12 52.1	14 6.0	12 34.7	1 7.9
53	9 3.3	12 30.0	15 39.8	13 31.8	1 33.8
54	8 39.7	12 5.8	17 23.7	14 33.0	2 1.3
55	8 14.3	11 39.1	19 19.2	15 38.7	2 30.7
56	7 47.0	11 9.7	21 28.3	16 49.5	3 2.2
57	7 17.5	10 36.9	23 53.2	18 6.2	3 36.0
58	6 45.3	10 0.1	26 36.4	19 29.6	4 12.7
59	6 10.0	9 18.6	29 40.7	21 0.8	4 52.5
60	5 31.0	8 31.2	♊ 3 9.6	22 41.3	5 36.2
61	4 47.7	7 36.7	7 6.4	24 32.6	6 24.5
62	3 58.9	6 33.1	11 34.7	26 37.1	7 18.4
63	3 3.2	5 17.7	16 37.3	28 57.8	8 19.4
64	1 58.3	3 46.7	22 15.9	♋ 1 39.1	9 29.8
65	0 40.6	1 54.1	28 29.7	4 47.4	10 53.3
66	♑29 3.1	♒29 29.8	♋ 5 14.8	8 33.2	12 36.8
S LAT	5	6	Descendant	8	9
		♋ 18° 27′ 56″			
7ʰ 20ᵐ 0ˢ		MC		110° 0′ 0″	

19ʰ 24ᵐ 0ˢ		MC		291° 0′ 0″	
		♑ 19° 24′ 4″			
N LAT	11	12	Ascendant	2	3
°	° ′	° ′	° ′	° ′	° ′
0	♒18 34.0	♓20 12.3	♈22 42.3	♉23 23.4	♊21 43.9
5	18 6.0	19 57.2	23 33.8	24 30.5	22 23.5
10	17 37.0	19 41.0	24 30.0	25 40.6	23 4.0
15	17 6.5	19 23.4	25 32.5	26 55.1	23 45.9
16	17 0.2	19 19.7	25 46.0	27 10.7	23 54.5
17	16 53.8	19 15.8	25 59.8	27 26.5	24 3.2
18	16 47.2	19 11.9	26 13.9	27 42.6	24 12.1
19	16 40.6	19 7.9	26 28.5	27 59.0	24 21.0
20	16 33.9	19 3.9	26 43.6	28 15.7	24 30.1
21	16 27.0	18 59.6	26 59.1	28 32.7	24 39.3
22	16 20.0	18 55.3	27 15.1	28 50.1	24 48.6
23	16 12.9	18 50.9	27 31.6	29 7.8	24 58.1
24	16 5.6	18 46.4	27 48.6	29 26.0	25 7.8
25	15 58.2	18 41.7	28 6.3	29 44.5	25 17.6
26	15 50.6	18 36.8	28 24.7	♊ 0 3.5	25 27.7
27	15 42.9	18 31.8	28 43.7	0 23.0	25 37.9
28	15 34.9	18 26.7	29 3.4	0 42.9	25 48.3
29	15 26.8	18 21.4	29 24.0	1 3.4	25 58.9
30	15 18.5	18 15.8	29 45.4	1 24.5	26 9.8
31	15 9.9	18 10.1	♉ 0 7.8	1 46.1	26 21.0
32	15 1.1	18 4.2	0 31.1	2 8.4	26 32.4
33	14 52.0	17 58.0	0 55.6	2 31.4	26 44.0
34	14 42.7	17 51.6	1 21.2	2 55.1	26 56.0
35	14 33.1	17 44.9	1 48.1	3 19.6	27 8.4
36	14 23.2	17 37.9	2 16.4	3 44.9	27 21.1
37	14 12.9	17 30.6	2 46.1	4 11.2	27 34.1
38	14 2.3	17 23.0	3 17.6	4 38.4	27 47.6
39	13 51.3	17 14.9	3 50.8	5 6.6	28 1.5
40	13 39.9	17 6.5	4 26.1	5 36.0	28 15.9
41	13 28.0	16 57.6	5 3.5	6 6.6	28 30.7
42	13 15.7	16 48.3	5 43.3	6 38.4	28 46.1
43	13 2.8	16 38.4	6 25.8	7 11.7	29 2.1
44	12 49.4	16 27.9	7 11.3	7 46.5	29 18.8
45	12 35.3	16 16.8	8 0.0	8 22.9	29 36.1
46	12 20.6	16 4.9	8 52.4	9 1.2	29 54.2
47	12 5.2	15 52.3	9 48.9	9 41.4	♋ 0 13.0
48	11 48.9	15 38.7	10 50.0	10 23.8	0 32.8
49	11 31.8	15 24.2	11 56.3	11 8.5	0 53.5
50	11 13.7	15 8.5	13 8.5	11 55.9	1 15.3
51	10 54.5	14 51.6	14 27.3	12 46.1	1 38.3
52	10 34.1	14 33.2	15 53.8	13 39.5	2 2.6
53	10 12.4	14 13.1	17 28.9	14 36.5	2 28.4
54	9 49.1	13 51.2	19 14.1	15 37.4	2 55.8
55	9 24.2	13 27.0	21 10.7	16 42.9	3 25.0
56	8 57.3	13 0.3	23 20.5	17 53.3	3 56.3
57	8 28.2	12 30.6	25 45.6	19 9.6	4 29.9
58	7 56.6	11 57.2	28 28.2	20 32.4	5 6.3
59	7 21.9	11 19.6	♊ 1 31.1	22 2.9	5 45.8
60	6 43.7	10 36.6	4 57.2	23 42.4	6 29.1
61	6 1.1	9 47.1	8 49.5	25 32.4	7 16.9
62	5 13.3	8 49.4	13 11.0	27 35.2	8 10.2
63	4 18.8	7 41.1	18 4.0	29 53.5	9 10.3
64	3 15.5	6 18.7	23 29.8	♋ 2 31.5	10 19.5
65	2 0.0	4 36.9	29 27.6	5 34.9	11 41.1
66	0 26.0	2 27.1	♋ 5 53.8	9 13.0	13 21.5
S LAT	5	6	Descendant	8	9
		♋ 19° 24′ 4″			
7ʰ 24ᵐ 0ˢ		MC		111° 0′ 0″	

19h 28m 0s		MC		292° 0′ 0″
		♑ 20° 20′ 19″		

N LAT	11	12	Ascendant	2	3
	° ′	° ′	° ′	° ′	° ′
0	♒19 35.0	♓21 17.5	♈23 46.0	♉24 22.0	♊22 39.2
5	19 7.5	21 4.0	24 39.8	25 29.8	23 18.9
10	18 39.0	20 49.6	25 38.3	26 40.7	23 59.3
15	18 9.1	20 33.9	26 43.4	27 55.9	24 41.3
16	18 2.9	20 30.5	26 57.4	28 11.6	24 49.9
17	17 56.5	20 27.1	27 11.8	28 27.5	24 58.6
18	17 50.1	20 23.7	27 26.5	28 43.8	25 7.5
19	17 43.6	20 20.1	27 41.7	29 0.3	25 16.4
20	17 37.0	20 16.5	27 57.3	29 17.1	25 25.5
21	17 30.3	20 12.7	28 13.4	29 34.3	25 34.7
22	17 23.4	20 8.9	28 30.0	29 51.8	25 44.1
23	17 16.4	20 4.9	28 47.2	♊ 0 9.7	25 53.6
24	17 9.3	20 0.9	29 4.9	0 27.9	26 3.2
25	17 2.0	19 56.7	29 23.3	0 46.6	26 13.1
26	16 54.5	19 52.4	29 42.3	1 5.7	26 23.1
27	16 46.9	19 47.9	♉ 0 2.0	1 25.3	26 33.3
28	16 39.1	19 43.3	0 22.5	1 45.4	26 43.7
29	16 31.1	19 38.6	0 43.9	2 6.0	26 54.3
30	16 22.9	19 33.6	1 6.1	2 27.2	27 5.2
31	16 14.5	19 28.5	1 29.2	2 49.0	27 16.3
32	16 5.8	19 23.3	1 53.4	3 11.4	27 27.7
33	15 56.9	19 17.7	2 18.7	3 34.5	27 39.4
34	15 47.8	19 12.0	2 45.2	3 58.3	27 51.4
35	15 38.3	19 6.1	3 13.0	4 22.9	28 3.7
36	15 28.5	18 59.8	3 42.2	4 48.4	28 16.4
37	15 18.5	18 53.3	4 13.0	5 14.7	28 29.4
38	15 8.0	18 46.5	4 45.5	5 42.1	28 42.9
39	14 57.2	18 39.3	5 19.8	6 10.4	28 56.7
40	14 46.0	18 31.8	5 56.1	6 39.9	29 11.1
41	14 34.3	18 23.9	6 34.6	7 10.5	29 25.9
42	14 22.2	18 15.5	7 15.6	7 42.5	29 41.3
43	14 9.5	18 6.7	7 59.3	8 15.8	29 57.3
44	13 56.3	17 57.3	8 46.0	8 50.7	♋ 0 13.9
45	13 42.5	17 47.4	9 36.0	9 27.2	0 31.1
46	13 28.0	17 36.8	10 29.7	10 5.5	0 49.1
47	13 12.8	17 25.5	11 27.5	10 45.7	1 8.0
48	12 56.9	17 13.4	12 30.0	11 28.1	1 27.7
49	12 40.0	17 0.4	13 37.6	12 12.8	1 48.3
50	12 22.2	16 46.4	14 51.2	13 0.1	2 10.0
51	12 3.3	16 31.2	16 11.3	13 50.3	2 32.9
52	11 43.3	16 14.8	17 39.0	14 43.6	2 57.1
53	11 21.9	15 56.9	19 15.3	15 40.4	3 22.7
54	10 59.1	15 37.2	21 1.5	16 41.1	3 50.0
55	10 34.5	15 15.6	22 58.8	17 46.3	4 19.0
56	10 8.1	14 51.7	25 9.0	18 56.4	4 50.1
57	9 39.5	14 25.1	27 33.9	20 12.1	5 23.5
58	9 8.4	13 55.3	♊ 0 15.8	21 34.4	5 59.6
59	8 34.4	13 21.5	3 16.9	23 4.1	6 38.8
60	7 56.9	12 43.1	6 40.0	24 42.5	7 21.7
61	7 15.2	11 58.7	10 27.6	26 31.2	8 9.0
62	6 28.3	11 7.1	14 42.3	28 32.2	9 1.6
63	5 35.0	10 5.9	19 26.1	♋ 0 48.2	10 0.9
64	4 33.3	8 52.3	24 39.9	3 22.8	11 8.9
65	3 20.0	7 21.5	♋ 0 22.9	6 21.5	12 28.7
66	1 49.4	5 26.2	6 31.8	9 52.3	14 6.2
S LAT	5	6	Descendant	8	9

		♋ 20° 20′ 19″		
7h 28m 0s		MC		112° 0′ 0″

19h 32m 0s		MC		293° 0′ 0″
		♑ 21° 16′ 40″		

N LAT	11	12	Ascendant	2	3
	° ′	° ′	° ′	° ′	° ′
0	♒20 36.1	♓22 22.6	♈24 49.7	♉25 20.5	♊23 34.4
5	20 9.2	22 10.8	25 45.6	26 29.0	24 14.1
10	19 41.3	21 58.2	26 46.5	27 40.5	24 54.6
15	19 11.9	21 44.4	27 54.1	28 56.4	25 36.6
16	19 5.8	21 41.5	28 8.6	29 12.2	25 45.2
17	18 59.6	21 38.5	28 23.5	29 28.3	25 53.9
18	18 53.3	21 35.5	28 38.8	29 44.7	26 2.8
19	18 46.9	21 32.4	28 54.5	♊ 0 1.3	26 11.7
20	18 40.4	21 29.2	29 10.8	0 18.3	26 20.8
21	18 33.8	21 25.9	29 27.4	0 35.6	26 30.0
22	18 27.0	21 22.5	29 44.7	0 53.2	26 39.4
23	18 20.1	21 19.0	♉ 0 2.4	1 11.2	26 48.9
24	18 13.1	21 15.5	0 20.8	1 29.6	26 58.5
25	18 6.0	21 11.8	0 39.8	1 48.4	27 8.4
26	17 58.7	21 8.0	0 59.5	2 7.6	27 18.4
27	17 51.2	21 4.1	1 19.9	2 27.3	27 28.6
28	17 43.5	21 0.1	1 41.2	2 47.5	27 39.0
29	17 35.7	20 55.9	2 3.2	3 8.3	27 49.6
30	17 27.6	20 51.6	2 26.2	3 29.5	28 0.5
31	17 19.4	20 47.1	2 50.1	3 51.4	28 11.6
32	17 10.9	20 42.5	3 15.1	4 14.0	28 23.0
33	17 2.1	20 37.7	3 41.2	4 37.2	28 34.7
34	16 53.1	20 32.6	4 8.5	5 1.1	28 46.6
35	16 43.8	20 27.4	4 37.2	5 25.9	28 58.9
36	16 34.2	20 21.9	5 7.4	5 51.4	29 11.6
37	16 24.3	20 16.2	5 39.1	6 17.9	29 24.6
38	16 14.1	20 10.2	6 12.5	6 45.3	29 38.0
39	16 3.4	20 3.9	6 47.8	7 13.7	29 51.9
40	15 52.4	19 57.3	7 25.1	7 43.2	♋ 0 6.2
41	15 41.0	19 50.4	8 4.7	8 14.0	0 21.0
42	15 29.0	19 43.0	8 46.8	8 46.0	0 36.3
43	15 16.6	19 35.2	9 31.6	9 19.4	0 52.2
44	15 3.6	19 27.0	10 19.4	9 54.3	1 8.8
45	14 50.0	19 18.3	11 10.6	10 30.8	1 26.0
46	14 35.8	19 9.0	12 5.5	11 9.2	1 44.0
47	14 20.9	18 59.0	13 4.5	11 49.4	2 2.7
48	14 5.2	18 48.4	14 8.2	12 31.8	2 22.3
49	13 48.6	18 37.0	15 17.0	13 16.5	2 42.9
50	13 31.1	18 24.7	16 31.8	14 3.7	3 4.6
51	13 12.6	18 11.3	17 53.1	14 53.8	3 27.4
52	12 52.9	17 56.9	19 21.8	15 46.9	3 51.4
53	12 31.9	17 41.1	20 59.0	16 43.6	4 16.9
54	12 9.4	17 23.8	22 45.9	17 44.1	4 44.0
55	11 45.4	17 4.8	24 43.7	18 48.9	5 12.9
56	11 19.4	16 43.8	26 54.0	19 58.6	5 43.7
57	10 51.4	16 20.3	29 18.5	21 13.8	6 16.9
58	10 20.8	15 54.1	♊ 1 59.2	22 35.4	6 52.7
59	9 47.4	15 24.4	4 58.4	24 4.3	7 31.6
60	9 10.7	14 50.5	8 18.2	25 41.7	8 14.1
61	8 29.8	14 11.4	12 1.1	27 29.0	9 0.8
62	7 44.0	13 25.9	16 9.1	29 28.3	9 52.8
63	6 51.9	12 32.1	20 44.1	♋ 1 41.9	10 51.2
64	5 51.8	11 27.3	25 46.6	4 13.2	11 58.0
65	4 40.6	10 7.7	♋ 1 15.8	7 7.3	13 16.1
66	3 13.3	8 26.8	7 9.0	10 31.2	14 50.6
S LAT	5	6	Descendant	8	9

		♋ 21° 16′ 40″		
7h 32m 0s		MC		113° 0′ 0″

19h 36m 0s — MC 294° 0′ 0″ — ♑ 22° 13′ 8″

11	12	Ascendant	2	3	N LAT
° ′	° ′	° ′	° ′	° ′	°
♒21 37.4	♓23 27.9	♈25 53.2	♉26 18.8	♊24 29.5	0
21 11.0	23 17.7	26 51.2	27 28.0	25 9.3	5
20 43.7	23 6.9	27 54.4	28 40.2	25 49.8	10
20 14.9	22 55.1	29 4.4	29 56.6	26 31.8	15
20 8.9	22 52.6	29 19.5	♊ 0 12.6	26 40.5	16
20 2.8	22 50.0	29 34.9	0 28.8	26 49.2	17
19 56.6	22 47.4	29 50.8	0 45.3	26 58.1	18
19 50.4	22 44.7	♉ 0 7.1	1 2.1	27 7.0	19
19 44.0	22 42.0	0 23.8	1 19.2	27 16.1	20
19 37.5	22 39.1	0 41.1	1 36.6	27 25.3	21
19 30.9	22 36.2	0 58.9	1 54.3	27 34.6	22
19 24.2	22 33.3	1 17.3	2 12.4	27 44.1	23
19 17.3	22 30.2	1 36.3	2 30.9	27 53.8	24
19 10.3	22 27.1	1 55.9	2 49.8	28 3.6	25
19 3.1	22 23.8	2 16.3	3 9.2	28 13.6	26
18 55.7	22 20.5	2 37.4	3 29.0	28 23.8	27
18 48.2	22 17.0	2 59.3	3 49.3	28 34.2	28
18 40.5	22 13.4	3 22.0	4 10.2	28 44.8	29
18 32.6	22 9.7	3 45.7	4 31.6	28 55.7	30
18 24.5	22 5.9	4 10.4	4 53.6	29 6.8	31
18 16.2	22 1.9	4 36.1	5 16.2	29 18.2	32
18 7.6	21 57.7	5 3.0	5 39.5	29 29.8	33
17 58.8	21 53.4	5 31.2	6 3.6	29 41.8	34
17 49.6	21 48.9	6 0.7	6 28.4	29 54.0	35
17 40.2	21 44.2	6 31.7	6 54.0	♋ 0 6.7	36
17 30.5	21 39.3	7 4.3	7 20.6	0 19.7	37
17 20.4	21 34.1	7 38.6	7 48.1	0 33.1	38
17 10.0	21 28.7	8 14.9	8 16.6	0 46.9	39
16 59.2	21 23.0	8 53.2	8 46.2	1 1.2	40
16 47.9	21 17.0	9 33.8	9 17.0	1 15.9	41
16 36.2	21 10.7	10 16.8	9 49.0	1 31.2	42
16 24.0	21 4.1	11 2.7	10 22.5	1 47.1	43
16 11.3	20 57.0	11 51.6	10 57.4	2 3.6	44
15 57.9	20 49.5	12 43.8	11 34.0	2 20.8	45
15 44.0	20 41.5	13 39.8	12 12.3	2 38.6	46
15 29.3	20 32.9	14 39.9	12 52.6	2 57.3	47
15 13.9	20 23.7	15 44.6	13 34.9	3 16.9	48
14 57.6	20 13.9	16 54.6	14 19.5	3 37.4	49
14 40.5	20 3.3	18 10.3	15 6.7	3 58.9	50
14 22.2	19 51.8	19 32.6	15 56.6	4 21.6	51
14 2.9	19 39.4	21 2.2	16 49.6	4 45.5	52
13 42.3	19 25.8	22 40.1	17 46.0	5 10.9	53
13 20.3	19 10.9	24 27.5	18 46.3	5 37.8	54
12 56.6	18 54.6	26 25.5	19 50.7	6 6.5	55
12 31.2	18 36.4	28 35.6	21 0.0	6 37.2	56
12 3.7	18 16.2	♊ 0 59.5	22 14.7	7 10.1	57
11 33.7	17 53.6	3 38.9	23 35.7	7 45.6	58
11 1.0	17 28.0	6 35.8	25 3.7	8 24.1	59
10 25.0	16 58.8	9 52.2	26 40.0	9 6.2	60
9 45.0	16 25.1	13 30.3	28 25.9	9 52.4	61
9 0.2	15 45.9	17 31.9	♋ 0 23.4	10 43.7	62
8 9.4	14 59.5	21 58.4	2 34.6	11 41.2	63
7 10.9	14 3.7	26 50.3	5 2.8	12 46.8	64
6 1.9	12 55.3	♋ 2 6.8	7 52.3	14 3.2	65
4 37.8	11 28.9	7 45.4	11 9.7	15 34.9	66
5	6	Descendant	8	9	S LAT

♋ 22° 13′ 8″ — 7h 36m 0s — MC 114° 0′ 0″

19h 40m 0s — MC 295° 0′ 0″ — ♑ 23° 9′ 44″

11	12	Ascendant	2	3	N LAT
° ′	° ′	° ′	° ′	° ′	°
♒22 38.9	♓24 33.2	♈26 56.5	♉27 17.0	♊25 24.7	0
22 13.1	24 24.7	27 56.7	28 26.8	26 4.5	5
21 46.3	24 15.7	29 2.0	29 39.6	26 45.0	10
21 18.1	24 5.8	♉ 0 14.5	♊ 0 56.7	27 27.0	15
21 12.2	24 3.7	0 30.1	1 12.7	27 35.7	16
21 6.3	24 1.6	0 46.0	1 29.1	27 44.4	17
21 0.2	23 59.4	1 2.4	1 45.7	27 53.2	18
20 54.1	23 57.1	1 19.2	2 2.6	28 2.2	19
20 47.8	23 54.8	1 36.6	2 19.8	28 11.3	20
20 41.5	23 52.5	1 54.4	2 37.3	28 20.5	21
20 35.0	23 50.1	2 12.8	2 55.1	28 29.8	22
20 28.4	23 47.6	2 31.8	3 13.3	28 39.3	23
20 21.7	23 45.0	2 51.3	3 32.0	28 49.0	24
20 14.8	23 42.4	3 11.6	3 51.0	28 58.8	25
20 7.8	23 39.7	3 32.6	4 10.4	29 8.8	26
20 0.6	23 36.9	3 54.3	4 30.4	29 19.0	27
19 53.2	23 34.0	4 16.9	4 50.8	29 29.3	28
19 45.6	23 31.0	4 40.3	5 11.7	29 40.0	29
19 37.9	23 27.9	5 4.7	5 33.2	29 50.8	30
19 29.9	23 24.7	5 30.1	5 55.3	♋ 0 1.9	31
19 21.8	23 21.4	5 56.6	6 18.1	0 13.2	32
19 13.4	23 17.9	6 24.2	6 41.5	0 24.9	33
19 4.7	23 14.3	6 53.2	7 5.6	0 36.8	34
18 55.8	23 10.5	7 23.5	7 30.5	0 49.1	35
18 46.5	23 6.6	7 55.3	7 56.2	1 1.7	36
18 37.0	23 2.5	8 28.7	8 22.8	1 14.6	37
18 27.1	22 58.2	9 3.9	8 50.4	1 28.0	38
18 16.9	22 53.7	9 41.0	9 19.0	1 41.8	39
18 6.3	22 48.9	10 20.2	9 48.6	1 56.0	40
17 55.2	22 43.9	11 1.7	10 19.5	2 10.7	41
17 43.8	22 38.6	11 45.8	10 51.6	2 26.0	42
17 31.8	22 33.1	12 32.6	11 25.1	2 41.8	43
17 19.3	22 27.1	13 22.4	12 0.0	2 58.3	44
17 6.2	22 20.9	14 15.6	12 36.6	3 15.4	45
16 52.5	22 14.2	15 12.6	13 14.9	3 33.2	46
16 38.1	22 7.0	16 13.7	13 55.1	3 51.8	47
16 23.0	21 59.4	17 19.4	14 37.4	4 11.3	48
16 7.1	21 51.1	18 30.2	15 22.0	4 31.7	49
15 50.2	21 42.3	19 46.8	16 9.0	4 53.1	50
15 32.3	21 32.7	21 9.9	16 58.8	5 15.7	51
15 13.4	21 22.3	22 40.2	17 51.6	5 39.5	52
14 53.2	21 10.9	24 18.6	18 47.8	6 4.7	53
14 31.6	20 58.4	26 6.3	19 47.8	6 31.5	54
14 8.4	20 44.7	28 4.3	20 51.9	6 59.9	55
13 43.5	20 29.6	♊ 0 14.1	22 0.7	7 30.4	56
13 16.5	20 12.7	2 37.1	23 14.9	8 3.1	57
12 47.2	19 53.7	5 14.9	24 35.1	8 38.3	58
12 15.1	19 32.2	8 9.4	26 2.2	9 16.4	59
11 39.9	19 7.8	11 22.4	27 37.4	9 58.0	60
11 0.8	18 39.6	14 55.8	29 22.0	10 43.7	61
10 17.1	18 6.8	18 51.0	♋ 1 17.6	11 34.3	62
9 27.6	17 28.0	23 9.4	3 26.5	12 31.0	63
8 30.6	16 41.3	27 51.3	5 51.5	13 35.4	64
7 23.8	15 44.1	♋ 2 55.9	8 36.7	14 50.0	65
6 2.8	14 32.1	8 21.1	11 47.8	16 19.2	66
5	6	Descendant	8	9	S LAT

♋ 23° 9′ 44″ — 7h 40m 0s — MC 115° 0′ 0″

19^h 44^m 0^s MC 296° 0′ 0″

♑ 24° 6′ 26″

11	12	Ascendant	2	3	N LAT
° ′	° ′	° ′	° ′	° ′	°
♒23 40.6	♓25 38.5	♈27 59.7	♉28 15.0	♊26 19.8	0
23 15.3	25 31.7	29 1.9	29 25.4	26 59.6	5
22 49.1	25 24.5	♉ 0 9.5	♊ 0 38.8	27 40.2	10
22 21.5	25 16.6	1 24.3	1 56.5	28 22.2	15
22 15.8	25 14.9	1 40.4	2 12.6	28 30.8	16
22 10.0	25 13.2	1 56.8	2 29.1	28 39.5	17
22 4.1	25 11.4	2 13.7	2 45.8	28 48.4	18
21 58.0	25 9.6	2 31.1	3 2.8	28 57.3	19
21 51.9	25 7.8	2 48.9	3 20.1	29 6.4	20
21 45.7	25 5.9	3 7.3	3 37.7	29 15.6	21
21 39.4	25 4.0	3 26.3	3 55.7	29 24.9	22
21 32.9	25 2.0	3 45.8	4 14.0	29 34.4	23
21 26.3	24 59.9	4 6.0	4 32.7	29 44.1	24
21 19.6	24 57.8	4 26.8	4 51.8	29 53.9	25
21 12.7	24 55.6	4 48.4	5 11.4	♋ 0 3.8	26
21 5.6	24 53.4	5 10.8	5 31.4	0 14.0	27
20 58.4	24 51.1	5 34.0	5 51.9	0 24.4	28
20 51.0	24 48.7	5 58.0	6 13.0	0 35.0	29
20 43.4	24 46.2	6 23.1	6 34.6	0 45.8	30
20 35.7	24 43.6	6 49.2	6 56.7	0 56.9	31
20 27.7	24 41.0	7 16.4	7 19.6	1 8.2	32
20 19.4	24 38.2	7 44.7	7 43.1	1 19.8	33
20 10.9	24 35.3	8 14.4	8 7.3	1 31.7	34
20 2.2	24 32.3	8 45.5	8 32.2	1 44.0	35
19 53.1	24 29.1	9 18.1	8 58.0	1 56.5	36
19 43.8	24 25.8	9 52.3	9 24.7	2 9.5	37
19 34.1	24 22.4	10 28.3	9 52.3	2 22.8	38
19 24.1	24 18.8	11 6.2	10 21.0	2 36.5	39
19 13.7	24 15.0	11 46.3	10 50.7	2 50.7	40
19 2.9	24 10.9	12 28.6	11 21.6	3 5.4	41
18 51.6	24 6.7	13 13.5	11 53.7	3 20.6	42
18 39.9	24 2.2	14 1.2	12 27.2	3 36.4	43
18 27.6	23 57.5	14 52.0	13 2.2	3 52.8	44
18 14.8	23 52.5	15 46.1	13 38.7	4 9.8	45
18 1.4	23 47.1	16 43.9	14 17.0	4 27.6	46
17 47.3	23 41.4	17 45.9	14 57.2	4 46.1	47
17 32.5	23 35.2	18 52.4	15 39.4	5 5.5	48
17 16.9	23 28.6	20 4.1	16 23.8	5 25.8	49
17 0.4	23 21.5	21 21.4	17 10.8	5 47.1	50
16 42.9	23 13.8	22 45.1	18 0.4	6 9.6	51
16 24.3	23 5.4	24 15.9	18 53.0	6 33.3	52
16 4.5	22 56.3	25 54.7	19 49.0	6 58.3	53
15 43.3	22 46.3	27 42.5	20 48.6	7 24.9	54
15 20.7	22 35.3	29 40.3	21 52.3	7 53.2	55
14 56.2	22 23.1	♊ 1 49.5	23 0.7	8 23.4	56
14 29.8	22 9.6	4 11.4	24 14.2	8 55.8	57
14 1.1	21 54.3	6 47.6	25 33.7	9 30.7	58
13 29.8	21 37.1	9 39.6	27 0.0	10 8.5	59
12 55.4	21 17.5	12 49.1	28 34.0	10 49.7	60
12 17.2	20 54.8	16 17.7	♋ 0 17.2	11 34.8	61
11 34.5	20 28.5	20 6.8	2 11.0	12 24.7	62
10 46.3	19 57.3	24 17.5	4 17.6	13 20.5	63
9 51.0	19 19.9	28 49.9	6 39.5	14 23.7	64
8 46.2	18 34.0	♋ 3 43.5	9 20.5	15 36.7	65
7 28.2	17 36.4	8 56.2	12 25.7	17 3.3	66
5	6	Descendant	8	9	S LAT

♋ 24° 6′ 26″

7^h 44^m 0^s MC 116° 0′ 0″

19^h 48^m 0^s MC 297° 0′ 0″

♑ 25° 3′ 17″

11	12	Ascendant	2	3	N LAT
° ′	° ′	° ′	° ′	° ′	°
♒24 42.5	♓26 43.8	♈29 2.8	♉29 12.8	♊27 14.8	0
24 17.8	26 38.8	♉ 0 7.0	♊ 0 23.9	27 54.7	5
23 52.2	26 33.3	1 16.7	1 37.8	28 35.3	10
23 25.2	26 27.4	2 33.8	2 56.0	29 17.3	15
23 19.6	26 26.1	2 50.4	3 12.3	29 25.9	16
23 13.9	26 24.8	3 7.3	3 28.8	29 34.6	17
23 8.1	26 23.5	3 24.7	3 45.7	29 43.5	18
23 2.2	26 22.2	3 42.6	4 2.8	29 52.4	19
22 56.2	26 20.8	4 1.0	4 20.2	♋ 0 1.5	20
22 50.2	26 19.4	4 19.9	4 37.9	0 10.7	21
22 44.0	26 17.9	4 39.3	4 55.9	0 20.0	22
22 37.6	26 16.4	4 59.4	5 14.4	0 29.5	23
22 31.2	26 14.9	5 20.2	5 33.2	0 39.1	24
22 24.6	26 13.3	5 41.6	5 52.4	0 48.9	25
22 17.9	26 11.7	6 3.8	6 12.0	0 58.9	26
22 11.0	26 10.0	6 26.7	6 32.2	1 9.0	27
22 3.9	26 8.3	6 50.5	6 52.8	1 19.4	28
21 56.7	26 6.5	7 15.2	7 13.9	1 30.0	29
21 49.3	26 4.6	7 40.9	7 35.6	1 40.8	30
21 41.6	26 2.7	8 7.7	7 57.8	1 51.8	31
21 33.8	26 0.6	8 35.5	8 20.7	2 3.1	32
21 25.8	25 58.6	9 4.6	8 44.3	2 14.7	33
21 17.4	25 56.4	9 34.9	9 8.6	2 26.6	34
21 8.9	25 54.1	10 6.7	9 33.6	2 38.8	35
21 0.0	25 51.8	10 40.0	9 59.5	2 51.3	36
20 50.9	25 49.3	11 15.0	10 26.2	3 4.2	37
20 41.4	25 46.7	11 51.8	10 53.8	3 17.5	38
20 31.6	25 44.0	12 30.5	11 22.5	3 31.2	39
20 21.4	25 41.1	13 11.3	11 52.3	3 45.4	40
20 10.9	25 38.1	13 54.5	12 23.2	4 0.0	41
19 59.8	25 34.9	14 40.2	12 55.3	4 15.2	42
19 48.4	25 31.6	15 28.7	13 28.8	4 30.9	43
19 36.4	25 28.0	16 20.2	14 3.8	4 47.2	44
19 23.8	25 24.2	17 15.1	14 40.3	5 4.2	45
19 10.7	25 20.2	18 13.8	15 18.6	5 21.9	46
18 56.9	25 15.9	19 16.5	15 58.7	5 40.3	47
18 42.4	25 11.2	20 23.8	16 40.8	5 59.6	48
18 27.1	25 6.3	21 36.1	17 25.1	6 19.8	49
18 10.9	25 0.9	22 54.0	18 11.9	6 41.0	50
17 53.8	24 55.1	24 18.2	19 1.4	7 3.4	51
17 35.6	24 48.9	25 49.4	19 53.8	7 26.9	52
17 16.2	24 42.0	27 28.4	20 49.4	7 51.8	53
16 55.6	24 34.5	29 16.1	21 48.7	8 18.2	54
16 33.4	24 26.2	♊ 1 13.6	22 52.0	8 46.3	55
16 9.5	24 17.0	3 22.1	23 59.9	9 16.3	56
15 43.7	24 6.8	5 42.8	25 12.8	9 48.4	57
15 15.6	23 55.4	8 17.1	26 31.6	10 23.0	58
14 45.0	23 42.4	11 6.5	27 56.9	11 0.4	59
14 11.4	23 27.6	14 12.4	29 29.9	11 41.1	60
13 34.1	23 10.6	17 36.4	♋ 1 11.6	12 25.7	61
12 52.6	22 50.8	21 19.6	3 3.6	13 14.9	62
12 5.6	22 27.3	25 22.9	5 7.9	14 9.8	63
11 11.9	21 59.2	29 46.4	7 26.8	15 11.9	64
10 9.3	21 24.8	♋ 4 29.7	10 3.8	16 23.2	65
8 54.2	20 41.6	9 30.8	13 3.3	17 47.4	66
5	6	Descendant	8	9	S LAT

♋ 25° 3′ 17″

7^h 48^m 0^s MC 117° 0′ 0″

19h 52m 0s MC 298° 0′ 0″

♑ 26° 0′ 15″

11	12	Ascendant	2	3	N LAT
° ′	° ′	° ′	° ′	° ′	°
♒25 44.5	♓27 49.2	♉ 0 5.7	♊ 0 10.5	♊28 9.9	0
25 20.4	27 45.8	1 11.9	1 22.1	28 49.8	5
24 55.4	27 42.2	2 23.7	2 36.6	29 30.4	10
24 29.1	27 38.2	3 43.1	3 55.3	♋ 0 12.3	15
24 23.6	27 37.4	4 0.1	4 11.7	0 21.0	16
24 18.0	27 36.5	4 17.5	4 28.4	0 29.7	17
24 12.4	27 35.7	4 35.4	4 45.3	0 38.5	18
24 6.6	27 34.8	4 53.7	5 2.5	0 47.4	19
24 0.8	27 33.8	5 12.6	5 20.0	0 56.5	20
23 54.9	27 32.9	5 32.0	5 37.8	1 5.7	21
23 48.8	27 31.9	5 52.0	5 55.9	1 15.0	22
23 42.6	27 30.9	6 12.6	6 14.5	1 24.5	23
23 36.3	27 29.9	6 33.9	6 33.4	1 34.1	24
23 29.9	27 28.8	6 55.9	6 52.7	1 43.8	25
23 23.3	27 27.8	7 18.6	7 12.4	1 53.8	26
23 16.6	27 26.6	7 42.2	7 32.6	2 3.9	27
23 9.7	27 25.5	8 6.6	7 53.3	2 14.3	28
23 2.6	27 24.3	8 31.9	8 14.5	2 24.8	29
22 55.4	27 23.0	8 58.2	8 36.2	2 35.6	30
22 47.9	27 21.7	9 25.5	8 58.5	2 46.7	31
22 40.3	27 20.4	9 54.0	9 21.5	2 57.9	32
22 32.4	27 19.0	10 23.7	9 45.1	3 9.5	33
22 24.3	27 17.6	10 54.8	10 9.5	3 21.3	34
22 15.9	27 16.0	11 27.2	10 34.6	3 33.5	35
22 7.2	27 14.5	12 1.2	11 0.5	3 46.0	36
21 58.3	27 12.8	12 36.9	11 27.3	3 58.9	37
21 49.0	27 11.1	13 14.4	11 55.0	4 12.1	38
21 39.5	27 9.3	13 53.8	12 23.7	4 25.8	39
21 29.5	27 7.4	14 35.4	12 53.4	4 39.9	40
21 19.2	27 5.3	15 19.3	13 24.3	4 54.5	41
21 8.4	27 3.2	16 5.7	13 56.5	5 9.6	42
20 57.2	27 1.0	16 54.9	14 30.0	5 25.3	43
20 45.4	26 58.6	17 47.2	15 4.9	5 41.5	44
20 33.2	26 56.1	18 42.8	15 41.4	5 58.4	45
20 20.3	26 53.4	19 42.2	16 19.6	6 16.0	46
20 6.8	26 50.5	20 45.5	16 59.6	6 34.4	47
19 52.7	26 47.4	21 53.4	17 41.7	6 53.6	48
19 37.7	26 44.1	23 6.3	18 25.9	7 13.7	49
19 21.9	26 40.5	24 24.7	19 12.5	7 34.8	50
19 5.2	26 36.7	25 49.3	20 1.8	7 57.0	51
18 47.4	26 32.5	27 20.7	20 53.9	8 20.4	52
18 28.4	26 27.9	28 59.8	21 49.3	8 45.1	53
18 8.2	26 22.9	♊ 0 47.3	22 48.2	9 11.3	54
17 46.5	26 17.3	2 44.3	23 51.1	9 39.2	55
17 23.2	26 11.2	4 51.9	24 58.5	10 8.9	56
16 58.0	26 4.4	7 11.3	26 10.8	10 40.8	57
16 30.6	25 56.7	9 43.6	27 28.8	11 15.1	58
16 0.7	25 48.1	12 30.4	28 53.2	11 52.1	59
15 27.9	25 38.2	15 32.8	♋ 0 24.9	12 32.3	60
14 51.6	25 26.8	18 52.2	2 5.2	13 16.3	61
14 11.2	25 13.6	22 29.6	3 55.5	14 4.9	62
13 25.5	24 57.9	26 25.8	5 57.5	14 58.9	63
12 33.4	24 39.1	♋ 0 41.0	8 13.4	15 59.8	64
11 32.9	24 16.2	5 14.6	10 46.5	17 9.6	65
10 20.7	23 47.4	10 4.9	13 40.6	18 31.4	66
5	6	Descendant	8	9	S LAT

♋ 26° 0′ 15″

7h 52m 0s MC 118° 0′ 0″

19h 56m 0s MC 299° 0′ 0″

♑ 26° 57′ 21″

11	12	Ascendant	2	3	N LAT
° ′	° ′	° ′	° ′	° ′	°
♒26 46.7	♓28 54.6	♉ 1 8.4	♊ 1 8.0	♊29 5.0	0
26 23.2	28 52.9	2 16.5	2 20.2	29 44.8	5
25 58.9	28 51.1	3 30.4	3 35.2	♋ 0 25.4	10
25 33.2	28 49.1	4 52.0	4 54.4	1 7.4	15
25 27.8	28 48.7	5 9.4	5 10.9	1 16.0	16
25 22.4	28 48.3	5 27.3	5 27.7	1 24.7	17
25 16.9	28 47.8	5 45.7	5 44.7	1 33.5	18
25 11.3	28 47.4	6 4.5	6 2.0	1 42.4	19
25 5.6	28 46.9	6 23.9	6 19.5	1 51.5	20
24 59.8	28 46.4	6 43.8	6 37.4	2 0.6	21
24 53.9	28 46.0	7 4.3	6 55.7	2 9.9	22
24 47.8	28 45.5	7 25.4	7 14.3	2 19.4	23
24 41.7	28 44.9	7 47.2	7 33.3	2 29.0	24
24 35.4	28 44.4	8 9.7	7 52.6	2 38.8	25
24 29.0	28 43.9	8 33.0	8 12.4	2 48.7	26
24 22.4	28 43.3	8 57.1	8 32.7	2 58.8	27
24 15.7	28 42.7	9 22.1	8 53.5	3 9.1	28
24 8.8	28 42.1	9 47.9	9 14.7	3 19.7	29
24 1.7	28 41.5	10 14.8	9 36.6	3 30.4	30
23 54.4	28 40.9	10 42.8	9 59.0	3 41.4	31
23 47.0	28 40.2	11 11.9	10 22.0	3 52.7	32
23 39.3	28 39.5	11 42.2	10 45.7	4 4.2	33
23 31.4	28 38.8	12 13.9	11 10.0	4 16.0	34
23 23.2	28 38.0	12 47.0	11 35.2	4 28.2	35
23 14.7	28 37.2	13 21.6	12 1.1	4 40.6	36
23 6.0	28 36.4	13 57.9	12 27.9	4 53.5	37
22 57.0	28 35.5	14 36.1	12 55.7	5 6.7	38
22 47.6	28 34.6	15 16.2	13 24.4	5 20.3	39
22 37.9	28 33.7	15 58.4	13 54.2	5 34.4	40
22 27.8	28 32.7	16 43.0	14 25.1	5 48.9	41
22 17.3	28 31.6	17 30.1	14 57.2	6 3.9	42
22 6.3	28 30.5	18 20.0	15 30.7	6 19.5	43
21 54.9	28 29.3	19 12.9	16 5.6	6 35.7	44
21 42.9	28 28.0	20 9.2	16 42.1	6 52.5	45
21 30.3	28 26.7	21 9.1	17 20.2	7 10.1	46
21 17.2	28 25.2	22 13.1	18 0.1	7 28.3	47
21 3.3	28 23.7	23 21.5	18 42.0	7 47.4	48
20 48.7	28 22.0	24 34.9	19 26.1	8 7.4	49
20 33.3	28 20.2	25 53.6	20 12.6	8 28.4	50
20 16.9	28 18.3	27 18.5	21 1.6	8 50.5	51
19 59.6	28 16.2	28 50.0	21 53.5	9 13.7	52
19 41.1	28 13.9	♊ 0 28.9	22 48.6	9 38.3	53
19 21.3	28 11.4	2 16.2	23 47.1	10 4.3	54
19 0.2	28 8.6	4 12.6	24 49.6	10 32.0	55
18 37.4	28 5.6	6 19.2	25 56.4	11 1.4	56
18 12.8	28 2.2	8 37.0	27 8.0	11 33.0	57
17 46.1	27 58.3	11 7.4	28 25.2	12 6.9	58
17 17.0	27 54.0	13 51.4	29 48.7	12 43.6	59
16 45.0	27 49.0	16 50.3	♋ 1 19.3	13 23.4	60
16 9.7	27 43.3	20 5.2	2 58.2	14 6.8	61
15 30.3	27 36.7	23 37.1	4 46.6	14 54.7	62
14 46.1	27 28.9	27 26.6	6 46.4	15 47.8	63
13 55.6	27 19.5	♋ 1 33.9	8 59.5	16 47.6	64
12 57.1	27 8.0	5 58.3	11 28.7	17 55.8	65
11 47.7	26 53.6	10 38.6	14 17.8	19 15.3	66
5	6	Descendant	8	9	S LAT

♋ 26° 57′ 21″

7h 56m 0s MC 119° 0′ 0″

20h 0m 0s — MC — 300° 0′ 0″

♑ 27° 54′ 36″

11	12	Ascendant	2	3	N LAT
° ′	° ′	° ′	° ′	° ′	°
♒27 49.1	♈ 0 0.0	♉ 2 10.9	♊ 2 5.4	♋ 0 0.0	0
27 26.2	0 0.0	3 21.0	3 18.1	0 39.9	5
27 2.5	0 0.0	4 36.9	4 33.7	1 20.4	10
26 37.5	0 0.0	6 0.6	5 53.3	2 2.3	15
26 32.3	0 0.0	6 18.5	6 9.9	2 11.0	16
26 27.0	0 0.0	6 36.8	6 26.7	2 19.7	17
26 21.6	0 0.0	6 55.6	6 43.8	2 28.5	18
26 16.1	0 0.0	7 14.9	7 1.2	2 37.4	19
26 10.6	0 0.0	7 34.8	7 18.9	2 46.4	20
26 4.9	0 0.0	7 55.2	7 36.8	2 55.6	21
25 59.2	0 0.0	8 16.2	7 55.2	3 4.9	22
25 53.3	0 0.0	8 37.8	8 13.8	3 14.3	23
25 47.3	0 0.0	9 0.1	8 32.9	3 23.9	24
25 41.2	0 0.0	9 23.1	8 52.3	3 33.6	25
25 34.9	0 0.0	9 46.9	9 12.2	3 43.5	26
25 28.5	0 0.0	10 11.5	9 32.6	3 53.6	27
25 22.0	0 0.0	10 37.0	9 53.4	4 3.9	28
25 15.2	0 0.0	11 3.4	10 14.7	4 14.4	29
25 8.3	0 0.0	11 30.9	10 36.6	4 25.2	30
25 1.3	0 0.0	11 59.4	10 59.0	4 36.1	31
24 54.0	0 0.0	12 29.1	11 22.1	4 47.4	32
24 46.5	0 0.0	13 0.0	11 45.8	4 58.9	33
24 38.7	0 0.0	13 32.2	12 10.3	5 10.7	34
24 30.8	0 0.0	14 5.9	12 35.4	5 22.8	35
24 22.5	0 0.0	14 41.2	13 1.4	5 35.2	36
24 14.0	0 0.0	15 18.1	13 28.3	5 48.0	37
24 5.2	0 0.0	15 56.9	13 56.0	6 1.1	38
23 56.1	0 0.0	16 37.6	14 24.7	6 14.7	39
23 46.6	0 0.0	17 20.5	14 54.5	6 28.7	40
23 36.7	0 0.0	18 5.7	15 25.4	6 43.2	41
23 26.5	0 0.0	18 53.4	15 57.6	6 58.2	42
23 15.8	0 0.0	19 43.9	16 31.0	7 13.7	43
23 4.6	0 0.0	20 37.4	17 5.8	7 29.8	44
22 52.9	0 0.0	21 34.2	17 42.2	7 46.6	45
22 40.7	0 0.0	22 34.7	18 20.2	8 4.0	46
22 27.9	0 0.0	23 39.2	19 0.1	8 22.2	47
22 14.3	0 0.0	24 48.0	19 41.9	8 41.2	48
22 0.1	0 0.0	26 1.7	20 25.8	9 1.0	49
21 45.0	0 0.0	27 20.8	21 12.1	9 21.9	50
21 29.1	0 0.0	28 45.7	22 0.9	9 43.8	51
21 12.2	0 0.0	♊ 0 17.2	22 52.5	10 6.9	52
20 54.2	0 0.0	1 56.0	23 47.2	10 31.3	53
20 34.9	0 0.0	3 42.8	24 45.4	10 57.1	54
20 14.3	0 0.0	5 38.4	25 47.4	11 24.6	55
19 52.1	0 0.0	7 43.9	26 53.6	11 53.8	56
19 28.1	0 0.0	10 0.3	28 4.6	12 25.1	57
19 2.1	0 0.0	12 28.5	29 21.0	12 58.7	58
18 33.8	0 0.0	15 9.8	♋ 0 43.5	13 34.9	59
18 2.7	0 0.0	18 5.2	2 13.0	14 14.2	60
17 28.3	0 0.0	21 15.8	3 50.4	14 57.1	61
16 50.1	0 0.0	24 42.3	5 37.1	15 44.2	62
16 7.1	0 0.0	28 25.4	7 34.7	16 36.5	63
15 18.3	0 0.0	♋ 2 25.2	9 44.9	17 35.2	64
14 21.8	0 0.0	6 41.1	12 10.5	18 41.8	65
13 15.2	0 0.0	11 12.0	14 54.8	19 59.2	66
5	6	Descendant	8	9	S LAT

♋ 27° 54′ 36″

8h 0m 0s — MC — 120° 0′ 0″

20h 4m 0s — MC — 301° 0′ 0″

♑ 28° 51′ 59″

11	12	Ascendant	2	3	N LAT
° ′	° ′	° ′	° ′	° ′	°
♒28 51.6	♈ 1 5.4	♉ 3 13.3	♊ 3 2.6	♋ 0 55.0	0
28 29.4	1 7.1	4 25.2	4 15.9	1 34.9	5
28 6.3	1 8.9	5 43.1	5 31.9	2 15.4	10
27 42.0	1 10.9	7 8.9	6 52.0	2 57.3	15
27 36.9	1 11.3	7 27.2	7 8.7	3 5.9	16
27 31.8	1 11.7	7 46.0	7 25.6	3 14.6	17
27 26.5	1 12.2	8 5.2	7 42.7	3 23.4	18
27 21.2	1 12.6	8 25.0	8 0.2	3 32.3	19
27 15.8	1 13.1	8 45.3	8 17.9	3 41.3	20
27 10.3	1 13.6	9 6.1	8 36.0	3 50.5	21
27 4.7	1 14.0	9 27.6	8 54.4	3 59.7	22
26 59.0	1 14.5	9 49.7	9 13.1	4 9.1	23
26 53.2	1 15.1	10 12.5	9 32.2	4 18.7	24
26 47.2	1 15.6	10 36.0	9 51.8	4 28.4	25
26 41.1	1 16.1	11 0.3	10 11.7	4 38.3	26
26 34.9	1 16.7	11 25.4	10 32.1	4 48.4	27
26 28.5	1 17.3	11 51.4	10 53.0	4 58.7	28
26 21.9	1 17.9	12 18.4	11 14.4	5 9.2	29
26 15.2	1 18.5	12 46.3	11 36.3	5 19.9	30
26 8.3	1 19.1	13 15.4	11 58.8	5 30.8	31
26 1.2	1 19.8	13 45.6	12 21.9	5 42.0	32
25 53.9	1 20.5	14 17.1	12 45.7	5 53.5	33
25 46.4	1 21.2	14 49.9	13 10.1	6 5.2	34
25 38.6	1 22.0	15 24.1	13 35.4	6 17.3	35
25 30.6	1 22.8	16 0.0	14 1.4	6 29.7	36
25 22.3	1 23.6	16 37.5	14 28.2	6 42.4	37
25 13.7	1 24.5	17 16.8	14 56.0	6 55.5	38
25 4.9	1 25.4	17 58.1	15 24.7	7 9.0	39
24 55.6	1 26.3	18 41.5	15 54.5	7 23.0	40
24 46.0	1 27.3	19 27.3	16 25.3	7 37.4	41
24 36.0	1 28.4	20 15.6	16 57.4	7 52.3	42
24 25.6	1 29.5	21 6.6	17 30.8	8 7.8	43
24 14.7	1 30.7	22 0.6	18 5.6	8 23.8	44
24 3.4	1 32.0	22 58.0	18 41.9	8 40.5	45
23 51.4	1 33.3	23 58.9	19 19.9	8 57.8	46
23 38.9	1 34.8	25 3.8	19 59.6	9 15.9	47
23 25.8	1 36.3	26 12.9	20 41.2	9 34.8	48
23 11.9	1 38.0	27 26.9	21 25.0	9 54.6	49
22 57.2	1 39.8	28 46.1	22 11.0	10 15.3	50
22 41.7	1 41.7	♊ 0 11.1	22 59.6	10 37.0	51
22 25.2	1 43.8	1 42.5	23 50.9	11 0.0	52
22 7.6	1 46.1	3 21.0	24 45.3	11 24.2	53
21 48.9	1 48.6	5 7.2	25 43.1	11 49.8	54
21 28.8	1 51.4	7 2.1	26 44.6	12 17.0	55
21 7.2	1 54.4	9 6.4	27 50.3	12 46.0	56
20 43.9	1 57.8	11 21.1	29 0.6	13 17.0	57
20 18.6	2 1.7	13 47.2	♋ 0 16.2	13 50.2	58
19 51.0	2 6.0	16 25.8	1 37.8	14 26.1	59
19 20.8	2 11.0	19 17.8	3 6.0	15 4.9	60
18 47.5	2 16.7	22 24.0	4 42.1	15 47.2	61
18 10.4	2 23.3	25 45.4	6 27.0	16 33.7	62
17 28.8	2 31.1	29 22.4	8 22.3	17 25.1	63
16 41.5	2 40.5	♋ 3 15.1	10 29.9	18 22.6	64
15 47.1	2 52.0	7 22.9	12 52.0	19 27.8	65
14 43.1	3 6.4	11 45.0	15 31.5	20 43.1	66
5	6	Descendant	8	9	S LAT

♋ 28° 51′ 59″

8h 4m 0s — MC — 121° 0′ 0″

20h 8m 0s MC 302° 0′ 0″

♑ 29° 49′ 31″

N LAT	11	12	Ascendant	2	3
°	° ′	° ′	° ′	° ′	° ′
0	♒29 54.3	♈ 2 10.8	♉ 4 15.5	♊ 3 59.8	♋ 1 50.1
5	29 32.8	2 14.2	5 29.3	5 13.5	2 30.0
10	29 10.3	2 17.8	6 49.1	6 29.9	3 10.4
15	28 46.7	2 21.8	8 16.9	7 50.5	3 52.3
16	28 41.8	2 22.6	8 35.6	8 7.2	4 0.8
17	28 36.8	2 23.5	8 54.8	8 24.2	4 9.5
18	28 31.7	2 24.3	9 14.5	8 41.4	4 18.3
19	28 26.5	2 25.2	9 34.7	8 59.0	4 27.2
20	28 21.3	2 26.2	9 55.4	9 16.8	4 36.2
21	28 15.9	2 27.1	10 16.7	9 34.9	4 45.3
22	28 10.5	2 28.1	10 38.6	9 53.4	4 54.6
23	28 4.9	2 29.1	11 1.2	10 12.2	5 3.9
24	27 59.2	2 30.1	11 24.4	10 31.3	5 13.5
25	27 53.5	2 31.2	11 48.4	10 50.9	5 23.2
26	27 47.5	2 32.2	12 13.2	11 10.9	5 33.1
27	27 41.5	2 33.4	12 38.8	11 31.4	5 43.1
28	27 35.3	2 34.5	13 5.3	11 52.3	5 53.4
29	27 28.9	2 35.7	13 32.7	12 13.7	6 3.8
30	27 22.4	2 37.0	14 1.2	12 35.7	6 14.5
31	27 15.7	2 38.3	14 30.8	12 58.3	6 25.4
32	27 8.8	2 39.6	15 1.5	13 21.4	6 36.6
33	27 1.7	2 41.0	15 33.5	13 45.2	6 48.0
34	26 54.3	2 42.4	16 6.8	14 9.7	6 59.7
35	26 46.8	2 44.0	16 41.6	14 34.9	7 11.7
36	26 39.0	2 45.5	17 17.9	15 0.9	7 24.1
37	26 30.9	2 47.2	17 56.0	15 27.8	7 36.8
38	26 22.6	2 48.9	18 35.8	15 55.5	7 49.8
39	26 13.9	2 50.7	19 17.7	16 24.3	8 3.3
40	26 4.9	2 52.6	20 1.6	16 54.0	8 17.2
41	25 55.6	2 54.7	20 47.9	17 24.9	8 31.5
42	25 45.9	2 56.8	21 36.7	17 56.9	8 46.4
43	25 35.8	2 59.0	22 28.2	18 30.3	9 1.8
44	25 25.2	3 1.4	23 22.7	19 5.0	9 17.7
45	25 14.1	3 3.9	24 20.4	19 41.2	9 34.3
46	25 2.5	3 6.6	25 21.7	20 19.0	9 51.6
47	24 50.3	3 9.5	26 26.9	20 58.6	10 9.5
48	24 37.5	3 12.6	27 36.4	21 40.1	10 28.3
49	24 24.0	3 15.9	28 50.5	22 23.7	10 47.9
50	24 9.8	3 19.5	♊ 0 9.8	23 9.5	11 8.5
51	23 54.7	3 23.3	1 34.8	23 57.8	11 30.2
52	23 38.6	3 27.5	3 6.0	24 48.9	11 52.9
53	23 21.6	3 32.1	4 44.0	25 42.9	12 16.9
54	23 3.3	3 37.1	6 29.6	26 40.2	12 42.4
55	22 43.8	3 42.7	8 23.5	27 41.2	13 9.3
56	22 22.8	3 48.8	10 26.6	28 46.3	13 38.0
57	22 0.2	3 55.6	12 39.7	29 56.0	14 8.7
58	21 35.6	4 3.3	15 3.6	♋ 1 10.8	14 41.6
59	21 8.8	4 11.9	17 39.5	2 31.3	15 17.1
60	20 39.5	4 21.8	20 28.0	3 58.5	15 55.4
61	20 7.2	4 33.2	23 30.2	5 33.1	16 37.2
62	19 31.3	4 46.4	26 46.6	7 16.3	17 22.9
63	18 51.0	5 2.1	♋ 0 17.7	9 9.5	18 13.5
64	18 5.4	5 20.9	4 3.6	11 14.4	19 9.9
65	17 12.9	5 43.8	8 4.0	13 33.0	20 13.6
66	16 11.6	6 12.6	12 17.7	16 8.2	21 26.9
S LAT	5	6	Descendant	8	9

♋ 29° 49′ 31″

8h 8m 0s MC 122° 0′ 0″

20h 12m 0s MC 303° 0′ 0″

♒ 0° 47′ 12″

N LAT	11	12	Ascendant	2	3
°	° ′	° ′	° ′	° ′	° ′
0	♓ 0 57.2	♈ 3 16.2	♉ 5 17.5	♊ 4 56.7	♋ 2 45.2
5	0 36.3	3 21.2	6 33.1	6 10.9	3 25.0
10	0 14.5	3 26.7	7 54.8	7 27.8	4 5.4
15	♒29 51.6	3 32.6	9 24.5	8 48.7	4 47.2
16	29 46.8	3 33.9	9 43.7	9 5.6	4 55.8
17	29 42.0	3 35.2	10 3.3	9 22.6	5 4.4
18	29 37.0	3 36.5	10 23.4	9 39.9	5 13.2
19	29 32.0	3 37.8	10 44.0	9 57.5	5 22.1
20	29 26.9	3 39.2	11 5.1	10 15.4	5 31.0
21	29 21.7	3 40.6	11 26.9	10 33.6	5 40.1
22	29 16.5	3 42.1	11 49.2	10 52.1	5 49.4
23	29 11.1	3 43.6	12 12.2	11 11.0	5 58.7
24	29 5.6	3 45.1	12 35.9	11 30.2	6 8.3
25	28 59.9	3 46.7	13 0.3	11 49.8	6 17.9
26	28 54.2	3 48.3	13 25.6	12 9.9	6 27.8
27	28 48.3	3 50.0	13 51.6	12 30.4	6 37.8
28	28 42.3	3 51.7	14 18.6	12 51.3	6 48.0
29	28 36.1	3 53.5	14 46.5	13 12.8	6 58.4
30	28 29.8	3 55.4	15 15.5	13 34.8	7 9.1
31	28 23.2	3 57.3	15 45.5	13 57.4	7 20.0
32	28 16.5	3 59.4	16 16.7	14 20.6	7 31.1
33	28 9.6	4 1.4	16 49.2	14 44.4	7 42.5
34	28 2.5	4 3.6	17 23.0	15 8.9	7 54.1
35	27 55.2	4 5.9	17 58.3	15 34.1	8 6.1
36	27 47.6	4 8.2	18 35.1	16 0.2	8 18.4
37	27 39.8	4 10.7	19 13.7	16 27.0	8 31.1
38	27 31.7	4 13.3	19 54.0	16 54.8	8 44.1
39	27 23.3	4 16.0	20 36.3	17 23.5	8 57.5
40	27 14.6	4 18.9	21 20.7	17 53.2	9 11.3
41	27 5.5	4 21.9	22 7.5	18 24.0	9 25.6
42	26 56.1	4 25.1	22 56.7	18 56.0	9 40.4
43	26 46.2	4 28.4	23 48.6	19 29.3	9 55.7
44	26 36.0	4 32.0	24 43.5	20 3.9	10 11.6
45	26 25.2	4 35.8	25 41.6	20 40.0	10 28.0
46	26 13.9	4 39.8	26 43.2	21 17.7	10 45.2
47	26 2.1	4 44.1	27 48.7	21 57.2	11 3.1
48	25 49.7	4 48.8	28 58.3	22 38.5	11 21.7
49	25 36.6	4 53.7	♊ 0 12.6	23 21.9	11 41.2
50	25 22.7	4 59.1	1 31.9	24 7.5	12 1.7
51	25 8.0	5 4.9	2 56.7	24 55.5	12 23.2
52	24 52.5	5 11.1	4 27.6	25 46.2	12 45.7
53	24 35.9	5 18.0	6 5.2	26 39.9	13 9.6
54	24 18.2	5 25.5	7 50.1	27 36.8	13 34.8
55	23 59.2	5 33.8	9 43.0	28 37.3	14 1.5
56	23 38.9	5 43.0	11 44.7	29 41.8	14 30.0
57	23 16.9	5 53.2	13 56.1	♋ 0 50.8	15 0.3
58	22 53.1	6 4.6	16 17.9	2 4.8	15 32.9
59	22 27.1	6 17.6	18 51.0	3 24.3	16 7.9
60	21 58.7	6 32.4	21 36.2	4 50.3	16 45.8
61	21 27.4	6 49.4	24 34.4	6 23.5	17 27.0
62	20 52.6	7 9.2	27 46.0	8 5.0	18 12.0
63	20 13.7	7 32.7	♋ 1 11.6	9 56.1	19 1.7
64	19 29.7	8 0.8	4 51.1	11 58.4	19 57.1
65	18 39.3	8 35.2	8 44.2	14 13.7	20 59.4
66	17 40.5	9 18.4	12 50.1	16 44.7	22 10.8
S LAT	5	6	Descendant	8	9

♌ 0° 47′ 12″

8h 12m 0s MC 123° 0′ 0″

20h 16m 0s — MC 304° 0′ 0″ — ♒ 1° 45′ 2″

N LAT	11	12	Ascendant	2	3
°	° ′	° ′	° ′	° ′	° ′
0	♓ 2 0.3	♈ 4 21.5	♉ 6 19.4	♊ 5 53.6	♋ 3 40.2
5	1 40.0	4 28.3	7 36.7	7 8.2	4 20.0
10	1 18.9	4 35.5	9 0.2	8 25.5	5 0.4
15	0 56.7	4 43.4	10 31.9	9 46.8	5 42.1
16	0 52.1	4 45.1	10 51.4	10 3.7	5 50.7
17	0 47.4	4 46.8	11 11.4	10 20.8	5 59.3
18	0 42.6	4 48.6	11 31.9	10 38.2	6 8.1
19	0 37.8	4 50.4	11 52.9	10 55.8	6 16.9
20	0 32.8	4 52.2	12 14.5	11 13.8	6 25.9
21	0 27.8	4 54.1	12 36.6	11 32.0	6 34.9
22	0 22.7	4 56.0	12 59.4	11 50.6	6 44.2
23	0 17.4	4 58.0	13 22.8	12 9.5	6 53.5
24	0 12.1	5 0.1	13 46.9	12 28.8	7 3.0
25	0 6.7	5 2.2	14 11.8	12 48.5	7 12.6
26	0 1.1	5 4.4	14 37.5	13 8.6	7 22.5
27	♒29 55.4	5 6.6	15 4.0	13 29.1	7 32.5
28	29 49.5	5 8.9	15 31.4	13 50.1	7 42.6
29	29 43.5	5 11.3	15 59.7	14 11.6	7 53.0
30	29 37.4	5 13.8	16 29.1	14 33.7	8 3.6
31	29 31.1	5 16.4	16 59.6	14 56.3	8 14.5
32	29 24.6	5 19.0	17 31.3	15 19.5	8 25.6
33	29 17.9	5 21.8	18 4.2	15 43.3	8 36.9
34	29 11.0	5 24.7	18 38.5	16 7.8	8 48.5
35	29 3.9	5 27.7	19 14.2	16 33.1	9 0.5
36	28 56.6	5 30.9	19 51.5	16 59.1	9 12.7
37	28 49.0	5 34.2	20 30.5	17 25.9	9 25.3
38	28 41.1	5 37.6	21 11.3	17 53.6	9 38.3
39	28 33.0	5 41.2	21 54.0	18 22.3	9 51.6
40	28 24.5	5 45.0	22 38.9	18 52.0	10 5.4
41	28 15.7	5 49.1	23 26.0	19 22.8	10 19.6
42	28 6.6	5 53.3	24 15.6	19 54.7	10 34.3
43	27 57.0	5 57.8	25 8.0	20 27.9	10 49.5
44	27 47.1	6 2.5	26 3.2	21 2.4	11 5.3
45	27 36.6	6 7.5	27 1.6	21 38.4	11 21.7
46	27 25.7	6 12.9	28 3.4	22 16.0	11 38.7
47	27 14.2	6 18.6	29 9.1	22 55.3	11 56.5
48	27 2.2	6 24.8	♊ 0 18.9	23 36.5	12 15.1
49	26 49.5	6 31.4	1 33.1	24 19.6	12 34.4
50	26 36.0	6 38.5	2 52.3	25 5.0	12 54.7
51	26 21.8	6 46.2	4 17.0	25 52.7	13 16.0
52	26 6.7	6 54.6	5 47.5	26 43.1	13 38.5
53	25 50.6	7 3.7	7 24.5	27 36.4	14 2.1
54	25 33.5	7 13.7	9 8.6	28 32.9	14 27.1
55	25 15.1	7 24.7	11 0.5	29 32.9	14 53.6
56	24 55.4	7 36.9	13 0.9	♋ 0 36.8	15 21.8
57	24 34.1	7 50.4	15 10.5	1 45.1	15 51.8
58	24 11.0	8 5.7	17 30.1	2 58.2	16 24.0
59	23 45.9	8 22.9	20 0.5	4 16.8	16 58.6
60	23 18.4	8 42.5	22 42.5	5 41.6	17 36.0
61	22 48.1	9 5.2	25 36.7	7 13.4	18 16.6
62	22 14.6	9 31.5	28 43.8	8 53.2	19 1.0
63	21 37.0	10 2.7	♋ 2 4.0	10 42.2	19 49.9
64	20 54.7	10 40.1	5 37.4	12 41.9	20 44.1
65	20 6.2	11 26.0	9 23.8	14 54.2	21 45.1
66	19 9.8	12 23.6	13 22.3	17 21.1	22 54.6
S LAT	5	6	Descendant	8	9

♌ 1° 45′ 2″ — 8h 16m 0s — MC 124° 0′ 0″

20h 20m 0s — MC 305° 0′ 0″ — ♒ 2° 43′ 2″

N LAT	11	12	Ascendant	2	3
°	° ′	° ′	° ′	° ′	° ′
0	♓ 3 3.5	♈ 5 26.8	♉ 7 21.1	♊ 6 50.3	♋ 4 35.3
5	2 43.9	5 35.3	8 40.1	8 5.4	5 15.1
10	2 23.5	5 44.3	10 5.4	9 23.0	5 55.4
15	2 2.0	5 54.2	11 38.9	10 44.7	6 37.0
16	1 57.5	5 56.3	11 58.8	11 1.6	6 45.6
17	1 53.0	5 58.4	12 19.2	11 18.8	6 54.2
18	1 48.4	6 0.6	12 40.1	11 36.2	7 2.9
19	1 43.7	6 2.9	13 1.5	11 53.9	7 11.7
20	1 38.9	6 5.2	13 23.4	12 11.9	7 20.7
21	1 34.0	6 7.5	13 46.0	12 30.2	7 29.7
22	1 29.1	6 9.9	14 9.1	12 48.9	7 38.9
23	1 24.0	6 12.4	14 33.0	13 7.8	7 48.2
24	1 18.9	6 15.0	14 57.5	13 27.2	7 57.7
25	1 13.6	6 17.6	15 22.8	13 46.9	8 7.3
26	1 8.2	6 20.3	15 48.8	14 7.0	8 17.1
27	1 2.7	6 23.1	16 15.8	14 27.6	8 27.1
28	0 57.0	6 26.0	16 43.6	14 48.6	8 37.2
29	0 51.2	6 29.0	17 12.4	15 10.2	8 47.6
30	0 45.3	6 32.1	17 42.2	15 32.2	8 58.2
31	0 39.2	6 35.3	18 13.1	15 54.8	9 9.0
32	0 32.9	6 38.6	18 45.2	16 18.0	9 20.0
33	0 26.4	6 42.1	19 18.6	16 41.9	9 31.3
34	0 19.8	6 45.7	19 53.3	17 6.4	9 42.9
35	0 12.9	6 49.5	20 29.4	17 31.6	9 54.8
36	0 5.8	6 53.4	21 7.2	17 57.6	10 7.0
37	♒29 58.4	6 57.5	21 46.5	18 24.5	10 19.5
38	29 50.8	7 1.8	22 27.7	18 52.2	10 32.4
39	29 42.9	7 6.3	23 10.9	19 20.8	10 45.7
40	29 34.7	7 11.1	23 56.1	19 50.4	10 59.4
41	29 26.2	7 16.1	24 43.6	20 21.2	11 13.5
42	29 17.4	7 21.4	25 33.6	20 53.0	11 28.1
43	29 8.1	7 26.9	26 26.2	21 26.1	11 43.3
44	28 58.5	7 32.9	27 21.7	22 0.6	11 59.0
45	28 48.4	7 39.1	28 20.3	22 36.5	12 15.3
46	28 37.8	7 45.8	29 22.4	23 13.9	12 32.2
47	28 26.7	7 53.0	♊ 0 28.2	23 53.0	12 49.9
48	28 15.0	8 0.6	1 38.0	24 34.0	13 8.3
49	28 2.7	8 8.9	2 52.2	25 16.9	13 27.5
50	27 49.7	8 17.7	4 11.3	26 2.0	13 47.7
51	27 36.0	8 27.3	5 35.6	26 49.5	14 8.8
52	27 21.3	8 37.7	7 5.7	27 39.6	14 31.1
53	27 5.8	8 49.1	8 42.1	28 32.5	14 54.5
54	26 49.2	9 1.6	10 25.4	29 28.5	15 19.3
55	26 31.4	9 15.3	12 16.2	♋ 0 27.9	15 45.6
56	26 12.3	9 30.4	14 15.2	1 31.2	16 13.4
57	25 51.7	9 47.3	16 23.0	2 38.8	16 43.2
58	25 29.4	10 6.3	18 40.5	3 51.1	17 15.0
59	25 5.1	10 27.8	21 8.2	5 8.7	17 49.2
60	24 38.6	10 52.2	23 47.0	6 32.3	18 26.1
61	24 9.4	11 20.4	26 37.4	8 2.7	19 6.2
62	23 37.0	11 53.2	29 40.0	9 40.9	19 49.8
63	23 0.8	12 32.0	♋ 2 55.1	11 27.9	20 37.9
64	22 20.1	13 18.7	6 22.8	13 25.1	21 31.1
65	21 33.6	14 15.9	10 2.7	15 34.3	22 30.7
66	20 39.7	15 27.9	13 54.3	17 57.4	23 38.5
S LAT	5	6	Descendant	8	9

♌ 2° 43′ 2″ — 8h 20m 0s — MC 125° 0′ 0″

20h 24m 0s		MC ♒ 3° 41′ 11″		306° 0′ 0″	
11	**12**	**Ascendant**	**2**	**3**	**N LAT**
° ′	° ′	° ′	° ′	° ′	°
♓ 4 6.8	♈ 6 32.1	♉ 8 22.6	♊ 7 46.9	♋ 5 30.5	**0**
3 47.9	6 42.3	9 43.3	9 2.4	6 10.2	**5**
3 28.3	6 53.1	11 10.3	10 20.4	6 50.4	**10**
3 7.5	7 4.9	12 45.6	11 42.4	7 31.9	**15**
3 3.2	7 7.4	13 5.9	11 59.4	7 40.5	**16**
2 58.8	7 10.0	13 26.6	12 16.6	7 49.1	**17**
2 54.3	7 12.6	13 47.9	12 34.1	7 57.8	**18**
2 49.8	7 15.3	14 9.7	12 51.8	8 6.6	**19**
2 45.2	7 18.0	14 32.0	13 9.9	8 15.5	**20**
2 40.5	7 20.9	14 54.9	13 28.2	8 24.5	**21**
2 35.7	7 23.8	15 18.5	13 46.9	8 33.7	**22**
2 30.8	7 26.7	15 42.7	14 5.9	8 43.0	**23**
2 25.9	7 29.8	16 7.6	14 25.3	8 52.4	**24**
2 20.8	7 32.9	16 33.3	14 45.0	9 2.0	**25**
2 15.6	7 36.2	16 59.7	15 5.2	9 11.7	**26**
2 10.2	7 39.5	17 27.0	15 25.8	9 21.7	**27**
2 4.8	7 43.0	17 55.3	15 46.9	9 31.8	**28**
1 59.2	7 46.6	18 24.5	16 8.4	9 42.1	**29**
1 53.4	7 50.3	18 54.7	16 30.5	9 52.6	**30**
1 47.5	7 54.1	19 26.0	16 53.1	10 3.4	**31**
1 41.4	7 58.1	19 58.5	17 16.3	10 14.4	**32**
1 35.2	8 2.3	20 32.2	17 40.2	10 25.7	**33**
1 28.8	8 6.6	21 7.4	18 4.7	10 37.2	**34**
1 22.1	8 11.1	21 43.9	18 29.9	10 49.0	**35**
1 15.2	8 15.8	22 22.0	18 55.9	11 1.2	**36**
1 8.1	8 20.7	23 1.8	19 22.7	11 13.6	**37**
1 0.8	8 25.9	23 43.3	19 50.4	11 26.5	**38**
0 53.2	8 31.3	24 26.8	20 19.0	11 39.7	**39**
0 45.3	8 37.0	25 12.4	20 48.5	11 53.3	**40**
0 37.0	8 43.0	26 0.2	21 19.2	12 7.4	**41**
0 28.5	8 49.3	26 50.5	21 51.0	12 21.9	**42**
0 19.5	8 55.9	27 43.4	22 24.0	12 37.0	**43**
0 10.2	9 3.0	28 39.1	22 58.3	12 52.6	**44**
0 0.5	9 10.5	29 37.9	23 34.1	13 8.8	**45**
♒29 50.2	9 18.5	♊ 0 40.1	24 11.4	13 25.6	**46**
29 39.5	9 27.1	1 45.9	24 50.3	13 43.2	**47**
29 28.2	9 36.3	2 55.8	25 31.1	14 1.5	**48**
29 16.3	9 46.1	4 9.9	26 13.8	14 20.6	**49**
29 3.8	9 56.7	5 28.8	26 58.6	14 40.6	**50**
28 50.5	10 8.2	6 52.8	27 45.8	15 1.6	**51**
28 36.4	10 20.6	8 22.4	28 35.5	15 23.6	**52**
28 21.3	10 34.2	9 58.1	29 28.0	15 46.9	**53**
28 5.3	10 49.1	11 40.5	♋ 0 23.6	16 11.4	**54**
27 48.1	11 5.4	13 30.2	1 22.5	16 37.4	**55**
27 29.7	11 23.6	15 27.7	2 25.2	17 5.0	**56**
27 9.8	11 43.8	17 33.7	3 32.0	17 34.4	**57**
26 48.3	12 6.4	19 49.0	4 43.4	18 5.9	**58**
26 24.9	12 32.0	22 14.2	6 0.1	18 39.7	**59**
25 59.2	13 1.2	24 49.8	7 22.5	19 16.1	**60**
25 31.1	13 34.9	27 36.6	8 51.6	19 55.6	**61**
24 59.9	14 14.1	♋ 0 34.9	10 28.1	20 38.6	**62**
24 25.2	15 0.5	3 45.1	12 13.1	21 25.8	**63**
23 46.0	15 56.3	7 7.3	14 8.0	22 18.0	**64**
23 1.4	17 4.7	10 41.1	16 14.2	23 16.3	**65**
22 9.9	18 31.1	14 26.1	18 33.6	24 22.3	**66**
5	**6**	**Descendant**	**8**	**9**	**S LAT**
8h 24m 0s		MC ♌ 3° 41′ 11″		126° 0′ 0″	

20h 28m 0s		MC ♒ 4° 39′ 30″		307° 0′ 0″	
N LAT	**11**	**12**	**Ascendant**	**2**	**3**
°	° ′	° ′	° ′	° ′	° ′
0	♓ 5 10.3	♈ 7 37.4	♉ 9 23.9	♊ 8 43.3	♋ 6 25.6
5	4 52.1	7 49.2	10 46.3	9 59.2	7 5.3
10	4 33.2	8 1.8	12 14.9	11 17.6	7 45.5
15	4 13.2	8 15.6	13 52.0	12 39.9	8 26.9
16	4 9.0	8 18.5	14 12.6	12 56.9	8 35.4
17	4 4.8	8 21.5	14 33.7	13 14.2	8 43.9
18	4 0.5	8 24.5	14 55.3	13 31.7	8 52.6
19	3 56.1	8 27.6	15 17.5	13 49.5	9 1.4
20	3 51.7	8 30.8	15 40.2	14 7.6	9 10.3
21	3 47.2	8 34.1	16 3.4	14 26.0	9 19.3
22	3 42.6	8 37.5	16 27.4	14 44.7	9 28.4
23	3 37.9	8 41.0	16 51.9	15 3.8	9 37.7
24	3 33.0	8 44.5	17 17.2	15 23.2	9 47.1
25	3 28.1	8 48.2	17 43.3	15 42.9	9 56.6
26	3 23.1	8 52.0	18 10.1	16 3.1	10 6.4
27	3 18.0	8 55.9	18 37.8	16 23.8	10 16.3
28	3 12.7	8 59.9	19 6.4	16 44.8	10 26.3
29	3 7.3	9 4.1	19 36.0	17 6.4	10 36.6
30	3 1.8	9 8.4	20 6.6	17 28.5	10 47.1
31	2 56.1	9 12.9	20 38.3	17 51.1	10 57.8
32	2 50.2	9 17.5	21 11.1	18 14.3	11 8.8
33	2 44.2	9 22.3	21 45.2	18 38.2	11 20.0
34	2 38.0	9 27.4	22 20.7	19 2.7	11 31.5
35	2 31.6	9 32.6	22 57.6	19 27.9	11 43.2
36	2 25.0	9 38.1	23 36.1	19 53.9	11 55.3
37	2 18.1	9 43.8	24 16.2	20 20.6	12 7.7
38	2 11.0	9 49.8	24 58.1	20 48.2	12 20.5
39	2 3.7	9 56.1	25 41.9	21 16.8	12 33.7
40	1 56.1	10 2.7	26 27.8	21 46.3	12 47.2
41	1 48.1	10 9.6	27 15.9	22 16.9	13 1.2
42	1 39.9	10 17.0	28 6.4	22 48.6	13 15.7
43	1 31.3	10 24.8	28 59.5	23 21.5	13 30.6
44	1 22.3	10 33.0	29 55.4	23 55.7	13 46.1
45	1 12.9	10 41.7	♊ 0 54.3	24 31.3	14 2.2
46	1 3.0	10 51.0	1 56.6	25 8.4	14 19.0
47	0 52.6	11 1.0	3 2.5	25 47.2	14 36.4
48	0 41.8	11 11.6	4 12.2	26 27.7	14 54.5
49	0 30.3	11 23.0	5 26.2	27 10.2	15 13.5
50	0 18.2	11 35.3	6 44.8	27 54.8	15 33.4
51	0 5.4	11 48.7	8 8.4	28 41.6	15 54.2
52	♒29 51.7	12 3.1	9 37.5	29 31.0	16 16.1
53	29 37.3	12 18.9	11 12.5	♋ 0 23.1	16 39.1
54	29 21.8	12 36.2	12 54.0	1 18.2	17 3.4
55	29 5.3	12 55.2	14 42.4	2 16.6	17 29.2
56	28 47.5	13 16.2	16 38.5	3 18.6	17 56.5
57	28 28.3	13 39.7	18 42.8	4 24.7	18 25.6
58	28 7.6	14 5.9	20 55.9	5 35.3	18 56.7
59	27 45.0	14 35.6	23 18.5	6 51.0	19 30.0
60	27 20.4	15 9.5	25 51.1	8 12.3	20 6.0
61	26 53.3	15 48.6	28 34.3	9 40.0	20 44.9
62	26 23.4	16 34.1	♋ 1 28.5	11 14.9	21 27.2
63	25 50.0	17 27.9	4 34.0	12 58.0	22 13.6
64	25 12.5	18 32.7	7 50.9	14 50.5	23 4.8
65	24 29.8	19 52.3	11 19.0	16 53.8	24 1.9
66	23 40.7	21 33.2	14 57.7	19 9.7	25 6.2
S LAT	**5**	**6**	**Descendant**	**8**	**9**
8h 28m 0s		MC ♌ 4° 39′ 30″		127° 0′ 0″	

20h 32m 0s		MC ♒ 5° 37′ 58″		308° 0′ 0″	N
11	**12**	**Ascendant**	**2**	**3**	**LAT**
° ′	° ′	° ′	° ′	° ′	°
♓ 6 14.0	♈ 8 42.5	♉ 10 25.0	♊ 9 39.7	♋ 7 20.8	**0**
5 56.5	8 56.0	11 49.0	10 55.9	8 0.4	**5**
5 38.3	9 10.4	13 19.3	12 14.6	8 40.5	**10**
5 19.1	9 26.1	14 58.0	13 37.2	9 21.8	**15**
5 15.0	9 29.5	15 19.0	13 54.3	9 30.3	**16**
5 11.0	9 32.9	15 40.5	14 11.6	9 38.8	**17**
5 6.9	9 36.3	16 2.4	14 29.2	9 47.5	**18**
5 2.7	9 39.9	16 24.9	14 47.0	9 56.2	**19**
4 58.4	9 43.5	16 47.9	15 5.2	10 5.1	**20**
4 54.0	9 47.3	17 11.6	15 23.6	10 14.1	**21**
4 49.6	9 51.1	17 35.8	15 42.3	10 23.2	**22**
4 45.1	9 55.1	18 0.8	16 1.4	10 32.4	**23**
4 40.4	9 59.1	18 26.4	16 20.8	10 41.8	**24**
4 35.7	10 3.3	18 52.8	16 40.6	10 51.3	**25**
4 30.9	10 7.6	19 20.0	17 0.8	11 1.0	**26**
4 26.0	10 12.1	19 48.0	17 21.5	11 10.8	**27**
4 20.9	10 16.7	20 17.0	17 42.6	11 20.9	**28**
4 15.7	10 21.4	20 46.9	18 4.2	11 31.1	**29**
4 10.4	10 26.4	21 17.9	18 26.2	11 41.6	**30**
4 4.9	10 31.5	21 49.9	18 48.9	11 52.2	**31**
3 59.3	10 36.7	22 23.1	19 12.1	12 3.1	**32**
3 53.5	10 42.3	22 57.6	19 35.9	12 14.3	**33**
3 47.5	10 48.0	23 33.4	20 0.4	12 25.7	**34**
3 41.3	10 53.9	24 10.6	20 25.6	12 37.4	**35**
3 34.9	11 0.2	24 49.4	20 51.5	12 49.5	**36**
3 28.4	11 6.7	25 29.8	21 18.2	13 1.8	**37**
3 21.5	11 13.5	26 12.0	21 45.8	13 14.5	**38**
3 14.5	11 20.7	26 56.1	22 14.3	13 27.6	**39**
3 7.1	11 28.2	27 42.2	22 43.7	13 41.1	**40**
2 59.5	11 36.1	28 30.6	23 14.2	13 55.0	**41**
2 51.5	11 44.5	29 21.3	23 45.8	14 9.4	**42**
2 43.3	11 53.3	♊ 0 14.6	24 18.6	14 24.2	**43**
2 34.6	12 2.7	1 10.6	24 52.7	14 39.6	**44**
2 25.5	12 12.6	2 9.7	25 28.2	14 55.6	**45**
2 16.1	12 23.2	3 12.0	26 5.1	15 12.3	**46**
2 6.1	12 34.5	4 17.8	26 43.7	15 29.5	**47**
1 55.6	12 46.6	5 27.4	27 24.0	15 47.6	**48**
1 44.6	12 59.6	6 41.2	28 6.3	16 6.4	**49**
1 32.9	13 13.6	7 59.5	28 50.5	16 26.1	**50**
1 20.6	13 28.8	9 22.6	29 37.1	16 46.7	**51**
1 7.5	13 45.2	10 51.1	♋ 0 26.1	17 8.4	**52**
0 53.6	14 3.1	12 25.4	1 17.7	17 31.2	**53**
0 38.7	14 22.8	14 5.9	2 12.3	17 55.3	**54**
0 22.8	14 44.4	15 53.1	3 10.2	18 20.8	**55**
0 5.7	15 8.3	17 47.7	4 11.6	18 47.9	**56**
♒29 47.3	15 34.9	19 50.2	5 17.0	19 16.6	**57**
29 27.3	16 4.7	22 1.2	6 26.7	19 47.4	**58**
29 5.7	16 38.5	24 21.2	7 41.4	20 20.3	**59**
28 42.0	17 16.9	26 50.9	9 1.6	20 55.7	**60**
28 16.0	18 1.3	29 30.6	10 27.9	21 34.1	**61**
27 47.3	18 52.9	♋ 2 20.9	12 1.2	22 15.7	**62**
27 15.3	19 54.1	5 21.9	13 42.4	23 1.3	**63**
26 39.4	21 7.7	8 33.8	15 32.7	23 51.5	**64**
25 58.7	22 38.5	11 56.3	17 33.2	24 47.4	**65**
25 11.8	24 33.8	15 29.1	19 45.7	25 50.1	**66**
5	**6**	**Descendant**	**8**	**9**	**S LAT**
8h 32m 0s		♌ 5° 37′ 58″ MC		128° 0′ 0″	

20h 36m 0s		MC ♒ 6° 36′ 37″		309° 0′ 0″	N
11	**12**	**Ascendant**	**2**	**3**	**LAT**
° ′	° ′	° ′	° ′	° ′	°
♓ 7 17.7	♈ 9 47.7	♉ 11 26.0	♊ 10 35.9	♋ 8 16.1	**0**
7 1.0	10 2.8	12 51.6	11 52.5	8 55.6	**5**
6 43.5	10 19.0	14 23.4	13 11.5	9 35.6	**10**
6 25.1	10 36.6	16 3.8	14 34.3	10 16.8	**15**
6 21.2	10 40.3	16 25.1	14 51.5	10 25.2	**16**
6 17.3	10 44.2	16 46.8	15 8.8	10 33.7	**17**
6 13.4	10 48.1	17 9.1	15 26.5	10 42.4	**18**
6 9.4	10 52.1	17 31.9	15 44.3	10 51.1	**19**
6 5.3	10 56.1	17 55.3	16 2.5	10 59.9	**20**
6 1.1	11 0.4	18 19.3	16 21.0	11 8.9	**21**
5 56.8	11 4.7	18 43.9	16 39.7	11 17.9	**22**
5 52.5	11 9.1	19 9.1	16 58.8	11 27.1	**23**
5 48.0	11 13.6	19 35.1	17 18.3	11 36.5	**24**
5 43.5	11 18.3	20 1.8	17 38.1	11 45.9	**25**
5 38.9	11 23.2	20 29.4	17 58.3	11 55.6	**26**
5 34.1	11 28.2	20 57.8	18 19.0	12 5.4	**27**
5 29.3	11 33.3	21 27.0	18 40.1	12 15.4	**28**
5 24.3	11 38.6	21 57.3	19 1.7	12 25.6	**29**
5 19.2	11 44.2	22 28.6	19 23.7	12 36.0	**30**
5 13.9	11 49.9	23 0.9	19 46.4	12 46.6	**31**
5 8.5	11 55.8	23 34.5	20 9.6	12 57.5	**32**
5 3.0	12 2.0	24 9.2	20 33.4	13 8.6	**33**
4 57.2	12 8.4	24 45.4	20 57.8	13 19.9	**34**
4 51.3	12 15.1	25 22.9	21 23.0	13 31.6	**35**
4 45.2	12 22.1	26 2.0	21 48.9	13 43.6	**36**
4 38.9	12 29.4	26 42.7	22 15.5	13 55.9	**37**
4 32.3	12 37.0	27 25.1	22 43.1	14 8.5	**38**
4 25.5	12 45.1	28 9.4	23 11.5	14 21.5	**39**
4 18.5	12 53.5	28 55.8	23 40.8	14 34.9	**40**
4 11.1	13 2.4	29 44.3	24 11.2	14 48.7	**41**
4 3.5	13 11.7	♊ 0 35.2	24 42.7	15 3.0	**42**
3 55.5	13 21.6	1 28.6	25 15.4	15 17.8	**43**
3 47.2	13 32.1	2 24.8	25 49.3	15 33.1	**44**
3 38.5	13 43.2	3 23.9	26 24.7	15 49.0	**45**
3 29.4	13 55.1	4 26.2	27 1.4	16 5.5	**46**
3 19.9	14 7.7	5 31.9	27 39.8	16 22.6	**47**
3 9.8	14 21.3	6 41.4	28 19.9	16 40.5	**48**
2 59.2	14 35.8	7 54.9	29 1.9	16 59.2	**49**
2 48.0	14 51.5	9 12.8	29 45.9	17 18.7	**50**
2 36.2	15 8.4	10 35.5	♋ 0 32.1	17 39.2	**51**
2 23.6	15 26.8	12 3.3	1 20.7	18 0.7	**52**
2 10.2	15 46.9	13 36.8	2 11.9	18 23.3	**53**
1 56.0	16 8.8	15 16.3	3 6.1	18 47.2	**54**
1 40.7	16 33.0	17 2.3	4 3.3	19 12.4	**55**
1 24.3	16 59.7	18 55.4	5 4.1	19 39.1	**56**
1 6.6	17 29.4	20 56.2	6 8.8	20 7.6	**57**
0 47.5	18 2.8	23 5.0	7 17.7	20 38.0	**58**
0 26.7	18 40.4	25 22.5	8 31.4	21 10.5	**59**
0 4.0	19 23.4	27 49.3	9 50.4	21 45.4	**60**
♒29 39.2	20 12.9	♋ 0 25.7	11 15.4	22 23.2	**61**
29 11.7	21 10.6	3 12.1	12 47.2	23 4.2	**62**
28 41.1	22 18.9	6 8.9	14 26.5	23 49.0	**63**
28 6.8	23 41.3	9 16.0	16 14.6	24 38.2	**64**
27 28.0	25 23.1	12 33.3	18 12.5	25 32.8	**65**
26 43.4	27 32.9	16 0.4	20 21.7	26 34.1	**66**
5	**6**	**Descendant**	**8**	**9**	**S LAT**
8h 36m 0s		♌ 6° 36′ 37″ MC		129° 0′ 0″	

20h 40m 0s — MC 310° 0′ 0″ — ♒ 7° 35′ 26″

11	12	Ascendant	2	3	N LAT
° ′	° ′	° ′	° ′	° ′	°
♓ 8 21.7	♈10 52.7	♉12 26.7	♊11 32.1	♋ 9 11.4	0
8 5.6	11 9.5	13 53.9	12 49.0	9 50.8	5
7 48.9	11 27.5	15 27.3	14 8.3	10 30.7	10
7 31.3	11 47.0	17 9.2	15 31.3	11 11.7	15
7 27.6	11 51.1	17 30.8	15 48.5	11 20.1	16
7 23.9	11 55.3	17 52.9	16 5.9	11 28.7	17
7 20.1	11 59.7	18 15.5	16 23.6	11 37.2	18
7 16.2	12 4.1	18 38.6	16 41.5	11 45.9	19
7 12.3	12 8.6	19 2.3	16 59.6	11 54.7	20
7 8.3	12 13.3	19 26.6	17 18.1	12 3.6	21
7 4.2	12 18.1	19 51.5	17 36.9	12 12.7	22
7 0.1	12 23.0	20 17.1	17 56.0	12 21.8	23
6 55.8	12 28.0	20 43.4	18 15.5	12 31.1	24
6 51.5	12 33.2	21 10.4	18 35.3	12 40.6	25
6 47.1	12 38.6	21 38.3	18 55.6	12 50.2	26
6 42.5	12 44.1	22 7.0	19 16.2	13 0.0	27
6 37.9	12 49.8	22 36.6	19 37.3	13 9.9	28
6 33.1	12 55.7	23 7.1	19 58.9	13 20.1	29
6 28.2	13 1.8	23 38.7	20 21.0	13 30.4	30
6 23.2	13 8.1	24 11.4	20 43.6	13 41.0	31
6 18.0	13 14.7	24 45.2	21 6.8	13 51.8	32
6 12.7	13 21.5	25 20.3	21 30.6	14 2.8	33
6 7.2	13 28.6	25 56.7	21 55.0	14 14.2	34
6 1.5	13 36.0	26 34.5	22 20.1	14 25.8	35
5 55.6	13 43.8	27 13.8	22 45.9	14 37.7	36
5 49.6	13 51.8	27 54.7	23 12.6	14 49.9	37
5 43.3	14 0.3	28 37.4	23 40.0	15 2.4	38
5 36.8	14 9.2	29 22.0	24 8.3	15 15.4	39
5 30.1	14 18.5	♊ 0 8.5	24 37.6	15 28.7	40
5 23.0	14 28.3	0 57.2	25 7.9	15 42.4	41
5 15.7	14 38.7	1 48.2	25 39.3	15 56.6	42
5 8.1	14 49.6	2 41.7	26 11.8	16 11.3	43
5 0.2	15 1.2	3 37.9	26 45.6	16 26.5	44
4 51.8	15 13.5	4 37.0	27 20.8	16 42.3	45
4 43.1	15 26.6	5 39.2	27 57.4	16 58.7	46
4 33.9	15 40.6	6 44.9	28 35.6	17 15.7	47
4 24.3	15 55.5	7 54.1	29 15.4	17 33.4	48
4 14.2	16 11.6	9 7.4	29 57.1	17 52.0	49
4 3.4	16 28.9	10 24.9	♋ 0 40.8	18 11.3	50
3 52.1	16 47.6	11 47.0	1 26.7	18 31.6	51
3 40.1	17 7.9	13 14.2	2 14.9	18 52.9	52
3 27.2	17 30.0	14 46.8	3 5.7	19 15.3	53
3 13.6	17 54.2	16 25.3	3 59.4	19 38.9	54
2 59.0	18 20.9	18 10.1	4 56.1	20 3.9	55
2 43.3	18 50.3	20 1.7	5 56.2	20 30.3	56
2 26.4	19 23.1	22 0.7	7 0.2	20 58.5	57
2 8.1	19 59.9	24 7.4	8 8.2	21 28.5	58
1 48.2	20 41.4	26 22.5	9 21.0	22 0.5	59
1 26.5	21 28.8	28 46.4	10 38.9	22 35.0	60
1 2.7	22 23.3	♋ 1 19.6	12 2.6	23 12.2	61
0 36.5	23 26.9	4 2.4	13 32.8	23 52.5	62
0 7.3	24 42.3	6 55.0	15 10.3	24 36.5	63
♒29 34.7	26 13.3	9 57.5	16 56.2	25 24.8	64
28 57.7	28 5.9	13 9.8	18 51.5	26 18.3	65
28 15.5	♉ 0 30.2	16 31.6	20 57.6	27 18.0	66
5	6	Descendant	8	9	S LAT

♌ 7° 35′ 26″ — 8h 40m 0s — MC 130° 0′ 0″

20h 44m 0s — MC 311° 0′ 0″ — ♒ 8° 34′ 25″

11	12	Ascendant	2	3	N LAT
° ′	° ′	° ′	° ′	° ′	°
♓ 9 25.7	♈11 57.7	♉13 27.3	♊12 28.1	♋10 6.7	0
9 10.4	12 16.1	14 55.9	13 45.3	10 46.0	5
8 54.5	12 35.8	16 30.9	15 4.9	11 25.8	10
8 37.7	12 57.2	18 14.2	16 28.1	12 6.7	15
8 34.2	13 1.8	18 36.2	16 45.3	12 15.1	16
8 30.6	13 6.4	18 58.6	17 2.8	12 23.6	17
8 27.0	13 11.2	19 21.5	17 20.5	12 32.2	18
8 23.3	13 16.0	19 44.9	17 38.4	12 40.8	19
8 19.5	13 21.0	20 8.9	17 56.6	12 49.6	20
8 15.7	13 26.1	20 33.5	18 15.1	12 58.5	21
8 11.8	13 31.3	20 58.7	18 33.9	13 7.5	22
8 7.9	13 36.7	21 24.6	18 53.1	13 16.6	23
8 3.8	13 42.2	21 51.2	19 12.5	13 25.8	24
7 59.7	13 47.9	22 18.5	19 32.4	13 35.2	25
7 55.4	13 53.8	22 46.7	19 52.6	13 44.8	26
7 51.1	13 59.8	23 15.6	20 13.3	13 54.5	27
7 46.7	14 6.1	23 45.5	20 34.4	14 4.5	28
7 42.1	14 12.6	24 16.4	20 55.9	14 14.6	29
7 37.4	14 19.3	24 48.2	21 18.0	14 24.9	30
7 32.6	14 26.2	25 21.2	21 40.6	14 35.4	31
7 27.7	14 33.4	25 55.3	22 3.8	14 46.1	32
7 22.6	14 40.9	26 30.6	22 27.5	14 57.1	33
7 17.3	14 48.6	27 7.3	22 51.9	15 8.4	34
7 11.9	14 56.8	27 45.3	23 17.0	15 19.9	35
7 6.3	15 5.2	28 24.9	23 42.7	15 31.7	36
7 0.6	15 14.1	29 6.1	24 9.3	15 43.9	37
6 54.6	15 23.3	29 48.9	24 36.7	15 56.4	38
6 48.4	15 33.0	♊ 0 33.7	25 4.9	16 9.2	39
6 41.9	15 43.2	1 20.4	25 34.1	16 22.4	40
6 35.2	15 54.0	2 9.2	26 4.3	16 36.1	41
6 28.2	16 5.3	3 0.3	26 35.6	16 50.2	42
6 20.9	16 17.3	3 53.9	27 8.0	17 4.8	43
6 13.3	16 29.9	4 50.1	27 41.6	17 19.9	44
6 5.4	16 43.4	5 49.1	28 16.6	17 35.5	45
5 57.1	16 57.7	6 51.3	28 53.0	17 51.8	46
5 48.3	17 13.0	7 56.7	29 31.0	18 8.7	47
5 39.1	17 29.4	9 5.7	♋ 0 10.6	18 26.3	48
5 29.4	17 46.9	10 18.6	0 52.0	18 44.7	49
5 19.2	18 5.8	11 35.7	1 35.4	19 3.9	50
5 8.3	18 26.3	12 57.3	2 20.9	19 24.0	51
4 56.8	18 48.4	14 23.8	3 8.8	19 45.1	52
4 44.6	19 12.6	15 55.5	3 59.1	20 7.2	53
4 31.6	19 39.0	17 33.0	4 52.3	20 30.6	54
4 17.6	20 8.1	19 16.5	5 48.4	20 55.3	55
4 2.6	20 40.2	21 6.6	6 47.9	21 21.5	56
3 46.5	21 16.0	23 3.8	7 51.1	21 49.3	57
3 29.0	21 56.1	25 8.5	8 58.4	22 18.9	58
3 10.1	22 41.3	27 21.3	10 10.1	22 50.5	59
2 49.4	23 32.9	29 42.4	11 26.9	23 24.5	60
2 26.8	24 32.4	♋ 2 12.4	12 49.3	24 1.2	61
2 1.8	25 41.8	4 51.6	14 18.0	24 40.8	62
1 34.0	27 4.0	7 40.3	15 53.8	25 24.1	63
1 3.0	28 43.4	10 38.4	17 37.5	26 11.4	64
0 27.9	♉ 0 46.7	13 46.0	19 30.4	27 3.7	65
♒29 47.9	3 25.7	17 2.6	21 33.5	28 2.0	66
5	6	Descendant	8	9	S LAT

♌ 8° 34′ 25″ — 8h 44m 0s — MC 131° 0′ 0″

20ʰ 48ᵐ 0ˢ		MC		312° 0′ 0″	
		♒ 9° 33′ 34″			
N LAT	**11**	**12**	**Ascendant**	**2**	**3**
0	♓10 29.9	♈13 2.6	♉14 27.7	♊13 24.0	♋11 2.1
5	10 15.3	13 22.7	15 57.8	14 41.6	11 41.3
10	10 0.2	13 44.1	17 34.2	16 1.4	12 21.0
15	9 44.2	14 7.4	19 19.0	17 24.8	13 1.8
16	9 40.8	14 12.3	19 41.2	17 42.0	13 10.1
17	9 37.5	14 17.3	20 3.9	17 59.5	13 18.6
18	9 34.0	14 22.5	20 27.1	18 17.2	13 27.1
19	9 30.5	14 27.8	20 50.8	18 35.2	13 35.7
20	9 26.9	14 33.2	21 15.1	18 53.4	13 44.4
21	9 23.3	14 38.7	21 40.0	19 11.9	13 53.3
22	9 19.6	14 44.4	22 5.5	19 30.7	14 2.2
23	9 15.8	14 50.3	22 31.6	19 49.9	14 11.3
24	9 12.0	14 56.3	22 58.5	20 9.4	14 20.5
25	9 8.0	15 2.5	23 26.1	20 29.2	14 29.9
26	9 4.0	15 8.8	23 54.6	20 49.5	14 39.4
27	8 59.9	15 15.4	24 23.8	21 10.1	14 49.1
28	8 55.7	15 22.2	24 54.0	21 31.2	14 59.0
29	8 51.3	15 29.2	25 25.1	21 52.7	15 9.0
30	8 46.9	15 36.5	25 57.2	22 14.8	15 19.3
31	8 42.3	15 44.0	26 30.4	22 37.4	15 29.7
32	8 37.6	15 51.9	27 4.8	23 0.5	15 40.4
33	8 32.7	16 0.0	27 40.4	23 24.2	15 51.4
34	8 27.7	16 8.4	28 17.2	23 48.5	16 2.6
35	8 22.6	16 17.2	28 55.5	24 13.6	16 14.0
36	8 17.3	16 26.4	29 35.3	24 39.3	16 25.8
37	8 11.8	16 36.0	♊ 0 16.6	25 5.8	16 37.9
38	8 6.1	16 46.1	0 59.7	25 33.1	16 50.3
39	8 0.1	16 56.6	1 44.6	26 1.2	17 3.0
40	7 54.0	17 7.7	2 31.4	26 30.3	17 16.2
41	7 47.6	17 19.3	3 20.3	27 0.4	17 29.7
42	7 41.0	17 31.6	4 11.5	27 31.5	17 43.8
43	7 34.0	17 44.6	5 5.1	28 3.8	17 58.2
44	7 26.8	17 58.3	6 1.2	28 37.3	18 13.2
45	7 19.2	18 12.9	7 0.2	29 12.1	18 28.8
46	7 11.3	18 28.4	8 2.2	29 48.3	18 44.9
47	7 3.0	18 45.0	9 7.5	♋ 0 26.0	19 1.6
48	6 54.2	19 2.7	10 16.2	1 5.4	19 19.1
49	6 45.0	19 21.7	11 28.8	1 46.6	19 37.3
50	6 35.2	19 42.2	12 45.4	2 29.6	19 56.4
51	6 24.9	20 4.3	14 6.4	3 14.8	20 16.3
52	6 14.0	20 28.3	15 32.1	4 2.2	20 37.2
53	6 2.3	20 54.4	17 3.0	4 52.2	20 59.1
54	5 49.9	21 23.0	18 39.4	5 44.8	21 22.3
55	5 36.6	21 54.5	20 21.6	6 40.4	21 46.7
56	5 22.3	22 29.2	22 10.3	7 39.2	22 12.5
57	5 7.0	23 7.9	24 5.7	8 41.7	22 40.0
58	4 50.4	23 51.2	26 8.4	9 48.1	23 9.2
59	4 32.4	24 40.1	28 18.8	10 58.9	23 40.5
60	4 12.7	25 35.9	♋ 0 37.3	12 14.6	24 14.0
61	3 51.2	26 40.1	3 4.3	13 35.7	24 50.0
62	3 27.5	27 55.0	5 40.0	15 2.9	25 29.1
63	3 1.1	29 23.9	8 24.9	16 37.0	26 11.5
64	2 31.7	♉ 1 11.5	11 18.8	18 18.7	26 58.0
65	1 58.5	3 25.5	14 21.8	20 9.1	27 49.2
66	1 20.7	6 19.2	17 33.5	22 9.3	28 46.1
S LAT	**5**	**6**	**Descendant**	**8**	**9**
		♌ 9° 33′ 34″			
8ʰ 48ᵐ 0ˢ		MC		132° 0′ 0″	

20ʰ 52ᵐ 0ˢ		MC		313° 0′ 0″	
		♒ 10° 32′ 55″			
N LAT	**11**	**12**	**Ascendant**	**2**	**3**
0	♓11 34.2	♈14 7.5	♉15 28.0	♊14 19.9	♋11 57.6
5	11 20.4	14 29.1	16 59.4	15 37.7	12 36.6
10	11 6.0	14 52.2	18 37.2	16 57.7	13 16.2
15	10 50.9	15 17.4	20 23.4	18 21.3	13 56.8
16	10 47.7	15 22.7	20 45.9	18 38.6	14 5.2
17	10 44.5	15 28.1	21 8.9	18 56.1	14 13.6
18	10 41.2	15 33.7	21 32.4	19 13.8	14 22.1
19	10 37.9	15 39.4	21 56.3	19 31.8	14 30.6
20	10 34.5	15 45.2	22 20.9	19 50.0	14 39.3
21	10 31.1	15 51.2	22 46.0	20 8.5	14 48.1
22	10 27.6	15 57.4	23 11.8	20 27.4	14 57.1
23	10 24.0	16 3.7	23 38.3	20 46.5	15 6.1
24	10 20.3	16 10.2	24 5.4	21 6.0	15 15.3
25	10 16.6	16 16.8	24 33.3	21 25.8	15 24.6
26	10 12.8	16 23.7	25 2.0	21 46.1	15 34.1
27	10 8.8	16 30.8	25 31.5	22 6.7	15 43.7
28	10 4.8	16 38.1	26 1.9	22 27.8	15 53.5
29	10 0.7	16 45.7	26 33.3	22 49.3	16 3.5
30	9 56.5	16 53.5	27 5.7	23 11.3	16 13.7
31	9 52.1	17 1.7	27 39.1	23 33.9	16 24.1
32	9 47.7	17 10.1	28 13.7	23 57.0	16 34.8
33	9 43.1	17 18.8	28 49.5	24 20.6	16 45.6
34	9 38.3	17 27.9	29 26.6	24 44.9	16 56.7
35	9 33.5	17 37.4	♊ 0 5.0	25 9.9	17 8.1
36	9 28.4	17 47.3	0 45.0	25 35.6	17 19.8
37	9 23.2	17 57.7	1 26.5	26 2.0	17 31.8
38	9 17.8	18 8.5	2 9.7	26 29.2	17 44.2
39	9 12.1	18 19.8	2 54.7	26 57.2	17 56.8
40	9 6.3	18 31.8	3 41.6	27 26.2	18 9.9
41	9 0.3	18 44.3	4 30.6	27 56.2	18 23.4
42	8 53.9	18 57.5	5 21.8	28 27.2	18 37.3
43	8 47.4	19 11.5	6 15.3	28 59.3	18 51.7
44	8 40.5	19 26.3	7 11.5	29 32.6	19 6.5
45	8 33.3	19 42.0	8 10.4	♋ 0 7.2	19 21.9
46	8 25.8	19 58.7	9 12.2	0 43.3	19 37.9
47	8 17.9	20 16.5	10 17.2	1 20.8	19 54.6
48	8 9.6	20 35.6	11 25.7	1 59.9	20 11.9
49	8 0.8	20 56.0	12 37.8	2 40.7	20 29.9
50	7 51.6	21 18.0	13 53.9	3 23.5	20 48.8
51	7 41.8	21 41.8	15 14.3	4 8.3	21 8.5
52	7 31.4	22 7.5	16 39.3	4 55.3	21 29.2
53	7 20.3	22 35.6	18 9.2	5 44.8	21 50.9
54	7 8.5	23 6.3	19 44.5	6 36.9	22 13.8
55	6 55.9	23 40.0	21 25.5	7 32.0	22 38.0
56	6 42.4	24 17.3	23 12.7	8 30.2	23 3.5
57	6 27.8	24 58.8	25 6.4	9 31.9	23 30.7
58	6 12.1	25 45.2	27 7.1	10 37.5	23 59.5
59	5 55.0	26 37.7	29 15.2	11 47.3	24 30.3
60	5 36.4	27 37.4	♋ 1 31.1	13 1.9	25 3.4
61	5 16.0	28 46.3	3 55.1	14 21.8	25 38.9
62	4 53.6	♉ 0 6.6	6 27.6	15 47.5	26 17.3
63	4 28.7	1 42.0	9 8.7	17 19.9	26 59.0
64	4 0.9	3 37.6	11 58.6	18 59.6	27 44.5
65	3 29.6	6 2.0	14 57.3	20 47.6	28 34.6
66	2 53.9	9 10.5	18 4.4	22 45.1	29 30.2
S LAT	**5**	**6**	**Descendant**	**8**	**9**
		♌ 10° 32′ 55″			
8ʰ 52ᵐ 0ˢ		MC		133° 0′ 0″	

20h 56m 0s — MC 314° 0′ 0″ — ♒ 11° 32′ 25″

11	12	Ascendant	2	3	N LAT
♓12 38.6	♈15 12.2	♉16 28.0	♊15 15.6	♋12 53.1	0
12 25.6	15 35.4	18 0.9	16 33.7	13 32.0	5
12 12.0	16 0.3	19 40.0	17 53.9	14 11.4	10
11 57.7	16 27.2	21 27.5	19 17.6	14 51.9	15
11 54.7	16 32.9	21 50.3	19 34.9	15 0.2	16
11 51.7	16 38.8	22 13.5	19 52.5	15 8.6	17
11 48.6	16 44.8	22 37.3	20 10.2	15 17.1	18
11 45.4	16 50.9	23 1.5	20 28.2	15 25.6	19
11 42.2	16 57.1	23 26.3	20 46.4	15 34.3	20
11 39.0	17 3.5	23 51.7	21 5.0	15 43.0	21
11 35.7	17 10.1	24 17.8	21 23.8	15 51.9	22
11 32.3	17 16.9	24 44.5	21 43.0	16 0.9	23
11 28.8	17 23.8	25 11.9	22 2.4	16 10.0	24
11 25.3	17 31.0	25 40.0	22 22.3	16 19.3	25
11 21.7	17 38.4	26 8.9	22 42.5	16 28.7	26
11 18.0	17 46.0	26 38.7	23 3.1	16 38.3	27
11 14.2	17 53.8	27 9.3	23 24.2	16 48.1	28
11 10.3	18 1.9	27 40.9	23 45.7	16 58.0	29
11 6.3	18 10.3	28 13.5	24 7.7	17 8.2	30
11 2.2	18 19.0	28 47.2	24 30.2	17 18.5	31
10 58.0	18 28.1	29 22.0	24 53.2	17 29.1	32
10 53.6	18 37.4	29 58.0	25 16.9	17 39.9	33
10 49.1	18 47.2	♊ 0 35.2	25 41.1	17 50.9	34
10 44.5	18 57.4	1 13.9	26 6.0	18 2.3	35
10 39.7	19 7.9	1 54.0	26 31.6	18 13.9	36
10 34.8	19 19.0	2 35.6	26 57.9	18 25.8	37
10 29.7	19 30.6	3 18.9	27 25.0	18 38.1	38
10 24.4	19 42.8	4 4.0	27 53.0	18 50.7	39
10 18.9	19 55.5	4 51.0	28 21.9	19 3.6	40
10 13.1	20 8.9	5 40.0	28 51.7	19 17.0	41
10 7.2	20 23.1	6 31.2	29 22.6	19 30.8	42
10 0.9	20 38.0	7 24.7	29 54.5	19 45.1	43
9 54.4	20 53.8	8 20.8	♋ 0 27.7	19 59.8	44
9 47.7	21 10.6	9 19.5	1 2.1	20 15.1	45
9 40.5	21 28.4	10 21.2	1 37.9	20 31.0	46
9 33.1	21 47.5	11 25.9	2 15.2	20 47.5	47
9 25.2	22 7.8	12 34.0	2 54.0	21 4.6	48
9 16.9	22 29.7	13 45.7	3 34.6	21 22.5	49
9 8.2	22 53.2	15 1.3	4 17.0	21 41.2	50
8 58.9	23 18.5	16 21.0	5 1.5	22 0.7	51
8 49.1	23 46.0	17 45.3	5 48.1	22 21.2	52
8 38.6	24 16.0	19 14.3	6 37.1	22 42.7	53
8 27.5	24 48.7	20 48.5	7 28.7	23 5.4	54
8 15.6	25 24.7	22 28.3	8 23.2	23 29.2	55
8 2.8	26 4.4	24 14.0	9 20.7	23 54.5	56
7 49.1	26 48.6	26 6.0	10 21.7	24 21.3	57
7 34.2	27 38.1	28 4.7	11 26.5	24 49.8	58
7 18.1	28 33.9	♋ 0 10.6	12 35.4	25 20.1	59
7 0.5	29 37.5	2 23.9	13 48.9	25 52.7	60
6 41.3	♉ 0 50.8	4 45.1	15 7.5	26 27.7	61
6 20.1	2 16.4	7 14.4	16 31.8	27 5.4	62
5 56.6	3 58.1	9 52.0	18 2.5	27 46.3	63
5 30.4	6 1.5	12 38.0	19 40.3	28 31.0	64
5 1.0	8 36.1	15 32.5	21 26.1	29 20.0	65
4 27.5	11 59.6	18 35.1	23 20.9	♌ 0 14.3	66
5	6	Descendant	8	9	S LAT

♌ 11° 32′ 25″ — 8h 56m 0s — MC 134° 0′ 0″

21h 0m 0s — MC 315° 0′ 0″ — ♒ 12° 32′ 6″

N LAT	11	12	Ascendant	2	3
0	♓13 43.2	♈16 16.8	♉17 27.9	♊16 11.3	♋13 48.7
5	13 30.9	16 41.7	19 2.1	17 29.6	14 27.5
10	13 18.1	17 8.2	20 42.5	18 50.0	15 6.8
15	13 4.6	17 36.9	22 31.3	20 13.9	15 47.1
16	13 1.8	17 43.0	22 54.4	20 31.2	15 55.3
17	12 59.0	17 49.3	23 17.8	20 48.7	16 3.7
18	12 56.1	17 55.6	23 41.8	21 6.5	16 12.1
19	12 53.1	18 2.2	24 6.3	21 24.5	16 20.6
20	12 50.1	18 8.8	24 31.4	21 42.7	16 29.2
21	12 47.0	18 15.7	24 57.0	22 1.3	16 37.9
22	12 43.9	18 22.7	25 23.3	22 20.1	16 46.8
23	12 40.7	18 29.9	25 50.2	22 39.2	16 55.7
24	12 37.5	18 37.3	26 17.9	22 58.7	17 4.8
25	12 34.1	18 45.0	26 46.3	23 18.6	17 14.0
26	12 30.7	18 52.8	27 15.4	23 38.8	17 23.4
27	12 27.3	19 0.9	27 45.4	23 59.4	17 33.0
28	12 23.7	19 9.3	28 16.3	24 20.4	17 42.7
29	12 20.0	19 17.9	28 48.1	24 41.9	17 52.6
30	12 16.3	19 26.9	29 20.9	25 3.8	18 2.6
31	12 12.4	19 36.2	29 54.7	25 26.3	18 12.9
32	12 8.4	19 45.8	♊ 0 29.7	25 49.3	18 23.4
33	12 4.3	19 55.8	1 5.8	26 12.8	18 34.2
34	12 0.1	20 6.1	1 43.3	26 37.0	18 45.1
35	11 55.8	20 17.0	2 22.0	27 1.8	18 56.4
36	11 51.3	20 28.3	3 2.3	27 27.4	19 7.9
37	11 46.6	20 40.0	3 44.0	27 53.6	19 19.8
38	11 41.8	20 52.4	4 27.4	28 20.6	19 31.9
39	11 36.8	21 5.3	5 12.6	28 48.5	19 44.5
40	11 31.6	21 18.9	5 59.6	29 17.2	19 57.3
41	11 26.2	21 33.1	6 48.6	29 46.9	20 10.6
42	11 20.6	21 48.2	7 39.8	♋ 0 17.7	20 24.3
43	11 14.7	22 4.1	8 33.3	0 49.5	20 38.5
44	11 8.6	22 20.9	9 29.2	1 22.4	20 53.1
45	11 2.2	22 38.7	10 27.8	1 56.7	21 8.3
46	10 55.5	22 57.7	11 29.2	2 32.2	21 24.0
47	10 48.5	23 17.9	12 33.7	3 9.3	21 40.3
48	10 41.1	23 39.5	13 41.4	3 47.8	21 57.3
49	10 33.3	24 2.7	14 52.6	4 28.1	22 15.1
50	10 25.0	24 27.7	16 7.7	5 10.2	22 33.6
51	10 16.3	24 54.6	17 26.7	5 54.3	22 52.9
52	10 7.1	25 23.8	18 50.2	6 40.5	23 13.2
53	9 57.3	25 55.5	20 18.3	7 29.1	23 34.5
54	9 46.8	26 30.3	21 51.4	8 20.2	23 56.8
55	9 35.6	27 8.4	23 29.9	9 14.1	24 20.4
56	9 23.5	27 50.5	25 14.1	10 11.0	24 45.4
57	9 10.6	28 37.3	27 4.5	11 11.2	25 11.8
58	8 56.6	29 29.7	29 1.3	12 15.2	25 39.9
59	8 41.4	♉ 0 28.8	♋ 1 4.9	13 23.1	26 9.9
60	8 24.9	1 36.2	3 15.8	14 35.6	26 42.0
61	8 6.8	2 53.8	5 34.2	15 53.0	27 16.4
62	7 46.9	4 24.4	8 0.4	17 15.9	27 53.5
63	7 24.9	6 12.0	10 34.6	18 44.9	28 33.7
64	7 0.3	8 23.0	13 17.0	20 20.8	29 17.5
65	6 32.8	11 7.7	16 7.4	22 4.4	♌ 0 5.5
66	6 1.5	14 46.4	19 5.7	23 56.6	0 58.5
S LAT	5	6	Descendant	8	9

♌ 12° 32′ 6″ — 9h 0m 0s — MC 135° 0′ 0″

21h 4m 0s MC 316° 0′ 0″

♒ 13° 31′ 59″

11	12	Ascendant	2	3	N LAT
° ′	° ′	° ′	° ′	° ′	°
♓14 47.8	♈17 21.4	♉18 27.6	♊17 6.9	♋14 44.4	0
14 36.3	17 47.8	20 3.0	18 25.4	15 23.1	5
14 24.4	18 15.9	21 44.7	19 46.0	16 2.1	10
14 11.7	18 46.5	23 34.8	21 10.0	16 42.3	15
14 9.1	18 53.0	23 58.1	21 27.3	16 50.5	16
14 6.4	18 59.6	24 21.8	21 44.8	16 58.8	17
14 3.7	19 6.4	24 46.0	22 2.6	17 7.2	18
14 0.9	19 13.3	25 10.8	22 20.6	17 15.7	19
13 58.1	19 20.4	25 36.1	22 38.9	17 24.2	20
13 55.2	19 27.6	26 1.9	22 57.4	17 32.9	21
13 52.3	19 35.1	26 28.4	23 16.2	17 41.7	22
13 49.3	19 42.7	26 55.6	23 35.4	17 50.6	23
13 46.3	19 50.6	27 23.5	23 54.8	17 59.6	24
13 43.2	19 58.7	27 52.0	24 14.6	18 8.8	25
13 40.0	20 7.0	28 21.4	24 34.8	18 18.1	26
13 36.7	20 15.6	28 51.6	24 55.4	18 27.6	27
13 33.4	20 24.5	29 22.7	25 16.4	18 37.3	28
13 29.9	20 33.7	29 54.7	25 37.8	18 47.1	29
13 26.4	20 43.2	♊ 0 27.6	25 59.7	18 57.1	30
13 22.8	20 53.0	1 1.7	26 22.2	19 7.3	31
13 19.1	21 3.2	1 36.8	26 45.1	19 17.8	32
13 15.2	21 13.8	2 13.1	27 8.6	19 28.4	33
13 11.3	21 24.8	2 50.7	27 32.7	19 39.4	34
13 7.2	21 36.3	3 29.6	27 57.5	19 50.5	35
13 3.0	21 48.2	4 9.9	28 22.9	20 2.0	36
12 58.6	22 0.7	4 51.8	28 49.1	20 13.8	37
12 54.1	22 13.8	5 35.2	29 16.0	20 25.8	38
12 49.4	22 27.5	6 20.4	29 43.7	20 38.3	39
12 44.6	22 41.8	7 7.5	♋ 0 12.4	20 51.0	40
12 39.5	22 56.9	7 56.5	0 41.9	21 4.2	41
12 34.2	23 12.9	8 47.6	1 12.5	21 17.8	42
12 28.8	23 29.7	9 41.0	1 44.1	21 31.8	43
12 23.0	23 47.5	10 36.8	2 16.9	21 46.4	44
12 17.0	24 6.4	11 35.1	2 50.9	22 1.4	45
12 10.7	24 26.4	12 36.3	3 26.3	22 17.0	46
12 4.2	24 47.8	13 40.5	4 3.1	22 33.2	47
11 57.2	25 10.7	14 47.8	4 41.4	22 50.0	48
11 49.9	25 35.2	15 58.6	5 21.3	23 7.6	49
11 42.2	26 1.5	17 13.0	6 3.1	23 25.9	50
11 34.0	26 30.0	18 31.4	6 46.8	23 45.1	51
11 25.4	27 0.8	19 54.1	7 32.6	24 5.1	52
11 16.1	27 34.3	21 21.2	8 20.7	24 26.2	53
11 6.3	28 10.9	22 53.3	9 11.3	24 48.3	54
10 55.8	28 51.1	24 30.5	10 4.6	25 11.6	55
10 44.5	29 35.5	26 13.3	11 0.9	25 36.3	56
10 32.4	♉ 0 24.8	28 1.9	12 0.4	26 2.4	57
10 19.3	1 20.0	29 56.8	13 3.5	26 30.1	58
10 5.1	2 22.3	♋ 1 58.4	14 10.5	26 59.6	59
9 49.7	3 33.2	4 6.8	15 21.9	27 31.2	60
9 32.8	4 54.9	6 22.6	16 38.1	28 5.1	61
9 14.1	6 30.3	8 45.8	17 59.7	28 41.6	62
8 53.6	8 23.8	11 16.7	19 27.1	29 21.1	63
8 30.6	10 42.1	13 55.5	21 1.2	♌ 0 4.0	64
8 4.9	13 36.6	16 42.1	22 42.6	0 51.0	65
7 35.8	17 30.6	19 36.3	24 32.3	1 42.7	66
5	6	Descendant	8	9	S LAT

♌ 13° 31′ 59″

9h 4m 0s MC 136° 0′ 0″

21h 8m 0s MC 317° 0′ 0″

♒ 14° 32′ 1″

N LAT	11	12	Ascendant	2	3
°	° ′	° ′	° ′	° ′	° ′
0	♓15 52.5	♈18 25.8	♉19 27.1	♊18 2.4	♋15 40.1
5	15 41.8	18 53.7	21 3.8	19 21.2	16 18.7
10	15 30.7	19 23.5	22 46.7	20 41.9	16 57.6
15	15 18.9	19 55.9	24 38.0	22 5.9	17 37.6
16	15 16.5	20 2.7	25 1.5	22 23.2	17 45.7
17	15 14.0	20 9.7	25 25.4	22 40.8	17 54.0
18	15 11.4	20 16.9	25 49.9	22 58.6	18 2.3
19	15 8.9	20 24.2	26 14.8	23 16.6	18 10.8
20	15 6.2	20 31.7	26 40.4	23 34.8	18 19.3
21	15 3.6	20 39.4	27 6.5	23 53.4	18 27.9
22	15 0.8	20 47.2	27 33.2	24 12.2	18 36.7
23	14 58.1	20 55.3	28 0.6	24 31.3	18 45.5
24	14 55.2	21 3.7	28 28.6	24 50.7	18 54.5
25	14 52.3	21 12.2	28 57.4	25 10.5	19 3.6
26	14 49.3	21 21.0	29 27.0	25 30.7	19 12.9
27	14 46.3	21 30.1	29 57.3	25 51.2	19 22.3
28	14 43.2	21 39.5	♊ 0 28.6	26 12.2	19 31.9
29	14 40.0	21 49.2	1 0.8	26 33.6	19 41.7
30	14 36.7	21 59.2	1 33.9	26 55.5	19 51.6
31	14 33.3	22 9.6	2 8.1	27 17.8	20 1.8
32	14 29.9	22 20.3	2 43.4	27 40.7	20 12.2
33	14 26.3	22 31.5	3 19.8	28 4.2	20 22.8
34	14 22.6	22 43.1	3 57.5	28 28.2	20 33.6
35	14 18.8	22 55.2	4 36.5	28 52.9	20 44.7
36	14 14.9	23 7.8	5 16.9	29 18.2	20 56.1
37	14 10.8	23 21.0	5 58.8	29 44.3	21 7.8
38	14 6.6	23 34.8	6 42.3	♋ 0 11.1	21 19.7
39	14 2.2	23 49.2	7 27.5	0 38.7	21 32.1
40	13 57.7	24 4.4	8 14.6	1 7.2	21 44.7
41	13 53.0	24 20.3	9 3.5	1 36.6	21 57.8
42	13 48.1	24 37.1	9 54.6	2 7.1	22 11.3
43	13 43.0	24 54.8	10 47.8	2 38.5	22 25.2
44	13 37.6	25 13.6	11 43.5	3 11.1	22 39.6
45	13 32.0	25 33.4	12 41.6	3 44.9	22 54.5
46	13 26.2	25 54.6	13 42.5	4 20.0	23 10.0
47	13 20.0	26 17.1	14 46.3	4 56.6	23 26.0
48	13 13.6	26 41.1	15 53.3	5 34.6	23 42.7
49	13 6.8	27 6.9	17 3.5	6 14.3	24 0.1
50	12 59.6	27 34.6	18 17.4	6 55.7	24 18.2
51	12 51.9	28 4.5	19 35.1	7 39.0	24 37.2
52	12 43.9	28 36.9	20 56.9	8 24.4	24 57.0
53	12 35.3	29 12.1	22 23.1	9 12.0	25 17.8
54	12 26.1	29 50.6	23 54.1	10 2.1	25 39.7
55	12 16.3	♉ 0 32.8	25 30.1	10 54.8	26 2.8
56	12 5.8	1 19.4	27 11.4	11 50.5	26 27.1
57	11 54.6	2 11.1	28 58.4	12 49.3	26 52.8
58	11 42.4	3 9.0	♋ 0 51.5	13 51.5	27 20.2
59	11 29.1	4 14.3	2 50.9	14 57.7	27 49.3
60	11 14.7	5 28.6	4 57.1	16 8.0	28 20.4
61	10 59.0	6 54.3	7 10.2	17 23.0	28 53.7
62	10 41.7	8 34.3	9 30.5	18 43.2	29 29.6
63	10 22.6	10 33.3	11 58.3	20 9.1	♌ 0 8.4
64	10 1.3	12 58.6	14 33.6	21 41.4	0 50.5
65	9 37.4	16 2.8	17 16.5	23 20.8	1 36.5
66	9 10.4	20 12.2	20 6.8	25 8.1	2 27.1
S LAT	5	6	Descendant	8	9

♌ 14° 32′ 1″

9h 8m 0s MC 137° 0′ 0″

21h 12m 0s — MC — 318° 0′ 0″

♒ 15° 32′ 15″

N LAT	11	12	Ascendant	2	3
°	° ′	° ′	° ′	° ′	° ′
0	♓16 57.4	♈19 30.1	♉20 26.4	♊18 57.9	♋16 36.0
5	16 47.5	19 59.6	22 4.4	20 16.8	17 14.4
10	16 37.2	20 31.0	23 48.4	21 37.7	17 53.1
15	16 26.3	21 5.1	25 40.9	23 1.7	18 32.9
16	16 24.0	21 12.3	26 4.6	23 19.1	18 41.0
17	16 21.7	21 19.7	26 28.7	23 36.6	18 49.2
18	16 19.3	21 27.2	26 53.4	23 54.4	18 57.5
19	16 16.9	21 34.9	27 18.6	24 12.4	19 5.9
20	16 14.5	21 42.8	27 44.3	24 30.7	19 14.4
21	16 12.0	21 50.9	28 10.6	24 49.2	19 23.0
22	16 9.5	21 59.2	28 37.5	25 8.0	19 31.7
23	16 6.9	22 7.7	29 5.1	25 27.1	19 40.5
24	16 4.3	22 16.5	29 33.3	25 46.5	19 49.4
25	16 1.6	22 25.5	♊ 0 2.3	26 6.3	19 58.5
26	15 58.9	22 34.7	0 32.1	26 26.4	20 7.7
27	15 56.0	22 44.3	1 2.6	26 46.9	20 17.0
28	15 53.2	22 54.2	1 34.0	27 7.9	20 26.6
29	15 50.2	23 4.4	2 6.3	27 29.2	20 36.3
30	15 47.2	23 14.9	2 39.6	27 51.0	20 46.2
31	15 44.0	23 25.8	3 14.0	28 13.3	20 56.3
32	15 40.8	23 37.1	3 49.4	28 36.2	21 6.6
33	15 37.5	23 48.9	4 25.9	28 59.5	21 17.1
34	15 34.1	24 1.1	5 3.7	29 23.5	21 27.9
35	15 30.6	24 13.8	5 42.8	29 48.1	21 38.9
36	15 26.9	24 27.1	6 23.3	♋ 0 13.3	21 50.2
37	15 23.2	24 40.9	7 5.2	0 39.3	22 1.8
38	15 19.3	24 55.4	7 48.8	1 6.0	22 13.7
39	15 15.2	25 10.6	8 34.0	1 33.5	22 25.9
40	15 11.0	25 26.5	9 21.0	2 1.9	22 38.5
41	15 6.7	25 43.2	10 9.9	2 31.1	22 51.4
42	15 2.1	26 0.8	11 0.8	3 1.4	23 4.8
43	14 57.4	26 19.4	11 53.9	3 32.7	23 18.6
44	14 52.4	26 39.1	12 49.4	4 5.1	23 32.9
45	14 47.3	27 0.0	13 47.3	4 38.7	23 47.6
46	14 41.8	27 22.1	14 47.9	5 13.5	24 2.9
47	14 36.1	27 45.7	15 51.3	5 49.8	24 18.8
48	14 30.1	28 10.9	16 57.8	6 27.6	24 35.4
49	14 23.8	28 37.9	18 7.6	7 6.9	24 52.6
50	14 17.2	29 7.0	19 20.8	7 48.0	25 10.5
51	14 10.1	29 38.3	20 37.8	8 30.9	25 29.3
52	14 2.6	♉ 0 12.2	21 58.8	9 15.9	25 48.9
53	13 54.7	0 49.0	23 24.1	10 3.1	26 9.5
54	13 46.2	1 29.3	24 53.9	10 52.6	26 31.1
55	13 37.1	2 13.4	26 28.7	11 44.7	26 53.9
56	13 27.4	3 2.1	28 8.6	12 39.7	27 17.9
57	13 17.0	3 56.2	29 54.0	13 37.8	27 43.3
58	13 5.7	4 56.6	♋ 1 45.3	14 39.3	28 10.3
59	12 53.4	6 4.8	3 42.7	15 44.5	28 39.0
60	12 40.1	7 22.4	5 46.5	16 53.8	29 9.6
61	12 25.6	8 51.7	7 57.1	18 7.6	29 42.4
62	12 9.6	10 36.1	10 14.7	19 26.5	♌ 0 17.6
63	11 51.9	12 40.5	12 39.4	20 50.9	0 55.7
64	11 32.2	15 12.5	15 11.4	22 21.4	1 37.0
65	11 10.2	18 26.0	17 50.7	23 58.8	2 22.0
66	10 45.3	22 51.0	20 37.3	25 43.8	3 11.4
S LAT	5	6	Descendant	8	9

♌ 15° 32′ 15″

9h 12m 0s — MC — 138° 0′ 0″

21h 16m 0s — MC — 319° 0′ 0″

♒ 16° 32′ 40″

N LAT	11	12	Ascendant	2	3
°	° ′	° ′	° ′	° ′	° ′
0	♓18 2.3	♈20 34.3	♉21 25.6	♊19 53.3	♋17 31.9
5	17 53.2	21 5.3	23 4.7	21 12.4	18 10.1
10	17 43.7	21 38.3	24 49.9	22 33.3	18 48.7
15	17 33.7	22 14.1	26 43.4	23 57.5	19 28.3
16	17 31.6	22 21.7	27 7.3	24 14.8	19 36.4
17	17 29.5	22 29.4	27 31.7	24 32.4	19 44.5
18	17 27.3	22 37.3	27 56.6	24 50.1	19 52.8
19	17 25.1	22 45.4	28 21.9	25 8.1	20 1.1
20	17 22.9	22 53.7	28 47.9	25 26.4	20 9.5
21	17 20.6	23 2.2	29 14.4	25 44.9	20 18.1
22	17 18.3	23 10.9	29 41.5	26 3.6	20 26.7
23	17 15.9	23 19.8	♊ 0 9.2	26 22.7	20 35.5
24	17 13.5	23 29.0	0 37.6	26 42.1	20 44.3
25	17 11.0	23 38.5	1 6.8	27 1.9	20 53.4
26	17 8.5	23 48.2	1 36.7	27 22.0	21 2.5
27	17 5.9	23 58.2	2 7.4	27 42.4	21 11.8
28	17 3.3	24 8.6	2 39.0	28 3.3	21 21.3
29	17 0.5	24 19.3	3 11.4	28 24.6	21 30.9
30	16 57.7	24 30.3	3 44.9	28 46.4	21 40.8
31	16 54.9	24 41.8	4 19.3	29 8.6	21 50.8
32	16 51.9	24 53.6	4 54.8	29 31.4	22 1.0
33	16 48.9	25 5.9	5 31.5	29 54.7	22 11.5
34	16 45.7	25 18.7	6 9.3	♋ 0 18.6	22 22.1
35	16 42.5	25 32.1	6 48.5	0 43.1	22 33.1
36	16 39.1	25 46.0	7 29.0	1 8.2	22 44.3
37	16 35.7	26 0.5	8 11.0	1 34.1	22 55.8
38	16 32.1	26 15.6	8 54.5	2 0.7	23 7.6
39	16 28.4	26 31.5	9 39.7	2 28.0	23 19.7
40	16 24.5	26 48.2	10 26.6	2 56.3	23 32.2
41	16 20.5	27 5.7	11 15.4	3 25.4	23 45.0
42	16 16.3	27 24.1	12 6.3	3 55.4	23 58.3
43	16 12.0	27 43.5	12 59.2	4 26.6	24 12.0
44	16 7.4	28 4.1	13 54.5	4 58.8	24 26.1
45	16 2.6	28 25.9	14 52.1	5 32.1	24 40.8
46	15 57.7	28 49.1	15 52.4	6 6.8	24 55.9
47	15 52.4	29 13.7	16 55.5	6 42.8	25 11.7
48	15 46.9	29 40.1	18 1.5	7 20.2	25 28.0
49	15 41.1	♉ 0 8.3	19 10.7	7 59.3	25 45.1
50	15 35.0	0 38.6	20 23.4	8 40.0	26 2.8
51	15 28.5	1 11.2	21 39.6	9 22.6	26 21.4
52	15 21.6	1 46.6	22 59.8	10 7.1	26 40.8
53	15 14.3	2 25.0	24 24.1	10 53.8	27 1.1
54	15 6.5	3 6.9	25 52.9	11 42.8	27 22.5
55	14 58.2	3 52.9	27 26.4	12 34.4	27 44.9
56	14 49.2	4 43.6	29 4.9	13 28.7	28 8.7
57	14 39.6	5 39.9	♋ 0 48.7	14 26.1	28 33.7
58	14 29.2	6 42.8	2 38.2	15 26.8	29 0.3
59	14 18.0	7 53.7	4 33.6	16 31.1	29 28.6
60	14 5.8	9 14.4	6 35.2	17 39.3	29 58.7
61	13 52.4	10 47.3	8 43.4	18 52.0	♌ 0 31.0
62	13 37.7	12 35.8	10 58.2	20 9.6	1 5.7
63	13 21.5	14 45.2	13 20.0	21 32.5	1 43.0
64	13 3.4	17 23.7	15 48.9	23 1.3	2 23.5
65	12 43.2	20 46.2	18 24.7	24 36.8	3 7.6
66	12 20.4	25 26.8	21 7.6	26 19.5	3 55.9
S LAT	5	6	Descendant	8	9

♌ 16° 32′ 40″

9h 16m 0s — MC — 139° 0′ 0″

21ʰ 20ᵐ 0ˢ MC 320° 0′ 0″

♒ 17° 33′ 15″

N LAT	11	12	Ascendant	2	3
°	° ′	° ′	° ′	° ′	° ′
0	♓19 7.3	♈21 38.3	♉22 24.6	♊20 48.6	♋18 27.9
5	18 59.0	22 10.8	24 4.8	22 7.9	19 6.0
10	18 50.3	22 45.4	25 51.1	23 28.9	19 44.4
15	18 41.2	23 22.9	27 45.7	24 53.1	20 23.7
16	18 39.3	23 30.8	28 9.8	25 10.4	20 31.8
17	18 37.4	23 38.9	28 34.3	25 27.9	20 39.9
18	18 35.4	23 47.2	28 59.4	25 45.7	20 48.1
19	18 33.4	23 55.7	29 25.0	26 3.7	20 56.4
20	18 31.4	24 4.4	29 51.1	26 21.9	21 4.8
21	18 29.3	24 13.3	♊ 0 17.7	26 40.4	21 13.2
22	18 27.2	24 22.4	0 45.0	26 59.2	21 21.8
23	18 25.0	24 31.7	1 12.9	27 18.2	21 30.5
24	18 22.8	24 41.3	1 41.5	27 37.6	21 39.3
25	18 20.6	24 51.2	2 10.8	27 57.3	21 48.3
26	18 18.3	25 1.4	2 40.9	28 17.4	21 57.4
27	18 15.9	25 11.9	3 11.7	28 37.8	22 6.6
28	18 13.5	25 22.7	3 43.4	28 58.6	22 16.0
29	18 11.0	25 33.9	4 16.0	29 19.9	22 25.6
30	18 8.4	25 45.4	4 49.6	29 41.6	22 35.4
31	18 5.8	25 57.4	5 24.1	♋ 0 3.8	22 45.3
32	18 3.1	26 9.8	5 59.7	0 26.4	22 55.5
33	18 0.4	26 22.6	6 36.4	0 49.7	23 5.8
34	17 57.5	26 36.0	7 14.4	1 13.5	23 16.5
35	17 54.5	26 49.9	7 53.6	1 37.9	23 27.3
36	17 51.5	27 4.4	8 34.1	2 2.9	23 38.4
37	17 48.3	27 19.6	9 16.1	2 28.6	23 49.8
38	17 45.1	27 35.4	9 59.6	2 55.1	24 1.5
39	17 41.7	27 52.0	10 44.8	3 22.4	24 13.6
40	17 38.2	28 9.3	11 31.6	3 50.4	24 25.9
41	17 34.5	28 27.6	12 20.3	4 19.4	24 38.7
42	17 30.7	28 46.8	13 11.0	4 49.3	24 51.8
43	17 26.7	29 7.1	14 3.8	5 20.2	25 5.4
44	17 22.6	29 28.5	14 58.8	5 52.2	25 19.4
45	17 18.2	29 51.3	15 56.2	6 25.4	25 33.9
46	17 13.7	♉ 0 15.4	16 56.1	6 59.8	25 48.9
47	17 8.9	0 41.1	17 58.8	7 35.5	26 4.5
48	17 3.9	1 8.5	19 4.4	8 12.7	26 20.7
49	16 58.6	1 37.8	20 13.0	8 51.4	26 37.5
50	16 53.0	2 9.4	21 25.0	9 31.7	26 55.1
51	16 47.1	2 43.3	22 40.6	10 13.9	27 13.4
52	16 40.8	3 20.1	23 59.9	10 58.0	27 32.6
53	16 34.1	4 0.0	25 23.3	11 44.2	27 52.7
54	16 27.0	4 43.6	26 51.0	12 32.8	28 13.8
55	16 19.4	5 31.3	28 23.2	13 23.7	28 36.0
56	16 11.3	6 23.9	♋ 0 0.3	14 17.4	28 59.4
57	16 2.5	7 22.3	1 42.6	15 14.1	29 24.1
58	15 53.1	8 27.6	3 30.3	16 14.0	29 50.4
59	15 42.8	9 41.0	5 23.8	17 17.4	♌ 0 18.2
60	15 31.7	11 4.6	7 23.3	18 24.6	0 47.9
61	15 19.5	12 40.8	9 29.0	19 36.2	1 19.6
62	15 6.1	14 33.3	11 41.3	20 52.5	1 53.7
63	14 51.4	16 47.4	14 0.2	22 13.9	2 30.4
64	14 34.9	19 32.0	16 26.0	23 41.1	3 10.0
65	14 16.6	23 3.1	18 58.6	25 14.6	3 53.2
66	13 55.9	27 59.4	21 38.0	26 55.2	4 40.4
S LAT	5	6	Descendant	8	9

♌ 17° 33′ 15″

9ʰ 20ᵐ 0ˢ MC 140° 0′ 0″

21ʰ 24ᵐ 0ˢ MC 321° 0′ 0″

♒ 18° 34′ 2″

N LAT	11	12	Ascendant	2	3
°	° ′	° ′	° ′	° ′	° ′
0	♓20 12.3	♈22 42.3	♉23 23.4	♊21 43.9	♋19 24.1
5	20 4.9	23 16.2	25 4.7	23 3.3	20 1.9
10	19 57.1	23 52.3	26 52.1	24 24.4	20 40.1
15	19 48.8	24 31.5	28 47.6	25 48.6	21 19.3
16	19 47.1	24 39.8	29 11.9	26 5.9	21 27.3
17	19 45.4	24 48.3	29 36.7	26 23.4	21 35.3
18	19 43.6	24 56.9	♊ 0 1.9	26 41.2	21 43.5
19	19 41.8	25 5.8	0 27.6	26 59.1	21 51.7
20	19 39.9	25 14.8	0 53.9	27 17.3	22 0.0
21	19 38.1	25 24.1	1 20.8	27 35.8	22 8.5
22	19 36.2	25 33.6	1 48.2	27 54.6	22 17.0
23	19 34.2	25 43.4	2 16.3	28 13.6	22 25.6
24	19 32.2	25 53.4	2 45.0	28 32.9	22 34.4
25	19 30.2	26 3.7	3 14.5	28 52.6	22 43.3
26	19 28.1	26 14.3	3 44.7	29 12.6	22 52.3
27	19 26.0	26 25.2	4 15.6	29 33.0	23 1.5
28	19 23.8	26 36.5	4 47.5	29 53.8	23 10.8
29	19 21.6	26 48.2	5 20.2	♋ 0 15.0	23 20.4
30	19 19.3	27 0.2	5 53.8	0 36.6	23 30.0
31	19 16.9	27 12.7	6 28.4	0 58.7	23 39.9
32	19 14.5	27 25.6	7 4.1	1 21.3	23 50.0
33	19 12.0	27 39.0	7 40.9	1 44.5	24 0.3
34	19 9.4	27 52.9	8 18.9	2 8.2	24 10.8
35	19 6.7	28 7.4	8 58.1	2 32.5	24 21.6
36	19 4.0	28 22.5	9 38.7	2 57.4	24 32.6
37	19 1.1	28 38.3	10 20.6	3 23.0	24 43.9
38	18 58.2	28 54.7	11 4.1	3 49.4	24 55.5
39	18 55.1	29 12.0	11 49.2	4 16.5	25 7.4
40	18 51.9	29 30.1	12 35.9	4 44.4	25 19.7
41	18 48.6	29 49.0	13 24.5	5 13.2	25 32.3
42	18 45.2	♉ 0 9.0	14 15.0	5 42.9	25 45.4
43	18 41.6	0 30.1	15 7.6	6 13.6	25 58.8
44	18 37.9	0 52.4	16 2.4	6 45.4	26 12.7
45	18 33.9	1 16.0	16 59.5	7 18.3	26 27.0
46	18 29.8	1 41.0	17 59.1	7 52.5	26 41.9
47	18 25.5	2 7.7	19 1.3	8 28.0	26 57.3
48	18 21.0	2 36.2	20 6.4	9 4.8	27 13.3
49	18 16.2	3 6.6	21 14.5	9 43.2	27 30.0
50	18 11.2	3 39.3	22 25.9	10 23.2	27 47.4
51	18 5.8	4 14.6	23 40.7	11 5.0	28 5.5
52	18 0.2	4 52.7	24 59.2	11 48.7	28 24.5
53	17 54.2	5 34.0	26 21.6	12 34.4	28 44.3
54	17 47.7	6 19.1	27 48.2	13 22.4	29 5.1
55	17 40.9	7 8.6	29 19.2	14 12.8	29 27.1
56	17 33.5	8 3.0	♋ 0 55.0	15 5.9	29 50.1
57	17 25.6	9 3.4	2 35.7	16 1.8	♌ 0 14.5
58	17 17.1	10 10.8	4 21.7	17 0.9	0 40.4
59	17 7.9	11 26.7	6 13.3	18 3.4	1 7.8
60	16 57.8	12 53.1	8 10.7	19 9.7	1 37.0
61	16 46.8	14 32.4	10 14.1	20 20.2	2 8.2
62	16 34.8	16 28.6	12 23.9	21 35.2	2 41.7
63	16 21.5	18 47.1	14 40.0	22 55.2	3 17.7
64	16 6.7	21 37.4	17 2.8	24 20.8	3 56.6
65	15 50.2	25 16.8	19 32.2	25 52.5	4 38.9
66	15 31.6	♊ 0 28.5	22 8.3	27 30.9	5 25.0
S LAT	5	6	Descendant	8	9

♌ 18° 34′ 2″

9ʰ 24ᵐ 0ˢ MC 141° 0′ 0″

21h 28m 0s MC 322° 0′ 0″

♒ 19° 34′ 59″

11	12	Ascendant	2	3	N LAT
° ′	° ′	° ′	° ′	° ′	°
♓21 17.5	♈23 46.0	♉24 22.0	♊22 39.2	♋20 20.3	0
21 10.8	24 21.4	26 4.4	23 58.6	20 58.0	5
21 3.9	24 59.1	27 52.8	25 19.8	21 36.0	10
20 56.5	25 39.9	29 49.3	26 44.0	22 14.9	15
20 55.0	25 48.6	♊ 0 13.7	27 1.3	22 22.8	16
20 53.4	25 57.4	0 38.7	27 18.8	22 30.8	17
20 51.9	26 6.4	1 4.1	27 36.5	22 38.9	18
20 50.3	26 15.6	1 30.0	27 54.5	22 47.1	19
20 48.6	26 25.0	1 56.4	28 12.7	22 55.4	20
20 47.0	26 34.7	2 23.4	28 31.1	23 3.8	21
20 45.3	26 44.6	2 51.0	28 49.8	23 12.2	22
20 43.5	26 54.7	3 19.2	29 8.8	23 20.8	23
20 41.7	27 5.1	3 48.1	29 28.1	23 29.5	24
20 39.9	27 15.9	4 17.7	29 47.7	23 38.4	25
20 38.1	27 26.9	4 48.0	♋ 0 7.7	23 47.3	26
20 36.2	27 38.3	5 19.1	0 28.0	23 56.4	27
20 34.2	27 50.0	5 51.0	0 48.8	24 5.7	28
20 32.2	28 2.1	6 23.8	1 9.9	24 15.1	29
20 30.2	28 14.6	6 57.5	1 31.5	24 24.7	30
20 28.1	28 27.6	7 32.2	1 53.5	24 34.5	31
20 25.9	28 41.0	8 8.0	2 16.0	24 44.5	32
20 23.7	28 54.9	8 44.8	2 39.1	24 54.8	33
20 21.4	29 9.4	9 22.8	3 2.7	25 5.2	34
20 19.0	29 24.5	10 2.0	3 26.9	25 15.9	35
20 16.6	29 40.2	10 42.6	3 51.7	25 26.8	36
20 14.0	29 56.5	11 24.5	4 17.2	25 38.0	37
20 11.4	♉ 0 13.6	12 8.0	4 43.4	25 49.5	38
20 8.7	0 31.5	12 52.9	5 10.4	26 1.4	39
20 5.8	0 50.3	13 39.6	5 38.1	26 13.5	40
20 2.9	1 10.0	14 28.0	6 6.8	26 26.0	41
19 59.8	1 30.7	15 18.4	6 36.3	26 38.9	42
19 56.6	1 52.5	16 10.7	7 6.8	26 52.2	43
19 53.3	2 15.6	17 5.2	7 38.4	27 6.0	44
19 49.8	2 40.1	18 2.0	8 11.1	27 20.2	45
19 46.1	3 6.0	19 1.2	8 45.0	27 34.9	46
19 42.3	3 33.7	20 3.1	9 20.2	27 50.1	47
19 38.3	4 3.1	21 7.7	9 56.8	28 6.0	48
19 34.0	4 34.6	22 15.2	10 34.8	28 22.5	49
19 29.5	5 8.5	23 25.9	11 14.5	28 39.7	50
19 24.8	5 44.9	24 40.0	11 55.9	28 57.6	51
19 19.7	6 24.3	25 57.6	12 39.1	29 16.3	52
19 14.4	7 7.0	27 19.1	13 24.4	29 35.9	53
19 8.6	7 53.6	28 44.6	14 11.8	29 56.5	54
19 2.5	8 44.6	♋ 0 14.5	15 1.7	♌ 0 18.1	55
18 56.0	9 40.8	1 48.9	15 54.1	0 40.9	56
18 49.0	10 43.1	3 28.1	16 49.3	1 4.9	57
18 41.4	11 52.6	5 12.4	17 47.6	1 30.4	58
18 33.1	13 10.8	7 2.1	18 49.3	1 57.4	59
18 24.2	14 39.7	8 57.5	19 54.6	2 26.1	60
18 14.4	16 22.0	10 58.7	21 3.9	2 56.8	61
18 3.7	18 21.5	13 5.9	22 17.7	3 29.7	62
17 51.9	20 44.2	15 19.5	23 36.3	4 5.0	63
17 38.7	23 39.8	17 39.4	25 0.3	4 43.2	64
17 24.0	27 27.0	20 5.7	26 30.2	5 24.5	65
17 7.5	♊ 2 54.0	22 38.5	28 6.6	6 9.6	66
5	6	Descendant	8	9	S LAT

♌ 19° 34′ 59″

9h 28m 0s MC 142° 0′ 0″

21h 32m 0s MC 323° 0′ 0″

♒ 20° 36′ 7″

N LAT	11	12	Ascendant	2	3
°	° ′	° ′	° ′	° ′	° ′
0	♓22 22.6	♈24 49.7	♉25 20.5	♊23 34.4	♋21 16.7
5	22 16.8	25 26.5	27 4.0	24 53.9	21 54.1
10	22 10.7	26 5.7	28 53.3	26 15.1	22 31.9
15	22 4.3	26 48.1	♊ 0 50.6	27 39.3	23 10.6
16	22 3.0	26 57.1	1 15.3	27 56.6	23 18.5
17	22 1.6	27 6.3	1 40.4	28 14.1	23 26.4
18	22 0.2	27 15.6	2 5.9	28 31.8	23 34.5
19	21 58.8	27 25.2	2 32.0	28 49.7	23 42.6
20	21 57.4	27 35.0	2 58.6	29 7.9	23 50.8
21	21 55.9	27 45.0	3 25.7	29 26.3	23 59.1
22	21 54.4	27 55.3	3 53.4	29 44.9	24 7.5
23	21 52.9	28 5.8	4 21.8	♋ 0 3.9	24 16.1
24	21 51.3	28 16.6	4 50.8	0 23.2	24 24.7
25	21 49.8	28 27.8	5 20.5	0 42.7	24 33.5
26	21 48.1	28 39.2	5 50.9	1 2.7	24 42.4
27	21 46.5	28 51.0	6 22.1	1 22.9	24 51.4
28	21 44.8	29 3.2	6 54.1	1 43.6	25 0.6
29	21 43.0	29 15.7	7 27.0	2 4.7	25 10.0
30	21 41.2	29 28.7	8 0.8	2 26.2	25 19.5
31	21 39.4	29 42.2	8 35.6	2 48.1	25 29.2
32	21 37.5	29 56.1	9 11.3	3 10.6	25 39.2
33	21 35.5	♉ 0 10.5	9 48.2	3 33.5	25 49.3
34	21 33.5	0 25.5	10 26.2	3 57.0	25 59.6
35	21 31.4	0 41.1	11 5.4	4 21.1	26 10.2
36	21 29.3	0 57.4	11 46.0	4 45.8	26 21.1
37	21 27.0	1 14.3	12 27.9	5 11.2	26 32.2
38	21 24.7	1 32.0	13 11.2	5 37.3	26 43.6
39	21 22.3	1 50.5	13 56.1	6 4.1	26 55.3
40	21 19.9	2 10.0	14 42.6	6 31.7	27 7.3
41	21 17.3	2 30.3	15 30.9	7 0.2	27 19.7
42	21 14.6	2 51.8	16 21.1	7 29.5	27 32.5
43	21 11.8	3 14.4	17 13.2	7 59.8	27 45.7
44	21 8.9	3 38.3	18 7.4	8 31.2	27 59.3
45	21 5.8	4 3.6	19 3.8	9 3.6	28 13.3
46	21 2.6	4 30.4	20 2.7	9 37.3	28 27.9
47	20 59.2	4 58.9	21 4.1	10 12.2	28 43.0
48	20 55.7	5 29.3	22 8.2	10 48.5	28 58.7
49	20 51.9	6 1.9	23 15.1	11 26.2	29 15.0
50	20 48.0	6 36.8	24 25.2	12 5.5	29 31.9
51	20 43.8	7 14.3	25 38.5	12 46.5	29 49.7
52	20 39.4	7 54.9	26 55.3	13 29.3	♌ 0 8.2
53	20 34.7	8 39.0	28 15.8	14 14.1	0 27.5
54	20 29.7	9 27.0	29 40.3	15 1.0	0 47.8
55	20 24.4	10 19.5	♋ 1 8.9	15 50.3	1 9.1
56	20 18.6	11 17.3	2 42.0	16 42.1	1 31.6
57	20 12.4	12 21.3	4 19.8	17 36.6	1 55.3
58	20 5.8	13 32.8	6 2.5	18 34.1	2 20.4
59	19 58.6	14 53.1	7 50.4	19 34.9	2 47.0
60	19 50.7	16 24.4	9 43.7	20 39.2	3 15.3
61	19 42.2	18 9.4	11 42.7	21 47.5	3 45.4
62	19 32.8	20 12.2	13 47.6	23 0.0	4 17.7
63	19 22.4	22 38.7	15 58.6	24 17.3	4 52.4
64	19 10.9	25 39.2	18 15.7	25 39.8	5 29.8
65	18 58.1	29 33.6	20 39.1	27 7.9	6 10.3
66	18 43.6	♊ 5 15.4	23 8.7	28 42.4	6 54.4
S LAT	5	6	Descendant	8	9

♌ 20° 36′ 7″

9h 32m 0s MC 143° 0′ 0″

21ʰ 36ᵐ 0ˢ MC 324° 0′ 0″

♒ 21° 37′ 27″

N LAT	11	12	Ascendant	2	3
°	° ′	° ′	° ′	° ′	° ′
0	♓23 27.9	♈25 53.2	♉26 18.8	♊24 29.5	♋22 13.1
5	23 22.9	26 31.4	28 3.3	25 49.2	22 50.4
10	23 17.7	27 12.1	29 53.5	27 10.4	23 27.9
15	23 12.1	27 56.1	♊ 1 51.7	28 34.5	24 6.3
16	23 11.0	28 5.4	2 16.5	28 51.8	24 14.2
17	23 9.8	28 14.9	2 41.7	29 9.3	24 22.1
18	23 8.6	28 24.6	3 7.5	29 26.9	24 30.1
19	23 7.4	28 34.5	3 33.7	29 44.8	24 38.2
20	23 6.2	28 44.6	4 0.4	♋ 0 3.0	24 46.3
21	23 4.9	28 55.0	4 27.7	0 21.3	24 54.6
22	23 3.6	29 5.7	4 55.5	0 40.0	25 2.9
23	23 2.3	29 16.6	5 24.0	0 58.9	25 11.4
24	23 1.0	29 27.8	5 53.1	1 18.1	25 20.0
25	22 59.6	29 39.3	6 22.9	1 37.6	25 28.7
26	22 58.3	29 51.2	6 53.4	1 57.5	25 37.5
27	22 56.8	♉ 0 3.4	7 24.7	2 17.7	25 46.5
28	22 55.4	0 16.0	7 56.8	2 38.3	25 55.6
29	22 53.9	0 29.0	8 29.7	2 59.3	26 4.9
30	22 52.3	0 42.5	9 3.6	3 20.7	26 14.3
31	22 50.7	0 56.4	9 38.4	3 42.6	26 24.0
32	22 49.1	1 10.8	10 14.2	4 5.0	26 33.8
33	22 47.4	1 25.7	10 51.1	4 27.8	26 43.8
34	22 45.7	1 41.2	11 29.1	4 51.2	26 54.1
35	22 43.9	1 57.3	12 8.3	5 15.2	27 4.6
36	22 42.1	2 14.1	12 48.8	5 39.8	27 15.4
37	22 40.1	2 31.6	13 30.6	6 5.0	27 26.4
38	22 38.2	2 49.9	14 13.9	6 31.0	27 37.7
39	22 36.1	3 9.1	14 58.7	6 57.6	27 49.3
40	22 34.0	3 29.1	15 45.1	7 25.1	28 1.2
41	22 31.8	3 50.2	16 33.2	7 53.3	28 13.5
42	22 29.5	4 12.3	17 23.1	8 22.5	28 26.1
43	22 27.0	4 35.6	18 15.0	8 52.6	28 39.1
44	22 24.5	5 0.3	19 8.9	9 23.8	28 52.6
45	22 21.9	5 26.3	20 5.0	9 56.0	29 6.5
46	22 19.1	5 54.0	21 3.4	10 29.4	29 20.9
47	22 16.3	6 23.4	22 4.4	11 4.0	29 35.9
48	22 13.2	6 54.7	23 8.0	11 39.9	29 51.4
49	22 10.0	7 28.2	24 14.3	12 17.3	♌ 0 7.5
50	22 6.6	8 4.2	25 23.7	12 56.2	0 24.2
51	22 3.0	8 42.9	26 36.3	13 36.8	0 41.7
52	21 59.2	9 24.6	27 52.3	14 19.2	1 0.0
53	21 55.2	10 9.9	29 11.8	15 3.5	1 19.1
54	21 50.9	10 59.2	♋ 0 35.2	15 50.0	1 39.2
55	21 46.3	11 53.1	2 2.7	16 38.7	2 0.2
56	21 41.4	12 52.5	3 34.5	17 29.8	2 22.3
57	21 36.1	13 58.2	5 10.8	18 23.6	2 45.7
58	21 30.4	15 11.4	6 51.9	19 20.4	3 10.4
59	21 24.2	16 33.8	8 38.0	20 20.3	3 36.6
60	21 17.5	18 7.3	10 29.4	21 23.7	4 4.4
61	21 10.1	19 54.9	12 26.3	22 30.9	4 34.0
62	21 2.1	22 0.5	14 28.9	23 42.2	5 5.8
63	20 53.2	24 30.5	16 37.3	24 58.2	5 39.8
64	20 43.3	27 35.5	18 51.8	26 19.1	6 16.4
65	20 32.3	♊ 1 36.5	21 12.3	27 45.6	6 56.1
66	20 19.9	7 32.4	23 38.9	29 18.1	7 39.2
S LAT	5	6	Descendant	8	9

♌ 21° 37′ 27″

9ʰ 36ᵐ 0ˢ MC 144° 0′ 0″

21ʰ 40ᵐ 0ˢ MC 325° 0′ 0″

♒ 22° 38′ 57″

N LAT	11	12	Ascendant	2	3
°	° ′	° ′	° ′	° ′	° ′
0	♓24 33.2	♈26 56.5	♉27 17.0	♊25 24.7	♋23 9.7
5	24 29.0	27 36.2	29 2.4	26 44.4	23 46.8
10	24 24.6	28 18.3	♊ 0 53.5	28 5.6	24 24.0
15	24 20.0	29 3.8	2 52.5	29 29.6	25 2.2
16	24 19.1	29 13.5	3 17.4	29 46.9	25 10.0
17	24 18.1	29 23.3	3 42.8	♋ 0 4.3	25 17.9
18	24 17.1	29 33.3	4 8.7	0 22.0	25 25.8
19	24 16.1	29 43.6	4 35.0	0 39.9	25 33.8
20	24 15.1	29 54.1	5 1.9	0 58.0	25 41.9
21	24 14.0	♉ 0 4.8	5 29.3	1 16.3	25 50.1
22	24 12.9	0 15.8	5 57.2	1 34.9	25 58.4
23	24 11.9	0 27.1	6 25.8	1 53.8	26 6.8
24	24 10.7	0 38.7	6 55.0	2 12.9	26 15.3
25	24 9.6	0 50.6	7 24.9	2 32.4	26 23.9
26	24 8.4	1 2.9	7 55.5	2 52.2	26 32.7
27	24 7.2	1 15.5	8 26.9	3 12.3	26 41.6
28	24 6.0	1 28.5	8 59.0	3 32.9	26 50.6
29	24 4.8	1 42.0	9 32.0	3 53.8	26 59.8
30	24 3.5	1 55.8	10 5.9	4 15.1	27 9.2
31	24 2.2	2 10.2	10 40.8	4 36.9	27 18.8
32	24 0.8	2 25.1	11 16.6	4 59.2	27 28.5
33	23 59.4	2 40.5	11 53.5	5 22.0	27 38.5
34	23 58.0	2 56.5	12 31.5	5 45.2	27 48.6
35	23 56.5	3 13.1	13 10.6	6 9.1	27 59.0
36	23 54.9	3 30.5	13 51.1	6 33.6	28 9.7
37	23 53.3	3 48.5	14 32.8	6 58.7	28 20.6
38	23 51.7	4 7.4	15 16.0	7 24.5	28 31.8
39	23 50.0	4 27.1	16 0.6	7 51.0	28 43.3
40	23 48.2	4 47.8	16 46.9	8 18.3	28 55.1
41	23 46.3	5 9.5	17 34.8	8 46.3	29 7.2
42	23 44.4	5 32.3	18 24.5	9 15.3	29 19.7
43	23 42.4	5 56.3	19 16.1	9 45.2	29 32.6
44	23 40.3	6 21.6	20 9.7	10 16.1	29 46.0
45	23 38.1	6 48.5	21 5.5	10 48.1	29 59.7
46	23 35.8	7 16.9	22 3.5	11 21.2	♌ 0 14.0
47	23 33.4	7 47.2	23 4.0	11 55.5	0 28.7
48	23 30.8	8 19.4	24 7.0	12 31.2	0 44.1
49	23 28.2	8 53.8	25 12.8	13 8.2	1 0.0
50	23 25.3	9 30.7	26 21.6	13 46.8	1 16.6
51	23 22.4	10 10.4	27 33.4	14 27.0	1 33.8
52	23 19.2	10 53.3	28 48.5	15 9.0	1 51.9
53	23 15.8	11 39.7	♋ 0 7.2	15 52.8	2 10.7
54	23 12.2	12 30.3	1 29.5	16 38.7	2 30.5
55	23 8.4	13 25.6	2 55.8	17 26.8	2 51.2
56	23 4.3	14 26.4	4 26.3	18 17.3	3 13.1
57	22 59.9	15 33.6	6 1.1	19 10.5	3 36.1
58	22 55.1	16 48.6	7 40.7	20 6.4	4 0.4
59	22 49.9	18 12.8	9 25.0	21 5.5	4 26.2
60	22 44.3	19 48.3	11 14.5	22 8.0	4 53.5
61	22 38.2	21 38.2	13 9.4	23 14.1	5 22.7
62	22 31.5	23 46.5	15 9.7	24 24.3	5 53.8
63	22 24.1	26 19.6	17 15.8	25 38.9	6 27.2
64	22 15.8	29 28.6	19 27.6	26 58.4	7 3.1
65	22 6.7	♊ 3 35.7	21 45.4	28 23.2	7 41.9
66	21 56.4	9 44.6	24 9.1	29 53.9	8 24.1
S LAT	5	6	Descendant	8	9

♌ 22° 38′ 57″

9ʰ 40ᵐ 0ˢ MC 145° 0′ 0″

21ʰ 44ᵐ 0ˢ — MC 326° 0′ 0″ — ♒ 23° 40′ 37″

N LAT °	11 ° ′	12 ° ′	Ascendant ° ′	2 ° ′	3 ° ′
0	♓25 38.5	♈27 59.7	♉28 15.0	♊26 19.8	♋24 6.4
5	25 35.1	28 40.7	♊ 0 1.3	27 39.5	24 43.2
10	25 31.7	29 24.3	1 53.3	29 0.7	25 20.3
15	25 28.0	♉ 0 11.4	3 53.0	♋ 0 24.7	25 58.2
16	25 27.2	0 21.3	4 18.1	0 41.9	26 5.9
17	25 26.4	0 31.4	4 43.6	0 59.3	26 13.7
18	25 25.6	0 41.8	5 9.6	1 17.0	26 21.6
19	25 24.8	0 52.4	5 36.1	1 34.8	26 29.5
20	25 24.0	1 3.2	6 3.0	1 52.8	26 37.6
21	25 23.2	1 14.3	6 30.5	2 11.1	26 45.7
22	25 22.3	1 25.7	6 58.6	2 29.7	26 53.9
23	25 21.4	1 37.3	7 27.3	2 48.5	27 2.2
24	25 20.5	1 49.3	7 56.6	3 7.6	27 10.7
25	25 19.6	2 1.6	8 26.6	3 27.0	27 19.2
26	25 18.7	2 14.2	8 57.2	3 46.8	27 27.9
27	25 17.7	2 27.3	9 28.7	4 6.9	27 36.8
28	25 16.8	2 40.7	10 0.9	4 27.3	27 45.7
29	25 15.7	2 54.5	10 33.9	4 48.2	27 54.9
30	25 14.7	3 8.8	11 7.8	5 9.4	28 4.2
31	25 13.7	3 23.6	11 42.7	5 31.1	28 13.6
32	25 12.6	3 39.0	12 18.5	5 53.3	28 23.3
33	25 11.4	3 54.8	12 55.4	6 15.9	28 33.2
34	25 10.3	4 11.3	13 33.3	6 39.1	28 43.2
35	25 9.1	4 28.5	14 12.5	7 2.9	28 53.5
36	25 7.9	4 46.3	14 52.8	7 27.2	29 4.1
37	25 6.6	5 4.9	15 34.5	7 52.2	29 14.9
38	25 5.3	5 24.3	16 17.5	8 17.8	29 26.0
39	25 3.9	5 44.6	17 2.0	8 44.2	29 37.3
40	25 2.5	6 5.9	17 48.1	9 11.3	29 49.0
41	25 1.0	6 28.2	18 35.8	9 39.2	♌ 0 1.0
42	24 59.4	6 51.6	19 25.3	10 7.9	0 13.4
43	24 57.8	7 16.3	20 16.6	10 37.6	0 26.2
44	24 56.1	7 42.4	21 9.9	11 8.3	0 39.4
45	24 54.4	8 9.9	22 5.3	11 40.0	0 53.0
46	24 52.5	8 39.1	23 2.9	12 12.9	1 7.1
47	24 50.6	9 10.2	24 2.9	12 46.9	1 21.7
48	24 48.6	9 43.2	25 5.5	13 22.3	1 36.8
49	24 46.4	10 18.5	26 10.7	13 59.0	1 52.5
50	24 44.2	10 56.4	27 18.7	14 37.2	2 8.9
51	24 41.8	11 37.1	28 29.8	15 17.0	2 25.9
52	24 39.2	12 21.0	29 44.1	15 58.5	2 43.7
53	24 36.5	13 8.5	♋ 1 1.8	16 41.8	3 2.4
54	24 33.7	14 0.3	2 23.1	17 27.2	3 21.9
55	24 30.6	14 56.8	3 48.2	18 14.7	3 42.3
56	24 27.3	15 58.9	5 17.4	19 4.6	4 3.8
57	24 23.8	17 7.6	6 50.9	19 57.1	4 26.5
58	24 19.9	18 24.1	8 28.9	20 52.3	4 50.4
59	24 15.8	19 50.1	10 11.6	21 50.5	5 15.8
60	24 11.3	21 27.5	11 59.2	22 52.1	5 42.7
61	24 6.4	23 19.4	13 52.0	23 57.2	6 11.3
62	24 1.0	25 30.1	15 50.2	25 6.2	6 41.9
63	23 55.1	28 6.1	17 53.9	26 19.5	7 14.7
64	23 48.5	♊ 1 18.6	20 3.3	27 37.6	7 49.9
65	23 41.2	5 30.9	22 18.3	29 0.8	8 27.9
66	23 32.9	11 51.4	24 39.2	♌ 0 29.7	9 9.1
S LAT	5	6	Descendant	8	9

♌ 23° 40′ 37″ — 9ʰ 44ᵐ 0ˢ — MC 146° 0′ 0″

21ʰ 48ᵐ 0ˢ — MC 327° 0′ 0″ — ♒ 24° 42′ 28″

N LAT °	11 ° ′	12 ° ′	Ascendant ° ′	2 ° ′	3 ° ′
0	♓26 43.8	♈29 2.8	♉29 12.8	♊27 14.8	♋25 3.3
5	26 41.3	29 45.1	♊ 1 0.0	28 34.6	25 39.8
10	26 38.7	♉ 0 30.1	2 52.8	29 55.8	26 16.6
15	26 35.9	1 18.6	4 53.2	♋ 1 19.6	26 54.2
16	26 35.4	1 28.9	5 18.5	1 36.9	27 1.9
17	26 34.8	1 39.3	5 44.1	1 54.3	27 9.6
18	26 34.2	1 50.0	6 10.2	2 11.8	27 17.5
19	26 33.6	2 0.9	6 36.8	2 29.6	27 25.3
20	26 33.0	2 12.1	7 3.9	2 47.7	27 33.3
21	26 32.3	2 23.5	7 31.5	3 5.9	27 41.4
22	26 31.7	2 35.2	7 59.6	3 24.4	27 49.5
23	26 31.0	2 47.3	8 28.4	3 43.2	27 57.8
24	26 30.4	2 59.6	8 57.8	4 2.2	28 6.2
25	26 29.7	3 12.2	9 27.8	4 21.6	28 14.6
26	26 29.0	3 25.3	9 58.6	4 41.2	28 23.3
27	26 28.3	3 38.7	10 30.0	5 1.3	28 32.0
28	26 27.5	3 52.5	11 2.3	5 21.6	28 40.9
29	26 26.8	4 6.7	11 35.4	5 42.4	28 50.0
30	26 26.0	4 21.5	12 9.3	6 3.6	28 59.2
31	26 25.2	4 36.7	12 44.1	6 25.2	29 8.6
32	26 24.4	4 52.5	13 20.0	6 47.2	29 18.1
33	26 23.5	5 8.8	13 56.8	7 9.8	29 27.9
34	26 22.7	5 25.7	14 34.7	7 32.9	29 37.9
35	26 21.8	5 43.4	15 13.8	7 56.5	29 48.1
36	26 20.8	6 1.7	15 54.1	8 20.7	29 58.5
37	26 19.9	6 20.8	16 35.6	8 45.5	♌ 0 9.2
38	26 18.9	6 40.7	17 18.5	9 11.0	0 20.2
39	26 17.9	7 1.6	18 2.9	9 37.2	0 31.4
40	26 16.8	7 23.4	18 48.8	10 4.1	0 43.0
41	26 15.7	7 46.3	19 36.3	10 31.8	0 54.9
42	26 14.5	8 10.4	20 25.5	11 0.4	1 7.2
43	26 13.3	8 35.7	21 16.5	11 29.9	1 19.8
44	26 12.1	9 2.4	22 9.5	12 0.3	1 32.8
45	26 10.7	9 30.7	23 4.5	12 31.8	1 46.3
46	26 9.3	10 0.6	24 1.7	13 4.4	2 0.2
47	26 7.9	10 32.4	25 1.3	13 38.1	2 14.6
48	26 6.4	11 6.3	26 3.2	14 13.1	2 29.6
49	26 4.8	11 42.4	27 7.8	14 49.5	2 45.1
50	26 3.1	12 21.2	28 15.2	15 27.3	3 1.3
51	26 1.3	13 2.8	29 25.5	16 6.7	3 18.1
52	25 59.4	13 47.7	♋ 0 39.0	16 47.8	3 35.6
53	25 57.3	14 36.3	1 55.8	17 30.7	3 54.0
54	25 55.2	15 29.1	3 16.1	18 15.5	4 13.2
55	25 52.9	16 26.8	4 40.1	19 2.5	4 33.4
56	25 50.4	17 30.2	6 8.0	19 51.8	4 54.6
57	25 47.7	18 40.2	7 40.1	20 43.5	5 16.9
58	25 44.9	19 58.2	9 16.6	21 38.0	5 40.5
59	25 41.8	21 25.6	10 57.6	22 35.4	6 5.4
60	25 38.4	23 4.8	12 43.5	23 36.0	6 31.9
61	25 34.7	24 58.6	14 34.3	24 40.1	7 0.0
62	25 30.7	27 11.3	16 30.4	25 48.0	7 30.0
63	25 26.2	29 49.8	18 31.8	27 0.1	8 2.1
64	25 21.3	♊ 3 5.4	20 38.7	28 16.7	8 36.6
65	25 15.8	7 22.1	22 51.2	29 38.4	9 13.8
66	25 9.6	13 52.5	25 9.3	♌ 1 5.5	9 54.1
S LAT	5	6	Descendant	8	9

♌ 24° 42′ 28″ — 9ʰ 48ᵐ 0ˢ — MC 147° 0′ 0″

21h 52m 0s — MC — 328° 0′ 0″

♒ 25° 44′ 30″

11	12	Ascendant	2	3	N LAT
° ′	° ′	° ′	° ′	° ′	°
♓27 49.2	♉ 0 5.7	♊ 0 10.5	♊28 9.9	♋26 0.2	0
27 47.5	0 49.3	1 58.6	29 29.7	26 36.6	5
27 45.8	1 35.6	3 52.1	♋ 0 50.8	27 13.1	10
27 43.9	2 25.7	5 53.2	2 14.6	27 50.4	15
27 43.6	2 36.2	6 18.6	2 31.7	27 58.0	16
27 43.2	2 47.0	6 44.3	2 49.1	28 5.7	17
27 42.8	2 58.0	7 10.5	3 6.7	28 13.4	18
27 42.4	3 9.2	7 37.2	3 24.4	28 21.2	19
27 42.0	3 20.7	8 4.4	3 42.4	28 29.2	20
27 41.5	3 32.5	8 32.1	4 0.6	28 37.1	21
27 41.1	3 44.5	9 0.3	4 19.0	28 45.2	22
27 40.7	3 56.9	9 29.2	4 37.7	28 53.4	23
27 40.2	4 9.5	9 58.6	4 56.7	29 1.7	24
27 39.8	4 22.6	10 28.7	5 16.0	29 10.1	25
27 39.3	4 36.0	10 59.5	5 35.6	29 18.7	26
27 38.8	4 49.7	11 31.0	5 55.6	29 27.3	27
27 38.3	5 3.9	12 3.3	6 15.8	29 36.2	28
27 37.8	5 18.6	12 36.4	6 36.5	29 45.1	29
27 37.3	5 33.7	13 10.3	6 57.6	29 54.3	30
27 36.8	5 49.3	13 45.2	7 19.1	♌ 0 3.6	31
27 36.2	6 5.5	14 21.0	7 41.1	0 13.0	32
27 35.7	6 22.3	14 57.8	8 3.5	0 22.7	33
27 35.1	6 39.7	15 35.6	8 26.5	0 32.6	34
27 34.5	6 57.8	16 14.6	8 49.9	0 42.7	35
27 33.9	7 16.6	16 54.8	9 14.0	0 53.0	36
27 33.2	7 36.2	17 36.2	9 38.7	1 3.6	37
27 32.6	7 56.7	18 19.0	10 4.0	1 14.5	38
27 31.9	8 18.0	19 3.2	10 30.0	1 25.6	39
27 31.2	8 40.4	19 48.9	10 56.8	1 37.0	40
27 30.4	9 3.9	20 36.2	11 24.3	1 48.8	41
27 29.7	9 28.5	21 25.1	11 52.7	2 0.9	42
27 28.8	9 54.5	22 15.9	12 21.9	2 13.4	43
27 28.0	10 21.8	23 8.5	12 52.1	2 26.3	44
27 27.1	10 50.8	24 3.1	13 23.4	2 39.6	45
27 26.2	11 21.4	24 59.9	13 55.7	2 53.3	46
27 25.2	11 53.9	25 59.0	14 29.1	3 7.6	47
27 24.2	12 28.5	27 0.4	15 3.8	3 22.3	48
27 23.1	13 5.5	28 4.4	15 39.8	3 37.7	49
27 22.0	13 45.0	29 11.1	16 17.3	3 53.6	50
27 20.8	14 27.5	♋ 0 20.7	16 56.3	4 10.2	51
27 19.5	15 13.3	1 33.3	17 36.9	4 27.6	52
27 18.2	16 2.9	2 49.1	18 19.3	4 45.7	53
27 16.7	16 56.7	4 8.4	19 3.6	5 4.6	54
27 15.2	17 55.5	5 31.3	19 50.0	5 24.5	55
27 13.6	19 0.1	6 58.0	20 38.7	5 45.4	56
27 11.8	20 11.3	8 28.7	21 29.8	6 7.3	57
27 9.9	21 30.6	10 3.7	22 23.5	6 30.5	58
27 7.8	22 59.5	11 43.2	23 20.1	6 55.1	59
27 5.6	24 40.2	13 27.2	24 19.8	7 21.1	60
27 3.1	26 35.6	15 16.2	25 22.9	7 48.7	61
27 0.4	28 50.3	17 10.2	26 29.7	8 18.2	62
26 57.4	♊ 1 30.8	19 9.4	27 40.5	8 49.7	63
26 54.2	4 49.1	21 14.0	28 55.8	9 23.5	64
26 50.5	9 9.2	23 23.9	♌ 0 15.9	9 59.9	65
26 46.4	15 47.4	25 39.4	1 41.3	10 39.3	66
5	6	Descendant	8	9	S LAT

♌ 25° 44′ 30″

9h 52m 0s — MC — 148° 0′ 0″

21h 56m 0s — MC — 329° 0′ 0″

♒ 26° 46′ 43″

11	12	Ascendant	2	3	N LAT
° ′	° ′	° ′	° ′	° ′	°
♓28 54.6	♉ 1 8.4	♊ 1 8.0	♊29 5.0	♋26 57.4	0
28 53.8	1 53.3	2 56.9	♋ 0 24.7	27 33.4	5
28 52.9	2 41.0	4 51.2	1 45.8	28 9.6	10
28 52.0	3 32.4	6 52.9	3 9.4	28 46.6	15
28 51.8	3 43.3	7 18.4	3 26.6	28 54.2	16
28 51.6	3 54.4	7 44.2	3 43.9	29 1.8	17
28 51.4	4 5.7	8 10.5	4 1.4	29 9.5	18
28 51.2	4 17.2	8 37.3	4 19.1	29 17.3	19
28 51.0	4 29.0	9 4.6	4 37.0	29 25.1	20
28 50.8	4 41.1	9 32.3	4 55.2	29 33.0	21
28 50.5	4 53.5	10 0.7	5 13.6	29 41.0	22
28 50.3	5 6.2	10 29.6	5 32.2	29 49.2	23
28 50.1	5 19.2	10 59.1	5 51.1	29 57.4	24
28 49.9	5 32.5	11 29.2	6 10.3	♌ 0 5.7	25
28 49.6	5 46.3	12 0.1	6 29.9	0 14.2	26
28 49.4	6 0.4	12 31.6	6 49.7	0 22.8	27
28 49.2	6 15.0	13 3.9	7 10.0	0 31.5	28
28 48.9	6 30.0	13 37.0	7 30.5	0 40.4	29
28 48.6	6 45.6	14 10.9	7 51.5	0 49.4	30
28 48.4	7 1.6	14 45.8	8 12.9	0 58.6	31
28 48.1	7 18.2	15 21.5	8 34.8	1 8.0	32
28 47.8	7 35.4	15 58.3	8 57.1	1 17.6	33
28 47.5	7 53.2	16 36.1	9 19.9	1 27.4	34
28 47.2	8 11.8	17 15.0	9 43.3	1 37.4	35
28 46.9	8 31.0	17 55.1	10 7.2	1 47.6	36
28 46.6	8 51.1	18 36.4	10 31.7	1 58.1	37
28 46.3	9 12.1	19 19.0	10 56.9	2 8.8	38
28 45.9	9 34.0	20 3.0	11 22.7	2 19.8	39
28 45.6	9 56.9	20 48.5	11 49.3	2 31.1	40
28 45.2	10 20.9	21 35.5	12 16.7	2 42.8	41
28 44.8	10 46.1	22 24.2	12 44.8	2 54.7	42
28 44.4	11 12.6	23 14.6	13 13.8	3 7.1	43
28 44.0	11 40.6	24 6.9	13 43.8	3 19.8	44
28 43.6	12 10.1	25 1.2	14 14.8	3 32.9	45
28 43.1	12 41.4	25 57.5	14 46.8	3 46.5	46
28 42.6	13 14.6	26 56.1	15 19.9	4 0.6	47
28 42.1	13 49.9	27 57.0	15 54.3	4 15.2	48
28 41.6	14 27.7	29 0.3	16 30.0	4 30.3	49
28 41.0	15 8.0	♋ 0 6.4	17 7.1	4 46.1	50
28 40.4	15 51.3	1 15.2	17 45.6	5 2.4	51
28 39.8	16 38.0	2 27.0	18 25.8	5 19.5	52
28 39.1	17 28.4	3 41.9	19 7.8	5 37.4	53
28 38.4	18 23.3	5 0.2	19 51.6	5 56.0	54
28 37.6	19 23.1	6 22.0	20 37.4	6 15.6	55
28 36.8	20 28.7	7 47.5	21 25.5	6 36.2	56
28 35.9	21 41.1	9 16.9	22 15.9	6 57.8	57
28 34.9	23 1.5	10 50.4	23 8.8	7 20.6	58
28 33.9	24 31.7	12 28.3	24 4.6	7 44.7	59
28 32.8	26 13.7	14 10.6	25 3.4	8 10.3	60
28 31.5	28 10.6	15 57.7	26 5.5	8 37.4	61
28 30.2	♊ 0 26.9	17 49.7	27 11.2	9 6.3	62
28 28.7	3 9.2	19 46.8	28 20.9	9 37.2	63
28 27.1	6 29.5	21 49.0	29 34.8	10 10.3	64
28 25.2	10 52.4	23 56.6	♌ 0 53.4	10 46.0	65
28 23.2	17 35.8	26 9.5	2 17.2	11 24.5	66
5	6	Descendant	8	9	S LAT

♌ 26° 46′ 43″

9h 56m 0s — MC — 149° 0′ 0″

22ʰ 0ᵐ 0ˢ — MC 330° 0′ 0″ — ♒ 27° 49′ 5″

N LAT	11	12	Ascendant	2	3
	° ′	° ′	° ′	° ′	° ′
0	♈ 0 0.0	♉ 2 10.9	♊ 2 5.4	♋ 0 0.0	♋27 54.6
5	0 0.0	2 57.1	3 55.1	1 19.8	28 30.4
10	0 0.0	3 46.1	5 50.1	2 40.7	29 6.3
15	0 0.0	4 39.0	7 52.4	4 4.2	29 43.0
16	0 0.0	4 50.1	8 17.9	4 21.3	29 50.5
17	0 0.0	5 1.5	8 43.9	4 38.6	29 58.1
18	0 0.0	5 13.1	9 10.3	4 56.1	♌ 0 5.7
19	0 0.0	5 24.9	9 37.1	5 13.7	0 13.4
20	0 0.0	5 37.0	10 4.5	5 31.6	0 21.1
21	0 0.0	5 49.4	10 32.3	5 49.7	0 29.0
22	0 0.0	6 2.1	11 0.7	6 8.0	0 36.9
23	0 0.0	6 15.1	11 29.7	6 26.6	0 45.0
24	0 0.0	6 28.5	11 59.2	6 45.4	0 53.1
25	0 0.0	6 42.2	12 29.4	7 4.6	1 1.4
26	0 0.0	6 56.3	13 0.3	7 24.0	1 9.8
27	0 0.0	7 10.8	13 31.8	7 43.8	1 18.3
28	0 0.0	7 25.7	14 4.1	8 4.0	1 26.9
29	0 0.0	7 41.1	14 37.2	8 24.4	1 35.7
30	0 0.0	7 57.0	15 11.1	8 45.3	1 44.6
31	0 0.0	8 13.4	15 45.9	9 6.6	1 53.8
32	0 0.0	8 30.4	16 21.7	9 28.4	2 3.1
33	0 0.0	8 48.0	16 58.3	9 50.6	2 12.5
34	0 0.0	9 6.3	17 36.1	10 13.3	2 22.2
35	0 0.0	9 25.3	18 14.9	10 36.5	2 32.1
36	0 0.0	9 45.0	18 54.8	11 0.3	2 42.2
37	0 0.0	10 5.5	19 36.0	11 24.6	2 52.6
38	0 0.0	10 26.9	20 18.5	11 49.6	3 3.2
39	0 0.0	10 49.3	21 2.3	12 15.3	3 14.1
40	0 0.0	11 12.7	21 47.6	12 41.7	3 25.3
41	0 0.0	11 37.3	22 34.3	13 8.8	3 36.8
42	0 0.0	12 3.0	23 22.7	13 36.8	3 48.6
43	0 0.0	12 30.1	24 12.9	14 5.6	4 0.8
44	0 0.0	12 58.6	25 4.8	14 35.3	4 13.4
45	0 0.0	13 28.8	25 58.6	15 6.0	4 26.3
46	0 0.0	14 0.7	26 54.5	15 37.7	4 39.8
47	0 0.0	14 34.6	27 52.6	16 10.6	4 53.6
48	0 0.0	15 10.6	28 52.9	16 44.6	5 8.0
49	0 0.0	15 49.0	29 55.7	17 20.0	5 23.0
50	0 0.0	16 30.1	♋ 1 1.1	17 56.7	5 38.5
51	0 0.0	17 14.1	2 9.1	18 34.8	5 54.7
52	0 0.0	18 1.6	3 20.1	19 14.6	6 11.5
53	0 0.0	18 52.9	4 34.1	19 56.1	6 29.1
54	0 0.0	19 48.6	5 51.4	20 39.4	6 47.5
55	0 0.0	20 49.4	7 12.1	21 24.6	7 6.8
56	0 0.0	21 55.9	8 36.4	22 12.1	7 27.0
57	0 0.0	23 9.4	10 4.5	23 1.8	7 48.3
58	0 0.0	24 30.9	11 36.6	23 54.0	8 10.7
59	0 0.0	26 2.2	13 12.9	24 49.0	8 34.4
60	0 0.0	27 45.4	14 53.6	25 46.9	8 59.5
61	0 0.0	29 43.6	16 38.9	26 48.1	9 26.2
62	0 0.0	♊ 2 1.2	18 29.0	27 52.7	9 54.5
63	0 0.0	4 44.9	20 23.9	29 1.2	10 24.8
64	0 0.0	8 6.7	22 24.0	♌ 0 13.8	10 57.3
65	0 0.0	12 31.4	24 29.2	1 31.0	11 32.2
66	0 0.0	19 17.5	26 39.6	2 53.1	12 9.8
S LAT	5	6	Descendant	8	9

♌ 27° 49′ 5″ — 10ʰ 0ᵐ 0ˢ — MC 150° 0′ 0″

22ʰ 4ᵐ 0ˢ — MC 331° 0′ 0″ — ♒ 28° 51′ 38″

N LAT	11	12	Ascendant	2	3
	° ′	° ′	° ′	° ′	° ′
0	♈ 1 5.4	♉ 3 13.3	♊ 3 2.6	♋ 0 55.0	♋28 52.0
5	1 6.2	4 0.8	4 53.1	2 14.8	29 27.5
10	1 7.1	4 51.0	6 48.7	3 35.6	♌ 0 3.1
15	1 8.0	5 45.2	8 51.6	4 58.9	0 39.5
16	1 8.2	5 56.7	9 17.2	5 16.0	0 46.9
17	1 8.4	6 8.3	9 43.3	5 33.3	0 54.4
18	1 8.6	6 20.2	10 9.7	5 50.7	1 2.0
19	1 8.8	6 32.4	10 36.7	6 8.3	1 9.6
20	1 9.0	6 44.8	11 4.1	6 26.1	1 17.3
21	1 9.2	6 57.5	11 32.0	6 44.1	1 25.1
22	1 9.5	7 10.5	12 0.4	7 2.4	1 32.9
23	1 9.7	7 23.8	12 29.4	7 20.9	1 40.9
24	1 9.9	7 37.5	12 59.0	7 39.7	1 49.0
25	1 10.1	7 51.5	13 29.2	7 58.8	1 57.1
26	1 10.4	8 5.9	14 0.1	8 18.1	2 5.4
27	1 10.6	8 20.8	14 31.7	8 37.8	2 13.9
28	1 10.8	8 36.1	15 4.0	8 57.9	2 22.4
29	1 11.1	8 51.8	15 37.1	9 18.3	2 31.1
30	1 11.4	9 8.1	16 11.0	9 39.0	2 40.0
31	1 11.6	9 24.9	16 45.7	10 0.2	2 49.0
32	1 11.9	9 42.3	17 21.4	10 21.8	2 58.2
33	1 12.2	10 0.3	17 58.0	10 43.9	3 7.6
34	1 12.5	10 18.9	18 35.6	11 6.5	3 17.1
35	1 12.8	10 38.3	19 14.3	11 29.6	3 26.9
36	1 13.1	10 58.4	19 54.2	11 53.2	3 36.9
37	1 13.4	11 19.4	20 35.2	12 17.4	3 47.2
38	1 13.7	11 41.3	21 17.5	12 42.3	3 57.7
39	1 14.1	12 4.1	22 1.1	13 7.7	4 8.4
40	1 14.4	12 28.0	22 46.1	13 33.9	4 19.5
41	1 14.8	12 53.0	23 32.7	14 0.9	4 30.8
42	1 15.2	13 19.3	24 20.8	14 28.6	4 42.5
43	1 15.6	13 46.9	25 10.6	14 57.2	4 54.6
44	1 16.0	14 16.0	26 2.1	15 26.7	5 7.0
45	1 16.4	14 46.7	26 55.6	15 57.1	5 19.8
46	1 16.9	15 19.3	27 51.0	16 28.5	5 33.0
47	1 17.4	15 53.7	28 48.6	17 1.1	5 46.7
48	1 17.9	16 30.4	29 48.4	17 34.8	6 0.9
49	1 18.4	17 9.4	♋ 0 50.5	18 9.8	6 15.7
50	1 19.0	17 51.2	1 55.2	18 46.1	6 31.0
51	1 19.6	18 36.0	3 2.5	19 23.9	6 46.9
52	1 20.2	19 24.2	4 12.7	20 3.2	7 3.5
53	1 20.9	20 16.3	5 25.8	20 44.2	7 20.9
54	1 21.6	21 12.8	6 42.1	21 27.0	7 39.0
55	1 22.4	22 14.4	8 1.7	22 11.7	7 58.0
56	1 23.2	23 21.9	9 24.8	22 58.5	8 17.9
57	1 24.1	24 36.2	10 51.7	23 47.6	8 38.8
58	1 25.1	25 58.7	12 22.4	24 39.1	9 0.9
59	1 26.1	27 31.0	13 57.2	25 33.3	9 24.2
60	1 27.2	29 15.3	15 36.3	26 30.3	9 48.8
61	1 28.5	♊ 1 14.5	17 19.8	27 30.5	10 15.0
62	1 29.8	3 33.2	19 7.9	28 34.1	10 42.8
63	1 31.3	6 18.0	21 0.9	29 41.4	11 12.5
64	1 32.9	9 40.9	22 58.7	♌ 0 52.7	11 44.3
65	1 34.8	14 6.4	25 1.7	2 8.5	12 18.4
66	1 36.8	20 52.5	27 9.7	3 29.0	12 55.2
S LAT	5	6	Descendant	8	9

♌ 28° 51′ 38″ — 10ʰ 4ᵐ 0ˢ — MC 151° 0′ 0″

22h 8m 0s — MC 332° 0′ 0″ — ♒ 29° 54′ 21″

N LAT °	11 ° ′	12 ° ′	Ascendant ° ′	2 ° ′	3 ° ′
0	♈ 2 10.8	♉ 4 15.5	♊ 3 59.8	♋ 1 50.1	♋ 29 49.5
5	2 12.5	5 4.2	5 50.9	3 9.8	♌ 0 24.7
10	2 14.2	5 55.7	7 47.2	4 30.5	1 0.1
15	2 16.1	6 51.3	9 50.5	5 53.6	1 36.1
16	2 16.4	7 3.0	10 16.3	6 10.7	1 43.5
17	2 16.8	7 14.9	10 42.4	6 27.9	1 50.9
18	2 17.2	7 27.1	11 8.9	6 45.2	1 58.4
19	2 17.6	7 39.5	11 35.9	7 2.8	2 5.9
20	2 18.0	7 52.2	12 3.4	7 20.5	2 13.5
21	2 18.5	8 5.2	12 31.4	7 38.5	2 21.2
22	2 18.9	8 18.5	12 59.8	7 56.7	2 29.0
23	2 19.3	8 32.1	13 28.9	8 15.1	2 36.9
24	2 19.8	8 46.1	13 58.5	8 33.9	2 44.9
25	2 20.2	9 0.5	14 28.7	8 52.8	2 53.0
26	2 20.7	9 15.2	14 59.6	9 12.1	3 1.2
27	2 21.2	9 30.4	15 31.2	9 31.7	3 9.5
28	2 21.7	9 46.0	16 3.5	9 51.7	3 18.0
29	2 22.2	10 2.1	16 36.5	10 12.0	3 26.6
30	2 22.7	10 18.7	17 10.4	10 32.6	3 35.4
31	2 23.2	10 35.9	17 45.1	10 53.7	3 44.3
32	2 23.8	10 53.7	18 20.7	11 15.2	3 53.4
33	2 24.3	11 12.0	18 57.2	11 37.2	4 2.7
34	2 24.9	11 31.1	19 34.8	11 59.6	4 12.1
35	2 25.5	11 50.9	20 13.3	12 22.5	4 21.8
36	2 26.1	12 11.4	20 53.0	12 46.0	4 31.7
37	2 26.8	12 32.8	21 33.9	13 10.1	4 41.8
38	2 27.4	12 55.1	22 16.0	13 34.7	4 52.2
39	2 28.1	13 18.4	22 59.4	14 0.1	5 2.8
40	2 28.8	13 42.7	23 44.2	14 26.1	5 13.7
41	2 29.6	14 8.2	24 30.5	14 52.8	5 25.0
42	2 30.3	14 35.0	25 18.3	15 20.3	5 36.5
43	2 31.2	15 3.1	26 7.8	15 48.6	5 48.4
44	2 32.0	15 32.7	26 59.0	16 17.9	6 0.6
45	2 32.9	16 4.0	27 52.0	16 48.0	6 13.3
46	2 33.8	16 37.1	28 47.0	17 19.2	6 26.4
47	2 34.8	17 12.1	29 44.0	17 51.4	6 39.9
48	2 35.8	17 49.4	♋ 0 43.3	18 24.8	6 53.9
49	2 36.9	18 29.1	1 44.8	18 59.5	7 8.4
50	2 38.0	19 11.5	2 48.8	19 35.4	7 23.5
51	2 39.2	19 56.9	3 55.4	20 12.8	7 39.2
52	2 40.5	20 45.9	5 4.7	20 51.6	7 55.6
53	2 41.8	21 38.7	6 17.0	21 32.1	8 12.6
54	2 43.3	22 35.9	7 32.3	22 14.4	8 30.5
55	2 44.8	23 38.3	8 50.9	22 58.6	8 49.2
56	2 46.4	24 46.5	10 12.8	23 44.8	9 8.8
57	2 48.2	26 1.7	11 38.4	24 33.2	9 29.4
58	2 50.1	27 25.1	13 7.7	25 24.0	9 51.0
59	2 52.2	28 58.2	14 41.1	26 17.4	10 13.9
60	2 54.4	♊ 0 43.3	16 18.5	27 13.6	10 38.1
61	2 56.9	2 43.5	18 0.3	28 12.8	11 3.8
62	2 59.6	5 2.9	19 46.6	29 15.4	11 31.1
63	3 2.6	7 48.5	21 37.6	♌ 0 21.5	12 0.2
64	3 5.8	11 11.9	23 33.4	1 31.6	12 31.3
65	3 9.5	15 37.4	25 34.1	2 46.0	13 4.7
66	3 13.6	22 20.9	27 39.7	4 5.0	13 40.7
S LAT	5	6	Descendant	8	9

♌ 29° 54′ 21″ — 10h 8m 0s — MC 152° 0′ 0″

22h 12m 0s — MC 333° 0′ 0″ — ♓ 0° 57′ 13″

N LAT °	11 ° ′	12 ° ′	Ascendant ° ′	2 ° ′	3 ° ′
0	♈ 3 16.2	♉ 5 17.5	♊ 4 56.7	♋ 2 45.2	♌ 0 47.2
5	3 18.7	6 7.4	6 48.6	4 4.8	1 22.1
10	3 21.3	7 0.2	8 45.4	5 25.4	1 57.1
15	3 24.1	7 57.0	10 49.2	6 48.3	2 32.8
16	3 24.6	8 9.0	11 15.0	7 5.3	2 40.1
17	3 25.2	8 21.2	11 41.2	7 22.4	2 47.5
18	3 25.8	8 33.6	12 7.9	7 39.7	2 54.9
19	3 26.4	8 46.4	12 34.9	7 57.2	3 2.3
20	3 27.0	8 59.3	13 2.4	8 14.9	3 9.9
21	3 27.7	9 12.6	13 30.4	8 32.8	3 17.5
22	3 28.3	9 26.2	13 59.0	8 50.9	3 25.2
23	3 29.0	9 40.2	14 28.0	9 9.3	3 33.0
24	3 29.6	9 54.4	14 57.7	9 27.9	3 40.9
25	3 30.3	10 9.1	15 27.9	9 46.9	3 48.9
26	3 31.0	10 24.2	15 58.8	10 6.1	3 57.1
27	3 31.7	10 39.7	16 30.4	10 25.6	4 5.3
28	3 32.5	10 55.6	17 2.6	10 45.4	4 13.7
29	3 33.2	11 12.0	17 35.6	11 5.6	4 22.2
30	3 34.0	11 29.0	18 9.4	11 26.2	4 30.9
31	3 34.8	11 46.5	18 44.1	11 47.1	4 39.7
32	3 35.6	12 4.6	19 19.6	12 8.5	4 48.7
33	3 36.5	12 23.4	19 56.0	12 30.3	4 57.8
34	3 37.3	12 42.8	20 33.5	12 52.6	5 7.2
35	3 38.2	13 2.9	21 11.9	13 15.4	5 16.8
36	3 39.2	13 23.9	21 51.5	13 38.7	5 26.5
37	3 40.1	13 45.7	22 32.2	14 2.6	5 36.5
38	3 41.1	14 8.4	23 14.1	14 27.1	5 46.8
39	3 42.1	14 32.1	23 57.3	14 52.2	5 57.3
40	3 43.2	14 56.9	24 41.8	15 18.0	6 8.1
41	3 44.3	15 22.8	25 27.8	15 44.6	6 19.2
42	3 45.5	15 50.1	26 15.3	16 11.9	6 30.5
43	3 46.7	16 18.7	27 4.5	16 40.0	6 42.3
44	3 47.9	16 48.8	27 55.3	17 8.9	6 54.4
45	3 49.3	17 20.6	28 47.9	17 38.8	7 6.9
46	3 50.7	17 54.1	29 42.4	18 9.7	7 19.7
47	3 52.1	18 29.7	♋ 0 39.0	18 41.6	7 33.1
48	3 53.6	19 7.5	1 37.6	19 14.7	7 46.9
49	3 55.2	19 47.8	2 38.6	19 49.0	8 1.2
50	3 56.9	20 30.8	3 41.9	20 24.5	8 16.1
51	3 58.7	21 16.9	4 47.8	21 1.5	8 31.6
52	4 0.6	22 6.5	5 56.3	21 39.9	8 47.7
53	4 2.7	23 0.0	7 7.7	22 20.0	9 4.5
54	4 4.8	23 57.9	8 22.0	23 1.7	9 22.0
55	4 7.1	25 0.9	9 39.5	23 45.3	9 40.4
56	4 9.6	26 9.9	11 0.3	24 30.9	9 59.7
57	4 12.3	27 25.8	12 24.7	25 18.7	10 19.9
58	4 15.1	28 49.9	13 52.7	26 8.8	10 41.3
59	4 18.2	♊ 0 23.8	15 24.6	27 1.4	11 3.7
60	4 21.6	2 9.6	17 0.5	27 56.7	11 27.5
61	4 25.3	4 10.4	18 40.6	28 55.0	11 52.7
62	4 29.3	6 30.5	20 25.1	29 56.6	12 19.5
63	4 33.8	9 16.5	22 14.2	♌ 1 1.6	12 48.0
64	4 38.7	12 40.0	24 7.9	2 10.5	13 18.5
65	4 44.2	17 4.6	26 6.4	3 23.4	13 51.1
66	4 50.4	23 43.1	28 9.8	4 41.0	14 26.3
S LAT	5	6	Descendant	8	9

♍ 0° 57′ 13″ — 10h 12m 0s — MC 153° 0′ 0″

22h 16m 0s MC 334° 0′ 0″

♓ 2° 0′ 16″

N LAT	11	12	Ascendant	2	3
°	° ′	° ′	° ′	° ′	° ′
0	♈ 4 21.5	♉ 6 19.4	♊ 5 53.6	♋ 3 40.2	♌ 1 45.0
5	4 24.9	7 10.4	7 46.1	4 59.8	2 19.6
10	4 28.3	8 4.4	9 43.5	6 20.2	2 54.3
15	4 32.0	9 2.5	11 47.7	7 42.9	3 29.7
16	4 32.8	9 14.8	12 13.6	7 59.9	3 36.9
17	4 33.6	9 27.2	12 39.9	8 16.9	3 44.2
18	4 34.4	9 39.9	13 6.5	8 34.2	3 51.5
19	4 35.2	9 52.9	13 33.6	8 51.6	3 58.9
20	4 36.0	10 6.2	14 1.2	9 9.2	4 6.3
21	4 36.8	10 19.7	14 29.2	9 27.1	4 13.9
22	4 37.7	10 33.6	14 57.8	9 45.1	4 21.5
23	4 38.6	10 47.8	15 26.9	10 3.4	4 29.2
24	4 39.5	11 2.4	15 56.5	10 22.0	4 37.1
25	4 40.4	11 17.4	16 26.8	10 40.8	4 45.0
26	4 41.3	11 32.7	16 57.6	10 59.9	4 53.0
27	4 42.3	11 48.5	17 29.2	11 19.3	5 1.2
28	4 43.2	12 4.8	18 1.4	11 39.1	5 9.5
29	4 44.3	12 21.6	18 34.4	11 59.1	5 17.9
30	4 45.3	12 38.8	19 8.1	12 19.6	5 26.4
31	4 46.3	12 56.7	19 42.7	12 40.4	5 35.2
32	4 47.4	13 15.1	20 18.1	13 1.7	5 44.0
33	4 48.6	13 34.2	20 54.5	13 23.4	5 53.1
34	4 49.7	13 54.0	21 31.8	13 45.5	6 2.3
35	4 50.9	14 14.5	22 10.1	14 8.2	6 11.8
36	4 52.1	14 35.9	22 49.5	14 31.3	6 21.4
37	4 53.4	14 58.0	23 30.0	14 55.1	6 31.3
38	4 54.7	15 21.2	24 11.7	15 19.4	6 41.4
39	4 56.1	15 45.3	24 54.7	15 44.3	6 51.8
40	4 57.5	16 10.5	25 39.0	16 9.9	7 2.5
41	4 59.0	16 36.9	26 24.7	16 36.2	7 13.4
42	5 0.6	17 4.5	27 11.9	17 3.3	7 24.7
43	5 2.2	17 33.6	28 0.7	17 31.1	7 36.2
44	5 3.9	18 4.1	28 51.1	17 59.9	7 48.2
45	5 5.6	18 36.4	29 43.3	18 29.5	8 0.5
46	5 7.5	19 10.5	♋ 0 37.4	19 0.1	8 13.2
47	5 9.4	19 46.6	1 33.4	19 31.7	8 26.3
48	5 11.4	20 24.9	2 31.5	20 4.4	8 39.9
49	5 13.6	21 5.7	3 31.9	20 38.4	8 54.1
50	5 15.8	21 49.3	4 34.5	21 13.5	9 8.7
51	5 18.2	22 36.0	5 39.6	21 50.1	9 24.0
52	5 20.8	23 26.1	6 47.4	22 28.1	9 39.8
53	5 23.5	24 20.2	7 57.9	23 7.6	9 56.4
54	5 26.3	25 18.7	9 11.3	23 48.9	10 13.6
55	5 29.4	26 22.4	10 27.7	24 31.9	10 31.7
56	5 32.7	27 32.0	11 47.5	25 16.9	10 50.7
57	5 36.2	28 48.5	13 10.6	26 4.0	11 10.6
58	5 40.1	♊ 0 13.2	14 37.3	26 53.4	11 31.5
59	5 44.2	1 47.7	16 7.7	27 45.3	11 53.6
60	5 48.7	3 34.2	17 42.1	28 39.8	12 16.9
61	5 53.6	5 35.5	19 20.6	29 37.2	12 41.6
62	5 59.0	7 55.9	21 3.3	♌ 0 37.7	13 7.9
63	6 4.9	10 42.0	22 50.5	1 41.6	13 35.8
64	6 11.5	14 5.1	24 42.3	2 49.3	14 5.6
65	6 18.8	18 28.1	26 38.7	4 0.9	14 37.6
66	6 27.1	24 59.5	28 39.9	5 17.0	15 12.0
S LAT	5	6	Descendant	8	9

♍ 2° 0′ 16″

10h 16m 0s MC 154° 0′ 0″

22h 20m 0s MC 335° 0′ 0″

♓ 3° 3′ 27″

N LAT	11	12	Ascendant	2	3
°	° ′	° ′	° ′	° ′	° ′
0	♈ 5 26.8	♉ 7 21.1	♊ 6 50.3	♋ 4 35.3	♌ 2 43.0
5	5 31.0	8 13.3	8 43.4	5 54.8	3 17.3
10	5 35.4	9 8.4	10 41.3	7 15.1	3 51.6
15	5 40.0	10 7.8	12 46.0	8 37.5	4 26.6
16	5 40.9	10 20.2	13 11.9	8 54.4	4 33.8
17	5 41.9	10 33.0	13 38.2	9 11.4	4 41.0
18	5 42.9	10 45.9	14 4.9	9 28.6	4 48.2
19	5 43.9	10 59.2	14 32.1	9 46.0	4 55.5
20	5 44.9	11 12.7	14 59.7	10 3.5	5 2.9
21	5 46.0	11 26.5	15 27.7	10 21.3	5 10.4
22	5 47.1	11 40.7	15 56.3	10 39.3	5 17.9
23	5 48.1	11 55.2	16 25.4	10 57.5	5 25.6
24	5 49.3	12 10.1	16 55.1	11 15.9	5 33.3
25	5 50.4	12 25.3	17 25.3	11 34.7	5 41.1
26	5 51.6	12 41.0	17 56.2	11 53.7	5 49.1
27	5 52.8	12 57.1	18 27.7	12 13.0	5 57.1
28	5 54.0	13 13.6	18 59.9	12 32.6	6 5.3
29	5 55.2	13 30.7	19 32.8	12 52.6	6 13.6
30	5 56.5	13 48.3	20 6.5	13 12.9	6 22.1
31	5 57.8	14 6.4	20 41.0	13 33.7	6 30.7
32	5 59.2	14 25.2	21 16.3	13 54.8	6 39.5
33	6 0.6	14 44.6	21 52.5	14 16.3	6 48.5
34	6 2.0	15 4.8	22 29.7	14 38.3	6 57.6
35	6 3.5	15 25.7	23 7.9	15 0.8	7 6.9
36	6 5.1	15 47.3	23 47.1	15 23.8	7 16.4
37	6 6.7	16 9.9	24 27.4	15 47.4	7 26.2
38	6 8.3	16 33.4	25 8.9	16 11.5	7 36.2
39	6 10.0	16 57.9	25 51.7	16 36.3	7 46.4
40	6 11.8	17 23.5	26 35.7	17 1.7	7 56.9
41	6 13.7	17 50.3	27 21.2	17 27.8	8 7.7
42	6 15.6	18 18.3	28 8.0	17 54.6	8 18.8
43	6 17.6	18 47.8	28 56.5	18 22.2	8 30.2
44	6 19.7	19 18.8	29 46.5	18 50.7	8 42.0
45	6 21.9	19 51.5	♋ 0 38.3	19 20.0	8 54.1
46	6 24.2	20 26.1	1 31.9	19 50.3	9 6.7
47	6 26.6	21 2.7	2 27.4	20 21.6	9 19.6
48	6 29.2	21 41.5	3 25.0	20 54.0	9 33.1
49	6 31.8	22 22.8	4 24.7	21 27.6	9 47.0
50	6 34.7	23 6.9	5 26.7	22 2.4	10 1.4
51	6 37.6	23 54.1	6 31.1	22 38.5	10 16.4
52	6 40.8	24 44.7	7 38.0	23 16.1	10 32.0
53	6 44.2	25 39.4	8 47.6	23 55.2	10 48.3
54	6 47.8	26 38.4	10 0.1	24 35.9	11 5.3
55	6 51.6	27 42.7	11 15.5	25 18.4	11 23.1
56	6 55.7	28 52.8	12 34.1	26 2.8	11 41.7
57	7 0.1	♊ 0 9.9	13 56.1	26 49.3	12 1.2
58	7 4.9	1 35.1	15 21.5	27 38.0	12 21.8
59	7 10.1	3 10.1	16 50.5	28 29.1	12 43.5
60	7 15.7	4 57.0	18 23.4	29 22.7	13 6.4
61	7 21.8	6 58.6	20 0.3	♌ 0 19.2	13 30.6
62	7 28.5	9 19.2	21 41.4	1 18.8	13 56.3
63	7 35.9	12 5.1	23 26.7	2 21.6	14 23.7
64	7 44.2	15 27.4	25 16.5	3 28.1	14 52.9
65	7 53.3	19 48.0	27 10.9	4 38.4	15 24.2
66	8 3.6	26 10.7	29 9.9	5 53.0	15 57.7
S LAT	5	6	Descendant	8	9

♍ 3° 3′ 27″

10h 20m 0s MC 155° 0′ 0″

22h 24m 0s MC 336° 0′ 0″
♓ 4° 6′ 48″

N LAT	11	12	Ascendant	2	3
°	° ′	° ′	° ′	° ′	° ′
0	♈ 6 32.1	♉ 8 22.6	♊ 7 46.9	♋ 5 30.5	♌ 3 41.2
5	6 37.1	9 15.9	9 40.6	6 49.8	4 15.1
10	6 42.3	10 12.2	11 39.0	8 9.9	4 49.1
15	6 47.9	11 12.7	13 44.0	9 32.1	5 23.7
16	6 49.0	11 25.4	14 10.0	9 48.9	5 30.8
17	6 50.2	11 38.4	14 36.3	10 5.9	5 37.9
18	6 51.4	11 51.6	15 3.1	10 23.0	5 45.1
19	6 52.6	12 5.1	15 30.3	10 40.3	5 52.3
20	6 53.8	12 18.9	15 57.9	10 57.8	5 59.6
21	6 55.1	12 33.0	16 26.0	11 15.5	6 7.0
22	6 56.4	12 47.4	16 54.6	11 33.4	6 14.5
23	6 57.7	13 2.2	17 23.7	11 51.5	6 22.0
24	6 59.0	13 17.3	17 53.3	12 9.9	6 29.6
25	7 0.4	13 32.9	18 23.5	12 28.5	6 37.4
26	7 1.7	13 48.8	18 54.4	12 47.4	6 45.2
27	7 3.2	14 5.2	19 25.8	13 6.6	6 53.2
28	7 4.6	14 22.1	19 58.0	13 26.1	7 1.3
29	7 6.1	14 39.4	20 30.8	13 46.0	7 9.5
30	7 7.7	14 57.3	21 4.5	14 6.2	7 17.9
31	7 9.3	15 15.8	21 38.9	14 26.8	7 26.4
32	7 10.9	15 34.9	22 14.1	14 47.8	7 35.1
33	7 12.6	15 54.6	22 50.2	15 9.2	7 43.9
34	7 14.3	16 15.1	23 27.3	15 31.1	7 52.9
35	7 16.1	16 36.3	24 5.3	15 53.4	8 2.1
36	7 17.9	16 58.3	24 44.3	16 16.2	8 11.5
37	7 19.9	17 21.2	25 24.5	16 39.6	8 21.2
38	7 21.8	17 45.1	26 5.7	17 3.6	8 31.0
39	7 23.9	18 9.9	26 48.2	17 28.1	8 41.1
40	7 26.0	18 35.9	27 32.0	17 53.3	8 51.5
41	7 28.2	19 3.1	28 17.2	18 19.2	9 2.1
42	7 30.5	19 31.5	29 3.7	18 45.8	9 13.1
43	7 33.0	20 1.4	29 51.8	19 13.2	9 24.3
44	7 35.5	20 32.8	♋ 0 41.4	19 41.4	9 35.9
45	7 38.1	21 6.0	1 32.8	20 10.5	9 47.9
46	7 40.9	21 40.9	2 25.9	20 40.5	10 0.2
47	7 43.7	22 18.0	3 20.9	21 11.5	10 13.0
48	7 46.8	22 57.2	4 17.9	21 43.5	10 26.2
49	7 50.0	23 39.0	5 17.0	22 16.7	10 39.9
50	7 53.4	24 23.6	6 18.3	22 51.2	10 54.1
51	7 57.0	25 11.2	7 22.0	23 26.9	11 8.9
52	8 0.8	26 2.4	8 28.2	24 4.0	11 24.2
53	8 4.8	26 57.5	9 36.9	24 42.6	11 40.3
54	8 9.1	27 57.1	10 48.5	25 22.8	11 57.0
55	8 13.7	29 1.8	12 2.9	26 4.8	12 14.4
56	8 18.6	♊ 0 12.4	13 20.5	26 48.6	12 32.7
57	8 23.9	1 30.0	14 41.2	27 34.4	12 52.0
58	8 29.6	2 55.6	16 5.4	28 22.4	13 12.1
59	8 35.8	4 31.0	17 33.1	29 12.7	13 33.4
60	8 42.5	6 18.1	19 4.5	♌ 0 5.6	13 55.9
61	8 49.9	8 19.9	20 39.8	1 1.2	14 19.6
62	8 57.9	10 40.4	22 19.2	1 59.8	14 44.8
63	9 6.8	13 25.9	24 2.8	3 1.6	15 11.6
64	9 16.7	16 46.9	25 50.7	4 6.9	15 40.2
65	9 27.7	21 4.5	27 43.1	5 15.9	16 10.8
66	9 40.1	27 17.2	29 40.0	6 29.2	16 43.6
S LAT	5	6	Descendant	8	9

♍ 4° 6′ 48″
10h 24m 0s MC 156° 0′ 0″

22h 28m 0s MC 337° 0′ 0″
♓ 5° 10′ 18″

N LAT	11	12	Ascendant	2	3
°	° ′	° ′	° ′	° ′	° ′
0	♈ 7 37.4	♉ 9 23.9	♊ 8 43.3	♋ 6 25.6	♌ 4 39.5
5	7 43.2	10 18.3	10 37.6	7 44.8	5 13.1
10	7 49.3	11 15.7	12 36.5	9 4.7	5 46.7
15	7 55.7	12 17.4	14 41.8	10 26.7	6 21.0
16	7 57.0	12 30.4	15 7.8	10 43.4	6 27.9
17	7 58.4	12 43.6	15 34.2	11 0.3	6 35.0
18	7 59.8	12 57.1	16 1.0	11 17.4	6 42.1
19	8 1.2	13 10.8	16 28.2	11 34.6	6 49.2
20	8 2.6	13 24.8	16 55.9	11 52.0	6 56.4
21	8 4.1	13 39.2	17 24.0	12 9.6	7 3.7
22	8 5.6	13 53.9	17 52.5	12 27.4	7 11.1
23	8 7.1	14 8.9	18 21.6	12 45.4	7 18.5
24	8 8.7	14 24.3	18 51.3	13 3.7	7 26.1
25	8 10.2	14 40.1	19 21.5	13 22.3	7 33.8
26	8 11.9	14 56.3	19 52.3	13 41.1	7 41.5
27	8 13.5	15 13.0	20 23.7	14 0.2	7 49.4
28	8 15.2	15 30.1	20 55.8	14 19.6	7 57.4
29	8 17.0	15 47.7	21 28.6	14 39.3	8 5.5
30	8 18.8	16 5.9	22 2.1	14 59.4	8 13.7
31	8 20.6	16 24.7	22 36.4	15 19.9	8 22.1
32	8 22.5	16 44.1	23 11.5	15 40.7	8 30.7
33	8 24.5	17 4.1	23 47.5	16 2.0	8 39.4
34	8 26.5	17 24.9	24 24.4	16 23.7	8 48.3
35	8 28.6	17 46.4	25 2.3	16 45.9	8 57.4
36	8 30.7	18 8.8	25 41.2	17 8.6	9 6.7
37	8 33.0	18 32.0	26 21.1	17 31.8	9 16.2
38	8 35.3	18 56.2	27 2.2	17 55.5	9 25.9
39	8 37.7	19 21.4	27 44.4	18 19.9	9 35.9
40	8 40.1	19 47.8	28 27.9	18 44.9	9 46.1
41	8 42.7	20 15.3	29 12.8	19 10.5	9 56.6
42	8 45.4	20 44.1	29 59.0	19 36.9	10 7.4
43	8 48.2	21 14.4	♋ 0 46.7	20 4.0	10 18.5
44	8 51.1	21 46.2	1 36.0	20 32.0	10 29.9
45	8 54.2	22 19.7	2 26.9	21 0.8	10 41.7
46	8 57.4	22 55.1	3 19.5	21 30.5	10 53.9
47	9 0.8	23 32.5	4 14.0	22 1.2	11 6.4
48	9 4.3	24 12.2	5 10.4	22 32.9	11 19.4
49	9 8.1	24 54.4	6 8.9	23 5.7	11 32.9
50	9 12.0	25 39.4	7 9.6	23 39.8	11 46.9
51	9 16.2	26 27.5	8 12.6	24 15.1	12 1.4
52	9 20.6	27 19.1	9 17.9	24 51.8	12 16.5
53	9 25.3	28 14.6	10 25.9	25 29.9	12 32.3
54	9 30.3	29 14.6	11 36.5	26 9.6	12 48.7
55	9 35.6	♊ 0 19.8	12 49.9	26 51.0	13 5.9
56	9 41.4	1 30.8	14 6.4	27 34.3	13 23.9
57	9 47.6	2 48.7	15 26.0	28 19.4	13 42.7
58	9 54.2	4 14.7	16 48.9	29 6.7	14 2.5
59	10 1.4	5 50.3	18 15.3	29 56.3	14 23.4
60	10 9.3	7 37.6	19 45.3	♌ 0 48.4	14 45.4
61	10 17.8	9 39.4	21 19.1	1 43.1	15 8.7
62	10 27.2	11 59.6	22 56.8	2 40.7	15 33.4
63	10 37.6	14 44.4	24 38.7	3 41.5	15 59.7
64	10 49.1	18 3.8	26 24.7	4 45.6	16 27.6
65	11 1.9	22 17.8	28 15.2	5 53.5	16 57.5
66	11 16.4	28 19.6	♌ 0 10.1	7 5.3	17 29.5
S LAT	5	6	Descendant	8	9

♍ 5° 10′ 18″
10h 28m 0s MC 157° 0′ 0″

22h 32m 0s — MC 338° 0′ 0″

♓ 6° 13′ 57″

11	12	Ascendant	2	3	N LAT
° ′	° ′	° ′	° ′	° ′	°
♈ 8 42.5	♉ 10 25.0	♊ 9 39.7	♋ 7 20.8	♌ 5 38.0	0
8 49.2	11 20.5	11 34.5	8 39.8	6 11.2	5
8 56.1	12 19.0	13 33.8	9 59.5	6 44.5	10
9 3.5	13 21.8	15 39.4	11 21.2	7 18.3	15
9 5.0	13 35.0	16 5.5	11 37.9	7 25.2	16
9 6.6	13 48.5	16 31.9	11 54.7	7 32.2	17
9 8.1	14 2.2	16 58.7	12 11.7	7 39.2	18
9 9.7	14 16.2	17 25.9	12 28.8	7 46.2	19
9 11.4	14 30.5	17 53.6	12 46.2	7 53.4	20
9 13.0	14 45.0	18 21.7	13 3.7	8 0.6	21
9 14.7	15 0.0	18 50.3	13 21.4	8 7.8	22
9 16.5	15 15.2	19 19.3	13 39.4	8 15.2	23
9 18.3	15 30.9	19 49.0	13 57.6	8 22.7	24
9 20.1	15 46.9	20 19.1	14 16.0	8 30.2	25
9 21.9	16 3.4	20 49.9	14 34.7	8 37.9	26
9 23.8	16 20.3	21 21.2	14 53.7	8 45.7	27
9 25.8	16 37.8	21 53.3	15 13.0	8 53.5	28
9 27.8	16 55.7	22 26.0	15 32.6	9 1.6	29
9 29.8	17 14.1	22 59.4	15 52.6	9 9.7	30
9 31.9	17 33.2	23 33.6	16 12.9	9 18.0	31
9 34.1	17 52.9	24 8.6	16 33.6	9 26.4	32
9 36.3	18 13.2	24 44.5	16 54.7	9 35.0	33
9 38.6	18 34.3	25 21.2	17 16.3	9 43.8	34
9 41.0	18 56.1	25 58.9	17 38.3	9 52.8	35
9 43.4	19 18.8	26 37.6	18 0.8	10 1.9	36
9 46.0	19 42.3	27 17.3	18 23.8	10 11.3	37
9 48.6	20 6.8	27 58.2	18 47.4	10 20.9	38
9 51.3	20 32.4	28 40.2	19 11.6	10 30.7	39
9 54.2	20 59.0	29 23.4	19 36.3	10 40.8	40
9 57.1	21 26.9	♋ 0 7.9	20 1.8	10 51.1	41
10 0.2	21 56.1	0 53.8	20 27.9	11 1.8	42
10 3.4	22 26.7	1 41.2	20 54.8	11 12.7	43
10 6.7	22 58.9	2 30.0	21 22.5	11 24.0	44
10 10.2	23 32.8	3 20.5	21 51.0	11 35.6	45
10 13.9	24 8.5	4 12.7	22 20.4	11 47.5	46
10 17.7	24 46.3	5 6.7	22 50.8	11 59.9	47
10 21.7	25 26.4	6 2.6	23 22.1	12 12.7	48
10 26.0	26 9.0	7 0.4	23 54.6	12 26.0	49
10 30.5	26 54.3	8 0.4	24 28.3	12 39.7	50
10 35.2	27 42.8	9 2.7	25 3.2	12 54.0	51
10 40.3	28 34.8	10 7.3	25 39.4	13 8.9	52
10 45.6	29 30.7	11 14.4	26 17.1	13 24.4	53
10 51.4	♊ 0 31.1	12 24.1	26 56.3	13 40.5	54
10 57.5	1 36.6	13 36.6	27 37.2	13 57.4	55
11 4.0	2 48.0	14 52.0	28 19.8	14 15.0	56
11 11.0	4 6.2	16 10.4	29 4.4	14 33.5	57
11 18.6	5 32.4	17 32.1	29 51.0	14 53.0	58
11 26.9	7 8.2	18 57.2	♌ 0 39.8	15 13.5	59
11 35.8	8 55.5	20 25.8	1 31.1	15 35.0	60
11 45.6	10 57.1	21 58.1	2 25.0	15 57.9	61
11 56.3	13 16.9	23 34.3	3 21.6	16 22.1	62
12 8.1	16 0.7	25 14.4	4 21.4	16 47.7	63
12 21.3	19 18.3	26 58.7	5 24.4	17 15.1	64
12 36.0	23 28.1	28 47.3	6 31.0	17 44.3	65
12 52.5	29 18.4	♌ 0 40.2	7 41.5	18 15.6	66
5	6	Descendant	8	9	S LAT

♍ 6° 13′ 57″

10h 32m 0s — MC 158° 0′ 0″

22h 36m 0s — MC 339° 0′ 0″

♓ 7° 17′ 44″

11	12	Ascendant	2	3	N LAT
° ′	° ′	° ′	° ′	° ′	°
♈ 9 47.7	♉ 11 26.0	♊ 10 35.9	♋ 8 16.1	♌ 6 36.6	0
9 55.1	12 22.5	12 31.3	9 34.9	7 9.5	5
10 2.9	13 22.1	14 30.9	10 54.4	7 42.3	10
10 11.2	14 26.0	16 36.8	12 15.7	8 15.8	15
10 12.9	14 39.4	17 2.9	12 32.4	8 22.6	16
10 14.6	14 53.1	17 29.3	12 49.1	8 29.5	17
10 16.4	15 7.0	17 56.2	13 6.0	8 36.4	18
10 18.2	15 21.2	18 23.4	13 23.1	8 43.4	19
10 20.1	15 35.8	18 51.1	13 40.3	8 50.4	20
10 21.9	15 50.6	19 19.2	13 57.8	8 57.5	21
10 23.8	16 5.7	19 47.7	14 15.4	9 4.7	22
10 25.8	16 21.3	20 16.8	14 33.3	9 12.0	23
10 27.8	16 37.2	20 46.4	14 51.3	9 19.4	24
10 29.8	16 53.5	21 16.5	15 9.7	9 26.8	25
10 31.9	17 10.2	21 47.2	15 28.3	9 34.4	26
10 34.0	17 27.4	22 18.5	15 47.2	9 42.0	27
10 36.2	17 45.0	22 50.5	16 6.3	9 49.8	28
10 38.4	18 3.2	23 23.1	16 25.8	9 57.7	29
10 40.7	18 21.9	23 56.4	16 45.7	10 5.8	30
10 43.1	18 41.3	24 30.5	17 5.9	10 14.0	31
10 45.5	19 1.2	25 5.4	17 26.4	10 22.3	32
10 48.0	19 21.8	25 41.1	17 47.4	10 30.8	33
10 50.6	19 43.2	26 17.7	18 8.8	10 39.4	34
10 53.3	20 5.3	26 55.2	18 30.6	10 48.3	35
10 56.0	20 28.3	27 33.7	18 53.0	10 57.3	36
10 58.9	20 52.1	28 13.2	19 15.8	11 6.5	37
11 1.8	21 16.9	28 53.8	19 39.2	11 16.0	38
11 4.9	21 42.8	29 35.6	20 3.1	11 25.7	39
11 8.1	22 9.8	♋ 0 18.5	20 27.7	11 35.6	40
11 11.4	22 37.9	1 2.7	20 52.9	11 45.8	41
11 14.8	23 7.5	1 48.3	21 18.8	11 56.2	42
11 18.4	23 38.4	2 35.3	21 45.4	12 7.0	43
11 22.1	24 10.9	3 23.7	22 12.8	12 18.1	44
11 26.1	24 45.1	4 13.8	22 41.1	12 29.5	45
11 30.2	25 21.2	5 5.5	23 10.2	12 41.3	46
11 34.5	25 59.4	5 58.9	23 40.2	12 53.5	47
11 39.0	26 39.8	6 54.3	24 11.3	13 6.1	48
11 43.8	27 22.7	7 51.5	24 43.4	13 19.1	49
11 48.8	28 8.4	8 50.9	25 16.7	13 32.6	50
11 54.2	28 57.3	9 52.4	25 51.2	13 46.7	51
11 59.8	29 49.6	10 56.2	26 27.0	14 1.3	52
12 5.8	♊ 0 45.9	12 2.5	27 4.2	14 16.5	53
12 12.3	1 46.6	13 11.3	27 42.9	14 32.3	54
12 19.1	2 52.4	14 22.9	28 23.2	14 48.9	55
12 26.5	4 4.0	15 37.2	29 5.3	15 6.2	56
12 34.4	5 22.4	16 54.6	29 49.2	15 24.4	57
12 42.9	6 48.8	18 15.1	♌ 0 35.2	15 43.5	58
12 52.1	8 24.6	19 38.9	1 23.3	16 3.6	59
13 2.2	10 11.9	21 6.1	2 13.7	16 24.7	60
13 13.2	12 13.2	22 36.9	3 6.7	16 47.1	61
13 25.2	14 32.3	24 11.5	4 2.5	17 10.8	62
13 38.5	17 15.0	25 50.0	5 1.2	17 35.9	63
13 53.3	20 30.3	27 32.6	6 3.1	18 2.6	64
14 9.8	24 35.7	29 19.3	7 8.5	18 31.2	65
14 28.4	♋ 0 14.1	♌ 1 10.3	8 17.7	19 1.7	66
5	6	Descendant	8	9	S LAT

♍ 7° 17′ 44″

10h 36m 0s — MC 159° 0′ 0″

22ʰ 40ᵐ 0ˢ		MC		340° 0′ 0″	
		♓ 8° 21′ 40″			
11	**12**	**Ascendant**	**2**	**3**	**N LAT**
♈10° 52.7′	♉12° 26.7′	♊11° 32.1′	♋ 9° 11.4′	♌ 7° 35.4′	**0°**
11 1.0	13 24.3	13 27.9	10 30.0	8 7.9	**5**
11 9.7	14 24.9	15 27.9	11 49.2	8 40.4	**10**
11 18.8	15 29.9	17 34.0	13 10.3	9 13.4	**15**
11 20.7	15 43.5	18 0.1	13 26.8	9 20.2	**16**
11 22.6	15 57.4	18 26.6	13 43.5	9 26.9	**17**
11 24.6	16 11.6	18 53.4	14 0.3	9 33.8	**18**
11 26.6	16 26.0	19 20.7	14 17.3	9 40.7	**19**
11 28.6	16 40.7	19 48.3	14 34.5	9 47.6	**20**
11 30.7	16 55.8	20 16.4	14 51.8	9 54.6	**21**
11 32.8	17 11.2	20 44.9	15 9.4	10 1.7	**22**
11 35.0	17 26.9	21 14.0	15 27.1	10 8.9	**23**
11 37.2	17 43.1	21 43.5	15 45.1	10 16.2	**24**
11 39.4	17 59.6	22 13.6	16 3.3	10 23.5	**25**
11 41.7	18 16.6	22 44.2	16 21.8	10 31.0	**26**
11 44.1	18 34.0	23 15.5	16 40.6	10 38.5	**27**
11 46.5	18 51.9	23 47.4	16 59.6	10 46.2	**28**
11 49.0	19 10.3	24 19.9	17 19.0	10 54.0	**29**
11 51.6	19 29.3	24 53.1	17 38.7	11 1.9	**30**
11 54.2	19 48.9	25 27.1	17 58.8	11 10.0	**31**
11 56.9	20 9.1	26 1.9	18 19.2	11 18.2	**32**
11 59.6	20 30.0	26 37.4	18 40.0	11 26.6	**33**
12 2.5	20 51.6	27 13.9	19 1.2	11 35.1	**34**
12 5.5	21 14.0	27 51.2	19 22.9	11 43.8	**35**
12 8.5	21 37.3	28 29.5	19 45.1	11 52.7	**36**
12 11.7	22 1.4	29 8.8	20 7.7	12 1.8	**37**
12 14.9	22 26.5	29 49.1	20 30.9	12 11.1	**38**
12 18.3	22 52.6	♋ 0 30.6	20 54.7	12 20.7	**39**
12 21.8	23 19.9	1 13.3	21 19.0	12 30.4	**40**
12 25.5	23 48.4	1 57.2	21 44.0	12 40.5	**41**
12 29.3	24 18.2	2 42.4	22 9.6	12 50.8	**42**
12 33.3	24 49.5	3 29.0	22 36.0	13 1.4	**43**
12 37.4	25 22.3	4 17.0	23 3.1	13 12.3	**44**
12 41.8	25 56.8	5 6.6	23 31.1	13 23.5	**45**
12 46.3	26 33.2	5 57.9	23 59.9	13 35.1	**46**
12 51.1	27 11.7	6 50.8	24 29.6	13 47.1	**47**
12 56.1	27 52.4	7 45.6	25 0.3	13 59.5	**48**
13 1.4	28 35.7	8 42.2	25 32.1	14 12.3	**49**
13 7.0	29 21.7	9 40.9	26 5.0	14 25.6	**50**
13 12.9	♊ 0 10.9	10 41.8	26 39.1	14 39.4	**51**
13 19.2	1 3.5	11 44.8	27 14.4	14 53.7	**52**
13 25.9	2 0.0	12 50.3	27 51.2	15 8.7	**53**
13 33.0	3 1.0	13 58.2	28 29.4	15 24.3	**54**
13 40.6	4 7.0	15 8.8	29 9.2	15 40.5	**55**
13 48.7	5 18.8	16 22.1	29 50.6	15 57.5	**56**
13 57.5	6 37.4	17 38.4	♌ 0 34.0	16 15.3	**57**
14 6.9	8 3.8	18 57.7	1 19.2	16 34.0	**58**
14 17.2	9 39.6	20 20.3	2 6.6	16 53.7	**59**
14 28.3	11 26.7	21 46.2	2 56.3	17 14.4	**60**
14 40.5	13 27.6	23 15.6	3 48.5	17 36.3	**61**
14 53.9	15 45.9	24 48.7	4 43.3	17 59.5	**62**
15 8.6	18 27.2	26 25.5	5 41.0	18 24.1	**63**
15 25.1	21 40.0	28 6.4	6 41.9	18 50.2	**64**
15 43.4	25 40.7	29 51.3	7 46.1	19 18.1	**65**
16 4.1	♋ 1 7.2	♌ 1 40.4	8 54.0	19 48.0	**66**
5	**6**	**Descendant**	**8**	**9**	**S LAT**
		♍ 8° 21′ 40″			
10ʰ 40ᵐ 0ˢ		MC		160° 0′ 0″	

22ʰ 44ᵐ 0ˢ		MC		341° 0′ 0″	
		♓ 9° 25′ 43″			
11	**12**	**Ascendant**	**2**	**3**	**N LAT**
♈11° 57.7′	♉13° 27.3′	♊12° 28.1′	♋10° 6.7′	♌ 8° 34.4′	**0°**
12 6.8	14 25.9	14 24.3	11 25.1	9 6.5	**5**
12 16.3	15 27.5	16 24.7	12 44.1	9 38.6	**10**
12 26.3	16 33.5	18 30.9	14 4.8	10 11.2	**15**
12 28.4	16 47.4	18 57.1	14 21.3	10 17.8	**16**
12 30.5	17 1.5	19 23.6	14 37.9	10 24.5	**17**
12 32.7	17 15.8	19 50.4	14 54.6	10 31.3	**18**
12 34.9	17 30.5	20 17.7	15 11.5	10 38.1	**19**
12 37.1	17 45.4	20 45.3	15 28.6	10 44.9	**20**
12 39.4	18 0.7	21 13.4	15 45.8	10 51.8	**21**
12 41.7	18 16.3	21 41.9	16 3.3	10 58.8	**22**
12 44.1	18 32.3	22 10.9	16 21.0	11 5.9	**23**
12 46.5	18 48.7	22 40.4	16 38.8	11 13.1	**24**
12 49.0	19 5.4	23 10.4	16 57.0	11 20.3	**25**
12 51.5	19 22.6	23 41.0	17 15.3	11 27.7	**26**
12 54.1	19 40.3	24 12.2	17 34.0	11 35.2	**27**
12 56.7	19 58.4	24 44.0	17 52.9	11 42.7	**28**
12 59.5	20 17.1	25 16.4	18 12.1	11 50.4	**29**
13 2.3	20 36.3	25 49.5	18 31.7	11 58.2	**30**
13 5.1	20 56.1	26 23.4	18 51.6	12 6.2	**31**
13 8.1	21 16.6	26 58.0	19 11.9	12 14.3	**32**
13 11.1	21 37.8	27 33.4	19 32.5	12 22.5	**33**
13 14.3	21 59.6	28 9.7	19 53.6	12 30.9	**34**
13 17.5	22 22.3	28 46.8	20 15.1	12 39.5	**35**
13 20.9	22 45.8	29 24.9	20 37.1	12 48.3	**36**
13 24.3	23 10.2	♋ 0 4.0	20 59.6	12 57.2	**37**
13 27.9	23 35.5	0 44.1	21 22.5	13 6.4	**38**
13 31.6	24 2.0	1 25.3	21 46.1	13 15.8	**39**
13 35.5	24 29.5	2 7.6	22 10.2	13 25.4	**40**
13 39.5	24 58.3	2 51.2	22 35.0	13 35.3	**41**
13 43.7	25 28.4	3 36.1	23 0.4	13 45.4	**42**
13 48.0	25 59.9	4 22.3	23 26.5	13 55.8	**43**
13 52.6	26 33.0	5 10.0	23 53.3	14 6.6	**44**
13 57.4	27 7.8	5 59.1	24 21.0	14 17.6	**45**
14 2.3	27 44.5	6 49.9	24 49.5	14 29.0	**46**
14 7.6	28 23.3	7 42.3	25 18.9	14 40.8	**47**
14 13.1	29 4.3	8 36.5	25 49.3	14 53.0	**48**
14 18.9	29 47.8	9 32.6	26 20.7	15 5.6	**49**
14 25.0	♊ 0 34.1	10 30.6	26 53.2	15 18.6	**50**
14 31.5	1 23.6	11 30.7	27 26.9	15 32.2	**51**
14 38.4	2 16.4	12 33.1	28 1.8	15 46.3	**52**
14 45.7	3 13.2	13 37.7	28 38.1	16 0.9	**53**
14 53.5	4 14.4	14 44.8	29 15.8	16 16.2	**54**
15 1.8	5 20.6	15 54.4	29 55.0	16 32.2	**55**
15 10.8	6 32.6	17 6.8	♌ 0 35.9	16 48.9	**56**
15 20.4	7 51.2	18 22.0	1 18.6	17 6.3	**57**
15 30.8	9 17.6	19 40.1	2 3.2	17 24.7	**58**
15 42.0	10 53.3	21 1.4	2 49.9	17 43.9	**59**
15 54.2	12 40.0	22 26.0	3 38.8	18 4.2	**60**
16 7.6	14 40.4	23 54.0	4 30.2	18 25.7	**61**
16 22.3	16 57.8	25 25.6	5 24.1	18 48.3	**62**
16 38.5	19 37.5	27 0.9	6 20.8	19 12.4	**63**
16 56.6	22 47.7	28 40.1	7 20.6	19 37.9	**64**
17 16.8	26 43.2	♌ 0 23.2	8 23.7	20 5.2	**65**
17 39.6	♋ 1 58.0	2 10.5	9 30.3	20 34.3	**66**
5	**6**	**Descendant**	**8**	**9**	**S LAT**
		♍ 9° 25′ 43″			
10ʰ 44ᵐ 0ˢ		MC		161° 0′ 0″	

22h 48m 0s MC 342° 0′ 0″

♓ 10° 29′ 54″

N LAT	11	12	Ascendant	2	3
°	° ′	° ′	° ′	° ′	° ′
0	♈13 2.6	♉14 27.7	♊13 24.0	♋11 2.1	♌ 9 33.6
5	13 12.5	15 27.3	15 20.7	12 20.3	10 5.2
10	13 22.8	16 29.9	17 21.3	13 39.0	10 36.9
15	13 33.7	17 36.9	19 27.7	14 59.4	11 9.1
16	13 36.0	17 50.9	19 53.9	15 15.7	11 15.7
17	13 38.3	18 5.2	20 20.4	15 32.3	11 22.3
18	13 40.7	18 19.8	20 47.3	15 48.9	11 28.9
19	13 43.1	18 34.7	21 14.5	16 5.7	11 35.6
20	13 45.5	18 49.8	21 42.1	16 22.7	11 42.4
21	13 48.0	19 5.3	22 10.2	16 39.9	11 49.2
22	13 50.5	19 21.1	22 38.6	16 57.2	11 56.1
23	13 53.1	19 37.3	23 7.6	17 14.8	12 3.1
24	13 55.7	19 53.9	23 37.0	17 32.5	12 10.1
25	13 58.4	20 10.9	24 7.0	17 50.5	12 17.3
26	14 1.1	20 28.3	24 37.5	18 8.8	12 24.5
27	14 4.0	20 46.2	25 8.6	18 27.3	12 31.9
28	14 6.8	21 4.5	25 40.3	18 46.1	12 39.3
29	14 9.8	21 23.4	26 12.6	19 5.2	12 46.9
30	14 12.8	21 42.9	26 45.7	19 24.7	12 54.6
31	14 16.0	22 3.0	27 19.4	19 44.4	13 2.4
32	14 19.2	22 23.7	27 53.8	20 4.5	13 10.4
33	14 22.5	22 45.0	28 29.1	20 25.0	13 18.5
34	14 25.9	23 7.2	29 5.2	20 45.9	13 26.8
35	14 29.4	23 30.1	29 42.1	21 7.3	13 35.2
36	14 33.1	23 53.8	♋ 0 20.0	21 29.1	13 43.9
37	14 36.8	24 18.4	0 58.8	21 51.3	13 52.7
38	14 40.7	24 44.1	1 38.7	22 14.1	14 1.7
39	14 44.8	25 10.7	2 19.6	22 37.5	14 10.9
40	14 49.0	25 38.5	3 1.7	23 1.4	14 20.4
41	14 53.3	26 7.6	3 44.9	23 25.9	14 30.1
42	14 57.9	26 37.9	4 29.4	23 51.0	14 40.1
43	15 2.6	27 9.7	5 15.3	24 16.9	14 50.4
44	15 7.6	27 43.1	6 2.5	24 43.5	15 0.9
45	15 12.7	28 18.2	6 51.2	25 10.9	15 11.8
46	15 18.2	28 55.1	7 41.5	25 39.1	15 23.0
47	15 23.9	29 34.1	8 33.4	26 8.1	15 34.6
48	15 29.9	♊ 0 15.4	9 27.1	26 38.2	15 46.5
49	15 36.2	0 59.2	10 22.6	27 9.2	15 58.9
50	15 42.8	1 45.8	11 20.0	27 41.3	16 11.7
51	15 49.9	2 35.4	12 19.4	28 14.6	16 25.0
52	15 57.4	3 28.5	13 21.0	28 49.1	16 38.9
53	16 5.3	4 25.5	14 24.8	29 24.9	16 53.2
54	16 13.8	5 26.8	15 31.0	♌ 0 2.1	17 8.2
55	16 22.9	6 33.1	16 39.7	0 40.8	17 23.9
56	16 32.6	7 45.2	17 51.1	1 21.1	17 40.2
57	16 43.0	9 3.8	19 5.2	2 3.2	17 57.4
58	16 54.3	10 30.1	20 22.3	2 47.2	18 15.3
59	17 6.6	12 5.6	21 42.4	3 33.1	18 34.2
60	17 19.9	13 51.9	23 5.7	4 21.3	18 54.1
61	17 34.4	15 51.6	24 32.3	5 11.8	19 15.1
62	17 50.4	18 8.0	26 2.4	6 4.8	19 37.2
63	18 8.1	20 45.9	27 36.2	7 0.6	20 0.7
64	18 27.8	23 53.3	29 13.7	7 59.4	20 25.7
65	18 49.8	27 43.6	♌ 0 55.2	9 1.3	20 52.3
66	19 14.7	♋ 2 46.9	2 40.6	10 6.7	21 20.7
S LAT	5	6	Descendant	8	9

♍ 10° 29′ 54″

10h 48m 0s MC 162° 0′ 0″

22h 52m 0s MC 343° 0′ 0″

♓ 11° 34′ 12″

N LAT	11	12	Ascendant	2	3
°	° ′	° ′	° ′	° ′	° ′
0	♈14 7.5	♉15 28.0	♊14 19.9	♋11 57.6	♌10 32.9
5	14 18.2	16 28.4	16 16.9	13 15.5	11 4.2
10	14 29.3	17 32.0	18 17.8	14 33.9	11 35.4
15	14 41.1	18 39.9	20 24.4	15 53.9	12 7.2
16	14 43.5	18 54.2	20 50.5	16 10.2	12 13.6
17	14 46.0	19 8.7	21 17.0	16 26.7	12 20.1
18	14 48.6	19 23.5	21 43.9	16 43.2	12 26.7
19	14 51.1	19 38.5	22 11.1	16 59.9	12 33.3
20	14 53.8	19 53.9	22 38.7	17 16.8	12 39.9
21	14 56.4	20 9.6	23 6.7	17 33.9	12 46.7
22	14 59.2	20 25.6	23 35.2	17 51.1	12 53.5
23	15 1.9	20 42.0	24 4.1	18 8.6	13 0.3
24	15 4.8	20 58.8	24 33.4	18 26.2	13 7.3
25	15 7.7	21 16.0	25 3.3	18 44.1	13 14.3
26	15 10.7	21 33.6	25 33.8	19 2.3	13 21.5
27	15 13.7	21 51.7	26 4.8	19 20.6	13 28.7
28	15 16.8	22 10.3	26 36.4	19 39.3	13 36.1
29	15 20.0	22 29.4	27 8.6	19 58.3	13 43.5
30	15 23.3	22 49.1	27 41.5	20 17.6	13 51.1
31	15 26.7	23 9.4	28 15.1	20 37.2	13 58.8
32	15 30.1	23 30.3	28 49.4	20 57.1	14 6.7
33	15 33.7	23 51.9	29 24.5	21 17.5	14 14.6
34	15 37.4	24 14.2	♋ 0 0.4	21 38.2	14 22.8
35	15 41.2	24 37.4	0 37.1	21 59.4	14 31.1
36	15 45.1	25 1.3	1 14.8	22 21.0	14 39.6
37	15 49.2	25 26.2	1 53.4	22 43.1	14 48.3
38	15 53.4	25 52.1	2 33.0	23 5.6	14 57.1
39	15 57.8	26 19.0	3 13.6	23 28.8	15 6.2
40	16 2.3	26 47.0	3 55.4	23 52.4	15 15.5
41	16 7.0	27 16.3	4 38.3	24 16.7	15 25.1
42	16 11.9	27 46.9	5 22.5	24 41.6	15 34.9
43	16 17.0	28 18.9	6 7.9	25 7.2	15 45.0
44	16 22.4	28 52.5	6 54.8	25 33.5	15 55.3
45	16 28.0	29 27.9	7 43.0	26 0.6	16 6.0
46	16 33.8	♊ 0 5.1	8 32.8	26 28.5	16 17.0
47	16 40.0	0 44.3	9 24.2	26 57.3	16 28.4
48	16 46.4	1 25.8	10 17.3	27 27.0	16 40.1
49	16 53.2	2 9.8	11 12.2	27 57.6	16 52.3
50	17 0.4	2 56.6	12 9.0	28 29.4	17 4.9
51	17 8.1	3 46.4	13 7.7	29 2.2	17 17.9
52	17 16.1	4 39.7	14 8.5	29 36.3	17 31.5
53	17 24.7	5 36.8	15 11.6	♌ 0 11.6	17 45.6
54	17 33.9	6 38.2	16 16.9	0 48.3	18 0.3
55	17 43.7	7 44.7	17 24.8	1 26.5	18 15.7
56	17 54.2	8 56.7	18 35.1	2 6.3	18 31.7
57	18 5.4	10 15.3	19 48.2	2 47.7	18 48.5
58	18 17.6	11 41.5	21 4.2	3 31.0	19 6.1
59	18 30.9	13 16.6	22 23.1	4 16.3	19 24.6
60	18 45.3	15 2.5	23 45.1	5 3.7	19 44.0
61	19 1.0	17 1.4	25 10.4	5 53.4	20 4.5
62	19 18.3	19 16.5	26 39.1	6 45.6	20 26.2
63	19 37.4	21 52.6	28 11.4	7 40.4	20 49.2
64	19 58.7	24 57.0	29 47.3	8 38.1	21 13.6
65	20 22.6	28 42.0	♌ 1 27.1	9 38.9	21 39.5
66	20 49.6	♋ 3 34.1	3 10.8	10 43.1	22 7.3
S LAT	5	6	Descendant	8	9

♍ 11° 34′ 12″

10h 52m 0s MC 163° 0′ 0″

22ʰ 56ᵐ 0ˢ MC 344° 0′ 0″

♓ 12° 38′ 37″

N LAT	11	12	Ascendant	2	3
°	° ′	° ′	° ′	° ′	° ′
0	♈15 12.2	♉16 28.0	♊15 15.6	♋12 53.1	♌11 32.4
5	15 23.7	17 29.4	17 13.1	14 10.8	12 3.2
10	15 35.6	18 33.9	19 14.2	15 28.9	12 34.0
15	15 48.3	19 42.8	21 20.8	16 48.5	13 5.3
16	15 50.9	19 57.2	21 47.0	17 4.7	13 11.7
17	15 53.6	20 11.9	22 13.5	17 21.1	13 18.1
18	15 56.3	20 26.8	22 40.3	17 37.5	13 24.6
19	15 59.1	20 42.1	23 7.5	17 54.2	13 31.1
20	16 1.9	20 57.7	23 35.1	18 10.9	13 37.6
21	16 4.8	21 13.5	24 3.0	18 27.9	13 44.3
22	16 7.7	21 29.8	24 31.4	18 45.0	13 51.0
23	16 10.7	21 46.4	25 0.3	19 2.4	13 57.7
24	16 13.7	22 3.3	25 29.6	19 19.9	14 4.6
25	16 16.8	22 20.7	25 59.4	19 37.7	14 11.5
26	16 20.0	22 38.6	26 29.8	19 55.7	14 18.6
27	16 23.3	22 56.9	27 0.7	20 14.0	14 25.7
28	16 26.6	23 15.6	27 32.2	20 32.5	14 32.9
29	16 30.1	23 35.0	28 4.3	20 51.3	14 40.3
30	16 33.6	23 54.9	28 37.0	21 10.5	14 47.7
31	16 37.2	24 15.4	29 10.5	21 29.9	14 55.3
32	16 40.9	24 36.5	29 44.6	21 49.7	15 3.0
33	16 44.8	24 58.3	♋ 0 19.6	22 9.9	15 10.9
34	16 48.7	25 20.9	0 55.3	22 30.5	15 18.9
35	16 52.8	25 44.2	1 31.8	22 51.4	15 27.1
36	16 57.0	26 8.4	2 9.2	23 12.8	15 35.4
37	17 1.4	26 33.5	2 47.6	23 34.7	15 43.9
38	17 5.9	26 59.6	3 26.9	23 57.1	15 52.7
39	17 10.6	27 26.7	4 7.3	24 20.0	16 1.6
40	17 15.4	27 55.0	4 48.8	24 43.5	16 10.7
41	17 20.5	28 24.5	5 31.4	25 7.5	16 20.1
42	17 25.8	28 55.3	6 15.2	25 32.2	16 29.8
43	17 31.2	29 27.5	7 0.2	25 57.5	16 39.7
44	17 37.0	♊ 0 1.4	7 46.7	26 23.5	16 49.8
45	17 43.0	0 36.9	8 34.5	26 50.3	17 0.3
46	17 49.3	1 14.3	9 23.8	27 17.9	17 11.1
47	17 55.8	1 53.8	10 14.7	27 46.3	17 22.3
48	18 2.8	2 35.5	11 7.2	28 15.7	17 33.8
49	18 10.1	3 19.7	12 1.5	28 46.0	17 45.8
50	18 17.8	4 6.6	12 57.6	29 17.3	17 58.1
51	18 26.0	4 56.6	13 55.7	29 49.8	18 10.9
52	18 34.6	5 50.0	14 55.8	♌ 0 23.4	18 24.2
53	18 43.9	6 47.2	15 58.1	0 58.2	18 38.1
54	18 53.7	7 48.7	17 2.6	1 34.5	18 52.5
55	19 4.2	8 55.2	18 9.5	2 12.1	19 7.5
56	19 15.5	10 7.2	19 18.9	2 51.3	19 23.2
57	19 27.6	11 25.7	20 31.0	3 32.2	19 39.6
58	19 40.7	12 51.6	21 45.8	4 14.9	19 56.9
59	19 54.9	14 26.4	23 3.6	4 59.4	20 15.0
60	20 10.3	16 11.7	24 24.4	5 46.1	20 34.0
61	20 27.2	18 9.7	25 48.4	6 35.0	20 54.0
62	20 45.9	20 23.6	27 15.7	7 26.3	21 15.2
63	21 6.4	22 57.7	28 46.5	8 20.2	21 37.7
64	21 29.4	25 58.9	♌ 0 20.8	9 16.9	22 1.5
65	21 55.1	29 38.6	1 59.0	10 16.6	22 26.8
66	22 24.2	♋ 4 19.9	3 40.9	11 19.6	22 53.9
S LAT	5	6	Descendant	8	9

♍ 12° 38′ 37″

10ʰ 56ᵐ 0ˢ MC 164° 0′ 0″

23ʰ 0ᵐ 0ˢ MC 345° 0′ 0″

♓ 13° 43′ 9″

N LAT	11	12	Ascendant	2	3
°	° ′	° ′	° ′	° ′	° ′
0	♈16 16.8	♉17 27.9	♊16 11.3	♋13 48.7	♌12 32.1
5	16 29.1	18 30.2	18 9.1	15 6.1	13 2.5
10	16 41.9	19 35.5	20 10.4	16 23.9	13 32.8
15	16 55.4	20 45.3	22 17.1	17 43.1	14 3.7
16	16 58.2	20 59.9	22 43.3	17 59.2	14 9.9
17	17 1.0	21 14.8	23 9.7	18 15.5	14 16.3
18	17 3.9	21 29.9	23 36.5	18 31.8	14 22.6
19	17 6.9	21 45.4	24 3.7	18 48.4	14 29.0
20	17 9.9	22 1.1	24 31.2	19 5.1	14 35.5
21	17 13.0	22 17.2	24 59.2	19 21.9	14 42.0
22	17 16.1	22 33.6	25 27.5	19 38.9	14 48.6
23	17 19.3	22 50.4	25 56.3	19 56.2	14 55.3
24	17 22.5	23 7.6	26 25.6	20 13.6	15 2.0
25	17 25.9	23 25.1	26 55.3	20 31.2	15 8.8
26	17 29.3	23 43.2	27 25.6	20 49.1	15 15.7
27	17 32.7	24 1.6	27 56.4	21 7.2	15 22.8
28	17 36.3	24 20.6	28 27.7	21 25.6	15 29.9
29	17 40.0	24 40.2	28 59.7	21 44.3	15 37.1
30	17 43.7	25 0.2	29 32.3	22 3.3	15 44.4
31	17 47.6	25 20.9	♋ 0 5.6	22 22.6	15 51.9
32	17 51.6	25 42.3	0 39.6	22 42.3	15 59.5
33	17 55.7	26 4.3	1 14.4	23 2.3	16 7.2
34	17 59.9	26 27.1	1 49.9	23 22.7	16 15.1
35	18 4.2	26 50.6	2 26.2	23 43.4	16 23.1
36	18 8.7	27 15.0	3 3.4	24 4.7	16 31.3
37	18 13.4	27 40.3	3 41.5	24 26.4	16 39.7
38	18 18.2	28 6.6	4 20.6	24 48.5	16 48.3
39	18 23.2	28 33.9	5 0.7	25 11.2	16 57.1
40	18 28.4	29 2.4	5 41.8	25 34.4	17 6.0
41	18 33.8	29 32.1	6 24.1	25 58.2	17 15.3
42	18 39.4	♊ 0 3.1	7 7.5	26 22.6	17 24.7
43	18 45.3	0 35.6	7 52.2	26 47.7	17 34.4
44	18 51.4	1 9.6	8 38.2	27 13.5	17 44.4
45	18 57.8	1 45.3	9 25.6	27 39.9	17 54.7
46	19 4.5	2 22.9	10 14.4	28 7.2	18 5.4
47	19 11.5	3 2.5	11 4.8	28 35.3	18 16.3
48	19 18.9	3 44.4	11 56.8	29 4.3	18 27.6
49	19 26.7	4 28.8	12 50.5	29 34.3	18 39.3
50	19 35.0	5 15.8	13 46.0	♌ 0 5.2	18 51.4
51	19 43.7	6 6.0	14 43.4	0 37.2	19 4.0
52	19 52.9	6 59.5	15 42.8	1 10.4	19 17.0
53	20 2.7	7 56.7	16 44.3	1 44.8	19 30.6
54	20 13.2	8 58.3	17 47.9	2 20.6	19 44.7
55	20 24.4	10 4.7	18 54.0	2 57.7	19 59.4
56	20 36.5	11 16.6	20 2.5	3 36.3	20 14.8
57	20 49.4	12 34.9	21 13.5	4 16.6	20 30.9
58	21 3.4	14 0.6	22 27.3	4 58.6	20 47.7
59	21 18.6	15 34.9	23 43.9	5 42.5	21 5.4
60	21 35.1	17 19.6	25 3.5	6 28.4	21 24.0
61	21 53.2	19 16.6	26 26.2	7 16.5	21 43.6
62	22 13.1	21 29.1	27 52.1	8 6.9	22 4.3
63	22 35.1	24 1.2	29 21.4	8 59.9	22 26.3
64	22 59.7	26 59.1	♌ 0 54.3	9 55.6	22 49.5
65	23 27.2	♋ 0 33.4	2 30.8	10 54.3	23 14.2
66	23 58.5	5 4.4	4 11.1	11 56.1	23 40.6
S LAT	5	6	Descendant	8	9

♍ 13° 43′ 9″

11ʰ 0ᵐ 0ˢ MC 165° 0′ 0″

23ʰ 4ᵐ 0ˢ	MC	346° 0′ 0″			
	♓ 14° 47′ 47″				
11	12	Ascendant	2	3	N LAT
° ′	° ′	° ′	° ′	° ′	°
♈17 21.4	♉18 27.6	♊17 6.9	♋14 44.4	♌13 32.0	0
17 34.4	19 30.7	19 5.0	16 1.5	14 1.9	5
17 48.0	20 36.9	21 6.5	17 18.9	14 31.8	10
18 2.3	21 47.6	23 13.2	18 37.7	15 2.2	15
18 5.3	22 2.4	23 39.4	18 53.7	15 8.3	16
18 8.3	22 17.4	24 5.8	19 9.9	15 14.5	17
18 11.4	22 32.8	24 32.6	19 26.2	15 20.8	18
18 14.6	22 48.4	24 59.7	19 42.6	15 27.1	19
18 17.8	23 4.3	25 27.2	19 59.2	15 33.5	20
18 21.0	23 20.5	25 55.1	20 15.9	15 39.9	21
18 24.3	23 37.1	26 23.4	20 32.8	15 46.4	22
18 27.7	23 54.1	26 52.1	20 49.9	15 52.9	23
18 31.2	24 11.4	27 21.3	21 7.2	15 59.5	24
18 34.7	24 29.2	27 50.9	21 24.8	16 6.3	25
18 38.3	24 47.4	28 21.1	21 42.5	16 13.1	26
18 42.0	25 6.1	28 51.8	22 0.5	16 20.0	27
18 45.8	25 25.3	29 23.1	22 18.8	16 26.9	28
18 49.7	25 45.0	29 54.9	22 37.3	16 34.0	29
18 53.7	26 5.2	♋ 0 27.4	22 56.1	16 41.3	30
18 57.8	26 26.1	1 0.5	23 15.3	16 48.6	31
19 2.0	26 47.6	1 34.4	23 34.8	16 56.1	32
19 6.4	27 9.9	2 8.9	23 54.6	17 3.6	33
19 10.9	27 32.8	2 44.2	24 14.8	17 11.4	34
19 15.5	27 56.5	3 20.3	24 35.4	17 19.3	35
19 20.3	28 21.1	3 57.3	24 56.5	17 27.3	36
19 25.2	28 46.6	4 35.2	25 17.9	17 35.6	37
19 30.3	29 13.1	5 14.0	25 39.9	17 44.0	38
19 35.6	29 40.6	5 53.8	26 2.3	17 52.6	39
19 41.1	♊ 0 9.3	6 34.6	26 25.3	18 1.4	40
19 46.9	0 39.1	7 16.5	26 48.9	18 10.5	41
19 52.8	1 10.3	7 59.6	27 13.0	18 19.8	42
19 59.1	1 43.0	8 43.9	27 37.8	18 29.3	43
20 5.6	2 17.2	9 29.5	28 3.3	18 39.1	44
20 12.3	2 53.1	10 16.4	28 29.5	18 49.2	45
20 19.5	3 30.8	11 4.8	28 56.5	18 59.6	46
20 26.9	4 10.6	11 54.7	29 24.3	19 10.4	47
20 34.8	4 52.7	12 46.1	29 52.9	19 21.5	48
20 43.1	5 37.1	13 39.2	♌ 0 22.5	19 32.9	49
20 51.8	6 24.3	14 34.1	0 53.0	19 44.8	50
21 1.1	7 14.5	15 30.8	1 24.7	19 57.1	51
21 10.9	8 8.1	16 29.4	1 57.4	20 9.9	52
21 21.4	9 5.4	17 30.2	2 31.4	20 23.2	53
21 32.5	10 7.0	18 33.0	3 6.6	20 37.0	54
21 44.4	11 13.3	19 38.2	3 43.2	20 51.4	55
21 57.2	12 25.1	20 45.7	4 21.3	21 6.5	56
22 10.9	13 43.2	21 55.8	5 1.0	21 22.2	57
22 25.8	15 8.5	23 8.5	5 42.3	21 38.7	58
22 41.9	16 42.3	24 24.0	6 25.5	21 56.0	59
22 59.5	18 26.2	25 42.4	7 10.7	22 14.1	60
23 18.7	20 22.3	27 3.8	7 58.0	22 33.3	61
23 39.9	22 33.3	28 28.4	8 47.6	22 53.5	62
24 3.4	25 3.2	29 56.4	9 39.7	23 14.9	63
24 29.6	27 57.8	♌ 1 27.7	10 34.4	23 37.6	64
24 59.0	♋ 1 26.8	3 2.7	11 32.0	24 1.7	65
25 32.5	5 47.8	4 41.3	12 32.7	24 27.4	66
5	6	Descendant	8	9	S LAT
	♍ 14° 47′ 47″				
11ʰ 4ᵐ 0ˢ	MC	166° 0′ 0″			

23ʰ 8ᵐ 0ˢ	MC	347° 0′ 0″			
	♓ 15° 52′ 32″				
N LAT	11	12	Ascendant	2	3
°	° ′	° ′	° ′	° ′	° ′
0	♈18 25.8	♉19 27.1	♊18 2.4	♋15 40.1	♌14 32.0
5	18 39.6	20 31.1	20 0.8	16 57.0	15 1.5
10	18 54.0	21 38.1	22 2.4	18 14.0	15 30.9
15	19 9.1	22 49.6	24 9.2	19 32.4	16 0.8
16	19 12.3	23 4.6	24 35.3	19 48.3	16 6.8
17	19 15.5	23 19.8	25 1.7	20 4.4	16 13.0
18	19 18.8	23 35.3	25 28.5	20 20.5	16 19.1
19	19 22.1	23 51.1	25 55.6	20 36.9	16 25.3
20	19 25.5	24 7.2	26 23.0	20 53.3	16 31.6
21	19 28.9	24 23.6	26 50.8	21 10.0	16 37.9
22	19 32.4	24 40.3	27 19.1	21 26.8	16 44.3
23	19 36.0	24 57.5	27 47.7	21 43.7	16 50.7
24	19 39.7	25 15.0	28 16.8	22 0.9	16 57.2
25	19 43.4	25 32.9	28 46.4	22 18.3	17 3.8
26	19 47.2	25 51.3	29 16.4	22 35.9	17 10.5
27	19 51.2	26 10.2	29 47.0	22 53.8	17 17.3
28	19 55.2	26 29.5	♋ 0 18.2	23 11.9	17 24.1
29	19 59.3	26 49.4	0 49.9	23 30.3	17 31.1
30	20 3.5	27 9.8	1 22.2	23 49.0	17 38.2
31	20 7.9	27 30.9	1 55.2	24 8.0	17 45.4
32	20 12.3	27 52.6	2 28.8	24 27.3	17 52.7
33	20 16.9	28 15.0	3 3.2	24 46.9	18 0.2
34	20 21.7	28 38.1	3 38.3	25 7.0	18 7.8
35	20 26.5	29 2.0	4 14.2	25 27.4	18 15.5
36	20 31.6	29 26.8	4 50.9	25 48.2	18 23.5
37	20 36.8	29 52.5	5 28.5	26 9.5	18 31.5
38	20 42.2	♊ 0 19.1	6 7.1	26 31.2	18 39.8
39	20 47.9	0 46.8	6 46.6	26 53.4	18 48.3
40	20 53.7	1 15.6	7 27.1	27 16.2	18 56.9
41	20 59.7	1 45.7	8 8.7	27 39.5	19 5.8
42	21 6.1	2 17.0	8 51.4	28 3.4	19 14.9
43	21 12.6	2 49.8	9 35.3	28 27.9	19 24.3
44	21 19.5	3 24.2	10 20.5	28 53.1	19 33.9
45	21 26.7	4 0.3	11 7.0	29 19.0	19 43.8
46	21 34.2	4 38.1	11 54.8	29 45.7	19 54.0
47	21 42.1	5 18.1	12 44.2	♌ 0 13.1	20 4.5
48	21 50.4	6 0.2	13 35.1	0 41.4	20 15.4
49	21 59.2	6 44.8	14 27.6	1 10.6	20 26.6
50	22 8.4	7 32.1	15 21.9	1 40.8	20 38.3
51	22 18.2	8 22.3	16 17.9	2 12.0	20 50.3
52	22 28.6	9 15.9	17 15.9	2 44.3	21 2.8
53	22 39.7	10 13.2	18 15.8	3 17.8	21 15.8
54	22 51.5	11 14.7	19 17.9	3 52.6	21 29.4
55	23 4.1	12 21.0	20 22.2	4 28.6	21 43.5
56	23 17.6	13 32.6	21 28.8	5 6.2	21 58.2
57	23 32.2	14 50.4	22 37.9	5 45.3	22 13.6
58	23 47.9	16 15.3	23 49.6	6 26.0	22 29.7
59	24 5.0	17 48.5	25 3.9	7 8.5	22 46.6
60	24 23.6	19 31.7	26 21.2	7 53.0	23 4.3
61	24 44.0	21 26.6	27 41.3	8 39.5	23 23.0
62	25 6.4	23 36.1	29 4.6	9 28.3	23 42.8
63	25 31.3	26 3.8	♌ 0 31.2	10 19.4	24 3.7
64	25 59.1	28 55.1	2 1.1	11 13.2	24 25.8
65	26 30.4	♋ 2 18.7	3 34.5	12 9.8	24 49.3
66	27 6.1	6 30.3	5 11.6	13 9.3	25 14.3
S LAT	5	6	Descendant	8	9
		♍ 15° 52′ 32″			
	11ʰ 8ᵐ 0ˢ	MC	167° 0′ 0″		

23h 12m 0s	MC 348° 0′ 0″	♓ 16° 57′ 21″			
N LAT	11	12	Ascendant	2	3
0	♈19 30.1	♉20 26.4	♊18 57.9	♋16 36.0	♌15 32.3
5	19 44.7	21 31.2	20 56.5	17 52.5	16 1.2
10	19 59.8	22 39.1	22 58.3	19 9.1	16 30.2
15	20 15.8	23 51.4	25 5.0	20 27.0	16 59.6
16	20 19.2	24 6.5	25 31.1	20 42.9	17 5.5
17	20 22.5	24 21.9	25 57.5	20 58.8	17 11.5
18	20 26.0	24 37.5	26 24.2	21 14.9	17 17.6
19	20 29.5	24 53.5	26 51.2	21 31.1	17 23.7
20	20 33.1	25 9.7	27 18.6	21 47.5	17 29.8
21	20 36.7	25 26.3	27 46.4	22 4.0	17 36.0
22	20 40.4	25 43.2	28 14.5	22 20.7	17 42.3
23	20 44.2	26 0.5	28 43.1	22 37.5	17 48.6
24	20 48.0	26 18.2	29 12.1	22 54.6	17 55.0
25	20 52.0	26 36.3	29 41.6	23 11.9	18 1.5
26	20 56.0	26 54.9	♋ 0 11.5	23 29.3	18 8.1
27	21 0.1	27 13.9	0 42.0	23 47.0	18 14.7
28	21 4.3	27 33.4	1 13.0	24 5.0	18 21.5
29	21 8.7	27 53.4	1 44.6	24 23.3	18 28.3
30	21 13.1	28 14.1	2 16.8	24 41.8	18 35.3
31	21 17.7	28 35.3	2 49.6	25 0.6	18 42.3
32	21 22.4	28 57.1	3 23.0	25 19.7	18 49.5
33	21 27.3	29 19.7	3 57.2	25 39.2	18 56.8
34	21 32.3	29 43.0	4 32.1	25 59.1	19 4.3
35	21 37.4	♊ 0 7.1	5 7.8	26 19.3	19 11.9
36	21 42.7	0 32.0	5 44.3	26 39.9	19 19.7
37	21 48.2	0 57.8	6 21.6	27 1.0	19 27.6
38	21 53.9	1 24.6	6 59.9	27 22.5	19 35.7
39	21 59.9	1 52.5	7 39.1	27 44.5	19 44.0
40	22 6.0	2 21.5	8 19.3	28 7.0	19 52.5
41	22 12.4	2 51.7	9 0.5	28 30.1	20 1.2
42	22 19.0	3 23.2	9 42.9	28 53.7	20 10.1
43	22 26.0	3 56.1	10 26.4	29 18.0	20 19.3
44	22 33.2	4 30.6	11 11.2	29 42.9	20 28.7
45	22 40.8	5 6.8	11 57.2	♌ 0 8.5	20 38.4
46	22 48.7	5 44.8	12 44.6	0 34.8	20 48.4
47	22 57.0	6 24.8	13 33.4	1 2.0	20 58.7
48	23 5.8	7 7.1	14 23.8	1 29.9	21 9.4
49	23 15.0	7 51.7	15 15.7	1 58.7	21 20.4
50	23 24.8	8 39.1	16 9.4	2 28.5	21 31.8
51	23 35.1	9 29.4	17 4.7	2 59.3	21 43.6
52	23 46.0	10 23.0	18 2.0	3 31.2	21 55.8
53	23 57.7	11 20.2	19 1.2	4 4.2	22 8.5
54	24 10.1	12 21.7	20 2.5	4 38.5	22 21.8
55	24 23.4	13 27.7	21 5.9	5 14.0	22 35.6
56	24 37.7	14 39.1	22 11.6	5 51.0	22 50.0
57	24 53.0	15 56.6	23 19.8	6 29.5	23 5.0
58	25 9.6	17 21.0	24 30.4	7 9.6	23 20.7
59	25 27.6	18 53.7	25 43.7	7 51.5	23 37.2
60	25 47.3	20 36.0	26 59.8	8 35.2	23 54.6
61	26 8.8	22 29.8	28 18.7	9 21.0	24 12.8
62	26 32.5	24 37.6	29 40.8	10 8.9	24 32.1
63	26 58.9	27 3.0	♌ 1 5.9	10 59.2	24 52.5
64	27 28.3	29 51.0	2 34.4	11 52.0	25 14.0
65	28 1.5	♋ 3 9.4	4 6.3	12 47.5	25 36.9
66	28 39.3	7 11.9	5 41.8	13 46.0	26 1.3
S LAT	5	6	Descendant	8	9
11h 12m 0s	MC 168° 0′ 0″	♍ 16° 57′ 21″			

23h 16m 0s	MC 349° 0′ 0″	♓ 18° 2′ 16″			
N LAT	11	12	Ascendant	2	3
0	♈20 34.3	♉21 25.6	♊19 53.3	♋17 31.9	♌16 32.7
5	20 49.6	22 31.2	21 52.1	18 48.1	17 1.2
10	21 5.5	23 39.8	23 54.0	20 4.3	17 29.6
15	21 22.3	24 52.9	26 0.6	21 21.8	17 58.5
16	21 25.8	25 8.2	26 26.7	21 37.5	18 4.3
17	21 29.4	25 23.7	26 53.1	21 53.4	18 10.2
18	21 33.0	25 39.5	27 19.7	22 9.3	18 16.2
19	21 36.7	25 55.6	27 46.7	22 25.4	18 22.2
20	21 40.5	26 12.0	28 14.1	22 41.7	18 28.2
21	21 44.3	26 28.7	28 41.8	22 58.1	18 34.3
22	21 48.2	26 45.8	29 9.8	23 14.6	18 40.4
23	21 52.1	27 3.3	29 38.3	23 31.4	18 46.6
24	21 56.2	27 21.1	♋ 0 7.2	23 48.3	18 52.9
25	22 0.3	27 39.4	0 36.6	24 5.4	18 59.3
26	22 4.6	27 58.1	1 6.4	24 22.7	19 5.7
27	22 8.9	28 17.3	1 36.8	24 40.3	19 12.3
28	22 13.3	28 36.9	2 7.7	24 58.1	19 18.9
29	22 17.9	28 57.1	2 39.1	25 16.2	19 25.6
30	22 22.6	29 17.9	3 11.1	25 34.6	19 32.4
31	22 27.4	29 39.3	3 43.7	25 53.2	19 39.4
32	22 32.3	♊ 0 1.3	4 17.0	26 12.2	19 46.4
33	22 37.4	0 24.0	4 51.0	26 31.5	19 53.6
34	22 42.7	0 47.4	5 25.7	26 51.2	20 0.9
35	22 48.1	1 11.7	6 1.1	27 11.2	20 8.4
36	22 53.7	1 36.8	6 37.4	27 31.6	20 16.0
37	22 59.4	2 2.7	7 14.5	27 52.5	20 23.8
38	23 5.4	2 29.7	7 52.4	28 13.8	20 31.7
39	23 11.6	2 57.7	8 31.3	28 35.6	20 39.8
40	23 18.1	3 26.8	9 11.2	28 57.8	20 48.2
41	23 24.8	3 57.1	9 52.1	29 20.6	20 56.7
42	23 31.8	4 28.8	10 34.1	29 44.0	21 5.4
43	23 39.1	5 1.9	11 17.2	♌ 0 8.0	21 14.4
44	23 46.7	5 36.5	12 1.6	0 32.6	21 23.7
45	23 54.6	6 12.8	12 47.2	0 57.9	21 33.2
46	24 2.9	6 50.9	13 34.1	1 23.9	21 43.0
47	24 11.7	7 31.0	14 22.4	1 50.7	21 53.1
48	24 20.9	8 13.3	15 12.2	2 18.3	22 3.5
49	24 30.6	8 58.0	16 3.6	2 46.8	22 14.2
50	24 40.8	9 45.4	16 56.6	3 16.2	22 25.4
51	24 51.7	10 35.7	17 51.3	3 46.6	22 36.9
52	25 3.2	11 29.2	18 47.9	4 18.0	22 48.9
53	25 15.4	12 26.4	19 46.4	4 50.6	23 1.3
54	25 28.4	13 27.7	20 46.8	5 24.3	23 14.3
55	25 42.4	14 33.6	21 49.4	5 59.4	23 27.8
56	25 57.4	15 44.7	22 54.3	6 35.8	23 41.8
57	26 13.5	17 1.8	24 1.4	7 13.7	23 56.5
58	26 31.0	18 25.8	25 11.1	7 53.2	24 11.9
59	26 49.9	19 57.8	26 23.3	8 34.4	24 28.0
60	27 10.6	21 39.1	27 38.3	9 17.5	24 44.9
61	27 33.2	23 31.7	28 56.0	10 2.5	25 2.7
62	27 58.2	25 37.9	♌ 0 16.8	10 49.6	25 21.5
63	28 26.0	28 1.0	1 40.6	11 39.0	25 41.3
64	28 57.0	♋ 0 45.6	3 7.7	12 30.8	26 2.4
65	29 32.1	3 58.9	4 38.2	13 25.4	26 24.7
66	♉ 0 12.1	7 52.9	6 12.1	14 22.8	26 48.4
S LAT	5	6	Descendant	8	9
11h 16m 0s	MC 169° 0′ 0″	♍ 18° 2′ 16″			

$23^h\ 20^m\ 0^s$ MC 350° 0′ 0″

♓ 19° 7′ 15″

N LAT	11	12	Ascendant	2	3
°	° ′	° ′	° ′	° ′	° ′
0	♈21 38.3	♉22 24.6	♊20 48.6	♋18 27.9	♌17 33.3
5	21 54.4	23 30.9	22 47.7	19 43.7	18 1.3
10	22 11.1	24 40.3	24 49.6	20 59.6	18 29.2
15	22 28.7	25 54.1	26 56.2	22 16.5	18 57.5
16	22 32.4	26 9.6	27 22.2	22 32.2	19 3.3
17	22 36.1	26 25.2	27 48.5	22 47.9	19 9.1
18	22 39.9	26 41.2	28 15.1	23 3.8	19 14.9
19	22 43.8	26 57.5	28 42.1	23 19.8	19 20.8
20	22 47.7	27 14.0	29 9.3	23 35.9	19 26.7
21	22 51.7	27 30.9	29 37.0	23 52.2	19 32.7
22	22 55.8	27 48.1	♋ 0 5.0	24 8.6	19 38.7
23	22 59.9	28 5.7	0 33.3	24 25.2	19 44.8
24	23 4.2	28 23.7	1 2.2	24 42.0	19 51.0
25	23 8.5	28 42.1	1 31.4	24 59.0	19 57.2
26	23 12.9	29 1.0	2 1.1	25 16.2	20 3.6
27	23 17.5	29 20.3	2 31.4	25 33.6	20 10.0
28	23 22.1	29 40.1	3 2.1	25 51.3	20 16.4
29	23 26.9	♊ 0 0.4	3 33.4	26 9.2	20 23.0
30	23 31.8	0 21.4	4 5.2	26 27.4	20 29.7
31	23 36.8	0 42.9	4 37.7	26 45.9	20 36.5
32	23 42.0	1 5.0	5 10.8	27 4.6	20 43.4
33	23 47.3	1 27.9	5 44.6	27 23.8	20 50.5
34	23 52.8	1 51.5	6 19.0	27 43.2	20 57.7
35	23 58.5	2 15.9	6 54.3	28 3.1	21 5.0
36	24 4.4	2 41.1	7 30.2	28 23.3	21 12.4
37	24 10.4	3 7.2	8 7.1	28 43.9	21 20.0
38	24 16.7	3 34.3	8 44.7	29 5.0	21 27.8
39	24 23.2	4 2.4	9 23.3	29 26.6	21 35.8
40	24 29.9	4 31.6	10 2.9	29 48.6	21 43.9
41	24 37.0	5 2.1	10 43.5	♌ 0 11.2	21 52.3
42	24 44.3	5 33.9	11 25.1	0 34.3	22 0.9
43	24 51.9	6 7.0	12 7.8	0 58.0	22 9.7
44	24 59.8	6 41.8	12 51.7	1 22.3	22 18.7
45	25 8.2	7 18.1	13 36.9	1 47.3	22 28.0
46	25 16.9	7 56.3	14 23.3	2 13.0	22 37.6
47	25 26.1	8 36.5	15 11.1	2 39.5	22 47.4
48	25 35.7	9 18.9	16 0.4	3 6.7	22 57.6
49	25 45.8	10 3.6	16 51.2	3 34.8	23 8.2
50	25 56.6	10 51.0	17 43.6	4 3.8	23 19.1
51	26 7.9	11 41.2	18 37.7	4 33.8	23 30.3
52	26 19.9	12 34.7	19 33.5	5 4.8	23 42.1
53	26 32.8	13 31.8	20 31.3	5 36.9	23 54.2
54	26 46.4	14 33.0	21 31.0	6 10.2	24 6.8
55	27 1.0	15 38.7	22 32.7	6 44.7	24 20.0
56	27 16.7	16 49.5	23 36.7	7 20.6	24 33.7
57	27 33.6	18 6.2	24 42.9	7 57.9	24 48.1
58	27 51.9	19 29.6	25 51.6	8 36.8	25 3.1
59	28 11.8	21 0.8	27 2.8	9 17.3	25 18.8
60	28 33.5	22 41.3	28 16.6	9 59.7	25 35.3
61	28 57.3	24 32.6	29 33.2	10 43.9	25 52.7
62	29 23.5	26 37.1	♌ 0 52.7	11 30.2	26 11.0
63	29 52.7	28 57.8	2 15.3	12 18.8	26 30.3
64	♉ 0 25.3	♋ 1 39.1	3 41.0	13 9.7	26 50.8
65	1 2.3	4 47.4	5 10.0	14 3.2	27 12.5
66	1 44.5	8 33.2	6 42.4	14 59.6	27 35.6
S LAT	5	6	Descendant	8	9

♍ 19° 7′ 15″

$11^h\ 20^m\ 0^s$ MC 170° 0′ 0″

$23^h\ 24^m\ 0^s$ MC 351° 0′ 0″

♓ 20° 12′ 19″

N LAT	11	12	Ascendant	2	3
°	° ′	° ′	° ′	° ′	° ′
0	♈22 42.3	♉23 23.4	♊21 43.9	♋19 24.1	♌18 34.0
5	22 59.0	24 30.5	23 43.1	20 39.5	19 1.5
10	23 16.5	25 40.6	25 45.1	21 54.9	19 29.0
15	23 34.9	26 55.1	27 51.5	23 11.4	19 56.8
16	23 38.8	27 10.7	28 17.5	23 26.9	20 2.4
17	23 42.7	27 26.5	28 43.8	23 42.5	20 8.1
18	23 46.6	27 42.6	29 10.4	23 58.3	20 13.8
19	23 50.6	27 59.0	29 37.2	24 14.1	20 19.6
20	23 54.7	28 15.7	♋ 0 4.4	24 30.1	20 25.4
21	23 58.9	28 32.7	0 32.0	24 46.3	20 31.2
22	24 3.2	28 50.1	0 59.9	25 2.6	20 37.1
23	24 7.5	29 7.8	1 28.2	25 19.0	20 43.1
24	24 12.0	29 26.0	1 56.9	25 35.7	20 49.2
25	24 16.5	29 44.5	2 26.1	25 52.5	20 55.3
26	24 21.1	♊ 0 3.5	2 55.7	26 9.6	21 1.5
27	24 25.9	0 23.0	3 25.7	26 26.9	21 7.8
28	24 30.7	0 42.9	3 56.3	26 44.4	21 14.1
29	24 35.7	1 3.4	4 27.5	27 2.1	21 20.6
30	24 40.8	1 24.5	4 59.1	27 20.2	21 27.1
31	24 46.1	1 46.1	5 31.4	27 38.5	21 33.8
32	24 51.5	2 8.4	6 4.3	27 57.1	21 40.6
33	24 57.0	2 31.4	6 37.9	28 16.0	21 47.5
34	25 2.8	2 55.1	7 12.1	28 35.3	21 54.5
35	25 8.7	3 19.6	7 47.1	28 54.9	22 1.6
36	25 14.8	3 44.9	8 22.9	29 15.0	22 9.0
37	25 21.1	4 11.2	8 59.4	29 35.4	22 16.4
38	25 27.7	4 38.4	9 36.8	29 56.2	22 24.0
39	25 34.5	5 6.6	10 15.1	♌ 0 17.5	22 31.8
40	25 41.5	5 36.0	10 54.3	0 39.3	22 39.8
41	25 48.9	6 6.6	11 34.5	1 1.6	22 48.0
42	25 56.5	6 38.4	12 15.8	1 24.5	22 56.4
43	26 4.5	7 11.7	12 58.1	1 47.9	23 5.0
44	26 12.8	7 46.5	13 41.6	2 12.0	23 13.8
45	26 21.5	8 22.9	14 26.3	2 36.6	23 22.9
46	26 30.6	9 1.2	15 12.3	3 2.0	23 32.3
47	26 40.1	9 41.4	15 59.6	3 28.1	23 41.9
48	26 50.2	10 23.8	16 48.4	3 55.0	23 51.9
49	27 0.8	11 8.5	17 38.6	4 22.8	24 2.2
50	27 12.0	11 55.9	18 30.4	4 51.4	24 12.8
51	27 23.8	12 46.1	19 23.8	5 21.0	24 23.8
52	27 36.4	13 39.5	20 19.0	5 51.5	24 35.3
53	27 49.8	14 36.5	21 16.0	6 23.2	24 47.1
54	28 4.0	15 37.4	22 14.9	6 56.0	24 59.5
55	28 19.3	16 42.9	23 15.9	7 30.0	25 12.3
56	28 35.7	17 53.3	24 18.9	8 5.3	25 25.7
57	28 53.4	19 9.6	25 24.3	8 42.1	25 39.7
58	29 12.5	20 32.4	26 31.9	9 20.3	25 54.4
59	29 33.3	22 2.9	27 42.1	10 0.2	26 9.7
60	29 56.0	23 42.4	28 54.8	10 41.9	26 25.8
61	♉ 0 20.8	25 32.4	♌ 0 10.3	11 25.3	26 42.7
62	0 48.3	27 35.2	1 28.6	12 10.9	27 0.5
63	1 18.9	29 53.5	2 49.8	12 58.5	27 19.4
64	1 53.2	♋ 2 31.5	4 14.2	13 48.6	27 39.3
65	2 32.0	5 34.9	5 41.8	14 41.1	28 0.4
66	3 16.6	9 13.0	7 12.7	15 36.4	28 22.8
S LAT	5	6	Descendant	8	9

♍ 20° 12′ 19″

$11^h\ 24^m\ 0^s$ MC 171° 0′ 0″

23h 28m 0s		MC		352° 0′ 0″	
		♓ 21° 17′ 27″			
11	12	Ascendant	2	3	N LAT
° ′	° ′	° ′	° ′	° ′	°
♈23 46.0	♉24 22.0	♊22 39.2	♋20 20.3	♌19 35.0	0
24 3.5	25 29.8	24 38.5	21 35.3	20 2.0	5
24 21.7	26 40.7	26 40.5	22 50.3	20 28.9	10
24 40.9	27 55.9	28 46.8	24 6.2	20 56.1	15
24 45.0	28 11.6	29 12.8	24 21.6	21 1.6	16
24 49.0	28 27.5	29 39.0	24 37.2	21 7.2	17
24 53.1	28 43.8	♋ 0 5.5	24 52.8	21 12.8	18
24 57.3	29 0.3	0 32.3	25 8.5	21 18.5	19
25 1.6	29 17.1	0 59.4	25 24.4	21 24.2	20
25 6.0	29 34.3	1 26.9	25 40.4	21 29.9	21
25 10.4	29 51.8	1 54.7	25 56.6	21 35.7	22
25 14.9	♊ 0 9.7	2 22.9	26 12.9	21 41.6	23
25 19.6	0 27.9	2 51.5	26 29.4	21 47.5	24
25 24.3	0 46.6	3 20.5	26 46.1	21 53.5	25
25 29.1	1 5.7	3 50.0	27 3.0	21 59.5	26
25 34.0	1 25.3	4 19.9	27 20.2	22 5.7	27
25 39.1	1 45.4	4 50.4	27 37.5	22 11.9	28
25 44.3	2 6.0	5 21.3	27 55.1	22 18.2	29
25 49.6	2 27.2	5 52.9	28 13.0	22 24.7	30
25 55.1	2 49.0	6 25.0	28 31.1	22 31.2	31
26 0.7	3 11.4	6 57.7	28 49.5	22 37.8	32
26 6.5	3 34.5	7 31.0	29 8.3	22 44.6	33
26 12.5	3 58.3	8 5.0	29 27.4	22 51.4	34
26 18.7	4 22.9	8 39.8	29 46.8	22 58.4	35
26 25.1	4 48.4	9 15.3	♌ 0 6.6	23 5.6	36
26 31.6	5 14.7	9 51.5	0 26.8	23 12.9	37
26 38.5	5 42.1	10 28.6	0 47.4	23 20.3	38
26 45.5	6 10.4	11 6.6	1 8.5	23 28.0	39
26 52.9	6 39.9	11 45.5	1 30.0	23 35.8	40
27 0.5	7 10.5	12 25.4	1 52.1	23 43.7	41
27 8.5	7 42.5	13 6.3	2 14.7	23 51.9	42
27 16.7	8 15.8	13 48.2	2 37.8	24 0.4	43
27 25.4	8 50.7	14 31.3	3 1.6	24 9.0	44
27 34.5	9 27.2	15 15.5	3 26.0	24 17.9	45
27 43.9	10 5.5	16 1.0	3 51.0	24 27.0	46
27 53.9	10 45.7	16 47.9	4 16.8	24 36.5	47
28 4.4	11 28.1	17 36.1	4 43.3	24 46.2	48
28 15.4	12 12.8	18 25.7	5 10.7	24 56.3	49
28 27.1	13 0.1	19 16.9	5 38.9	25 6.7	50
28 39.4	13 50.3	20 9.7	6 8.1	25 17.4	51
28 52.5	14 43.6	21 4.2	6 38.2	25 28.6	52
29 6.4	15 40.4	22 0.5	7 9.4	25 40.2	53
29 21.3	16 41.1	22 58.6	7 41.7	25 52.2	54
29 37.2	17 46.3	23 58.8	8 15.2	26 4.7	55
29 54.3	18 56.4	25 1.0	8 50.0	26 17.8	56
♉ 0 12.7	20 12.1	26 5.4	9 26.2	26 31.4	57
0 32.7	21 34.4	27 12.1	10 3.9	26 45.7	58
0 54.3	23 4.1	28 21.3	10 43.1	27 0.6	59
1 18.0	24 42.5	29 32.9	11 24.0	27 16.3	60
1 44.0	26 31.2	♌ 0 47.2	12 6.8	27 32.8	61
2 12.7	28 32.2	2 4.3	12 51.5	27 50.1	62
2 44.7	♋ 0 48.2	3 24.3	13 38.4	28 8.5	63
3 20.6	3 22.8	4 47.4	14 27.5	28 27.8	64
4 1.3	6 21.5	6 13.6	15 19.1	28 48.4	65
4 48.2	9 52.3	7 43.1	16 13.3	29 10.2	66
5	6	Descendant	8	9	S LAT
		♍ 21° 17′ 27″			
11h 28m 0s		MC		172° 0′ 0″	

23h 32m 0s		MC		353° 0′ 0″	
		♓ 22° 22′ 38″			
N LAT	11	12	Ascendant	2	3
°	° ′	° ′	° ′	° ′	° ′
0	♈24 49.7	♉25 20.5	♊23 34.4	♋21 16.7	♌20 36.1
5	25 7.9	26 29.0	25 33.8	22 31.3	21 2.6
10	25 26.8	27 40.5	27 35.8	23 45.7	21 28.9
15	25 46.8	28 56.4	29 41.9	25 1.2	21 55.6
16	25 51.0	29 12.2	♋ 0 7.8	25 16.5	22 1.0
17	25 55.2	29 28.3	0 34.0	25 31.9	22 6.5
18	25 59.5	29 44.7	1 0.4	25 47.4	22 12.0
19	26 3.9	♊ 0 1.3	1 27.2	26 3.0	22 17.5
20	26 8.3	0 18.3	1 54.2	26 18.7	22 23.1
21	26 12.8	0 35.6	2 21.6	26 34.6	22 28.7
22	26 17.4	0 53.2	2 49.3	26 50.6	22 34.4
23	26 22.1	1 11.2	3 17.4	27 6.8	22 40.1
24	26 27.0	1 29.6	3 45.9	27 23.2	22 45.9
25	26 31.9	1 48.4	4 14.8	27 39.8	22 51.8
26	26 36.9	2 7.6	4 44.1	27 56.5	22 57.7
27	26 42.0	2 27.3	5 13.9	28 13.5	23 3.7
28	26 47.3	2 47.5	5 44.2	28 30.7	23 9.8
29	26 52.7	3 8.3	6 15.0	28 48.1	23 16.0
30	26 58.2	3 29.5	6 46.4	29 5.8	23 22.3
31	27 3.9	3 51.4	7 18.3	29 23.7	23 28.7
32	27 9.8	4 14.0	7 50.8	29 42.0	23 35.2
33	27 15.8	4 37.2	8 23.9	♌ 0 0.5	23 41.8
34	27 22.0	5 1.1	8 57.7	0 19.4	23 48.5
35	27 28.4	5 25.9	9 32.2	0 38.6	23 55.3
36	27 35.0	5 51.4	10 7.4	0 58.2	24 2.3
37	27 41.9	6 17.9	10 43.4	1 18.2	24 9.4
38	27 49.0	6 45.3	11 20.3	1 38.6	24 16.7
39	27 56.3	7 13.7	11 57.9	1 59.4	24 24.2
40	28 3.9	7 43.2	12 36.5	2 20.7	24 31.8
41	28 11.9	8 14.0	13 16.0	2 42.5	24 39.6
42	28 20.1	8 46.0	13 56.5	3 4.8	24 47.6
43	28 28.7	9 19.4	14 38.1	3 27.7	24 55.8
44	28 37.7	9 54.3	15 20.7	3 51.2	25 4.3
45	28 47.1	10 30.8	16 4.5	4 15.2	25 13.0
46	28 57.0	11 9.2	16 49.5	4 40.0	25 21.9
47	29 7.4	11 49.4	17 35.9	5 5.4	25 31.1
48	29 18.2	12 31.8	18 23.5	5 31.6	25 40.6
49	29 29.7	13 16.5	19 12.6	5 58.6	25 50.4
50	29 41.8	14 3.7	20 3.2	6 26.4	26 0.6
51	29 54.6	14 53.8	20 55.4	6 55.2	26 11.1
52	♉ 0 8.3	15 46.9	21 49.2	7 24.9	26 21.9
53	0 22.7	16 43.6	22 44.8	7 55.6	26 33.2
54	0 38.2	17 44.1	23 42.2	8 27.4	26 45.0
55	0 54.7	18 48.9	24 41.5	9 0.5	26 57.2
56	1 12.5	19 58.6	25 42.9	9 34.7	27 9.9
57	1 31.7	21 13.8	26 46.4	10 10.3	27 23.2
58	1 52.4	22 35.4	27 52.2	10 47.4	27 37.1
59	2 15.0	24 4.3	29 0.3	11 25.9	27 51.7
60	2 39.6	25 41.7	♌ 0 10.9	12 6.2	28 6.9
61	3 6.7	27 29.0	1 24.1	12 48.2	28 23.0
62	3 36.6	29 28.2	2 40.0	13 32.2	28 39.8
63	4 10.0	♋ 1 41.9	3 58.8	14 18.2	28 57.7
64	4 47.5	4 13.2	5 20.5	15 6.4	29 16.5
65	5 30.2	7 7.3	6 45.4	15 57.0	29 36.5
66	6 19.3	10 31.2	8 13.4	16 50.3	29 57.6
S LAT	5	6	Descendant	8	9
			♍ 22° 22′ 38″		
	11h 32m 0s		MC		173° 0′ 0″

23ʰ 36ᵐ 0ˢ — MC 354° 0′ 0″ — ♓ 23° 27′ 53″

N LAT	11	12	Ascendant	2	3
°	° ′	° ′	° ′	° ′	° ′
0	♈25 53.2	♉26 18.8	♊24 29.5	♋22 13.1	♌21 37.4
5	26 12.1	27 28.0	26 29.1	23 27.3	22 3.3
10	26 31.7	28 40.2	28 31.0	24 41.3	22 29.2
15	26 52.5	29 56.6	♋ 0 37.0	25 56.1	22 55.3
16	26 56.8	♊ 0 12.6	1 2.8	26 11.3	23 0.6
17	27 1.2	0 28.8	1 28.9	26 26.6	23 5.9
18	27 5.7	0 45.3	1 55.3	26 42.0	23 11.3
19	27 10.2	1 2.1	2 21.9	26 57.5	23 16.7
20	27 14.8	1 19.2	2 48.9	27 13.1	23 22.1
21	27 19.5	1 36.6	3 16.2	27 28.8	23 27.6
22	27 24.3	1 54.3	3 43.8	27 44.7	23 33.2
23	27 29.2	2 12.4	4 11.8	28 0.8	23 38.8
24	27 34.1	2 30.9	4 40.1	28 17.0	23 44.5
25	27 39.2	2 49.8	5 8.9	28 33.4	23 50.2
26	27 44.4	3 9.2	5 38.1	28 50.0	23 56.0
27	27 49.8	3 29.0	6 7.8	29 6.8	24 1.9
28	27 55.2	3 49.3	6 37.9	29 23.8	24 7.9
29	28 0.8	4 10.2	7 8.5	29 41.1	24 13.9
30	28 6.6	4 31.6	7 39.7	29 58.6	24 20.0
31	28 12.5	4 53.6	8 11.4	♌ 0 16.4	24 26.3
32	28 18.6	5 16.2	8 43.7	0 34.4	24 32.6
33	28 24.8	5 39.5	9 16.6	0 52.8	24 39.1
34	28 31.2	6 3.6	9 50.2	1 11.5	24 45.6
35	28 37.9	6 28.4	10 24.4	1 30.5	24 52.3
36	28 44.8	6 54.0	10 59.4	1 49.9	24 59.2
37	28 51.9	7 20.6	11 35.1	2 9.6	25 6.1
38	28 59.2	7 48.1	12 11.7	2 29.8	25 13.2
39	29 6.8	8 16.6	12 49.0	2 50.4	25 20.5
40	29 14.7	8 46.2	13 27.2	3 11.4	25 28.0
41	29 23.0	9 17.0	14 6.4	3 33.0	25 35.6
42	29 31.5	9 49.0	14 46.5	3 55.0	25 43.4
43	29 40.5	10 22.5	15 27.7	4 17.6	25 51.4
44	29 49.8	10 57.4	16 9.9	4 40.7	25 59.7
45	29 59.5	11 34.0	16 53.3	5 4.5	26 8.1
46	♉ 0 9.8	12 12.3	17 37.8	5 28.9	26 16.9
47	0 20.5	12 52.6	18 23.6	5 54.0	26 25.8
48	0 31.8	13 34.9	19 10.8	6 19.9	26 35.1
49	0 43.7	14 19.5	19 59.3	6 46.5	26 44.7
50	0 56.2	15 6.7	20 49.3	7 13.9	26 54.6
51	1 9.5	15 56.6	21 40.8	7 42.2	27 4.8
52	1 23.6	16 49.6	22 34.0	8 11.5	27 15.4
53	1 38.7	17 46.0	23 28.9	8 41.8	27 26.4
54	1 54.7	18 46.3	24 25.5	9 13.1	27 37.8
55	2 11.9	19 50.7	25 24.1	9 45.6	27 49.7
56	2 30.3	21 0.0	26 24.6	10 19.4	28 2.1
57	2 50.2	22 14.7	27 27.2	10 54.4	28 15.1
58	3 11.7	23 35.7	28 32.1	11 30.8	28 28.6
59	3 35.1	25 3.7	29 39.2	12 8.8	28 42.8
60	4 0.8	26 40.0	♌ 0 48.8	12 48.4	28 57.6
61	4 28.9	28 25.9	2 0.9	13 29.7	29 13.2
62	5 0.1	♋ 0 23.4	3 15.7	14 12.8	29 29.6
63	5 34.8	2 34.6	4 33.2	14 58.0	29 46.9
64	6 14.0	5 2.8	5 53.7	15 45.4	♍ 0 5.2
65	6 58.6	7 52.3	7 17.2	16 35.1	0 24.6
66	7 50.1	11 9.7	8 43.8	17 27.3	0 45.2
S LAT	5	6	Descendant	8	9

♍ 23° 27′ 53″ — 11ʰ 36ᵐ 0ˢ — MC 174° 0′ 0″

23ʰ 40ᵐ 0ˢ — MC 355° 0′ 0″ — ♓ 24° 33′ 10″

N LAT	11	12	Ascendant	2	3
°	° ′	° ′	° ′	° ′	° ′
0	♈26 56.5	♉27 17.0	♊25 24.7	♋23 9.7	♌22 38.9
5	27 16.1	28 26.8	27 24.3	24 23.4	23 4.3
10	27 36.5	29 39.6	29 26.1	25 36.9	23 29.5
15	27 58.0	♊ 0 56.7	♋ 1 31.9	26 51.2	23 55.1
16	28 2.5	1 12.7	1 57.7	27 6.3	24 0.3
17	28 7.0	1 29.1	2 23.7	27 21.4	24 5.5
18	28 11.6	1 45.7	2 50.0	27 36.7	24 10.7
19	28 16.3	2 2.6	3 16.6	27 52.0	24 16.0
20	28 21.1	2 19.8	3 43.4	28 7.5	24 21.3
21	28 26.0	2 37.3	4 10.6	28 23.1	24 26.7
22	28 30.9	2 55.1	4 38.1	28 38.9	24 32.2
23	28 36.0	3 13.3	5 6.0	28 54.8	24 37.6
24	28 41.1	3 32.0	5 34.2	29 10.8	24 43.2
25	28 46.4	3 51.0	6 2.9	29 27.1	24 48.8
26	28 51.8	4 10.4	6 31.9	29 43.5	24 54.4
27	28 57.3	4 30.4	7 1.4	♌ 0 0.2	25 0.2
28	29 3.0	4 50.8	7 31.4	0 17.0	25 6.0
29	29 8.8	5 11.7	8 1.9	0 34.1	25 11.9
30	29 14.7	5 33.2	8 32.8	0 51.4	25 17.9
31	29 20.8	5 55.3	9 4.4	1 9.0	25 24.0
32	29 27.1	6 18.1	9 36.5	1 26.9	25 30.2
33	29 33.6	6 41.5	10 9.2	1 45.1	25 36.5
34	29 40.2	7 5.6	10 42.5	2 3.5	25 42.9
35	29 47.1	7 30.5	11 16.5	2 22.3	25 49.4
36	29 54.2	7 56.2	11 51.2	2 41.5	25 56.1
37	♉ 0 1.6	8 22.8	12 26.6	3 1.0	26 2.9
38	0 9.2	8 50.4	13 2.8	3 21.0	26 9.8
39	0 17.1	9 19.0	13 39.9	3 41.3	26 16.9
40	0 25.3	9 48.6	14 17.8	4 2.1	26 24.2
41	0 33.8	10 19.5	14 56.6	4 23.4	26 31.6
42	0 42.6	10 51.6	15 36.3	4 45.1	26 39.3
43	0 51.9	11 25.1	16 17.1	5 7.4	26 47.1
44	1 1.5	12 0.0	16 58.9	5 30.3	26 55.1
45	1 11.6	12 36.6	17 41.8	5 53.7	27 3.4
46	1 22.2	13 14.9	18 25.9	6 17.8	27 11.9
47	1 33.3	13 55.1	19 11.2	6 42.6	27 20.6
48	1 45.0	14 37.4	19 57.8	7 8.1	27 29.7
49	1 57.3	15 22.0	20 45.8	7 34.3	27 39.0
50	2 10.3	16 9.0	21 35.2	8 1.4	27 48.6
51	2 24.0	16 58.8	22 26.1	8 29.3	27 58.6
52	2 38.7	17 51.6	23 18.6	8 58.1	28 8.9
53	2 54.2	18 47.8	24 12.8	9 28.0	28 19.6
54	3 10.8	19 47.8	25 8.7	9 58.8	28 30.8
55	3 28.6	20 51.9	26 6.5	10 30.8	28 42.4
56	3 47.7	22 0.7	27 6.2	11 4.0	28 54.4
57	4 8.3	23 14.9	28 7.9	11 38.5	29 7.0
58	4 30.6	24 35.1	29 11.9	12 14.3	29 20.2
59	4 54.9	26 2.2	♌ 0 18.0	12 51.6	29 33.9
60	5 21.4	27 37.4	1 26.6	13 30.5	29 48.4
61	5 50.6	29 22.0	2 37.6	14 11.1	♍ 0 3.5
62	6 23.0	♋ 1 17.6	3 51.2	14 53.5	0 19.5
63	6 59.2	3 26.5	5 7.6	15 37.9	0 36.3
64	7 39.9	5 51.5	6 26.8	16 24.4	0 54.0
65	8 26.4	8 36.7	7 49.0	17 13.1	1 12.9
66	9 20.3	11 47.8	9 14.3	18 4.3	1 32.8
S LAT	5	6	Descendant	8	9

♍ 24° 33′ 10″ — 11ʰ 40ᵐ 0ˢ — MC 175° 0′ 0″

23h 44m 0s — MC 356° 0′ 0″ — ♓ 25° 38′ 29″

11	12	Ascendant	2	3	N LAT
° ′	° ′	° ′	° ′	° ′	°
♈27 59.7	♉28 15.0	♊26 19.8	♋24 6.4	♌23 40.6	**0**
28 20.0	29 25.4	28 19.4	25 19.7	24 5.4	**5**
28 41.1	♊ 0 38.8	♋ 0 21.2	26 32.6	24 30.1	**10**
29 3.3	1 56.5	2 26.7	27 46.3	24 55.0	**15**
29 7.9	2 12.6	2 52.4	28 1.3	25 0.1	**16**
29 12.6	2 29.1	3 18.4	28 16.3	25 5.2	**17**
29 17.4	2 45.8	3 44.6	28 31.4	25 10.3	**18**
29 22.2	3 2.8	4 11.1	28 46.6	25 15.5	**19**
29 27.2	3 20.1	4 37.8	29 2.0	25 20.7	**20**
29 32.2	3 37.7	5 4.9	29 17.4	25 25.9	**21**
29 37.3	3 55.7	5 32.3	29 33.0	25 31.2	**22**
29 42.6	4 14.0	6 0.1	29 48.8	25 36.6	**23**
29 47.9	4 32.7	6 28.2	♌ 0 4.7	25 42.0	**24**
29 53.3	4 51.8	6 56.7	0 20.8	25 47.5	**25**
29 58.9	5 11.4	7 25.6	0 37.1	25 53.0	**26**
♉ 0 4.6	5 31.4	7 55.0	0 53.6	25 58.6	**27**
0 10.5	5 51.9	8 24.8	1 10.2	26 4.3	**28**
0 16.5	6 13.0	8 55.0	1 27.1	26 10.1	**29**
0 22.6	6 34.6	9 25.8	1 44.3	26 15.9	**30**
0 28.9	6 56.7	9 57.1	2 1.7	26 21.8	**31**
0 35.4	7 19.6	10 29.0	2 19.4	26 27.9	**32**
0 42.1	7 43.1	11 1.5	2 37.3	26 34.0	**33**
0 49.0	8 7.3	11 34.6	2 55.6	26 40.3	**34**
0 56.1	8 32.2	12 8.3	3 14.2	26 46.6	**35**
1 3.4	8 58.0	12 42.8	3 33.1	26 53.1	**36**
1 11.0	9 24.7	13 17.9	3 52.4	26 59.8	**37**
1 18.9	9 52.3	13 53.8	4 12.1	27 6.5	**38**
1 27.0	10 21.0	14 30.5	4 32.2	27 13.5	**39**
1 35.5	10 50.7	15 8.1	4 52.8	27 20.5	**40**
1 44.3	11 21.6	15 46.6	5 13.8	27 27.8	**41**
1 53.4	11 53.7	16 25.9	5 35.3	27 35.2	**42**
2 3.0	12 27.2	17 6.3	5 57.3	27 42.8	**43**
2 12.9	13 2.2	17 47.7	6 19.8	27 50.7	**44**
2 23.4	13 38.7	18 30.2	6 43.0	27 58.7	**45**
2 34.3	14 17.0	19 13.8	7 6.7	28 7.0	**46**
2 45.8	14 57.2	19 58.6	7 31.1	28 15.5	**47**
2 57.8	15 39.4	20 44.7	7 56.3	28 24.3	**48**
3 10.5	16 23.8	21 32.1	8 22.1	28 33.4	**49**
3 24.0	17 10.8	22 20.9	8 48.8	28 42.8	**50**
3 38.2	18 0.4	23 11.2	9 16.3	28 52.5	**51**
3 53.3	18 53.0	24 3.0	9 44.7	29 2.5	**52**
4 9.4	19 49.0	24 56.5	10 14.1	29 12.9	**53**
4 26.5	20 48.6	25 51.7	10 44.5	29 23.8	**54**
4 44.9	21 52.3	26 48.7	11 16.0	29 35.0	**55**
5 4.6	23 0.7	27 47.6	11 48.6	29 46.8	**56**
5 25.9	24 14.2	28 48.5	12 22.5	29 59.0	**57**
5 49.0	25 33.7	29 51.5	12 57.8	♍ 0 11.8	**58**
6 14.1	27 0.0	♌ 0 56.7	13 34.5	0 25.2	**59**
6 41.6	28 34.0	2 4.2	14 12.7	0 39.2	**60**
7 11.9	♋ 0 17.2	3 14.2	14 52.5	0 53.9	**61**
7 45.4	2 11.0	4 26.7	15 34.2	1 9.4	**62**
8 23.0	4 17.6	5 41.9	16 17.8	1 25.7	**63**
9 5.3	6 39.5	6 59.9	17 3.4	1 42.9	**64**
9 53.8	9 20.5	8 20.8	17 51.2	2 1.2	**65**
10 50.2	12 25.7	9 44.7	18 41.5	2 20.5	**66**
5	6	Descendant	8	9	S LAT

♍ 25° 38′ 29″ — 11h 44m 0s — MC 176° 0′ 0″

23h 48m 0s — MC 357° 0′ 0″ — ♓ 26° 43′ 50″

11	12	Ascendant	2	3	N LAT
° ′	° ′	° ′	° ′	° ′	°
♈29 2.8	♉29 12.8	♊27 14.8	♋25 3.3	♌24 42.5	**0**
29 23.7	♊ 0 23.9	29 14.5	26 16.0	25 6.7	**5**
29 45.5	1 37.8	♋ 1 16.2	27 28.4	25 30.8	**10**
♉ 0 8.4	2 56.0	3 21.4	28 41.5	25 55.1	**15**
0 13.2	3 12.3	3 47.1	28 56.3	26 0.1	**16**
0 18.0	3 28.8	4 13.0	29 11.2	26 5.0	**17**
0 23.0	3 45.7	4 39.1	29 26.2	26 10.0	**18**
0 28.0	4 2.8	5 5.5	29 41.3	26 15.1	**19**
0 33.1	4 20.2	5 32.1	29 56.5	26 20.2	**20**
0 38.3	4 37.9	5 59.1	♌ 0 11.8	26 25.3	**21**
0 43.5	4 55.9	6 26.4	0 27.3	26 30.4	**22**
0 48.9	5 14.4	6 54.0	0 42.9	26 35.7	**23**
0 54.4	5 33.2	7 22.0	0 58.6	26 40.9	**24**
1 0.1	5 52.4	7 50.4	1 14.6	26 46.3	**25**
1 5.8	6 12.0	8 19.1	1 30.7	26 51.7	**26**
1 11.7	6 32.2	8 48.3	1 47.0	26 57.1	**27**
1 17.7	6 52.8	9 17.9	2 3.5	27 2.7	**28**
1 23.9	7 13.9	9 48.0	2 20.2	27 8.3	**29**
1 30.2	7 35.6	10 18.6	2 37.2	27 14.0	**30**
1 36.8	7 57.8	10 49.7	2 54.4	27 19.8	**31**
1 43.5	8 20.7	11 21.4	3 11.9	27 25.7	**32**
1 50.4	8 44.3	11 53.6	3 29.6	27 31.7	**33**
1 57.5	9 8.6	12 26.5	3 47.7	27 37.8	**34**
2 4.8	9 33.6	13 0.0	4 6.1	27 44.0	**35**
2 12.4	9 59.5	13 34.1	4 24.8	27 50.3	**36**
2 20.2	10 26.2	14 9.0	4 43.8	27 56.7	**37**
2 28.3	10 53.8	14 44.6	5 3.3	28 3.3	**38**
2 36.7	11 22.5	15 21.0	5 23.2	28 10.1	**39**
2 45.4	11 52.3	15 58.2	5 43.4	28 17.0	**40**
2 54.5	12 23.2	16 36.3	6 4.2	28 24.0	**41**
3 3.9	12 55.3	17 15.3	6 25.4	28 31.3	**42**
3 13.8	13 28.8	17 55.3	6 47.1	28 38.7	**43**
3 24.0	14 3.8	18 36.3	7 9.3	28 46.3	**44**
3 34.8	14 40.3	19 18.3	7 32.2	28 54.1	**45**
3 46.1	15 18.6	20 1.4	7 55.6	29 2.2	**46**
3 57.9	15 58.7	20 45.8	8 19.7	29 10.5	**47**
4 10.3	16 40.8	21 31.3	8 44.5	29 19.0	**48**
4 23.4	17 25.1	22 18.2	9 10.0	29 27.9	**49**
4 37.3	18 11.9	23 6.4	9 36.2	29 37.0	**50**
4 52.0	19 1.4	23 56.1	10 3.3	29 46.4	**51**
5 7.5	19 53.8	24 47.3	10 31.3	29 56.2	**52**
5 24.1	20 49.4	25 40.1	11 0.2	♍ 0 6.3	**53**
5 41.8	21 48.7	26 34.5	11 30.1	0 16.9	**54**
6 0.8	22 52.0	27 30.8	12 1.1	0 27.8	**55**
6 21.1	23 59.9	28 28.9	12 33.2	0 39.2	**56**
6 43.1	25 12.8	29 28.9	13 6.6	0 51.1	**57**
7 6.9	26 31.6	♌ 0 31.0	13 41.2	1 3.5	**58**
7 32.9	27 56.9	1 35.3	14 17.3	1 16.5	**59**
8 1.3	29 29.9	2 41.8	14 54.8	1 30.1	**60**
8 32.6	♋ 1 11.6	3 50.8	15 34.0	1 44.4	**61**
9 7.4	3 3.6	5 2.2	16 14.9	1 59.4	**62**
9 46.3	5 7.9	6 16.2	16 57.7	2 15.2	**63**
10 30.3	7 26.8	7 33.0	17 42.5	2 31.9	**64**
11 20.7	10 3.8	8 52.6	18 29.4	2 49.6	**65**
12 19.5	13 3.3	10 15.2	19 18.7	3 8.3	**66**
5	6	Descendant	8	9	S LAT

♍ 26° 43′ 50″ — 11h 48m 0s — MC 177° 0′ 0″

23ʰ 52ᵐ 0ˢ MC 358° 0′ 0″

♓ 27° 49′ 13″

N LAT	11	12	Ascendant	2	3
°	° ′	° ′	° ′	° ′	° ′
0	♉ 0 5.7	♊ 0 10.5	♊28 9.9	♋26 0.2	♌25 44.5
5	0 27.2	1 22.1	♋ 0 9.6	27 12.5	26 8.1
10	0 49.7	2 36.6	2 11.1	28 24.3	26 31.6
15	1 13.3	3 55.3	4 16.1	29 36.8	26 55.4
16	1 18.2	4 11.7	4 41.6	29 51.5	27 0.2
17	1 23.2	4 28.4	5 7.4	♌ 0 6.2	27 5.0
18	1 28.3	4 45.3	5 33.5	0 21.1	27 9.9
19	1 33.5	5 2.5	5 59.7	0 36.0	27 14.8
20	1 38.7	5 20.0	6 26.3	0 51.1	27 19.8
21	1 44.1	5 37.8	6 53.2	1 6.2	27 24.7
22	1 49.5	5 55.9	7 20.3	1 21.5	27 29.8
23	1 55.1	6 14.5	7 47.8	1 37.0	27 34.9
24	2 0.8	6 33.4	8 15.7	1 52.6	27 40.0
25	2 6.5	6 52.7	8 43.9	2 8.4	27 45.2
26	2 12.5	7 12.4	9 12.5	2 24.3	27 50.5
27	2 18.5	7 32.6	9 41.5	2 40.4	27 55.8
28	2 24.7	7 53.3	10 11.0	2 56.8	28 1.2
29	2 31.1	8 14.5	10 40.9	3 13.3	28 6.7
30	2 37.6	8 36.2	11 11.3	3 30.1	28 12.2
31	2 44.3	8 58.5	11 42.2	3 47.1	28 17.8
32	2 51.2	9 21.5	12 13.6	4 4.4	28 23.6
33	2 58.3	9 45.1	12 45.6	4 21.9	28 29.4
34	3 5.7	10 9.5	13 18.2	4 39.8	28 35.3
35	3 13.2	10 34.6	13 51.5	4 57.9	28 41.4
36	3 21.0	11 0.5	14 25.4	5 16.4	28 47.5
37	3 29.1	11 27.3	14 59.9	5 35.3	28 53.8
38	3 37.4	11 55.0	15 35.2	5 54.5	29 0.2
39	3 46.1	12 23.7	16 11.3	6 14.1	29 6.8
40	3 55.1	12 53.4	16 48.2	6 34.1	29 13.5
41	4 4.4	13 24.3	17 25.9	6 54.6	29 20.4
42	4 14.1	13 56.5	18 4.6	7 15.5	29 27.4
43	4 24.2	14 30.0	18 44.1	7 36.9	29 34.6
44	4 34.8	15 4.9	19 24.7	7 58.9	29 42.0
45	4 45.9	15 41.4	20 6.3	8 21.4	29 49.6
46	4 57.5	16 19.6	20 48.9	8 44.5	29 57.5
47	5 9.7	16 59.6	21 32.8	9 8.2	♍ 0 5.5
48	5 22.5	17 41.7	22 17.8	9 32.6	0 13.8
49	5 36.0	18 25.9	23 4.1	9 57.7	0 22.4
50	5 50.2	19 12.5	23 51.8	10 23.6	0 31.3
51	6 5.3	20 1.8	24 40.8	10 50.3	0 40.5
52	6 21.4	20 53.9	25 31.4	11 17.9	0 49.9
53	6 38.4	21 49.3	26 23.5	11 46.3	0 59.8
54	6 56.7	22 48.2	27 17.2	12 15.8	1 10.0
55	7 16.2	23 51.1	28 12.7	12 46.2	1 20.6
56	7 37.2	24 58.5	29 10.0	13 17.8	1 31.7
57	7 59.8	26 10.8	♌ 0 9.2	13 50.6	1 43.2
58	8 24.4	27 28.8	1 10.5	14 24.7	1 55.3
59	8 51.2	28 53.2	2 13.8	15 0.1	2 7.8
60	9 20.5	♋ 0 24.9	3 19.4	15 37.0	2 21.0
61	9 52.8	2 5.2	4 27.2	16 15.5	2 34.9
62	10 28.7	3 55.5	5 37.6	16 55.6	2 49.4
63	11 9.0	5 57.5	6 50.5	17 37.6	3 4.8
64	11 54.6	8 13.4	8 6.1	18 21.5	3 20.9
65	12 47.1	10 46.5	9 24.5	19 7.6	3 38.1
66	13 48.4	13 40.6	10 45.8	19 55.9	3 56.2
S LAT	5	6	Descendant	8	9

♍ 27° 49′ 13″

11ʰ 52ᵐ 0ˢ MC 178° 0′ 0″

23ʰ 56ᵐ 0ˢ MC 359° 0′ 0″

♓ 28° 54′ 36″

N LAT	11	12	Ascendant	2	3
°	° ′	° ′	° ′	° ′	° ′
0	♉ 1 8.4	♊ 1 8.0	♊29 5.0	♋26 57.4	♌26 46.7
5	1 30.6	2 20.2	♋ 1 4.6	28 9.0	27 9.7
10	1 53.7	3 35.2	3 6.0	29 20.3	27 32.6
15	2 18.0	4 54.4	5 10.6	♌ 0 32.1	27 55.7
16	2 23.1	5 10.9	5 36.1	0 46.7	28 0.4
17	2 28.2	5 27.7	6 1.8	1 1.3	28 5.1
18	2 33.5	5 44.7	6 27.7	1 16.0	28 9.9
19	2 38.8	6 2.0	6 53.9	1 30.8	28 14.7
20	2 44.2	6 19.5	7 20.4	1 45.7	28 19.5
21	2 49.7	6 37.4	7 47.1	2 0.7	28 24.3
22	2 55.3	6 55.7	8 14.2	2 15.9	28 29.2
23	3 1.0	7 14.3	8 41.5	2 31.2	28 34.2
24	3 6.8	7 33.3	9 9.2	2 46.6	28 39.2
25	3 12.8	7 52.6	9 37.3	3 2.2	28 44.3
26	3 18.9	8 12.4	10 5.7	3 18.0	28 49.4
27	3 25.1	8 32.7	10 34.6	3 33.9	28 54.6
28	3 31.5	8 53.5	11 3.9	3 50.1	28 59.8
29	3 38.1	9 14.7	11 33.6	4 6.4	29 5.1
30	3 44.8	9 36.6	12 3.8	4 23.0	29 10.5
31	3 51.7	9 59.0	12 34.5	4 39.8	29 16.0
32	3 58.8	10 22.0	13 5.7	4 56.9	29 21.6
33	4 6.1	10 45.7	13 37.5	5 14.3	29 27.3
34	4 13.6	11 10.0	14 9.8	5 31.9	29 33.0
35	4 21.4	11 35.2	14 42.8	5 49.8	29 38.9
36	4 29.4	12 1.1	15 16.4	6 8.1	29 44.9
37	4 37.7	12 27.9	15 50.7	6 26.7	29 51.0
38	4 46.3	12 55.7	16 25.7	6 45.7	29 57.2
39	4 55.1	13 24.4	17 1.4	7 5.0	♍ 0 3.6
40	5 4.4	13 54.2	17 38.0	7 24.8	0 10.1
41	5 14.0	14 25.1	18 15.4	7 44.9	0 16.8
42	5 24.0	14 57.2	18 53.6	8 5.6	0 23.6
43	5 34.4	15 30.7	19 32.8	8 26.7	0 30.6
44	5 45.3	16 5.6	20 12.9	8 48.4	0 37.8
45	5 56.6	16 42.1	20 54.0	9 10.6	0 45.2
46	6 8.6	17 20.2	21 36.3	9 33.3	0 52.8
47	6 21.1	18 0.1	22 19.6	9 56.7	1 0.7
48	6 34.2	18 42.0	23 4.1	10 20.8	1 8.7
49	6 48.1	19 26.1	23 49.9	10 45.5	1 17.1
50	7 2.8	20 12.6	24 37.0	11 11.0	1 25.7
51	7 18.3	21 1.6	25 25.4	11 37.3	1 34.6
52	7 34.8	21 53.5	26 15.3	12 4.4	1 43.8
53	7 52.4	22 48.6	27 6.8	12 32.4	1 53.3
54	8 11.1	23 47.1	27 59.8	13 1.4	2 3.2
55	8 31.2	24 49.6	28 54.5	13 31.4	2 13.5
56	8 52.8	25 56.4	29 51.0	14 2.4	2 24.2
57	9 16.1	27 8.0	♌ 0 49.4	14 34.7	2 35.4
58	9 41.4	28 25.2	1 49.8	15 8.1	2 47.1
59	10 9.0	29 48.7	2 52.2	15 42.9	2 59.3
60	10 39.2	♋ 1 19.3	3 56.8	16 19.2	3 12.1
61	11 12.5	2 58.2	5 3.7	16 57.0	3 25.5
62	11 49.6	4 46.6	6 12.9	17 36.4	3 39.6
63	12 31.2	6 46.4	7 24.7	18 17.6	3 54.4
64	13 18.5	8 59.5	8 39.1	19 0.7	4 10.1
65	14 12.9	11 28.7	9 56.3	19 45.8	4 26.6
66	15 16.9	14 17.8	11 16.3	20 33.2	4 44.2
S LAT	5	6	Descendant	8	9

♍ 28° 54′ 36″

11ʰ 56ᵐ 0ˢ MC 179° 0′ 0″